Andrew F. Nagy • André Balogh • Thomas E. Cravens •
Michael Mendillo • Ingo Mueller-Wodarg
Editors

Comparative Aeronomy

Previously published in *Space Science Reviews* Volume 139,
Issues 1–4, 2008

Springer

Andrew F. Nagy
Dept. of Atmospheric, Oceanic and Space Sciences
University of Michigan
Ann Arbor, MI, USA

André Balogh
Internation Space Science Institute
Bern, Switzerland

Thomas E. Cravens
Dept. of Physics and Astronomy
University of Kansas
Lawrence, KS, USA

Michael Mendillo
Department of Astronomy
Boston University
Boston, MA, USA

Ingo Mueller-Wodarg
Space and Atmospheric Physics Group
The Blackett Laboratory
Imperial College London
London, UK

Cover illustration: Courtesy NASA/JPL-Caltech

Library of Congress Control Number: 2008939139

ISBN-978-0-387-87824-9 e-ISBN-978-0-387-87825-6

Printed on acid-free paper.

1

springer.com

Contents

Preface

Andrew F. Nagy

Originally published in the journal Space Science Reviews, Volume 139, Nos 1–4.
DOI: 10.1007/s11214-008-9353-0 © Springer Science+Business Media B.V. 2008

Keywords Aeronomy

The term "aeronomy" has been used widely for many decades, but its origin has mostly been lost over the years. It was introduced by Sydney Chapman in a Letter to the Editor, entitled "Some Thoughts on Nomenclature", in *Nature* in 1946 (Chapman 1946). In that letter he suggested that aeronomy should replace meteorology, writing that the word "meteor is now irrelevant and misleading". This proposal was apparently not received with much support so in a short note in Weather in 1953 Chapman (1953) wrote:

> "If, despite its obvious convenience of brevity in itself and its derivatives, it does not commend itself to aeronomers, I think there is a case for modifying my proposal so that instead of the word being used to signify the study of the atmosphere in general, it should be adopted with the restricted sense of the science of the *upper* atmosphere, for which there is no convenient short word."

In a chapter, he wrote in a 1960 book (Chapman 1960), he give his final and definitive definition, by stating that "Aeronomy is the science of the upper region of the atmosphere, where dissociation and ionization are important".

The Workshop on "Comparative Aeronomy" was held at ISSI during the week of June 25–29, 2007. Participation of this workshop was by invitation only due to space limitations at the available meeting facility. The structure of the meeting was such that each of the 32 selected topic was allocated a 30 minute presentation, which was then followed by 20 minutes of open discussion. This book is based on those presentations, but is not a collection of the talks, but is a synthesis, presented as 11 chapters.

This was the second week-long conference dealing with this topic; the first with the same title was held as a Chapman Conference at the Yosemite National Park during February

A.F. Nagy (✉)
Department of Atmospheric, Oceanic and Space Sciences, University of Michigan, Ann Arbor, MI 48109, USA
e-mail: anagy@umich.edu

8–11, 2000 (Mendillo et al. 2002). A full-day symposium on comparative aeronomy was subsequently sponsored by the Royal Astronomical Society in London in January 2003. These meetings, and some preceding CEDAR Workshops, have clearly established that there is a great deal of similarity in the physical and chemical processes controlling the various upper atmospheres and ionospheres in the solar system. Yet, there are significant differences as well due to such factors as distance from the Sun, different neutral atmospheres, roles of intrinsic and induced magnetic fields, and the presence of a surface or regolith of the object under study. Therefore it has become quite clear that there is a great deal to be learned by discussions among scientists working on different aeronomical problems in diverse settings in our Solar System and, increasingly, on extra-solar-system planets. This ISSI workshop addressed all of these topics in a venue that fostered comparisons and syntheses.

References

S. Chapman, Some thoughts on nomenclature. Nature **157**, 105 (1946)

S. Chapman, Nomenclature in meteorology. Weather **7–8**, 62 (1953)

S. Chapman, The thermosphere – The Earth's outermost atmosphere, in *Physics of the Upper Atmosphere*, ed. by J.A. Ratcliffe (Academic Press, 1960)

M. Mendillo, A. Nagy, J.H. Waite (eds.), *Atmospheres in the Solar System: Comparative Aeronomy* (American Geophysical Union, 2002)

Energy Deposition in Planetary Atmospheres by Charged Particles and Solar Photons

Jane L. Fox · Marina I. Galand · Robert E. Johnson

Originally published in the journal Space Science Reviews, Volume 139, Nos 1–4.
DOI: 10.1007/s11214-008-9403-7

Abstract We discuss here the energy deposition of solar FUV, EUV and X-ray photons, energetic auroral particles, and pickup ions. Photons and the photoelectrons that they produce may interact with thermospheric neutral species producing dissociation, ionization, excitation, and heating. The interaction of X-rays or keV electrons with atmospheric neutrals may produce core-ionized species, which may decay by the production of characteristic X-rays or Auger electrons. Energetic particles may precipitate into the atmosphere, and their collisions with atmospheric particles also produce ionization, excitation, and heating, and auroral emissions. Auroral energetic particles, like photoelectrons, interact with the atmospheric species through discrete collisions that produce ionization, excitation, and heating of the ambient electron population. Auroral particles are, however, not restricted to the sunlit regions. They originate outside the atmosphere and are more energetic than photoelectrons, especially at magnetized planets. The spectroscopic analysis of auroral emissions is discussed here, along with its relevance to precipitating particle diagnostics. Atmospheres can also be modified by the energy deposited by the incident pickup ions with energies of eV's to MeV's; these particles may be of solar wind origin, or from a magnetospheric plasma. When the modeling of the energy deposition of the plasma is calculated, the subsequent modeling of the atmospheric processes, such as chemistry, emission, and the fate of hot recoil particles produced is roughly independent of the exciting radiation. However, calculating the spatial distribution of the energy deposition versus depth into the atmosphere produced by an incident plasma is much more complex than is the calculation of the solar excitation profile. Here, the nature of the energy deposition processes by the incident plasma are described as is the fate of the hot recoil particles produced by exothermic chemistry and by knock-on collisions by the incident ions.

J.L. Fox (✉)
Wright State University, Dayton, USA
e-mail: jane.fox@wright.edu

M.I. Galand
Imperial College, London, England

R.E. Johnson
University of Virginia, Charlottesville, USA

Keywords Energy deposition · Photon absorption · Chapman functions · Heating efficiencies · Auger electrons · Characteristic X-rays · X-ray absorption · Doubly charged ions · Core-excited ions · Auroral electrons · Auroral emissions · Color ratios · Auroral particles · Heavy ions · Electron transport · Pickup ions · Range of energetic particles · Knock-on · Recoil particles · Energy loss per ion pair

1 Introduction

The source for nearly all atmospheric processes is ultimately the interaction of solar photons and energetic particles of solar or magnetospheric origin with the atmosphere. We discuss first the energy deposition of solar FUV, EUV and X-ray photons. These photons and the photoelectrons that they produce may interact with thermospheric neutral species producing dissociation, ionization, excitation, and heating. The interaction of X-rays or keV electrons with atmospheric neutrals may produce ejection of inner shell electrons. The resulting core-ionized species may decay by the production of characteristic X-rays or Auger electrons.

Energetic particles may precipitate into the atmosphere, producing auroral emissions. Auroral energetic particles, like photoelectrons, interact with the atmospheric species through collisions that produce ionization, excitation, and heating of the ambient electron population. Auroral particles are, however, not restricted to the sunlit regions. They originate outside the atmosphere and are more energetic than photoelectrons, especially at the magnetized planets. Their spectral shape is very different from those of photoelectrons. The focus here is on the spectroscopic analysis of auroral emissions, their relevance to precipitating particle diagnostics, and the similarities and differences of such an approach applied to different planetary atmospheres.

Neutral atmospheres and ionospheres can be affected by coupling to an external plasma, such as the solar wind or a magnetospheric plasma. Therefore, in addition to being modified by the solar photon flux they can be modified by the energy deposited by the incident plasma ions and electrons. This flux might be associated with the solar wind plasma or, for satellites orbiting in a planet's magnetosphere, a trapped, magnetospheric plasma. The energy range of the incident ions and electrons can be large, extending from eV's to many MeV's. When the energy deposition by the plasma can be calculated, then the subsequent modeling of the atmospheric processes, which is discussed in a number of associated reviews in this issue, is roughly independent of the exciting radiation. This is the case both for the quasi-thermal chemistry and the emission processes, but is also the case for the fate of the hot recoil particles produced by the incident ions or by exothermic chemistry. However, calculating the spatial distribution of the energy deposition versus depth into the atmosphere produced by an incident plasma is much more complex than is the calculation of the solar excitation profile, as discussed briefly below, but in detail by Ledvina et al. (2008). The complexity is due not only to the feedback processes that control the interaction of an ionized upper atmosphere with the ambient fields, but also because of the flux of locally produced pick-up ions. In this paper, the nature of the energy deposition processes by the incident plasma are described as is the fate of the hot recoils produced by exothermic chemistry and by knock-on collisions by the incident ions. This involves not only describing the molecular physics initiated by the incident radiations but also the transport process that occurs in the energized regions of the atmosphere.

2 Solar Photon and Photoelectron Energy Deposition

Most of the solar energy flux is in the visible and infrared regions of the spectrum, which are characterized by photons with wavelengths in the range ~ 4000–50,000 Å. The photons in the visible (4000–8000 Å) and near ultraviolet (NUV) (~ 2000–4000 Å) region of the solar spectrum arise from the photosphere, which is characterized approximately by a black-body spectrum with a temperature of ~ 6000 K. Solar photons with wavelengths in the far ultraviolet (FUV) (~ 1000–2000 Å) and the extreme ultraviolet (EUV) (~ 100–1000 Å), originate in the chromosphere and the transition region to the corona, where the temperatures are in the range 10^4–10^6 K. Soft X-rays (~ 10–100 Å) arise from the solar corona. Some authors refer to an XUV region, which comprises the wavelength range of about from about 10 to 250 Å (e.g., Solomon et al. 2001). Harder X-rays ($1 < \lambda < 10$ Å), which arise from solar active regions, are absorbed in the mesosphere and the mesosphere/thermosphere boundary. Only about 2% of the solar energy flux is carried by photons in the ultraviolet and X-ray regions of the spectrum. The regions of the solar spectrum that are absorbed in the thermospheres and upper mesospheres of the planets are generally characterized by wavelengths less than 2000 Å.

2.1 Photoabsorption and Scattering of Visible Photons

In the visible portion of the spectrum, the photoabsorption cross sections for the major atmospheric species at high altitudes in solar system bodies are negligible. Exceptions to this generalization include some trace species, such as ozone, which absorbs weakly in the Chappuis bands from ~ 4500–8500 Å in the terrestrial and Martian atmospheres. Some hydrocarbon radicals, which may be found in small abundances in the middle atmospheres of the outer planets, theoretically may undergo photodissociation in the visible region to produce an energetic H atom, such as $C_2H_5 + h\nu \rightarrow C_2H_4 + H$, for which the threshold dissociation energy (DE) is 1.65 eV (7535 Å) (e.g., Gilbert et al. 1999) and the 2-propyl radical $C_3H_7 + h\nu \rightarrow C_3H_6 + H$ for which the DE is 1.536 eV (8071.9 Å) (e.g., Noller and Fischer 2007). Visible photons therefore either penetrate to the surfaces or are scattered by cloud and haze particles on all the planets and satellites that have significant atmospheres.

The surface of the Earth is only partially obscured by water and water-ice clouds, which cover about 40% of the planet. On Mars, visible photons are partly attenuated by highly temporally and spatially variable airborne dust (e.g., Kahn et al. 1992) and water-ice hazes, such as the those that surround the poles during winter, and those that form during the afternoon over the Tharsis and Elysium uplands (e.g., Jakosky and Haberle 1992; Zurek et al. 1992).

On Venus, Titan and Triton, layers of clouds and hazes scatter visible photons and prevent them from penetrating to the surface. On Venus, the multi-layered cloud deck, which is composed of mostly sulfuric acid particles, water ices, and chlorine-containing species, extends from about 45 to 65 km. In the lower cloud layer there is evidence for a phosphorus-containing species such as phosphoric acid (H_3PO_4). Haze layers form above and below the main cloud layers (e.g., Esposito et al. 1983; Prinn 1985; Chamberlain and Hunten 1987; Esposito et al. 1997; de Pater and Lissauer 2001).

The ubiquitous hazes on Titan are probably composed of C_2 and higher hydrocarbons and large nitriles. Nitriles are organic species containing a triple CN bond, such as HCN (hydrogen cyanide), HC_3N (cyanoacetylene), CH_3CN (methyl cyanide) or C_2N_2 (cyanogen). These species are formed by ionizing and dissociating interactions of ultraviolet photons or energetic electrons with the major constituents CH_4 and N_2, followed by a rich and complicated photochemistry. The Titan surface is completely obscured by yellow-orange haze particles that are probably composed of tholins, which are nitrogen-rich organic compounds or polymers. The haze particles are believed to contain a solid organic core (e.g., Baines et al. 1995; Israel et al. 2005; Tomasko et al. 2005; Lavvas et al. 2008). The cold surface of Triton, the temperature of which is of the order of 40 K, is also obscured partly by hazes that are probably composed of condensed hydrocarbons, and partly by patchy clouds of condensed N_2 (e.g., Gurrola et al. 1992; Stevens et al. 1992; Yelle et al. 1995).

The middle atmospheres of Jupiter and Saturn are characterized by various aerosols, hazes and multiple cloud layers that scatter visible radiation and limit its penetration to the lower atmospheres (e.g., Atreya et al. 1999; Moses 2000; West et al. 2004; Kim et al. 2006). Although NH_3 is an important minor component of the lower atmospheres of these planets (e.g., (Taylor et al. 2004); de Pater and Massie 1985; de Pater and Lissauer 2001), ammonia and NH_4SH condense to form cloud layers in the tropospheres of Jupiter and on Saturn. At higher altitudes, in the stratospheres of these planets, hydrocarbon haze layers are also present (e.g., Moses 2000; Kim et al. 2006). In the colder atmospheres of Uranus and Neptune, photochemical hydrocarbon hazes and ices may form in the stratospheres (e.g., Moses et al. 1992), and methane ices probably form tropospheric clouds (e.g., Baines et al. 1995).

2.2 Absorption of Ultraviolet Photons

The major thermospheric species of the terrestrial planets with oxidizing atmospheres (Venus, Earth and Mars) include O_2, N_2, CO_2, and Ar, with small admixtures of He, H_2, and the photolysis products, O, N, NO, CO and H. CO_2 is the major species in the lower thermospheres of Mars and Venus, but is a minor species in the terrestrial thermosphere. Although CO_2 condenses onto the surface of Mars none of the major or minor species condenses in the thermospheres of the planets, which are heated by EUV and FUV radiation (e.g., Roble et al. 1987; Fox and Dalgarno 1979, 1981; Fox 1988; Fox et al. 1995).

In the reducing thermospheres of the outer planets (Jupiter, Saturn, Neptune and Uranus), H_2, He, H, CH_4 and its photolysis products, CH_3, CH_2, and CH, along with photochemically produced higher hydrocarbons, are the dominant constituents (Gladstone et al. 1996; Yelle and Miller 2002).

The atmospheres of Titan and Triton are composed mostly N_2 and are of intermediate oxidation state; the thermospheres also contain small amounts of CH_4, H_2, and small radical species formed by chemistry initiated by photon or energetic electron impact, such as N, C and H. The abundances of C_2H_2, C_2H_6, C_2H_4, higher hydrocarbons, and nitriles such as HCN and HC_3N, and CH_3CN are significant in the lower thermospheres and the middle atmospheres (e.g., Krasnopolsky et al. 1993; Krasnopolsky and Cruikshank 1995; Keller et al. 1992; Marten et al. 2002; Yelle et al. 2006).

The photoabsorption cross sections for the small molecules and atoms that make up planetary thermospheres maximize in the extreme ultraviolet (EUV), with values of $\sim 10^{-17} - 10^{-15}\,\mathrm{cm}^2$, and optical depth unity is reached for column densities in the range $(1-100) \times 10^{15}\,\mathrm{cm}^{-2}$. Therefore, any solar system body with a substantial atmosphere has a thermosphere and an ionosphere.

We here limit our discussion to the effects of absorption of FUV, EUV and soft X-ray photons. We specifically ignore the effects of the absorption of solar near infrared photons, which, however, play an important role in heating the lower thermospheres of the planets (e.g., Bougher et al. 1990; López-Valverde et al. 1998; Roldán et al. 2000; Bougher et al. 2008).

The absorption of photons in the far UV, EUV, and soft X-ray regions of the spectrum may lead to dissociation, ionization, or, in some cases, fluorescence of thermospheric species. The rate of absorption of photons that are characterized by wavelength λ at altitude z, $q_\lambda^a(z)$, is given by

$$q_\lambda^a(z) = F_\lambda(z)\sigma_\lambda^a n(z) \tag{1}$$

where σ_λ^a is the absorption cross section, $n(z)$ is the local number density, $F_\lambda(z) = F_\lambda^\infty \exp(-\tau_\lambda(z))$ is the local solar photon flux, and F_λ^∞ is the photon flux at the top of the atmosphere. The optical depth in a plane parallel atmosphere at altitude z is $\tau_\lambda(z,\chi) = \int_z^\infty \sigma_\lambda^a n(z') \sec(\chi) dz'$, where χ is the solar zenith angle. It can be easily shown that the maximum absorption rate in such an atmosphere for solar zenith angle χ is found at the altitude at which $\tau_\lambda(\chi)$ is unity. In order to compute the total photoabsorption rate, (1) must of course be integrated over all wavelengths. In any realistic multi-constituent atmosphere, the rate of photoabsorption must also be summed over all major species.

The column density above an altitude z is defined as $N(z) = \int_z^\infty n(z')dz'$. Note that here we have denoted the local number density $n(z)$, and we have reserved $N(z)$ to indicate the column density. For an atmosphere in hydrostatic equilibrium, the variation of pressure P with altitude is given by

$$\frac{dP(z)}{dz} = -\rho(z)g(z) \tag{2}$$

where the $\rho(z)$ is the mass density and $g(z)$ is the acceleration of gravity.

In the following equations, most of the variables are altitude dependent, but for the purpose of compactness, we have suppressed the variable z. The pressure P at a given altitude is just the force (or weight) per unit area of the atmosphere above that altitude, which is simply given by $P = Nm_a g$. The pressure can also be expressed by the ideal gas law:

$$P = nkT = Nm_a g \tag{3}$$

where k is Boltzmann's constant. Combining the ideal gas law with the definition of $\rho = nm_a$, where m_a is the average mass of the atmospheric constituents, we can derive an expression for the mass density, $\rho = (Pm_a/kT)$. Substituting this expression into (2) and rearranging, we obtain

$$dP/P = \frac{-m_a g}{kT} dz = -1/H_p, \tag{4}$$

where the pressure scale height H_p is defined as kT/mg. Equation (4) can be integrated to give the barometric formula

$$P = P_0 \exp\left(- \int_{z_0}^{z} 1/H_p \right). \tag{5}$$

In this equation, the subscript 0 indicates an arbitrary reference level.

Equations (3) can be rearranged to give a simple expression for the vertical column density

$$N = n\frac{kT}{m_a g} = nH_p. \tag{6}$$

This rather general expression is valid if we assume that the acceleration of gravity and m_a are constant; these assumptions are valid over altitude ranges of the order of a scale height in the mixed region of the atmosphere, that is, below the homopause. For $\chi \neq 0$, the altitude of unit optical depth occurs where the column density along the line of sight to the sun $N(\chi) = N(0)\sec(\chi)$ is the inverse of the absorption cross section, i.e., $N(\chi) = {\sigma_a}^{-1}$. The photon fluxes F_λ^∞ at the top of the atmosphere depend on the distance of the planet from the Sun at a given place in its orbit, solar activity (usually as measured by a suitable proxy), and the solar flux model that is used.

2.3 Solar Flux Models

Among the early versions of the solar spectra used in modeling are those of Hinteregger et al. (e.g., 1981), which were based on measurements of the EUVS experiment on the Atmosphere Explorer satellites. The measurements were normalized and extended outside the wavelength range of the satellite measurements, 142–1850 Å using data from rocket experiments (e.g., Heroux and Hinteregger 1978). Often the solar flux models are denoted by the last two digits of the year and the 3 digit ordinal day of year that they apply to, such as 74113, 76200 (which is also known as SC#21REFW, the successor to the F76REF), and 79050. The first two spectra pertain to low solar activities and the latter to high solar activity of solar cycle 21. In "Hinteregger-style" spectra, the solar fluxes are given at 1 Å resolution in the continua, and as delta functions at the central wavelength of the strong solar lines, for a total of more than 1800 wavelengths from 18 to 2000 Å. Sometimes larger spectral ranges are used, for a total of 37 intervals from 50 to 1050 Å (e.g., Torr et al. 1979). The Hinteregger spectra are also known as the SERF1 solar flux models (e.g., Tobiska 1991).

More recently, the solar fluxes that have been used in thermosphere/ionosphere modeling have been derived from the SOLAR 2000 (S2K) models of Tobiska (e.g., Tobiska 2004; Tobiska and Bouwer 2006). The S2K v2.2x spectra are normalized to the measurements from the Solar EUV Experiment (SEE) on the Thermosphere Ionosphere Mesosphere Energetics and Dynamics (TIMED) spacecraft. The SEE instrument has measured solar irradiances in the range 1 to 1940 Å in 10 Å intervals from 2002 to the present. (e.g., Woods et al. 2005; see also the instrument website at lasp.colorado.edu/see). Other commonly used models include the S2K v1.24 spectra, which are normalized to the data from the Student Nitric Oxide Explorer (SNOE) spacecraft (e.g., Bailey et al. 2000). The latter spectra yield larger peak electron densities that are in better agreement with the measured electron density profiles for Mars and Venus. In Fig. 1a, we present the photon fluxes of the S2K 2.22 model for day 76200 (low solar activity) from 18 to 2000 Å; in Fig. 1b, we show the ratio of the S2K 2.22 photon fluxes to those of the S2K v1.24 spectrum over the range 18 to 1100 Å.

2.4 Chapman Layer Theory

The interaction of photons with atmospheres was first described in a simple but insightful way by Chapman (1931a). The details of Chapman layer theory have been described in many textbooks (e.g., Rishbeth and Garriott 1969; Bauer 1973; Banks and Kockarts 1973; Schunk and Nagy 2000; Bauer and Lammer 2004), and will not be repeated here. Although a Chapman layer need not be an ion/electron layer, we will confine ourselves to a brief presentation of the salient features and most important equations for ionospheric ion and electron density profiles.

In Chapman layer theory for ions, the thermosphere is assumed to be composed of one molecular constituent, XY, which is ionized by the absorption of monochromatic solar photons

$$\mathrm{XY} + h\nu \rightarrow \mathrm{XY}^+ + \mathrm{e}. \tag{7}$$

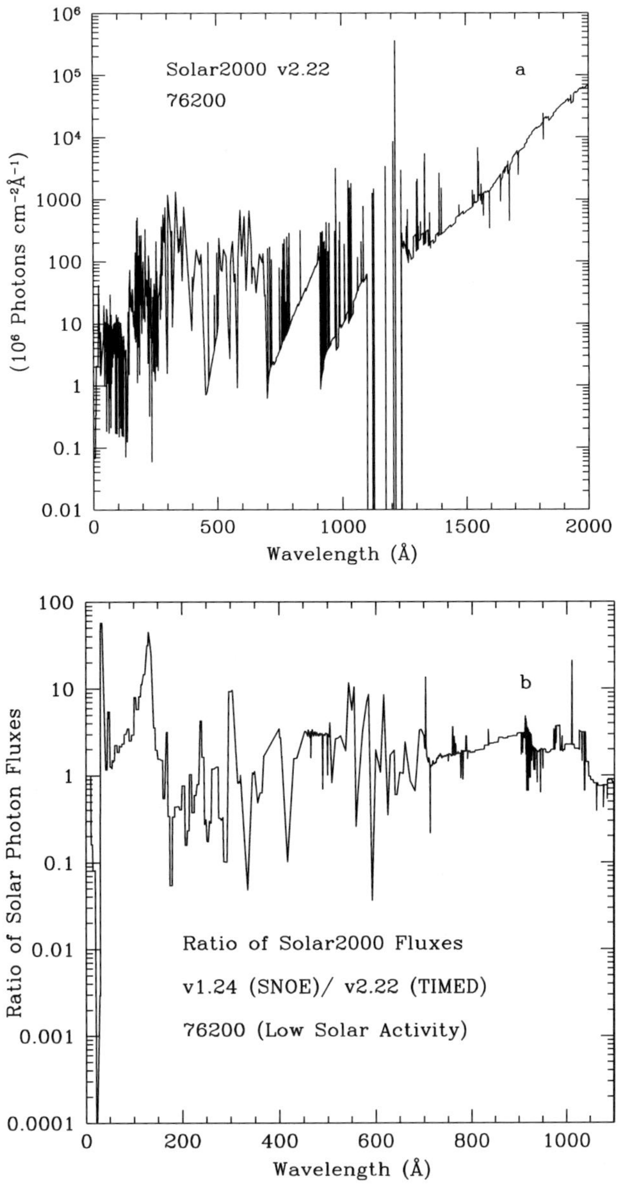

Fig. 1 (**a**) Photon fluxes from 18 to 2000 Å for the 76200 S2K v2.22 solar flux model of Tobiska (2004). (**b**) Ratio of S2K 2.22 v1.24 photon fluxes to the S2K 2.22 photon fluxes from 18 to 1100 Å

The ion XY^+ is destroyed locally by dissociative recombination with a rate coefficient α_{dr}:

$$XY^+ + e \rightarrow X + Y. \tag{8}$$

The production rate of XY^+ by photoionization is given by a general equation similar to that for photoabsorption above (see (1)):

$$q^i = F\sigma^i n_{XY}, \tag{9}$$

where n_{XY} is again the number density of the neutral molecule XY, and σ^i is the ionization cross section. The loss rate for dissociative recombination is given by $L = \alpha_{DR} n_i n_e$, where n_i is the ion density, which, because of charge neutrality, is equal to n_e, the electron density. In Chapman theory photochemical equilibrium (PCE) is assumed, which means that at steady-state, the photochemical production and loss rates are equal. The electron and ion densities are given by

$$n_e = n_i = \left[\frac{F\sigma^i n_{XY}}{\alpha_{dr}}\right]^{1/2}. \tag{10}$$

In Chapman theory the temperature T and mass of XY are assumed to be constant; furthermore, if the altitude dependence of the acceleration of gravity g is ignored, $H_p = kT/mg$ is a constant and the barometric formula (5) can be expressed in terms of the number density $n_{XY}(z) = n_{XY}^0 \exp(-z/H)$, where n_{XY}^0 is the number density at an arbitrary reference altitude, which may be defined as $z = 0$. Under these conditions the pressure and number density scale heights are equal. If the production rate of ions in a Chapman layer for solar zenith angle (χ) is given by (9), and the maximum ionization rate occurs at $z = 0$ where $\tau = 1$:

$$q^i_{\max,\chi} = \frac{F^\infty}{e} \frac{\sigma^i}{\sigma^a H \sec\chi} = \frac{q^i_{\max,0}}{\sec\chi}, \tag{11}$$

The ionization rate q^i_χ at altitude z can be expressed in terms of the maximum ionization rate for overhead sun as

$$q^i_\chi(z) = q^i_{\max,0} \exp\left[1 - \frac{z}{H} - \sec\chi\, e^{-z/H}\right]. \tag{12}$$

Combining (10) and (12), the ion density profile in Chapman theory for a plane parallel atmosphere is given by

$$n_{\mathrm{i}}(z) = \left[\frac{q^i(z)}{\alpha_{\mathrm{dr}}}\right]^{1/2} = \left[\frac{q^i_{\max,0}}{\alpha_{\mathrm{dr}}}\right]^{1/2} \exp\left[\frac{1}{2} - \frac{z}{2H} - \frac{1}{2}\sec\chi\, e^{-z/H}\right], \tag{13}$$

and the maximum ion or electron density as a function of solar zenith angle is then

$$n^i_{\max,\chi} = n^i_{\max,0}(\cos\chi)^{0.5}. \tag{14}$$

The mathematically elegant theory described above is confined to the plane parallel approximation, which becomes increasingly invalid as the terminator is approached. For near terminator region, the sphericity of the atmosphere must be taken into account. In these cases, $\sec\chi$ is often replaced by the Chapman Function, $Ch(x,\chi)$, where $x = R/H_n$, R is the distance from the center of the planet, and H_n is the (constant) neutral scale height. The Chapman Function, which is the ratio of the number density along the line of sight to the Sun in spherical geometry to the vertical column density, has been approximated by various combinations of analytical functions (e.g., Chapman 1931b; Rishbeth and Garriott 1969; Smith and Smith 1972; Bauer 1973). Huestis (2001) has reviewed the various approximations, and described a new analytical evaluation of the Chapman function that is accurate for a large range of χ and for small values of x. Since the advent of fast computers, however, the use of the Chapman function has become unnecessary.

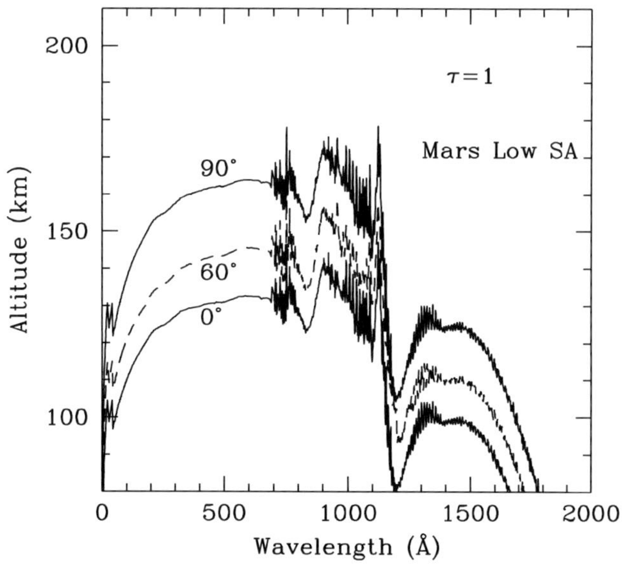

Fig. 2 Altitude of optical depth unity for a low solar activity model of Mars for 0, 60 and 90° SZA

It is fairly easy to compute the optical depth $\tau(\lambda, z, \chi)$ along the line of sight to the Sun in spherical geometry numerically, as described, for example by Rees (1989). For solar zenith angles χ less than 90°,

$$\tau(\lambda, z, \chi) = \sum_j \int_z^\infty n_j(z')\sigma_j^a(\lambda)\left[1 - \left(\frac{R+z}{R+z'}\right)^2 \sin^2\chi\right]^{-0.5} dz', \tag{15}$$

where R is the planetary radius, and the sum over species j is shown explicitly. For χ greater than 90°, the optical depth can be computed as

$$\tau(\lambda, z, \chi) = \sum_j \left\{ 2\int_{z_s}^\infty n_j(z')\sigma_j^a(\lambda)\left[1 - \left(\frac{R+z_s}{R+z'}\right)^2 \sin^2 90°\right]^{-0.5} dz' \right.$$
$$\left. - \int_z^\infty n_j(z')\sigma_j^a(\lambda)\left[1 - \left(\frac{R+z}{R+z'}\right)^2 \sin^2\chi\right]^{-0.5} dz' \right\} \tag{16}$$

where z_s is the tangent altitude. It is of course unnecessary to include the $\sin^2 90° = 1$ factor explicitly in the first term on the right of (16), but we include it here in order to clarify the origin of the formula for the optical depth as twice the total horizontal optical depth along the line of sight to the sun minus that beyond the solar zenith angle χ.

Figure 2 shows optical depth unity as a function of wavelength from the soft X-ray region to the mid FUV region for a low solar activity model of Mars for solar zenith angles 0, 60 and 90°. There is a small increase in the penetration depth as the solar zenith angle increases from 0 to 60°, but a somewhat larger increase occurs from 60 to 90° SZA.

Because of the simplifying assumptions built into the Chapman layer theory, there is no reason to believe that real ionospheric profiles are even quasi-Chapman. Although thermospheric temperatures approach a constant value, T_∞, at high altitudes in stationary atmospheres, thermospheric temperatures increase rapidly near and above the ion peak, where much of the solar energy is deposited.

Table 1 Ionization potentials (I_P) of Some Species of relevance to Planetary Atmospheres[a]. Units are eV

High I_P		Medium I_P		Ionized by Ly α	
Species	I_P	Species	I_P	Species	I_P
He	24.59	H_2O	12.61	C_4H_2	10.18
Ne	21.56	CH_4	12.51	NH_3	10.16
Ar	15.76	SO_2	12.32	CH_3	9.84
N_2	15.58	CH_3CN	12.194	C_3H_6	9.73
H_2	15.43	O_2	12.07	NO	9.264
N	14.53	HC_3N	11.64	C_6H_6	9.246
CO	14.01	C_2H_6	11.52	Si	8.152
CO_2	13.77	C_2H_2	11.40	C_2H_5	8.13
O	13.618	C	11.26	HCO	8.10
H	13.598	C_3H_8	10.95	C_3H_7	8.09
HCN	13.60	CH	10.64	Fe	7.87
OH	13.00	C_2H_4	10.51	Mg	7.65
		H_2S	10.45	trans-HCNH	7.0[b]
		CH_2	10.4	cis-HCNH	6.8[b]
		S	10.35	Ca	6.11
				Na	5.139

[a]Computed with data from Lias et al. (1988), except as noted

[b]From Nesbitt et al. (1991)

Thermospheres also have multiple neutral constituents, and ionospheres are composed of many ions; they are not in PCE at high altitudes. In fact, the largest peaks on the Earth and outer planets are composed of atomic ions O^+ and H^+, respectively, and are F_2 peaks (e.g., Banks and Kockarts 1973; Hinson et al. 1998; Waite and Cravens 1987). F_2 peaks are formed where the time constant for loss by chemical reactions, $\tau_c = 1/\mathcal{L}$, where $\mathcal{L} = L/n$ is the specific loss rate, is equal to the time constant for diffusion, $\tau_d = H^2/D$, where D is the ambipolar diffusion coefficient. Thus PCE breaks down as the F_2 peaks are approached from below.

In addition, the ionizing fluxes are not monochromatic, but cover a range from the ionization potentials of the species to the soft X-ray region of the solar spectrum. Ionization potentials of the common atmospheric species in the atmospheres of the planets are shown in Table 1. The photoionization cross sections, $\sigma^i(\lambda, j)$ and the photoabsorption cross sections $\sigma^a(\lambda, j)$ are functions of wavelength λ and are species dependent. The dissociative recombination coefficient α_{dr} depends on the identity of the ion and has a dependence on the electron temperature that is usually expressed as $(300/T_e)^b$. In this formula, the exponent b has a theoretical value of 0.5, but is found experimentally to be in the range 0.2–0.7, and may itself be temperature dependent. Most important, ionization by solar photons is supplemented by that of photoelectrons, which in general deposit their maximum energy below that of photons.

On the topsides, the ion and electron density profiles are determined to varying extents by the electron (T_e) and ion (T_i) temperatures which are equal to the neutral temperatures only in the lower ionosphere where collision rates between neutrals, ions and electrons are high. The plasma temperatures are larger than T_n at higher altitudes. Some examples of neutral, ion and electron temperature profiles are given in Fig. 3 of Witasse et al. (2008).

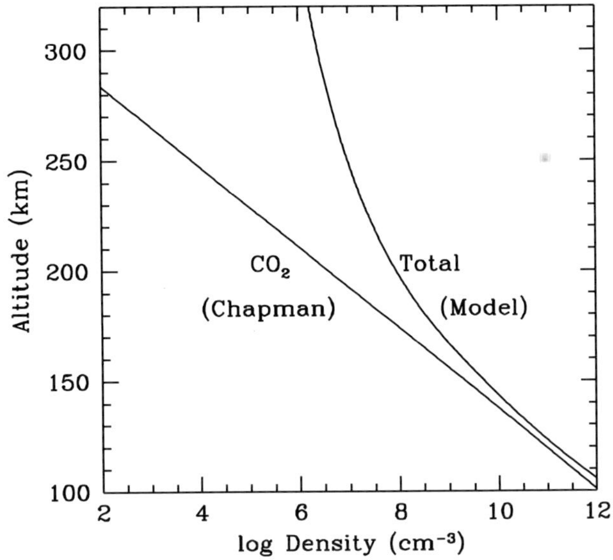

Fig. 3 Altitude profiles of total neutral densities for models based on Chapman profiles and a realistic profile from a model of Mars. The Chapman neutral model profile consists of pure CO_2, while the realistic model profile is made up of 12 neutral species. It is clear that the scale height increases with altitude in the realistic model. This is a result of increasing neutral temperatures, and increased abundances of lighter species

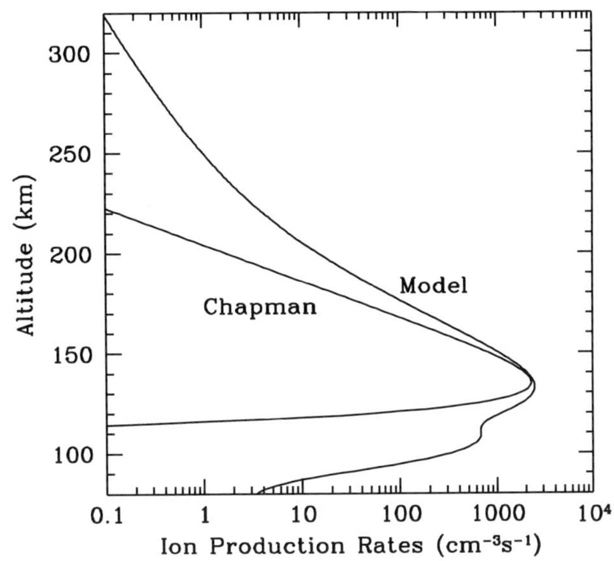

Fig. 4 Ion production rates for the models shown in Fig. 3. Neither the topside nor the bottomsides of the Chapman profiles fit the model profiles

In order to compare Chapman profiles with detailed, realistic numerical models, we have constructed a 60° SZA low solar activity model of the Martian thermosphere/ionosphere similar to those of Fox (2004). In order to fit the Chapman profile, we have determined the CO_2 number density profile for which the ionospheric peak magnitude and altitude reproduce the F_1 peak of the realistic model. This CO_2 profile is compared to the total neutral number density profile of the realistic model in Fig. 3. The resulting ionization profile from 100 to 320 km for the realistic model is presented in Fig. 4 where it is compared with that of the Chapman layer production profile. It is obvious that the realistic ionization profile does not fit the Chapman model at either low or high altitudes. At high altitudes the temperature increases above that at the peak, and the abundance of light atoms increase. This results in an increasing neutral scale height above the peak, rather than a constant scale height as required by Chapman theory.

Photoionization of the various atmospheric species by EUV photons occurs over a broad wavelength range, which in turn causes broadening of the upper F_1 electron density peak. The major mechanism for production of ions in this region is absorption of EUV photons. In fact, this is how we define the F_1 peak here (Banks and Kockarts 1973; Bauer 1973; Bauer and Lammer 2004). Other workers have defined it differently. The photoelectrons produced near the F_1 peak may cause further ionization. The maximum in the ion production rate profile for photoelectrons is slightly below that of photons in the F_1 region. Solar soft X-rays penetrate to lower altitudes in the thermosphere, and produce very high energy photoelectrons. These photoelectrons may produce multiple ionizations at altitudes below the lower E-region peak in the photoionization profile. For example, in a high solar activity model of the Martian ionosphere, the average energy of the photoelectrons produced near the F_1 peak at 135 km is 25.1 eV, and that near the E peak at 117 km is 161 eV.

In fact, ionization at the lower (E-region) peak is caused mainly by impact of photoelectrons and secondary electrons:

$$CO_2 + e^* \rightarrow CO_2^+ + e^* + e, \tag{17a}$$

$$CO_2 + e^* \rightarrow CO^+ + O + e^* + e, \tag{17b}$$

$$CO_2 + e^* \rightarrow O^+ + CO + e^* + e, \tag{17c}$$

where the asterisk denotes an electron with enough energy to potentially produce further ionization. Equations (17b) and (17c) illustrate dissociative ionization of CO_2, which is the main source of the fragment ions O^+ and CO^+ ions at the E-region peak. A comparison of the sources of ionization caused by photons and photoelectrons is shown in Figs. 5a and 5b.

The model electron density profile between the altitudes of 100 and 320 km is compared to that of a Chapman layer in Fig. 6. It is easily seen that the model profile and the Chapman layer are very different both above and below the peak. The Chapman layer is characterized by a constant scale height above the peak, and the electron densities fall off rapidly below the peak. The model electron densities are, however, larger and are characterized by scale heights that vary above the peak. Above about 180 km the scale height increases rapidly, so that near the 300 km, the difference in densities is nearly 3 orders of magnitude. The ion number density scale height H_n^i differs from the plasma pressure scale height $H_p^i = kT_p/(m^i g)$, where m^i is the ion mass and $T_p = T_i + T_e$ is the plasma temperature according to the equation

$$\frac{1}{H_n^i} = \frac{1}{H_p^i} + \frac{1}{T_p}\frac{dT_p}{dz}. \tag{18}$$

At high altitudes where the plasma temperatures increase rapidly in the model, the second term on the right becomes larger than the first, and all the ions are characterized by the same scale height. In addition, PCE breaks down for O_2^+ in the Mars models above an altitude of about ~ 184 km at low solar activity and near ~ 216 km at high solar activity.

As Fig. 6 illustrates, in the region below the F_1 peak, where the Martian ion production rate is dominated by absorption of soft X-rays and the concomitant ionization by high energy photoelectrons and secondary electrons, the model densities are much larger than those of the single Chapman layer model, in which the ionizing photons are monochromatic. The total electron content (TEC) of the model is $\sim 4.9 \times 10^{11}$ cm^{-2}, whereas the TEC for the Chapman profile is 2.7×10^{11} cm^{-2}, which is smaller by almost a factor of two.

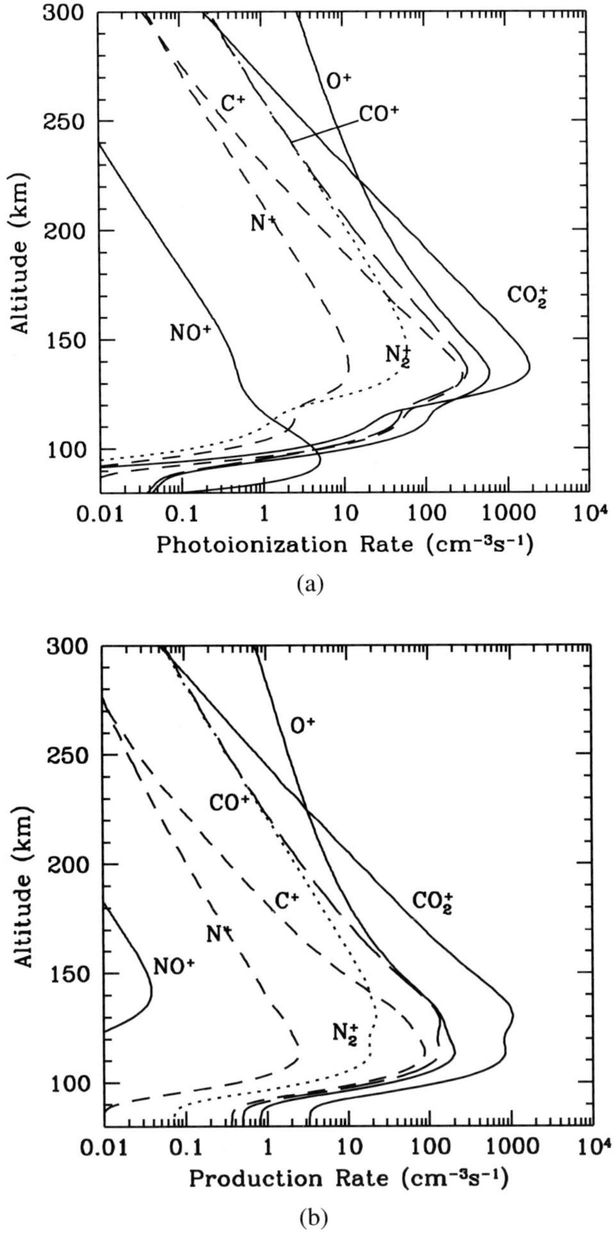

Fig. 5 Computed ionization rate profiles from 80 to 300 km for 7 ions for a high solar activity 60° SZA model of Mars. (**a**) Production rates by photoionization. (**b**) Production rates by energetic photoelectron and secondary electron impact. Near the F_1 peak, the ionization is mostly by EUV photons. Near the E-region peak, however, ionization by very energetic photoelectrons and further electrons produced by electron impact ionization dominate

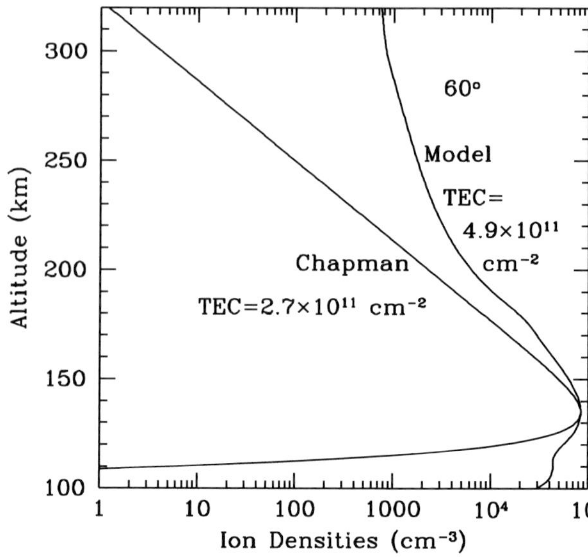

Fig. 6 Total ion or electron density profiles for the Chapman model and the realistic 60° SZA Mars models. These densities correspond to the neutral density profiles and production rates in Figs. 3 and 4. The total electron contents (TEC) are compared

2.5 Photodissociation

Photodissociation is a major source of thermal, and translationally or electronically excited atoms and small fragments in the thermospheres and mesospheres of the planets. Photodissociation can be represented as

$$\mathrm{XY} + \mathrm{h}\nu \rightarrow \mathrm{X} + \mathrm{Y}, \tag{19}$$

where XY is a molecule, X and Y are fragments; the rate of dissociation, q^d, can theoretically be determined similarly to that of ionization (see (9))

$$q_\lambda^d = F_\lambda \sigma_\lambda^d n_{XY}, \tag{20}$$

where σ_λ^d is the wavelength dependent photodissociation cross section, and n_{XY} is the number density of the molecule. As always, q^d, F and n_{XY} are functions of altitude. To a first approximation, the photoabsorption cross section is the sum of the photoionization and photodissociation cross sections. To obtain the total photodissociation rate, q_λ^d must be summed over wavelengths in the solar spectrum from the photodissociation threshold to the point where the photodissociation cross section is zero, that is, where photoabsorption cross section is equal to the ionization cross section. For simple molecules, such as H_2, N_2, O_2, CO, and CO_2, that wavelength is the range ~600–750 Å (e.g., Berkowitz 2002). Model calculations of photodissociation rates in atmospheres may appear to be simple, but in practice, they are complicated by several factors. Measured photoionization cross sections in the continuum shortward of about 600 Å, where photodissociation does not compete with ionization are fairly accurate. Immediately shortward of the photoionization threshold, however, the photoabsorption cross sections usually are highly structured. In this region the photodissociation cross sections are calculated as the difference between the photoabsorption cross sections and the photoionization cross sections, which may be the difference between two large numbers. In addition, if the photoionization and photoabsorption cross sections are not

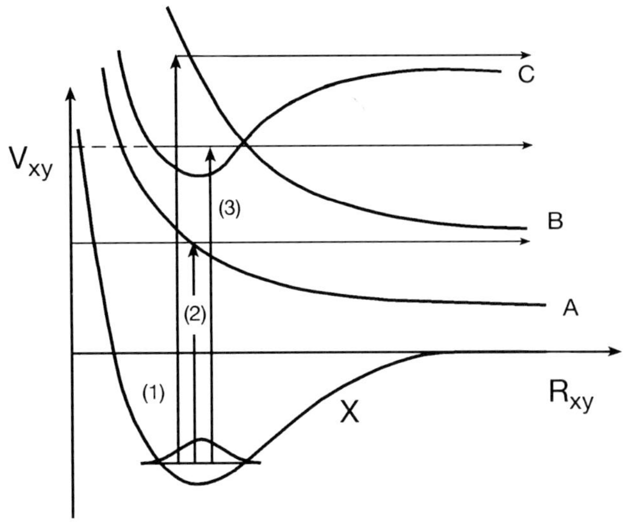

Fig. 7 Schematic representation of the three major mechanisms for photodissociation. (*1*) Excitation to the continuum of bound state (C) (*2*) Excitation to the purely repulsive state (A). Discrete excitation to a bound state (C) that is predissociated by a radiationless transition to the repulsive state (B)

adopted from a single source, there is a large potential for error in the computation of the photodissociation cross sections.

Longward of the photoionization threshold, the photoabsorption cross sections are usually taken to be equal to the photodissociation cross sections. In principle, the absorption of some photons may lead directly to fluorescence, but the fraction is generally small. For example, weak visible fluorescence in the wavelength range of 5500–7500 Å has been detected in the photoabsorption of O_2 at 1162 Å by Lee and Nee (2000), and provisionally attributed to the O_2 $D^3\Sigma_u^+(v' = 6) \rightarrow C^3\Pi_g(v'')$ transition.

Photodissociation can proceed via three possible mechanisms, which are illustrated schematically in Fig. 7. Absorption of a photon may excite a molecule into the continuum of an excited state (process 1), or to a purely repulsive state (process 2). The photodissociation cross sections for both of these processes are fairly smooth as a function of wavelength, and the photodissociation rates may usually be modeled with relatively low resolution cross sections and solar fluxes, of the order of 0.5–1 Å. Alternatively, photons may be absorbed into discrete excited states of the molecule, followed by predissociation (process 3). In order to model the rates of photodissociation proceeding via this mechanism, the individual bands must be resolved and the predissociation probabilities must be known. The photoabsorption rate must be modeled using very high resolution (of the order of 10^{-3} Å) photodissociation cross sections and a similarly high resolution solar spectrum. In addition, the photoabsorption cross sections are usually temperature dependent, and therefore the cross sections must be measured and calculations carried out at temperatures relevant to the part of the atmosphere where the absorption takes place.

We will illustrate these processes by describing the photoabsorption characteristics of some atmospherically important molecules. In order to illustrate the wealth of possible processes, we will focus on the details for O_2, which, because of its importance in the terrestrial atmosphere, has been the subject of many investigations. Photoabsorption by other relevant planetary thermospheric molecules will be discussed briefly.

2.6 Photoabsorption of O_2

In the wavelength region between the O_2 photoabsorption threshold at 2424 Å (5.11 eV) and about 2050 Å, most to the photodissociation is via the dipole forbidden absorption directly into the continuum of the $A^3\Sigma_u^+$ state (process 1). This is the upper state of the Herzberg I system:

$$O_2(X^3\Sigma_g^-) + h\nu \rightarrow O_2(A^3\Sigma_u^+) \tag{21a}$$

$$\rightarrow O(^3P) + O(^3P). \tag{21b}$$

Excitation into the continua of the O_2 $(A'^3\Delta_u)$ state (the upper state of the Herzberg III band system) and the O_2 $(c^1\Sigma_u^-)$ state (the upper state of the Herzberg II band system) also contribute to a lesser extent (e.g., Saxon and Slanger 1986). From 2050 to 2400 Å the cross sections decrease from about 7×10^{-24} to 1×10^{-24} cm^2 (e.g., Yoshino et al. 1988). Because the cross sections are very small, absorption by O_2 in the Herzberg continuum takes place in the terrestrial stratosphere where it competes with absorption by O_3. In the wavelength region between 192 and 205 nm, the absorption into the Herzberg continuum is much less efficient than that into the first Schumann-Runge (S-R) bands (e.g., Coquert et al. 1990; Yoshino et al. 1992).

Photoabsorption into the discrete states of the S-R band system of O_2 followed by predissociation dominates the absorption in the 1750–2050 Å range (process 3). The S-R bands arise from the dipole allowed photoabsorption process into discrete levels of the O_2 $(B^3\Sigma_u^-;\, v')$ state, followed by predissociation via a radiationless transfer to the repulsive $1^1\Pi_u$, $1^3\Pi_u$, $^5\Pi_u$, or $2^3\Sigma_u^+$ states:

$$O_2(X^3\Sigma_g^-;\, v'') + h\nu \rightarrow O_2(B^3\Sigma_u^-;\, v') \rightarrow O(^3P) + O(^3P) \tag{22}$$

(e.g., Julienne et al. 1997; Allison et al. 1986; Lin et al. 1996; Balakrishnan et al. 2000). As shown above, the product O atoms are in the ground ^3P states. Since the absorption cross sections in the region of the S-R bands vary greatly over wavelengths intervals of $\sim 10^{-4}$ nm, the cross sections must be measured with high resolution (e.g., Yoshino et al. 1984; Cheung et al. 1996; Matsui et al. 2003). High resolution absorption cross sections in the O_2 S-R Bands can be found on the CFA website (cfa-www.harvard.edu/amp/ampdata). Because the cross sections in the Schumann-Runge bands are of the order of $10^{-21} - 10^{-19}$ cm^2, photons in this region of the absorption spectrum penetrate into the terrestrial mesosphere and stratosphere, where their photoabsorption is the principal O_2 dissociation process. They do not, however, affect the terrestrial thermosphere. The thermospheres of Mars and Venus do not contain enough O_2 to make photodissociation of O_2 an important as a source of O atoms.

Photoabsorption by O_2 in the wavelength range (130–175 nm) is mostly by direct absorption into the continuum of the $B^3\Sigma_u^-$ state, the S-R continuum (process 1):

$$O_2(X^3\Sigma_g^-;\, v'') + h\nu \rightarrow O_2(B^3\Sigma_u^-) \rightarrow O(^3P) + O(^1D). \tag{23}$$

The maximum cross section at 295 K is about 1.44×10^{-18} cm^2 near 1400 Å (e.g., Yoshino et al. 2005), and the products of this dissociation process are mostly $O(^3P)$ + $O(^1D)$. Absorption of solar radiation by O_2 in the S-R continuum is the main source of atomic O in the terrestrial thermosphere.

Direct absorption from the ground state of O_2 to the $1^3\Pi_u$ state of O_2 is an approximate example of absorption into an excited state that has no bound state (process 2). Except for a shallow minimum that is found in the 1.05–1.27 Å region, the potential curve is mostly repulsive (e.g., Allison et al. 1982). The photoabsorption cross sections into that state are small and smoothly varying at long wavelengths, but exhibit a sharp increase near 1358 Å, with considerable structure shortward of that wavelength (e.g., Allison et al. 1986; Balakrishnan et al. 2000).

In the wavelength range 1030 to 1300 Å, photodissociation of O_2 takes place via dipole allowed excitation of the $O_2(X^3\Sigma_g^-)$ ground state to discrete Rydberg states, such as the $E, E'^3\Sigma_u^-$ and the $F, F'^3\Pi_u$ states, which are strongly predissociated (process 3). This effect leads to highly structured cross sections, and an accidental, but important minimum appears at Lyman alpha (1216 Å). This is the "window" that allows penetration of solar Lyman alpha photons to low altitudes in the terrestrial atmosphere (e.g., Lee and Nee 2000, 2001). La Coursiere et al. (1999) have computed the relative yields of $O(^3P) + O(^3P)$ and $O(^3P) + O(^1D)$ over the solar Lyman alpha line, and have found that the $O(^1D)$ yield is about 0.58. Most of the rest of the yield in this region is to the products $O(^3P) + O(^3P)$. Only a small fraction of the dissociation produces $O(^3P) + O(^1S)$, with an upper limit less than 0.02.

Since the ionization potential of O_2 is 12.07 eV, photons shortward of 1027 Å can ionize or dissociate O_2. Jones et al. (1996) have shown that photons in the wavelength range 750 to 850 Å can excite O_2 to the I, I' and I'' Rydberg states that converge to the $O_2^+(a^4\Pi_u)$ state. In this wavelength range predissociation competes with autoionization, but, in general, predissociation is more important. Carlson (1974) has shown that absorption into these Rydberg states leads to production of one ground state $O(^3P)$ atom plus one excited state $O(^3S^o)$, which radiates to the ground state leading to the OI triplet at 1302, 1304, and 1306 Å.

2.7 Photodissociation of CO

Photodissociation of CO is important as a source of C on Venus and Mars. In the wavelength region 885 to threshold at 1118 Å, the photodissociation of CO takes place via discrete absorptions to a number of predissociating states (e.g., van Dishoeck and Black 1988). The line spacings in some of the bands are of the order of 10^{-4} Å. Fox and Black (1989) constructed high resolution cross sections for excitation to in six bands in the range 885–912 Å, and 33 bands considered by Black and van Dishoeck (1987) in the range 912–1118 Å, for a total of 39 bands to several electronic states. Fox and Black combined the computed high resolution cross sections with a model high resolution solar flux spectrum, which included measured or estimated lineshapes for the strong solar lines in this region of the spectrum. The photodissociation rates for the various bands in the Venus thermosphere were computed, and the total photodissociation rate was found to be smaller by a factor of two than those computed using the low resolution cross sections. The correction was small because CO is not the primary absorber in this region of the spectrum in the Venus atmosphere.

More recent measurements of high resolution photoabsorption cross sections for CO have been carried out for the wavelength regions 967–988 Å, and 925–974 Å by Stark et al. (1993) and Yoshino et al. (1995), respectively.

2.8 Photodissociation of N_2

Photodissociation of N_2 is potentially important for production of translationally and electronically excited N atoms in the planetary thermospheres. Low resolution cross sections for photoabsorption and photoionization of N_2 from 1 Å to the experimental threshold at 1021 Å (12.14 eV) are shown in Fig. 8. Photodissociation of N_2 in the region longward of the IP at 796 Å (15.58 eV) to 1021 Å takes place via line absorptions into discrete valence and

Rydberg states in the singlet manifold, which may decay by emission or preferably by predissociation (e.g., Helm and Cosby 1989). Although the energetic threshold for production of two ground state N atoms:

$$N_2 + h\nu \rightarrow N(^4S) + N(^4S) \tag{24a}$$

is 1270 Å (9.76 eV), the photodissociation cross sections longward of 1021 Å ($E <$ 12.14 eV) are found to be negligible. The possible channels for photodissociation of N_2 thus include

$$N_2 + h\nu(E > 12.14\ \text{eV}) \rightarrow N(^4S) + N(^2D), \tag{24b}$$

$$N_2 + h\nu(E > 13.34\ \text{eV}) \rightarrow N(^4S) + N(^2P), \tag{24c}$$

and

$$N_2 + h\nu(E > 14.52\ \text{eV}) \rightarrow N(^2D) + N(^2D). \tag{24d}$$

For photon energies smaller than 14.52 eV, the dissociation produces one ground state and one excited state N atom. At energies larger than 14.52 eV (wavelengths shortward of 854 Å), in addition to line absorptions, there may also be some photodissociation directly into the continua of singlet *ungerade* repulsive states that result in the production of channel (24d) above (Michels 1981; Walter et al. 1993). Shortward of the ionization threshold at 796 Å ($E > 15.58$ eV) the photoabsorption cross sections are highly structured, and autoionization competes with predissociation. Photodissociations via the channels

$$N_2 + h\nu(E > 15.7\ \text{eV}) \rightarrow N(^2D) + N(^2P) \tag{24e}$$

$$N_2 + h\nu(E > 16.9\ \text{eV}) \rightarrow N(^2P) + N(^2P) \tag{24f}$$

are also possible, although the photoabsorption and photodissociation cross sections are difficult to measure in this spectral region (G. Stark, private communication, 2008).

In the highly structured regions of the N_2 photoabsorption spectrum, ultrahigh resolution cross sections and predissociation probabilities are required in order to quantitatively model the production of N atoms. Cross sections with resolutions of the order of $\sim 6 \times 10^{-3}$ Å are currently being measured by, for example, Stark et al. (2005, and references therein) and

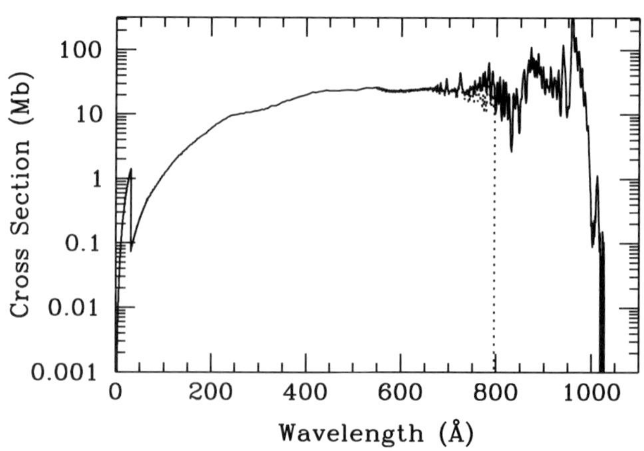

Fig. 8 Low resolution N_2 photoabsorption cross sections (*solid curve*). The photoionization cross sections are represented by the dotted curve. The experimental threshold for photodissociation is about 1021 Å (12.14 eV). In the highly structured region shortward of the ionization threshold at 796 Å predissociation and autoionization compete. Below about 650 Å the photoabsorption and photoionization cross sections are equal

Sprengers et al. (2005, and references therein). Lewis et al. (2005, and references therein) have computed predissociation lifetimes. Calculations carried out with lower resolution cross sections and low resolution solar fluxes can be considered to be accurate to a factor of only a few.

In addition, branching ratios to the various possible channels (24b–24f) must be known over the entire range of photodissociation. These data are not available, except for some information at energies less than 14.52 eV (see Fox 2007 for a review of this subject). Although progress is being made in determining high resolution cross sections and product yields of N_2, not enough information is currently available to accurately compute the production rates of various states of N atoms or the total photodissociation rate of N_2 in thermospheres/ionospheres of the earth, Venus, Mars, Titan or Triton.

2.9 Photodissociation of CO_2

The photoabsorption cross sections of CO_2 are also highly structured. Temperature dependent cross sections have been measured by Stark et al. (2007) from 1061 to 1187 Å with a moderate resolution of 0.05–0.1 Å. The values of the cross sections range from 5×10^{-16} to 2×10^{-20} cm^2 at 195 K, and thus photons in this region of the spectrum are absorbed over a wide altitude range in the atmospheres of Mars and Venus. In the wavelength region 1187 to 1755 Å, the CO_2 photoabsorption cross sections exhibit considerable structure. Photoabsorption cross sections in the 1178.08 to 1633.99 Å range have been measured by Yoshino et al. (1996) at moderate (~ 0.05 Å) resolution, and are available on the CFA website referenced above. Anbar et al. (1993) showed that the use of moderate resolution temperature dependent cross sections in the 1225 to 1970 Å region changed the CO_2 photolysis rates in the Martian lower atmosphere by about 33%, and those of H_2O by 950% as compared to those arising from the use of cross sections averaged over 50 Å bins.

2.10 Photodissociation of H_2

Dissociation of H_2 has a thermodynamic threshold of 4.48 eV (2769 Å), but the photoabsorption cross sections longward of 1116 Å are negligible. Photodissociation in the wavelength region 845–1116 Å proceeds largely by dipole-allowed absorption from the ground $X^1\Sigma_g^+(v'')$ state into discrete states, including $B^1\Sigma_u^+(v')$, $C^1\Pi_u(v')$, $B'^1\Sigma_u^+(v')$, and $D^1\Pi_u(v')$, which may then radiate either to the discrete levels of the ground state, producing emission, or to the continuum of the ground state, producing dissociation to form two H(1s) atoms (e.g., Dalgarno et al. 1970; Abgrall et al. 1997). The $B^1\Sigma_u^+(v')$ and $C^1\Pi_u(v')$ states are the upper states of the Lyman and Werner band systems, respectively, of H_2. Black and van Dishoeck (1987) have investigated these processes as they relate to the interstellar medium, where only radiation longward of the H ionization threshold of 912 Å is important. They found that about 10–15% of the initial line absorptions fluoresce to the continuum of the $X^1\Sigma_g^+(v'')$ state.

Except for a small contribution from predissociation of the $D^1\Pi_u$; v' state for $v \geq 3$ (e.g, Mentall and Gentieu 1970), these excited states are not predissociated in the usual sense because no suitable crossings to repulsive states are available. The lowest purely repulsive state of H_2 is the $b^3\Sigma_u^+$ state (Herzberg 1950), which, because of dipole selection rules, cannot be significantly populated by photoabsorption. Shortward of 845 Å, direct absorption into the continua of the $B^1\Sigma_u^+$, $C^1\Pi_u$, $B'^1\Sigma_u^+$, and $D^1\Pi_u$ states dominates the photodissociation. The cross sections for these processes are substantial, and the products of dissociation are one H(1s) atom and one H(2s, 2p) (e.g., Glass-Maujean 1986).

Autoionization of H_2 begins to compete with dissociation shortward of the ionization threshold at 803 Å, and the cross sections exhibit complicated structure in this region, which has not been fully analyzed (e.g., Yan et al. 1998). For example, dissociation of the $B''\bar{B}^1\Sigma_g^+$ double-welled state proceeds by radiation to the continuum of the $B'^1\Sigma_u^+; v'$ state, with a rate of the order of 5×10^{10} s^{-1} (e.g, Glass-Maujean et al. 2007).

In environments in which H_2 is the major absorber, such as the thermosphere of Jupiter, radiation shortward of 912 Å is important. Significant effects of discrete absorption on the atmospheric absorption profiles are predicted. Kim and Fox (1991, 1994) used the compilations of oscillator strengths, transition probabilities, and fluorescent dissociation fractions of Black and van Dishoeck (1987) and some additional lines provided by J. H. Black (private communication, 1990) shortward of 912 Å to construct high resolution H_2 photoabsorption cross sections. Kim and Fox combined these cross sections with a similarly high resolution solar flux spectrum, and found that, in the Jovian atmosphere, solar radiation in the line centers penetrates to only 420 km above the ammonia cloud tops, but in the wings of the photoabsorption lines, two strong solar lines, CIII at 977.02 Å and OVI at 1031.91 Å, and about 30% of the continuum, penetrate 100 km further to below the methane homopause, where the production of a layer of hydrocarbon ions was predicted.

2.11 Photodissociation of Hydrocarbons

The photoabsorption cross sections for methane fall off rapidly longward of about 1450 Å (e.g., Lee and Chiang 1983), but other hydrocarbons expected to be found below the methane homopauses of the giant planets, such as acetylene, ethane, and ethylene, absorb at longer wavelengths. The cross sections are found, however, to be highly structured and temperature dependent. Photoabsorption cross sections for acetylene have been measured, for example, by Smith et al. (1991) at a resolution of $\sim 0.1-0.5$ Å. Temperature dependent cross sections for ethylene have recently been measured with a resolution of 0.6 Å by Wu et al. (2004). For other hydrocarbons, the reader is referred to the *science-softCon UV/Vis Spectra Data Base*, in which the available data on photoabsorption of hydrocarbons and other molecules of atmospheric interest are summarized and presented (Noelle et al. 2007; see also Huestis et al. 2008).

2.12 Heating by Absorption of Solar Photons and Heating Efficiencies

The heating efficiency is usually defined as the fraction of solar energy absorbed that is deposited locally as heat. Solar energy is transformed into heat in photodissociation and photoelectron-impact dissociation of molecules, and in exothermic reactions, including ion-molecule reactions, neutral-neutral reactions, and dissociative recombination of ions with electrons. Quenching (or collisional deactivation) of metastable ions, such as $O^+(^2D)$, or neutrals, such as $O(^1D)$ or $N(^2D)$, is a particularly important class of reactions that lead to heating. A major uncertainty in modeling heating efficiencies is determining the fraction, f_v, of the exothermicities in these processes that appears as vibrational excitation of molecular products. Vibrational excitation usually leads to cooling either by direct radiation to space for heteronuclear diatomics or polyatomics, or by vibrational energy transfer from homonuclear diatomics to heteronuclear species, and subsequent radiation. By contrast rotational and translationally excited products are thermalized rapidly.

In photodissociation, the amount of energy that appears as kinetic energy is the difference between the energy of the photon and the dissociation energy (which may include some electronic excitation of the fragments). The energy that appears as vibrational excitation in

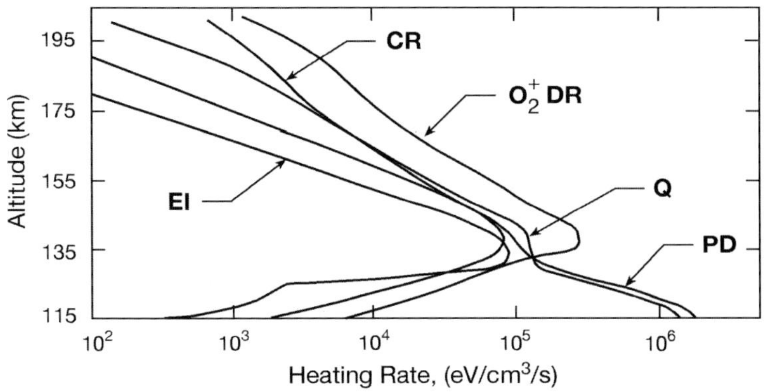

Fig. 9 Altitude profiles of computed heating rates in the Venus thermosphere. The *curve labeled CR* represents the heating rate due to exothermic chemical reactions; the *curve labeled* O_2^+ *DR* is that due to dissociative recombination of O_2^+; the *curve labeled Q* represents the heating due to quenching reactions of metastable species; the *curve labeled PD* is the heating rate due to photodissociation; the *curve labeled EI* is that due to electron impact processes. Adapted from Fox (1988)

photodissociation is found to be small, of the order of 25%. A half-collision model suggests that this fraction is particularly small if one of the fragments is light, such as H or H_2. Vibrational excitation fractions of 10–15% are indicated for this case (cf., Fox 1988, and references therein). In suprathermal electron-impact dissociation, most of the energy is carried away by the electron, and the energy that appears as translation has been found to be of the order of 1 eV (cf., Fox and Dalgarno 1979).

Exothermic ion-neutral and neutral-neutral reactions can be a significant source of heating for the ions and neutrals. The energies of the atomic products are determined by conservation of momentum and energy, but if there are molecular products, some of the available energy can appear as vibrational excitation. That fraction depends on the mechanism of the reaction, and is generally greater for reactions that proceed via a collision complex than those that proceed via a direct insertion/decomposition mechanism. Energy tends also to be deposited in vibration when a new bond is formed. In quenching of metastable species, such as $O(^1D)$, especially those that proceed via the formation of a collision complex, a significant (usually statistical) proportion of the energy can appear as vibrational excitation of the molecular products.

The most important reactions for heating on Venus and Mars are generally dissociative recombination reactions (see (8)). DR reactions tend to be very exothermic, and are the main loss process for ions whose parent neutrals have low ionization potentials. For DR of diatomic molecules, all the exothermicity that does not appear as electronic excitation appears as heat.

Fox (1988) computed the heating rates and efficiencies for a high solar activity model of Venus. The heating rates due to various processes are shown in Fig. 9. It can be seen that the most important source of heating is the DR reaction of the major molecular ion, O_2^+, above about 130 km. Below that altitude the most important sources of heat are almost equally photodissociation and quenching of metastable species. Electron impact dissociation and chemical reactions other than DR or quenching of metastable species are unimportant. Altitude profiles of resulting heating efficiencies are shown in Fig. 10a. The heating efficiency curve labeled “A” is for the standard model, and that labeled “B” is a lower limit. The lower limit model is based on extreme assumptions about the fraction of energy being deposited

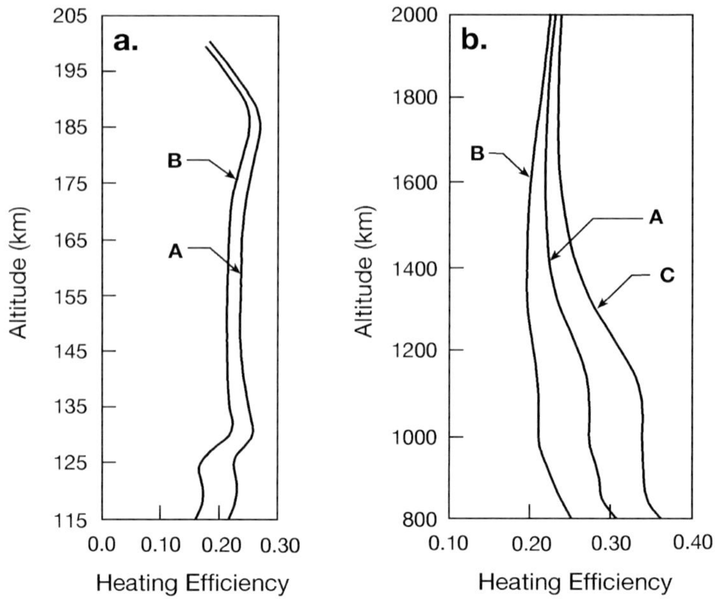

Fig. 10 Altitude profiles of heating efficiencies in the atmospheres of (**a**) Venus (adapted from Fox 1988) and (**b**) Titan (Fox and Yelle, unpublished calculations). The *curves labeled A* are from the standard models; those *labeled B* are lower limits. The *curve labeled C* in (**b**) is the upper limit for the Titan model (see text)

as vibrational excitation. The heating efficiencies range from 16 to 22% at altitudes near 100 km. A similar model for Mars exhibits heating rates and efficiencies that are comparable to those obtained for Venus (Fox et al. 1995).

By contrast, the heating rates on Titan are not dominated by DR reactions, partly because the dominant ions have been "processed more", that is, they have been transformed via many ion–molecule reactions before they can recombine. In addition, since most of the DR reactions are of polyatomics, some of the exothermicity may appear as vibration of the fragments produced. DR of N_2^+ is not an important heat source because N_2 has a high ionization potential, and therefore in the region near and for a significant distance above the main ion peak, N_2^+ tends to destroyed by ion–molecule reactions, rather than by DR. The main sources of heat are found to be photodissociation of N_2 and CH_4, and neutral-neutral chemical reactions. The standard "best guess" and upper and lower limits for the heating efficiencies on Titan are presented in Fig. 10b. These heating efficiency profiles correspond to different assumptions about the fraction of the exothermicities that appears as vibrational excitation in different processes. For example, in neutral-neutral and ion-neutral reactions, fractions of 60%, 40% and 80% are assumed for the standard, upper limit and lower limit models. The heating efficiencies are found to be in the range 25–35% at the lower boundary, and decrease with altitude to values near 22% at the top boundary of the model near 2000 km.

Roble et al. (1987) computed the heating efficiencies in the terrestrial thermosphere, and reported values that increased from 30% at about 100 km to $\sim$ 55% near the F_1 peak ($\sim$ 175 km) and then decreased to $\sim$ 30% near the exobase at 400–500 km. The main sources of heat for the earth are similar to those for Titan, and are photodissociation of O_2 in the Schumann-Runge continuum, and exothermic neutral–neutral reactions. The heating rates are significantly larger than those for Venus and Mars. The reasons for this are various, but may be related to the fact that the major metastable species in the terrestrial ionosphere ($N(^2D)$ and $O(^1D)$) have long radiative lifetimes and are therefore quenched before they can radiate. The major metastable species produced on Venus and Mars, $CO(a^3\Pi)$, is characterized by a short radiative lifetime that exhibits a strong dependence on vibrational and rotational levels, and is in the range $\sim 3-150$ ms (e.g., Jongma et al. 1997). The $CO(a^3\Pi)$ state therefore mostly emits to the ground $X^1\Sigma$ state, producing the Cameron bands, before it can be quenched.

Also, the major species on Venus and Mars, CO_2, is a triatomic molecule. Its interaction with photons and electrons, and its chemical reactions, produce more molecular species for which the exothermicity can be taken up as vibrational excitation, than do reactions of O_2 and N_2 in the terrestrial atmosphere.

2.13 Auger Electrons and Characteristic X-rays

Measurements of doubly charge ions in the terrestrial thermosphere began with the Atmosphere Explorer in the mid 1970's, and in the Venus ionosphere by Pioneer Venus beginning in 1978. Since then, many models of altitude profiles of doubly charged ions densities have been constructed. The first studies were of O^{++} in the terrestrial and Venus atmospheres. More recently, studies of O^{++} and doubly charged molecular ions on other bodies have been carried out. The mechanisms for the production and loss of these ions are uncertain. The production processes include double valence shell ionization and Auger ionization, which may be produced by absorption of X-rays or by impact of very energetic particles. Signatures of Auger ionization have been identified in suprathermal electron flux measurements or predicted by models. Little is known about the loss processes for doubly charged ions, which complicates their modeling.

In addition, measurements of X-rays from various solar system bodies have been made, and their sources have been modeled. We describe the ways in which X-rays interact with atmospheric species, the cross sections for various competing processes, and the mechanisms for emission of X-rays that have been identified for various solar system bodies. We begin by describing the Auger effects and the production of characteristic X-rays.

Most atmospheric molecules are made up of atoms with atomic numbers $Z < 10$, and therefore only the K-and L-shells, defined as those characterized by principal quantum numbers $n = 1$, and $n = 2$, respectively, are populated in the ground states. The ground state of the noble gas He ($Z = 2$) has a filled K-shell with electron configuration $1s^2$; Ne ($Z = 10$) has filled K and L shells with electron configuration $1s^2 2s^2 2p^6$. Ar ($Z = 18$), a minor constituent in planetary atmospheres, has electrons in the ($n = 3$) M-shell; its ground state electron configuration is $1s^2 2s^2 2p^6 3s^2 3p^6$. Metals and other atoms that are formed from ablation of meteors in the mesospheres/thermospheres of planets also have electrons in levels with principal quantum numbers greater than 2. These atoms include, for example, Na (11), Mg (12), Si (14), Ca (20), and Fe (26), where the atomic numbers Z are shown in parentheses. Meteoric ion layers are present in all planetary atmospheres with substantial neutral densities.

The first ionization potentials of atoms and molecules, which are listed in Table 1, pertain to the ejection of outer shell electrons, which are fairly loosely bound. An inner K-shell or "core" ($1s$) electron can be ejected from an atom by absorption of an energetic photon or via a collision with an energetic electron. For atoms with atomic numbers greater than 4, the core-ionizing photons must be in the X-ray region of the spectrum, and the impinging electrons must be characterized by energies of the order of kilovolts.

After the ejection of the core electron, an electron in the L-shell may then make a transition to the $1s$ orbital. Selection rules for one-electron jumps (i.e., $\Delta\ell = \pm 1$) require that

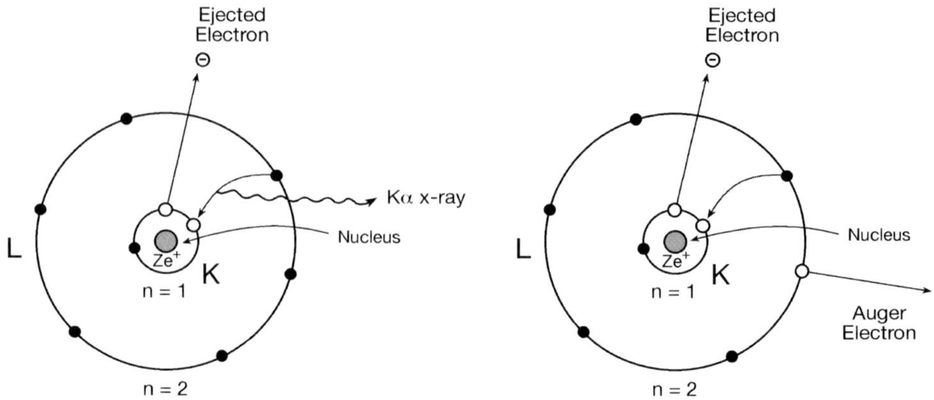

Fig. 11 Cartoon that represents the production of characteristic X-rays and Auger electrons. Only the K and L shells are shown. (*Left*) Characteristic X-rays are produced when absorption of an X-ray photon or energetic charged particle causes the ejection of a K-shell electron (*labeled "Ejected electron"*). An outer shell electron the makes transition to fill the hole in the inner shell. In this process, the excess energy is carried away by a $K\alpha$ X-ray. The ion is left in a singly ionized state. (*Right*) Auger electrons are produced by a similar sequence, except that the excess energy released in reorganization of the ion is carried away by an energetic "Auger" electron, which leaves the ion in the ground or an excited doubly ionized state

this electron be a $2p$ electron. The energy released in the decay of the core ionized state of the ion may be carried away by a characteristic X-ray of approximate energy,

$$E = 13.6(Z-1)^2\left(\frac{1}{{n_1}^2} - \frac{1}{{n_2}^2}\right)\mathrm{eV}. \quad (25)$$

In this equation n_1 is the principal quantum number of the ejected electron and n_2 is the principal quantum number of the valence electron that replaces the inner electron. The characteristic X-ray is designated as "Kα" for $n_1 = 1$ and $n_2 = 2$. The English physicist H. G. C. Moseley studied this effect, which allowed him to determine the atomic numbers of many elements. This simple formula (25) yields energies of 255, 367, 499 and 2947 eV for the Kα X-rays for the atmospherically important atoms C, N, O, and Ar, respectively. If an atom has electrons in the shells with $n \geq 3$, it may, after ejection of an electron from the L-shell, produce a characteristic $L\alpha$ X-ray, which corresponds to $n_1 = 2$ and $n_2 = 3$ in (25). Kβ X-rays may also be emitted, if, after core ionization, an electron from the $n = 3$ (M-shell) drops down to fill the core hole; this corresponds to $n_1 = 1$ and $n_2 = 3$ in (25).

Alternatively, as the valence electron drops down to fill the vacancy in the K-shell, the excess energy may be carried away by the ejection of an outer shell electron, leaving the ion in a doubly ionized state. This ejected "Auger electron" is named after the physicist Pierre Auger, who discovered the phenomenon in the 1920's. These alternatives for relaxation of a core-hole ion are illustrated in Fig. 11. Alternatively, a related phenomenon, the radiative Auger process, may take place; here the excess energy is shared by the Auger electron and a simultaneously emitted photon (e.g., Mühleisen et al. 1996; Penent et al. 2005). For light elements, such as those that are found in the atmospheres of the planets, the Auger process is more important than the emission of characteristic X-rays, while the reverse is true for heavy elements. This is due to the increasing importance of photon emission as the atomic number increases, rather than to a decrease in the probability of ejection of an Auger electron (e.g., Condon and Shortley 1964). Krause (1979) has summarized the X-ray emission and Auger yields for K and L shells for atoms with $5 \leq Z \leq 110$, and showed that they are equally probable for an atomic number of about 31. The Auger yields are 0.997, 0.995, and 0.992, and 0.882 for C, N, O and Ar, respectively. The thresholds for core ($1s^{-1}$) ionization of C, N, O and Ar are 291 eV (42.6 Å), 404.8 eV (30.6 Å), 538 eV (23.0 Å), and 3203 eV (3.87 Å), respectively, (e.g., Verner and Yakovlev 1995) and the Kα X-rays are characterized by wavelengths of 43.7, 30.9, 23.3, and 3.87 Å, respectively (Lide 2008).

We shall describe the details of these processes, specifically for atomic oxygen in the text that follows. The K-shell ionization of O by an X-ray or keV electron can be represented by

$$\mathrm{O}(1s^2s^22p^4) + (h\nu, e^*, [E > 538\ \mathrm{eV}]) \rightarrow \mathrm{O}^{+*}(1s2s^22p^4) + (e, 2e), \tag{26}$$

where O^{+*} represents a core-hole ion. In (26) the electron configurations of the ground state of the neutral and of the core-ionized ion are shown explicitly. After the ejection of the $1s$ electron, an outer L-shell electron may make a transition to the inner shell to fill the vacancy left by the ejected electron; this may be followed by the emission of a characteristic $K\alpha$ X-ray photon, e.g.,

$$\mathrm{O}^{+*}(1s2s^22p^4) \rightarrow \mathrm{O}^{+}(1s^22s^22p^3) + h\nu\ (\sim 23.6\,\text{Å}). \tag{27}$$

In the competing Auger process, after the ejection of the core electron, as the outer shell electron makes a transition to fill the core hole, the excess energy may be carried away by emission of an Auger electron from an outer shell, e.g.,

$$\mathrm{O}^{+*}(1s2s^22p^4) \rightarrow \mathrm{O}^{++}(1s^22s^22p^2) + e\ (\mathrm{Auger}). \tag{28}$$

The energy of the Auger electron may be approximated as that of the $K\alpha$ X-ray minus the binding energy of the $n = 2$ electron. For O^+, the binding energy of the outer electron is approximately equal to the second ionization potential, ~ 35 eV, so that the Auger electrons are expected to have energies of about 500 eV.

This energy is, however, only an approximation. In fact, the core excited O^{+*} ion with electron configuration $(1s2s^22p^4)$ ion actually corresponds to any of four electronic states with term symbols $^4P_{5/2,3/2,1/2}$, $^2P_{3/2,1/2}$, $^2D_{5/2,3/2}$ and $^2S_{1/2}$. Only the ^4P and ^2P states, however, have been observed experimentally from the $\mathrm{O}(^3P)$ ground state (e.g., Petrini and de Araújo 1994). The energies of the $^2P_{3/2}$ and $^2P_{1/2}$ states have been computed by Lohmann and Fritzsche (1996) to lie 4.675 and 4.708 eV, respectively, above the lowest $^4P_{5/2}$ state. The $^4P_{3/2}$ and $^4P_{1/2}$ fine structure levels lie 0.04 eV and 0.053 eV, respectively, above the $^4P_{5/2}$ level.

The $K\alpha$ X-rays may arise from any of ten dipole allowed transitions from the two observed core-excited O^+ states to the three electronic states characterized by the ground state electron configuration $1s^22s^22p^3$, including the ground $\mathrm{O}^+(^4S^{\mathrm{o}}_{3/2})$, and the excited $\mathrm{O}^+(^2D^{\mathrm{o}}_{5/2,3/2})$, and $\mathrm{O}^+(^2P^{\mathrm{o}}_{3/2,1/2})$ states; the energies of the latter two states are about 3.32 and 5.02 eV, respectively, above the ground state. The allowed radiative transitions between the core hole upper O^+ states and the lower states with the ground state electron configuration obey the dipole selection rules for radiative transitions: $\Delta S = 0$, $\Delta L = 0, \pm 1$; $\Delta J = 0, \pm 1$; $L = 0 \nleftrightarrow L = 0$; $J = 0 \nleftrightarrow J = 0$; e $\leftrightarrow$ o). Thus the X-rays produced in dipole-allowed transitions will be characterized by slightly different energies and intensities, and the $K\alpha$ "line" actually represents a series of closely spaced lines.

In the Auger process (see (28)) a radiationless transition takes place between the core hole ^4P and ^2P states of O^+ and the final O^{++} states. The end state with electron configuration $\mathrm{O}^{++}(1s^22s^22p^2)$ comprises three electronic states ^3P, ^1D, and ^1S with Auger electron energies in the range 492–498 eV; six states comprise the O^{++} final state with electron configuration $1s^22s2p^3$: $^5S^o$, $^3D^o$, $^3P^o$, $^1D^o$, $^3S^o$, and $^1P^o$ with Auger energies in the range 470–491.6 eV; three states of O^{++} are associated with the electron configuration $1s^22p^4$: ^3P, ^1D, and ^1S with Auger energies in the range 452–461 eV above the ground state. The total angular momentum quantum numbers J for these terms have been suppressed for compactness. The more stringent selection rules for Auger transitions apply to this process ($\Delta L = \Delta J = 0$; $\Delta S = 0$; $\Delta M_L = \Delta M_S = 0$) (Craseman 2006). Figure 12 illustrates some of the processes that may occur after ejection of a $1s$ electron from oxygen by absorption of an X-ray photon or a collision with a keV electron.

Ten oxygen Auger transitions were observed for Auger electron energies from 465 to 505 eV by Caldwell and Krause (1993). Figure 13 shows an Auger electron spectrum for atomic oxygen. Lohmann and Fritzsche (1996) have computed the energies and the rates of 59 transitions between the core-hole O^{+*} $^4P_{5/2,3/2,1/2}$ and $^2P_{3/2,1/2}$ states and the fine structure levels of the O^{++} states listed above. In general, the dominant Auger transitions are those that terminate on the O^{++} ($1s^22s^22p^2\,^3P$) ground state configuration, but several other transitions are predicted to occur with significant probabilities.

If the O^{++} ion is left in an excited state after ejection of the Auger electron, photons may be emitted that arise from dipole allowed transitions to a lower state. Figure 12 illustrates one example of a radiative transition from the excited $O^{++}(1s^22s2p^3)$ $^3P^o$ state to the ground $O^{++}(1s^22s^22p^2)^3P$ state. This transition results in the appearance of a multiplet which is characterized by the emission of photons in the 702–704 Å range.

2.14 The Auger Effect and Double Valence Shell Ionization in Molecules

In atmospheric molecules, such as N_2, CO, NO, CO_2, and O_2, the inner shell electrons are tightly held around the atomic nuclei, and have large binding energies. By contrast, the valence electrons are in more diffuse molecular orbitals that surround all the nuclei. These electrons are more loosely held and are characterized by much smaller binding energies, which differ, however, from those for the constituent atoms. In addition, the Auger spectra of molecules are more complex than those of atoms. For example, 22 peaks have been identified in the Auger spectrum of O_2 compared to 10 peaks for that of O (e.g., Larsson et al. 1990).

In general, the thermodynamic limit for dissociation of the doubly charged molecular to fragment ions lies below that for production of the doubly excited ion. (A noteworthy exception to this rule is the $CaCl^{++}$ ion, for which the $^2\Pi_{3/2,1/2}$ states are stable by 0.87–0.95 eV (Wright et al. 1997)). Almost all states of doubly charged ions are inherently unstable or metastable against dissociation to two ionized fragments in a "Coulomb explosion". This process can take place via predissociation or through tunneling through the Coulomb barrier. This fragmentation is an important effect that differentiates atomic and molecular Auger processes.

The identification of electronic states of doubly ionized molecules, their potential energy curves, and decay processes have been the subject of numerous investigations over the last ~ 40 years. Mathur (1993, 2004) has reviewed the subject of multiply charged molecules. In many cases, the identification of the ground state is not yet secure. It is difficult to determine the second ionization potential (IP) of a molecule, which is not necessarily the same as the appearance potential for production of doubly ionized species by photons or charged particles. The appearance potential is the minimum energy for double ionization by a vertical (Franck-Condon) transition; the doubly ionized molecules produced may then fragment into ions with substantial kinetic energy.

Theoretical studies of doubly ionized molecules have included calculation of potential energy curves, the Coulomb barriers and identification of states responsible for predissociation of these ions. Many experimental studies of doubly ionized molecules have focused

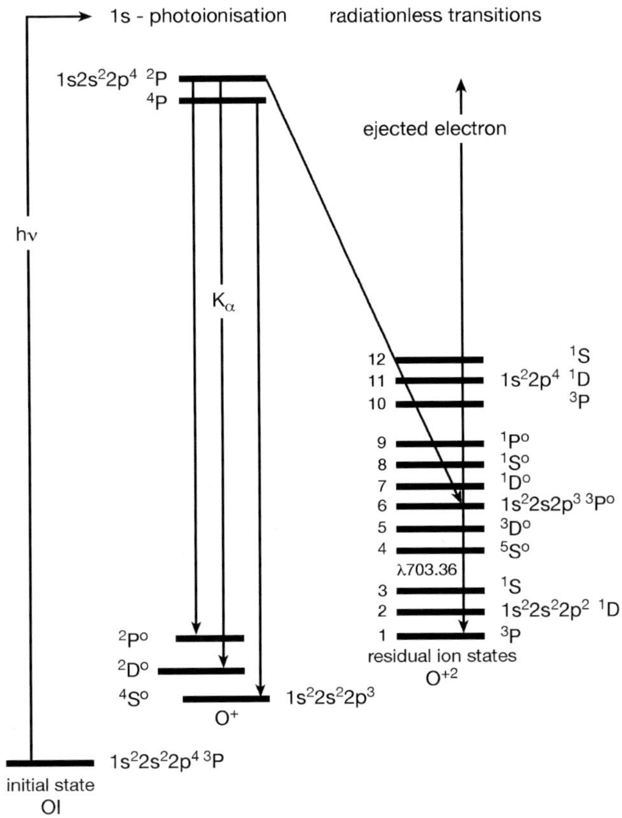

Fig. 12 Diagram of the states involved in the production of $K\alpha$ X-rays and Auger electrons from atomic oxygen (not to scale). (*Left*) An X-ray (or energetic particle) causes the ejection of a "core" ($1s^{-1}$) electron (*middle*) One of the core-ionized states of O^+ (^4P or ^2P) is produced. A $K\alpha$ photon is emitted in the transition from one of these states to any of the three states of O^+ with ground state electron configuration, $1s^2 2s^2 2p^3$. (*Right*) Alternatively, a radiationless transition to one of the states shown on the right may occur (although selection rules must be obeyed). For each transition, an Auger electron is produced with a different energy. The figure also shows a radiative transition between the O^{++} $^3P^o$ state, which may be produced in Auger decay, and the ground ^3P state. A photon with wavelength of 702–704 Å is produced. (Adapted from Petrini and de Araújo 1994)

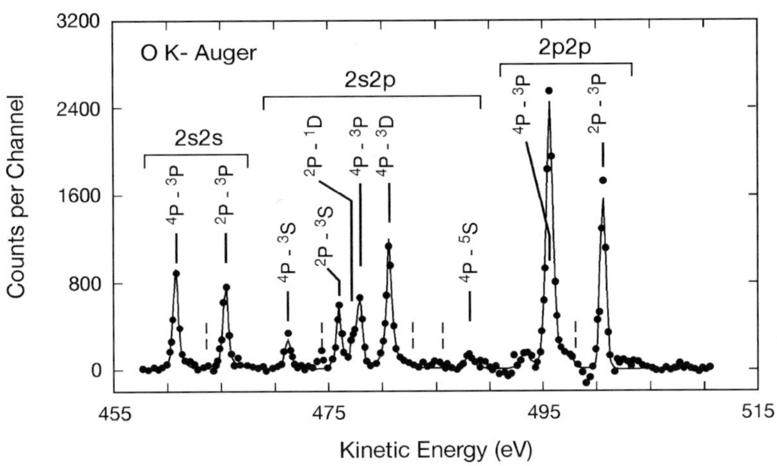

Fig. 13 K-shell Auger spectrum of atomic oxygen, showing 10 discrete transitions. The experimental points are shown; the *solid curve* is a fit to the data. The vertical *dashed lines* indicate transitions from the ^2P state that are too weak to measure. Adapted from Caldwell and Krause (1993). Copyright 1993 by the American Physical Society

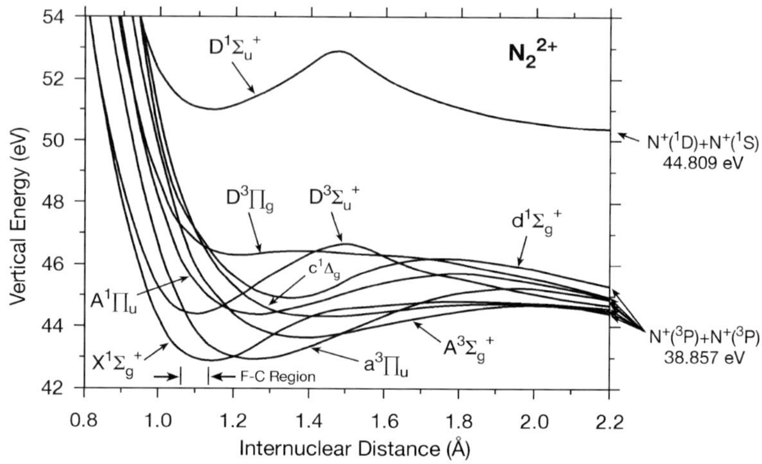

Fig. 14 Potential curves of some low-lying states of N_2^{++}. The energy scale on the *vertical axis* is relative to the $v = 0$ ground state of N_2, and the Franck-Condon region from this state is indicated at the *bottom*. The energy of the $N_2^{++}\ ^1\Sigma_g^+(v'' = 0)$ state relative to the ground state is 43.0 eV. See text for a discussion of the various potential curves. From Lundqvist et al. (1996). Reprinted by permission of IOP Publishing, Ltd

on the Auger electron spectra (e.g., Moddeman et al. 1971), which is an important tool to investigate the properties of doubly charged ions. Recently a more powerful tool, the photoelectron-Auger electron coincidence method has been developed (Penent et al. 2005). In Kinetic Energy Release (KER) experiments, the distributions of fragment ions that arise from decay of doubly ionized molecules are determined. In order to study the KER spectrum of a doubly charged ion, the lifetime of the ion must be in the "experimental window", which is generally less than of 100's of ns.

A wide range of lifetimes of doubly ionized molecules has been found, which complicates studies of their properties. The lifetime of the doubly ionized molecule depends strongly on its electronic state, and the details of its potential curve, including the availability of curve crossings which can predissociate it, and the width of the Coulomb barrier, which varies with the vibrational energy level of the doubly charged molecular ion. Lifetimes range from many seconds to several μs for low vibrational levels of discrete "quasi-bound" states with large Coulomb barriers, to nanoseconds for states that are observed in Kinetic Energy Release (KER) spectroscopy to picoseconds for states whose potential curves are characterized by very shallow minima. Often the higher vibrational levels of a state may be characterized by lifetimes that are too short to be detectable, while those of lower vibrational levels have been inferred to be as long as μs (e.g., Curtis and Boyd 1984). In a seminal study, Mathur et al. (1995) investigated the lifetimes of multiply charged molecular ions using an ion storage ring. They found that all the doubly charged ions studied had components with lifetimes that were of the order of seconds or more.

We will illustrate the molecular Auger process and the related process of double valence ionization specifically for N_2, which, has been the subject of many experimental investigations in the last several decades, including Auger electron spectroscopy (e.g, Moddeman et al. 1971; Svensson et al. 1992, and references therein), X-ray absorption spectroscopy, (e.g., Shigemasa et al. 2002; Svensson 2005 and references therein), photofragment spectroscopy (e.g., Martin et al. 1994; Larsson et al. 1992; Cosby et al. 1983; Sundström et al. 1994), ion coincidence Kinetic Energy Release experiments (e.g, Stockdale 1977; Brehm and de Frênes 1978; Lundqvist et al. 1996), electron–ion coincidence (e.g, Benndorf et al. 1998;

Hsieh and Eland 1996; Fainelli et al. 2002), and translational energy spectroscopy (e.g., Mathur et al. 1995). Double ionization of N_2 by electron impact has been studied by Märk (1975) from threshold ($\sim$ 43 eV) to 170 eV. Ab initio calculations include, for example, those by Wetmore and Boyd (1986), Taylor and Partridge (1987), Koslowski et al. (1991), and Mathur et al. (1995). Figure 14 from Lundqvist et al. (1996) shows some of the low-lying potential curves of N_2^{++}, and indicates the Franck-Condon region from the ground state of N_2. We will discuss some of the features of N_2^{++} by reference to this figure.

In a molecule, as in an atom, an inner shell electron may be ejected by absorption of an X-ray photon, a keV electron, or a heavy charged particle (e.g., Edwards and Wood 1982)

$$N_2 + (h\nu, e^*) \rightarrow N_2^{+*} + (e, 2e), \tag{29}$$

where the asterisk represents an excited core hole state of the molecular ion. The K-shell ionization threshold of N_2 is 409.9 eV, compared to 404.8 eV for an atomic N (e.g., Shigemasa et al. 2002). Just as for core-ionized atoms, a valence electron may make a transition to fill the core hole, and the excess energy may be carried away by ejection of an outer shell Auger electron:

$$N_2^{+*} \rightarrow N_2^{++} + e^* \text{ (Auger)}. \tag{30}$$

This process occurs on a femtosecond time scale (Benndorf et al. 1998). As with small atoms, small core hole molecular ions may emit characteristic X-rays, but Auger processes are more probable. Subsequent fragmentation of the doubly charged N_2^{++} ion produces N^+ ions with kinetic energies of the order of a few electron volts:

$$N_2^{++*} \rightarrow N^+ + N^+ + \Delta E. \tag{31}$$

The states shown in Fig. 14, except for the $D^1\Sigma_u^+(v')$ state dissociate to two ground state N^+ ions.

The $v'' = 0$ level of the ground $^1\Sigma_g^+$ state of N_2^{++} lies about 43 eV above the $^1\Sigma_g^+(v = 0)$ ground state of N_2 (Benndorf et al. 1998; Lundqvist et al. 1996), whereas the minimum energy for production of two ground state $N^+(^3P)$ fragment ions with zero kinetic energies from N_2 is about 38.9 eV, about 4 eV below that for production of the doubly charged ion. The ground state of N_2^{++} shown in Fig. 14 exhibits a deep well, and supports a large number of long-lived vibrotational levels. Higher excited states of N_2^{++} may, however, be produced in the Auger process.

A wide range of heights of the Coulomb barriers are apparent for all the states. Highly vibrationally excited levels in the "quasi-bound" states may tunnel through the Coulomb barrier, while lower vibrationally excited levels may predissociate via a repulsive curve crossing. The ground $X^1\Sigma_g^+$ and $a^3\Pi_u$ states can be seen to have large Coulomb barriers for low vibrational levels, and significant lifetimes are predicted for these states. Mathur et al. (1995) found that a significant N_2^{++} component exists with lifetimes of the order of 3 s.

Hsieh and Eland (1996) identified eight electronic states of N_2^{++} in the Franck-Condon region, although the ground $X^1\Sigma_g^+$ and the $c^3\Sigma_u^+$ (identified in Fig. 14 as $D^3\Sigma_u^+$) states were not observed, apparently because the vibrational states accessible are too low for tunneling through the Coulomb barriers. The energies of the N^+ ions produced by the absorption of 48.4 eV photons were observed to range from 5 to 10 eV. In this experiment, the lifetimes of the doubly ionized N_2^{++} states observed in KER distributions of fragment ions produced from decay of doubly ionized states were on the order of 100's of ns.

Excited doubly charged ions may be optically active, and they may be studied by photofragment spectroscopy (e.g., Cosby et al. 1983; Larsson et al. 1992). The N_2^{++}

$A^1\Pi_u(v')$ and $D^1\Sigma_u^+(v')$ states are connected to the ground $X^1\Sigma_g^+(v'')$ states by optically allowed transitions. These states were studied by Lundqvist et al. (1995, 1996), who observed KER spectra with 19 vibrationally resolved peaks. The $A^1\Pi_u$ state supports 9 or 10 metastable bound states; vibrational levels with $v \geq 7$ were observed, with lifetimes ranging from 1 ns to $\sim$ 300 ns. Larsson et al. did not observe the vibrational levels for $v < 7$ in their experiment; they were inferred to have lifetimes greater than 10 μs. The method of decay for the high vibrationally excited states was assumed to be tunneling or predissociation. A broad peak in the KER spectrum may correspond to excitation to the purely repulsive part of the $A^1\Pi_u$ state. Because the potential curves of the $N_2^{++}(A^1\Pi_u)$ state and the ground $X^1\Sigma_g^+$ state are significantly offset, the emission from this state is expected to be diffuse, and has not been observed.

By contrast, low vibrational levels are populated in the Franck-Condon transitions to the $D^1\Sigma_u^+(v')$ state. The large Coulomb barrier for this state can be seen in Fig. 14. The decay of the $v = 0$ state is dominated by optical transitions to the ground state. The (0,0) and (0,1) bands emit photons that are characterized by wavelengths 1587 and 1594 Å, respectively (e.g, Ehresmann et al. 2000). Higher vibrational levels decay by predissociation. The $D^3\Pi_g$ state shown in Fig. 14 exhibits a shallow minimum which supports two vibrational levels; the lifetimes of these states against tunneling are less than 70 ps (Mullin et al. 1992).

Since the fragment ions produced in the decay of doubly ionized molecules have kinetic energies of the order of several eV, it is not easy to estimate the energies of the Auger electrons. In order to include Auger processes of molecules in atmospheric modeling, it is necessary to use data from Auger spectroscopy, calculations of potential curves and the transitions between them, and/or measurements of the energies of fragment ions produced from X-ray absorption spectroscopy. We should note here that the contribution of $1s$ ejection and Auger decay is very small for the production of doubly ionized species and energetic ion pairs for atmospherically important molecules. Just as for atomic oxygen in the atmosphere, we expect double ionization of valence electrons in molecules to be more important than Auger ionization. If the double ionization in the Franck-Condon region of a molecule proceeds via a transition to a purely repulsive state, or to the repulsive part of a "quasi-bound" potential energy curve, two energetic ionic fragments may be produced directly. The relative rates of the processes depend on the details of the potential energy curves and by the vibrational level (if any) of the doubly charged ion, as well as the energy of the impinging photon or charged particle.

2.15 X-ray Cross Sections for Atoms and Molecules

Berkowitz (2002) has summarized the photoabsorption cross sections for several atoms, and diatomic, triatomic, and polyatomic molecules for a wide range of wavelengths from threshold to the X-ray regions. Although Berkowitz does not differentiate photodissociation and photoionization cross sections, the references in this work cover a nearly complete range of photoabsorption data from threshold to very high energies to which the reader is referred. Svensson (2005) has reviewed the soft X-ray photoionization cross sections of atoms and molecules as determined with synchrotron radiation.

Henke et al. (1993; cf., 1982) have tabulated the cross sections for absorption and scattering of X-rays for the elements with atomic numbers from 1–92, for photon energies from $\sim$ 50 to 30 keV. The cross sections for core ionization of atomic species can be estimated fairly accurately by the fitting functions of Verner and Yakovlev (1995) or Verner et al. (1996). X-ray absorption cross sections of Ar, N_2, O_2, CH_4, He and H_2 have been measured by Denne (1970) in the wavelength range 23.7–82.1 Å. We discuss below the X-ray absorption cross sections of a number of atmospherically relevant species.

Atomic hydrogen has only one electron, for which the ground state is ($1s\ ^2S$), and the ionization potential is 13.598 eV (911.8 Å). There is obviously no Auger effect for H, and the cross sections in the soft X-ray region are very small. For example, at 30 Å, the H ionization cross section is about 2×10^{-22} cm^2, compared to those for C and N of 4.5×10^{-19} and 6.8×10^{-19} cm^2, respectively. At very high energies the Compton ionization (or scattering) cross sections become larger than the photoionization cross sections (e.g, Hubbell et al. 1975). In Compton ionization the photon is scattered rather than absorbed, and the photon and the ejected electron share the energy and momentum. The atomic hydrogen Compton ionization cross sections exceed the photoionization cross sections for energies greater than $\sim$ 2.8 keV (4.4 Å) (Yan et al. 1998). The maximum Compton cross section for H falls in the energy range of 10–15 keV (0.8–1.2 Å), and has a value of about 6×10^{-25} cm^2.

The ionization potential of H_2 is 15.43 eV ($\sim$ 803 Å), and only one state of the H_2^+ ion, $X^2\Sigma_g^+$, is bound. Samson and Haddad (1994) have measured the H_2 photoionization cross sections from 18 to 113 eV (688–110 Å); Yan et al. (1998) have extended the experimental cross sections with theoretical cross sections to the soft X-ray region, from 113 eV to 300 eV (41.3 Å), and they have provided a formula for the cross sections at higher photon energies. The maximum Compton cross section for H_2 is about 1.2×10^{-24} cm^2 for incident photons with energies of about 15 keV (0.8 Å) (Hubbell et al. 1975). The H_2 Compton ionization cross sections are larger than those for photoionization at about 3.1 keV (4 Å) (e.g., Yan et al. 1998). The ratio of the Compton cross sections for H_2 to those for H are about 2.

Since there are no bound states of H_2^{++}, there is no true "threshold" for double ionization. The energy required to produce two protons with zero kinetic energy from H_2 is about 31.7 eV. Double ionization proceeds, however, via excitation from the H_2 ground state to a repulsive $H_2{}^{++}$ state for which the vertical (Franck-Condon) transition energy is about 50.5 eV. Double ionization is followed by a "Coulomb explosion", in which two protons are produced with combined energies of 14–26 eV, with a most probable energy of 18 eV (e.g., McCulloh 1968; Brehm and de Frênes 1978).

The first double photoionization cross sections of H_2 were measured by Dujardin et al. (1987) from 50 to 140 eV (88.5–248 Å). Sadeghpour and Dalgarno (1993) computed the high energy cross sections for double ionization of H_2 and showed that asymptotic ratio of the double-to-single photoionization is about 0.0225. The maximum cross section for double photoionization peaks near 64 eV, and the maximum ratio is about 0.058 near 132 eV photon energy (94 Å).

The inert gas He has two K-shell electrons in the ground electronic state ($1s^2\ ^1S$), for which the first and second ionization potentials are 24.59 and 54.4 eV, respectively. The thresholds for single and double photoionization are therefore 504.2 and 157 Å, respectively. Although He does not exhibit Auger behavior, double ionization can take place to produce an α-particle for photons with energies greater than 79 eV (157 Å). At high energies, the ratio of the cross sections for double to single ionization tend to a limiting value of 0.0164 (e.g., Dalgarno and Sadeghpour 1992). The photoionization cross sections of He at soft X-ray wavelengths are, however, small compared with those of heavier elements. For example, the cross section at 40 Å is about 1.5×10^{-20} cm^2. Photoionization cross sections of He that are accurate for X-ray wavelengths have been computed by Yan et al. (1998). Compton ionization becomes more important than photoionization at a photon energy of 6.5 keV (1.9 Å).

Because the K-shell electrons are localized around the constituent nuclei, the cross sections for K-shell ionization of most molecules can be approximated as the sum of the cross sections for the constituent atoms. For H, the core electron is also the valence electron, but X-ray photoabsorption cross sections for molecules containing H can be estimated as the sum of the constituent atoms, if only because the cross sections for H are very small compared to other species. For example, the H photoabsorption cross section at 42.6 Å, the K-shell threshold for C, is $\sim 6.6 \times 10^{-22}\,cm^2$, compared to that for C of about $10^{-18}\,cm^2$. The CH_4 K-shell ionization cross sections shortward of about 200 Å are approximately equal to that of C (e.g., Denne 1970; Carter and Kelly 1976; Jannitti et al. 1990) with only a small correction for the presence of the H atoms. For $CH_4{}^{++}$ and $NH_3{}^{++}$, the hydrogen atoms are embedded in the electron clouds of the central atoms and the positive charges are located further apart, which leads to greater stability of the doubly charged ion. Similarly, for CO_2^{++}, the larger volume the positive charges occupy leads to a smaller Coulomb repulsion and longer lifetimes. In addition, the two outer electrons in ground state CO_2 are in non-bonding orbitals, so their removal does not decrease the stability of the ion, and leads to ground state lifetimes of about 4 s (Mathur 2004).

It should be noted that the inner shell binding energies for molecules may be shifted by a few eV due to chemical effects (e.g., Svensson 2005). Carbon Kα X-ray emission rates and Auger transition rates from ethane, ethylene, and acetylene have been studied by Kimura (1992), who showed that the X-ray emission rates decrease as the C–C bond order increases (or as the bond length decreases). Uda et al. (1979) showed that K-shell X-ray emission from an atom within a molecule increases as the electronegativity of the neighboring atom increases. Svensson (2005) showed that the core hole ionization energy of alkanes depends on the location of the C atom in each molecule.

2.16 Applications of the Auger Process, Double Ionization and Characteristic X-rays to Ionospheres

The effect of Auger electrons on the terrestrial ionosphere was explored by Avakyan (1978), who suggested it as the major source of O^{++}. O^{++} was detected by the mass spectrometers on the Atmosphere Explorer (AE) Satellites (e.g., Breig et al. 1977). The measurement of rate coefficients of O^{++} with atmospheric neutrals has enabled modeling of O^{++} density profiles. Rate coefficients for reaction of O^{++} with N_2, O_2, and Ar were measured by Howorka et al. (1979); with CO_2 by Johnsen and Biondi (private communication to Fox and Victor, 1980); and with H by Honvault et al. (1995). Victor and Constantinides (1979) showed that double ionization of valence electrons was more important than the Auger effect for the production of the doubly ionized species. Avakyan (1980) came to the same conclusion, but pointed out that during solar flares when the X-ray fluxes were elevated, the Auger process could be more important. Breig et al. (1982) analyzed four AE-C orbits, with the assumption that the most important source of O^{++} was double ionization of valence electrons. The measured O^{++} densities exceeded those that were modeled particularly below 200 km. Breig et al. (1982) attributed this phenomenon to a spacecraft effect, but considered other possibilities for this source, including production of O^{++} by the Auger effect, or general X-ray ionization. They concluded that there was no evidence for enhanced X-ray fluxes for these orbits, and these sources were not likely to account for the data for the orbits considered. Kudryashev and Avakyan (2000) later explored in more detail the effect of solar flares on the ionization and excitation of the terrestrial upper atmosphere, and showed that Auger electrons made a substantial contribution to the high energy electron fluxes and to increased intensities of emissions produced by electron impact on atmospheric species. Simon et al. (2005) modeled doubly charged ions in the terrestrial ionosphere. They fitted the O^{++} densities to those derived from AE-C and obtained excellent agreement with AE-C O^{++} density profiles.

Fox and Victor (1981) modeled the Venusian O^{++} profiles that were returned by the Pioneer Venus ion mass spectrometer, (e.g., Taylor et al. 1980), and showed that double valence shell ionization of O could explain the measured densities; the Auger process was found to be a minor effect. The solar fluxes, in the X-ray region were not, however, known to a great deal of accuracy at that time. Since no measurements of the rate coefficient for the charge transfer reaction of O^{++} with O are available, the rate coefficient in all the models was varied to fit the altitude profiles of O^{++}. Values of $1.5 \times 10^{-10}\,cm^3\,s^{-1}$, $6.6 \times 10^{-11}\,cm^3\,s^{-1}$, and $1.06 \times 10^{-10}\,cm^3\,s^{-1}$, were derived by Fox and Victor (1981), Breig et al. (1982) and Simon et al. (2005), respectively. Note that the rate coefficients quoted by Gronoff et al. (2007) for these reactions are too large by a factor 10 in the text (but not in the table).

Recently, there have been several measurements of the dissociative recombination of molecular doubly charged ions in their vibrational ground states using ion storage rings. For example, Safvan et al. (1999) measured the DR coefficient for CO^{++} with electrons, and reported a value of $0.9 \times 10^{-7}\,cm^3\,s^{-1}$ at 300 K, with a quoted accuracy of a factor of 2. Seiersen et al. (2003a, 2003b) remeasured the DR coefficient for CO^{++} along with that for CO_2^{++} and N_2^{++} and obtained values of $3.0 \times 10^{-7}(300/T)^{0.5}$, $6.2 \times 10^{-7}(300/T)^{0.5}$, and $5.8 \times 10^{-7}(300/T)^{0.5}\,cm^3\,s^{-1}$, respectively.

Experimental studies of the reactions of doubly charged molecular ions with atmospheric species are sparse. The rate coefficients for reactions of O_2^{++} with O_2, N_2, CO_2, NO, and Ne were measured by Chatterjee and Johnsen (1989) using a drift-tube/mass spectrometer experiment. In this experiment the doubly charged ions were formed in vibrationally excited states by the electron impact ion source. Vibrational excitation may reduce the lifetime of a doubly charged ion if its decay process is by tunneling through a Coulomb barrier.

The products of all the reactions of O_2^{++} measured by Chatterjee and Johnsen except that with NO were observed to be simple charge transfer reactions. The dominance of charge transfer reactions that lead to charge separation of the products is a general feature of reactions of multiply ionized species with neutrals, although bond rearrangements of the products may occur (e.g., Mathur 2004). Chemical reactions of CO_2^{++} with neutrals have, however, been observed to compete with simple charge transfer reactions (e.g., Mrázek et al. 2000; Franceschi et al. 2003). In particular, dissociative charge transfer may be the most important channel for these very exothermic reactions.

Witasse et al. (2002) measured a rate coefficient for the reaction of CO_2^{++} with CO_2 and reported a value of $2.13 \times 10^{-10}\,(T/300)^{0.5}\,cm^3\,s^{-1}$. Rate coefficient for reactions of N_2^{++} with N_2 and CH_4 of 2.7×10^{-9} and $1.8 \times 10^{-9}\,cm^3\,s^{-1}$, respectively, were measured by Lilensten et al. (2005). Simon et al. (2005) reported measurements of the rate coefficients for the reaction of N_2^{++} with O_2, and obtained a value of $2.8 \times 10^{-9}\,cm^3\,s^{-1}$.

By employing these measured rate coefficients, models have been constructed of CO_2^{++} in the Martian ionosphere (Witasse et al. 2002), N_2^{++} in the Titan ionosphere (Lilensten et al. 2005), N_2^{++} and O_2^{++} in the terrestrial atmosphere (Simon et al. 2005), and CO_2^{++} and N_2^{++} in the Venusian ionosphere (Gronoff et al. 2007). With the exception of CO_2^{++} at high solar activity, the computed doubly charged molecular ion density profiles exhibited maxima that were of the order of or less than of than 1 cm^{-3}. In addition, because the doubly charged ions are not necessarily formed in the long-lived states, and many potentially important reactions were neglected, the computed density profiles are probably upper limits.

Winningham et al. (1989) used data from the low altitude plasma instrument on the Dynamics Explorer Spacecraft and found a population of high energy photoelectrons, which were shown by models to be produced by solar soft X-rays; the peaks in the electron spectrum were ascribed to Auger electrons emitted from oxygen and nitrogen. Siskind et al. (1989) included the Auger effect for the first time in their models of thermospheric NO densities. Along with several other changes to their model, they found that the new computed NO profiles better fit the more recent data.

Recently, Mitchell et al. (2000) have interpreted a feature in the Martian electron fluxes at ~ 500 eV measured by the Electron Reflectometer on board the MGS spacecraft as oxygen Auger electrons. Cravens et al. (2004) modeled the photoelectron flux at an altitude of 1220 km on Titan, and predicted peaks near the 284 and 402 eV arising from the C and N K-shell Auger electrons.

X-rays have been detected from many solar system objects, including Venus, Earth, Mars, Jupiter, Saturn, the moon, Io, Europa, Ganymede, comets, the Io plasma torus, the rings of Saturn, and the coronas of Earth and Mars. This subject has been recently reviewed (Bhardwaj et al. 2007a). For example, the X-ray spectrum from Venus was observed with the Chandra X-ray observatory, and discrete emissions were seen from the unresolved O-$K\alpha$ emission lines near 530 eV, from the C-$K\alpha$ lines near 280 eV, with a marginal detection of the N-$K\alpha$ lines near 400 eV (Dennerl et al. 2002). A similar spectrum of Mars exhibited only O-$K\alpha$ lines (Dennerl 2002) Additional emission was attributed to scattering of solar X-rays and bremmstralhung emission, and charge exchange between highly stripped heavy ions in the solar wind and H and O atoms in the coronas.

Bhardwaj et al. (2005a) have reported the discovery of oxygen Kα X-rays at 530 eV from the rings of Saturn with the Chandra X-ray Observatory. The rings are known to be made up of mostly water ice. They also proposed that fluorescent scattering of solar X-rays from the icy H_2O ring material plays a role.

Maurellis et al. (2000) modeled the X-ray emission spectrum from Jupiter and predicted that the C-$K\alpha$ lines contributed 8–12% of the spectrum. Bhardwaj et al. (2005b) found that Jovian low-latitude X-ray emissions varied as the solar X-ray flux, and suggested that resonance scattering of solar radiation dominated, with a 10% contribution from characteristic X-rays. They argued against the theory of Waite et al. (1997) that the X-rays were produced by precipitation of ring current particles.

On the other hand, the X-ray emission from comets is proposed to arise mainly from charge-exchange of highly stripped solar wind ions with cometary neutrals (e.g., Cravens 2002). A similar mechanism may be responsible for X-rays from the Jovian auroral regions. For example, Kharchenko et al. (2006) modeled the precipitation of O^+ and S^+ ions with energies of 1–2 MeV/amu and found that stripping produced very high charge state O^{q+} and S^{q+} ions, which emit X-rays when they charge exchange with neutral species, such as H, He and H_2, in the Jovian auroral regions.

When characteristic X-rays are detected from the atmospheres of the solar system objects, the dominance of Auger electron ejection over that of these X-rays for small atoms dictates that there must be many more Auger electrons produced. When these electrons are ejected from molecules, it is likely that ion fragments are produced with substantial kinetic energy. The production of energetic ion fragments from the molecular Auger effect could play a role in heating in the lower ionospheres of the planets, especially during solar flares. This effect and others due to Auger processes, and the emission of characteristic X-rays on various planetary bodies remain to be modeled. In any case, the primary uncertainty in modeling the effects of X-rays on planetary atmospheres is determining the solar fluxes in this region of the spectrum, which are not well known and are quite variable.

3 Auroral Particle Deposition

3.1 Introduction

Auroral particles are energetic, extra-atmospheric particles, which arise from the magnetized environment of a planet or satellite and precipitate into the upper atmosphere, where they deposit their energy. The particles can be electrons, light ions, such as protons, or heavy ions, such as O^+ or S^+. Precipitating ions may charge transfer to atmospheric species to produce energetic neutrals. For magnetized planets, where magnetospheric acceleration processes are effective, the auroral particles have energies ranging from a few hundred eV to a few hundred keV for electrons, and from a few keV up to a few MeV/amu for heavy ions (Kharchenko et al. 2006). Auroral heavy particles at Jupiter mainly originate from the moon Io. At Earth, auroral electrons and heavier particles come originally from the ionosphere, as shown by the presence of oxygen ions in the magnetospheric plasma. There is also a contribution from the solar wind, especially in the cusp regions, where the particles can enter the atmosphere directly. At Venus and Mars, which have induced magnetospheres, the auroral particles originate from the shocked solar wind population and from the planetary atmosphere. The energization processes are weaker at these planets than at magnetized planets. The auroral particles at Venus and Mars are therefore softer than at Earth, Jupiter, and Saturn, with energies ranging from a few tens of eV to a few hundred eV.

Particle precipitation represents a significant source of energy for the auroral regions of magnetized planets and on the nightsides of non-magnetized planets. The total energy input from auroral particles and Joule heating is also often comparable to or exceeds that due to solar photons (e.g., Strobel 2002). Auroral particles deposit their energy in a planetary atmosphere through impacts with the atmospheric neutrals via ionization, excitation, dissociation, and elastic scattering. In addition, suprathermal electrons heat the ambient electrons via Coulomb collisions. All these processes cause the upper atmosphere to respond through changes in its electrodynamical and dynamical processes, thermal structure and constituent distributions. The focus in this section is on one aspect of auroral particle deposition, namely, the auroral emissions. First, the rationale for analyzing auroral emissions is discussed, followed by a short description of the modeling of transport and energy deposition of auroral particles. Then different approaches for the spectroscopic analysis of auroral emissions which are used for identifying and retrieving the spectral characteristics of auroral particles are presented. Finally, some concluding remarks on the spectroscopic analyses are given. The specific examples of aurorae given here are taken from the X-ray, FUV, and visible spectral regions. Discussion of infrared aurora as well as complementary information on the X-ray aurora can be found in Slanger et al. (2008).

3.2 Relevance of Auroral Emission Analysis

The term "aurora" refers to the observed emission spectra (from γ-rays and X-rays to ultraviolet, visible and infrared) arising from the interaction of these extra-atmospheric, energetic particles with an atmosphere (Galand and Chakrabarti 2002). Auroral emissions are associated with an excited state of an atmospheric species that is either (1) the result of a direct impact of an auroral particle with an atmospheric species (e.g., the N_2 Lyman-Birge-Hopfield (LBH) bands) or (2) the result of a chain of chemical processes initially originated with the impact of an auroral particle with an atmospheric gas. Auroral emissions may also be produced by the energetic particles themselves, as is the case of the Doppler-shifted H emissions produced in an energetic H^+/H beam, the K-shell lines produced by charge transfer from highly stripped heavy ions to atmospheric gases, or Bremsstrahlung continuum radiation produced by very energetic electrons.

While the auroral emissions were first observed from the ground in the high latitude regions of Earth, the capability to image aurorae from space opened a new era. The entire terrestrial auroral oval has been imaged from several Earth radii by orbiting spacecraft, including Dynamics Explorer 1 (DE1), Polar, and IMAGE, in the far ultraviolet (FUV), visible, and X-ray spectral regions. This approach is the only way to obtain snapshots of the particle input over entire auroral regions and permits inference of the time variability of the total hemispheric auroral particle power (e.g., Lummerzheim et al. 1997; Germany et al. 1997; Frey et al. 2002; Østgaard et al. 2000, 2001; Liou et al. 2001). *In situ* particle measurements acquired over many years have been used to derive statistical patterns of the particle energy input (e.g., Fuller-Rowell and Evans 1987; Hardy et al. 1989; Brautigam et al. 1991). While such an approach is of significant relevance for morphological trend studies, it is not appropriate for studies of a particular day or geomagnetic event (e.g., Østgaard et al. 2000). Spectroscopic analyses of global auroral images have been used for a wide range of purposes, such as assessing the atmospheric response to particle precipitation (Rees et al. 1995; Doe et al. 1997; Aksnes et al. 2004, 2006; Sætre et al. 2007), estimating the overall budget during a magnetic cloud event (Lu et al. 1998), inferring the ionospheric contribution to magnetosphere-ionosphere coupling (Liou et al. 1997; Meng et al. 2001), and inferring the magnetospheric source regions of auroral precipitation during substorms (Lu et al. 2000).

In addition to the observations of auroral emissions produced in the terrestrial atmosphere, access to space has allowed the discovery and analysis of aurorae throughout the solar system. This has been made possible due to observations from flyby spacecraft (e.g., Voyager, Rosetta, New Horizons); from orbiting planetary probes, such as Galileo (Jupiter), Cassini (Saturn), Pioneer Venus, Venus Express and Mars Express; and from Earth-orbiting observatories, such as the International Ultraviolet Explorer (IUE), the Hubble Space Telescope (HST), Chandra, and XMM-Newton. Discovering and analyzing auroral emissions have led us to a greater understanding of the physical processes taking place in the upper atmospheres and magnetospheres of different solar system bodies. Investigations of aurorae have been used to determine magnetic field configuration, to trace plasma interactions, and to identify atmospheric constituents and auroral particle sources. Reviews of the modeling, observations, and analysis of auroral emissions can be found in Galand and Chakrabarti (2002) for bodies throughout the solar system, Bhardwaj and Gladstone (2000) and Waite and Lummerzheim (2002) for the outer planets and Earth and Bhardwaj et al. (2007a) for X-rays.

3.3 Modeling of Auroral Energy Deposition

Unlike solar photons which are absorbed in the atmosphere, auroral particles are scattered (and heavy ions can also charge exchange) and lose their energy in finite steps along their path through the atmosphere. As a consequence, describing the transport and energy deposition of auroral particles is more complicated than for solar photons. In addition, the contribution of secondary, tertiary (and so on) electrons produced through ionization of atmospheric species by particle-impact must be included in the calculation, as they play an important role as a further source of ionization, excitation or heating. Unlike the thermal ionospheric particles which can be described by macroscopic quantities, suprathermal, auroral particles, similar to photoelectrons, need to be treated through a kinetic approach. Their energy distribution changes as they propagate through the atmosphere. A comprehensive description of the modeling of auroral particle transport and induced emission brightnesses is presented by Galand and Chakrabarti (2002) and hence in the following is only briefly outlined.

Different approaches have been used to describe the transport and energy deposition of auroral electrons. One approach is based on the continuous slowing down approximation (e.g., Rees 1963; Fox and Stewart 1991; Régo et al. 1999; Ågren et al. 2007). This method is restricted to the estimation of ionization and excitation rates. In addition to these quantities, the following three methods provide suprathermal electron intensities and thermal electron heating rates. The first two methods are based on the analytical or numerical solution of the Boltzmann equation. First, the two-stream approximation describes the upward and downward fluxes of electrons propagating along a magnetic field line, as discussed, for example by Gan et al. (1992), Grodent et al. (2001), and Cravens et al. (2005). All of the latter studies are based on the approach developed by Nagy and Banks (1970). Second, the multi-stream approach describes the particle fluxes for a large number of pitch angles—typically 8 or 16—(e.g., Strickland et al. 1976; Lummerzheim et al. 1989; Perry et al. 1999; Galand et al. 1999; Leblanc et al. 2006). While the former method is suitable for estimating integrated quantities, such as excitation and heating rates, the latter method is required for comparison with observed particle fluxes or for studying anisotropy. The final type of calculation is the Monte Carlo approach, which is a stochastic method based on the tracking of numerous individual particles (e.g., Onda et al. 1999; Solomon 1993).

Calculations of the transport and energy deposition of auroral ions are even more complex than those of auroral electrons due to the production of neutrals and ions of different charge states through charge-changing reactions. Each charge-state population needs to be described individually while their distributions are coupled through the charge-changing reactions (e.g., Cravens et al. 1995; Galand et al. 1997; Kallio et al. 1997; Basu et al. 2001; Kallio and Barabash 2001; Kharchenko et al. 2006). In addition, when energetic neutral atoms are produced, the particle beam spreads over space, reducing the energy deposition at the center of the beam (e.g., Lorentzen 2000; Fang et al. 2005).

The heating efficiency is defined as the local rate at which the gas (e.g, electrons or neutrals) is heated divided by the total local energy deposition rate. Such heating efficiencies have been calculated using all the above mentioned approaches except the first. The neutral heating efficiency is sensitive to the composition, density and temperature of the neutral atmosphere, but this dependency can be minimized by adopting a pressure coordinate system instead of an altitude-based coordinate system (Rees et al. 1983). The vertical profile of the heating efficiency was found to be independent of the auroral electron spectrum (Rees et al. 1983), but dependent on the auroral proton spectrum (Srivastava and Singh 1988).

Transport and energy deposition models are used to estimate the excitation rate associated with auroral emissions. If the excited state has no sources other than auroral particle impact, and its radiative lifetime is very short compared with those of other loss processes, the auroral emission brightness can be easily estimated. The calculation of the emission brightness requires the inclusion of photoabsorption, when the emission undergoes "true" absorption by atmospheric species, as is the case for the emissions commonly used for auroral diagnostics, such as the N_2 Lyman-Birge-Hopfield and the H_2 Werner and Lyman bands. For excited states which have long lifetimes against radiation, the continuity equation for the excited state must be solved in order to estimate the emission brightness (Lummerzheim et al. 2001). Finally, radiative transfer calculations are required when the emitted photons undergo scattering, such as resonance scattering, through their paths in the atmosphere. This is the case for modeling the OI 1304 Å resonant triplet in the Earth's auroral regions (Gladstone 1992) and the H Lyman α spectral profile at the giant planets (Régo et al. 1999).

3.4 Spectroscopic Analysis of Auroral Emissions

Spectroscopic analysis of auroral emissions using comprehensive modeling tools, such as particle kinetic codes, can aide in identifying the type (e.g., electrons, light or heavy ions) of the precipitating particles, as well as in determining their mean energy, spectral shape, and total energy flux. Two approaches that have been widely used include the determination of color ratios, and spectral line profiles. Here, rather than providing an exhaustive review of the subject, a few examples are presented for illustration.

3.5 Color Ratios

A "color ratio" is a ratio of total brightness in two wavelength regions, each of which includes one or more strong auroral emissions. Variants of this definition are based on ratios which are a function of the total brightnesses in two (or more) wavelength regions (Régo et al. 1999). The choice of the wavelength regions is determined by the dependence of the color ratio on the altitude of maximum emission. If one assumes an atmospheric model, i.e., altitude profiles of neutral densities, the ratio can be related to the energy of the incident auroral particles. More energetic particles deposit their energy deeper in the atmosphere, and this is manifested by a lower altitude of maximum emission. Once the mean energy of the incident particles is inferred, the total energy flux can be retrieved from the brightness in one of the spectral windows. Such an approach may be used after removal of any non-auroral contributions, such as scattered sunlight and airglow. The solar contribution has been successfully removed from FUV images by, for example, Liou et al. (1997) and Lummerzheim et al. (1997). Since visible radiation is scattered more effectively than UV radiation, high resolution auroral spectroscopy from the ground under sunlit conditions has only recently begun (e.g., Pallamraju and Chakrabarti 2006 and references therein). In addition FUV auroral emissions are prompt and originate from direct energetic particle impact compared with the visible emissions that can arise from chemical reactions of the particles that are produced. Therefore their theoretical interpretation, when multiple scattering can be neglected, is easier.

There are different ways to interpret the color ratio as an indicator of the particle energy. The "colors" associated with emissions may be produced by different mechanisms or by interaction of the particles with different neutral species whose relative density varies with altitude. Examples include the following ratios widely-used for the analysis of the terrestrial aurora: N_2 LBH/OI (1356 Å), where the wavelength range of the LBH bands is 1273–2555 Å (Frey et al. 2002), OI(6300 Å)/N_2^+ 1NG (4278 Å) (also called red-to-blue ratio) (Strickland et al. 1989; Waite and Lummerzheim 2002), OI(8446 Å)/OI(7774 Å) (Waite and Lummerzheim 2002). The analysis of the OI(1304 Å)/OI(1356 Å) intensity ratio derived from Pioneer Venus observations has provided evidence for precipitation of very soft electrons into the nightside atmosphere of Venus (Fox and Stewart 1991). At Mars, the brightness ratio of the nightside CO Cameron band emission and CO_2^+ ultraviolet doublet emission has been derived from Mars Express observations. Based on the analysis of these emissions, Leblanc et al. (2006) inferred that they are consistent with precipitation of electrons characterized by a few tens of eV, that is, of lower energies but with larger fluxes than those of the typical "auroral" electron spectra. This conclusion indicates that the emissions observed may not be auroral in origin, but rather may be induced by photoelectrons produced in a sunlit region that is magnetically connected to the region where emissions were observed.

One of the limitations of using these color ratios is their sensitivity to the ratio of the column densities of the atmospheric constituents involved. This is particularly the case for color ratios based on atomic oxygen and molecular nitrogen emissions, since the [O]/[N_2] ratio is very sensitive to magnetospheric activity and is difficult to estimate (e.g., Strickland et al. 1999; Drob et al. 1999; Meier et al. 2005). As a result, the use of more than two "colors" is required for providing additional constraints on the atmospheric composition

Table 2 Auroral spectroscopic analysis based on the use of FUV color ratios assuming pure electron precipitation in the auroral regions of Earth and Jupiter

Planet:	Earth[a]	Jupiter[b]
Auroral emission band system:	N_2 LBH	H_2 Lyman and Werner
Centered λ of selected bands[c]:	∼ 150 nm & ∼ 170 nm	∼ 125 nm & ∼ 160 nm
Absorber:	O_2	CH_4
Strong photo-absorption below:	∼ 160 nm	∼ 140 nm
Electron energy range retrieved by this technique:	∼ 0.2–20 keV	∼ 10–200 keV
Type of electron aurora identified:	Discrete aurora[d]	Diffuse and discrete aurora[e]

[a]Lummerzheim et al. (1997), Germany et al. (1997)

[b]Régo et al. (1999) and references therein

[c]The wavelength bands selected for the analysis are about 10 nm or less wide (FWHM)

[d]Primarily discrete aurora (Liou et al. 1997), with contribution of diffuse aurora, especially at storm-time (Chen et al. 2005)

[e]Hard electron component only (Grodent et al. 2001; Ajello et al. 2001, 2005)

with four key channels recently proposed by Hecht et al. (2006) for ground-based auroral analysis at Earth.

Another type of color ratio is based on the use of a band system that is attenuated strongly in one part of the spectrum and significantly less in the other. The relative shape of the spectrum of this system, and thus the associated color ratio, can be used as an indicator of the total column density of the absorber above the auroral emitting layer; if an atmospheric model is assumed, the ratio can be related to the energy of the incident auroral particles. Such an approach, which is used at Earth is based on observations of the FUV N_2 LBH band system, for which the shortward spectral component is strongly absorbed by O_2 in the Schumann-Runge continuum (e.g., Germany et al. 1997; Lummerzheim et al. 1997). The energy flux of the incident particles is inferred from the total brightness in the longward spectral window, which is not strongly dependent on the energy of the incident particles (e.g., Liou et al. 1998 and references therein). At the outer planets, FUV H_2 Lyman and Werner band emissions are used in a similar manner with the hydrocarbon layer (primarily methane) as the wavelength-dependent absorber (Régo et al. 1999, and references therein; Grodent et al. 2001; Gérard et al. 2004; Ajello et al. 2005 and references therein). In that case, the color ratio provides the altitude of the auroral emitting layer relative to the methane homopause. A comparison of the technique used at Earth and Jupiter is summarized in Table 2 and color ratios as a function of the mean energy of the incident auroral particles are presented in Fig. 15. The strong dependence of the color ratios on electron mean energy shows clear evidence of the relevance of such ratios for auroral electron diagnostics.

Analysis of auroral N_2 LBH observations allows the retrieval of the spectral characteristics of electrons ranging from a few 100's of eV to less than 20 keV. For electrons of energies below ∼0.2 keV, the altitude of maximum emission is high enough that there is no significant O_2 absorption over the whole N_2 LBH spectrum; the color ratio becomes independent of the energy of the incident electrons. For electrons with energies greater than 15–20 keV, the altitude of maximum emission is low enough that O_2 absorbs significantly over the whole N_2 LBH spectrum. Not only does the color ratio becomes insensitive to the energy of the

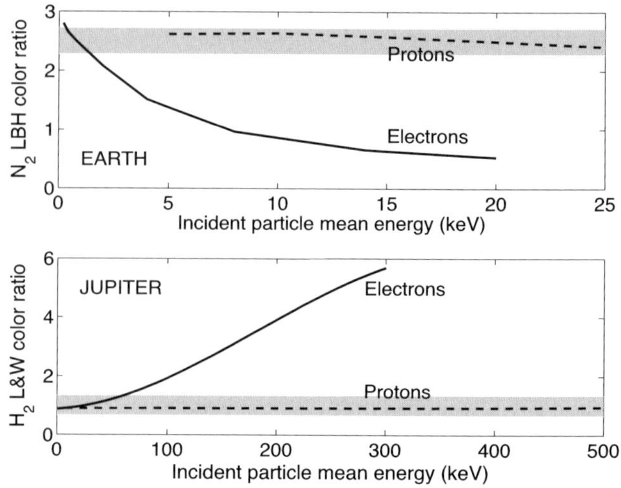

Fig. 15 Modeled FUV color ratios plotted as a function of the mean energy of the incident, auroral particles. Since they are strongly dependent on the energy of the incident electrons, they are used to infer the electron energy from FUV images. The *shaded area* highlights the similar values in color ratios between protons and low energy electrons. (*Top*) Color ratio associated with the N_2 LBH bands for Polar/UVI imaging the terrestrial auroral regions (after Galand and Lummerzheim 2004). (*Bottom*) Color ratio associated with the H_2 Lyman and Werner bands observed by IUE and HST in the auroral regions of Jupiter (after Régo et al. 1999)

incident electrons, but the energy flux cannot be derived from the total brightness of the less attenuated color. At the outer planets where the absorber is the hydrocarbon layer, the technique can be used for inferring the mean energy and energy flux of ~ 10–200 keV electrons. The lower energy boundary is determined by the energy required to reach the hydrocarbon layers. At Jupiter, the FUV H_2 color ratio has been used to retrieve the hard component of the electron precipitation (~10 keV to ~100–200 keV) (e.g., Morrissey et al. 1997; Grodent et al. 2001; Ajello et al. 2001; Pallier and Prangé 2004; Gustin et al. 2004; Ajello et al. 2005; Gladstone et al. 2007). Another technique is required to retrieve the soft component, such as the analysis of EUV observations (e.g., Ajello et al. 2001, 2005). At Saturn, typical auroral electrons are expected to range from a few keV to a few tens of keV. Even though the FUV H_2 color ratio, based on Jovian calculations, has been used (Gérard et al. 2004), an alternative approach, such as the analysis of the EUV emissions or the H Lyman α profile, seems more suitable for Saturn.

At Earth and Jupiter, electron aurorae may be "diffuse" or "discrete". These are the two major types of electron aurora that are identified in or close to the main auroral oval. The diffuse aurora is unstructured, at least on large spatial scales, and is located at the equatorward edge of the main oval. At Jupiter, the associated electron precipitation is induced by pitch angle scattering as a result of the interactions of a warm plasma from the inner plasma sheet with the cold plasma escaping from Io and diffusing outward. At Earth, except during storm-time conditions (Chen et al. 2005), diffuse aurora are characterized by precipitating electron energies of a few tens of keV and energy fluxes less than 1 mW m^{-2}. At Jupiter, the hard component of the diffuse aurora is associated with precipitation of electrons characterized by energies of the order of a few tens of keV. Discrete aurorae, which are brighter than diffuse aurorae, are associated with upward field-aligned currents (i.e., field-aligned precipitating electrons), which are characterized by energies of 1–10 keV at Earth and 30–100 keV at Jupiter. At Earth, discrete aurorae are very dynamic over time scales ranging from fractions of a second to hours, and are highly structured over horizontal scales smaller than a kilometer. It should also be noted that a soft electron component is often added to the electron distribution in both the diffuse and discrete aurorae in order to meet the temperature constraints (Grodent et al. 2001) or EUV observations of Jovian aurorae (e.g., Ajello et al. 2001, 2005).

There is one interesting consequence of the limited range of electron mean energies which can be recovered by the color ratio from the approaches presented in Table 1. At Earth, the approach based on N_2 LBH ratio is primarily useful for the analysis of discrete electron aurora (Liou et al. 1997). Diffuse terrestrial aurorae are primarily induced by electrons that are too hard to be diagnosed by this technique, and are not intense enough to be detected by the UV imagers from high Earth orbit (e.g., Polar), except during storm-time conditions (Chen et al. 2005). At Jupiter this technique can, however, be used to retrieve the characteristics of the harder components of both discrete and diffuse aurorae (Règo et al. 1999). Therefore, while similar techniques can be used for quantitative diagnostics of the electron aurora at different solar system bodies, different limitations apply to the derived physical quantities and to the type of aurora which can be analyzed.

When the emissions in the two colors originate from the same process and the same molecule, uncertainties in the identity of the atmospheric absorber(s) are limiting factors in the analyses, especially at Jupiter where the location of the hydrocarbon layer is variable. Pallier and Prangé (2004) used the H_2-FUV color ratio technique to analyze an atypically bright auroral region observed with HST/STIS. They identified this region as a transient bright cusp. The color ratio obtained implies that the electrons are characterized by energies as large as 200 keV, or that the hydrocarbon layer has been significantly uplifted by the power input.

Spectroscopic analyses based on the use of color ratios are commonly undertaken by assuming pure electron precipitation. Auroral particles of different types, however, such as protons and heavy ions (Règo et al. 1999, and references therein; Galand et al. 2002; Galand and Lummerzheim 2004), may also contribute to the FUV emissions. Under these conditions the color ratios may be misinterpreted. At Earth, the contribution of proton precipitation to the total energy flux averaged over the entire auroral oval is about 15% that of electrons (Hardy et al. 1989). In addition, at given locations and times, particularly in the cusp and at the equatorward boundary of the evening sector of the auroral oval, proton precipitation can become the dominant energy source for the upper atmosphere. Galand and Lummerzheim (2004) have shown that the presence of proton precipitation with an energy flux even as modest as 10% of the total can yield an underestimate of the electron mean energy and energy flux, especially in regions of hard electron precipitation. The color ratio produced by a proton aurora has values similar to that induced by soft electrons (Fig. 15a). In addition, auroral keV protons are more efficient at ionization than are soft electrons (Galand et al. 1999, 2001). The differences between proton and electron aurora can lead to misinterpretation when brightness ratios are used to derive ionospheric conductances, if pure electron precipitation is assumed. It is crucial to separate the electron and proton components of the precipitation in order to improve the auroral diagnostics (Galand et al. 2002).

Figure 15a shows that unlike electron aurorae, the color ratio for proton aurorae is weakly dependent on the mean energy of the incident particles. The decrease in brightness of the emission induced by proton/H atom impact with increasing energy of the incident protons is compensated by the increase of the secondary electron contribution. Molecular oxygen absorption is a rather secondary effect in proton aurora (Galand and Lummerzheim 2004).

Similar to the terrestrial case, auroral protons at Jupiter give the same results as soft electrons for commonly-used color ratios (Fig. 15b). The limitation of the color ratio analysis based on the assumption of pure electron precipitation is however difficult to estimate at Jupiter. To date the relative importance of proton precipitation in terms of incident energy flux is still unknown, even though it is not dominant (Règo et al. 2001). As for heavy ions, constraints on the spectral characteristics derived from the analysis of X-ray spectra could be used to estimate the induced FUV brightness.

Table 3 Analysis of auroral spectral profiles at Earth and Jupiter

H spectral profile	Planet	Earth	Jupiter
	Emission line	H Balmer	H Lyman α
	Emission characteristics (dominant process)	Doppler-shifted as emitted by energetic H atoms[a]	Produced by auroral particles with a central reversal due to H absorption[b]
	Type of auroral particles characterized	Protons	Electrons[c]
X-ray spectral profile[d]	Spectral range	(1) ≥ 3 keV (2) 0.1–2 keV	(1) > 2 keV (2) < 1 keV
	Emission characteristics	(1) Bremstrahlung emission by electrons (2) K-shell lines emitted by the atmospheric species excited by energetic electrons	(1) Bremstrahlung emission by electrons (2) K-shell lines emitted by the precipitating heavy ions
	Type of auroral particles characterized	Electrons (>3 keV) (including diffuse aurora)	(1) Electrons (2) Heavy ions (S+, O+)

[a]Galand and Chakrabarti (2006) and references therein

[b]Prangé et al. (1997), Régo et al. (1999), Régo et al. (2001)

[c]Assuming the electrons are the dominant particle source

[d]Bhardwaj et al. (2007a) and references therein

3.6 Spectral Profiles

Another approach for identifying the type of precipitating particles and for retrieving their spectral characteristics is based on the analysis of spectral profiles resulting from the interaction of the auroral particles with the atmosphere. Depending on the atmospheric composition, the type of auroral particles, and the wavelength range, the spectral profiles are either emitted by the atmospheric species or by the auroral particles. The shape of these auroral profiles may be used to determine the spectral shape or the mean energy of the impinging particles, and their total brightness may be used to infer the total energy flux input. We have chosen to focus on two types of spectral profiles: that associated with the emission of hydrogen atoms, either atmospheric or precipitating, and that associated with X-ray emissions. The dominant production mechanism and relevant physical parameters retrieved from the analysis of such profiles observed in the auroral regions at Earth and at Jupiter are summarized in Table 3.

3.7 H spectral profiles

At Earth, the H Lyman and Balmer series emissions are unique signatures of proton precipitation (e.g., Eather 1967). The high spectral resolution required for a comprehensive analysis of the H emission profiles, however, has been achieved in the visible region of the spectrum from ground-based observations (Galand and Chakrabarti 2006, and references therein), but not yet from space (Galand and Lummerzheim 2004). While multi-spectral analysis of space-based observations has been performed for retrieving both electron and proton components of the precipitation, assumptions about the energies of the incident protons had to be made due to the low resolution of the H profiles (Strickland et al. 2001; Galand et al. 2002). IMAGE/SI12 has returned the first global images of the proton aurora over the entire oval. These images have provided crucial information about the morphology and dynamics of such aurorae and have considerably improved our understanding of them (e.g., Frey 2007, and references therein). The SI12 channel is aimed at the Doppler-shifted portion of the H Lyman α line profile induced by protons and hydrogen atoms with energies larger than 1 keV. It effectively excludes geocoronal Lyman α. Only the integrated brightness over the profiles is, however, provided, which includes the contamination by the nearby NI 1200 Å emission. No information on the spectral shape of the H line is available, which limits quantitative analyses (Frey et al. 2002; Immel et al. 2002). It is surprising that no high resolution spectral imaging has been done of the H Lyman alpha in the terrestrial atmosphere, despite the fact that such imaging has been carried out at other planetary bodies, such as Jupiter.

At Jupiter, the shape and brightness of the H Lyman α profile could be analyzed to derive the mean energy and energy flux of the incident precipitating particles (Régo et al. 1999). The "cold" or atmospheric component of the hydrogen profile is the result of the dissociative excitation of atmospheric H_2 by the auroral particles:

$$\mathrm{H}_2 + (e, p, \mathrm{H}) \rightarrow \mathrm{H}(1s) + \mathrm{H}(2p) + (e, p, \mathrm{H}). \tag{32}$$

$$\mathrm{H}(2p) \rightarrow \mathrm{H}(1s) + h\nu \ (1216\text{Å}). \tag{33}$$

Absorption of the photons emitted by the hydrogen atoms is followed by re-emission in the wings of the profile. The depth and width of the central self reversal as seen on HST observations have been associated with the H column density above the emission region (Prangé et al. 1997). If an atmospheric model is assumed, the energy of the incident particles can be retrieved. These H Lyman α profiles do not depend upon the identity of the particles (electrons versus protons) for a given penetration depth; they only constrain the atmospheric H column density above the emitting layer (Régo et al. 1999). Similar to the terrestrial case, the presence of proton precipitation in the Jovian auroral regions is indicated by the presence of a strong Doppler-shifted (or "hot") component in the H profile (Régo et al. 1999). The absence of a clear detection of this component in the HST observations rules out auroral protons as a major component of the precipitating flux at Jupiter (Régo et al. 2001). The limitations of this technique lies in the uncertainties in the atmospheric density profiles of atomic and molecular hydrogen as well as the possible extinction by methane, which would uniformly affect the profile. A multi-spectral approach in which an FUV color ratio technique is combined with an H profile analysis would aid in better constraining the problem by reducing the number of uncertainties (Régo et al. 2001).

3.8 X-ray Spectral Profiles

In the hard X-ray region, which we define here as photons with energies greater than 2 keV, the auroral emission is produced by the energetic electrons themselves through continuum Bremsstrahlung radiation. This phenomenon has been extensively observed at Earth for deriving the spectral characteristics of hard electron precipitation. For more than a decade, global 2D imaging has been possible from the PIXIE instrument onboard the Polar satellite, which is sensitive to X-rays in the 2–60 keV range. Such images have allowed the retrieval of the characteristics of the energetic electrons and their morphology and variability in time (e.g., Østgaard et al. 2000). Although its presence at Jupiter had been predicted, the Bremsstrahlung component in the aurorae has only recently been identified from analysis of XMM-Newton observations by Branduardi-Raymont et al. (2007).

In the soft X-ray region ($E \leq 2$ keV), the Jovian aurora has been extensively observed since the 1980s, while there have been few observations of X-rays in the terrestrial aurora. At Earth, the first imaging of the auroral region in the soft X-ray region was undertaken by Bhardwaj et al. (2007b). The analysis of the terrestrial spectra has provided evidence of $K\alpha$ lines for nitrogen and oxygen near 0.4–0.5 keV overlying the Bremsstrahlung emission spectrum. $K\alpha$ X-rays are initiated by core-ionization of atmospheric species by X-rays or energetic particles, followed by the filling of the core hole by a valence electron. The excess energy must be carried away by a characteristic X-ray or an Auger electron. Another possible source of soft X-rays is the interaction of solar wind heavy ions with the terrestrial upper atmosphere in the cusp regions (Bhardwaj et al. 2007b). At Jupiter, high spectral resolution observations have clearly identified the presence of X-ray lines in the spectrum which are consistent with charge exchange by precipitating highly stripped oxygen and possibly sulfur ions (Elsner et al. 2005; Kharchenko et al. 2006).

3.9 Concluding Remarks on the Comparative Spectroscopic Analysis of Auroral Emissions

Spectroscopic analysis of auroral emissions allows a quantitative diagnostic of the precipitating particles. When applied to a sequence of auroral images, it provides information about the morphology and variability in time of the source(s) of the particles. For example, the analysis may provide information about the role of the solar wind in producing the emissions (e.g., Liou et al. 1998; Fillingim et al. 2005; Pallier and Prangé 2004; Frey 2007).

Such analyses, however, have limitations, which should be kept in mind. While similar techniques can be applied at different solar system bodies, the energy range covered by the analysis, the type of aurora which can be probed, or the physical quantities derived from the analysis vary from one body to another (see Tables 1 and 2). Among the limitations are the physical parameters required for spectroscopic analyses, including particle-impact cross sections (cf., Huestis et al. 2008), atmospheric neutral densities (cf., Müller-Wodarg et al. 2008), and pitch angle and energy distributions of the incident, auroral particles. For instance, recent updates in the electron impact cross sections of the N_2 LBH excitation (Johnson et al. 2005) potentially have implications for auroral FUV analyses at Earth and other bodies with nitrogen-dominated atmosphere, such as Titan. In addition, large uncertainties lie in determining the $[O]/[N_2]$ ratio at Earth and the hydrocarbon layer and H column density at Jupiter, which are all very variable in the auroral regions.

One method of addressing these limitations is to employ a multi-spectral approach, which includes the use of more than one color ratio, or combinations of different spectral regions (e.g., EUV, FUV, soft/hard X-rays, IR). Combining the analysis of auroral emissions from different origins provides complementary information to further constrain the auroral diagnostic. Such an approach has been used: (1) to assess characteristics of different particle types. At Jupiter, the regions of bright FUV auroral emissions and hard X-ray spectra have been found to be coincident and consistent in brightness with predicted emissions from a population of energetic electrons, while soft X-ray spectra from heavy precipitating ions are located at the periphery (Branduardi-Raymont et al. 2008). (2) to cover a spectral range for the auroral particles that is larger than that covered with only one technique. The analyses of FUV color ratios and hard X-ray profiles have been combined to derive spectral characteristics of the auroral electrons from a few 100's of eV to a few 100's of keV (e.g., Østgaard et al. 2001). Such an approach is critical for a better understanding of the energy budget and of the response of the atmosphere to particle precipitation (Aksnes et al. 2004, 2006; Sætre et al. 2007). (3) to constrain the atmospheric model used for the analysis (Hecht et al. 2006; Régo et al. 2001).

A cross body comparison, illustrated here through a comparison between Earth and Jupiter, yields a synthetic, and thus more critical view of auroral analysis and ultimately of interactions taking place at different solar system bodies, including solar wind-magnetosphere-ionosphere coupling. Such an approach is extremely relevant to planetary aeronomy (Galand et al. 2006). The optimum cross body comparison is based on multi-spectral observations of two or more solar system bodies responding to similar forcing. Prangé et al. (2004) have given a preview of such an analysis by the comparison of the global auroral response of Earth, Jupiter, and Saturn to an interplanetary shock triggered by a violent solar eruption. Another example, even though not auroral, is the observation of the response of the terrestrial and martian ionospheres to the same solar flares (Mendillo et al. 2006; Mendillo and Withers 2008).

4 Energy Deposition of Pick-up Ions

Heating and loss of the atmosphere due to plasma bombardment occur at Mars, Venus, Pluto and Titan (when it is outside of Saturn's magnetopause), due to impacting solar wind ions and pick-up ions. Such processes also occur at Titan, Triton, Io, and Europa due to impacting magnetospheric plasma and pick-up ions. Atmospheric heating also occurs at Earth, primarily due to magnetospheric O^+ ions that are formed in the corona and re-impact the thermosphere as described below. The incident plasma ions typically have energies of the order of 10's of eV to tens of keV and lose their energy in a very small column of atmosphere of the order of 10^{15} to 10^{17} atoms cm^{-2}. Therefore, even a relatively small energy flux that is deposited close to the nominal exobase ($\sim 10^{15}$ atoms cm^{-2}) (cf. Johnson et al. 2008) can cause heating and escape, processes often lumped together as "atmospheric sputtering".

Atmospheric sputtering affects the long term evolution of these atmospheres. The increased exobase temperatures can enhance Jeans escape by the lightest components, typically H or H_2. In addition, the energy transfer to the atoms in the thermosphere by momentum transfer collisions, often referred to as knock-on collisions, is much less sensitive to the mass of the escaping species and, therefore, can lead to enhanced loss of all atmospheric constituents, which is the atmospheric sputtering process (Johnson 1990, 1994). To estimate this contribution to atmospheric loss, sputtering is often divided into two components: those incident ions which penetrate the exobase and enter the thermosphere and those which only pass through the corona. For the penetrating particles, a cascade of momentum transfer collisions is initiated in the thermosphere by the incident particles. This forms a population of "hot" atoms and molecules. For the other contribution, ejection can occur in single collisions of the incident ion with a coronal molecule and by charge exchange in which the fast ion is replaced by a fast escaping neutral. The relative importance depends on the extent of the corona, the escape energy, and the escape depth from the thermosphere (cf. Johnson et al. 2008).

The composition of the thermosphere near the exobase is often atomic: e.g., O, N, or H atoms. When the composition is molecular, as it the case at Titan and Io for example, an energetic plasma ion transfers energy in close collisions with an individual atom in a molecule. These collisions cause dissociation and produce energetic fragments. Therefore, in both atomic and molecular thermospheres, energetic "secondary atoms" are set in motion. These hot atoms either populate the corona or make further collisions with the constituents of the thermosphere. As the energies of these recoil atoms decrease, whole molecules are eventually set in motion in a molecular thermosphere (Johnson and Liu 1998). This leads to molecular escape, a process which dominates at Io (McGrath and Johnson 1987; Pospieszalska and Johnson 1996; McGrath et al. 2004).

Solar wind ions can have access to the atmospheres of bodies like Venus, Mars and Pluto that are unprotected by a magnetosphere. These are predominantly light ions that cause ionization and charge exchange but are otherwise inefficient sputtering agents. Therefore, atmospheric sputtering is dominated by locally formed ions referred to as pick-up ions (Luhmann et al. 1992; Johnson et al. 2008; Lammer et al. 2008). These ions are formed in the extended corona by electron and photon impact and by charge exchange and are accelerated by the local fields. They can either be swept away, a loss process, or can re-impact the atmospheric exobase causing sputtering. In the case of bodies trapped in a planetary magnetosphere, an additional enhancement can occur. Atoms and molecules that have escaped from the satellites or ring particles and are ionized in the rotating magnetosphere become trapped and accumulate, forming a toroidal plasma in the magnetosphere. This accumulated plasma can enhance the sputtering effect, but also provides an impacting plasma that consists of both light and heavy ions representative of the composition of the atmosphere of a satellite embedded in the magnetosphere.

Such ion acceleration processes also occur in an object's atmosphere. That is, neutrals are similarly ionized by electrons, photons or charge exchange in the object's ionosphere. If the local fields penetrate into the corona, or even below the exobase as seen at Titan and Mars, the newly formed ions can be accelerated by local electric fields. At the Earth the fields are associated with its rotating magnetosphere. At Venus, Mars and Pluto they are the induced fields due to interaction of the ionosphere with the solar wind, and on Io, Titan and Triton, they are due to the induced fields associated with the interaction of the atmosphere with the giant planet's magnetosphere (Ma et al. 2008).

The ions formed in the corona and those formed below the exobase clearly have a composition characteristic of the atmosphere and are called locally produced pick-up ions. Therefore, their detection in the ambient plasma can lead to information on the composition in the exobase region. The ionospheric outflow and pick-up in the corona are typically artificially separated, although the spatial transition from one to the other is smooth. Both sets of ions can result in loss by being swept away down the tail of the interaction region, and they can cause atmospheric sputtering by momentum transfer collisions both in the corona or below the exobase. To date the focus has been on the ions formed and accelerated in the corona. However, the ions dragged out of the atmosphere, often called "ionospheric outflow" (e.g., Fox 1997; Ma and Nagy 2007) are critically important to atmospheric loss and heating at Io (Wilson et al. 2002; Mendillo et al. 2007) and, likely, at Titan (Johnson 2008). Ions that are swept down the tail of the interaction region are an important atmospheric loss process. In addition, depending on their path length in the atmosphere's corona they can cause further ionization by charge exchange. Pick-up ions also make momentum transfer collisions in the atmosphere, both those re-impacting ions formed in the corona and the ionospheric outflow. This populates and expands the corona (Michael and Johnson 2005) producing a complex feedback process. That is, pick-up ions formed in the corona and in the atmosphere can enhance the population of the corona and, thereby, enhance the pick-up ion production and atmospheric loss (Johnson and Luhmann 1998). Approximate models for this complex interaction have been made in which the coronal processes and the ionospheric outflow are separated, but complete simulations are only recently available (Chaufray et al. 2007).

The energy deposited produces hot atoms and molecules both by dissociation or by direct momentum transfer collisions with incident ions and the energetic neutral atoms (ENAs) formed by charge exchange. In this way energized molecular fragments are seen in emission, as discussed earlier, and can initiate a set of collision processes often referred to as a collision cascade. Of particular interest is the energy that goes into producing atmospheric escape and the hot corona (e.g., Michael et al. 2005; Shematovich et al. 2003; Cipriani et al. 2007). The cascades can be described by direct solution of the transport equations, by analytic approximations to the transport equation, or by Monte Carlo simulation. Johnson et al. (2000) have compared such models for a simple O thermosphere energized by an incident O^+ pick-up ion plasma.

4.1 Transport Equations

Consistent with what was described above, a set of kinetic, time-independent Boltzmann equations for the incident particles and the atmospheric particles can in principal be written and solved. There is one such equation for each species i,

$$\mathbf{v}\partial f_i/\partial \mathbf{r} + \mathbf{g}\partial f_i/\partial \mathbf{v} = Q_i - L_i + \sum_j J_{ij}(f_i, f_j). \tag{34}$$

Here the $f_i(\mathbf{r}, \mathbf{v})$ are the distribution functions for the translational and internal degrees of freedom for particles i in the atmosphere, $\mathbf{g}$ is the gravitational acceleration, Q_i and L_i are source and loss functions for species i, and the J_{ij} describe the collisions in the atmosphere (momentum transfer excitations, dissociation, ionization, and charge transfer collisions) between particles i and all other particles j. In 1-D, single component atmospheres, such equations have been solved. Although such equations are often written down for the purpose of discussing the importance of the various processes, they are almost never solved when the spatial distribution of hot recoils is required.

4.2 Analytic Models

A useful approximation to the Boltzmann equation gives the energy distribution of recoils for a single component atmosphere. This is obtained by integrating over the spatial dimensions and ignoring gravity. It has been shown that the total number of recoils with energy between E and $E + dE$ produced by a hot atom of energy E_o colliding in a background atmosphere of identical atoms is

$$G(E_o, E_i) \approx \beta_n E_o/E^2. \tag{35}$$

This is obtained analytically as the lead term in the distribution by assuming an interaction potential which is a power law with power n where β_n is a constant that varies slowly with n. This remarkably simple expression can be used to give the energy distribution of recoil particles moving locally in a planetary thermosphere:

$$f(E) \approx \phi(E_o)G(E_o, E)/\nu_m(E). \tag{36}$$

Here $\phi(E_o)$ is the local rate at which atoms of energy E_o are initially produced by dissociation or momentum transfer collisions by the incident ions and $\nu_m(E)$ is momentum transfer collision frequency at the local density for atoms of energy E. This result can be used to describe the hot atoms in the exobase region and, thereby, give simple expressions for the rates of production of sputtering and of the production of the hot corona. In addition, if the spatial morphology of incident radiation is known such expressions can be used to obtain the spatial structure of the hot corona.

These simple expressions were compared to atmospheric loss simulations for an O thermosphere and shown to be accurate (Johnson et al. 2000). Cipriani et al. (2007) simulated a more complex Martian atmosphere, involving both atoms and molecules excited by dissociative recombination and ion sputtering, using Monte Carlo simulations described below. They compared their steady-state recoil distribution to the expression for the total number of recoils produced, $G(E_o, E)$ rather than $f(E)$, and concluded the analytic model was only approximate. However, their steady-state recoil distributions produced by either dissociation or ion-induced sputtering compare well with the correct expression, $f(E)$ (Johnson 2008). Remarkably, the analytic expression for $f(E)$ compares well even for the steady-state distribution of molecular recoils found by Cipriani et al. (2007).

4.3 Monte Carlo Simulations

Since the above transport equations are usually not solved, simulations that are equivalent to such solutions are carried out. There are two principal types of simulations and sometimes a combination of these is used (Leblanc and Johnson 2001). 1-D, 2-D and 3-D versions have been implemented depending on the aspect being described. The most common model is a test particle simulation in which the incident radiation sets representative atoms and molecules in motion. These hot atoms and molecules, also called primary recoils, are tracked as they move under the influence of gravity in a background atmosphere obtained either from observations or a model. The hot recoils interact with the background gas causing additional excitations and momentum transfer collisions producing additional recoils. In this cascade of collisions the deposited energy can be transported away from the initial energy deposition site and a distribution of hot particles is produced. In this way the collisional energy deposited by each incident ion is rapidly dispersed. Whereas the transport equations and other approximations above typically assume some steady state for a given incoming flux, in the Monte Carlo simulations the fate of the energy deposited by sample ions, called test particles, is tracked (e.g., Shematovich et al. 1994).

The test particles are assigned a weight such that the net production rate represents the hot particle source profiles described earlier. The initially produced hot particles and their recoils are tracked until they fall below some energy cut-off after which they are assumed to be part of the background atmosphere. The size of this energy cut-off depends on what aspect of the atmosphere is being modeled, the escape rate, the structure of the corona, or the local heating, in which case the energy must be tracked to some value related to the ambient temperature.

The representative particles are tracked ballistically for time steps which are short enough that the collision probability is small. At the end of the time step the velocities of the particles are changed to account for the outcome of those collisions that occurred. These new velocities are used as the initial velocities for the next time step. Because the collision probability along the path of a hot particle and the outcome of each collision are typically chosen using a Monte Carlo procedure, these are referred to as Monte Carlo simulations.

Whereas only hot particles are tracked in the test particle method, in the second type of simulation the background atmosphere is allowed to evolve in response to the energy deposition by hot particle production (e.g., Michael and Johnson 2005). Therefore, both representative hot particles and representative atmospheric molecules are tracked. In such simulations collisions are allowed to occur between hot particles, between hot particles and atmospheric particles, and between atmospheric particles, each having weights that allow the representative particles to describe both the source distribution and the atmospheric structure and temperature. Because such simulations are computationally intensive, the computational zone has a lower boundary at which temperatures and densities are assigned. These are made to be consistent with the properties found by continuum models of the atmosphere at that depth. The lower boundary typically occurs at a depth at which the energy deposition can be neglected or treated by a continuous slowing down (CSDA) model. The upper boundary is typically where collisions are improbable and the hot particles are in ballistic orbits. Particles reaching the upper boundary are tested and either escape and are removed or are allowed to re-enter the atmosphere as appropriate to their ballistic trajectory.

Since the atmosphere is allowed to respond to the energy deposition in such simulations, care must be taken to account for energy loss processes other than conduction to the lower boundary which is automatically included. Following Bird (1994) these are often referred to as Direct Simulation Monte Carlo (DSMC) models.

In both types of simulations, the weights for different species in a multi-component atmosphere can be different, so that trace species can be accurately accounted for. In addition, the atmosphere is divided into cells in which its properties do not change significantly, so that random particles in each cell can be used to describe the outcome of the collisions. Most simulations use the method in Bird (1994) (e.g., Leblanc and Johnson 2001; Michael and Johnson 2005) to describe the collision probabilities based on the weights and densities of each type of particle. The accuracy and computational time are optimized by the choice of collision model, weights, time step, cell sizes and the simulation type used in each region of the atmosphere. For instance, in describing the transition from the thermosphere to the exosphere, weights that are different in different regions are often used as well as different simulation models which are coupled at the boundary between the regions. Such models have been used to directly produce the corona and to test the simpler analytic models (cf., Johnson et al. 2008; Lammer et al. 2008).

4.4 Energetic Particle Deposition

Based on the discussions above, there are a number of effects produced by the incident radiation: auroral emissions, an ionosphere, thermospheric heating, sputter loss, etc. These effects typically peak at very different depths in the atmosphere; therefore the relevance of the various radiation types depends both on the slant depth into the atmosphere as well as the energy flux for each type of incident radiation. The quantity of interest is the energy deposition per unit volume per unit time. This is the radiation flux, $\phi(z)$ at each altitude, z, times the energy deposited per unit path length by the incident particles. The latter quantity is called the stopping power of the medium and is typically written as (dE/dx) where x is along the path of the incident particles. The dependence of ϕ on depth and the redistribution of the deposited energy is determined by transport processes. Here the focus is on the energy deposition by different charged particle types via (dE/dx) and the penetration depths. Comparisons with the effect of the photon flux (e.g., Michael and Johnson 2005; Johnson 2008) are hampered by the fact that dE/dx for energetic ions and electrons are not very sensitive to the atmospheric composition, but the energy deposition by the solar flux UV and EUV flux, as well as the low energy electrons (≤ 100 eV) are very specific to the composition.

The energy lost per unit path length depends on the density of the atmosphere, and thus one often writes $dE/dx = \sum_i (n_i S_i)$, where n_i is the local number density for atmospheric molecules i and S_i is called the stopping cross section for the incident particle by molecules of type i. For incident ions S is divided into qualitatively different types of energy

Table 4 W-values (in eV) and ratios W/I

Gas	w	W/I
He	44	1.8
Ar	27	1.7
H_2	36	2.3
N_2	37	2.4
O_2	31	2.6
CO	34	2.4
CO_2	34	2.5
H_2O	30	2.4
NH_3	26	2.5
CH_4	31	2.5
C_2H_6	25	2.1
C_2H_4	26	2.5
C_2H_2	26	2.3
SO_2	32	2.6

transfer, as mentioned earlier: momentum transfer to the atmospheric atoms and molecules, often referred to as the elastic nuclear component, S_n, which also scatters the incident ions, and energy transfer to excitations and ionizations of the atmospheric atoms and molecules, S_e. Incident electrons are also scattered, especially at low energies, but the direct energy/momentum transfer to atomic motion is typically negligible.

Because the amount of energy transferred depends on the nuclear charge of the molecules, which is nearly proportional to its mass and equals the number of electrons available for excitation, the stopping power is often divided by the mass density, ρ. This gives a quantity that for fast ions and electrons is nearly material independent: $(dE/dx)/\rho$. Therefore, in Fig. 16 we compare the stopping powers of an electron, proton and O^+ ion in H_2O as a function of the incident velocity, v, given as the kinetic energy per unit mass. Such results can be obtained for a variety of incident ion/target combinations using freeware programs such as SRIM (http://www.srim.org/). It is seen that at velocities above the Bohr velocity (2.18×10^8 cm s^{-1}), the variation with velocity for each incident particle is similar, until one reaches very high velocities (e.g., $\sim$ a fraction of c or a few MeV/u) where radiation losses come into play. That is, the amount of energy deposited in the target electrons and ions is determined primarily by the speed of the charged particles over a broad range of relevant energies (e.g., Johnson 1990). Also shown in Fig. 17 is the projected range, or penetration depth, of these particles versus velocity, again given as their initial energy per unit mass. Because the atmospheric density versus altitude depends on the local gravity and temperature, the range is also multiplied by the mass density, giving the atmospheric mass per unit area penetrated by the particles. This is a quantity that is again nearly independent of atmospheric type.

There have been numerous studies of the fraction of energy deposited by fast ions and electrons in electronic excitations and ionizations (e.g., Paretzke 1989). The latter process, like photoionization, produces secondary electrons (Long and Paretzke 1991). As a rule, fast incident charged particles expend $\sim 50-60$ eV per ionization, so that the secondary electrons carry off a considerable fraction of the deposited energy. Since these electrons can also ionize and excite atmospheric molecules, the average energy deposited per ionization produced by either the incident particle or a secondary electron is called the W-value. Quite

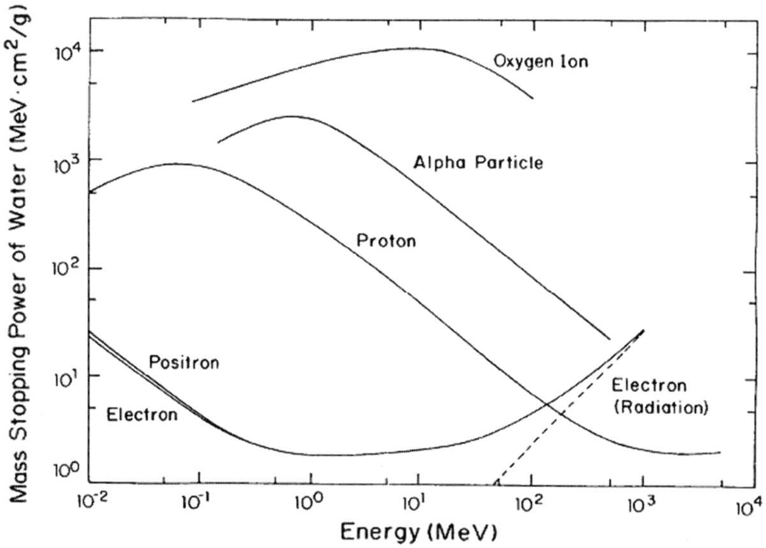

Fig. 16 Stopping power (dE/dx) of a material versus ion energy. Here water vapor is used and the stopping power is given as an energy loss per mass column density penetrated. For the energy range shown this quantity is nearly independent of material, i.e., multiply by the mass density of the material to obtain energy loss per unit path length (from Johnson 1990)

Fig. 17 Range (distance an incident ion travels in a material) versus ion energy. Range is given as a mass column density penetrated; the material is again water vapor, but is similar for most atmospheric compositions, i.e., divide by mass density to get distance traveled

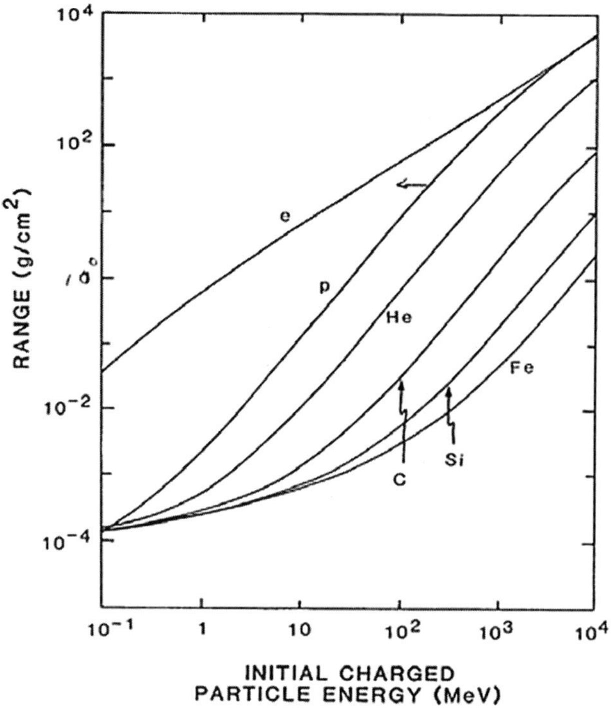

remarkably this is nearly constant for fast ions and electrons, as is the fraction of the deposited energy going into excitations. Values for relevant gases are given in Table 4. It is seen that the W for a molecular atmosphere is $\sim 2.5\,I$, so the fast ions and electrons expend, on the average, more energy per ionization produced than is typical for EUV-UV photons described earlier. In a molecular atmosphere, the fraction of the deposited energy going into electronic excited states is $\sim 10\%$ and the energy dissipated by the secondary electrons in elastic collisions and vibrational excitations is $\sim 50\%$ of the total energy deposited. In addition, the ionizations produced by the incident plasma ions and electrons are distributed spatially very differently than are those produced by incident UV and EUV photons.

5 Summary and Conclusions

We have discussed here the energy deposition of solar photons and charged particles of various origins, with emphasis on problems of current interest. For photons, these include determining EUV and FUV heating efficiencies, for which the principal unknown is the fraction of energy that appears as vibrational excitation of molecules in various processes. These heating processes occur in competition with other external perturbations such as meteoroid bombardment. We have discussed the validity of Chapman layer theory. By comparison of a Chapman layer to a detailed, realistic model, we have shown that more elaborately modeled electron density profiles bear little resemblance to Chapman layers.

We have shown that modeling the photodissociation rates of many thermospheric molecules is difficult for several reasons. For photodissociations that proceed by excitation to discrete states followed by predissociation, the cross sections must be known to very high resolution, of the order of 10^{-3} Å. For most atmospheric molecules, there are wavelength regions in which the cross sections are highly oscillatory on such wavelength scales. Branching ratios for the production of various electronically excited states of the products of dissociation are also required, and are not generally known. Immediately shortward of the ionization thresholds of molecules, the photoabsorption cross sections are usually highly structured, and autoionization competes with predissociation. In this region, the photodissociation cross sections need to be determined as the difference between the photoabsorption cross sections and the photoionization cross sections, a process that is difficult and usually produces only approximations.

We have also discussed the interaction of X-rays with atmospheric species, in particular the Auger effect, and the production of characteristic X-rays from atoms and molecules. We have described the Auger effect in detail for atoms for O and that for molecules for N_2. Auger electrons have been observed or predicted in photoelectron flux spectra from several planetary bodies, as have X-rays of various origins. The Auger process and double ionization of valence shell electrons produce doubly ionized species of uncertain lifetime. Recent measurements have provided a basis for the construction of approximate models of such species. Much more data is needed, however, in order to accurately model their density profiles, especially for doubly ionized molecules.

Analyses of emissions are valuable for remote-sensing of auroral particles, which precipitate into the upper atmosphere of a solar system body, where they deposit their energy. Spectroscopic analyses of auroral emissions based on color ratios and spectral profiles have been used to identify the type of precipitating particles and to determine their characteristics in terms of total energy flux and mean energy or spectral shape. This information is crucial for identifying the origin of the auroral particles and the associated forcing, for assessing the global response of a planetary upper atmosphere to auroral particles, and to infer the ionospheric contribution to magnetosphere–ionosphere coupling. Based on a cross body comparison, we have highlighted the differences in applying similar spectroscopic techniques at different solar system bodies. The auroral particle energy range covered by the analysis, the type of aurora which can be probed, and the physical quantities derived from the analysis vary from one solar system body to another. Multi-spectral analyses of auroral emissions have been employed for increasing the auroral particle energy range covered by only one technique, for determining the characteristics of more than one auroral particle type, and for constraining the atmospheric model used for the analysis.

The upper atmospheres of planets and planetary satellites are often affected by energetic ions and electrons, either locally created or from an ambient plasma. Since these ions and electrons are in turn affected by the local fields, there are complex feedback processes that affect the atmospheric heating and loss rates. Whereas the effect of solar photons on the upper atmosphere has been studied for years, only recently has accurate modeling of the plasma interactions been carried out, as discussed in the other articles of this volume. In addition, it is only very recently that we have had good spacecraft measurements of thermospheres other than that of the Earth to test such models.

Assuming one knows the local plasma flux, in this article we described briefly how one calculates the energy deposition rate. As is the case for photon-energy deposition, our ability to describe the effect of the incident and locally produced plasma particles is limited by the available atomic and molecular data. However, as pointed out in the text, for fast ions and electrons the energy deposition rates are much less sensitive to the atmospheric composition than is the case for incident photon or low energy incident or secondary electrons. When the energy deposition by the energetic plasma particles can be calculated, then the subsequent modeling of the atmospheric processes is roughly independent of the exciting radiation. That is, the subsequent processes are very similar to those initiated by photo-absorption. We also pointed out the similarity in the effect of hot recoils on atmospheric heating, whether the recoils are produced by incident ions or exothermic chemistry. Therefore, the principal limiting factor in our ability to describe the heating of the upper atmosphere by the incident plasma is our knowledge, from spacecraft measurements or modeling efforts, of the morphology of the plasma flow into the atmosphere and our knowledge of the fields that penetrate the atmosphere, issues discussed in later articles in this volume.

Acknowledgements JLF acknowledges partial support from the NASA Planetary Atmospheres Program under grant NNG06GF12G and from the NASA Mars Fundamental Research Program under grant NNG05GL82G to Wright State University. REJ acknowledges support from NASA's Planetary Atmospheres Program under grant NNG06GC09G to the University of Virginia.

References

H. Abgrall, E. Roueff, X. Liu, D.E. Shemansky, Astrophys. J. **481**, 557–566 (1997)

K. Ågren, J.-E. Wahlund, R. Modolo, D. Lummerzheim, M. Galand, I. Müller-Wodarg et al., Ann. Geophys. **25**, 2359–2369 (2007)

J.M. Ajello, D.E. Shemansky, W.R. Pryor, A.I. Stewart, K.E. Simmons, T. Majeed, J.H. Waite, G.R. Gladstone, D. Grodent, Icarus **152**, 151–171 (2001). doi:10.1006/icar.2001.6619

J.M. Ajello, W. Pryor, L. Esposito, A.I. Stewart, W. McClintock, J. Gustin, D. Grodent, J.-C. Gerard, T. Clarke, Icarus **178**, 327 (2005).

A. Aksnes, J. Stadsnes, G. Lu, N. Østgaard, R.R. Vondrak, D.L. Detrick, T.J. Rosenberg, G.A. Germany, M. Schulz, Ann. Geophys. **22**, 475–496 (2004)

A. Aksnes, J. Stadsnes, N. Østgaard, G.A. Germany, K. Oksavik, R.R. Vondrak, A. Brekke, U.P. Løvhaug, J. Geophys. Res. **111**, A02301 (2006). doi:10.1029/2005JA011331

A.C. Allison, S.L. Guberman, A. Dalgarno, J. Geophys. Res. **87**, 923–925 (1982)

A.C. Allison, S.L. Guberman, A. Dalgarno, J. Geophys. Res. **91**, 10,193–10,198 (1986)
A.D. Anbar, M. Allen, H.A. Nair, J. Geophys. Res. **98**, 10,925–10,931 (1993)
S.K. Atreya, M.H. Wong, T.C. Owen, P.R. Mahaffy, H.B. Niemann, I. de Pater, P. Drossart, Th. Encrenaz, Planet. Space Sci. **47**, 1243–1262 (1999)
S.V. Avakyan, Kosm. Issl. **16**, 144–148 (1978)
S.V. Avakyan, Cosmic Res. **17**, 786–789 (1980)
S.M. Bailey, T.N. Woods, C.A. Barth, S.C. Solomon, L.R. Canfield, R. Korde, J. Geophys. Res. **105**, 27,179–27,193 (2000)
K.H. Baines, H.B. Hammel, K.A. Rages, P.N. Romani, R.S. Samuelson, in *Neptune and Triton*, ed. by D.P. Cruikshank (University of Arizona Press, Tucson, 1995)
N. Balakrishnan, M.J. Jamieson, A. Dalgarno, Y. Li, R.J. Buenker, J. Chem. Phys. **112**, 1255–1259 (2000)
P.M. Banks, G. Kockarts, *Aeronomy* (Academic Press, New York, 1973)
B. Basu, D.T. Decker, J.R. Jasperse, J. Geophys. Res. **106**, 93–105 (2001)
S.J. Bauer, *Physics of Planetary Ionospheres* (Springer, New York, 1973)
S.J. Bauer, H. Lammer, *Planetary Aeronomy* (Springer, Berlin, 2004)
M. Benndorf, W.B. Westerveld, J.V. Eck, J.V.D. Weg, H.G.M. Heideman, Chem. Phys. Lett. **286**, 321–328 (1998)
J. Berkowitz, *Atomic and Molecular Photoabsorption* (Academic Press, San Diego, 2002)
A. Bhardwaj, G.R. Gladstone, Rev. Geophys. **38**, 295–353 (2000)
A. Bhardwaj, G. Branduardi-Raymont, R.F. Elsner, G.R. Gladstone, G. Ramsay, P. Rodriguez, R. Soria, J.H. Waite Jr., T.E. Cravens, Geophys. Res. Lett. **32**, L03S08 (2005a). doi:10.1029/2004GL021497
A. Bhardwaj, R.F. Elsner, J.H. Waite Jr., G.R. Gladstone, T.E. Cravens, P.F. Ford, Astrophys. J. **627**, L73–L76 (2005b)
A. Bhardwaj, R. Gladstone, R. Elsner, N. Østgaard, H. Waite, T. Cravens, S.-W. Chang, T. Majeed, A. Metzger, J. Atmos. Sol.-Terr. Phys. **69**, 179–187 (2007b). doi:10.1016/j.jastp.2006.07.011
A. Bhardwaj et al., Planet. Space Sci. **55**, 1135–1189 (2007a). doi:10.1016/j.Planet.SpaceSci.2006.11.009
G.A. Bird, *Molecular Gas Dynamics and the Direct Simulation of Gas Flows* (Clarendon, Oxford, England, 1994)
J.H. Black, E.F. van Dishoeck, Astrophys. J. **322**, 412–449 (1987)
S.W. Bougher, R.G. Roble, E.C. Ridley, R.E. Dickinson, J. Geophys. Res. **95**, 14,811–14,827 (1990)
S.W. Bougher, P.-L. Blelly, M. Combi, J.L. Fox, I. Mueller-Wodarg, A. Ridley, R.G. Roble, Space Sci. Rev. (2008, this issue)
G. Branduardi-Raymont, A. Bhardwaj, R.F. Elsner, G.R. Gladstone, G. Ramsay, P. Rodriguez, R. Soria, J.H. Waite Jr., T.E. Cravens, Astron. Astrophys. **463**, 761–774 (2007). doi:10.1051/00046361:20066406
G. Branduardi-Raymont, R.F. Elsner, M. Galand, D. Grodent, T.E. Cravens, P. Ford, G.R. Gladstone, J.H. Waite Jr., J. Geophys. Res. **113**, A02202 (2008). doi:10.1029/2007JA012600
D.H. Brautigam, M.S. Gussenhoven, D.A. Hardy, J. Geophys. Res. **96**(A4), 5525–5538 (1991)
B. Brehm, G. de Frênes, Int. J. Mass Spectrom. Ion Proc., 251–266 (1978)
E.L. Breig, M.R. Torr, D.G. Torr, W.B. Hanson, J.H. Hoffman, J.C.G. Walker, A.O. Nier, J. Geophys. Res. **82**, 1008–1012 (1977)
E.L. Breig, M.R. Torr, D.C. Kayser, J. Geophys. Res. **87**, 7653–7665 (1982)
C.D. Caldwell, M.O. Krause, Phys. Rev. A **47**, R759–R762 (1993)
R.W. Carlson, J. Chem. Phys. **60**, 2350–2352 (1974)
S.L. Carter, H.P. Kelly, J. Phys. B **9**, 1887–1897 (1976)
J.W. Chamberlain, D.M. Hunten, *Theory of Planetary Atmospheres* (Academic Press, Orlando, 1987)
S. Chapman, Proc. Phys. Soc. Lond. **43**, 26–45 (1931a)
S. Chapman, Proc. Phys. Soc. Lond. **43**, 483–501 (1931b)
B.K. Chatterjee, R. Johnsen, J. Chem. Phys. **91**, 1378–1379 (1989)
J.Y. Chaufray, R. Modolo, F. Leblanc, G. Chanteur, R.E. Johnson, J.G. Luhmann, J. Geophys. Res. **112**, E09009 (2007). doi:10.1029/2007JE002915
M.W. Chen, M. Schulz, P.C. Anderson, G. Lu, G. Germany, M. Wüest, J. Geophys. Res. **110**, A03210 (2005). doi:10.1029/2004JA010725
A.S.-C. Cheung, K. Yoshino, J.R. Esmond, W.H. Parkinson, J. Mol. Spectrosc. **178**, 66–77 (1996)
F. Cipriani, F. Leblanc, J.J. Berthelier, J. Geophys. Res. **E07001**, (2007). doi:10.1029/2006JE00281
E.U. Condon, G.H. Shortley, *Theory of Atomic Spectra* (Cambridge University Press, Cambridge, 1964)
B. Coquert, M.F. Merienne, A. Jenouvrier, Planet. Space. Sci. **38**, 287–300 (1990)
P.C. Cosby, R. Möller, H. Helm, Phys. Rev. A **28**, 766–722 (1983)
B. Craseman, in *Atomic, Molecular and Optical Physics Handbook*, 2nd edn. ed. by G.W.F. Drake (American Institute of Physics Press, Woodbury, 2006), pp. 915–928
T.E. Cravens, Science **296**, 1042–1045 (2002)
T.E. Cravens, E. Howell, J.H. Waite Jr., G.R. Gladstone, J. Geophys. Res. **100**, 17,153–17,161 (1995)

T.E. Cravens, J. Vann, J. Clark, J. Yu, C.N. Keller, C. Brull, Adv. Space Res. **33**, 212–215 (2004)
T.E. Cravens, T.E. Cravens, I.P. Robertson, J. Clark, J.-E. Wahlund, J.H. Waite Jr., S.A. Ledvina, H.B. Niemann, R.V. Yelle, W.T. Kasprzak, J.G. Luhmann, R.L. McNutt, W.H. Ip, V. De La Haye, I. Mueller-Wodarg, D.T. Young, A.J. Coates, Geophys. Res. Lett. **32**, L12108 (2005). doi:10.1029/2005GL023249
J.M. Curtis, R.K. Boyd, J. Chem. Phys. **80**, 1150–1161 (1984)
A. Dalgarno, G.H. Herzberg, T.L. Stevens, Astrophys J. **162**, L47 (1970)
I. de Pater, J.J. Lissauer, *Planetary Sciences* (Cambridge University Press, Cambridge, 2001)
I. de Pater, S.T. Massie, Icarus **62**, 143–171 (1985)
A. Dalgarno, H.R. Sadeghpour, Phys. Rev. A **46**, R3591–R3593 (1992)
D.R. Denne, J. Phys. D **3**, 1392–1398 (1970)
K. Dennerl, V. Astron. Astrophys. **386**, 319–330 (2002)
K. Dennerl, V. Burwitz, J. Englhauser, C. Lisse, S. Wolke, Astron. Astrophys. **386**, 319–330 (2002)
R.A. Doe, J.D. Kelly, D. Lummerzheim, G.K. Parks, M.J. Brittnacher, G.A. Germany, J. Spann, Geophys. Res. Lett. **24**(8), 999–1002 (1997)
D.P. Drob, R.R. Meier, J.M. Picone, D.J. Strickland, R.J. Cox, A.C. Nicholas, J. Geophys. Res. **104**, 4267 (1999)
G. Dujardin, M.J. Besnard, L. Heller, Y. Malinovich, Phys. Rev. A **35**, 5012–5019 (1987)
R.H. Eather, Rev. Geophys. **5**, 207–285 (1967)
A.K. Edwards, R.M. Wood, J. Chem. Phys. **76**, 2938–2942 (1982)
A. Ehresmann, S. Machida, M. Kitayama, M. Ukai, K. Kameta, N. Kouichi, Y. Hatano, E. Shigemasa, T. Hayashi, J. Phys. B **33**, 473–490 (2000)
R.F. Elsner et al., J. Geophys. Res. **110**, A10207 (2005). doi:10.1029/2004JA010717
L.W. Esposito, R.G. Knollenberg, M. Ya Marov, O.B. Toon, R.P. Turco, in *Venus* (University of Arizona Press, Tucson, 1983)
L.W. Esposito, J.-L. Bertaux, V. Krasnopolsky, V.I. Moroz, L.V. Zasova, in *Venus II*, ed. by S.W. Bougher, D.M. Hunten, R.J. Phillips (University of Arizona Press, Tucson, 1997)
E. Fainelli, R. Maracci, L. Avaldi, J. Electron. Spectrosc. Rad. Transf. **123**, 277–286 (2002)
X. Fang, M.W. Liemohn, J.U. Kozyra, S.C. Solomon, J. Geophys. Res. **110**, A07302 (2005). doi:10.1029/2004JA010915
M.O. Fillingim, G.K. Parks, H.U. Frey, T.J. Immel, S.B. Mende, Geophys. Res. Lett. **32**, L03113 (2005). doi:10.1029/2004GL021635
J.L. Fox, Planet. Space Sci. **36**, 37–46 (1988)
J.L. Fox, Geophys. Res. Lett. **24**, 2901–2904 (1997)
J.L. Fox, J. Geophys. Res. **109**, A11310 (2004). doi:10.1029/2004JA010380
J.L. Fox, Icarus 192 (2007). doi:10.1016/j.icarus2007.05.02
J.L. Fox, J.H. Black, Geophys. Res. Lett. **16**, 291–294 (1989)
J.L. Fox, A. Dalgarno, J. Geophys. Res. **84**, 7315–7333 (1979)
J.L. Fox, A. Dalgarno, J. Geophys. Res. **86**, 629–639 (1981)
J.L. Fox, A.I.F. Stewart, Geophys. Res. **96**(A6), 9821–9828 (1991)
J.L. Fox, G.A. Victor, J. Geophys. Res. **86**, 2438–2442 (1981)
J.L. Fox, P. Zhou, S.W. Bougher, Adv. Space Res. **17**, (11)203–(11)218 (1995)
R. Franceschi, R. Thissen, J. Žabka, J. Roithová, Z. Herman, O. Dutuit, Int. J. Mass Spectrom. Ion Proc. **228**, 507–516 (2003)
H.U. Frey, Rev. Geophys. **45**, RG1003 (2007). doi:10.1029/2005RG000174
H.U. Frey, S.B. Mende, T.J. Immel, S.A. Fuselier, E.S. Claflin, J.-C. Gérard, B. Hubert, J. Geophys. Res. **107**(A7), 1091 (2002). doi:10.1029/2001JA900161
T.J. Fuller-Rowell, D.S. Evans, J. Geophys. Res. **92**, 7606 (1987)
M. Galand, S. Chakrabarti, J. Atmos. Terr. Phys. **68**, 1488–1501 (2006). doi:10.1016/jastp.2005.04.013
M. Galand, S. Chakrabarti, in *Atmospheres in the Solar System: Comparative Aeronomy*, ed. by M. Mendillo, A. Nagy, H. Waite, AGU Monograph, vol. 130 (AGU Press, 2002), pp. 55–76
M. Galand, D. Lummerzheim, J. Geophys. Res. **109**(A3), A03307 (2004). doi:10.1029/2003JA010321
M. Galand, J. Lilensten, W. Kofman, R.B. Sidje, J. Geophys. Res. **102**, 22,261–22,272 (1997)
M. Galand, R. Roble, D. Lummerzheim, J. Geophys. Res. **104**, 27,973–27,990 (1999)
M. Galand, T.J. Fuller-Rowell, M.V. Codrescu, J. Geophys. Res. **106**, 127–139 (2001)
M. Galand, D. Lummerzheim, A.W. Stephan, B.C. Bush, S. Chakrabarti, J. Geophys. Res. **107**(A7) (2002). doi:10.1029/2001JA000235
M. Galand, A. Bhardwaj, S. Chakrabarti, in *Advances in Geosciences*, vol. 2: Solar Terrestrial (ST) Science. Editor-in-Chief Wing-Huen Ip, Volume Editor-in-Chief Marc Duldig (2006), pp. 239–248
L. Gan, C.N. Keller, T.E. Cravens, Geophys. Res. **97**, 12,137–12,151 (1992)
J.-C. Gérard, D. Grodent, J. Gustin, A. Saglam, J.T. Clarke, J.T. Trauger, J. Geophys. Res. **109**, A09207 (2004). doi:10.1029/2004JA010513

G.A. Germany, G.K. Parks, M. Brittnacher, J. Cumnock, D. Lummerzheim, J.F. Spann, L. Chen, P.G. Richards, F.J. Rich, Geophys. Res. Lett. **24**, 995 (1997)
T. Gilbert, T.L. Grebner, I. Fischer, P. Chen, J. Chem. Phys. **110**, 5485–5488 (1999)
G.R. Gladstone, J. Geophys. Res. **97**, 1377–1387 (1992)
G.R. Gladstone, M. Allen, Y.L. Yung, Icarus **119**, 1–52 (1996)
G.R. Gladstone, S.A. Stern, D.C. Slater, M. Versteeg, M.W. Davis, K.D. Retherford, L.A. Young, A.J. Steffl, H. Throop, J.W. Parker, H.A. Weaver, A.F. Cheng, G.S. Orton, J.T. Clarke, J.D. Nichols, Science **318**, 229 (2007). doi:10.1126/science.1147613
M. Glass-Maujean, Phys. Rev. A **33**, 342–345 (1986)
M. Glass-Maujean, S. Klumpp, L. Werner, A. Ehresmann, H. Schmoranzer, J. Chem. Phys. **126**, 144303 (2007)
D. Grodent, J.H. Waite, J.-C. Gérard, J. Geophys. Res. **106**, 12,933 (2001)
G. Gronoff, J. Lilensten, C. Simon, O. Witasse, R. Thissen, O. Dutuit, C. Alcaraz, Astron. Astrophys. **465**, 641–645 (2007)
E.M. Gurrola, E.A. Marouf, V.R. Eshleman, G.L. Tyler, P.A. Rosen, in *Abstract Book* (AZ, Tucson, 1992), p. 28
J. Gustin, J.-C. Gerard, D. Grodent, S.W.H. Cowley, J.T. Clarke, A. Grard, J. Geophys. Res. **109**, A10205 (2004). doi:10.1029/2003JA010365
D.A. Hardy, M.S. Gussenhoven, D. Brautigam, J. Geophys. Res. **94**, 370–392 (1989)
J.H. Hecht, D.J. Strickland, M.G. Conde, J. Atmos. Sol.-Terr. Phys. **68**, 1502–1519 (2006). doi:10.1016/j.jastp.2005.06.022
H. Helm, P.C. Cosby, J. Chem. Phys. **90**, 4208–4215 (1989)
B.L. Henke, P. Lee, J. Tanaka, R.L. Shimabukuro, B.K. Fujikawa, Atomic Data Nucl. Data Tables **27**, 1–144 (1982)
B.L. Henke, E.M. Gullikson, J.C. Davis, Atomic Data Nucl. Data Tables **54**, 181–342 (1993)
L. Heroux, H.E. Hinteregger, J. Geophys. Res. **83**, 5305–5308 (1978)
G. Herzberg, *Spectra of Diatomic Molecules* (van Nostrand Reinhold, New York, 1950)
D.P. Hinson, J.D. Twicken, E.T. Karayel, J. Geophys. Res. **103**, 9505–9520 (1998)
H.E. Hinteregger, K. Fukui, B.R. Gibson, Geophys. Res. Lett. **8**, 1147–1150 (1981)
P. Honvault, M. Gargaud, M.C. Bacchus-Montabonel, R. McCarroll, Astron. Astrophys. **302**, 931–934 (1995)
F. Howorka, A.A. Viggiano, D.L. Albritton, E.E. Ferguson, F.C. Fehsenfeld, J. Geophys. Res. **84**, 5941–5942 (1979)
S. Hsieh, J.H.D. Eland, J. Phys. B **29**, 5795–5809 (1996)
J.H. Hubbell, W.J. Veigele, E.A. Briggs, R.T. Brown, D.T. Cromer, R.J. Howerron, J. Phys. Chem. Ref. Data **4**, 471–538 (1975)
D.L. Huestis, J. Quant. Spectrosc. Rad. Trans. **69**, 709–721 (2001)
D.L. Huestis et al., Space Sci. Rev. (2008, this issue)
T.J. Immel, S.B. Mende, H.U. Frey, L.M. Peticolas, C.W. Carlson, J.-C. Gérard, B. Hubert, S.A. Fuselier, J.L. Burch, Geophys. Res. Lett. **29**(11), 1519 (2002). doi:10.1029/2001GL013847
G. Israel et al., Nature **438** (2005), doi:10.1038/nature04349
B.M. Jakosky, R.M. Haberle, in *Mars*, ed. by H.H. Kieffer, B.M. Jakosky, C.W. Snyder, M.S. Matthews (University of Arizona Press, Tucson, 1992), pp. 969–1016
E. Jannitti, P. Nicoloso, G. Tondello, Physica Scripta **41**, 458–463 (1990)
R.E. Johnson, *Energetic Charged Particle Interactions with Atmospheres and Surfaces* (Springer, Berlin, 1990)
R.E. Johnson, Space Sci. Revs. **69**, 215–253 (1994)
R.E. Johnson, Proc. Royal Soc. (Lond.) (2008, in press)
R.E. Johnson, M. Liu, J. Geophys. Res. **103**, 3639–3647 (1998)
R.E. Johnson, J.G. Luhmann, J. Geophys. Res. **103**, 3649–3653 (1998)
P.V. Johnson, D. Schnellenberger, M.C. Wong, J. Geophys. Res. **105**, 1659–1670 (2000)
P.V. Johnson, C.P. Malone, I. Kanik, K. Tran, M.A. Khakoo, J. Geophys. Res. **110**, A11311 (2005). doi:10.1029/2005JA011295
R.E. Johnson, M.R. Combi, J.L. Fox, W.-H. Ip, F. Leblanc, M.A. McGrath, V.I. Shematovich, D.F. Strobel, J.H. Waite, Space Sci. Rev. (2008, this issue)
A.L. Jones, A.J. Blake, L. Torop, D.G. McCoy, Chem. Phys. **211**, 291–297 (1996)
R.T. Jongma, G. Berden, G. Meijer, J. Chem. Phys. **107**, 7034–7040 (1997)
P.S. Julienne, D. Neumann, M. Krauss, J. Chem. Phys. **64**, 2990–2996 (1976)
R.A. Kahn, T.Z. Martin, R.W. Zurek, S.W. Lee, in *Mars*, ed. by H.H. Kieffer, B.M. Jakosky, C.W. Snyder, M.S. Matthews (University of Arizona Press, Tucson, 1992), pp. 1017–1053
E. Kallio, S. Barabash, J. Geophys. Res. **106**, 165–177 (2001)

E. Kallio, J.G. Luhmann, S. Barabash, Charge exchange near Mars. J. Geophys. Res. **102**, 22,183–22,198 (1997)
C.N. Keller, T.E. Cravens, L. Gan, J. Geophys. Res. **97**, 12,117–12,135 (1992)
V. Kharchenko, A. Dalgarno, D.R. Schultz, P.C. Stancil, Geophys. Res. Lett. **33**, L11105 (2006). doi:10.1029/2006GL026039
Y.H. Kim, J.L. Fox, Geophys. Res. Lett. **18**, 123 (1991)
Y.H. Kim, J.L. Fox, Icarus **112**, 310–325 (1994)
J.H. Kim, S.J. Kim, T.R. Geballe, S.S. Kim, L.R. Brown, Icarus **185**, 476–486 (2006)
M. Kimura, Chem. Phys. Lett. **200**, 524–527 (1992)
H.R. Koslowski, H. Lebuis, V. Staemmler, R. Fink, J. Phys. B **24**, 5023–5034 (1991)
V.A. Krasnopolsky, D.P. Cruikshank, J. Geophys. Res. **100**, 21,271–21,286 (1995)
V.A. Krasnopolsky, B.R. Sandel, R. Herbert, R.J. Vervack, J. Geophys. Res. **98**, 3065–3078 (1993)
M.O. Krause, J. Phys. Chem. Ref. Data **8**, 307–327 (1979)
G.S. Kudryashev, S.V. Avakyan, Phys. Chem. Earth (C) **25**, 511–514 (2000)
J. La Coursiere, S.A. Meyer, G.W. Faris, T.G. Slanger, B.R. Lewis, S.T. Gibson, J. Chem. Phys. **110**, 1949–1958 (1999)
H. Lammer, J.F. Kasting, E. Chassefiére, R.E. Johnson, Y.N. Kulikov, F. Tian, Space Sci. Rev. (2008, this issue)
M. Larsson, P. Baltzer, S. Svensson, B. Wannberg, N. Martensson, A. Naves de Brito, N. Correia, M.P. Keane, M. Carlsson-Göthe, L. Karlsson, J. Phys. B **23**, 1175–1195 (1990)
M. Larsson, G. Sundström, L. Broström, S. Mannervik, J. Chem. Phys. **97**, 1750–1756 (1992)
P.P. Lavvas, A. Keystones, I.M. Vardavas, Planet. Space Sci. **56**, 67–99 (2008)
F. Leblanc, R.E. Johnson, Planet. Space Sci. **49**, 645–656 (2001)
F. Leblanc, O. Witasse, J. Winningham, D. Brain, J. Lilensten, P.-L. Blelly, R.A. Frahm, J.S. Halekas, J.L. Bertaux, J. Geophys. Res. **111**, A09313 (2006). doi:10.1029/2006JA011763
Ledvina et al., Space Sci. Rev. (2008, this issue)
L.C. Lee, C.C. Chiang, J. Chem. Phys. **78**, 688–691 (1983)
P.C. Lee, J.B. Nee, J. Chem. Phys. **112**, 1763–1768 (2000)
P.C. Lee, J.B. Nee, J. Chem. Phys. **114**, 792 (2001)
B.R. Lewis, S.T. Gibson, J.P. Sprengers, W. Ubachs, A. Johansson, C.G. Wahlströom, J. Chem. Phys. **123**, 23,601 (2005)
S.G. Lias, J.E. Bartmess, J.F. Liebman, J.L. Holmes, R.D. Levin, W.G. Mallard, Gas phase ion and neutral thermochemistry. J. Phys. Chem. Ref. Data **17**, 1–861 (1988)
D.R. Lide, *Editor in Chief, Handbook of Chemistry and Physics*, 88th edn. (Chemical Rubber Company Press, Boca Raton, 2008)
J. Lilensten, O. Witasse, C. Simon, H. Soldi-Lose, O. Dutuit, R. Thissen, C. Alcaraz, Geophys. Res. Lett. L03203 (2005). doi:10.1029/2004GL021432
J.J. Lin, D.W. Hwang, Y.T. Lee, X. Yang, J. Chem. Phys. **109**, 1758–1762 (1996)
K. Liou, P.T. Newell, C.-I. Meng, A.T.Y. Lui, M. Brittnacher, G. Parks, J. Geophys. Res. **102**, 27,197 (1997)
K. Liou, P.T. Newell, C.-I. Meng, M. Brittnacher, G. Parks, J. Geophys. Res. **103**(A8), 17,543–17,557 (1998)
K. Liou, P.T. Newell, C.-I. Meng, J. Geophys. Res. **106**(A4), 5531–5542 (2001)
B. Lohmann, S. Fritzsche, J. Phys. B **29**, 5711–5723 (1996)
K.A. Long, H.G. Paretzke, J. Chem. Phys. **95**, 1049 (1991)
M.A. López-Valverde, D.P. Edwards, M. López-Puertas, C. Roldan, J. Geophys. Res. **103**, 16,799–16,811 (1998)
D.A. Lorentzen, Ann. Geophys. **18**, 81–89 (2000)
G. Lu, M. Brittnacher, G. Parks, D. Lummerzheim, J. Geophys. Res. **105**, 18,483 (2000)
G. Lu, D.N. Baker, R.L. McPherron, C.J. Farrugia, D. Lummerzheim, J.M. Ruohoniemi et al., J. Geophys. Res. **103**, 11,685 (1998)
J.G. Luhmann, R.E. Johnson, M.H.G. Zhang, Geophys. Res. Lett. **19**, 2151–2154 (1992)
D. Lummerzheim, M.H. Rees, H.R. Anderson, Planet. Space Sci. **37**, 109–129 (1989)
D. Lummerzheim, M. Brittnacher, D. Evans, G.A. Germany, G.K. Parks, M.H. Rees, J.F. Spann, Geophys. Res. Lett. **24**, 987 (1997)
D. Lummerzheim, M. Galand, J. Semeter, M.J. Mendillo, M.H. Rees, F.J. Rich, J. Geophys. Res. **106**, 141–148 (2001)
M. Lundqvist, P. Baltzer, D. Edvardsson, L. Karlsson, B. Wannberg, Phys. Rev. Lett. **75**, 1058–1061 (1995)
M. Lundqvist, D. Edvardsson, P. Baltzer, B. Wannberg, J. Phys. B **29**, 1489–1499 (1996)
Y.-J. Ma, A.F. Nagy, Geophys. Res. Lett. **34** (2007). doi:10.1029/2006GL029208
Y.-J. Ma, K. Altwegg, T. Breus, M.R. Combi, T.E. Cravens, E. Kallio, S.A. Ledvina, J.G. Luhmann, S. Miller, A.F. Nagy, A.J. Ridley, D.F. Strobel, Space Sci. Rev. (2008, this issue)
T.D. Märk, J. Chem. Phys. **63**, 3731–3736 (1975)

A. Marten, T. Hidayat, Y. Biraud, R. Moreno, Icarus **158**, 532–544 (2002)
P.A. Martin, F.R. Bennett, J.P. Maier, J. Chem. Phys. **100**, 4766–4774 (1994)
D. Mathur, Phys. Rep. **225**, 193–272 (1993)
D. Mathur, Phys. Rep. **391**, 1–118 (2004)
D. Mathur, L.H. Andersen, P. Hvelplund, D. Kella, C.P. Safvan, J. Phys. B **28**, 3415–3426 (1995)
T. Matsui, A.S.-C. Cheung, W.W.-S. Leung, K. Yoshino, W.H. Parkinson, A.P. Thorne, J.E. Murray, K. Ito, T. Imajo, J. Mol. Spectrosc. **219**, 45–57 (2003)
A. Maurellis, T.E. Cravens, G. Randall Gladstone, J.H. Waite, L.W. Acton, Geophys. Res. Lett. **27**, 1339–1342 (2000)
K.E. McCulloh, J. Chem. Phys. **48**, 2090–2093 (1968)
M.A. McGrath et al., in *Jupiter: Planet, Atmosphere, Magnetosphere*, ed. by F. Bagenal et al. (2004), Chap. 19
M.A. McGrath, R.E. Johnson, Icarus **69**, 519 (1987)
R. Meier, G. Crowley, D.J. Strickland, A.B. Christensen, L.J. Paxton, D. Morrison, C.L. Hackert, J. Geophys. Res. **110**, A09S41 (2005). doi:10.1029/2004JA010990
M. Mendillo, P. Withers, in *Radio Sounding and Plasma Physics*, ed. by P. Song, J. Foster, M. Mendillo, D. Bilitza (Am. Inst. Physics, 2008), #978–0–7354–0493–9
M. Mendillo, P. Withers, D. Hinson, H. Rishbeth, B. Reinisch, Science **311**, 1135–1138 (2006)
M. Mendillo, S. Laurent, J. Wilson, J. Baumgardner, J. Konrad, W.C. Karl, Nature **448**, 330–332 (2007)
C.-I. Meng, K. Liou, P.T. Newell, Phys. Chem. Earth (C) **26**, 43–47 (2001)
J.E. Mentall, E.P. Gentieu, J. Chem. Phys. **52**, 5641–5645 (1970)
M. Michael, R.E. Johnson, Planet. Space Sci. **53**, 1510–1514 (2005)
M. Michael, R.E. Johnson, F. Leblanc, M. Liu, J.G. Luhmann, V.I. Shematovich, Icarus **175**, 263–267 (2005)
H.H. Michels, in *The Excited State in Chemical Physics* (Wiley, New York, 1981) pp. 241–265
D.L. Mitchell, R.P. Lin, H. Rème, D.H. Crider, P.A. Cloutier, J.E.P. Connerney, M.H. Acuña, N.F. Ness, Geophys. Res. Lett. **27**, 1871–1874 (2000)
W.E. Moddeman, T.A. Carlson, M.O. Krause, B.P. Pullen, W.E. Bull, G.K. Schweitzer, J. Chem. Phys. **55**, 2317–2336 (1971)
P.F. Morrissey, P.D. Feldman, J.T. Clarke, B.C. Wolven, D.F. Strobel, S.T. Durrance, J.T. Trauger, Astrophys. J. **476**, 918 (1997)
J.I. Moses, Icarus **143**, 244–298 (2000)
J.I. Moses, M. Allen, Y.L. Yung, Icarus **99**, 318–346 (1992)
L. Mrázek, J. Žabka, Z. Dolejšek, J. Hrušák, Z. Herman, J. Phys. Chem. A **104**, 7294–7303 (2000)
A. Mühleisen, M. Budnar, J.-Cl. Dousse, Phys. Rev. A **54**, 3852–3858 (1996)
Müller-Wodarg et al., Space Sci. Rev. (2008, this issue)
A.S. Mullin, D.M. Szaflarski, K. Yokoyama, G. Gerber, W.C. Lineberger, J. Chem. Phys. **96**, 3636–3648 (1992)
A.F. Nagy, P.M. Banks, J. Geophys. Res. **75**(31), 6260–6270 (1970)
F.L. Nesbitt, G. Marston, L.J. Stief, J. Phys. Chem. **95**, 7613–7617 (1991)
A. Noelle, G.K. Hartman, D. Lary, S. Le Calve, S. Trick, A.D. Vandaele, R.P. Wayne, C.Y.R. Wu, *UV/Vis Spectra Data Base, Science-softCon*, 5th edn (2007)
B. Noller, I. Fischer, J. Chem. Phys. **126**, 144302 (2007)
K. Onda, M. Ejiri, Y. Itikawa, J. Geophys. Res. **104**, 27,991–28,001 (1999)
N. Østgaard, J. Stadsnes, J. Bjordal, R.R. Vondrak, S.A. Cummer, D.L. Chenette, M. Schulz, J. Pronko, J. Geophys. Res. **105**, 20,869 (2000)
N. Østgaard, J. Stadsnes, J. Bjordal, G.A. Germany, R.R. Vondrak, G.K. Parks, S.A. Cummer, D.L. Chenette, J.G. Pronko, J. Geophys. Res. **106**(A11), 26,081–26,089 (2001)
D. Pallamraju, S. Chakrabarti, J. Atmos. Sol.-Terr. Phys. **68**, 1459–1471 (2006). doi:10.1016/j.jastp.2005.05.013
L. Pallier, R. Prangé, Geophys. Res. Lett. **31**(6), L06701 (2004). doi:10.1029/2003GL018041
H.G. Paretzke, Adv. Space Res. **9**, 15–20 (1989)
F. Penent, P. Lablanquie, R.I. Hall, J. Palaudoux, K. Ito, Y. Hikosaka, T. Aoto, J.H.D. Eland, J. Electron. Spectrosc. Rel. Phenom. **7–11**, 144–147 (2005)
J.J. Perry, Y.H. Kim, J.L. Fox, H.S. Porter, J. Geophys. Res. **104**, 16,541–16,565 (1999)
D. Petrini, F.X. de Araújo, Astron. Astrophys. **282**, 315–317 (1994)
M. Pospieszalska, R.E. Johnson, J. Geophys. Res. **101**, 7565–7573 (1996)
R. Prangé, D. Régo, L. Pallier, L. Ben Jaffel, C. Emerich, J. Ajello, J.T. Clarke, G.E. Ballester, Astrophys. J. **484**, L169–L173 (1997)
R. Prangé, L. Pallier, K.C. Hansen, R. Howard, A. Vourlidas, R. Courtin, C. Parkinson, Nature **432**, 78–81 (2004). doi:10.1038/nature02986

R.G. Prinn, in *The Photochemistry of Atmospheres: Earth, the Other Planets, and Comets*, ed. by J.S. Levine (Academic Press, Orlando, 1985)
M.H. Rees, Planet. Space Sci. **11**, 1209–1218 (1963). doi:10.1016/0032-0633(63)90252-6
M.H. Rees, *Physics and Chemistry of the Upper Atmosphere* (Cambridge University Press, New York, 1989)
M.H. Rees, B.A. Emery, R.G. Roble, K. Stamnes, J. Geophys. Res. **88**, 6289–6300 (1983)
M.H. Rees, D. Lummerzheim, R.G. Roble, Space Sci. Rev. **71**, 691 (1995)
D. Régo, R. Prangé, L. Ben Jaffel, J. Geophys. Res. **104**, 5939–5954 (1999)
D. Régo, J.T. Clarke, L. Ben Jaffel, G.E. Ballester, R. Prangé, J. McConnell, Icarus **150**, 234–243 (2001). doi:10.1006/icar.2000.6583
H. Rishbeth, O.K. Garriott, *Introduction to Atmospheric Physics* (Academic Press, New York, 1969)
R.G. Roble, E.C. Ridley, R.E. Dickinson, J. Geophys. Res. **92**, 8745–8758 (1987)
C. Roldán, M.A. Lopez-Valverde, M. Lopez-Puertas, D.P. Edwards, Icarus **147**, 11–25 (2000)
H.R. Sadeghpour, A. Dalgarno, Phys. Rev. A. **47**, R2458–R2459 (1993)
C. Sætre, C.A. Barth, J. Stadsnes, N. Østgaard, S.M. Bailey, D.N. Baker, G.A. Germany, J.W. Gjerloev, J. Geophys. Res. **112**, A08306 (2007). doi:10.1029/2006JA012203
C.P. Safvan, M.J. Jensen, H.B. Pedersen, L.H. Andersen, Phys. Rev. A **60**, R3361–R3364 (1999). Erratum in Phys. Rev. A, **60**, R3361 (1999)
J.A.R. Samson, G.N. Haddad, J. Opt. Soc. Am. B **11**, 277–279 (1994)
R.P. Saxon, T.G. Slanger, J. Geophys. Res. **91**, 9877 (1986)
R.W. Schunk, A.F. Nagy, *Ionospheres: Physics, Plasma Physics, and Chemistry* (Cambridge University Press, New York, 2000)
K. Seiersen, A. Al-Khalili, O. Heber, M.J. Jensen, I.B. Pedersen, C.P. Safvan, L.H. Andersen, Phys. Rev. A **68**, 022708 (2003a)
K. Seiersen, O. Heber, M.J. Jensen, C.P. Safvan, L.H. Andersen, J. Chem. Phys. **119**, 839–843 (2003b)
V.I. Shematovich, D.V. Bisikalo, J.-C. Gérard, J. Geophys. Res. **99**, 23,217 (1994)
V.I. Shematovich, R.E. Johnson, M. Michael, J.G. Luhmann, J. Geophys. Res. **108**, 5087 (2003). doi:10.1029/2003JE002094
E. Shigemasa, T. Gejo, M. Nagasono, T. Hatsui, N. Kosugi, Phys. Rev. A **66**, 022508 (2002)
C. Simon, J. Lilensten, O. Dutuit, R. Thissen, O. Witasse, C. Alcaraz, H. Soldi-Lose, Ann. Geophys. **23**, 781–797 (2005)
D.E. Siskind, D.J. Strickland, R.R. Meier, T. Majeed, F.G. Eparvier, J. Geophys. Res. **100**, 19,687–19,694 (1989)
T.G. Slanger et al., Space Sci. Rev. (2008, this issue)
F.L. Smith, C. Smith, J. Geophys. Res. **77**, 3592–3597 (1972)
P.F. Smith, K. Yoshino, W.H. Parkinson, K. Ito, G. Stark, J. Geophys. Res. **96**, 17,529–17,533 (1991)
S.C. Solomon, Geophys. Res. Lett. **20**, 185–188 (1993)
J.P. Sprengers, W. Ubachs, K.G.H. Baldwin, J. Chem. Phys. **122**, 144301 (2005)
S.C. Solomon, S.M. Bailey, T.N. Woods, Geophys. Res. Lett. **28**, 2149–2152 (2001)
V. Srivastava, V. Singh, J. Geophys. Res. **93**, 5845–5854 (1988)
G. Stark, K. Yoshino, P.L. Smith, J.R. Esmond, K. Ito, M.H. Stevens, Astrophys. J. **410**, 837–842 (1993)
G. Stark, K.P. Huber, K. Yoshino, P.L. Smith, K. Ito, J. Chem. Phys. **123**, 214303 (2005)
G. Stark, K. Yoshino, P.L. Smith, K. Ito, J. Quant. Spectrosc. Rad. Transf. **103**, 67–73 (2007)
M.H. Stevens, D.F. Strobel, M.E. Summers, R.V. Yelle, Geophys. Res. Lett. **19**, 669–672 (1992)
J.A.D. Stockdale, J. Chem. Phys. **66**, 1792–1794 (1977)
D.J. Strickland, D.L. Book, T.P. Coffey, J.A. Fedder, J. Geophys. Res. **81**, 2755–2764 (1976)
D.J. Strickland, R.R. Meier, J.H. Hecht, A.B. Christensen, J. Geophys. Res. **94**(A10), 13,527–13,539 (1989)
D.J. Strickland, R.J. Cox, R.R. Meier, D.P. Drob, J. Geophys. Res. **104**, 4251 (1999)
D.J. Strickland, J. Bishop, J.S. Evans, T. Majeed, R.J. Cox, D. Morrison, G.J. Romick, J.F. Carbary, L.J. Paxton, C.-I. Meng, J. Geophys. Res. **106**, 65 (2001)
D.F. Strobel, in *Atmospheres in the Solar System: Comparative Aeronomy*, ed. by M. Mendillo, A. Nagy, H. Waite, AGU Monograph, vol. 130 (AGU Press, 2002), pp. 7–22
G. Sundström, M. Carlson, M. Larsson, Broström, Chem. Phys. Lett. **218**, 17–20 (1994)
S. Svensson, J. Phys. B **38**, S821–S835 (2005)
S. Svensson, A. Naves de Brito, M.P. Keane, N. Correia, L. Karllson, C.-M. Liegenner, A. Ågren, J. Phys. B **25**, 135–144 (1992)
P.R. Taylor, H. Partridge, J. Phys. Chem. **91**, 6148–6151 (1987)
H.A. Taylor, H.C. Brinton, S.J. Bauer, R.E. Hartle, P.A. Cloutier, R.E. Daniell, J. Geophys. Res. **85**, 7765–7777 (1980)
H.A. Taylor, Th. Encrenaz, D.M. Hunten, P.G.J. Irwin, T. Cowan, in *Jupiter: The Planet, Satellites, and Magnetospheres*, ed. by F. Bagenal, T.E. Dowling, W.B. McKinnon (2004)
W.K. Tobiska, J. Atmos. Terr. Phys. **53**, 1005–1018 (1991)

W.K. Tobiska, Adv. Space. Res. **34**(8), 1736–1746 (2004)
W.K. Tobiska, S.D. Bouwer, Adv. Space Res. **37**(2), 347–358 (2006). doi:10.1016/j.asr.2005.08.015
M.G. Tomasko et al., Nature **438**, 765–778 (2005)
M.R. Torr, D.G. Torr, R.A. Ong, H.E. Hinteregger, Geophys. Res. Lett. **6**, 771–774 (1979)
M. Uda, H. Endo, K. Maeda, Y. Awaya, M. Kobayashi, Y. Sasa, H. Kumagi, T. Tomma, Phys. Rev. Lett. **42**, 1257–1269 (1979)
E.F. van Dishoeck, J.H. Black, Astrophys. J. **334**, 771–802 (1988)
D.A. Verner, D.G. Yakovlev, Astron. Astrophys. Suppl. Ser. **109**, 125–133 (1995)
D.A. Verner, G.J. Ferland, K.T. Korista, D.G. Yakovlev, Astrophys. J. **465**, 487–498 (1996)
G.A. Victor, E.R. Constantinides, Geophys. Res. Lett. **6**, 519–522 (1979)
J.H. Waite, T.E. Cravens, Adv. Space Res. **7**(12), 119–134 (1987)
J.H. Waite, D. Lummerzheim, in *Atmospheres in the Solar System: Comparative Aeronomy*, ed. by M. Mendillo, A. Nagy, H. Waite, AGU Monograph, vol. 130 (AGU Press, 2002), pp. 115–139
J.H. Waite, G.R. Gladstone, W.S. Lewis, P. Drossart, T.E. Cravens, A.N. Maurellis, B.H. Mauk, S. Miller, Science **276**, 104–108 (1997)
C.W. Walter, P.C. Cosby, H. Helm, J. Chem. Phys. **99**, 3553–3561 (1993)
R.A. West, K.H. Baines, A.J. Friedson, D. Banfield, B. Ragent, F.W. Taylor, in *Jupiter: The Planet, Satellites, and Magnetosphere*, ed. by F. Bagenal, T.E. Dowling, W. McKinnon (Cambridge University Press, Cambridge, 2004)
R.W. Wetmore, R.K. Boyd, J. Phys. Chem. **90**, 5540–5551 (1986)
J.K. Wilson, M. Mendillo, J. Baumgardner, N.M. Schneider, J.T. Trauger, B. Flynn, Icarus **157**, 476–489 (2002)
J.D. Winningham, D.T. Decker, J.U. Kozyra, J. Geophys. Res. **94**, 15,335–15,348 (1989)
O. Witasse et al., Space Sci. Rev. (2008, this issue)
O. Witasse, O. Dutuit, J. Lilensten, R. Thissen, J. Zabka, C. Alcaraz, P.-L. Blelly, S.W. Bougher, S. Engel, L.H. Andersen, K. Seiersen, Geophys. Res. Lett. **29**, 1263 (2002). doi:10.1029/2002GL014781
T.N. Woods, F.G. Eparvier, S.M. Bailey, P.C. Charberlin, J. Lean, G.J. Rottman, S.C. Solomon, W.K. Tobiska, D.L. Woodraska, J. Geophys. Res. **110**, A01312 (2005). doi:10.1029/2004JA010765
J.S. Wright, D.J. Carpenter, A.B. Alekseyev, H.-P. Lieberman, R. Lingott, R.J. Buenker, Chem. Phys. Lett. **266**, 391–396 (1997)
C.Y.R. Wu, F.Z. Chen, D.L. Judge, J. Geophys. Res. **109**, E07S15 (2004). doi:10.1029/2003JE002180
M. Yan, H.R. Sadeghpour, A. Dalgarno, Astrophys. J. **496**, 1044–1050 (1998)
Yelle, Miller, in *Jupiter, the Planet, Satellites, and Magnetosphere*, ed. by F. Bagenal et al. (Cambridge University Press, Cambridge, 2002)
R.V. Yelle, J.I. Lunine, J.B. Pollack, R.W. Brown, in *Neptune and Triton*, ed. by D.P. Cruikshank (University of Arizona Press, Tucson, 1995), pp. 1031–1106
R.V. Yelle, N. Borggren, V. de la Haye, W.T. Kasprzak, H.B. Niemann, I. Müller-Wodard, J.H. Waite, Icarus **182**, 567–576 (2006)
K. Yoshino, D.E. Freedman, W.H. Parkinson, J. Phys. Chem. Ref. Data **13**, 207–227 (1984)
K. Yoshino, A.S.-C. Cheung, J.R. Esmond, W.H. Parkinson, D.E. Freedman, S.L. Guberman, A. Jenvrier, B. Coquart, M.F. Merienne, Planet. Space Sci. **36**, 1469–1475 (1988)
K. Yoshino, J.R. Esmond, A.S.-C. Cheung, D.E. Freedman, W.H. Parkinson, Planet. Space Sci. **40**, 185–192 (1992)
K. Yoshino, G. Stark, J.R. Esmond, P.L. Smith, K. Ito, T. Matsui, Astrophys. J. **438**, 1013–1016 (1995)
K. Yoshino, J.R. Esmond, Y. Sun, W.H. Parkinson, K. Ito, T. Matsui, J. Quant. Spectrosc. Rad. Trans. **55**, 53–60 (1996)
K. Yoshino, W.H. Parkinson, K. Ito, T. Matsui, J. Mol. Spectrosc. **229**, 238–243 (2005)
R.W. Zurek, J.R. Barnes, R.M. Haberle, J.B. Pollack, J.E. Tillman, C.B. Leovy, in *Mars*, ed. by H.H. Kieffer, B.M. Jakosky, C.W. Snyder, M.S. Matthews (University of Arizona Press, Tucson, 1992), pp. 835–933

Cross Sections and Reaction Rates for Comparative Planetary Aeronomy

David L. Huestis · Stephen W. Bougher · Jane L. Fox · Marina Galand · Robert E. Johnson · Julianne I. Moses · Juliet C. Pickering

Originally published in the journal Space Science Reviews, Volume 139, Nos 1–4.
DOI: 10.1007/s11214-008-9383-7

Abstract In this chapter we describe the current knowledge of a selection of collision processes and chemical reactions of importance to planetary aeronomy. Emphasis is placed on critical evaluation of what we know and what we wish we knew about fundamental processes required for interpretation, explanation, and modeling of atmospheric observations.

Keywords Collisions · Photoabsorption · Chemical reactions · Critical evaluation · Laboratory measurement · First-principles theory · Needs in atmospheric models

1 Introduction, Objectives, and Outline

In this chapter we highlight some of the optical, chemical, and collisional processes involving atoms and molecules of importance in the upper atmospheres, ionospheres, and magne-

---

Prepared for publication in Space Science Reviews (Springer) and in the International Space Science Institute (ISSI Bern) book series, Volume 29.

D.L. Huestis (✉)
Molecular Physics Laboratory, SRI International, Menlo Park, CA 94025, USA
e-mail: david.huestis@sri.com

S.W. Bougher
Atmospheric, Oceanic and Space Sciences, University of Michigan, Ann Arbor, MI 48109, USA

J.L. Fox
Department of Physics, Wright State University, Dayton, OH 45435, USA

M. Galand · J.C. Pickering
Space and Atmospheric Physics, Imperial College London, London, SW7 2AZ, UK

R.E. Johnson
Engineering Physics and Materials Science, University of Virginia, Charlottesville, VA 22904, USA

J.I. Moses
Planetary Science, Lunar and Planetary Institute, Houston, TX 77058, USA

tospheres of the planets, moons, and comets in the solar system. The atmospheric domain extends roughly from the lowest altitudes at which dissociation and ionization are important, through the corona and into the ambient plasma where ion-neutral collisions are important. The range of chemical species extends from photons, electrons, and protons, through moderate molecular-weight organic molecules. The collision energies range from several meV, corresponding to the local temperature, up to 100's of keV, appropriate for particles in the solar wind and planetary magnetospheres.

As a companion with the other chapters in this volume, the discussion here must necessarily be selective, emphasizing processes that are important in more than one atmosphere, or at least are similar to those in other atmospheres. The limitations of space imply that we can include only a very small fraction of the microscopic chemical and collisional processes that are included in current atmospheric chemistry models. The obvious bias is for processes that are "important." In addition, we emphasize processes that are currently not well quantified, have been a topic of recent controversy, or have been the subject of recent experimental or theoretical investigations.

The optical, chemical, and collision processes to be addressed are divided into the following seven categories:

1. Chemical reactions of neutral atoms and molecules in their ground electronic states
2. Ion–molecule reactions
3. Chemistry, relaxation, and radiation of electronically excited atoms and molecules
4. Vibrational and rotational excitation, relaxation, and radiative emission
5. Photoabsorption, photodissociation, and photoionization
6. Electron-impact excitation, dissociation, ionization, and recombination
7. Energetic heavy particle excitation and charge exchange

Below we will briefly review how cross sections and reaction rates are used in data interpretation and modeling. The main component of this chapter is a limited critical evaluation of a few processes in each of the seven categories listed above. We will conclude with an extensive list of cited references, highlighting key laboratory measurements, first-principles calculations, the atmospheric models in which they are incorporated, and the atmospheric observations they help to characterize and explain.

2 Fundamental Processes in Planetary Atmospheres

In this section we will provide several examples of how cross sections and reaction rates are used in observational data interpretation and modeling, by atmospheric type, region, event, or observable. The key is in identifying what is important and what we wish we knew.

2.1 Inner Planets

The recent Mars and Venus Express missions have stimulated renewed interest in comparative investigations of the atmospheres of the Earth and its near neighbors. In this Section we will review interconnections between microscopic collisional processes, atmospheric motion, thermal balance, and remotely observable signatures.

2.1.1 Nightglow

Direct wind measurements of the Venus and Mars upper atmospheres are currently lacking. Such wind retrievals, in conjunction with simultaneous density and temperature measurements, would provide a means to constrain General Circulation Models (GCMs) and thereby characterize the self-consistent thermal, wind, and compositional structure of these upper atmospheres over solar cycle, seasonal, and diurnal timescales. This ideal situation is presently far from being realized in our study of the Venus and Mars upper atmospheres (e.g. Bougher et al. 1997, 2006a, 2006b).

Instead, creative application of available nightglow observations at both Venus and Mars, in conjunction with GCMs, is being used to infer global circulation patterns, wind magnitudes, and their variations over solar cycle and seasonal timescales (e.g. Bougher et al. 1990, 1997, 2006a, 2006b; Bougher and Borucki 1994; Bertaux et al. 2005). At Venus, NO UV nightglow, $O_2(^1\Delta_g)$ 1.27-μm nightglow, and O_2 400–800 nm nightglow intensity distributions and their temporal variability are being used to constrain GCMs and uncover the thermospheric general circulation patterns and wind magnitude responsible. Pioneer Venus, Venus Express, and ground based nightglow observations have all contributed to this process of tracing Venus thermospheric wind patterns (see reviews by Lellouch et al. 1997; Bougher et al. 1997, 2006a; Gerard et al. 2008). At Mars, recent Mars Express SPICAM observations of NO nightglow emissions at high latitudes (Bertaux et al. 2005) plus several winter polar warming measurements from aerobraking missions (e.g. Bougher et al. 2006b) are motivating new GCM simulations that seek to determine the seasonal and solar cycle variations in the Mars thermospheric circulation.

The effectiveness of these nightglow studies to constrain the Venus and Mars thermospheric circulation patterns is predicated upon a firm understanding of the physical processes responsible for the creation of the nightglow emission features. A few key reactions and rates are poorly constrained, leading to ambiguity in nightglow calculations for comparison to observations. Specifically, 3-body recombination ($O+O+CO_2$) is assumed to be the major contributor to the production of the $O_2(a^1\Delta_g)$ state and the observed Venus 1.27-μm nightglow emission. Quenching of this $O_2(a^1\Delta_g)$ state, notably by CO_2, competes with this production. Firm measurements of this 3-body reaction rate (and ~100–250 K temperature dependence) and the corresponding CO_2 quenching rate are needed to refine these nightglow calculations within model simulations. Furthermore, the effective yield of the $O_2(a^1\Delta_g)$ state from 3-body recombination is assumed to be ~0.75–1.0 (e.g. Gerard et al. 2008). A measurement of the absolute value of this yield in CO_2 is also greatly needed to further constrain model simulations. The interpretation of recent Venus Express $O_2(^1\Delta_g)$ 1.27-μm nightglow distributions (Gerard et al. 2008) will be greatly advanced by such new laboratory measurements.

See also Slanger et al. (2008b) in this Volume and Sects. 3.1.2 and 3.3.2 of this chapter.

2.1.2 O–CO_2 Cooling

The role of CO_2 15-μm cooling in the thermospheric heat budgets of Venus, Earth, and Mars is described to be profoundly different (see reviews by Bougher et al. 1999, 2000, 2002). However, a self-consistent and definitive treatment of these differences requires the application of a realistic and common O–CO_2 vibrational re-excitation rate. This rate has recently been measured in the laboratory at room temperatures and found to be $\sim 1.5 \times 10^{-12}$ cm^3/sec (see Khvorostovskaya et al. 2002; Akmaev 2003; Sect. 3.4.2 of this chapter), in contrast to values of $\sim 3.0 \times 10^{-12}$ cm^3/sec (or larger) commonly used in recent GCM model simulations that seek to compare the heat budgets of Venus, Earth, and Mars using a common modeling framework (e.g. Bougher et al. 1999, 2000, 2002). This discrepancy may be due in part to uncertainties in the upper atmosphere atomic O abundances for Venus and Mars. In particular, O abundances have never been directly measured for the Mars upper atmosphere, only inferred from UV airglow measurements (e.g. Stewart et al. 1992;

Alexander et al. 1993) and ionospheric calculations (e.g. Hanson et al. 1977; Fox and Dalgarno 1979). Values should vary considerably over the solar cycle, seasons, and latitudes on Mars. A proper evaluation of the impacts of CO_2 15-µm cooling on the Mars thermosphere, even assuming the modern laboratory O–CO_2 vibrational re-excitation rate, must await new Martian in-situ composition measurements. Similarly, Venus atomic O abundances have been measured by the Pioneer Venus ONMS instrument near the equator above ~135 km for solar maximum conditions (e.g. Niemann et al. 1979, 1980; Hedin et al. 1983). However, abundance measurements throughout the rest of the solar cycle and over wide latitude regions are missing.

In addition, our present understanding of both Venus and Mars thermospheric heat budgets is predicated upon detailed EUV-UV heating efficiency calculations conducted most recently for Venus (Fox 1988) and Mars (Fox et al. 1995). Efficiencies of $21 \pm 2\%$ are reasonable and should be applicable to the Mars thermosphere; profiles of 20–22% are acceptable for the Venus thermosphere. Uncertainties in these efficiencies, although small, still provide some flexibility when applying the recently measured O–CO_2 rate ($\sim 1.5 \times 10^{-12}$ cm^3/sec) in model calculations designed to reproduce measured temperatures and their variations over the solar cycle and seasons.

In short, CO_2 15-µm cooling rates and their impacts upon the Venus and Mars thermospheres are best quantified and compared once: (1) atomic O abundances are properly characterized for both planets, (2) an appropriate EUV-UV heating efficiency is assumed, and (3) a common (and modern) O–CO_2 vibrational re-excitation rate ($\sim 1.5 \times 10^{-12}$ cm^3/sec near 300 K) is applied. See also Sect. 3.4.2 in this chapter.

2.2 Outer Planets

Solar-radiation-driven photochemical processes are very important on the giant planets despite the relatively large distances of these planets from the Sun. In the uppermost portions of giant-planet atmospheres, extreme ultraviolet radiation and X-rays interact with hydrogen and other constituents, leading to photoionization (for reviews of ionospheric processes on the giant planets, see Witasse et al. (2008) in this issue, Nagy and Cravens (2002), and Majeed et al. (2004). The dominant primary ion, H_2^+, reacts quickly with H_2 to produce H_3^+ or charge exchanges with H to form H^+. Radiative recombination of H^+ is slow, so H^+ can become the dominant ion in giant-planet ionospheres, especially in regions where the daily average flux of solar radiation is highest (e.g., in the summer, at low latitudes, at solar maximum) and at night at all latitudes. The H_3^+ ions can dominate during the afternoon at middle and high latitudes (e.g. Moore et al. 2004), but dissociative recombination is fast enough that H_3^+ can recombine at night, and a significant diurnal variation of H_3^+ is expected.

Ion–molecule reactions are prevalent, especially in the lower ionosphere, where neutral hydrocarbons can interact with ions. Rate coefficients for ion–molecule reactions relevant to the giant planets (and Titan) have been reviewed by Anicich and McEwan (1997). Although the rate coefficients for the most important reactions are reasonably well known, the product distributions are less well known. In addition, the rate coefficients and products for important electron recombination reactions need further characterization. Interaction of charged particles from the magnetosphere also instigates upper-atmospheric chemistry on the giant planets, especially in the auroral regions (see Wong et al. 2000, 2003; Friedson et al. 2002), and many of the details (including rate coefficients, pathways, role of excited states, etc.) of this process remain to be worked out.

In the middle atmospheres of the giant planets, methane photolysis (primarily at Lyman alpha wavelengths) drives a complex hydrocarbon photochemistry that results in the production of species like CH_3, C_2H_2, C_2H_4, C_2H_6, C_3H_4, C_3H_8, C_4H_2, and C_6H_6, which have all been observed on one or more of the giant planets. Details of hydrocarbon photochemistry on the giant planets can be found in Moses et al. (2004, 2005); Yung and DeMore (1999); and Gladstone et al. (1996). Many uncertainties in this hydrocarbon chemistry remain, largely due to a lack of relevant laboratory and theoretical data on photodissociation cross sections, photodissociation quantum yields and branching ratios, reaction rate coefficients, and reaction pathways at low temperatures and pressures. Although much progress in determining the photoabsorption cross sections at relevant temperatures has been made in recent years (e.g. Wu et al. 2001, 2004; Chen and Wu 2004; Fahr and Nayak 1994, 1996; Bénilan et al. 2000; Chen et al. 1991, 2000; Smith et al. 1991, 1998), critical information about the *photodissociation* cross sections and product yields is missing. One particularly relevant case in point is that for methane photolysis at Lyman alpha. Only in recent years have the primary products and H and H_2 quantum yields been well characterized (e.g. Cook et al. 2001; Wang et al. 2000; Smith and Raulin 1999; Brownsword et al. 1997; Heck et al. 1996; Mordaunt et al. 1993); however, the quantum yield of CH, which is of critical importance for the production of unsaturated hydrocarbons, is still uncertain. In addition, the branching ratios from photolysis of other major hydrocarbons (e.g., C_2H_2, C_2H_4, C_2H_6, CH_3C_2H, C_3H_8) need to be worked out in better detail.

Other critical missing information is the low-pressure limiting behavior and/or falloff-region rate coefficients for termolecular reactions at low temperatures that are relevant to giant-planet atmospheres. Much of the action in terms of hydrocarbon photochemistry occurs in the 0.001–10 mbar pressure region, and typical temperatures range from 80–170 K. Termolecular reactions are of great importance on the giant planets despite the low pressures involved. Given the prevalence of atomic hydrogen in the photochemically active regions, low-pressure rate coefficients for $H+C_2H_x$, $H+C_3H_x$, and $H+C_4H_x$ are particularly needed, as are the details of benzene production and loss under appropriate conditions for giant-planet stratospheres. In Table 6 of Moses et al. (2005), some specific chemical kinetics needs for giant-planet hydrocarbon photochemical modeling are described.

See also Sects. 3.1.1, 3.2.1, and 3.2.2 of this chapter.

2.3 Ionospheres

The non-auroral ionospheres of the Earth and planets are produced by photoionization and photoelectron impact ionization of thermospheric atoms and molecules.

2.3.1 Inner Planet Ionospheres

Earth, Venus and Mars all have oxidizing atmospheres, but the major constituents of the atmospheres are N_2 and O_2 for Earth, and CO_2 and N_2 for Venus and Mars. In all three atmospheres, there are small admixtures of the stable species Ar, He, and H_2, and small chemically unstable atomic and molecular radicals, such as O, NO, N, H, CO, and C, which are produced photochemically.

In the F_1 regions of these ionospheres, the ions are produced by absorption of solar EUV photons, whose wavelengths range from about 100 to 1000 Å. The major ions produced in this region, including CO_2^+ and N_2^+, and most of the minor ions are rarely the terminal ions. In the presence of sufficient densities of neutrals, ions whose parent neutrals are characterized by high ionization potentials (IP's) are transformed to ions whose parent neutrals are characterized by low IP's. Thus N_2^+ and CO_2^+ are transformed to O_2^+ and NO^+ by charge transfer and other ion-molecule reactions (cf. Fox 2002, 2006). The terminal ions in the lower ionospheres of the terrestrial planets are O_2^+ and NO^+, which are destroyed mainly by dissociative recombination (DR), e.g.,

$$NO^+ + e \rightarrow N + O. \quad (1)$$

In fact, NO has the lowest IP (9.26 eV) among the major and minor thermospheric species; it is thus an important ion on all three planets, even in the absence of significant densities of NO. At higher altitudes, where the fractional ionization is large and neutral densities are small, the molecular ions produced by direct ionization and by photochemical processes may be destroyed by DR. Atomic ions, such as O^+, N^+, H^+, and C^+ cannot undergo DR, and radiative recombination, e.g.,

$$O^+ + e \rightarrow O + h\nu \quad (2)$$

is slow, with rate coefficients on the order of 10^{-12} cm^3/s (Escalante and Victor 1992; Slanger et al. 2004). As a result, the fractional ionization would be expected to increase indefinitely as the altitude increases, but eventually the major loss process becomes transport downward by ambipolar diffusion. Atomic ion density profiles thus form an F_2 peak at the altitude where the time constant for loss by chemistry is equal to that for loss by transport. On Earth, O^+ forms a prominent F_2 peak near 300 km. On Venus, the O^+ peak near 200 km is not as prominent, and is generally not visible in electron density profiles. On Mars, the Viking RPA data showed that the O^+ density peaks near 230 km, but it is everywhere less than that of the major ion, O_2^+ (e.g. Hanson et al. 1977). In the electron density profiles of the terrestrial planets, a secondary, but visible E-region peak is seen below the F_1 peak. This peak is formed from ionization by solar soft X-rays, with wavelengths in the interval ~10–100 Å, and by the concomitant high energy photoelectrons and secondary electrons. On earth, ionization of O_2 by solar Lyman β at 1026 Å is an additional source of the E-region peak (e.g., Bauer 1973).

2.3.2 Outer Planet Ionospheres

The outer planets, including Jupiter, Saturn, Uranus and Neptune, are composed primarily of H_2, He, and CH_4. There are also small atmospheric abundances of NH_3, C_2H_6, and C_2H_2 that vary among the four planets. The latter two species are formed by photochemical processes that originate ultimately from CH_4. Jupiter and Saturn also have small abundances of PH_3. Because ammonia condenses to form clouds in the middle atmospheres, NH_3 is not a major component of the thermospheres of any of the outer planets. PH_3 and large hydrocarbons may also be incorporated into aerosols and hazes. In the thermospheres, the densities of higher hydrocarbons fall off rapidly above the methane homopauses, and as a result above the upper thermospheres are composed of mostly of H_2, H and He.

In reducing atmospheres, where hydrogen is abundant, and neutral densities are large, ionization flows from ions whose (unprotonated) parent species have low proton affinities (PA's) to those whose parent neutrals have high PA's. Thus the principle ion produced in the main part of these atmospheres, H_2^+, is never the terminal ion; the PA of H is only 2.69 eV, while that of H_2 is 4.39 eV, and those of hydrocarbons are even larger. Thus the terminal major molecular ion in the F_1 region is predicted to be H_3^+, which is produced by reaction of H_2^+ with H_2, and the major atomic ion at the F_2 peaks is H^+. Modelers have, however, encountered difficulty in fitting the model electron density profiles, for which the F_2 peak density is about 10^6 cm^{-3}, compared to the smaller measured electron density peak of about 10^5 cm^{-3}, without postulating some chemical loss process for H^+, such as charge transfer of H^+ to vibrationally excited H_2^+ (e.g., McElroy 1973; Cravens 1987). For Saturn, influx of water (PA = 7.24 eV) from the rings has been postulated to reduce the H^+ densities (Waite et al. 1997; Moore et al. 2006).

However, Kim and Fox (1994) showed that solar radiation in the wings of the H_2 absorption lines penetrates to the Jovian hydrocarbon layer, producing one or more layers of hydrocarbon ions that peak in the altitude range 300–400 km above the ammonia cloud tops near 0.6 bar. This is likely to be the case for Saturn as well, although it has not been modeled.

2.3.3 Ionospheres on Titan and Triton

The atmospheres of the satellites Titan and Triton are of intermediate oxidation state and are composed mostly of N_2, with small amounts of CH_4 and H_2 (e.g., Krasnopolsky and Cruikshank 1995; Yelle et al. 2006). The surface pressure (13.5 μbar vs. 1.5 bar) and temperature (~38 vs. 94 K) are much lower on Triton than on Titan, and the major species on Triton condense at the surface. In the atmospheres of Triton and Titan, hydrocarbons may be produced ultimately from methane photochemically or by processes the follow the impact of energetic electrons from the magnetospheres of Neptune and Saturn, respectively. On Triton, however, hydrocarbons seem to be formed mostly below the ionospheric peak, while on Titan, production of higher hydrocarbons takes place below the methane homopause, which has been found by the Cassini Huygens probe to be unexpectedly high in the thermosphere (e.g., Strobel and Summers 1995; Yelle et al. 2006).

Predictions for the composition of the ionospheres of the two satellites thus differ substantially. On Titan the major ion was predicted and found to be $HCNH^+$, with large quantitites of hydrocarbon ions, such as $C_2H_5^+$, CH_3^+, CH_5^+, $C_3H_5^+$, and $C_4H_3^+$, and nitrile ions, $C_nN_kH_m^+$ (e.g. Fox and Yelle 1997; Keller et al. 1998; Cravens et al. 2006; Vuitton et al. 2007). N_2^+ (IP = 15.58 eV, PA = 5.13) is the dominant ion formed at high altitudes, but it is transformed photochemically in regions of sufficient neutral densities to species with either lower IP's, such as CH_4^+ (IP_{CH_4} = 12.51 eV) or those with higher PA's, such as CH_5^+ (PA_{CH_4} = 5.72 eV) and $HCNH^+$ (PA_{HCN} = 7.46 eV). The transformation of protonated hydrocarbon and nitrile ions on Titan continues, as in reducing atmospheres, from species whose parent (unprotonated) neutrals are characterized by low PA's to those whose parent neutrals have higher PA's.

On Triton, the ionosphere was found by the Voyager RSS to be quite robust (Tyler et al. 1989), with a maximum ion density near 350 km of (2.6–4.3) $\times 10^4$ cm^{-3}, and a secondary peak near 100 km with a density of about 3000 cm^{-3}. The major ion is predicted by most models to be C^+ (IP_C = 11.3 eV) (e.g. Lyons et al. 1992; Krasnopolsky and Cruikshank 1995; Strobel and Summers 1995), although N^+ (IP_N = 14.5 eV) was suggested by early models (e.g. Majeed et al. 1990). The ionosphere thus consists primarily of an F_2 peak, although the secondary peak may be an F_1 or E peak, depending on its formation mechanisms. The details of chemistry of C and C^+ are currently, however, uncertain (in both reducing and oxidizing atmospheres) and measurements of rate coefficients are needed for reactions involving both species such as charge transfer from N_2^+, N^+, O^+, and O_2^+ to C.

2.3.4 Ionosphere Summary

For all the planets, the data required to model these ionospheres include photoabsorption, photoionization, and photodissociation cross sections; the cross sections for interaction of solar photoelectrons with atmospheric gases; the rate coefficients for ion–molecule reactions that transform the ions, and the diffusion coefficients for ions and minor neutrals.

Little information is available on the diffusion coefficients for metastable species, such as $O(^1D)$ and $O(^1S)$. Of particular interest are the rate coefficients for charge transfer of atomic ions to atoms. The reactions of He^+ and O^{++} with molecules may result in fragmentation and the product branching ratios are often unknown. Significant progress has been make in determining the chemistry of the important metastable species $O(^1D)$, $O(^1S)$, $O^+(^2D)$, $O^+(^2P)$, and $N_2(A^3\Sigma_u^+)$, but their sources and sinks are not currently as well known as those of stable species.

For the reducing atmospheres, the rate coefficients for many reactions of hydrocarbons and their ions are unknown. Although it is doubtful that the chemistry of C_3 or higher hydrocarbons and their ions will be completely understood, some progress can certainly be made for the smaller species.

See also Sects. 3.2.1, 3.2.2, 3.3.1, 3.6.1, and 3.7.1 of this chapter.

2.4 Aurora

Energetic particle precipitation in auroral regions is an important energy source upon a planetary atmosphere. It affects the atmospheric composition and dynamics, the thermal structure and the electrodynamical properties of the upper atmosphere. Auroral emissions observed from ground-based and Earth-orbiting observatories and from space probes offer us an extremely valuable remote-sensing of the auroral particle source. Its quantitative analysis allows the identification of the type of precipitating particles and the assessment of the incident particle characteristics in terms of mean energy and energy flux (e.g., Galand and Chakrabarti 2002; Fox et al. 2008; Slanger et al. 2008b).

Assessing the response of an upper atmosphere to auroral particles and analyzing auroral emissions quantitatively require comprehensive modelling tools, describing the transport of these energetic particles in an atmosphere. One of the key inputs of such tools is the collision cross section set between the energetic particles—electrons, ions, or neutrals—and the atmospheric species (e.g., H_2, H, He, CO_2, CO, CH_4, N_2, N, O_2, O). The type of collisions includes ionization, dissociation, and excitation of the neutral species, as well as scattering and excitation of the energetic particles and charge-changing collisions in the case of energetic ions or neutrals. The auroral particle energies extend to larger values than that of the photoelectrons, reaching a few tens of keV in the auroral regions at Earth, up to a few tens of MeV at Jupiter. The minimum energy to consider for describing the transport of auroral particles in a planetary atmosphere is the ionospheric thermal energy for electrons ($\sim$0.1 eV) and the minimum collision threshold energy for ions and neutrals when cross section data are available. The assessment of physical quantities from auroral analysis is significantly limited by the uncertainties in impact cross sections. For illustration, the energy flux and mean energy of the incident particles derived from auroral emission analysis are changed by up to 16% and 23%, respectively, for 10-keV incident electrons, as a result of the mere use of different N_2 ionization cross section sets (Germany et al. 2001).

Apart from the particle impact cross sections, reaction rates are also required for analyzing the auroral emissions which are associated with excited states produced or lost through chemical reactions, such as the OI 630.0 nm red line (e.g., Lummerzheim et al. 2001) and the OI 557.7 nm green line (e.g., Jones et al. 2006). Photo-absorption cross sections are needed when the auroral emissions undergo 'true' absorption by atmospheric neutrals. This is the case of the following emissions widely used in auroral particle diagnostic: N_2 Lyman-Birge-Hopfield (LBH) band emissions partially absorbed by O_2 in the terrestrial atmosphere (e.g. Galand and Lummerzheim 2004) and H_2 Lyman and Werner band emissions partially absorbed by the hydrocarbon layer at Jupiter (e.g. Régo et al. 1999). Finally, in addition to photo-absorption cross sections, scattering cross sections (e.g., resonance) are required when the auroral photons experience significant scattering in the atmosphere. Radiative transfer needs to be taken into account for modeling the OI 130.4 nm resonant triplet in the Earth's auroral regions (e.g. Gladstone 1992) and the H Lyman α spectral profile at the giant planets (e.g. Régo et al. 1999).

Thanks to recent laboratory measurements or theoretical derivations, more accurate cross section sets, such as electron impact ionization cross sections of N_2 (Shemansky and Liu 2005) and CH_4 (Liu and Shemansky 2006), excitation cross sections associated with N_2 LBH band emissions (Johnson et al. 2005), excitation cross sections of O_2 (Kanik et al. 2003; Jones et al. 2006) and proton charge-transfer cross section with O (Pandey et al. 2007), are now available. They will help towards more reliable auroral transport modelling results at planets and moons in the Solar System, with possible implication on auroral diagnostic (e.g., Johnson et al. 2005). An illustration of recent progress in auroral physics, made possible thanks to the availability of new cross section data, is proposed by Kharchenko et al. (2006). They used new S^{q+} and updated O^{q+} cross sections for a more reliable modeling of the X-ray spectra induced by precipitating oxygen and sulfur ions. The satisfactory agreement obtained by these authors between the modeled and observed X-ray spectra in the auroral regions of Jupiter provides a reconciliation between both datasets. It also allows the identification of the types of precipitating particles—an equal mixture of oxygen and sulfur ions—and an estimation of their energies, which is consistent with a magnetospheric origin and acceleration processes up to a few MeV/amu.

Despite the new cross sections made available the past years, there is still a true need for additional measurements or modelling due to lack of data or dataset with too restricted energy coverage, too large uncertainness or too low spectral resolution (e.g., Lindsay and Stebbings 2005; Galand and Chakrabarti 2006; Karwasz et al. 2006). Finally, the applicability of measurements obtained under laboratory conditions to space environment should be borne in mind (Lindsay and Stebbings 2005).

2.5 Tenuous Atmospheres

On a planetary body with a significant atmosphere, the density decreases with increasing altitude until the atmospheric molecules move very large distances in ballistic trajectories and collisions between atmospheric species are improbable. This region of the atmosphere, called the corona or the exosphere, directly interacts with the space environment, as described by Johnson et al. (2008) and Ledvina et al. (2008). Similarly the 'airless' bodies, such as the Moon, Mercury, Saturn's main rings, Europa and many of the other large satellites have a nearly collisionless, gravitationally bound ambient gas forming a tenuous atmosphere. A fraction of the atoms and molecules in these atmospheres escape and enter the plasma environment. The escaping or gravitationally bound atoms and molecules can be ionized by photon and electron impact, the same processes which form the ionosphere. Because the interaction between the planetary corona and the ambient fields is complex, the principal problem has been lack of accurate the electron temperatures and densities rather than the lack of cross sections. In this region additional processes become important: charge exchange and knock-on collisions with the ions in the incident plasma, either the background solar wind or magnetospheric ions or the locally formed pick-up ions.

Since tenuous atmospheres are simulated by Monte Carlo models, the primary collision processes dominate so that one needs to know the ionization, charge exchange and angular differential cross sections. For most atomic ion-neutral collisions either data is available, there are good models for extrapolating data (Johnson 1990), or there are freeware programs (e.g., the SRIM package, Ziegler 2008) to estimate the cross sections and the deflections produced. Having said that, detailed angular-differential cross section for elastic collisions between open-shell atoms have been calculated for only a few systems (e.g., O + O: Kharchenko et al. 2000; Tully and Johnson 2001). Such cross sections are critical in describing the escape depths (Johnson et al. 2008). If the collisions in the exobase region are molecular, it is inappropriate to use the hard-sphere approximation to derive the differential cross section from the known total cross section, because that method assumes isotropic scattering.

If one needs to know the emission spectra produced by an incident ion, much less data is available. For fast ions, ionization typically dominates dissociation and the production of secondary electrons has again been well studied. Also, the ambient electron flux and the secondary electrons produced typically dominate the emission spectra.

However, many of interesting tenuous atmospheres have a molecular component. Therefore, the locally produced pick-up ions can also be molecular, so that impacts of energetic ($\geq$10's of eV) molecular ions with molecules can be the dominant heating process for the corona (Michael and Johnson 2005). For fast atomic ions colliding with molecules, again the ionization, total charge exchange and the secondary electron production are either measured or can be estimated from good models. However, the angular differential cross sections and dissociation channels are known for only a limited number of collision pairs and over limited energy range.

For charge exchange by very fast ions, which is dominated by distant collisions, one can often use dissociation product energies obtained for electron impact ionization and the deflections are often small. However, this is not the case for lower energy ($\leq$1 keV/u) processes. Recent measurements for H^+ and O^+ on O_2 (Luna et al. 2005) and H^+ and N^+ on N_2 (Luna et al. 2003) have improved the situation at Europa and Titan for a limited range of ion energies. However, energetic atomic or molecular ion interactions with H_2O, SO_2, and CO_2 at Europa, SO_2 at Io, CH_4 at Titan, Triton and Pluto, CO_2 and CO at Mars, etc. are not well described and simulations have relied on simple classical collision cross sections (Johnson et al. 2002). The data base for relevant molecular ions colliding with molecules, a critical process in Titan's corona, is very limited, except at very low, quasi-thermal energies ($\leq$1 eV) where ion-molecule reaction expressions can be used, as discussed earlier. The incident ions not only produce hot electrons but can, by charge exchange or by knock-on collisions, produce energetic neutrals. Total and angular cross sections and product distributions for collisions of energetic ($>$10 eV) neutrals with molecules are essentially not readily available. Recently, collisions of a hot ($>$10 eV) N with N_2 were calculated (Tully and Johnson 2002, 2003).

See also Sects. 3.6.2 and 3.7.1 of this chapter.

3 Representative Optical, Chemical, and Collisional Processes

3.1 Chemical Reactions of Neutrals

3.1.1 Recombination of Methyl Radicals

Planetary emissions of the methyl radical were observed for the first time in 1998 on Saturn and Neptune by the ISO mission satellite (Bezard et al. 1998, 1999). Concentrations were derived for valuable comparisons to models. CH_3 is produced by VUV photolysis of CH_4 and is the key photochemical intermediate leading to complex organic molecules on the giant planets and moons. Thus correct model predictions are required for the correct simulation of the photochemical synthesis. These observations form a very sensitive test of the mechanism, and as a reactive intermediate methyl is also a good marker for the transport parameterizations employed.

A very sensitive parameter controlling methyl concentrations is the loss process by recombination,

$$CH_3 + CH_3 + H_2 \rightarrow C_2H_6 + H_2 \tag{3}$$

which is also the main ethane formation step. At low upper atmosphere pressures, collisional deactivation of hot ethane intermediates by hydrogen bath gas is rate determining. Data is only available for temperatures above the 115–140 K needed, and only at much higher pressures. Extrapolation of existing rate expressions beyond their intended ranges gives differences of over two orders of magnitude (MacPherson et al. 1983, 1985; Slagle et al. 1988).

New theoretical master equation calculations were therefore performed, guided by the higher temperature pressure dependent rate constant data and theoretical calculations of transition state structures, to provide sound rate constants for this important reaction (Smith 2003). The results are intermediate to the prior extrapolations, and somewhat lower than used in recent model studies that match the methyl data (Lee et al. 2000; Moses et al. 2005). In regions of high H atom concentration, the

$$H + CH_3 \rightarrow CH_4 \tag{4}$$

recombination reaction is also important in determining methyl concentrations, so a similar calculation was performed for this step (Smith 2003).

These low pressure limit stabilization rate constants can increase rapidly at low temperatures, and for larger molecules any pressure dependence or falloff must also be considered. So it is important to determine the proper temperature dependence, and this often will need to be done theoretically because experimental results at low temperature and pressure are rare and difficult to obtain. Some other recombination reactions have small energy barriers but are still important, such as

$$H + C_2H_2 \leftrightarrow C_2H_3. \tag{5}$$

Tunneling has profound effects here for increasing low pressure and temperature rate constants orders of magnitude above normal expectations and extrapolations (Knyazev and Slagle 1996).

3.1.2 Three-Body Recombination of Oxygen Atoms

Dayside photoabsorption of solar ultraviolet (UV) radiation by molecular oxygen in the upper atmosphere of Earth, and by carbon dioxide in the upper atmospheres of Venus and Mars, results in production of atomic oxygen

$$O_2 + h\nu \rightarrow O + O \quad \text{Earth} \tag{6}$$

$$CO_2 + h\nu \rightarrow O + CO \quad \text{Venus and Mars} \tag{7}$$

that eventually undergoes three-body recombination (Huestis 2002)

$$O + O + M \rightarrow O_2(^5\Pi_g, A^3\Sigma_u^+, A'^3\Delta_u, c^1\Sigma_u^-, b^1\Sigma_g^+, a^1\Delta_g, X^3\Sigma_g^-) + M. \tag{8}$$

On all three planets, the competition between photodissociation, recombination, and diffusive vertical transport controls the transition between the homosphere and the heterosphere, and thus the atomic and molecular composition of the mesosphere and lower thermosphere. After transport to the nightside by planetary rotation or by high-altitude atmospheric winds, the chemical energy stored in O atoms is converted by three-body recombination into electronic, vibrational, and rotational energy, eventually appearing as ultraviolet, visible, and infrared nightglow emissions (250–1270 nm range), which have been the subject of many laboratory and interpretive investigations. Oxygen atom recombination is the only source for O_2 nightglow and the resulting emissions of electronically excited O_2 are key tracers for photochemical and wave activity near the mesopause. Moreover, O-atom recombination contributes significantly to the total thermospheric heating rate below about 100 km (Roble 1995). Knowledge of the temperature-dependent rate coefficient for recombination of atomic oxygen is thus essential for accurate modeling of the atmospheric composition and energetics. Equally important is a detailed understanding of the electronic states of O_2 produced in oxygen atom recombination, and of what can be learned from their nightglow emissions.

Until recently, the most modern measurement of the rate coefficient for O-atom three-body recombination was 35 years old (Campbell and Gray 1973). The most recent comprehensive review (Baulch et al. 1976) is also over 30 years old and shows that the absolute rate coefficients for recombination and the reverse process, collision-induced dissociation, as well as the dependence on temperature and collider, were poorly determined, in spite of the relatively narrow error bars reported in individual studies. The available information is illustrated in Fig. 1.

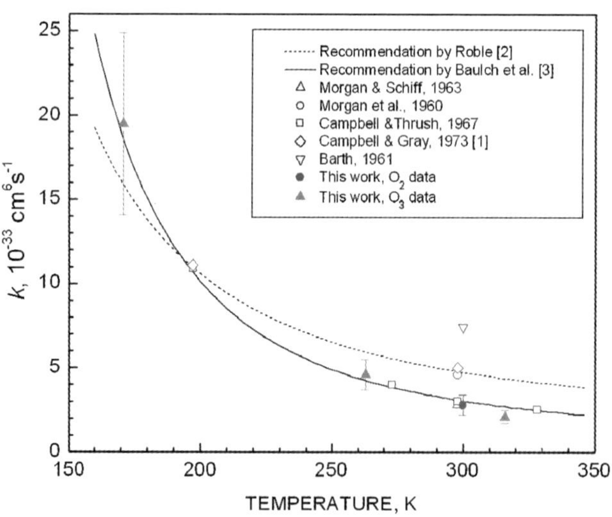

Fig. 1 Comparison of recommendations and measurements of the rate coefficient for oxygen atom three-body recombination in nitrogen. *Solid line*: recommendation by Baulch et al. (1976), adopted by the combustion modeling community. *Dashed line*: recommendation by Roble (1995), adopted by the atmospheric modeling community. *Open circle*: measurement by Morgan et al. (1960). *Open down triangle*: measurement by Barth (1961). *Open up triangle*: measurement by Morgan and Schiff (1963). *Open squares*: measurements by Campbell and Thrush (1967). *Open diamonds*: measurements by Campbell and Gray (1973). *Filled up triangles*: measurements using laser photodissociation of O_3 by Smith and Robertson (2008). *Filled circle*: measurement using laser photodissociation of O_2 by Pejaković et al. (2005, 2008)

Two recent laser-based experimental investigations have measured rate coefficients for O-atom three-body recombination in nitrogen (Pejaković et al. 2005, 2008; Smith and Robertson 2008). One experiment employed the pulsed output of a fluorine laser at 157.6 nm to achieve high degrees of photodissociation of molecular oxygen. In a high-pressure (1 atm) background of N_2, the produced oxygen atoms recombine in a time scale of several milliseconds. The O-atom population is monitored by two-photon laser-induced fluorescence at 845 nm, using a second laser with output near 226 nm. The measured value of the rate coefficient at room temperature is $(2.7 \pm 0.3) \times 10^{-33}$ cm^{6}s^{-1} (1-σ uncertainty), a value a factor of 2 lower than that currently adopted by the atmospheric modeling community. The second experiment employed a 248 nm KrF laser to achieve 100% photodissociation of molecular ozone. The time scale for O-atom recombination was followed by the same two-photon laser-induce-fluorescence technique. Recombination rate coefficients were measured at 170, 260, 300, and 315 K. The results of these two experiments are displayed in Fig. 1. These modern measurements are clearly consistent with the 30-year-old recommendations in use by the combustion modeling community (Baulch et al. 1976) and clearly inconsistent with the values favored by the atmospheric modeling community (Roble 1995).

The rate of recombination in CO_2 has not been measured, but has been estimated at 200 K to be about a factor of 2.5 faster than that in N_2 (Nair et al. 1994; Slanger et al. 2006). In both N_2 and CO_2 the overall yields of $O_2(a^1\Delta_g)$ and 1.27 μm radiation are believed to be close to unity, after accounting for collisional relaxation of the higher O_2 excited states produced initially by recombination (Huestis 2002; Slanger et al. 2006).

3.2 Ion Molecule Reactions

The diversity of the ionized molecules in planetary ionospheres was introduced above in Sects. 2.2 and 2.3 of this chapter. Correspondingly, complex sequences of chemical reaction transform the primary products of ionization by photon, electron, or heavy particle impact into the species that are eventually neutralized by recombination with electrons. In this section we review ion–molecule reactions that control the ionospheric electron density on the giant planets. Other important ion–molecule reactions are described in Sects. 3.3.1, 3.4.3, and 3.6.1 of the chapter. Also see the extensive collections of rate coefficients and recommendations by Anicich (2003).

3.2.1 $H_2^+ + H_2 \rightarrow H_3^+(v) + H$

H_3^+ made a surprise appearance in the very first molecular mass spectrum (Thompson 1911). Observation of M/Z of 3 instead of the expected M/Z of 2 is now understood as resulting from the fast reaction

$$H_2^+ + H_2 \rightarrow H_3^+(v) + H. \tag{9}$$

The H_3^+ laboratory infrared spectrum was first reported 69 years later (Oka 1980), observed in outer planet infrared aurorae about a decade thereafter (Drossart et al. 1989; Oka and Geballe 1990; Geballe et al. 1993; Trafton et al. 1993), and a widely used diagnostic of and initiator of isotopic and heavy-atom chemistry in interstellar clouds (Gerlich et al. 2002). The literature of experimental and theoretical investigations of the spectroscopy, electronic structure, potential energy surfaces, and chemical reactions of H_3^+ is far too vast for an adequate review here. Suffice it to say that H_3^+ is now better known than almost any other polyatomic molecule. Correspondingly, Reaction (9) is plausibly the most important ion–molecule reaction in outer planet ionospheres, and in the universe as a whole. The rate coefficient is known. The product H_3^+ ion is known to contain vibrational energy, but the vibrational population distribution is currently unknown (Kim et al. 1974; Huestis and Bowman 2007; Huestis 2008).

3.2.2 $H^+ + H_2(v) \rightarrow H + H_2^+$ *or* $H^+ + H_2(v')$

The Pioneer and Voyager radio occultation experiments found electron densities in the ionospheres of the giant planets that were about an order of magnitude smaller than expected. Subsequent measurements from the Galileo and Cassini spacecraft have confirmed these observations. This has been one of the major puzzles in understanding planetary ionospheres.

The technical problem is illustrated by the following simple model:

$$H_2 + h\nu \rightarrow e + H_2^+ \text{ or } e + H^+ + H \tag{10}$$

$$H_2^+ + H_2 \rightarrow H_3^+ + H \qquad \text{(fast)} \tag{11}$$

$$H_3^+ + e \rightarrow H_2 + H \text{ or } 3H \qquad \text{(fast)} \tag{12}$$

$$H^+ + 2H_2 \rightarrow H_3^+ + H_2 \qquad \text{(slow)} \tag{13}$$

$$H^+ + e \rightarrow H + h\nu \qquad \text{(slow).} \tag{14}$$

Unless some new reaction is found to convert H^+ into H_2^+, H_3^+, or some other species that rapidly leads to electron–ion recombination, models predict that the protons (and thus the electrons) will reach greater densities than is consistent with observations.

As a result, modelers seized on the suggestion by McElroy (1973) that the endothermic charge transfer reaction

$$H^+ + H_2(v) \rightarrow H + H_2^+ \tag{15}$$

becomes exothermic for vibrational levels $v \geq 4$, and therefore might be expected to be fast. That point is illustrated in Fig. 2, which shows that the two asymptotic channels $H^+ + H_2$ and $H + H_2^+$ have the same energy at one value of the H_2 or H_2^+ internuclear distance, $R_{vib} \approx 2.5$ bohr or 1.32 Å. A number of modeling studies (Cravens 1974, 1987; Atreya et al. 1979; McConnell et al. 1982; Moses and Bass 2000; Hallett et al. 2004, 2005a, 2005b; Moore et al. 2004; Majeed et al. 2004) have followed the McElroy suggestion. Given a rate coefficient for Reaction (15), the key piece of missing information would be the H_2 vibrational distribution. A few studies explored reactions producing vibrationally excited hydrogen. Others used parameterized models of the vibrational distributions in the giant planet ionospheres, represented as altitude dependent rates for Reaction (15) or Reaction (14). Reaction (15) would be an important loss process for magnetic confinement fusion. A mitigating companion process is the vibrational relaxation reaction

$$H^+ + H_2(v_i) \rightarrow H^+ + H_2(v_f < v_i) \tag{16}$$

that had not been considered in previous ionospheric modeling studies.

We can use publications from the plasma fusion quantum theory community (Ichihara et al. 2000; Krstic 2002; Krstic et al. 2002; Krstic and Schultz 2003; Janev et al. 2003) to estimate rates of Reactions (15) and (16) at atmospheric temperatures (Huestis 2005b, 2008). Two studies (Ichihara et al. 2000; Krstic et al. 2002) investigated the charge transfer Reaction (15). By extrapolating to lower temperatures the results from the earlier study (Ichihara et al. 2000, Table 2) we estimate that Reaction (15) has a rate coefficient at 600 K

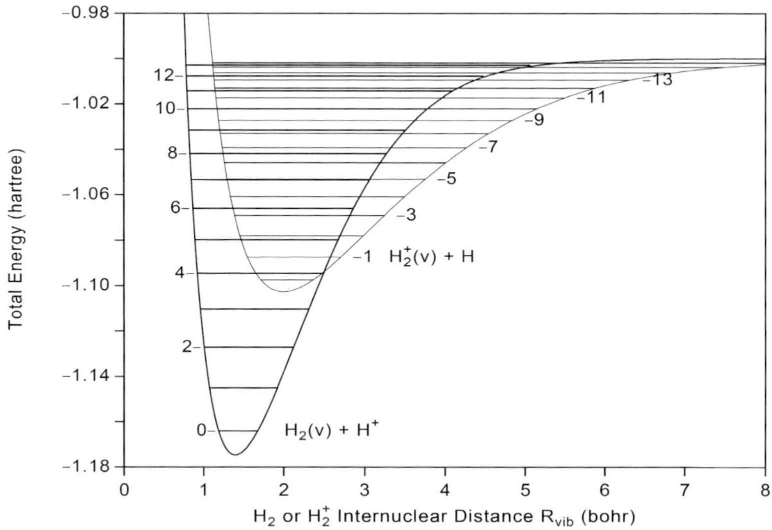

Fig. 2 H_2 and H_2^+ potential energy curves and vibrational energies

Table 1 Recommended rate coefficients at 600 K for $H^+ + H_2(v)$ charge transfer (k_{CT}) and vibrational relaxation (k_{VR}) in units of 10^{-9} cm^3/s

	v								
	0	1	2	3	4	5	6	7	8
k_{CT}	0	0	0	0	0.6	1.3	1.3	1.3	1.3
k_{VR}	0	1.2	1.8	1.8	1.5	1.2	1.2	1.2	1.2

of approximately 1.3×10^{-9} cm^3/s, for $v \geq 4$, consistent with numbers in current models. However, the later study (Krstic et al. 2002) (using more capable theory) found cross sections at thermal energies for $v = 4$ that are much smaller than those for $v \geq 5$. As a result of this disagreement, we have arbitrarily reduced the "recommended" rate coefficient for $v = 4$ shown in Table 1 by 50% compared to those for $v = 5$–8.

Another fusion-motivated investigation (Krstic 2002, Fig. 11) suggested that the vibrational relaxation Reaction (16) should be fast. From graphical analysis of this work we recommend thermal-energy rate coefficients of between 1.2×10^{-9} cm^3/s and 1.8×10^{-9} cm^3/s as indicated in Table 1. New quantum theory calculations are underway to reduce the uncertainties in these recommendations at low energy (Quemener et al. 2008).

Inclusion of Reaction (16) will significantly reduce calculated vibrational temperatures in ionospheric models. Protons are less abundant than neutral hydrogen atoms by a factor of about 10,000 and the proton rate coefficients for vibrational relaxation are about a factor of 10,000 larger. Just as important, Reaction (16) depletes excited vibrational population in vibrational levels 1, 2, and 3, which will contain the vast majority of vibrational energy for plausible vibrational distributions. The results of our earlier analysis (Huestis 2005b) have been adopted in a recent ionospheric modeling study (Moore et al. 2006), which suggests that the high influx of water indicated by Cassini observations will lead to a different mechanism for reducing the ionospheric electron density on Saturn (Connerney and Waite 1984;

Waite et al. 1997; Maurellis and Cravens 2001):

$$H^+ + H_2O \rightarrow H + H_2O^+. \quad (17)$$

3.3 Collisions of Excited Electronic States

3.3.1 Relaxation of $O^+(^2D)$ *and* $O^+(^2P)$ *in Collisions with* N_2 *and* $O(^3P)$

Atomic oxygen ions are primary charge carriers in the Earth's ionosphere. Optical emissions from the metastable excited states $O^+(^2D)$ and $O^+(^2P)$ provide useful diagnostics of energy deposition processes. The $O^+(^2P \rightarrow {}^2D,{}^4S)$ emissions near 732 and 247 nm are common features of the daytime and nighttime airglow. Previously, atmospheric $O^+(^2D \rightarrow {}^4S)$ emissions near 373 nm were known only during polar cusp aurorae (Sivjee 1983, 1991).

New observational information has been provided by recent analysis of sky spectra from the VLT (Very Large Telescope) in Chile during periods of large solar storms: 6–7 April 2000, 6–7 Nov. 2001, and 28 Oct.–1 Nov. 2003 (O'Neill et al. 2006; Slanger and Cosby 2007). $O^+(^2D \rightarrow {}^4S)$ 373 nm emissions are prominent and show a strong correlation with the Dst (disturbance storm time) index. Interpretation of the relative strengths of the $O^+(^2P \rightarrow {}^2D)$ 732 nm and $O^+(^2D \rightarrow {}^4S)$ 373 nm emissions requires reliable values for the rates of relaxation of the excited states in collisions with the principal components of the neutral ionosphere, N_2 and $O(^3P)$.

The current status of laboratory measurements and atmospheric modeling inferences is summarized in Table 2, along with recommended rate coefficients.

Of the four laboratory studies of charge-transfer, electronic-deexcitation, and ion-molecule reactions in collisions of $O^+(^4S,^2D,^2P)$ with N_2, only the most recent actually resolved the composition of the reactants, i.e., the relative abundance of $O^+(^4S)$, $O^+(^2D)$,

Table 2 Literature and recommended values of rate coefficients for collisional removal of $O^+(^2P)$ and $O^+(^2D)$, in units of 1.0×10^{-10} cm^3/s, taken from Huestis et al. (2007)

Reference	Method	$O^+(^2P) + N_2$	$O^+(^2P) + O(^3P)$	$O^+(^2D) + N_2$	$O^+(^2D) + O(^3P)$
WTH75	airglow	0.5 or 5	2 or 0		
RTH77	airglow	4.8	0.52		
GRT78	laboratory	1.5 ± 0.45^a		1.5 ± 0.45^a	
JB80a-b	laboratory	8 ± 2^a		8 ± 2^a	
RFF80	laboratory	8.3 ± 3.4^a		8.3 ± 3.4^a	
ATR84	N_2^+ model	4.8^b	0.52^b	8^b	<0.05
CTR93	airglow	3.4 ± 1.5	4.0 ± 1.9		
LHF97	laboratory	2.0 ± 0.5		1.5 ± 0.4	
SMD03	airglow	1.8 ± 0.3	0.50 ± 0.34		
Recommendations		2.0 ± 0.5	0.4 ± 0.2	1.5 ± 0.4	0.6 ± 0.3

[a]Unknown mixture of $O^+(^2P) + O^+(^2D)$

[b]Values selected from works above

ATR84 = Abdou et al. 1984; CTR93 = Chang et al. 1993; GRT78 = Glosik et al. 1978; JB80a-b = Johnsen and Biondi 1980a, 1980b; LHF97 = Li et al. 1997; RFF80 = Rowe et al. 1980; RTH77 = Rusch et al. 1977; SMD03 = Stephan et al. 2003; WTK75 = Walker et al. 1975 and $O^+(^2P)$, as well as the identity of the product ion, e.g., N_2^+, NO^+, or N^+. The most recent laboratory and airglow modeling numbers agree for $O^+(^2P)$, providing strong support for each other and for the laboratory value for $O^+(^2D)$. There are no published laboratory or theoretical constraints for collisions with $O(^3P)$. The airglow modeling numbers similarly provide only very broad limits. Qualitative theoretical analysis (Huestis et al. 2007) based on the best-available O_2^+ potential energy curves by Beebe et al. (1976), suggests that we expect numbers around $5 \times 10^{-11}\ \mathrm{cm^3/s}$ for both $O^+(^2D)$ and $O^+(^2P)$, with the rate coefficient for $O^+(^2D)$ likely to be about twice as large as that for $O^+(^2P)$.

3.3.2 *Rates and Products of Collisional Relaxation of* $O_2(a^1\Delta_g)$ *and* $O_2(b^1\Sigma_g^+)$

The low-lying electronic excited states of molecular oxygen, $O_2(a^1\Delta_g)$ and $O_2(b^1\Sigma_g^+)$ are produced directly or indirectly after absorption of ultraviolet solar radiation in the atmospheres of Venus, Earth, and Mars. For example

$$O_3 + h\nu \rightarrow O(^1D) + O_2(a^1\Delta_g) \quad (18)$$

$$O(^1D) + O_2(X^3\Sigma_g^-) \rightarrow O(^3P) + O_2(b^1\Sigma_g^+) \quad (19)$$

$$O(^3P) + O(^3P) + M \rightarrow O_2(a^1\Delta_g) \quad \text{and} \quad O_2(b^1\Sigma_g^+). \quad (20)$$

At high altitudes, both $O_2(a^1\Delta_g)$ and $O_2(b^1\Sigma_g^+)$ are weakly quenched and produce strong airglow emission at 1270 nm (Venus, Earth, and Mars) and 762 nm (Earth), respectively.

At lower altitudes, oxygen molecules in either electronic state undergo collisional relaxation. The question to be addressed here is what can be said about the rates and products of this relaxation in the terrestrial atmosphere. More specifically, we wish to know how much of the electronic energy eventually ends up in vibrational excitation in $O_2(X^3\Sigma_g^-)$ which may contribute to observable infrared emission after vibrational energy transfer to H_2O or CO_2. A portion of the following analysis has already been published (Huestis 2005a). The reactions in question are

$$O_2(b^1\Sigma_g^+, v=0) + O_2(X^3\Sigma_g^-, v=0) \rightarrow O_2(X^3\Sigma_g^-, v') \text{ or } O_2(a^1\Delta_g, v') + O_2(X^3\Sigma_g^-, v'') \quad (21)$$

$$O_2(a^1\Delta_g, v=0) + O_2(X^3\Sigma_g^-, v=0) \rightarrow O_2(X^3\Sigma_g^-, v') + O_2(X^3\Sigma_g^-, v''). \quad (22)$$

While no direct measurements exist of the product distributions in either reaction, enough indirect information is available for construction of relatively reliable estimates.

Combining the data from low temperature (Seidl et al. 1991), room temperature (Sander et al. 2006), and high temperature shock tube studies (Borrell et al. 1979, 1982) we find that the temperature dependence of overall rate of electronic relaxation of $O_2(b^1\Sigma_g^+,\ v=0)$ in collisions with $^{16}O_2(X^3\Sigma_g^-,\ v=0)$ can be adequately represented from 90 to 1800 K by the formula

$$k_{21}(T) = 3 \times 10^{-18} + 4.5 \times 10^{-11} \exp(-91/T^{1/3})\ \mathrm{cm^3/s}. \quad (23)$$

The form of this expression is a generalization of the Landau and Teller (1936) representation of vibrational and rotational relaxation (Huestis 2006a, 2008).

The overall rate of electronic relaxation of $O_2(a^1\Delta_g,\ v=0)$ in Reaction (22) has been measured near room temperature in a number of investigations (Sander et al. 2006), in liquid oxygen (Huestis et al. 1974; Protz and Maier 1980; Faltermeier et al. 1981; Klingshirn et al.

1982; Wild et al. 1982, 1984), and more recently in gaseous oxygen between 100 and 500 K (Billingham and Borrell 1986; Chatelet et al. 1986; Seidl et al. 1991). The gas phase data in normal oxygen can be represented adequately by the formula

$$k_{22}(T) = 0.14 \times 10^{-18} + 1.43 \times 10^{-18}\,(T/298). \tag{24}$$

Many investigators have reported high yields of $O_2(a^1\Delta_g)$ in Reaction (21) in the gas phase, including relaxation in collisions with H_2O, CO_2, and other collision partners (Singh and Setser 1985; Knickelbein et al. 1987; Wildt et al. 1988; Sander et al. 2006). Experiments performed in natural and isotopic liquid oxygen provide additional information about the resulting vibrational distribution (Klingshirn and Maier 1985). In natural liquid oxygen $O_2(b^1\Sigma_g^+)$ is quenched by $O_2(X^3\Sigma_g^-)$ with a rate coefficient of 5.5×10^{-18} cm^3/s, while in liquid $^{18}O_2$, the rate coefficient is only 3×10^{-19} cm^3/s. Subsequent experiments in low temperature gaseous oxygen also showed that $O_2(b^1\Sigma_g^+)$ is quenched about 10 times more slowly in $^{18}O_2$ (Seidl et al. 1991). The slower quenching in $^{18}O_2$ is interpreted as indicating that vibrational energy resonance is a critical factor, corresponding to the fact that the translational/rotational energy released in the reaction

$$O_2(b^1\Sigma_g^+, v=0) + O_2(X^3\Sigma_g^-, v=0) \rightarrow O_2(a^1\Delta_g, v') + O_2(X^3\Sigma_g^-, 3-v') \tag{25}$$

is about 150 cm^{-1} greater for $^{18}O_2$ than for $^{16}O_2$, implying that quenching of $O_2(b^1\Sigma_g^+)$ produces $O_2(a^1\Delta_g)$ plus three quanta of vibration. A similar discussion of vibrational energy resonance is used to explain why electronic relaxation of $O_2(b^1\Sigma_g^+,\ v=0)$ is so much faster at room temperature when the collider is N_2, producing $O_2(a^1\Delta_g, v=2) + N_2(v=1)$.

A similar argument applies for identification of the vibrational product distribution for relaxation of $O_2(a^1\Delta_g)$. Investigations in liquid $^{18}O_2$ showed much slower relaxation (Klingshirn et al. 1982; Wild et al. 1984; Seidl et al. 1991), with a rate coefficient of 2×10^{-21} cm^3/s compared to 1×10^{-18} cm^3/s in $^{16}O_2$. In low temperature gaseous oxygen, relaxation in $^{18}O_2$ is also slower by a similar factor (Seidl et al. 1991). This was explained by the observation that the translational/rotational energy released in the reaction

$$O_2(a^1\Delta_g) + O_2(X^3\Sigma_g^-) \rightarrow O_2(X^3\Sigma_g^-, v) + O_2(X^3\Sigma_g^-, 5-v) \tag{26}$$

is more than 400 cm^{-1} greater for $^{18}O_2$ than for $^{16}O_2$, implying that quenching of $O_2(a^1\Delta_g)$ produces $O_2(X^3\Sigma_g^-)$ plus five quanta of vibration. This conclusion is supported by one of the liquid oxygen studies (Wild et al. 1982) in which anti-Stokes Raman scattering was used to infer production of 3.8 ± 0.8 vibrational quanta in $O_2(X^3\Sigma_g^-, v)$.

3.4 Relaxation and Excitation of Rotational and Vibrational Levels

3.4.1 Vibrational Energy Transfer and Relaxation in O_2 and H_2O

Vibrational energy transfer from oxygen molecules to water molecules helps control the local temperature in the Earth's mesosphere through radiative cooling. Infrared emissions from water molecules provide remote diagnostics for the altitude profile of water. Modeling these emissions is made more complicated by the near degeneracy between the first excited vibrational levels of the water and oxygen molecules. The rate of vibrational energy exchange has been of interest to atmospheric scientists, combustion modelers, and developers of chemical lasers. It has been recently inferred from analysis of unpublished laser-based laboratory experiments (Huestis 2006a). The reactions in question are

$$O_2(1) + O_2(0) \rightarrow O_2(0) + O_2(0) \quad (27)$$

$$O_2(1) + H_2O(000) \rightarrow O_2(0) + H_2O(000) \quad (28)$$

$$O_2(1) + H_2O(000) \leftrightarrow O_2(0) + H_2O(010) \quad (29)$$

$$H_2O(010) + H_2O(000) \rightarrow H_2O(000) + H_2O(000) \quad (30)$$

$$H_2O(010) + O_2(0) \rightarrow H_2O(000) + O_2(0). \quad (31)$$

The chemical kinetics and sound absorption literatures provide reliable values for the rates of Reactions (27) and (30) and strong evidence that Reactions (28) and (31) are slow in comparison with Reaction (29). Our analytical solution to the chemical reaction system above shows that the rate of Reaction (29) can only be measured with water mole fractions higher than 1%. The only measurement that satisfies this requirement is reported in a Ph.D. thesis (Diskin 1997) and a conference presentation (Diskin et al. 1996) from the combustion community. Reanalysis of that data yields our recommended value of $(5.5 \pm 0.4) \times 10^{-13}$ cm^3/s at 300 K, between the values favored by the atmospheric and laser modeling communities.

3.4.2 O–CO_2 *Cooling*

As indicated in Sect. 2.1.2 above, infrared emissions from bending-mode excited carbon dioxide, $CO_2(010)$, is a primary regulator of thermal balance in the atmospheres of Venus, Earth, and Mars. At higher altitudes the time scale for infrared fluorescence is shorter than the collisional excitation time. Oxygen atoms are expected to be the principle collision partner in the vibrational reexcitation. The rate coefficient remains a matter of some dispute, with laboratory experiments providing values that are factors of 2 or more lower than those preferred by some modelers. Here we will review the current status of laboratory experiments and theory, which the former continue to confirm the preference for rate coefficients in the low range.

The key reaction in question is

$$CO_2(v_2 = 0) + O(^3P_{J'}) \leftrightarrow CO_2(v_2 = 1) + O(^3P_J). \quad (32)$$

While the rate in the excitation direction, as written, is the key parameter for modeling atmospheric infrared emission, laboratory measurements are usually reported in the de-excitation or relaxation direction, which atmospheric scientists have also adopted for purposes of comparison.

The results from laboratory investigations at room temperature are summarized in Table 3, along with the summary recommendation provided here and adopted above in Sect. 2.1.2. The several diverse laboratory approaches are all returning mutually consistent rate coefficients for the presumed key reaction that are all well below the inferences of atmospheric modelers, which suggests that uncertainty in rate coefficient is not the critical issue. As discussed below, this conclusion is supported by theoretical studies and the experimentally observed temperature dependence.

Three rather old theoretical studies (Bass 1974; Schatz and Redmon 1981; Harvey 1982) of excitation of $CO_2(010)$ in collisions with oxygen atoms all ignored the fact that the open-shell $O(^3P)$ atom actually generated three electronic states, one of symmetry $^3A'$ and two of

Table 3 Laboratory measurements of the rate coefficient for relaxation of $CO_2(010)$ in collision with $O(^3P)$ at 300 K

Year	k (in units if 10^{-12} cm^3/sec)	Reference
1991	1.5 ± 0.5	Shved et al. 1991
1993	1.2 ± 0.2	Pollock et al. 1993
1994	0.5 ± 0.2	Lilenfeld 1994
2002	1.4 ± 0.2	Khvorostovskaya et al. 2002
2006	1.8 ± 0.3	Castle et al. 2006
2007	1.6 ± 0.2	Dodd et al. 2007
2008	1.5 ± 0.2	Recommendation, this work

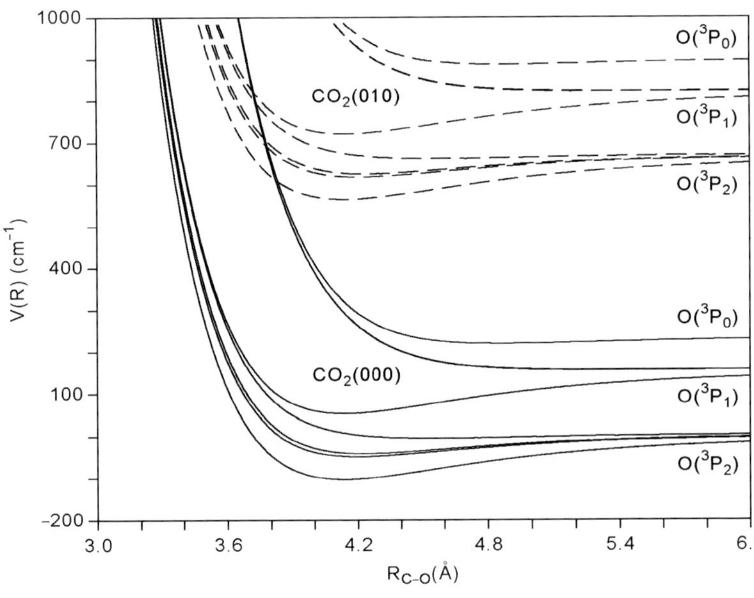

Fig. 3 Potential surfaces for $O(^3P) + CO_2(\nu_2 = 0)$, *solid lines*, and $O(^3P) + CO_2(\nu_2 = 1)$, *dashed lines*, with spin–orbit coupling, for 22.5° approach, from Huestis et al. (2002)

symmetry $^3A''$. Scott et al. (1993) provided the first curve-crossing model, following Nikitin and Umanski (1972), which enabled them to show that rate coefficients on the order of $1\text{–}2 \times 10^{-12}$ cm^3/s are indeed plausible.

Huestis et al. (2002) constructed the first set of model potential energy surfaces, including the three electronic states, calibrated by *ab initio* calculations. When spin–orbit coupling is included, a total of nine potential surfaces result, as shown in Fig. 3. The more recent calculations by de Lara-Castells et al. (2006, 2007) of CO_2+ $O(^3P)$, also included a representation of spin–orbit coupling. The two independent sets of potential energy surfaces provide strong support for the concept that the $O(^3P_J)$ fine structure states play critical roles in $CO_2(010)$ excitation and relaxation, as illustrated by the reaction

$$CO_2(000) + O(^3P_{0,1}) \leftrightarrow CO_2(010) + O(^3P_{1,2}) \tag{33}$$

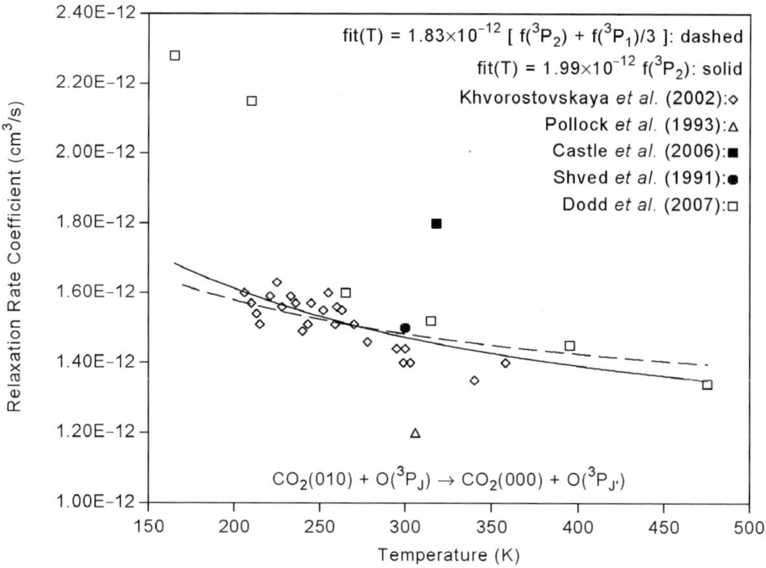

Fig. 4 Experimental rate coefficients and model fits for vibrational relaxation of $CO_2(010)$ in collisions with $O(^3P)$

which will be discussed further below. Comprehensive accurate chemical dynamics calculations have not been completed for this very complex system with four active atoms, nine active electronic states, and multiple potential surface crossings.

Additional theoretical guidance comes from the recent studies of vibrational relaxation at low temperature by Nikitin and Troe (Dashevskaya and Nikitin 2000; Dashevskaya et al. 2003, 2006, 2007; Nikitin and Troe 2006). They point out that quantum mechanics (Bethe 1937; Wigner 1948; Dashevskaya et al. 2003) implies the surprising conclusion that exoergic deexcitation reactions can have non-zero rate coefficients as the temperature approaches zero, in contrast to the normally expected $T^{1/2}$ dependence, as long as the interaction potential energy surface is attractive. They have verified this conclusion in accurate calculations of some simple vibrational relaxation collisions at low collision velocities (Dashevskaya et al. 2003).

Figure 3 shows that several of the potential surfaces arising from $CO_2(010) + O(^3P_{1,2})$ are attractive and intersect repulsive potential surfaces leading to curve-crossing deexcitation to $CO_2(000) + O(^3P_{0,1})$ at thermally accessible energies, even at low temperature. More detailed analysis of the potential surfaces in Fig. 3 suggests that all five of the potential surfaces arising from $O(^3P_2)$ and one arising from $O(^3P_1)$ are candidates.

Figure 4 shows the measured temperature-dependent $CO_2(010)$ deexcitation rate coefficients. The solid and dashed lines correspond to models in which we assume that the rate coefficient depends only on the temperature-dependent fraction of the oxygen atom population in the $O(^3P_2)$ and $O(^3P_1)$ spin–orbit sub-levels:

$$f(^3P_2) = 5/[5 + 3\exp(-225/T) + \exp(-326/T)] \tag{34}$$

$$f(^3P_1) = 3\exp(-225/T)/[5 + 3\exp(-225/T) + \exp(-326/T)]. \tag{35}$$

The quantitative adequacy of these simple models suggests that we are on the right conceptual track and that there is little reason to doubt the validity and accuracy of the laboratory measurements. The single-measurement outlying values around room temperature, $(1.2 \pm 0.2) \times 10^{-12}$ cm^3/s (Pollock et al. 1993) and $(1.8 \pm 0.3) \times 10^{-12}$ cm^3/s (Castle et al. 2006) are both statistically consistent with the nominal value of $(1.5 \pm 0.2) \times 10^{-12}$ cm^3/s. The other two outlying values, at 165 and 210 K, from the most recent experiments of Dodd et al. (2007), might suggest a possible importance of long-range attraction due to interaction of the CO_2 vibrational dipole moment with the O-atom quadrupole moment. Given the atmospheric importance of this temperature range, and the divergence from the low temperature values from Khvorostovskaya et al. (2002), additional low temperature experiments are needed.

3.4.3 Vibrationally Excited H_3^+ *in Outer Planet Atmospheres*

The observed temperatures of planetary upper atmospheres, ionospheres, and thermospheres depend to a significant extent on the balance between absorption of ultraviolet solar radiation and infrared emission. On the outer planets the primary emitter is H_3^+ on the ν_2 vibrational fundamental and overtone transitions between 2 and 4 μm bands. At higher altitudes the low atmospheric densities imply that the reduced rates of collisional excitation of the emitting levels will eventually fall below the rates of spontaneous emission, thus limiting the efficiency of infrared emission. See Sect. 3.2.1 of the chapter above for additional discussion of production, ion–molecule reactions, and electron-ion recombination of atomic and molecular hydrogen ions.

The primary sources of vibrationally excited $H_3^+(v_2 > 0)$ are the H_3^+ formation reaction (Bowers et al. 1973; Theard and Huntress 1974; Kim et al. 1974) (note that the initial $H_3^+(\nu_2)$ vibrational distribution is only weakly constrained)

$$H_2^+ + H_2 \rightarrow H_3^+(v_2 > 0) + H \tag{36}$$

and $T \rightarrow V$ and $V \rightarrow V$ excitation in collisions with ambient H_2

$$H_3^+(v_2) + H_2(v) \rightarrow H_3^+(v_2') + H_2(v'). \tag{37}$$

Observations of H_3^+ emissions on Jupiter (Lellouch 2006) suggest that the $H_3^+(\nu_2)$ vibrational distribution is non-thermal and cooler than the ambient translational temperature as inferred from the observed $H_3^+(J)$ rotational distribution. Departures from local thermodynamic equilibrium (LTE) of H_3^+ population distributions are also observed in interstellar clouds (Oka and Epp 2004) and have been the subject of previous modeling studies of the Jovian ionosphere (McConnell and Majeed 1987; Kim et al. 1992; Miller et al. 2000, 2006; Grodent et al. 2001; Melin et al. 2005, 2006). In addition, the potential extent of vibrational excitation of H_3^+ is important in modeling gas discharges (Phelps 2001) and fusion plasmas (Janev et al. 2003), from which we can derive guidance for modeling the ionospheres of the outer planets.

Key missing pieces of information include the rate coefficients for redistribution of proton labeling and vibrational and rotational energy in the H_5^+ rearrangement collision

$$H_3^+(v_1, v_2, J, G) + H_2(v, J) \rightarrow H_5^+(v'', J'') \rightarrow H_3^+(v_1', v_2', J', G') + H_2(v', J'). \tag{38}$$

Characterizing this reaction experimentally is overwhelmingly difficult. The first challenge would be to prepare reactants with well-defined initial quantum numbers. The second challenge is measurement of the product quantum numbers. The third challenge is presented by the fact that each of the five protons in the H_5^+ reaction intermediate complex could have come from either of the H_3^+ and H_2 reactants and could end up in either of the H_3^+ and H_2 products.

The third challenge could be considered an advantage if our actual interest were proton scrambling that leads to *ortho–para* conversion (Uy et al. 1997; Cordonnier et al. 2000; Oka 2004; Park and Light 2007). Alternatively, one could consider experiments and quantum theory calculations in which one or more of the H atoms are isotopically substituted by D atoms (Terao and Back 1969; Huntress and Anicich 1976; McMahon et al. 1976; Smith and Futrell 1976; Gerlich et al. 2002; Moyano and Collins 2003). In that case we have proton or deuteron transfer or more complicated rearrangement reactions such as

$$H_3^+ + D_2 \leftrightarrow (H_3D_2)^+ \leftrightarrow H_2 + HD_2^+ \text{ or } HD + H_2D^+ \tag{39}$$

$$D_3^3 + H_2 \leftrightarrow (H_2D_3)^+ \leftrightarrow D_2 + H_2D+ \text{ or } HD^+HD_2^+ \tag{40}$$

$$H_3^+ + HD \leftrightarrow (H_4D)^+ \leftrightarrow H_2 + H_2D^+. \tag{41}$$

In these cases the differences between the zero-point vibrational energies means that the various product channels have different non-zero reaction exothermicities or endothermicities. In addition, proton and deuteron transfer reactions should have different rates because of the mass-dependent kinetic isotope effect. Furthermore, *ortho*/*para* selection rules and nuclear-spin degeneracies imply that the various entrance and exit channels may have different statistical weights. Finally, well-defined experiments would normally begin with vibrationally relaxed H_3^+ or D_3^+ reactants and thus would not provide direct information about vibrational deexcitation.

In spite of these complexities, isotopic substitution studies provide important information about the extent to which internal vibrational and rotational energy are statistically redistributed in formation and decomposition of the $H_5^+(v'', J'')$ reaction intermediate in Reaction (38). Stated in another way, can we use the probability of proton transfer (Janev et al. 2003) as a measure of rate of vibrational energy exchange, excitation, and deexcitation of H_3^+ with H_2 in Reaction (38)? In addition, Reactions (39)–(41) are important in their own right in modeling isotope fractionation in planetary ionospheres and interstellar clouds.

Table 4 contains a summary of the available kinetic information on isotopic forms of Reaction (38). As we can see from Table 4, the reported rate coefficients cluster into two groups:

(a) Small values in the range of 2.0–5.2 $\times$ 10^{-10} cm^3/s
(b) Large values in the range of 6.6–14.5 $\times$ 10^{-10} cm^3/s.

In some cases, numbers are reported that differ by more than a factor of three for the same reaction at the same temperature. The earlier room temperature experiments (Terao and Back 1969; Kim et al. 1974; Huntress and Anicich 1976; McMahon et al. 1976; Smith and Futrell 1976), the one 10 K experiment (Gerlich et al. 2002), and the very recent theoretical study (Park and Light 2007) of hyperfine spin scrambling are in the "small numbers" group, while the most comprehensive experiment (Giles et al. 1992) and theory (Moyano and Collins 2003) investigations are in the "large numbers" group.

As indicated by the divergences shown in Table 4, the current state of knowledge is most unsatisfactory. The factor-of-three scatter means that we have no plausible basis for an appropriate estimate for the rate coefficient for our target process, vibrational excitation and relaxation in Reaction (38). Further confusion comes from the fact that proton and deuteron exchange reactions should always lead to some V–V and V–T equilibration and thus could

Table 4 Summary of literature values of rate coefficients for selected $H_3^+ + H_2$ isotopomeric ion molecule reactions. Total reaction rate coefficients (to all products) are listed in units of 10^{-10} cm^3/s

Reaction						T (K)	Ref.	Note
(38)	(41)	(39)		(40)				
$H_3^+ + H_2$	$H_3^+ + HD$	$H_3^+ + D_2$	$H_2D^+ + HD$	$HD_2^+ + HD$	$D_3^+ + H_2$			
	3.5		2.6	2.0		10	GHR02	(a)
	12.1					10	MC03	(b)
0.0						10	PL07	(c)
4.9						10	PL07	(d)
	12.0	14.5	8.5	8.0	5.2	80	GAS92	(a)
	11.3	10.4				80	MC03	(b)
3.5						80	PL07	(c)
6.6						80	PL07	(d)
		3.3			3.3	300	TB69	(a)
2.7						300	KTH74	(e)
	3.0					300	HA76	(a)
			2.6	3.5		300	MMB76	(a)
					4.0	300	SF76	(a)
	9.6	12.6	5.0	4.5	8.2	300	GAS92	(a)

(a) Experiment
(b) Theory
(c) Theory: hyperfine scrambling for *para* H_2
(d) Theory: hyperfine scrambling for normal H_2
(e) Experiment: vibrational relaxation

GAS92 = Giles et al. 1992; GHR = Gerlich et al. 2002; HA76 = Huntress and Anicich 1976; MMB76 = McMahon et al. 1976; KTH74 = Kim et al. 1974; MC03 = Moyano and Collins 2003; PL07 = Park and Light 2007; SF76 = Smith and Futrell 1976; TB69 = Terao and Back 1969

be thought to provide lower limits for vibrational relaxation. In contrast, almost all the numbers in Table 4 are significantly larger than our most direct estimate for vibrational relaxation of $(2.7 \pm 0.6) \times 10^{-10}$ cm^3/s (Kim et al. 1974).

Thus we are left with the following unanswered science questions

- What is the mechanistic origin and cause of the sub-thermal vibrational distributions of H_3^+ (v_2) observed in the upper atmospheres of the outer planets?
- What are the key reactions and temperature-dependent rate coefficients that should be included in atmospheric models?
- What are the origin and resolution of the divergence between the "small values" and "large values" of the rate coefficients reported in the laboratory and quantum theory literature?

To answer these questions new theoretical investigations are planned for quantum electronic structure and chemical dynamics calculations for Reactions (36) and (38) (Xie et al. 2005; Huestis and Bowman 2007).

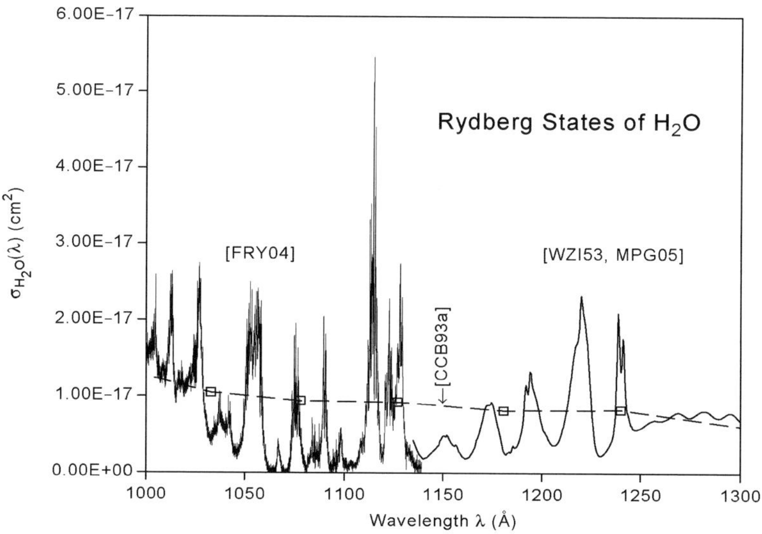

Fig. 5 A portion of the absorption cross section for water, H_2O, from Huestis and Berkowitz (2007). [CCB93a] = Chan et al. 1993; [FRY04] = Fillion et al. 2004; [MPG05] = R. Mota et al. 2005; [WZI63] = Watanabe et al. 1953

3.5 Photoabsorption

3.5.1 Absorption Cross Sections and Excited Photofragment Yields for CO_2 *and* H_2O

The Mariner flyby missions found strong ultraviolet emissions from CO_2^+(B,A), CO(a,A), and O(^{1}S) in the Mars dayglow. These observations, which have been confirmed by the recent Mars Express Mission, led to a number of laboratory quantum yield measurements.

Comet observers have supposed that the relative strengths of the O(^{1}S $\rightarrow$ ^{1}D) and O(^{1}D $\rightarrow$ ^{3}P) green (557.7 nm) and red (630.0 nm) lines can be used to infer the relative abundance of water and carbon dioxide in cometary coma (Delsemme 1980; Festou 1981; Festou and Feldman 1981; Huebner 1985; Cochran and Cochran 2001; Capria et al. 2005, 2008). Laboratory measurements have determined the yield of O(^{1}D) from photodissociation of water, but none of these experiments would have been capable of detecting O(^{1}S). In contrast, the wavelength dependent yield of O(^{1}S) from CO_2 has been investigated, but the yield of O(^{1}D) has been determined only at longer wavelengths, where the CO_2 absorption cross section is low (Huestis and Slanger 2006).

A portion of the water absorption spectrum is shown in Fig. 5.

For solar UV photodissociation Delsemme (1980) suggests yields of 12% for O(^{1}D) + H_2 and 1% for $O^* + H_2^*$ (other). On the other hand, Festou (1981) recommends 6.22% for O(^{1}D) + H_2 and 0.6% for O(^{1}S) + H_2. Interestingly, near Earth comets show a green/red ratio of about 0.1 (Cochran and Cochran 2001; Capria et al. 2005, 2008), hinting that Delsemme and Festou somehow got it right! What is actually known from previous laboratory experiments is the following:

Solar Lyman-α is by far the most important wavelength

$$H_2O + h\nu\ (\lambda < 242\ \text{nm}) \rightarrow OH(X^2\Pi) + H \quad \text{known} \tag{42}$$

$$H_2O + h\nu\ (\lambda < 177\ \text{nm}) \rightarrow O(^1D) + H_2 \quad \text{known} \tag{43}$$

$$H_2O + h\nu\ (\lambda < 136\ \text{nm}) \rightarrow OH(A^2\Sigma^+) + H \quad \text{known} \tag{44}$$

$$H_2O + h\nu\ (\lambda < 134\ \text{nm}) \rightarrow O(^1S) + H_2 \quad \text{unknown} \tag{45}$$

$$H_2O + h\nu\ (\lambda < 130\ \text{nm}) \rightarrow O(^3P) + H + H \quad \text{inferred.} \tag{46}$$

The yield of $O(^1D)$ was measured through its chemical reaction with H_2 or by 130 nm fluorescence scattering from $O(^3P)$, produced by relaxation in collisions with added N_2. No experiments performed to date would have been capable of identifying production of $O(^1S)$, a point made explicitly by Stief et al. (1975). McNesby et al. (1962) mention $O(^1S)$, not because it was detected, but only to show that it could not have been the source of observed OH fluorescence. Misinterpretation of this work is perhaps the source of Festou's estimate of the $O(^1S)$ yield.

A portion of the carbon dioxide absorption spectrum is shown in Fig. 6, along with a version of the solar UV spectrum, and the product of the two. Delsemme (1980) suggests yields from CO_2 67% for $O(^1D)$ and 22% for $CO^* + O^*$ (other). He never mentions $O(^1S)$, nor does he cite any of the papers reporting its production. He also does not cite the source of the high yield of $O(^1D)$. Similarly, Huebner (1985) presented an extensive table of photophysical processes (with no references to the primary literature), in which $O(^1S)$ is never mentioned as a CO_2 photodissociative product, in spite of the fact that in the same volume, Barth (1985) describes $O(^1S)$ production from CO_2 photodissociation at Mars. In fact, it was precisely the Mars observations (Barth et al. 1971) that led to the burst of quantum yield measurements in the 1970s.

The UV spectroscopy (Cossart-Magos et al. 1982, 1987) and absorption cross section (Nakata et al. 1965; Ogawa 1971; Slanger et al. 1974; Lewis and Carver 1983; Shaw et al. 1995; Yoshino et al. 1996; Berkowitz 2002; Huestis 2006b; Huestis and Slanger 2006; Stark et al. 2007; Huestis and Berkowitz 2007; Keller-Rudek and Moortgat 2008) of CO_2 have been extensively investigated. Several studies have determined the quantum yields for production of a wide variety of dissociation and ionization products, including $O(^3P)$ (Slanger and Black 1971; Zhu and Gordon 1990), $O(^1D)$ (Slanger and Black 1971; Zhu and Gordon 1990), $O(^1S)$ (Lawrence 1972; Ridley et al. 1973; Koyano et al. 1973; Slanger et al. 1977; Bibinov et al. 1979), as well as $CO(A^1\Pi)$, $CO(a^3\Pi$, and higher triplets), $CO_2^+(A^2\Pi$ and $B^2\Sigma^+)$, and $CO^+(B^2\Sigma^+)$. Astronomical observations or space missions motivated many of these studies. Okabe (1978) reviewed the earlier work. The data are not of uniform quality. There are wavelength gaps in the measured yields and neither the solar spectrum nor the absorption cross section is known with sufficient wavelength resolution.

Because of spin-conservation selection rules, it is widely believed (Schiff 1965; McElroy and Hunten 1970; Slanger and Black 1971, 1978; Slanger et al. 1974; Delsemme 1980; Zhu and Gordon 1990; Miller et al. 1992) that $O(^1D)$ is the primary product in the region from the energetic threshold at 7.5 eV up to the $O(^1S)$ threshold. However, there is little in the literature about direct detection of $O(^1D)$ at higher energy or shorter wavelength (Welge and Gilpin 1971). Thus we are unable to confirm the usual assumption that if the known quantum yield is less than unity, and if production of $O(^1D)$ is spin allowed, then it must have been the "dark" product. This uncertainty is important because the CO_2 absorption cross section is increasing rapidly at higher energy (shorter wavelength).

$O(^1S)$ is known to be produced from threshold at 9.6 eV (129 nm) to beyond the ionization limit at 13.8 eV (90 nm) (Lawrence 1972; Koyano et al. 1973; Slanger et al. 1977; Bibinov et al. 1979). In this spectral region, the absorption cross-section is quite strongly structured. Some absorption features are known to have unit yields of $O(^1S)$, while some others appear to produce purely $O(^1D)$. This variation has been difficult to quantify because

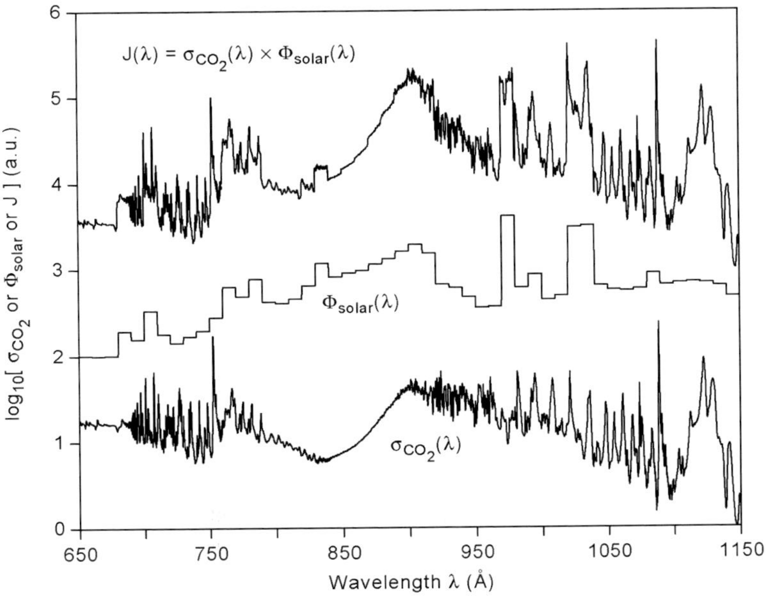

Fig. 6 Absorption cross section for carbon dioxide [σ_{CO_2}], the intensity of solar radiation [Φ_{solar}] and their product [J], from Huestis and Berkowitz (2007)

previous experiments generally used broadband light sources and made measurements at only a few widely spaced wavelengths.

The critical evaluation summarized here formed the basis for a new NASA-funded research program to measure the yields of $O(^1S)$ and $O(^1D)$ from photodissociation of H_2O and CO_2 using high spectral resolution VUV radiation from the Advanced Light Source at Lawrence Berkeley National Laboratory (Slanger et al. 2008a).

3.5.2 *High Resolution Photoabsorption Cross Sections for* SO_2

SO_2 is a known constituent of the atmosphere of Io (Ballester et al. 1994), but atmospheric studies of SO_2 using spectra acquired from the HST Faint Object Spectrograph (McGrath et al. 2000) have shown that the lack of laboratory measured high resolution cross sections limited the reliability of estimates of the SO_2 column density on Io. The use of low-resolution SO_2 data had led to the modelling of SO_2 ultraviolet (UV) absorption as a continuum when in reality it is a dense line spectrum. The spectrographs on board the current Venus Express also require high resolution cross sections for SO_2. It has been found that saturation of the very sharp SO_2 line features can lead to large underestimates of the SO_2 column density when instrumentally broadened absorption spectra are analyzed with low-resolution laboratory-derived cross sections (Stark et al. 1999). For very narrow spectral absorption features it is important to understand the difference between high- and low-resolution cross-section measurements (Hudson 1971). With inadequate instrumental resolution, measured cross sections at the centre of narrow absorption features are consistently underestimated and the cross sections in the "wing" regions between narrow features are consistently overestimated; the magnitude of the error being a function of the ratio of line width to instrument profile width, with the largest errors being associated with the lowest resolution. These systematic errors can be surprisingly large, as is seen in the case of SO_2 (see Fig. 7).

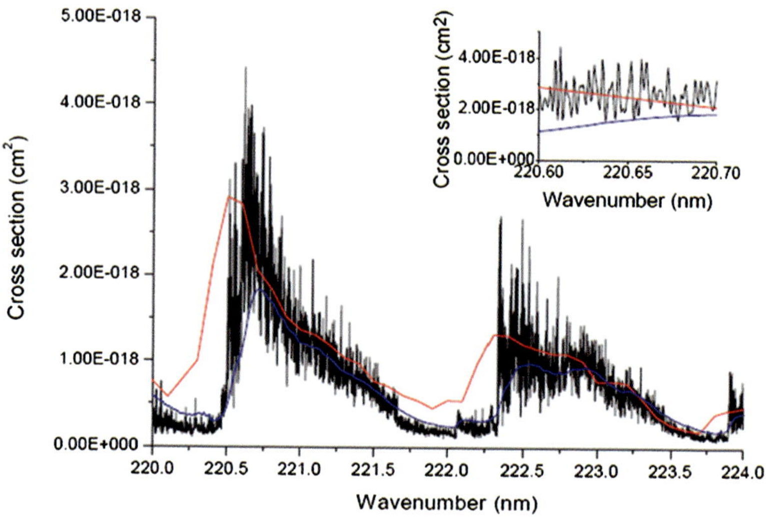

Fig. 7 Comparison of room temperature SO_2 absorption measurements (*4 nm segment, and inset 0.1 nm section*) carried out at different spectral resolution: high resolution 0.0004 nm (Rufus et al. 2003), low resolution compilation (0.1 nm) of Manatt and Lane (1993) (*red line*) and resolution 0.05 nm Wu et al. (2000) (*blue line*)

The UV spectrum of SO_2 has two main regions of significant absorption: a stronger absorbing region 175–230 nm, and a weaker region 250–320 nm. The spectrum is extremely complex at room temperature, and it is not possible to calculate the spectrum to sufficient accuracy for applications in planetary atmospheres (Stark et al. 1999). High resolution laboratory measurements of the SO_2 spectrum are required, and at a range of temperatures relevant to planetary atmospheres. Outlined here are the state-of-the-art measurements of high resolution photoabsorption cross sections undertaken at Imperial College in order to provide cross sections for SO_2 of sufficient accuracy for planetary atmosphere applications.

The basic components for experimental measurement of photoabsorption cross sections σ are: spectrometer, continuum light source, and absorption cell. The spectrometer used in this study was the Imperial College visible-UV Fourier Transform (FT) Spectrometer (Thorne et al. 1987), and resolving powers up to 550,000 were chosen to resolve the majority of the narrow SO_2 line features. The FT Spectrometer has advantages of high resolution, smoothly varying spectral response, linear wavenumber scale, and simultaneous observation of a wide spectral range. The continuum light sources were: a positive column hydrogen discharge, and high power deuterium lamp for the shorter wavelength region, and a 300 W xenon arc for the longer wavelength. The absorption cell contained 99.9% pure SO_2 with column densities ranging from 2.3×10^{16} cm^{-2} to 1.0×10^{19} cm^{-2} depending on the spectral region and range in σ magnitude. For each region several measurements at different pressures were made as checks to ensure that saturation effects were not present. The spectrum was measured in wavelength sections (10–20 nm bandwidth) by use of a novel zero-deviation zero-dispersion pre-monochromator, built at Imperial College (Murray 1992). This meant that good signal-to-noise ratio (SNR > 50) could be achieved within a reasonable spectrum acquisition time (4–8 hours).

Commonly the experimental method involves recording the spectrum of a continuum source with (I_T) and without (I_o) the sample gas in the absorption cell. Although the continuum light sources are typically stable in intensity for an hour, over 8 hours, their intensity has been observed to vary by a few percent. If the usual method of measurement with an empty gas cell, followed by filled cell, and a final empty cell, is used there will be errors in the resulting photoabsorption cross sections, as the continuum light source will have varied in intensity over the measurement time. For long acquisition times it was therefore essential to use the "dual beam technique". This makes use of the two outputs of the FT spectrometer (Rufus et al. 2003; Thorne et al. 1999; Davis et al. 2001). One output detector continuously measures the continuum light source, and the other measures the continuum light source with empty absorption cell, filled cell and then empty cell. Analysis of these spectra yields I_o and I_T, without errors arising from variations in the continuum light source.

However, care must be taken with choice of detectors. The response of photomultiplier tube detectors may also vary over the measurement timescales, and this variation can differ between detectors themselves. The photomultiplier detectors (Hamamatsu R166 and 1P28) were selected to be matched pairs from the manufacturer, and care was also taken to ensure their temperature remained constant during the day, again avoiding drifts in detector response. Wavelength calibration was carried out using iron standard lines (Learner and Thorne 1988), and wavelength accuracy is better than 10 mÅ.

Room temperature SO_2 measurements at Imperial College are completed (Stark et al. 1999; Rufus et al. 2003), and are the highest resolution SO_2 photoabsorption cross sections measurements undertaken to date. Uncertainties for σ in region 198–220 nm are estimated to vary from 10% for larger σ (10^{-17} cm^2) to 50% for lower σ ($<10^{-17}$ cm^2). In the longer wavelength region 220–325 nm uncertainties are typically 5% in regions of higher σ. In comparisons with previous photoabsorption cross section measurements in the literature (Rufus et al. 2003) there were differences in up to a factor of 2 in some cases arising in the main part from the resolution effect, and also shifts arising from wavelength errors. Figure 7 shows an example of the improvement in cross section data.

High resolution, low temperature, measurements are also required, to match atmospheric temperatures on Io and Venus. The Imperial College group, using the techniques described above, and a cooled gas cell, are undertaking measurements at 160 K in the range 190–220 nm, and at 200 K in the range 220–325 nm (Blackie et al. 2007). Some preliminary results are illustrated in Fig. 8.

These high resolution measurements over a range of temperatures have immediate applications in atmospheres where SO_2 is important, for example Io (Jessup et al. 2007). The use of low resolution laboratory photoabsorption cross sections leads to large errors in column densities estimates. Observations of planetary atmospheres with modern high resolution spectrographs require high resolution laboratory photoabsorption cross section data, at a range of temperatures, for full and reliable analysis.

3.6 Electron Collisions

As illustrated in previous sections of this chapter, electrons represent a primary source of excitation, dissociation, and ionization in planetary atmospheres and ionospheres. In addition, after thermalization in collisions with the ambient neutral atmosphere, electrons act like a chemical species, that is eventually consumed by dissociative recombination. Here we will review one specific case of dissociative recombination that contributes to atomic oxygen nightglow emissions. A second topic is creation of the atomic oxygen airglow on Europa and Ganymede by dissociative electron-impact excitation of molecular oxygen.

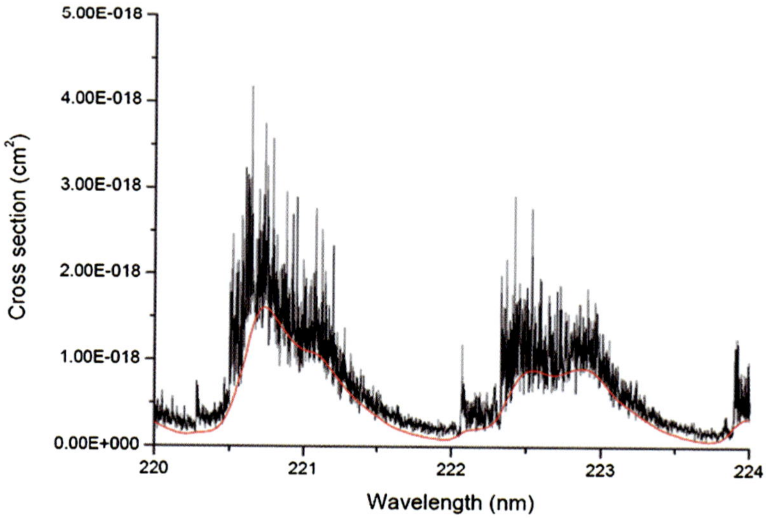

Fig. 8 Comparison of new high resolution 198 K temperature SO_2 photoabsorption cross sections (highly structured *black line*, Blackie et al. 2007) with the lower resolution results of Wu et al. (2000) (*red line*)

3.6.1 Dissociative Recombination of Electrons with Vibrationally Excited O_2^+

Dissociative recombination (DR) is the primary mechanism for electron loss in ionized, low-pressure molecular gases and plasmas, such as planetary ionospheres (Mitchell and Guberman 1989). Dissociative recombination occurs through a reaction written as

$$AB^+ + e \rightarrow A + B + \Delta E \tag{47}$$

for a diatomic molecular ion AB^+, atomic fragment products A and B, and kinetic energy release ΔE. Because the initial potential energy of the ion is usually 4 to 9 eV above the lowest dissociation limit of the neutral AB molecule, ample energy is available to leave some or all of the products A and B in electronically excited states (or in rotationally or vibrationally excited states if they are molecules instead of atoms). The excess energy ΔE appears as the center-of-mass kinetic energy given to the dissociation fragments. These two product characteristics, electronic excitation and high translation energy (which are connected by energy conservation), amplify the importance of dissociative recombination in the plasma by contributing to atmospheric heating (Torr et al. 1980; Fox 1988), planetary escape (Hunten 1982; Fox and Dalgarno 1983), optical emissions (Bates 1990), and the opening of new reaction channels in the ion and neutral chemistries.

The overall rates for dissociative recombination have been measured for many important species (Mitchell and Guberman 1989) and are generally considered to be reliably known, at least to within a factor of 2. Much less is known about the dependence of the rate on the initial vibrational state of the molecular ion, or about the branching ratios for production of specific excited states of the fragments. A previous study (Kella et al. 1997) of O_2^+ DR showed that vibrational excitation doubled the yield of $O(^1S)$. Although that experiment highlighted the effects of vibrational excitation, it offered no way of measuring the actual vibrational distribution. The expected influence of vibrational excitation on yields of atomic oxygen excited states is illustrated schematically in Fig. 9. With increasing vibrational excitation, the vibrational wavefunction is able to reach curve crossings between the initial

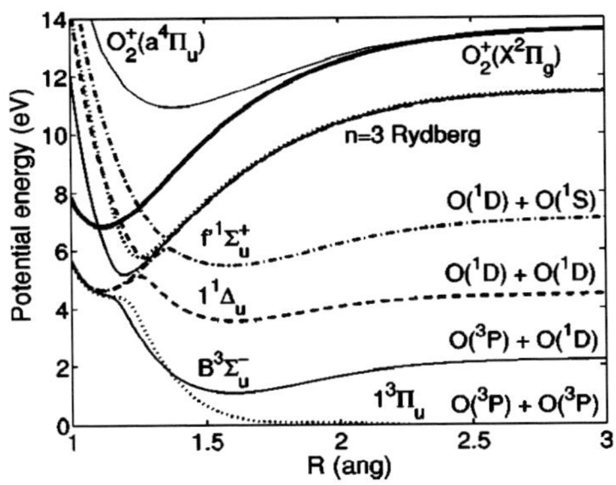

Fig. 9 Schematic potential curves relevant for the DR process with O_2^+, taken from Petrignani et al. (2005). The bound electronic ground $X^2\Pi_g$ and metastable excited $a^4\Pi_u$ states of the molecular ion are shown, along with the bound 3s Rydberg states of the neutral oxygen molecule (parallel to the $X^2\Pi_g$ ion state). The repulsive neutral electron-capture states, correlating with the various atomic asymptotes, illustrate the non-adiabatic interactions with the intermediate 3s Rydberg states

$O_2^+(X^2\Pi_g)$ potential curve and higher repulsive potential curves of neutral O_2 that correlate with more highly excited atomic fragments.

In the ionospheres of both Venus and Mars, O_2^+ is the most abundant molecular ion, followed by CO_2^+ and NO^+ (Hanson et al. 1977; Fox 2006). This result was initially surprising, given that there is relatively little O_2 in either atmosphere. O_2^+ is formed in two fast exothermic reactions:

$$O + CO_2^+ \rightarrow O_2^+ + CO \quad (48)$$

and

$$O^+ + CO_2 \rightarrow O_2^+ + CO. \quad (49)$$

As described by Fox (1985), Reaction (48) is more important below 150 km and Reaction (49) above 150 km. Reaction (49) has been studied in the laboratory (Walter et al. 1993) and has been shown to produce O_2^+ that is rotationally hot and vibrationally excited, with about 40% of the reaction exothermicity distributed statistically as internal excitation of the O_2^+ product. The resulting vibrational distribution is 38% in $v = 0$, 30% in $v = 1$, 18% in $v = 2$, and 15% in $v \geq 3$. Fox (1985) had earlier proposed that, above 150 km on Venus, dissociative recombination of excited vibrational levels of O_2^+ must be considered, because vibrational relaxation in collisions with CO_2 would be relatively slow. On Earth, O_2^+ is vibrationally relaxed by fast symmetric charge transfer collisions with O_2, which is improbable in the atmospheres of Mars and Venus.

Recently DR experiments have been performed at the CRYRING heavy-ion storage ring facility (Petrignani et al. 2005) using vibrationally defined molecular oxygen ions. The O_2^+ vibrational distribution was modified by varying the gas pressure, electron-beam voltage, and residence time in the ion source and measured by dissociative charge exchange in cesium vapor (Cosby et al. 2003; Petrignani et al. 2005). The results of these experiments are summarized in Table 5, from which we can see that both the DR cross section and the distribution of products are indeed strong functions of the vibrational quantum number.

Table 5 Partial relative cross sections, σ_v, quantum yields, and branching fractions for dissociative recombination of O_2^+ ($X^2\Pi_g$, $v = 0$–2), from Petrignani et al. (2005)

v	σ_v	Quantum yields			Blanching fractions			
		$O(^1S)$	$O(^1D)$	$O(^3P)$	$O(^1D)+O(^1S)$	$O(^1D)+O(^1D)$	$O(^3P)+O(^1D)$	$O(^3P)+O(^3P)$
0	1	0.06	0.94	1.00	5.8 ± 0.5	20.4 ± 0.3	47.3 ± 0.8	26.5 ± 0.8
1	0.31 ± 0.13	0.14	1.44	0.42	13.9 ± 3.1	51.0 ± 5.4	27.8 ± 5.1	7.3 ± 7.5
2	0.52 ± 0.16	0.21	1.02	0.76	21.1 ± 2.5	2.5 ± 2.1	76.4 ± 2.2	0.02 ± 0.03

3.6.2 Electron Impact Dissociative Excitation of O_2 and the Oxygen Airglow on Europa and Ganymede

Europa and Ganymede were observed to have an unsuspected ultraviolet dayglow dominated by the oxygen atomic emissions near 130.4 and 135.6 nm (Hall et al. 1995, 1998; Feldman et al. 2001). These observations confirmed the existence of a tenuous atmosphere (Broadfoot et al. 1979; Kumar and Hunten 1982; Hunten 1995), far less dense than thought earlier (Carlson et al. 1973; Yung and McElroy 1977). Using these emissions to infer information about the neutral atmosphere required quantitative information on relative yields of $O(2p^33s\ ^3P)$ and $O(2p^33s\ ^5P)$ from plausible oxygen sources, such as H_2O, O_2, and $O(^3P)$, excited by photons, electrons, or heavy particles. Water is an unlikely choice because its vapor pressure is extremely low at the relevant temperature. Hall et al. (1995) used the then available information on the collisions

$$e + O \rightarrow e + O^* \tag{50}$$

and

$$e + O_2 \rightarrow e + O + O^* \tag{51}$$

to calculate an expected ratio of emission strengths. Collision (51) gave an emission ratio $I(135.6)/I(130.4) = 1.9$, within the estimated uncertainty in the observed ratio, while Collision (50) gave an emission ratio 0.1.

These observations stimulated renewed interest in modeling these atmospheres and the surface impact processes that could produce a molecular oxygen atmosphere (Shematovich and Johnson 2001; Shematovich et al. 2005; Marconi 2007). Yung and McElroy (1977) had correctly indicated that any stream of energetic particles (photons, electrons, protons, etc., from the solar spectrum, solar wind, or jovian radiation environment) incident on an ice surface will eventually produce some H_2, which will quickly evaporate. The surface left behind will then have an oxygen excess, which will gradually produce some O_2, which will also evaporate. The UV dayglow observations also motivated new laboratory measurements (Noren et al. 2001a, 2001b; Kanik et al. 2003) and first-principles calculations (Zatsarinny and Tayal 2002).

3.7 Energetic Atomic/Molecular Collisions

As suggested in Sects. 2.4 and 2.5 above in this chapter, collisions of energetic heavy particles, from the solar wind or the local magnetized planetary environment, can deposit energy that may be observable as optical emissions, produce ionospheric modifications, or facilitate planetary escape. One good planetary science example is generation of X-ray emission from comets resulting from charge exchange electron capture by multicharged ions in the solar wind (Otranto and Olson 2008). Here we critically review the available information on the specific case of ion-atom charge exchange collisions of hydrogen and oxygen atoms. See also the review by Lindsay and Stebbings (2005).

3.7.1 Degenerate Charge Exchange Collisions of H^+ *and* O^+ *with* H *and* O

Hydrogen and oxygen are among the most abundant elements in the universe. Correspondingly, their neutral atomic and ionized forms are primary components of interstellar clouds, the solar wind, and planetary magnetospheres and ionospheres. The fact that the ionization energies of H and O are nearly identical implies that the charge exchange collisions

$$H^+ + H \leftrightarrow H + H^+ \tag{52}$$

$$O^+ + O \leftrightarrow O + O^+ \tag{53}$$

$$O^+ + H \leftrightarrow O + H^+ \tag{54}$$

all have large reaction cross sections on the order of 10^{-15} cm^2 or greater from meV to keV collision energies.

H^+ from the solar wind, or precipitating from planetary magnetospheres, collides with H and O atoms in the upper atmospheres of Venus, Earth, and Mars. On Venus and Mars this is the origin of about 10% of exospheric ionization (Zhang et al. 1993). The sudden buildup of the magnetic field in the Martian bowshock is attributed to these charge exchange collisions in the Martian exosphere (Chen et al. 2001). Other effects of Reactions (52), (53), and (54) include the production of hot H and O atoms; facilitating the upward flow of H, H^+, O, and O^+, as well as interhemispheric transport of hydrogen; transferring kinetic energy to O and O^+; mediating magnetosphere-ionosphere energetic coupling; modifying the $[O^+]/[O]$ and $[H^+]/[H]$ ratios versus altitude; and limiting the lifetimes of O^+ and H^+ ions in the ring current colliding with hydrogen atoms in the geocorona.

The current knowledge of the cross section for collision (52) is summarized in Fig. 10, following the analysis of Huestis (2008). From the numerous laboratory experimental studies in the literature we have selected data for subsequent analysis from only two of the most recent (but still very old: McClure 1966; Wittkower et al. 1966) as the only that appear to be quantitatively reliable because on experiment-to-experiment consistency and agreement with theory. They cover the range from 2 to 250 keV. We lack high quality laboratory data at lower energies. From the various theoretical investigations at low collision energy we have also selected only a subset (Brinkman and Kramers 1930; Dalgarno and Yadav 1953; Jackson and Schiff 1953; Bates and Boyd 1962; Smith 1967; Hunter and Kuriyan 1977; Shakeshaft 1978; Olson 1983; Hodges and Breig 1991) for analysis. In this case, we have excluded some of the most reliable recent information (Davis and Thorson 1978; Krstic and Schultz 1999; Krstic et al. 2004; Furlanetto and Furlanetto 2007) because it presents more fine detail than our analysis can reproduce, and because it agrees quantitatively with the best low-energy data included, when degraded to the same energy resolution. Also shown in Fig. 10 is a piece-wise polynomial representation versus collision energy (Huestis 2008) that can be used for modeling studies. The formula is given in Table 6.

For Reaction (53) at low energies, three quantum theory investigations (Stallcop et al. 1991; Pesnell et al. 1993; Hickman et al. 1997a, 1997b) attempted to resolve disagreements between calculated collision frequencies at thermal energies with those inferred from ionospheric observations. The latter investigation, which included spin–orbit interactions,

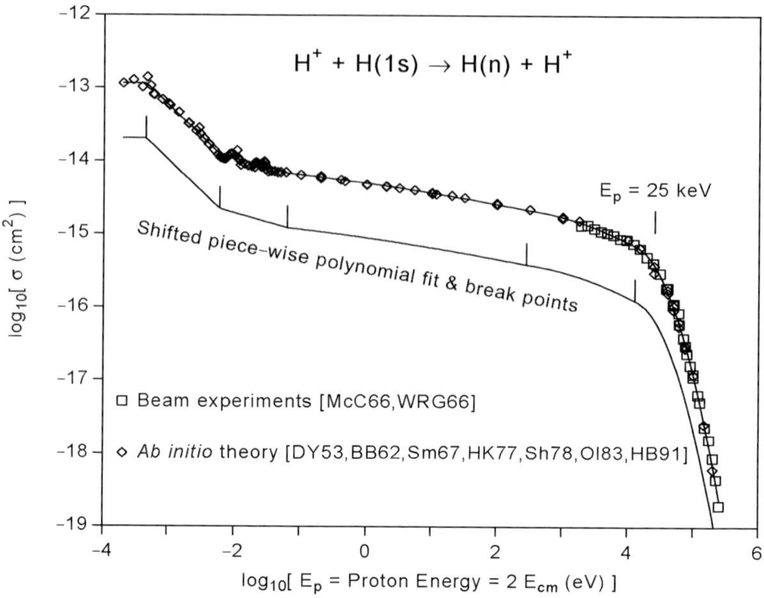

Fig. 10 Fitting the $H^+ + H$ Charge Transfer Cross Section, from Huestis (2008). The formula for the piece-wise polynomial fit is given in Table 6.
DY53 = Dalgarno and Yadav 1953; BB62 = Bates and Boyd 1962; McC66 = McClure 1966; WRG66 = Wittkower et al. 1966; Sm67 = Smith 1967; HK77 = Hunter and Kuriyan 1977; Sh78 = Shakeshaft 1978; Ol83 = Olson 1983; HB91 = Hodges and Breig 1991

gave results that were consistent with the earlier work for energies above 0.03 eV. At lower energies, spin–orbit interactions had a significant effect and the charge exchange and momentum transfer cross sections were no longer proportional.

For Reaction (54) at low energies, three quantum theory investigations (Chambaud et al. 1980; Stancil et al. 1999; Spirko et al. 2003) all included spin–orbit interactions. At energies above 0.03 eV, they agree with each other and calculations ignoring spin–orbit interactions agree with appropriate averages of spin–orbit-resolved cross sections. Below this energy, there are substantial differences, that are presumably due to different representations of the low-lying potential curves, spin–orbit couplings, charge-transfer matrix elements of the OH^+ molecule. Stancil et al. (1999) have provided explicit mathematical expressions that adequately represent the rate coefficients and cross sections for Reactions (54) from 0.1 to 10^6 eV.

4 Summary, Conclusions, and Recommendations

In this chapter we have followed the spirit expressed in the classic texts by Banks and Kockarts (1973), who supply the definitions

- Aeronomy is the scientific discipline devoted to the study of the composition, movement, and thermal balance of planetary atmospheres.
- As a field of research, aeronomy demands understanding of the basic concepts of both chemistry and physics as applied to a highly rarefied medium composed of neutral and charged particles.

Table 6 FORTRAN code fragment that returns values of the piece-wise linear fit for the cross section shown in Fig. 10 for the process $H^+ + H(1s) \rightarrow H(n\ell) + H^+$. Given the proton initial energy, `Ep`, in units of eV, `sigma( Ep )` returns the charge transfer cross section in units of cm^2

```
real function sigma( Ep )
x = alog10( Ep )
if( x .le. -3.3396 ) then
  y = -12.9451
else if( x .le. -2.2227 ) then
  y = -15.8262 -0.86272 * x
else if( x .le. -1.202 ) then
  y = -14.4664 -0.25094 * x
else if( x .le. 2.4513 ) then
  y = -14.3008 + ( -0.12455 -0.0094356 * x ) * x
else if( x .le. 4.1059 ) then
  y = -14.9562 + ( +0.36821 -0.10138 * x ) * x
else
  y = -43.9887 + ( +14.4120 -1.79965 * x ) * x
end if
sigma = 10.0**y
return
end
```

Here we have followed their emphasis on what we know, or wish we knew, about the fundamental underlining microscopic chemical and collisional processes, based on, or confirmed by, laboratory measurements and first-principles theory calculations, that are documented in the published peer-reviewed literature. Correspondingly we make the following recommendations,

- Laboratory experiments and first-principles theory calculations are essential components of planetary aeronomy research programs, comparable in importance to observational and modeling efforts.
- Critical evaluation and documentation of the current state of knowledge or ignorance of the microscopic chemical and collisional processes are of comparable importance to "original research" because they provide essential guidance and constraints on interpretative and modeling attempts to "explain" surprising or unexpected observations.

References

W.A. Abdou, D.G. Torr, P.G. Richards, M.R. Torr, E.L. Breig, J. Geophys. Res. **89**(A10), 9069–9079 (1984)

R.A. Akmaev, J. Geophys. Res. **108**(A7), 1292 (2003). doi:10.1029/2003JA009896

M.J. Alexander, A.I.F. Stewart, S.C. Solomon, S.W. Bougher, J. Geophys. Res. **98**(E6), 10,849–10,871 (1993)

V.G. Anicich, An index of the literature for bimolecular gas phase cation-molecule. Reaction kinetics. JPL Publication 03-19 (2003). trs-new.jpl.nasa.gov/dspace/bitstream/2014/7981/1/03-2964.pdf

V.G. Anicich, M.J. McEwan, Planet. Space Sci. **45**, 897–921 (1997)

S.K. Atreya, T.M. Donahue, J.H. Waite Jr., Nature **280**, 795–796 (1979)

G.E. Ballester, M.A. McGrath, D.F. Strobel, X. Zhu, P.D. Feldman, H.W. Moos, Icarus **111**, 2–17 (1994)

P.M. Banks, G. Kockarts, *Aeronomy, Parts A and B* (Academic Press, New York, 1973)

C.A. Barth, JPL Res. Summ. 36–39 **1**, 64 (1961)

C.A. Barth, The photochemistry of the atmosphere of Mars, in *The Photochemistry of Atmospheres*, ed. by J.S. Levine (Academic Press, New York, 1985)
C.A. Barth, C.W. Hord, J.B. Pearce, K.K. Kelly, G.P. Anderson, A.I. Stewart, J. Geophys. Res. **76**(10), 2213–2227 (1971)
J.N. Bass, J. Chem. Phys. **60**, 2913–2921 (1974)
D.R. Bates, Planet. Space Sci. **38**, 889–902 (1990)
D.R. Bates, A.H. Boyd, Proc. Phys. Soc. **80**, 1301–1307 (1962)
S.J. Bauer, *Physics of Planetary Ionospheres* (Springer, Berlin, 1973)
D.L. Baulch, D.D. Drysdale, J. Duxbury, S.J. Grant, *Evaluated Kinetic Data for High Temperature Reactions*, vol. 3 (Butterworths, London, 1976)
N.H.F. Beebe, E.W. Thulstrup, A. Andersen, J. Chem. Phys. **64**, 2080–2093 (1976). Supplementary Data Tables (private communication)
Y. Bénilan, N. Smith, A. Jolly, F. Raulin, Planet. Space Sci. **48**, 463–471 (2000)
J. Berkowitz, *Atomic and Molecular Photoabsorption – Absolute Total Cross Sections* (Academic Press, New York, 2002)
J.-L. Bertaux, F. Leblanc, S. Perrier, E. Quemerais, O. Korablev, E. Dimarellis, A. Reberac, F. Forget, P.C. Simon, S.A. Stern, B. Sandel, the SPICAM team, Science **307**, 566–569 (2005)
H.A. Bethe, Rev. Mod. Phys. **9**, 69–249 (1937)
B. Bezard, H. Feuchtgruber, J.I. Moses, T. Encrenaz, Astron. Astrophys. **334**, L41–L44 (1998)
B. Bezard, P.N. Romani, H. Feuchtgruber, T. Encrenaz, Astrophys. J. **515**, 868–872 (1999)
N.K. Bibinov, F.I. Vilesov, I.P. Vinogradov, L.D. Mineev, A.M. Pravilov, Sov. J. Quantum Electron. **9**, 838–844 (1979)
A.P. Billingham, P. Borrell, J. Chem. Soc. Faraday Trans. 2 **82**, 963–970 (1986)
D. Blackie, R. Blackwell-Whitehead, G. Stark, J.C. Pickering, J. Rufus, A. Thorne, P. Smith, Sulphur dioxide: high resolution ultra-violet photoabsorption cross section measurements at 200 K. Eos Trans. AGU **88**(52). Fall Meet. Suppl., Abstract P21A-0226 (2007)
P. Borrell, P.M. Borrell, M.D. Pedley, K.R. Grant, Proc. R. Soc. Lond. **367**, 395–410 (1979)
P.M. Borrell, P. Borrell, K.R. Grant, M.D. Pedley, J. Phys. Chem. **86**, 700–703 (1982)
S.W. Bougher, W.J. Borucki, J. Geophys. Res. **99**(E2), 3759–3776 (1994)
S.W. Bougher, J.C. Gerard, A.I.F. Stewart, C.G. Fesen, J. Geophys. Res. **95**(A5), 6271–6284 (1990)
S.W. Bougher, M.J. Alexander, H.G. Mayr, Upper atmosphere dynamics: Global circulation and gravity waves, in *Venus II*, ed. by S.W. Bougher, D.M. Hunten, P.J. Phillips (U. Arizona Press, Tucson, 1997), pp. 259–292
S.W. Bougher, S. Engel, R.G. Roble, B. Foster, J. Geophys. Res. **104**(E7), 16,591–16,611 (1999)
S.W. Bougher, S. Engel, R.G. Roble, B. Foster, J. Geophys. Res. **105**(E7), 17669–17692 (2000)
S.W. Bougher, R.G. Roble, T.J. Fuller-Rowell, Simulations of the upper atmospheres of the terrestrial planets, in *Atmospheres in the Solar System – Comparative Aeronomy*, M. Mendillo, A. Nagy, J.H. Waite, eds., AGU Monograph **130**, 261–288 (2002)
S.W. Bougher, S. Rafkin, P. Drossart, Planet. Space Sci. **54**, 1371–1380 (2006a)
S.W. Bougher, J.M. Bell, J.R. Murphy, M.A. Lopez-Valverde, P.G. Withers, Geophys. Res. Lett. **33**, L02203 (2006b)
M.T. Bowers, W.H. Chesnavich, W.T. Huntress Jr., Int. J. Mass Spectrom. Ion Phys. **12**, 357–382 (1973)
H.C. Brinkman, H.A. Kramers, Proc. Acad. Sci. Amst. **33**, 973–984 (1930)
A.L. Broadfoot et al., Science **204**, 979–982 (1979)
R.A. Brownsword, M. Hillenkamp, T. Laurent, R.K. Vasta, H.-R. Volpp, J. Wolfrum, Chem. Phys. Lett. **266**, 259–266 (1997)
I.M. Campbell, C.N. Gray, Chem. Phys. Lett. **18**, 607–609 (1973)
I.M. Campbell, B.A. Thrush, Proc. R. Soc. A **296**, 222–232 (1967)
M.T. Capria, G. Cremonese, A. Bhardwaj, M.C. de Sanctis, Astron. Astrophys. **442**, 1121–1126 (2005)
M.T. Capria, G. Cremonese, A. Bhardwaj, M.C. de Sanctis, E. Mazzotta Epifani, Astron. Astrophys. **479**, 257–263 (2008)
R.W. Carlson et al., Science **182**, 53–55 (1973)
K.J. Castle, K.M. Kleissas, J.M. Rhinehart, E.S. Hwang, J.A. Dodd, J. Geophys. Res. **111**, A09303 (2006)
G. Chambaud, J.M. Launay, B. Lefy, P. Millie, E. Roueff, F. Tran Minh, J. Phys. B **13**, 4255–4216 (1980)
W.F. Chan, G. Cooper, C.E. Brion, Chem. Phys. **178**, 387–400 (1993)
T. Chang, D.G. Torr, P.G. Richards, S.C. Solomon, J. Geophys. Res. **98**(A9), 15,589–15,597 (1993)
M. Chatelet, A. Tardieu, W. Spreitzer, M. Maier, Chem. Phys. **102**, 387–394 (1986)
F.Z. Chen, C.Y.R. Wu, J. Quant. Spectrosc. Radiat. Transfer **85**, 195–209 (2004)
F.Z. Chen, D.L. Judge, C.Y.R. Wu, J. Caldwell, H.P. White, R. Wagener, J. Geophys. Res. **96**, 17519–17527 (1991)
F.Z. Chen, D.L. Judge, C.Y.R. Wu, Chem. Phys. **260**, 215–223 (2000)

Y. Chen, P.A. Cloutier, D.H. Crider, C. Mazelle, H. Reme, J. Geophys. Res. **106**(A12), 29,387–29,399 (2001)
A.L. Cochran, W.D. Cochran, Icarus **154**, 381–390 (2001)
J.E.P. Connerney, J.H. Waite, Nature **312**, 136–138 (1984)
P.A. Cook, M.N.R. Ashfold, Y.J. Jee, K.H. Jung, S. Harich, X.M. Yang, Phys. Chem. Chem. Phys. **3**, 1848–1860 (2001)
M. Cordonnier, D. Uy, R.M. Dickson, K.E. Kerr, Y. Zhang, T. Oka, J. Chem. Phys. **113**, 3181–3193 (2000)
P.C. Cosby, J.R. Peterson, D.L. Huestis, Dissociative recombination of vibrationally excited levels in oxygen molecular ions, in *Dissociative Recombination of Molecular Ions with Electrons*, ed. by S.L. Guberman (Kluwer/Plenum, New York, 2003), pp. 101–108
C. Cossart-Magos, S. Leach, M. Eidelsberg, F. Launay, R. Rostas, J. Chem. Soc. Faraday Trans. **78**(2), 1477–1487 (1982)
C. Cossart-Magos, M. Jungen, F. Launay, Molec. Phys. **61**, 1077–1117 (1987)
T.E. Cravens, Astrophysical applications for electron energy deposition in molecular hydrogen. Ph.D. thesis, Harvard University, Cambridge, MA (1974)
T.E. Cravens, J. Geophys. Res. **92**(A10), 11,083–11,100 (1987)
T.E. Cravens et al., Geophys. Res. Lett. **33**, L07105 (2006)
A. Dalgarno, H.N. Yadav, Proc. Phys. Soc. **A66**, 173–177 (1953)
E.I. Dashevskaya, E.E. Nikitin, Chem. Phys. Lett. **328**, 119–123 (2000)
E.I. Dashevskaya, J.A. Kunc, E.E. Nikitin, I. Oref, J. Chem. Phys. **118**, 3141–3147 (2003)
E.I. Dashevskaya, I. Litvin, E.E. Nikitin, J. Troe, J. Chem. Phys. **125**, 154315 (2006)
E.I. Dashevskaya, I. Litvin, E.E. Nikitin, J. Troe, J. Chem. Phys. **127**, 114317 (2007)
J.P. Davis, W.R. Thorson, Can. J. Phys. **56**, 996–1020 (1978)
S.P. Davis, M.C. Abrams, J.W. Brault, *Fourier Transform Spectrometry* (Academic Press, New York, 2001)
A.H. Delsemme, Photodissociation of CO_2 into CO + $O(^1D)$, in *Les Spectres des Molécules Simples au Laboratorie et en Astrophysique* (Liège Institut d'Astrophysique, Université de Liège, Liège, 1980), pp. 515–523
G.S. Diskin, Experimental and theoretical investigation of the physical processes important to the RELIEF flow tagging diagnostic. Ph.D. Thesis, Princeton University (1997)
G.S. Diskin, W.R. Lempert, R.B. Miles, Observation of vibrational dynamics in $X3\Sigma_g^-$ oxygen following stimulated Raman excitation to the $v = 1$ Level: Implications for the RELIEF Flow Tagging Technique. AIAA 96-3001, 34th Aerospace Sciences Meeting and Exhibit (Reno, NV, 15–18 January 1996)
J.A. Dodd, E.S. Hwang, M. Simione, K.J. Castle, Laboratory measurement of $CO_2(v_2)$ + O temperature-dependent vibrational energy transfer. Eos Trans. AGU **88**(52), Fall Meet. Suppl., Abstract SA41A-0278 (2007); private communication; K.J. Castle, L.A. Black, M.W. Simione, E.S. Hwang, J.A. Dodd (manuscript in preparation)
P. Drossart et al., Nature **340**, 539–542 (1989)
V. Escalante, G.A. Victor, Planet. Space Sci. **40**, 1705–1718 (1992)
A. Fahr, A.K. Nayak, Chem. Phys. **189**, 725–731 (1994)
A. Fahr, A. Nayak, Chem. Phys. **203**, 351–358 (1996)
B. Faltermeier, R. Protz, M. Maier, Chem. Phys. **62**, 377–385 (1981)
P.D. Feldman, M.A. McGrath, D.F. Strobel, H.W. Moos, K.D. Retherford, B.C. Wolven, Astrophys. J. **535**, 1085–1090 (2001)
M.C. Festou, Astron. Astrophys. **96**, 52–57 (1981)
M.C. Festou, P.D. Feldman, Astron. Astrophys. **103**, 154–159 (1981)
J.-H. Fillion, J. Ruiz, X.-F. Yang, M. Castillejo, F. Rostas, J.-L. Lemaire, J. Chem. Phys. **120**, 6531–6541 (2004)
J.L. Fox, Adv. Space Res. **5**(9), 165–169 (1985)
J.L. Fox, Planet. Space Sci. **36**, 37–46 (1988)
J.L. Fox, Chemistry of the atmosphere: ion chemistry, in *Encyclopedia of Atmospheric Sciences*, ed. by J.R. Holton, J. Pyle, J.A. Curry (Elsevier, London, 2002), pp. 359–375
J.L. Fox, Aeronomy, in *Springer Handbook of Atomic, Molecular, and Optical Physics*, ed. by G.W.F. Drake (Springer, New York, 2006), pp. 1259–1292, Chapter 84
J.L. Fox, A. Dalgarno, J. Geophys. Res. **84**(A12), 7315–7331 (1979)
J.L. Fox, A. Dalgarno, J. Geophys. Res. **88**, 9027–9032 (1983)
J.L. Fox, R.V. Yelle, Geophys. Res. Lett. **24**, 2179–2182 (1997)
J.L. Fox, P. Zhou, S.W. Bougher, Adv. Space Res. **17**, (11)203–(11)218 (1995)
J.L. Fox et al., Energy deposition, Space Sci. Rev. (2008, this issue). Chapter 1
A.H. Friedson, A.-S. Wong, Y.L. Yung, Icarus **158**, 389–400 (2002)
S.R. Furlanetto, M.R. Furlanetto, Mon. Not. R. Astron. Soc. **374**, 547–555 (2007)
M. Galand, S. Chakrabarti, Auroral processes in the solar system, in *Atmospheres in the Solar System – Comparative Aeronomy*, ed. by M. Mendillo, A. Nagy, J.H. Waite, AGU Monograph **130**, 55–76 (2002)

M. Galand, S. Chakrabarti, J. Atmos. Solar Terr. Phys. **68**, 1488–1501 (2006)
M. Galand, D. Lummerzheim, J. Geophys. Res. **109**, A03307 (2004)
T.R. Geballe, M.F. Jagod, T. Oka, Astrophys. J. **408**, L109–L112 (1993)
J. Gerard, A. Saglam, G. Piccioni, P. Drossart, C. Cox, S. Erard, R. Hueso, A. Sanchez-Lavega, The distribution of the O_2 infrared nightglow observed with VIRTIS on board Venus Express. Geophys Res. Lett. (2008, in press). doi:10.1029/2007GL032021
D. Gerlich, E. Herbst, E. Roueff, Planet. Space Sci. **50**, 1275–1285 (2002)
G.A. Germany, D. Lummerzheim, P.G. Richards, J. Geophys. Res. **106**, 12,837–12,843 (2001)
K. Giles, N.G. Adams, D. Smith, J. Phys. Chem. **96**, 7645–7650 (1992)
G.R. Gladstone, J. Geophys. Res. **97**, 1377–1387 (1992)
G.R. Gladstone, M. Allen, Y.L. Yung, Icarus **119**, 1–52 (1996)
J. Glosik, A.B. Rakshit, N.C. Twiddy, N.G. Adams, D. Smith, J. Phys. B **11**, 3365–3379 (1978)
D. Grodent, J.H. Waite Jr., J.-C. Gerard, J. Geophys. Res. **106**(A7), 12,933–12,952 (2001)
D.T. Hall, D.F. Strobel, P.D. Feldman, M.A. McGrath, H.A. Weaver, Nature **373**, 677–679 (1995)
D.T. Hall, P.D. Feldman, M.A. McGrath, D.F. Strobel, Astrophys. J. **499**, 475–481 (1998)
J.T. Hallett, D.E. Shemansky, X. Liu, The Cassini UVIS Team, Cassini UVIS observations of the Saturn H_2 dayglow emission. Eos Trans. AGU **85**(47), F1268 (2004)
J.T. Hallett, D.E. Shemansky, X. Liu, Geophys. Res. Lett. **32**, L02204 (2005a)
J.T. Hallett, D.E. Shemansky, X. Liu, Astrophys. J. **624**, 448–461 (2005b)
W.B. Hanson, S. Sanatani, D.R. Zuccaro, J. Geophys. Res. **82**, 4351–4363 (1977)
N.M. Harvey, Chem. Phys. Lett. **88**, 553–558 (1982)
A.J.R. Heck, R.N. Zare, D.W. Chandler, J. Chem. Phys. **104**, 4019–4030 (1996)
A.E. Hedin, H.B. Niemann, W.T. Kasprzak, A. Seiff, J. Geophys. Res. **88**, 73–83 (1983)
A.P. Hickman, M. Medikeri-Naphade, C.D. Chapin, D.L. Huestis, Geophys. Res. Lett. **24**, 119–122 (1997a)
A.P. Hickman, M. Medikeri-Naphade, C.D. Chapin, D.L. Huestis, Phys. Rev. A **56**, 4633–4643 (1997b)
R.R. Hodges Jr., E.L. Breig, J. Geophys. Res. **96**(A5), 7697–7708 (1991)
R.D. Hudson, Rev. Geophys. Space Phys. **9**, 305–406 (1971)
W.F. Huebner, "The photochemistry of comets" and Appendix "Unattenuated solar photo rate coefficients at 1 AU heliocentric distance", in *The Photochemistry of Atmospheres*, ed. by J.S. Levine (Academic Press, New York, 1985)
D.L. Huestis, Current laboratory experiments for planetary aeronomy, in *Atmospheres in the Solar System – Comparative Aeronomy*, M. Mendillo, A. Nagy, J.H. Waite, eds., AGU Monograph **130**, 245–258 (2002)
D.L. Huestis, Chem. Phys. Lett. **411**, 108–110 (2005a)
D.L. Huestis, $H^+ + H_2$ ion-molecule reactions in the ionospheres of the outer planets. Bull. Am. Astron. Soc. **37**, 757 (2005b)
D.L. Huestis, J. Phys. Chem. **110**, 6638–6642 (2006a)
D.L. Huestis, Radiative transition probabilities, in *Springer Handbook of Atomic, Molecular, and Optical Physics*, ed. by G.W.F. Drake (Springer, New York, 2006b), pp. 515–533, Chapter 33
D.L. Huestis, Hydrogen collisions in planetary atmospheres, ionospheres, and magnetospheres. AOGS 2007 Proceedings, Planet. Space Sci. (2008, in press)
D.L. Huestis, J. Berkowitz, Photoabsorption cross sections for polyatomic molecules in planetary atmospheres. SRI International proposal PYU 07-091 submitted to NASA Planetary Atmospheres (June 2007)
D.L. Huestis, J.M. Bowman, Vibrationally excited H_3^+ in outer planet atmospheres. SRI International proposal PYU 07-173 submitted to NASA Outer Planets Research (November 2007)
D.L. Huestis, T.G. Slanger, Cross sections and yields of $O(^1S)$ and $O(^1D)$ in photodissociation of H_2O and CO_2. Bull. Am. Astron. Soc. **38**, 609 (2006)
D.L. Huestis, G. Black, S.A. Edelstein, R.L. Sharpless, J. Chem. Phys. **60**, 4471–4474 (1974)
D.L. Huestis, J. Marschall, G.D. Billing, R. Maclagan, Theoretical and experimental studies of O–CO_2 collisions. Eos Trans. AGU **83**, F1106 (2002)
D.L. Huestis, B.D. Sharpee, T.G. Slanger, Eos Trans. AGU **88**(52) (2007). Fall Meet. Suppl., Abstract SA41-0261
D.M. Hunten, Planet. Space Sci. **30**, 773–783 (1982)
D.M. Hunten, Nature **373**, 654 (1995)
G. Hunter, M. Kuriyan, Proc. R. Soc. Lond. A **353**, 575–588 (1977)
W.T. Huntress Jr., V.G. Anicich, Astrophys. J. **208**, 237–244 (1976)
A. Ichihara, O. Iwamoto, R.K. Janev, J. Phys. B **33**, 4747–4758 (2000)
J.D. Jackson, H. Schiff, Phys. Rev. **89**, 359–365 (1953)
R.K. Janev, D. Reiter, U. Samm, Collision processes in low-temperature hydrogen plasmas (2003). http://www.eirene.de/reports/report_4105.pdf

K.L. Jessup, J. Spencer, R. Yelle, Icarus **192**, 24–40 (2007)
R. Johnsen, M.A. Biondi, Geophys. Res. Lett. **7**, 401–403 (1980a)
R. Johnsen, M.A. Biondi, J. Chem. Phys. **73**, 190–193 (1980b)
R.E. Johnson, *Energetic Charged Particle Interactions with Atmospheres and Surfaces* (Springer, Berlin, 1990)
R.E. Johnson, M. Liu, C. Tully, Planet. Space Sci. **50**, 123–128 (2002)
P.V. Johnson, C.P. Malone, I. Kanik, K. Tran, M.A. Khakoo, J. Geophys. Res. **110**, A11311 (2005)
R.E. Johnson et al., Exospheres. Space Sci. Rev. (2008, this issue). Chapter 9
D.B. Jones, L. Campbell, M.J. Bottema, P.J.O. Teubner, D.C. Cartwright, W.R. Newell, M.J. Brunger, Planet. Space Sci. **54**, 45–59 (2006)
I. Kanik, C. Noren, O.P. Vattipalle, J.M. Ajello, D.E. Shemansky, J. Geophys. Res. **108**(E11), 5126 (2003)
G.P. Karwasz, T. Wroblewski, R.S. Brusa, E. Illenberger, Jpn. J. Appl. Phys. **45**, 8192–8196 (2006)
D. Kella, L. Vejby-Christensen, P.J. Johnson, H.B. Pederson, L.H. Andersen, Science **276**, 1530–1533 (1997)
C.N. Keller, V.G. Anicich, T.E. Cravens, Planet. Space Sci. **46**, 1157–1174 (1998)
H. Keller-Rudek, G.K. Moortgat, MPI-Mainz-UV-VIS Spectral Atlas of Gaseous Molecules (2008). http://www.atmosphere.mpg.de/enid/2295
V. Kharchenko, A. Dalgarno, B. Zygelman, J.-H. Yee, J. Geophys. Res. **103**, 24,899–24,906 (2000)
V. Kharchenko, A. Dalgarno, D.R. Schultz, P.C. Stancil, Geophys. Res. Lett. **33**, L11105 (2006)
L.E. Khvorostovskaya, I.Yu. Potekhin, G.M. Shved, V.P. Ogibalov, T.V. Uzyukova, Izvestiya Atmos. Ocean. Phys. **38**, 613–624 (2002)
Y.H. Kim, J.L. Fox, Icarus **112**, 310–325 (1994)
J.K. Kim, L.P. Heard, W.T. Huntress Jr., Int. J. Mass Spectr. Ion Phys. **15**, 223–244 (1974)
Y.H. Kim, J.L. Fox, H.S. Porter, J. Geophys. Res. **97**(E4), 6093–6101 (1992)
H. Klingshirn, B. Faltermeier, W. Hengl, M. Maier, Chem. Phys. Lett. **93**, 485–489 (1982)
H. Klingshirn, M. Maier, J. Chem. Phys. **82**, 714–719 (1985)
M.B. Knickelbein, K.L. Marsh, O.E. Ulrich, G.E. Busch, J. Chem. Phys. **87**, 2392–2393 (1987)
V.D. Knyazev, I.R. Slagle, J. Phys. Chem. **100**, 16899–16911 (1996)
I. Koyano, T.S. Wauchop, K.H. Welge, J. Chem. Phys. **63**, 110–112 (1973)
V.A. Krasnopolsky, D.P. Cruikshank, J. Geophys. Res. **100**(E10), 21,271–21,286 (1995)
P.S. Krstic, Phys. Rev. A **66**, 042717 (2002)
P.S. Krstic, D.R. Schultz, J. Phys. B **32**, 3485–3509 (1999)
P.S. Krstic, D.R. Schultz, J. Phys. B **36**, 385–398 (2003)
P.S. Krstic, D.R. Schultz, R.K. Janev, Phys. Scripta **T96**, 61–71 (2002)
P.S. Krstic, J.H. Macek, S.Yu. Ovchinnikov, D.R. Schulz, Phys. Rev. A **70**, 042711 (2004)
S. Kumar, D.M. Hunten, The atmospheres of Io and other satellites, in *Satellites of Jupiter*, ed. by D. Morrison (U. Arizona Press, Tucson, 1982)
L. Landau, E. Teller, Zur Theorie der Schalldispersion. Phys. Z. Sowjet. **10**, 34–43 (1936) [trans. D. Ter Haar, *Collected Papers of L.D. Landau* (Gordon and Breach, New York, 1965)]
M.P. de Lara-Castells, M.I. Hedrnandez, G. Delgado-Barrio, P. Villarreal, M. Lopez-Puertas, J. Chem. Phys. **124**, 164302 (2006)
M.P. de Lara-Castells, M.I. Hedrnandez, G. Delgado-Barrio, P. Villarreal, M. Lopez-Puertas, Molec. Phys. **105**, 1171–1181 (2007)
G.M. Lawrence, J. Chem. Phys. **57**, 5616–5617 (1972)
R.C.M. Learner, A.P. Thorne, J. Opt. Soc. Am. B **5**(10), 2045–2059 (1988)
S.A. Ledvina et al., Modelling of plasma flows and related phenomena. Space Sci. Rev. (2008, this issue). Chapter 4
A.Y.T. Lee, Y.L. Yung, J. Moses, J. Geophys. Res. **105**(E8), 20,207–20,225 (2000)
E. Lellouch, Phil. Trans. R. Soc. Lond. A **364**, 3139–3146 (2006)
E. Lellouch, T. Clancy, D. Crisp, A. Kliore, D. Titov, S.W. Bougher, Monitoring of mesospheric structure and dynamics, in *Venus II*, ed. by S.W. Bougher, D.M. Hunten, P.J. Phillips (U. Arizona Press, Tucson, 1997), pp. 295–324
B.R. Lewis, J.H. Carver, J. Quant. Spectrosc. Radiat. Transfer **30**, 297–309 (1983)
X. Li, Y.-L. Huang, G.D. Flesch, C.Y. Ng, J. Chem. Phys. **104**, 1373–1381 (1997)
H.V. Lilenfeld, Deactivation of vibrationally excited NO and CO_2 by O-atoms. Final report PL-TR-94-2180 (McDonnell Douglas Corp., St. Louis, June 1994)
B.G. Lindsay, R.F. Stebbings, J. Geophys. Res. **110**, A12213 (2005)
X. Liu, D.E. Shemansky, J. Geophys. Res. **111**, A04303 (2006)
D. Lummerzheim, M. Galand, J. Semeter, M.J. Mendillo, M.H. Rees, F.J. Rich, J. Geophys. Res. **106**, 141–148 (2001)
H. Luna, M. Michael, M.B. Shah, R.E. Johnson, C.J. Latimer, J.W. McConkey, J. Geophys. Res. **108**(E4) (2003). doi:10.1029/2002JE001950

H. Luna, C. McGrath, M.B. Shah, R.E. Johnson, M. Liu, C.J. Latimer, E.C. Montenegro, Astrophys. J. **628**, 1086–1096 (2005)
J.R. Lyons, Y.L. Yung, M. Allen, Science **256**, 204–206 (1992)
M.T. MacPherson, M.J. Pilling, M.J.C. Smith, Chem. Phys. Lett. **94**, 430–433 (1983)
M.T. MacPherson, M.J. Pilling, M.J.C. Smith, J. Phys. Chem. **89**, 2268–2274 (1985)
T. Majeed, J.C. McConnell, D.F. Strobel, M.E. Summers, Geophys. Res. Lett. **17**, 1721–1724 (1990)
T. Majeed, J.H. Waite Jr., S.W. Bougher, R.V. Yelle, G.R. Gladstone, J.C. McConnell, A. Bhardwaj, Adv. Space Res. **33**, 197–211 (2004)
M.L. Marconi, Icarus **190**, 155–174 (2007)
S.L. Manatt, A.L. Lane, J. Quant. Spectrosc. Radiat. Transf. **50**, 267–276 (1993)
A.N. Maurellis, T.E. Cravens, Icarus **154**, 350–371 (2001)
G.E. McClure, Phys. Rev. **148**, 47–54 (1966)
J.C. McConnell, T. Majeed, J. Geophys. Res. **92**(A8), 8570–8578 (1987)
J.C. McConnell, J.B. Holdberg, G.R. Smith, B.R. Sandel, D.E. Shemansky, A.L. Broadfoot, Planet. Space Sci. **30**, 151–167 (1982)
M.B. McElroy, Space Sci. Rev. **14**, 460–473 (1973)
M.B. McElroy, D.M. Hunten, J. Geophys. Res. **75**, 1188–1201 (1970)
M.A. McGrath, M.J.S. Belton, J.R. Spencer, P. Sartoretti, Icarus **146**, 476–493 (2000)
T.B. McMahon, P.G. Miasek, J.I. Beauchamp, Int. J. Mass Spectrom. Ion Phys. **21**, 63–71 (1976)
J.R. McNesby, I. Tanaka, H. Okabe, J. Chem. Phys. **36**, 605–607 (1962)
H. Melin, S. Miller, T. Stallard, D. Grodent, Icarus **178**, 97–103 (2005)
H. Melin, S. Miller, T. Stallard, C. Smith, D. Grodent, Icarus **181**, 256–265 (2006)
M. Michael, R.E. Johnson, Planet. Space Sci. **53**, 1510–1514 (2005)
R.L. Miller, S.H. Kable, P.L. Houston, I. Burak, J. Chem. Phys. **96**, 332–338 (1992)
S. Miller et al., Phil. Trans. R. Soc. Lond. A **358**, 2485–2512 (2000)
S. Miller, T. Stallard, C. Smith, G. Millward, H. Melin, M. Lystrup, A. Aylward, Phil. Trans. R. Soc. Lond. A **364**, 3121–3137 (2006)
J.B.A. Mitchell, S.L. Guberman (eds.), *Dissociative Recombination: Theory, Experiment, and Application* (World Scientific, NJ, 1989)
L.E. Moore, M. Mendillo, I.C.F. Müller-Wodarg, D.L. Murr, Icarus **172**, 503–520 (2004)
L. Moore, A.F. Nagy, A.J. Kliore, I. Müller-Wodarg, J.D. Richardson, M. Mendillo, Geophys. Res. Lett. **33**, L22202 (2006)
D.H. Mordaunt, I.R. Lambert, G.P. Morley, M.N.R. Ashford, R.N. Dixon, C.M. Western, J. Chem. Phys. **98**, 2054–2065 (1993)
J.E. Morgan, H.I. Schiff, J. Chem. Phys. **38**, 1495–1500 (1963)
J.E. Morgan, L. Elias, H.I. Schiff, J. Chem. Phys. **33**, 930–931 (1960)
J.I. Moses, S.F. Bass, J. Geophys. Res. **105**(E3), 7013–7052 (2000)
J.I. Moses, T. Fouchet, R.V. Yelle, A.J. Friedson, G.S. Orton, B. Bézard, P. Drossart, G.R. Gladstone, T. Kostiuk, T.A. Livengood, The stratosphere of Jupiter, in *Jupiter: Planet, Satellites and Magnetosphere*, ed. by F. Bagenal, T.E. Dowling, W.B. McKinnon (Cambridge Univ. Press, New York, 2004), pp. 129–157
J.I. Moses, T. Fouchet, B. Bézard, G.R. Gladstone, E. Lellouch, H. Feuchtgruber, J. Geophys. Res. **110**, E08001 (2005)
R. Mota, R. Parafita, A. Giuliani, M.-J. Hubin-Franskin, J.M.C. Lourenco, G. Garcia, S.V. Hoffman, N.J. Mason, P.A. Ribeiro, M. Paposo, P. Limao-Vieira, Chem. Phys. Lett. **416**, 152–159 (2005)
G.E. Moyano, M.A. Collins, J. Chem. Phys. **119**, 5510–5517 (2003)
J.E. Murray, High resolution spectrometry of neutral chromium using a Fourier transform spectrometer. Ph.D Thesis, Imperial College, London University (1992)
A.F. Nagy, T.E. Cravens, Solar system ionospheres, in *Atmospheres in the Solar System – Comparative Aeronomy*, ed. by M. Mendillo, A. Nagy, J.H. Waite, AGU Monograph **130**, 39–54 (2002)
H. Nair, M. Allen, A.D. Anbar, Y.L. Yung, R.T. Clancy, Icarus **111**, 124–150 (1994)
R.S. Nakata, K. Watanabe, F.M. Matsunaga, Sci. Light **14**(1), 54–71 (1965)
H.B. Niemann, R.E. Hartle, A.E. Hedin, W.T. Kasprzak, N.W. Spencer, D.M. Hunten, G.R. Carignan, Science **205**, 54–56 (1979)
H.B. Niemann, W.T. Kasprzak, A.E. Hedin, D.M. Hunten, N.W. Spencer, J. Geophys. Res. **85**, 7817–7827 (1980)
E.E. Nikitin, J. Troe, Phys. Chem. Chem. Phys. **125**, 154315 (2006)
E.E. Nikitin, S.Ya. Umanski, Faraday Disc. Chem. Soc. **53**, 7–17 (1972)
C. Noren, I. Kanik, J.M. Ajello, P. McCartney, O.P. Makarov, W.E. McClintock, V.A. Drake, Geophys. Res. Lett. **28**, 1379–1392 (2001a)
C. Noren, I. Kanik, P.V. Johnson, P. McCartney, G.K. James, J.M. Ajello, J. Phys. B **34**, 2667–2677 (2001b)

M. Ogawa, J. Chem. Phys. **54**, 2550–2556 (1971)
T. Oka, Phys. Rev. Lett. **45**, 531–534 (1980)
T. Oka, J. Molec. Spectrosc. **228**, 635–639 (2004)
T. Oka, E. Epp, Astrophys. J. **613**, 349–354 (2004)
T. Oka, T.R. Geballe, Astrophys. J. **351**, L53–L56 (1990)
H. Okabe, *Photochemistry of Small Molecules* (Wiley, New York, 1978)
R.E. Olson, Phys. Rev. A **27**, 1871–1878 (1983)
E.R. O'Neill, B.D. Sharpee, T.G. Slanger, The nightglow content of astronomical spectra taken during the October/November 2003 solar storm. Eos Trans. AGU **87**(52). Fall Meet. Suppl., Abstract SA13B-0277 (2006)
S. Otranto, R.E. Olson, Phys. Rev. A **77**, 022709 (2008)
M.K. Pandey, R.K. Dubey, D.N. Tripathi, Eur. Phys. J. D **45**, 273–277 (2007)
K. Park, J.C. Light, J. Chem. Phys. **126**, 044305 (2007)
D.A. Pejaković, K.S. Kalogerakis, R.A. Copeland, D.L. Huestis, R.M. Robertson, G.P. Smith, Rate coefficients for O-atom three-body recombination in N_2 at temperatures in the range 170–320 K. Eos Trans. AGU **86**(52) Fall Meet. Suppl. Abstract SA53B-1172 (2005)
D.A. Pejaković, K.S. Kalogerakis, R.A. Copeland, D.L. Huestis, J. Geophys. Res. **113**, A04303 (2008)
W.D. Pesnell, K. Omidvar, W.R. Hogey, Geophys. Res. Lett. **20**, 1343–1356 (1993)
A. Petrignani, W.J. van der Zande, P.C. Cosby, F. Hellberg, R.D. Thomas, M. Larsson, J. Chem. Phys. **122**, 014302 (2005)
A.V. Phelps, $H_3^+ + H_2$ cross sections (unpublished report dated 19 August 2001) (private communication)
D.S. Pollock, G.B.I. Scott, L.F. Phillips, Geophys. Res. Lett. **20**, 727–729 (1993)
R. Protz, M. Maier, J. Chem. Phys. **73**, 5464–5467 (1980)
G. Quemener, B. Naduvalath, D.L. Huestis, Charge exchange and vibrational relaxation in collisions of H^+ with $H_2(v)$. Quant. Chem. Calc. (2008, in progress)
D. Régo, R. Prangé, L. Ben Jaffel, J. Geophys. Res. **104**, 5939–5954 (1999)
B.A. Ridley, R. Atkinson, K.H. Welge, J. Chem. Phys. **58**, 3878–3880 (1973)
R.G. Roble, Energetics of the mesosphere and thermosphere, in *The Upper Mesosphere and Lower Thermosphere: A Review of Experiment and Theory*, R.M. Johnson, T.L. Killeen, eds., AGU Monograph **87**, 1–21 (1995)
B.R. Rowe, D.W. Fahey, F.C. Fehsenfeld, D.L. Albrighton, J. Chem. Phys. **73**, 194–205 (1980)
J. Rufus, G. Stark, P.L. Smith, J.C. Pickering, A.P. Thorne, J. Geophys. Res. **108**(E2) (2003). doi:10.1029/2002JE001931
D.W. Rusch, D.G. Torr, P.B. Hays, J.C.G. Walker, J. Geophys. Res. **82**(4), 719–722 (1977)
S.P. Sander, V.L. Orkin, M.J. Kurylo, D.M. Golden, R.E. Huie, C.E. Kolb, B.J. Finlayson-Pitts, M.J. Molina, R.R. Friedl, A.R. Ravishankara, G.K. Moortgat, H. Keller-Rudek, P.H. Pine, Chemical kinetics and photochemical data for use in atmospheric studies, evaluation number 15. JPL Publication 06-2 (Jet Propulsion Laboratory, California Institute of Technology, Pasadena, CA, November 20, 2006). http://jpldataeval.jpl.nasa.gov/
G.C. Schatz, M.J. Redmon, Chem. Phys. **58**, 195–201 (1981)
H.I. Schiff, The photolysis of CO_2 at 1470 Å. Final report on grant No. DA-ARO(D)-31-124-0507, McGill University, Montreal Canada, November 1965
G.B.I. Scott, D.S. Pollock, L.F. Phillips, J. Chem. Soc. Faraday Trans. **98**, 1183–1188 (1993)
M. Seidl, J. Kaa, M. Maier, Chem. Phys. **157**, 279–285 (1991)
R. Shakeshaft, Phys. Rev. A **18**, 1930–1934 (1978)
D.A. Shaw, D.M.P. Holland, M.A. Hayes, M.A. MacDonald, A. Hopkirk, S.M. McSweeney, Chem. Phys. **198**, 381–396 (1995)
D.E. Shemansky, X. Liu, J. Geophys. Res. **110**, A07307 (2005)
V.I. Shematovich, R.E. Johnson, Adv. Space Res. **27**, 1881–1888 (2001)
V.I. Shematovich, R.E. Johnson, J.F. Cooper, M.C. Wong, Icarus **173**, 480–498 (2005)
G.M. Shved, L.E. Khvorostovskaya, I.Yu. Potekhin, A.I. Dem'yanikov, A.A. Kutepov, V.I. Fomichev, Izvestiya. Atmos. Ocean. Phys. **27**, 295–299 (1991)
J.P. Singh, D.W. Setser, J. Phys. Chem. **89**, 5353–5358 (1985)
G.G. Sivjee, J. Geophys. Res. **88**(A1), 435–441 (1983)
G.G. Sivjee, Planet. Space Sci. **39**, 777–784 (1991)
I.R. Slagle, D. Gutman, J.W. Davies, M.J. Pilling, J. Phys. Chem. **92**, 2455–2462 (1988)
T.G. Slanger, G. Black, J. Chem. Phys. **54**, 1889–1898 (1971)
T.G. Slanger, G. Black, J. Chem. Phys. **68**, 1844–1949 (1978)
T.G. Slanger, P.C. Cosby, Ground-based optical signatures of solar storms interacting with the ionosphere. SRI International proposal PYU 07-016, submitted in response to solicitation NSF 07-520 (January 2007)

T.G. Slanger, R.L. Sharpless, G. Black, S.V. Filseth, J. Chem. Phys. **61**, 5022–5027 (1974)
T.G. Slanger, R.L. Sharpless, G. Black, J. Chem. Phys. **67**, 5317–5323 (1977)
T.G. Slanger, P.C. Cosby, D.L. Huestis, R.R. Meier, J. Geophys. Res. **109**, A10309 (2004)
T.G. Slanger, D.L. Huestis, P.C. Cosby, N.J. Chanover, T.A. Bida, Icarus **182**, 1–9 (2006)
T.G. Slanger, K.S. Kalogerakis, L.C. Lee, Excited state photofragment yields from CO_2 and H_2O (2008a, experiments in progress)
T.G. Slanger et al., Photoemission phenomena in the solar system. Space Sci. Rev. (2008b, this issue). Chapter 7
F.J. Smith, Proc. Phys. Soc. **92**, 866–870 (1967)
G.P. Smith, Chem. Phys. Lett. **376**, 381–388 (2003)
D.L. Smith, J.H. Futrell, Chem. Phys. Lett. **40**, 229–232 (1976)
N. Smith, F. Raulin, J. Geophys. Res. **104**(E1), 1873–1876 (1999)
G.P. Smith, R. Robertson, Chem. Phys. Lett. **458**, 6–10 (2008). doi:10.1016/j.cplett.2008.04.074
P.L. Smith, K. Yoshino, W.H. Parkinson, K. Ito, G. Stark, J. Geophys. Res. **96**(E2), 17,529–17,533 (1991)
N.S. Smith, Y. Bénilan, P. Bruston, Planet. Space Sci. **46**, 1215–1220 (1998)
J.A. Spirko, J.J. Zirbel, A.P. Hickman, J. Phys. B **36**, 1645–1662 (2003)
J.R. Stallcop, H. Partridge, E. Levin, J. Chem. Phys. **95**, 6429–6439 (1991)
P.C. Stancil, D.R. Schultz, M. Kimura, J.-P. Gu, G. Hirsch, R.J. Buenker, Astron. Astrophys. Suppl. Ser. **140**, 225–236 (1999)
G. Stark, P.L. Smith, J. Rufus, A.P. Thorne, J.C. Pickering, G. Cox, J. Geophys. Res. **104**(E4), 16,585–16,590 (1999)
G. Stark, K. Yoshino, P.L. Smith, K. Ito, J. Quant. Spectrosc. Radiat. Transf. **103**, 67–73 (2007)
A.W. Stephan, R.R. Meier, K.F. Dymond, S.A. Budzien, R.P. McCoy, J. Geophys. Res. **108**(A1), 1034 (2003)
A.I.F. Stewart, M.J. Alexander, R.R. Meier, J.J. Paxton, S.W. Bougher, C.G. Fesen, J. Geophys. Res. **97**, 91–102 (1992)
L.J. Stief, W.A. Payne, R.B. Klemm, J. Chem. Phys. **62**, 4000–4008 (1975)
D.F. Strobel, M.E. Summers, Triton's upper atmosphere and ionosphere, in *Neptune and Triton*, ed. by D.P. Cruikshank (U. Arizona Press, Tucscon, 1995), pp. 1107–1148
T. Terao, R.A. Back, J. Phys. Chem. **73**, 3884–3890 (1969)
L.P. Theard, W.T. Huntress Jr., J. Chem. Phys. **60**, 2840–2848 (1974)
J.J. Thompson, Philos. Mag. **21**, 225 (1911)
A.P. Thorne, C.J. Harris, I. Wynne-Jones, R.C.M. Learner, G. Cox, J. Phys. E Sci. Instrum. **20**, 54–60 (1987)
A.P. Thorne, U. Litzen, S.E. Johannson, *Spectrophysics: Principles and Applications* (Springer, Berlin, 1999)
M.R. Torr, P.G. Richards, D.G. Torr, J. Geophys. Res. **85**, 6819–6826 (1980)
L.M. Trafton, T.R. Geballe, S. Miller, J. Tennyson, G.E. Ballester, Astrophys. J. **405**, 761–766 (1993)
C. Tully, R.E. Johnson, Planet. Space Sci. **49**, 533–537 (2001)
C. Tully, R.E. Johnson, J. Chem. Phys. **117** 6556–6561 (2002)
C. Tully, R.E. Johnson, J. Chem. Phys. **119**(E) 10452–10453 (2003)
G.L. Tyler et al., Science **246**, 1466–1473 (1989)
D. Uy, M. Cordonnier, T. Oka, Phys. Rev. Lett. **78**, 3844–3846 (1997)
V. Vuitton, R.V. Yelle, M.J. McEwan, Icarus **191**, 722–742 (2007)
J.H. Waite Jr., W.S. Lewis, G.R. Gladstone, T.E. Cravens, A.N. Maurellis, P. Drossart, J.E.P. Connerney, S. Miller, H.A. Lam, Adv. Space Res. **20**, 243–252 (1997)
J.C.G. Walker, D.G. Torr, P.B. Hays, D.W. Rusch, K. Docken, G. Victor, M. Oppenheimer, J. Geophys. Res. **82**(4), 719–722 (1975)
C.W. Walter, P.C. Cosby, J.R. Peterson, J. Chem. Phys. **98**, 2860 (1993)
J.-H. Wang, K. Liu, Z. Min, H. Su, R. Bersohn, J. Preses, J. Larese, J. Chem. Phys. **113**, 4146–4152 (2000)
K. Watanabe, M. Zelikoff, E.C.Y. Inn, Absorption coefficients of several atmospheric gases. Air Force Cambridge Research Labs, Hanscom AFB, MA (1953) [DTIC AD0019700]
K.H. Welge, R. Gilpin, J. Chem. Phys. **54**, 4224–4227 (1971)
E.P. Wigner, Phys. Rev. **73**, 1002–1009 (1948)
E. Wild, H. Klingshirn, B. Faltermeier, M. Maier, Chem. Phys Lett. **93**, 490–494 (1982)
E. Wild, H. Klingshirn, M. Maier, J. Photochem. **25**, 131–143 (1984)
J. Wildt, G. Bednarek, E.H. Fink, R.P. Wayne, Chem. Phys. **122**, 463–470 (1988)
O. Witasse et al., Ionospheres. Space Sci. Rev. (2008, this issue). Chapter 6
A.B. Wittkower, G. Ryding, H.B. Gilbody, Proc. Phys. Soc. **89**, 541–546 (1966)
A.-S. Wong, A.Y.T. Lee, Y.L. Yung, J.M. Ajello, Astrophys. J. **534**, L215–L217 (2000)
A.-S. Wong, Y.L. Yung, A.J. Friedson, Geophys. Res. Lett. **30**, 1447 (2003). doi:10.1029/2002GL016661
C.Y.R. Wu, B.W. Yang, F.Z. Chen, D.L. Judge, J. Caldwell, L.M. Trafton, Icarus **145**, 289–296 (2000)
C.Y.R. Wu, F.Z. Chen, D.L. Judge, J. Geophys. Res. **106**(E4), 7629–7636 (2001)
C.Y.R. Wu, F.Z. Chen, D.L. Judge, J. Geophys. Res. **109**, E07S15 (2004)

Z. Xie, B.J. Braams, J.M. Bowman, J. Chem. Phys. **122**, 224307 (2005)

R.V. Yelle, N. Borggren, V. de la Haye, W.T. Kasprzak, H.B. Niemann, I. Müller-Wodard, J.H. Waite, Icarus **182**, 567–675 (2006)

K. Yoshino, J.R. Esmond, Y. Sun, W.H. Parkinson, K. Ito, T. Matsui, J. Quant. Spectrosc. Radiat. Transf. **55**, 53 (1996)

Y.L. Yung, W.B. DeMore, *Photochemistry of Planetary Atmospheres* (Oxford University Press, New York, 1999)

Y.L. Yung, M.B. McElroy, Icarus **30**, 97–103 (1977)

O. Zatsarinny, S.S. Tayal, J. Phys. B **35**, 241–253 (2002)

M.H.G. Zhang, J.G. Luhmann, A.F. Nagy, J.R. Spreiter, S.S. Stahara, J. Geophys. Res. **98**(E2), 311–3318 (1993)

Y.-F. Zhu, R.J. Gordon, J. Chem. Phys. **92**, 2897–2901 (1990)

J.F. Ziegler, SRIM – The stopping and range of ions in matter (2008). http://www.srim.org/

Neutral Upper Atmosphere and Ionosphere Modeling

Stephen W. Bougher · Pierre-Louis Blelly · Michael Combi · Jane L. Fox · Ingo Mueller-Wodarg · Aaron Ridley · Raymond G. Roble

Originally published in the journal Space Science Reviews, Volume 139, Nos 1–4.
DOI: 10.1007/s11214-008-9401-9

Abstract Numerical modeling tools can be used for a number of reasons yielding many benefits in their application to planetary upper atmosphere and ionosphere environments. These tools are commonly used to predict upper atmosphere and ionosphere characteristics and to interpret measurements once they are obtained. Additional applications of these tools include conducting diagnostic balance studies, converting raw measurements into useful physical parameters, and comparing features and processes of different planetary atmospheres. This chapter focuses upon various classes of upper atmosphere and ionosphere numerical modeling tools, the equations solved and key assumptions made, specified inputs and tunable parameters, their common applications, and finally their notable strengths and weaknesses. Examples of these model classes and their specific applications to individual planetary environments will be described.

Keywords Planets · Thermospheres · Ionospheres · Numerical modeling

S.W. Bougher (✉) · M. Combi · A. Ridley
Atmospheric, Oceanic and Space Sciences Department, University of Michigan, Ann Arbor, MI 48109-2143, USA
e-mail: bougher@umich.edu

P.-L. Blelly
CESR, Toulouse, France

J.L. Fox
Wright State University, Dayton, OH 45435, USA

I. Mueller-Wodarg
Imperial College London, London, UK

R.G. Roble
National Center for Atmospheric Research, Boulder, CO 80309, USA

1 Introduction and Scope

The arrival of new measurements of the neutral upper atmospheres and ionospheres of various solar system planets and moons over the past 4-decades from various spacecraft missions has been astounding (e.g., see Mueller-Wodarg et al. 2008; Witasse et al. 2008; Johnson et al. 2008, and other chapters from this book). These measurements have been used to characterize the structure and dynamics of these atmospheric environments and to compare them to one another. A corresponding evolution of modeling tools, from simple to complex frameworks, has occurred over the same timeframe. These tools are commonly used to predict upper atmosphere and ionosphere characteristics and to interpret measurements once they are obtained. This chapter focuses upon various classes of upper atmosphere and ionosphere numerical modeling tools, the equations solved and assumptions, specified inputs and tunable parameters, their applications, and finally their notable strengths and weaknesses. Examples of these model classes and their specific applications to individual planetary environments will be described.

1.1 General Uses/Benefits of Modeling Tools

Numerical modeling tools can be used for a number of reasons yielding many benefits in their application to planetary upper atmosphere and ionosphere environments. First and foremost, modeling tools are commonly utilized to understand the processes that maintain observed atmospheric structures and drive their variations over various timescales (e.g., solar cycle, seasonal, diurnal, etc.). Time varying inputs (e.g., solar) are often specified to drive model simulations and monitor the resulting variations among the simulated fields. These same model simulations can also be examined to provide a diagnostic analysis of the individual terms of the solved equations; e.g. thermal and momentum balances. Such diagnostic studies provide valuable insight into the underlying processes that are responsible for the time variable features of the atmosphere. Spatially or temporally limited measurements are commonly used to constrain model simulations in an effort to construct reasonable predictions outside available dataset domains and/or time periods. This effort places available observations in a more general/global context. Such model predictions can also be used to motivate new measurements and/or conduct more thorough data analysis studies combining existing datasets.

Model simulations can also be utilized in data processing to facilitate the conversion of raw measurements into useful physical parameters. A good example involves the analysis of aerobraking accelerometer measurements (e.g. Withers 2006; Tolson et al. 2007). Raw accelerations are typically calibrated to yield mass densities and corresponding scale heights. The estimation of temperatures from these density scale heights requires independent information about the composition of the thermosphere (e.g. relative abundance of atomic and molecular species). Global thermospheric general circulation model (TGCM) simulations for Mars are available to provide a first estimate of these global abundances, enabling neutral temperatures to be estimated from scale heights (see Sect. 4.2.5 below).

The comparative approach to investigating planetary upper atmospheres is becoming increasingly fruitful as new information from various planet atmospheres is assimilated using state-of-the-art modeling tools (e.g., Bougher et al. 2002). A comparison of the basic features of the structure and dynamics of planetary upper atmospheres and ionospheres can often be understood by examining the implications of their fundamental planetary parameters (e.g., Bougher et al. 1999a, 2000, 2002; Rishbeth et al. 2000b). Such analysis can be used to guide new model simulations (e.g., to estimate the relative importance of individual processes), and to subsequently interpret completed model simulations. Recent studies have also shown that substantial advances in our understanding can be realized by investigating common aeronomic processes across various planetary environments. A common modeling framework, modified to incorporate planet specific fundamental parameters, provides a useful platform for examining the relative importance of these common physical processes. Finally, model predictions for planetary upper atmospheres and ionospheres with limited or no measurements are often made based upon our experience with previously successful model frameworks encompassing similar physical processes (see Sect. 1.2).

1.2 Usefulness and Shortcomings of the Earth Paradigm

Terrestrial modeling frameworks and their assumptions have typically been used to launch new simulations of other planetary upper atmospheres and ionospheres. This terrestrial paradigm is both useful and hazardous at the same time. The primary benefit of the Earth paradigm can be realized for planetary upper atmospheres having similarities in their fundamental planetary parameters, basic processes, and vertical domains (atmospheric regions). Simulations across these similar planetary environments can be effectively used to examine the relative importance of common aeronomic processes. A good example is the determination of the relative importance of CO_2 15-micron emission as a cooling agent in the thermospheres of Venus, Earth, and Mars (e.g., Bougher et al. 1999a, 2000, 2002). However, planet specific assumptions are often applied when casting the model equations to be solved and the physical formulations employed. Furthermore, fundamental planetary parameters may be so different that application of a terrestrial model framework may no longer be appropriate. A good example of the latter is the application of traditional Earth thermospheric models to the upper atmosphere of Saturn's moon Titan (see Sect. 4.2.3). Here, the assumption of constant gravity over the Titan thermospheric domain ($\sim$600–1500 km) is not sufficient to characterize the extended atmosphere associated with this small body.

In short, care must be taken when applying an existing modeling framework to a new planetary environment. A review of the key equations to be solved and all supporting assumptions must be made in light of the important processes to be incorporated and the vertical domain to be addressed (see Sect. 2).

1.3 Roadmap for Chapter

This chapter describes various numerical model classes (and representative model tools) that are typically used in simulations of the upper atmospheres and ionospheres of planets. Assumptions about the model equations to be solved by the different model frameworks can be classified by the number of moments carried in the solution of the Boltzmann equation (see Table 1 and Sect. 2). Both 1D and multi-dimensional model frameworks are employed, both for the upper atmosphere (UATM) and whole atmosphere (WATM) environment (see Table 1). One-dimensional models (see Sect. 3) are commonly used to thoroughly test detailed aeronomic processes (e.g., thermal, diffusion, and chemical) before the addition of global winds in a multi-dimensional model framework (see Sect. 4). This progression from 1D to multi-dimensional models follows a development strategy involving increasing complexity, internal self-consistency, and expanded temporal plus spatial coverage. Finally, modeling frontiers and key problems for further research are described in Sect. 5.

Table 1 Classes and Examples of Upper Atmosphere Modeling Tools

Dimensions	5-Moment	8-Moment	13-Moment
1D (UATM)	Earth (Roble)	Earth GITM	Mars
	Venus (Fox)	Mars GITM	
	Mars (Fox)	Titan GITM	
3D (UATM)	Earth TIE-GCM	Earth GITM	
	Earth TIME-GCM	Mars GITM	
	Earth CTIM,CTIP, CMAT	Titan GITM	
	Venus TGCM		
	Mars TGCM		
	Jupiter JIM		
	Jupiter TGCM		
	Saturn STIM		
	Titan TTIM		
3D (WATM)	Earth WACCM	Mars GITM (MWACM)	
	Mars MGCM-MTGCM		
	Mars LMD-MGCM		

TGCM (Thermosphere General Circulation Model); GITM (Global Ionosphere Thermosphere Model); WACCM (Whole Atmosphere Community Climate Model); TIE-GCM (Thermosphere Ionosphere Electrodynamics General Circulation Model); TIME-GCM (Thermosphere Ionosphere Mesosphere Electrodynamics General Circulation Model); CTIM (Coupled Thermosphere Ionosphere Model); CTIP (Coupled Thermosphere Ionosphere Plasmasphere Model); CMAT (Coupled Middle Atmosphere Thermosphere Model); JIM (Jupiter Ionosphere Model); STIM (Saturn Thermosphere Ionosphere Model); TTIM (Titan Thermosphere Ionosphere Model); MWACM (Mars Whole Atmosphere Climate Model); MGCM-MTGCM (Mars General Circulation Model – Mars Thermosphere General Circulation Model); LMD-MGCM (Laboratoire de Meteorologie Dynamique Mars General Circulation Model)

2 Moment Solutions of the Boltzman Equation: Applicability Concerns for Upper Atmosphere Models

There are numerous different ways to obtain the transport equations needed to model neutral atmosphere and ionospheric behavior. The one which starts from basic physical principles uses the so called moments of the Boltzmann equation. In the Boltzmann approach one is not interested in the behavior of individual gas particles, but the gas is described by the velocity distribution function $f_s(\mathbf{r}, \mathbf{v}, t)$, where the subscript s, denotes a given species, s, $\mathbf{r}_s$ is the spatial location, $\mathbf{v}_s$ is the velocity and t is time. It can be easily shown (e.g., Schunk and Nagy 2000) that the integro-differential equation for f_s is:

$$\frac{\partial f_s}{\partial t} + \mathbf{v}_s.\nabla f_s + \mathbf{a}_s.\nabla_v f_s = \frac{\delta f_s}{\delta t} \tag{1}$$

where ∇_s is the velocity gradient and $\delta f_s/\delta t$ is a short hand way to write the change in f_s due to collisions. In order to deduce the transport equations of macroscopic parameters one first defines velocity moments of this distribution function. The most commonly used macroscopic parameters of interest are:

$$n_s(r,t) = \int_{-\infty}^{\infty} d^3 v_s f_s, \tag{2}$$

$$\mathbf{u}_s(\mathbf{r}, t) = \int_{-\infty}^{\infty} d^3 v_s f_s \mathbf{v}_s / n_s, \tag{3}$$

$$T_s = \frac{m_s}{3kn_s} d^3 v_s f_s (\mathbf{v}_s - \mathbf{u}_s)^2, \tag{4}$$

$$\mathbf{q}_s = \frac{m_s}{2} \int_{-\infty}^{\infty} d^3 v_s f_s (\mathbf{v}_s - \mathbf{u}_s)^2 (\mathbf{v}_s - \mathbf{u}_s), \tag{5}$$

$$p_s = \frac{m_s}{3} \int_{-\infty}^{\infty} d^3 v_s f_s (\mathbf{v}_s - \mathbf{u}_s)^2, \tag{6}$$

$$\mathbf{P}_s = m_s \int_{-\infty}^{\infty} d^3 v_s f_s (\mathbf{v}_s - \mathbf{u}_s)(\mathbf{v}_s - \mathbf{u}_s), \tag{7}$$

$$\tau_s = \mathbf{P}_s - p_s \mathbf{I}, \tag{8}$$

where n_s is the number density of species s, $\mathbf{u}_s$ is the drift (mean) velocity, T_s is the temperature, $\mathbf{q}_s$ is the heat flow vector, p_s is the scalar pressure, $\mathbf{P}_s$ is the pressure tensor and τ_s is the stress tensor.

In order to obtain the appropriate transport equations for these macroscopic parameters of interest one multiplies the Boltzmann equation with the appropriate function of velocity and then integrates over all velocities. Now a very important point should be noted that the resulting, so called "transport equations", do not lead to a closed system of equations. Namely an equation governing the moment of order k contains the moment of order $(k+1)$. As an example the equation for density contains the drift velocity. A number of approaches have been suggested in order to achieve closure. The most commonly used method finds an approximate expression for the distribution function which allows closure and the evaluation of the collision term.

The so-called 13 moment approximation is the most complex set of equations that have been used so far in neutral atmosphere and ionosphere modeling. In this approach, f_s is approximated as a truncated expansion about the Maxwell Boltzmann distribution function in terms of density, drift velocity, temperature, stress tensor and heat flow vector. The name 13-moment approximation comes from the fact that the gas is described in terms of 13 parameters ($n_s = 1, \mathbf{u}_s = 3, T_s = 1, \mathbf{q}_s = 3, \tau_s = 5$). In most cases the 13 moment equations (see Schunk and Nagy 2000) are too complicated to be able to be solved in a comprehensive global model. The 8-moment equation neglects the stress tensor and the 5 moment equations neglect both stress and heat flow. The Navier-Stokes equation are obtained from the 13-moment equations by assuming that the collision frequency is very high and dropping all $\mathbf{q}_s$ and $\boldsymbol{\tau}_s$ terms, except those that are multiplied by the collision frequency. In this approximation $\mathbf{q}_s$ and $\boldsymbol{\tau}_s$ can be expressed in term of n_s, $\mathbf{u}_s$ and T_s. The most common set of equations used in global models are these Navier-Stokes ones. However, it should be remembered that in this approximation the distribution function must be close to a Maxwellian.

3 Representative 1-D Neutral and/or Ion Models

3.1 Earth

A global mean model of a planetary atmosphere is useful for the development of a self-consistent aeronomical scheme, determining the vertical resolution necessary to determine the basic atmospheric and ionospheric structure, time constants of physical and chemical processes and a host of other properties of the atmosphere and ionosphere. Such a model is numerically fast for long-time integrations, and easy to test and analyze the sensitivity of the atmosphere to physical and chemical processes, such as eddy diffusion, chemical reactions, branching ratios, radiation to space and many other parameters. One can easily conduct a large number of numerical simulations to develop an understanding of the important processes responsible for atmospheric structure and thus make it easier to transfer the important processes to other higher dimensional models.

A 1D model that was important for the development of the series of National Center for Atmospheric Research (NCAR) TGCMs was developed by Roble et al. (1987) for the thermosphere and ionosphere. It was further extended to include the mesosphere and upper stratosphere by Roble (1995). A thorough description of the model has been given in the 1995 paper. The model was designed to be fully consistent internally with only specification of temperature and composition at the lower boundary at 10 mb (30 km) and the upper boundary in thermal and diffusive equilibrium (500–700 km). It is a time dependent model generally run to steady state for global average forcing from solar EUV radiation, the aurora, and specified eddy diffusion. It was primarily used to develop the aeronomical scheme for the series of TGCMs and to test ideas and parameters. It has been used for numerous studies by colleagues and students to test various ideas. One example of a simulation for solar medium conditions is shown in Fig. 1.

This terrestrial 1D code has also been used to study global change in the upper atmosphere by Roble and Dickinson (1989) and more recently been used by Qian et al. (2006, 2008) to study decadal changes in satellite drag and ionospheric structure.

3.2 Venus and Mars

One-dimensional models have been constructed of the thermospheres and ionospheres of Mars and Venus since the first radio occultation measurements of the electron density profiles were reported from early flybys and orbiters. Lodders and Fegley (1998) have reviewed missions to these planets up to 1998.

There are two basic types of one-dimensional models: photochemical equilibrium (PCE) models, in which transport is ignored, and those that include (usually) vertical transport of species. While in the former type of model, the densities at each altitude can be computed independently, in the latter type of model numerical coupling of each altitude to those above and below it must be taken into account. Altitude profiles of neutral species in thermospheres cannot be modeled with the PCE approximation, although ion density profiles in the high density lower peak regions may be.

All of the early (and most of the subsequent) radio occultation electron density profiles measured by radio science experiments of both Mars and Venus have exhibited two peaks: an upper F1 peak, which is produced by absorption of the main portion of the EUV ($\sim$150–1000 Å), and a lower E peak which is produced by the absorption of soft X-rays (e.g., Kliore et al. 1967; Stewart 1971). Curiously, however, the ion density profiles derived from the Viking RPA measurements showed no lower peak (e.g., Hanson et al. 1977).

Most of the early models of the ionospheres of Mars and Venus that were designed to fit radio occultation electron density profiles, assumed PCE (e.g., McElroy 1967, 1968a, 1969; Stewart 1968, 1971). The Venus ionospheric model of Kumar and Hunten (1974), however, was a hybrid model, in which the heavy ions were assumed to be in PCE, while the lighter ions were subject to transport. Shimazaki and Shimizu (1970) constructed a number of models of the Martian ionosphere that included transport by diffusion and eddy diffusion, which were compared to the electron density profile measured by Mariner 4. Interestingly,

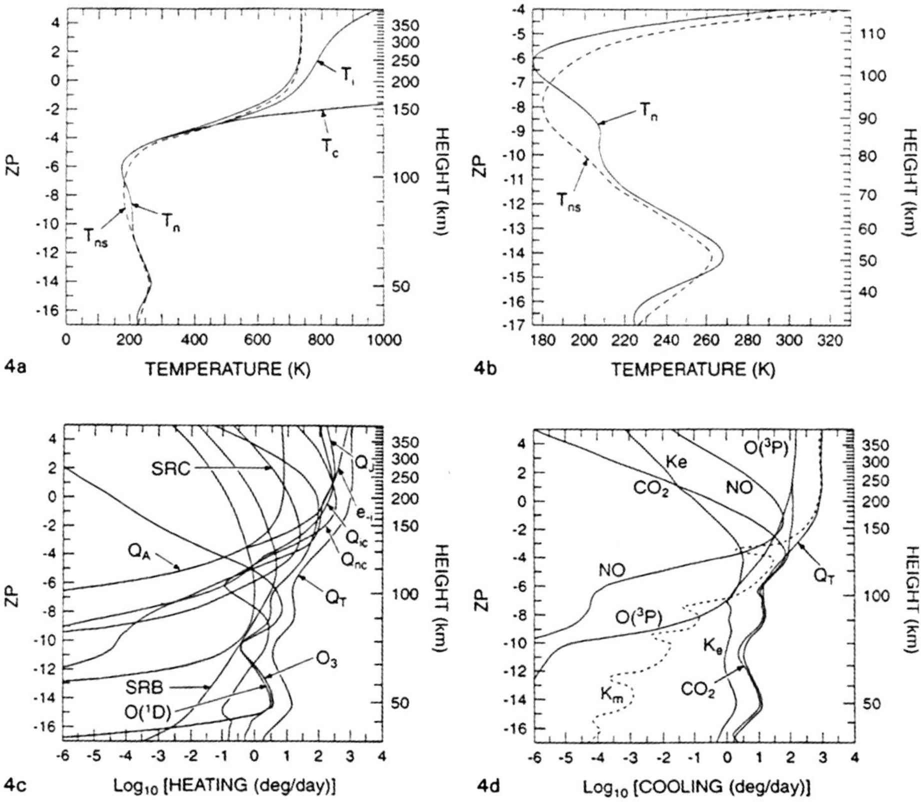

Fig. 1 NCAR 1D global mean model simulations for solar medium conditions. (**a**) Neutral, ion, and electron temperatures (~30–400 km); (**b**) Neutral temperatures (40–120 km); (**c**) Heating terms of thermal equation (~30–400 km); and (**d**) Cooling terms of the thermal equation (~30–400 km). Individual curves for (**a**): Tn, Te, Ti: neutral, electron and ion temperatures; T_{ns}: neutral temperature from MSIS-90. Individual curves for (**c**): SRC: Schumann-Runge Continuum; SRB: Schumann-Runge Bands; Q_T: total heating; Q_J: Joule heating; Q_A: auroral heating; O_3: O_3 heating; $O(^1D)$: heating from $O(^1D)$ quenching; Q_{NC}: neutral chemistry heating; Q_{IC}: ion chemistry heating; Q_{EI}: ion–electron heating. Individual curves for (**d**): Q_T: total heating; K_e: cooling rates from eddy thermal conduction; CO_2: 15-micron cooling; NO: 5.3-micron cooling; K_m: cooling rate from downward molecular thermal conduction; $O(^3P)$: oxygen fine structure cooling. ZP (left vertical axis, log pressure scale used by the NCAR codes)

McElroy (1968b) noted the difficulty of constructing model ionospheres when the only information was in the form of radio occultation electron density profiles, without the benefit of *in situ* measurements of ion and neutral densities. The lack of *in situ* measurements has also plagued interpretation of the electron density profiles of the outer planets for many years.

The PCE approximation becomes inaccurate above the boundary where the lifetime of a species due to chemistry ($\tau_c = 1/L$), where $L = L/n$ is the specific loss rate, L is the total chemical loss rate and n is the number density of the species, becomes longer than that due to vertical transport. If the main transport process is by diffusion, the lifetime of a species due to transport is given approximately as $\tau_D \sim H^2/D$, where D is the diffusion coefficient of that species and $H = kT/mg$ is the scale height. In this expression, k is Boltzmann's constant, while T, g, and m are the altitude dependent temperature, acceleration of gravity, the mass of the atmospheric species, respectively. For ions, the scale height is $H_i = k(T_i + T_e)/m_i g$, where T_i is the ion temperature, and T_e is the electron temperature.

If the major transport process (for neutrals) is mixing, the lifetime of a species against transport is given by $\tau_K \sim H_{avg}^2/K$, where K is the eddy diffusion coefficient, $H_{avg} = kT/m_{avg}g$, and m_{avg} is the average mass of the constituents. It is remarkable that the eddy diffusion coefficient is on the order of $1.0 \times 10^{13}/n^{0.5}$ cm^2 s^{-1}, where n is the total number density in the lower thermospheres for both Mars and Venus (cf., Krasnopolsky 1982; von Zahn et al. 1980).

The boundary where diffusion of a neutral species becomes more important than mixing is known as the homopause. In fact, however, that altitude is different for each species. In early models of the thermospheres of Venus and Mars, a single homopause altitude was assumed (e.g., McElroy 1967, 1969; Kumar and Hunten 1974; Chen and Nagy 1978). Below the homopause the thermosphere is considered to be completely mixed for chemical tracers/inert species; above the homopause, these species density profiles are assumed to be distributed according to their own scale heights. Species that are formed photochemically do not exhibit this behavior. Because of the availability of *in situ* measurements of the neutral densities, and the computing power that is available today, even in one-dimensional models, the homopause approximation is neither necessary nor used widely.

While PCE approximations are almost never used to compute the density profiles of minor neutral thermospheric species, such models of ion density profiles continue to be used for specific purposes, including studies focused on the electron density peak regions (e.g., Cravens et al. 1981; Fox and Dalgarno 1981; Kim et al. 1989; Martinis et al. 2003), or for airglow calculations of processes that originate near the electron density peaks (see, for example Fox 1992, and references therein).

The Viking I and II probes carried neutral mass spectrometers through the Martian atmosphere, and so the major neutral densities in the thermosphere at low solar activity were measured in situ for the first time (e.g., Nier and McElroy 1976). Early ionospheric models based on these measurements included, for example, those of McElroy et al. (1976), Fox and Dalgarno (1979), and Chen et al. (1978). While the latter model included vertical transport. The former two were photochemical equilibrium models. The *in situ* measurements of the Pioneer Venus orbiter and probes enabled more accurate models of the Venusian ionosphere (e.g., Chen and Nagy 1978; Nagy et al. 1980; Fox 1982).

In the terrestrial ionosphere, the absolute maximum in the electron density profile is an F_2 peak, which appears near 300 km. At this altitude, the chemical lifetime of the major ion (O^+) is approximately equal to that of transport. At high altitudes in the Venus ionosphere, O^+ becomes the most important species, yet models show that it forms a peak that is not (or is barely) visible in the electron density profile. On Mars, the O^+ density forms a peak at high altitudes, but thus far measurements indicate that densities of O^+ are everywhere smaller than those of O_2^+ (e.g., Hanson et al. 1977). Where the major loss is by transport, photochemical equilibrium models do not reproduce the profiles of ions, such as O^+ and other (mostly) atomic ions.

More sophisticated one-dimensional thermosphere/ionosphere models have included both chemistry and transport by molecular and eddy diffusion (for neutrals), and ambipolar diffusion (for ions), and do not assume a fixed homopause (e.g., Nagy et al. 1980; Fox 1982, 2004; Krasnopolsky 2002). The one-dimensional thermosphere-ionosphere models of Venus and Mars of Shinagawa and Cravens (1988, 1989) have also included magnetic fields.

Although one-dimensional models have limitations, mainly that horizontal transport by convection is ignored, they are simple enough that many species and reactions among those

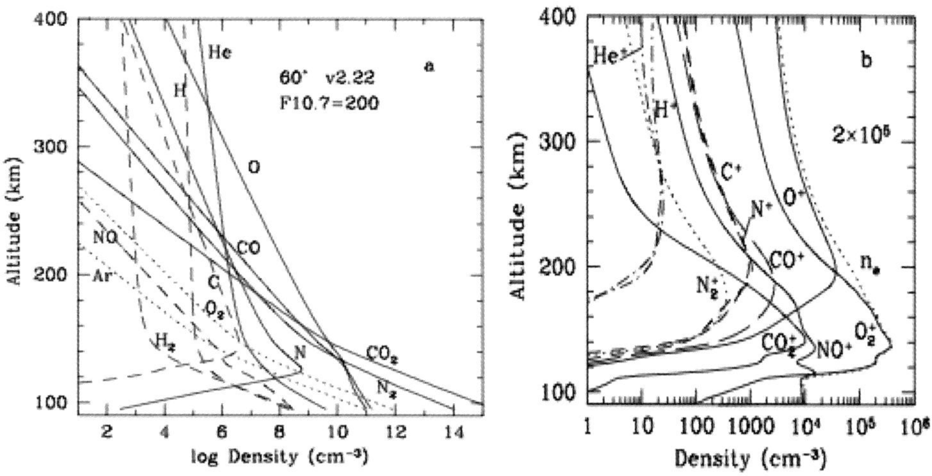

Fig. 2 Venus 1D simulations for solar maximum conditions: (**a**) Neutral densities; (**b**) ion densities and electron density. From Fox (unpublished)

species can be included in them. For example, the current Mars and Venus models of Fox and coworkers (e.g., Fox and Sung 2001; Fox 2004; Fox and Yeager 2006) contain 23 species and more than 200 reactions. These models include photoionization and excitation, photoelectron impact ionization and excitation, and electron-impact excitation of twelve neutral background species, and photodissociation, photodissociative excitation, and photodissociative ionization of all of the molecular neutral background species. Cross sections for all these processes are required input for the models.

An example of a one-dimensional Venus model for high solar activity is shown in Fig. 2. Figure 2a illustrates the background density profiles of 12 species; Fig. 2b shows the computed ion density profiles. In this model an upward velocity boundary condition of 2×10^5 cm s^{-1} was imposed on most ions, since the ionosphere of Venus is eroded by day-to-night ion transport. O^{++} ions are assumed to be in photochemical equilibrium. Using these models, density profiles of metastable species, such as $O^+(^2D)$, $O^+(^2P)$, $N(^2D)$, $N(^2P)$, $O(^1S)$, and $O(^1D)$ can be computed. These species can participate in reactions that are not available to ground state species, and are important to the energy balance in the thermosphere/ionosphere. Quenching of these species produces local heating, and radiation produces cooling. The radiation rates of shorter-lived excited species may be computed to produce profiles of airglow intensities.

Using one-dimensional models, the altitude-dependent heating efficiencies can be readily computed. The heating efficiency is defined as the ratio of the local heating rate to the solar energy absorbed at a given altitude. Computations show that for Mars and Venus (Fox 1988; Fox et al. 1995), the most important heat source near the ion peak is dissociative recombination of O_2^+,

$$O_2^+ + e \rightarrow O + O + \Delta E \tag{9}$$

where the product O atoms may be in various combinations of electronically excited states and the energy released Δ/E varies with the branching ratios (e.g., Kella et al. 1997). At low altitudes, photodissociation and quenching of metastable species are more important. Exothermic chemical reactions are of secondary importance, and electron impact is of minor importance over the entire thermosphere. For Venus, the altitude dependent heating efficiencies range from about 16% at low altitudes for the "lower limit" model to 22% for the "best guess" model. At high altitudes the heating efficiencies increase slightly with altitude up to 185 km. On Mars, the "best guess model" yields heating efficiencies of about $21 \pm 2\%$ from 100 to 200 km, although at low altitudes the lower limit model shows heating efficiencies of about 16%.

4 Representative Multidimensional Models

4.1 Model Classes to be Addressed

Table 1 summarizes the classes of multi-dimensional models that we will consider in Sects. 4 and 5. Notice that multi-dimensional model development is following two important trends. First, existing upper atmosphere models are being extended to encompass the entire atmosphere domain (ground to exosphere) for Earth (NCAR WACCM) and Mars (Michigan-MWACM). The LMD-MGCM code for the Mars lower atmosphere has also be extended upward into the thermosphere. These efforts reflect the availability of both lower and upper atmosphere datasets for these planets. Coupling processes (thermal, chemical, dynamical) linking these atmospheric regions are important to address with these "whole atmosphere" modeling tools. Second, model frameworks are being developed and exercised using 8-moment and 13-moment solutions of the Boltzman equation. The motivations here are at least twofold: (a) to incorporate a non-hydrostatic treatment for improvement of the simulation of vertical velocities (Ridley et al. 2006; Deng et al. 2008), and (b) to capture the physical processes that bridge the collisional and non-collisional regions near the exobase (Boqueho and Blelly 2005).

Under hydrostatic equilibrium, a typical assumption used in most global planetary models, the pressure gradient in the vertical direction is exactly balanced by the gravity force (Deng et al. 2008). However, for large vertical velocities, acceleration terms in the vertical momentum equation cannot be ignored, and the basic balance is no longer hydrostatic. Two likely examples of these conditions are realized for the sudden intense enhancement of high latitude Joule heating for the Earth's thermosphere-ionosphere (Deng et al. 2008), and that of Jupiter as well. If the hydrostatic assumption is relaxed, the vertical momentum equation can be expanded to include additional acceleration terms: (1) forces due to ion-neutral and neutral-neutral friction (when each constituent is solved independently), (2) centrifugal and Coriolis forces, and (3) non-linear advection terms. The altitude variation of gravity is easily accommodated in this framework. Vertical propagation of a non-hydrostatic "disturbance" results in acoustic waves and non-hydrostatic gravity waves (Deng et al. 2008). Care must be taken to either damp or accommodate these waves in non-hydrostatic models. In short, global non-hydrostatic models are needed to address phenomenon associated with large vertical winds in planetary upper atmospheres. New planet specific global thermosphere–ionosphere models are being developed and validated for this purpose, based upon the Global Thermosphere–Ionosphere Model (GITM) for Earth (Ridley et al. 2006; Deng et al. 2008).

4.2 Representative GCM Model Descriptions/Results

4.2.1 Earth (NCAR TGCMs)

A series of TGCMs have been developed at NCAR over the past 30 years with each model incorporating new processes of the coupled thermosphere-ionosphere system. The historical development of the TGCM suite of models is discussed in Bougher et al. (2002). Self-consistent temperatures, neutral-ion densities, neutral dynamics, and self-consistent electrodynamics are contained in the TIE-GCM (Richmond et al. 1992). This code was then extended to include the mesosphere and upper stratosphere to become the TIME-GCM (Roble and Ridley 1994). These two codes are now the basic upper atmosphere models at NCAR. The TIE-GCM is used to explore thermosphere-ionosphere-electrodynamic interactions and the TIME-GCM has the same processes but extended to include aeronomic processes associated with the mesosphere and upper stratosphere and to examine physical and chemical interactions between upper atmospheric regions.

The TIME-GCM is a self-consistent model of the upper atmosphere extending between 30 km and 500 km altitude. It has been used for comparison and interpretation of satellite, rocket and ground-based data for many years by a wide variety of scientific colleagues, post doctoral fellows and graduate students. The TIME-GCM was initially designed for a 5° latitude and longitude grid in the horizontal and 2 grid points per scale height in the vertical with a 5 minute time step. This "coarse" horizontal and vertical resolution was later refined for specific model applications (see below). The TIME-GCM solves for the neutral gas temperature, winds and constituents of the thermosphere, mesosphere and upper stratosphere both major and minor species self-consistently. It also solves for the ionospheric plasma properties of electron and ion temperature, electron density and ion composition, and the electric field, plasma drift and resulting magnetic perturbations. The most recent description of the model is given by Roble (2000).

In the late 1990s, the TIME-GCM was extended to use lower boundary data at 10 mb (30 km) to force the variability propagating upward from the lower atmosphere and to study couplings between the lower and upper atmospheres (Roble 2000). This included specification of tides, gravity waves, planetary waves and other disturbances on a daily basis so that continuous simulations of the upper atmosphere could be made for realistic daily simulations that would be used to compare with observational time averaged campaign or satellite orbital tracking data.

In order to examine the feasibility of developing a model that extended from the ground-to-exosphere the TIME-GCM was flux coupled to the NCAR community climate model CCM3 (Khiel et al. 1998) at the boundary between the models near 10 mb. This allowed information between the upper and lower atmospheres to be exchanged simulating the entire atmosphere, troposphere, stratosphere, mesosphere and thermosphere/ionosphere. This coupled model simulated a strong stratospheric warming, mesospheric cooling, thermospheric warming at high latitudes that was generated spontaneously from planetary waves propagating upward from the troposphere (Liu and Roble 2002, 2005). These studies indicated that a continuous model from the ground to exosphere could be developed to examine coupling aspects between regions of the whole atmosphere. This was a precursor for the development of the WACCM that is discussed in Sect. 5.1.2.

The TIME-GCM resolution was limited by computer power in the 1990s and early 2000s. While this was sufficient for a number of upper atmosphere studies, it was inadequate to represent the shorter wave migrating and non-migrating tides and planetary wave propagation and dissipation. One temporary solution was to double the amplitudes of the Global Scale Wave Model (GSWM) forcing at the lower boundary to get agreement with observational data. Subsequent simulations showed that it was not necessary to double the amplitudes but rather to improve both the vertical (0.25 scale height) and horizontal (2.5° × 2.5°) resolution. These comparisons are shown in Fig. 3. With a double resolution, large tides developed without the need to double GSWM amplitudes. This exercise clearly illustrates the need to

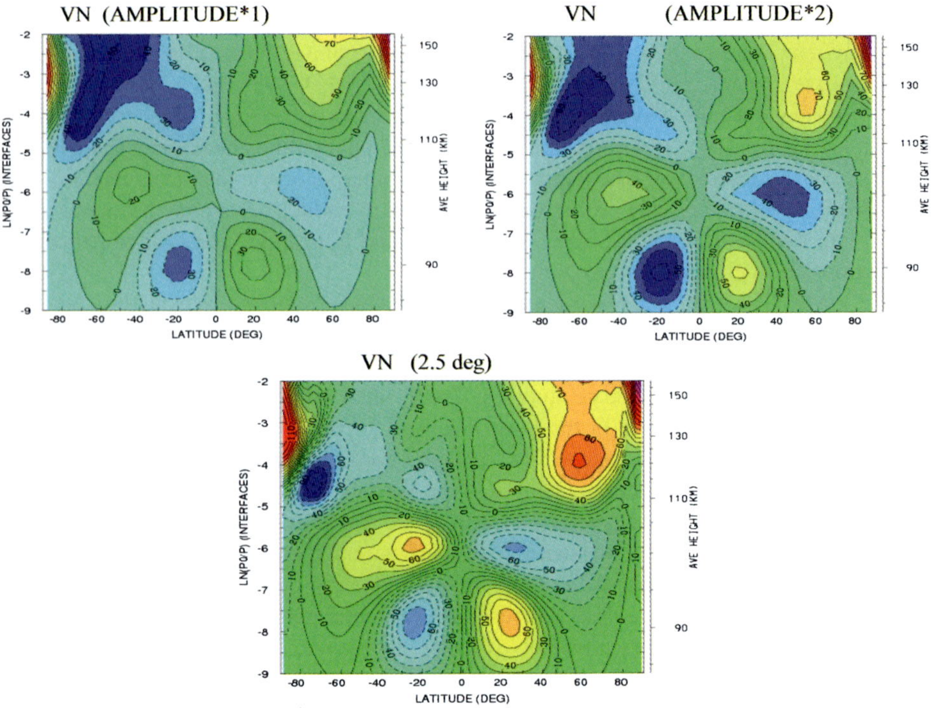

Fig. 3 NCAR TIME-GCM model simulations for solar moderate conditions. Meridional tidal winds (VN) in the ~80–150 km region at local noon for: (**a**) standard GSWM forcing, (**b**) doubled GSWM forcing, and (**c**) standard GSWM forcing but with 2.5° horizontal resolution

match the model resolution (both vertical and horizontal) with the phenomena being examined. This new high-resolution model resolved a number of problems with simulation/data comparisons.

Work is continuing on evaluating the impact of the new resolution on a number of previous studies, such as the migrating and non-migrating tides, auroral dynamics, thermospheric densities and satellite drag and ionospheric F-region dynamics as well as a number of airglow studies. A web site that describes the model and some important simulations is: http://www.hao.ucar.edu/modelling/tgcm/tgcm.html.

Validation of other planetary GCM codes (see Sects. 4.2.2–4.2.9) may also require sensitivity tests to confirm the model resolution necessary to obtain converged solutions; i.e. finer vertical (e.g. ~0.25 scale height) and horizontal (e.g. 2.5° × 2.5°) resolution may also be required to obtain converged solutions for important desired applications.

4.2.2 Earth (CTIM, CTIP, CMAT codes)

Another major atmosphere general circulation model for Earth is the CTIP model developed jointly by groups in the UK and US. The code originates from two initially separate models for the thermosphere (Fuller-Rowell and Rees 1980, 1983) and high latitude ionosphere (Quegan et al. 1982) which were later coupled (Fuller-Rowell et al. 1987, 1996) for two-way coupling between the neutral and ionized regimes of the upper atmosphere, a version of the model often referred to as CTIM. A self-consistent plasmasphere model for the regions equatorward of around 30° geomagnetic latitude was added by Millward et al. (1996a) to form CTIP. More recently, the original bottom boundary (80 km, 0.01 mb) of the CTIM model was lowered by Harris (2001) into the stratosphere (30 km, 10 mb) in order to include the stratospheric and mesospheric chemistry and full vertical dynamical and chemical coupling. This extended version of the model is referred to as the CMAT model.

What distinguishes these from other thermosphere/ionosphere models is primarily the fact that ionospheric calculations are carried out along magnetic field lines rather than being treated on the same spherical grid as the neutral gases. This has proven to be a most useful approach since one key issue is how to treat upper boundary conditions for ions and electrons, in particular the plasma fluxes along field lines in the topside ionosphere where field-aligned transport forms a dominant process affecting the distribution of O^+ and H^+ ions and electrons. The plasma flux boundary condition becomes important for calculations of ionospheric densities when the boundary is located within a few scale heights above the density peak. The particular choice of flux tube coordinates for plasma in CTIP/CTIM eliminates this difficulty. At high latitudes field lines are open and extend to around 10000 km altitude in the model, while at low latitudes flux tubes in CTIP are closed. With plasma densities near 10000 km being negligible compared with ionospheric densities, a zero flux boundary condition can safely be assumed, while in the regime of closed flux tubes both ends of a field line are within the photochemical domain of the ionosphere (in opposite hemispheres), allowing the simple boundary condition of photochemical equilibrium.

The CTIP, CTIM and CMAT models have been used extensively over the past decades to understand the global morphology of the thermosphere and ionosphere both through purely theoretical studies and in comparisons with observations. Studies have investigated (amongst others) the morphology of responses to geomagnetic storms (Field et al. 1998; Fuller-Rowell et al. 2002), thermospheric composition and dynamics (Fuller-Rowell 1998; Rishbeth and Mueller-Wodarg 1999), semiannual variations in the ionosphere (Millward et al. 1996b; Rishbeth et al. 2000a), effects of tidal forcing (Millward et al. 2001; Mueller-Wodarg et al. 2003) and NO chemistry (Dobbin et al. 2006).

4.2.3 Michigan GITM Codes

Earth GITM. The GITM code (Ridley et al. 2006) was designed from the bottom up with flexibility in mind for every aspect of modeling upper atmospheres of planetary systems. The grid system within GITM is fully parallel and is quite versatile. Users can run 1D cases at a specified latitude and longitude (which is set at run-time in the input file) or 3D cases with almost any latitudinal and longitudinal resolution that the user wants (once again, set at run-time). GITM can run on a single processor machine or multi-processor machines. It has been run on 256 processors resolving the upper atmosphere with a 1.25° latitudinal by 2.5° longitudinal resolution. For testing, GITM has been run with 10° by 20° resolution. This flexibility in the resolution allows rapid development of the model and facilitates testing of new physics within the code.

The main feature that differentiates GITM from all other coupled ionosphere thermosphere models is the easing of the hydrostatic assumption within the vertical momentum equation—the pressure does not have to balance with gravity (although it almost always does). In regions in which there are non-hydrostatic forces (e.g., the auroral zone), large vertical winds can develop (Deng et al. 2008). In addition, the GITM vertical momentum equation allows gravity to be dependent on altitude, instead of constant, which is crucial for small bodies with extended atmospheres, such as Titan. GITM utilizes an altitude grid, which is also different than other upper atmosphere models. The resolution in the vertical direction is stretched such that it is approximately $1/3$ of a scale height at code initiation.

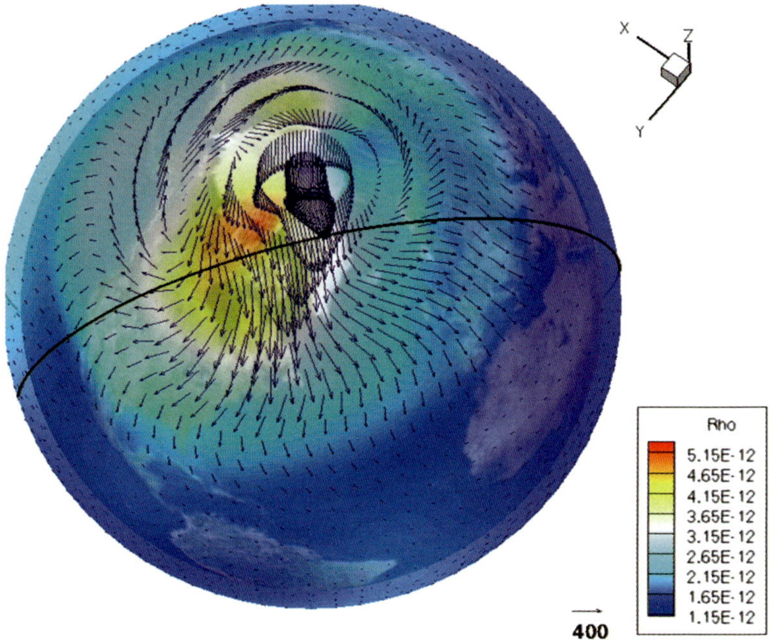

Fig. 4 Earth GITM simulation at 400 km. Horizontal wind vectors are superimposed upon mass density color contours (1.15 to 5.15 $\times$ 10^{-12} g/m^3)

An additional feature of GITM is the ability to turn on and off physics through the input file. Different source terms (such as Coriolis, Joule heating, solar EUV heating, and thermal conduction), can be turned off through the input file. This allows users to conduct numerical experiments in which they self-consistently determine the effects of different source terms on the coupled nonlinear system.

In order to allow GITM to be utilized for more than a single body, very little was hard-coded into the core of GITM. Each atmospheric and ionospheric constituent is specified in a planet specific module, such that the advective core need only loop over the number of species to determine the hydrodynamic behavior of the atmosphere. Other common source terms, such as solar EUV inputs, and inter-species frictional drag in the vertical direction, are handled in a similar manner, allowing the code to be adapted to another planet very easily. More specific source terms, such as radiative cooling, need to be coded for the particular problem and are easily linked to GITM through hooks.

At Earth, GITM has an extremely flexible high-latitude energy input module. This allows users to try different electric fields and particle inputs to drive GITM. At other planets, this can be easily adapted to different types of forcing. Models of electron precipitation and electric potential can be added with little difficulty. These will then be utilized to calculate electric fields and ion and electron velocities. Joule heating and ion-neutral momentum coupling are then self-consistently calculated.

Figure 4 shows results from the Earth-based GITM at 400 km altitude. The vectors show the thermospheric neutral winds, while the coloring is the thermospheric mass density. The neutral winds roughly follow a two-cell convection pattern at high latitudes, due to the strong forcing by the ion convection and aurora. On the dusk-side (left), the neutrals are able to form a completely closed cell, while on the dawn-side (right), the cell is less well defined.

This is because on the dusk-side, the Coriolis force is in the same direction as the circulation pattern, so the closed cell is accentuated, while on the dawn-side, the flow is inhibited, since the Coriolis force opposes it. At very high latitudes and at the tail end of the convection cells on the night side (i.e., just over Canada), there is a large enhancement in the thermospheric mass density. This is caused by Joule heating in the auroral zone, where there are extremely large electric fields and strong conductivities. During this strong driving period, the density peaks at the poles, while during quieter times, the mass density peaks at lower latitudes.

Titan GITM. A new application of the GITM framework was recently developed (Bell et al. 2006; Bell 2008) in an effort to interpret and place in a global context new Cassini INMS neutral and ion density and inferred temperature datasets of the Titan upper atmosphere (e.g., Waite et al. 2005). The GITM framework was chosen to capture the unique physics of the Titan upper atmosphere that requires: (a) variable gravity, (b) calculated fluxes of major species out the top of the atmosphere, and (c) variable Saturn magnetospheric forcing. Density gradients (yielding fluxes) were specified at the upper boundary in order to simulate measured Cassini INMS CH_4 density profiles; the self-consistent feedback of these fluxes upon temperatures was also included. The Titan GITM code was designed to span $\sim$500 to 1500 km in the Titan upper atmosphere, covering the region below the homopause ($\sim$800 km) to just above the exobase ($\sim$1300 km). Solar EUV forcing (heating, photo-dissociation, photo-ionization) is dominated by N_2 and CH_4 solar absorption between 1.6–170.0 nm. Hydrogen cyanide (HCN) rotational band infrared radiative cooling is incorporated and represents the dominant IR cooling agent in the thermosphere. A self-consistent treatment of chemical production and loss processes for 4-major, 7-minor, several isotopes, and thermally active (e.g. HCN) constituents is incorporated, based upon the scheme outlined by DeLa Haye (2005). Major photochemical ions are limited to key species: N_2^+, N^+, CH_3^+, H_2CN^+, and $C_2H_5^+$. Finally, a differentially, super-rotating lower boundary is specified at $\sim$500 km, in accord with ground-based observations (Hubbard et al. 1993). Horizontal and vertical distributions of simulated temperatures and densities have been successfully compared to specific Cassini orbit measurements, especially profiles of key isotopes (e.g., Bell 2008).

Mars GITM. The need for a ground-to-exobase GCM at Mars has motivated another new application of the GITM framework. Existing Mars lower and upper atmosphere datasets need to be interpreted and connected using such a "whole atmosphere" model framework (e.g., Bougher et al. 2006b). This is the first extension of the GITM framework over a wide range of altitudes encompassing both upper and lower atmosphere processes.

A prototype MWACM code using the GITM framework has been developed, enabling initial 1D and 3D simulations to be conducted over 0–300 km for specific solar cycle, seasonal and dust conditions at Mars. Specifically, the terrestrial GITM code (Ridley et al. 2006) was adapted to include Mars fundamental parameters, constants, and key radiative processes in order to capture the basic observed features of the thermal and dynamical structure of the Mars atmosphere from the ground to 300 km. For the Mars lower atmosphere (0–80 km), an efficient (fast) radiation code was adapted from the NASA Ames MGCM code to the framework of the MWACM. This now provides MWACM solar heating (long and short wavelength), aerosol heating, and CO_2 15-micron cooling in the LTE region of the Mars atmosphere below $\sim$80 km. For the Mars upper atmosphere ($\sim$80 to 300 km), a fast formulation for NLTE CO_2 15-micron cooling was implemented into the MWACM code, along with a correction for non-LTE (NLTE) near-IR heating rates ($\sim$80–120 km). In addition, a thermospheric EUV-UV heating routine (based upon a CO_2 dominated atmosphere) was adapted to the MWACM framework. Finally, detailed neutral-ion chemistry was recently incorporated above 80 km, based upon Mars TGCM reactions and rates (see Sect. 4.2.5).

For the entire atmosphere, the MWACM dynamical core solver was modified to work with the new terrain following coordinate system. The Martian terrain is now being incorporated into the MWACM code making use of Mars Global Surveyor MOLA topographic data files. NASA Ames MGCM CO_2 condensation, and boundary layer routines will be added below 80 km. At the surface, global empirical maps of albedo and thermal inertia will be supplied to the radiation calculations. Initial simulations indicate this extended model is stable and captures the basic observed temperatures and expected wind structures throughout the Mars atmosphere.

4.2.4 Venus (NCAR VTGCM)

The large-scale circulation of the Venus upper atmosphere from ~90 to ~200 km (upper mesosphere and thermosphere) is a combination of two distinct flow patterns: (1) a relatively stable subsolar-to-antisolar (SS-AS) circulation cell driven by solar (EUV-UV-IR) heating, and (2) a highly variable retrograde superrotating zonal (RSZ) flow (see reviews by Bougher et al. 1997, 2006a, 2006b; Schubert et al. 2007). GCMs have proven useful for synthesizing the available Pioneer Venus, Magellan, Venus Express, and ground-based density, temperature, and/or airglow datasets and thereby extracting these upper atmosphere wind components (see reviews by Bougher et al. 1997, 2006a; Schubert et al. 2007).

The Venus TGCM (VTGCM) is a 3D finite difference hydrodynamic model of the Venus upper atmosphere that is based on the NCAR terrestrial TGCM. The VTGCM has been documented in detail as revisions and improvements have been made over nearly 2-decades (see Bougher et al. 1988, 1990, 1997, 1999a, 2002; Bougher and Borucki 1994; Zhang et al. 1996).

The modern VTGCM code (e.g., Brecht et al. 2007; Rafkin et al. 2007; Bougher et al. 2008) calculates global distributions of major species (CO_2, CO, O, and N_2), minor species (e.g. O_2, NO, $N(^4S)$, $N(^2D)$), and dayside photochemical ions (CO_2^+, O_2^+, O^+, and NO^+). These constituent fields are all consistent with the simulated 3-D temperature structure and the corresponding 3-component neutral winds. The VTGCM model covers a 5° by 5° latitude-longitude grid, with 46 evenly spaced log-pressure levels in the vertical, extending from approximately ~80 to 200 km at local noon. Dayside O and CO sources arise primarily from CO_2 net dissociation and ion-neutral chemistry; the latter utilizes the ion-neutral chemical reactions and rates of Fox and Sung (2001). Simplified catalytic ClOx and HOx reactions can also employed to specifically improve the chemical sources and sinks for O and CO below 120 km (e.g., Bougher and Borucki 1994).

Formulations for CO_2 15-micron cooling, wave drag, and eddy diffusion are incorporated into the VTGCM (see Bougher et al. 1999a; Brecht et al. 2007). In particular, CO_2 15-micron emission is known to be enhanced by collisions with O atoms, providing increased cooling in NLTE regions of the upper atmosphere (see Bougher et al. 1994; Kasprzak et al. 1997). VTGCM CO_2 15-micron cooling is parameterized as described by Bougher et al. (1986), making use Roldan et al. (2000) exact cooling profiles at reference temperatures and atomic oxygen abundances. The collisional O–CO_2 relaxation rate adopted for simulated 15-micron cooling is $\sim 3 \times 10^{-12}$ cm^3/s. In addition, near-IR heating rates are incorporated using modern offline look-up tables from Roldan et al. (2000). These parameterizations provide strong CO_2 15-micron cooling that is consistent with the use of EUV-UV heating efficiencies of ~20–22%, which are in agreement with detailed offline heating efficiency calculations of Fox (1988).

The VTGCM is typically run to examine Venus thermospheric structure and winds for solar maximum, moderate, and minimum EUV-UV flux conditions, corresponding to terrestrial F10.7-cm indices of 200, 110-130, and 68-80, respectively. In addition, the VTGCM

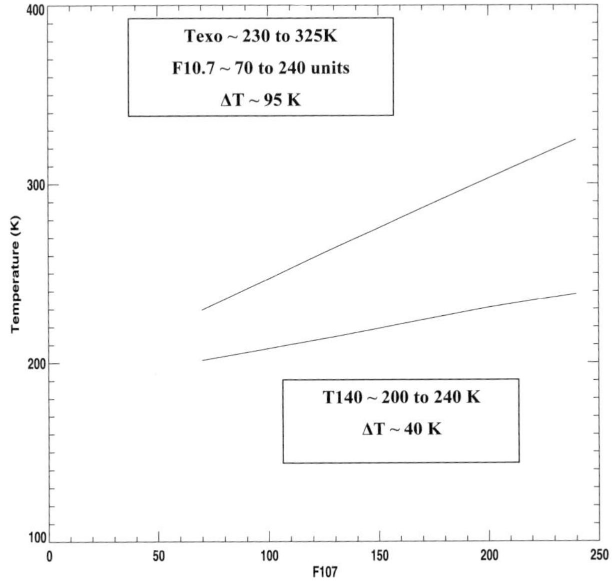

Fig. 5 VTGCM predicted exospheric temperature variations for dayside, equatorial conditions. $\Delta Texo \sim 95$ K for ΔF 10.7~170 units. $Tmax = 325$ K; $Tmin = 230$ K. Temperatures at 140 km are also plotted: ~200 to 240 K

is designed to calculate O_2 visible (400–800 nm), O_2 IR (1.27 microns) and NO ultraviolet (198.0 nm) nightglow distributions for comparison with various Venera, Pioneer Venus, Venus Express, and/or ground-based measurements (e.g. Bougher et al. 1997, 2006a, 2008; Brecht et al. 2007). These night airglow layers, and the controlling global circulation patterns being traced, span ~90 to 150 km. Figure 5 illustrates the VTGCM predicted dayside, equatorial exospheric temperature variation, in substantial agreement with available Pioneer Venus and Magellan observations (e.g., Keating et al. 1980; Keating and Hsu 1993; Kasprzak et al. 1997). Figure 6 illustrates the VTGCM exobase circulation pattern and the underlying asymmetric thermospheric temperature structure for solar moderate conditions. Night airglow distributions respond to this mean circulation pattern, with maxima that occur on average near (or just beyond) midnight and close to the equator (e.g., Bougher et al. 2006a, 2008).

However, existing GCMs are presently unable to reproduce the significant variations in observed diurnal density, temperature, and airglow fields utilizing a unique set of wind fields, eddy diffusion coefficients, and wave drag parameters (Bougher et al. 1997, 2006a). This problem may reflect missing physical processes or inputs, e.g., exospheric transport above 180–200 km, upward propagating planetary waves, and limited gravity wave constraints for formulating wave breaking (Bougher et al. 2006a).

4.2.5 Mars (NCAR MTGCM)

The Mars TGCM (MTGCM) is a finite difference primitive equation model that self-consistently solves for time-dependent neutral temperatures, neutral-ion densities, and three component neutral winds over the Mars globe (e.g., Bougher et al. 1999a, 1999b, 2000, 2002, 2004, 2006b; Bell et al. 2007). The MTGCM code is adapted from the NCAR TGCM framework (see Sect. 4.2.1).

The modern MTGCM code contains prognostic equations for the major neutral species (CO_2, CO, N_2, and O), selected minor neutral species (Ar, NO, $N(^4S)$, O_2), and several photochemically produced ions (e.g. O_2^+, CO_2^+, O^+, and NO^+ below 180 km). All fields are

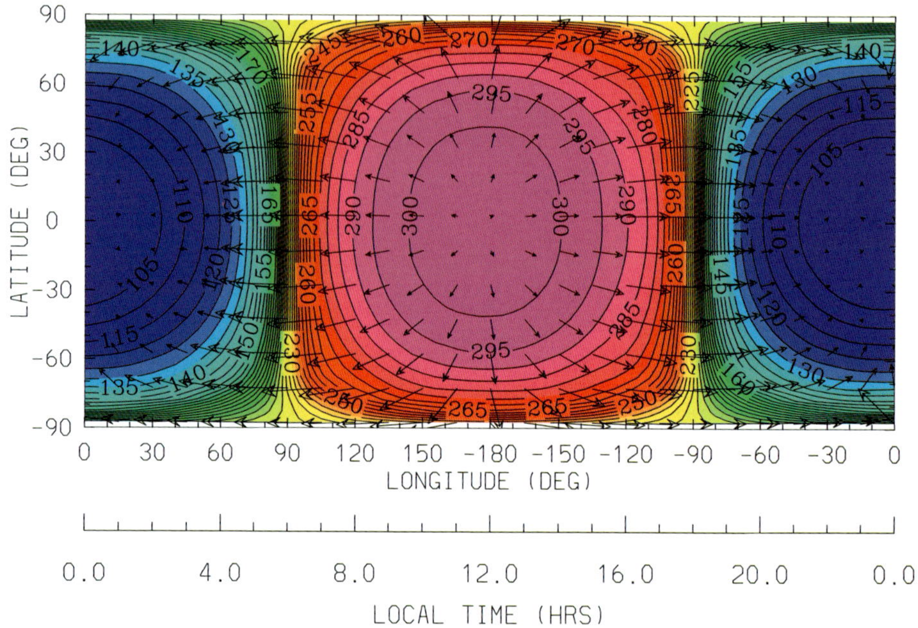

Fig. 6 VTGCM simulated exospheric temperatures and superimposed horizontal neutral winds for Equinox, $F10.7 = 200$ conditions (similar to early Pioneer Venus observations). Dayside equatorial temperatures reach ~305 K. Nightside minimum temperatures drop to ~100 K. Maximum horizontal winds reach ~220 m/sec (~70–90% of the sound speed in the upper thermosphere) across the evening terminator. The average altitude for this slice is ~180 km

calculated on 33 pressure levels above 1.32 μbar, corresponding to altitudes from roughly 70 to 300 km (at solar maximum conditions), with a 5° resolution in latitude and longitude. The vertical coordinate is log pressure, with a vertical spacing of 0.5 scale heights. Key adjustable parameters which can be varied for MTGCM cases include the $F10.7$-cm index (solar EUV/UV flux variation), the heliocentric distance and solar declination corresponding to Mars seasons. A fast NLTE 15-micron cooling scheme is implemented in the MTGCM, along with corresponding near-IR heating rates (Bougher et al. 2006b). These inputs are based upon detailed 1D NLTE model calculations for the Mars atmosphere (e.g., López-Valverde et al. 1998).

A simple dayside photochemical ionosphere is formulated for the MTGCM, including the major ions. Key ion-neutral reactions and rates are taken from Fox and Sung (2001); empirical electron and ion temperatures are adapted from the Viking mission for various solar conditions. The ionization rates required for the production rates are calculated self-consistently, making use of specified solar EUV fluxes. Nightside ions are not yet simulated, but will require either day-to-night ion drifts from modern magnetohydrodynamic models (e.g., Ma et al. 2004), or energetic electron precipitation sources of nightside ionization (e.g., Fillingim et al. 2007).

The MTGCM is driven from below by the NASA Ames Mars MGCM code (Haberle et al. 1999) at the 1.32-μbar level (near 60–80 km). This coupling allows both the migrating and non-migrating tides to cross the MTGCM lower boundary and the effects of the expansion and contraction of the Mars lower atmosphere to extend to the thermosphere. The entire atmospheric response to simulated dust storms can also be calculated using these coupled

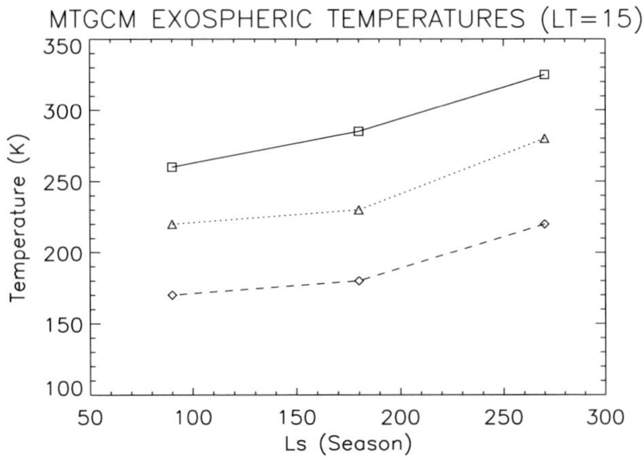

Fig. 7 MTGCM exospheric temperatures as a function of *Ls* (season) and solar cycle ($F10.7$-cm index). Dayside ($LT = 1500$) equatorial conditions are displayed. Curves indicated: $F10.7 = 175$–200 (*top*), 110–130 (*middle*), and 70–80 (*bottom*)

models. Key prognostic variables are passed upward from the MGCM to the MTGCM at the 1.32-μbar level at every MTGCM grid point: temperatures, zonal and meridional winds, and geopotential heights. These two climate models are each run with a 2-minute time step, with the MGCM exchanging fields with the MTGCM at this frequency.

This coupled configuration has been validated using an assortment of spacecraft observations, including Mars Global Surveyor, 2001 Mars Odyssey, and Mars Reconnaissance Orbiter thermosphere and/or ionosphere data sets (Bougher et al. 1999b, 2000, 2004, 2006b). Figure 7 shows dayside, equatorial exospheric temperature variations predicted by the MTGCM over the solar cycle and Mars seasons. These variations are in reasonably good agreement with available aerobraking and orbital drag measurements (e.g., Keating et al. 1998, 2003; Forbes et al. 2008; Mazarico et al. 2007). Exobase neutral horizontal winds and temperatures for solar maximum, southern summer solstice conditions are also illustrated in Fig. 8. In general, the global wind patterns simulated by the MTGCM reveal very strong summer-to-winter inter-hemispheric Hadley circulations that are consistent with observed NO nightglow (Bertaux et al. 2005) and winter polar warming (Bougher et al. 2006b) features. Finally, no downward coupling is presently activated between the MGCM and the MTGCM. However, the impacts of lower atmosphere dynamics upon the upper atmosphere are dominant (see Sect. 5.1.3).

4.2.6 Jupiter (JIM)

The JIM code is a general circulation model calculating the global thermosphere and ionosphere on Jupiter above the 2 μbar level (set to 357 km above the 1 bar level). This model, the first of its kind for Jupiter, was developed in the UK by Achilleos et al. (1998) and based largely on the thermosphere model of CTIM (see Sect. 4.2.2). Solar heating and photoionization are calculated self-consistently by solving the Beer–Lambert law. The thermospheric dynamics are calculated assuming all non-linear terms in the momentum equation, including ion drag, and ion dynamic calculations include diffusion and neutral drag. Calculations of the distribution of major neutral species H, H_2 and He include transport by winds and diffusion as well as simple ion-neutral photochemistry above the homopause. Unlike CTIM, ions in JIM are calculated on the same spherical pressure grid as the neutrals, which assumes a 10° horizontal resolution. The principal ions considered are H^+, H_2^+ and H_3^+, with more complex organic molecules which occur mainly below the homopause being

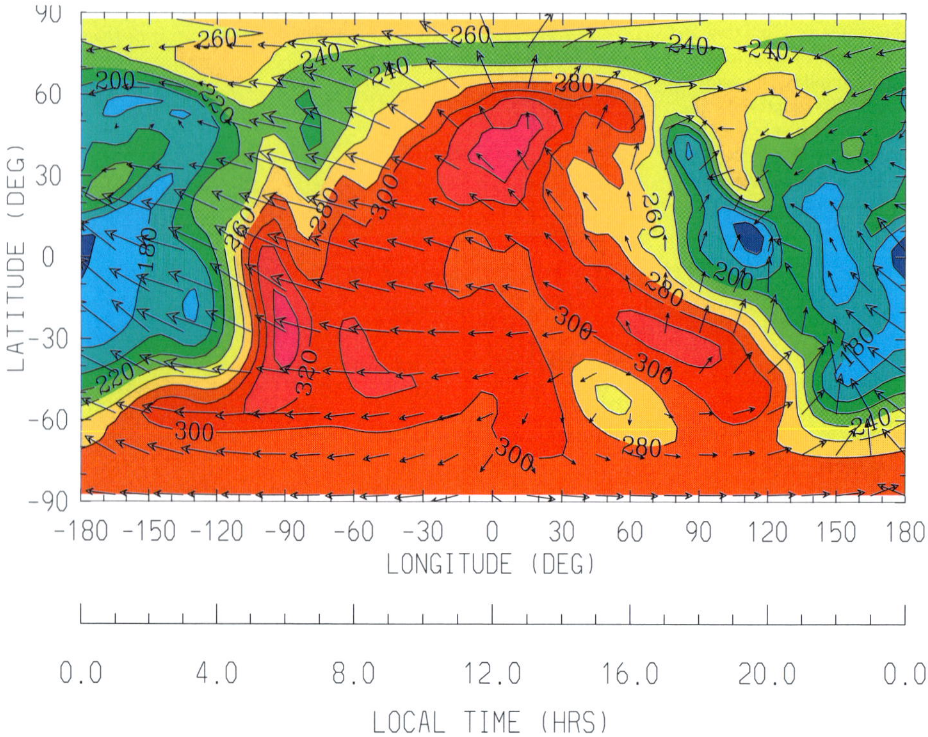

Fig. 8 MTGCM simulated exospheric temperatures and superimposed horizontal neutral winds for $Ls = 270$, $F10.7 = 175$ conditions (similar to late Mars Odyssey aerobraking observations). Dayside subsolar latitude (~25 S) temperatures reach ~320 K, nightside minimum temperatures drop to ~145 K. Maximum horizontal winds reach ~550 m sec^{-1} (slightly in excess of the sound speed). The average altitude for this slice is ~215 km

treated through a generic single molecule in the ion recombination reactions. A magnetic field is included for the calculations of ion dynamics and transport. In the auroral zones JIM allows for the inclusion of particle precipitation from the magnetosphere to calculate ionization and energy deposition. At high latitudes an electric field is included to simulate the auroral electrojet (Achilleos et al. 2001). Studies with JIM have helped investigate the morphology of thermosphere ionosphere coupling on Jupiter, in particular the dynamics at high latitudes (Millward et al. 2005), electrodynamic coupling between the thermosphere and the auroral ionosphere (Achilleos et al. 2001) and ionospheric conductivities (Millward et al. 2002a, 2002b).

4.2.7 Jupiter (NCAR JTGCM)

The proper characterization of Jupiter's upper atmosphere, embedded ionosphere, and auroral features requires the examination of underlying processes, including the feedbacks of energetics, neutral-ion dynamics, composition, and magnetospheric coupling. Coupled thermosphere and ionosphere GCM models (with magnetospheric inputs) can be used to address these feedbacks.

A Jupiter TGCM (JTGCM) code has now been developed and exercised to address global temperatures, three-component neutral winds, and neutral-ion species distributions

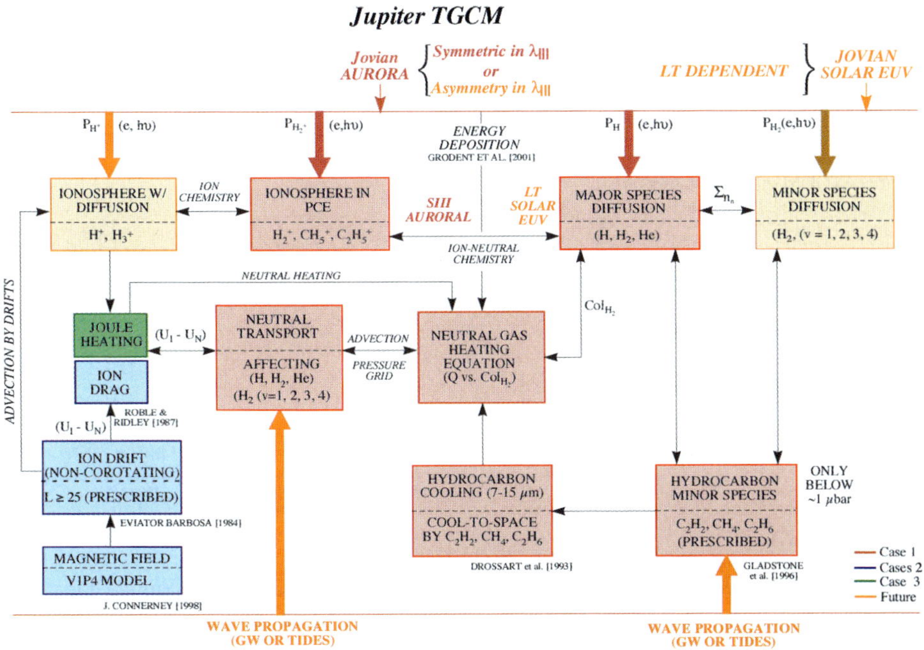

Fig. 9 JTGCM. This schematic summarizes the various thermosphere, ionosphere, magnetosphere processes and their couplings that are presently incorporated into the JTGCM code. Case 1 (auroral forcing alone), Case 2 (auroral forcing plus moderate ion drag/Joule heating), and Case 3 (auroral forcing plus strong ion drag/Joule heating) processes are identified corresponding to JTGCM simulations presented in Bougher et al. (2005). Future planned upgrades and processes are also indicated (*orange shading*)

(Bougher et al. 2005; Majeed et al. 2005, 2008). This code is based upon the NCAR TIE-GCM framework. Neutral temperatures, 3-component neutral winds, major neutral species (H, H_2, He), and major ion species (H^+, H_2^+, H_3^+) are simulated. The domain of this JTGCM framework extends from 20-mbar (capturing hydrocarbon cooling) to $\sim 1.0 \times 10^{-4}$ nbar (including auroral and Joule heating processes). A $5 \times 5°$ horizontal grid is utilized, along with a 0.5 scale height vertical resolution. Auroral and Joule heating processes are incorporated, both of which contribute to maintaining the neutral temperatures and driving the global winds. Joule heating is produced by the action of a prescribed plasma drift pattern that is consistent with the convection electric field derived from the Voyager constrained magnetospheric model of Eviatar and Barbosa (1984). Simulated auroral electrojet ion wind magnitudes approach $\sim$1.5–3.0 km sec^{-1}. The benchmark JTGCM was fully spun-up and integrated for >50 Jupiter rotations, thereby achieving steady state solutions above the $\sim$1.0-μbar level.

Initial results from three JTGCM cases incorporating moderate realistic auroral heating, ion drag, and moderate to strong Joule heating processes were presented by Bougher et al. (2005). Figure 9 summarizes the various thermosphere-ionosphere-magnetosphere processes and their couplings that are presently incorporated into the JTGCM code. The neutral horizontal winds at ionospheric heights vary from $\sim$0.5 km s^{-1} to 1.2 km s^{-1} (approaching the sound speed), atomic hydrogen is transported equatorward, and auroral exospheric temperatures range from $\sim$1200–1300 K to above 3000 K, depending on the magnitude of Joule heating. Figure 10 shows JTGCM simulated equatorial temperature profiles, and comparisons with multispectral and Galileo ASI observations (Majeed et al. 2005). The

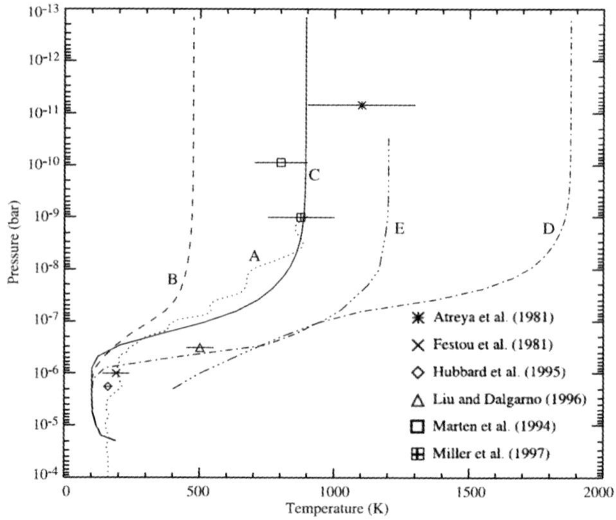

Fig. 10 JTGCM equatorial temperature profiles are shown in comparison with corresponding temperature profiles from JIM (*curve E*) and in-situ measurements from the Galileo ASI probe (*curve A*). *Curve B* assumes JTGCM auroral heating alone (no Joule heating). *Curve C* is from the "best" JTGCM simulation that incorporates auroral plus 15% joule heating conditions. *Curve D* is from the JTGCM simulation that assumes auroral plus 30% joule heating conditions. Remotely sensed (multi-spectral) temperature observations are also displayed (see key). From Majeed et al. (2005)

best fit to the Galileo data implies that the major energy source for maintaining the equatorial temperatures is due to dynamical heating induced by the low-latitude convergence of the high-latitude-driven thermospheric circulation. Simulated Joule heating can be quite large, requiring a scaling factor (~15%) to enable both observed equatorial and auroral oval temperatures to be simulated (Majeed et al. 2005, 2008). Overall, the Jupiter thermosphere/ionosphere system is highly variable and is shown to be strongly dependent on magnetospheric coupling that regulates Joule heating.

Diagnostic heat balance studies utilizing the JTGCM have been employed to quantify in detail the thermal balance processes required to maintain thermospheric temperatures consistent with multi-spectral observations of Jupiter's equatorial (Majeed et al. 2005), auroral oval and polar cap regions (Majeed et al. 2008). It is found that upwelling/divergent winds in the auroral oval regions provide local cooling (reducing temperatures from values otherwise expected using 1D models), while downwelling/convergent winds in the equatorial region provide local heating largely responsible for maintaining the warm temperatures measured. In general, the significant Jovian auroral plus Joule heating processes appear sufficient to drive a strong equatorward (meridional) flow that is adequate to overcome Coriolis forces and to dynamically produce warm equatorial temperatures. This ability of Jovian equatorward winds to overcome Coriolis forces is apparently different than for Saturn (Mueller-Wodarg et al. 2006), where the combined auroral plus Joule heating magnitudes at polar latitudes are reduced from those at Jupiter (Strobel 2002). Correspondingly, Jovian auroral and Joule heating at polar latitudes combine to drive local heating and corresponding dynamical cooling, together controlling the thermal structure of the ovals and the polar cap regions. These results are beginning to address the "heat budget problem" of the Jovian upper atmosphere that has been debated for decades (c.f. Yelle and Miller 2004).

4.2.8 Saturn (STIM)

The first general circulation model of Saturn's thermosphere and ionosphere was presented by Mueller-Wodarg et al. (2006) and is referred to as the STIM code. The model forms part of a collaborative project between Boston University, Imperial College London and

University College London. It calculates the response of Saturn's upper atmosphere to solar heating and ionization, including global dynamics, composition and the thermal structure. The coupled non-linear equation of momentum, energy and continuity are solved on a global spherical pressure level grid above the 100 nbar level (800 km above the 1 bar level). Horizontal and vertical resolutions are flexible, allowing them to be optimized for specific problems under investigation. A simplified scheme of ion and neutral photochemistry is included and recent additions include ion diffusion and full ion-neutral dynamical coupling. The model has been used to investigate the thermal balance on Saturn in order to investigate the origin of abnormally large thermospheric temperatures on Saturn and other gas giants. It was found that the fast rotation of Saturn and strong Coriolis forces prevented energy deposited at auroral latitudes in the form of Joule heating from propagating equatorward and explain the observed low latitude thermosphere temperatures (Mueller-Wodarg et al. 2006; Smith et al. 2005), as shown in Fig. 11. The importance of global dynamics for understand-

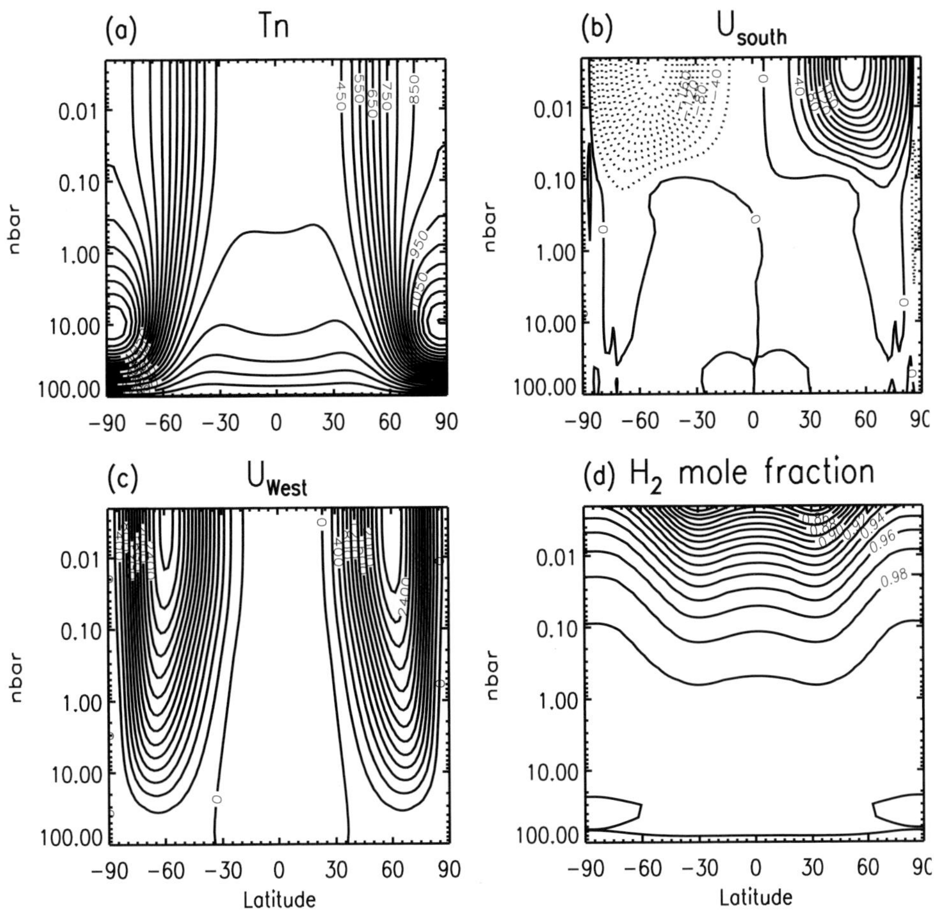

Fig. 11 Diurnally averaged temperatures, horizontal winds and H_2 mixing ratios versus latitude and height for equinox and solar maximum conditions, as calculated by the STIM GCM. The simulation considers Joule heating in the auroral regions. U_{South} and U_{West} are meridional and zonal winds, respectively, defined as positive southward and westward

ing the thermal balance on gas giants such as Saturn makes the use of general circulation models particularly relevant there. Other studies with the ionospheric module of STIM investigated the global structure of Saturn's highly variable ionosphere, considering shadowing by Saturn's rings (Moore et al. 2004) and effects of water precipitating into Saturn's ionosphere from the rings (Moore et al. 2006). These calculations found the presence of water to be important to reproduce the dawn dusk asymmetries in electron densities observed by the Cassini Radio Science experiment (Nagy et al. 2006). A recent study by Moore and Mendillo (2007) proposed variable water influx rates to be responsible for the high variability of Saturn's ionospheric densities.

4.2.9 Titan (TTGCM)

To understand the global structure and dynamics of Titan's thermosphere, Mueller-Wodarg et al. (2000) presented the first general circulation model for Titan's atmosphere above 600 km altitude. While this model, a collaborative project between University College London, the University of Arizona and Boston University, originally was based on the terrestrial thermosphere model of Fuller-Rowell and Rees (1980), it turned out that approximations frequently made for Earth are no longer valid on Titan, and ultimately a new model was developed from scratch. The Titan GCM predicted solar driven day-night temperature differences on Titan of up to 20 K which drive vigorous thermospheric dynamics. The extended nature of the thermosphere leads to effects such as continuous solar illumination also on the nightside at sufficiently polar latitudes. Subsequent studies showed that such winds would effectively redistribute constituents in Titan's thermosphere, particularly CH_4, causing large local time and hemispheric asymmetries in the CH_4 densities (Mueller-Wodarg and Yelle 2002; Mueller-Wodarg et al. 2003), as shown in Fig. 12. These calculations showed that dynamics and the distribution of CH_4 in Titan's thermosphere are intimately coupled. The first in-situ observations by the Cassini Ion Neutral Mass Spectrometer (INMS) showed that Titan's real thermosphere was more complex that suggested by the simple solar-driven calculations. Current developments of the model include adding the effects of more realistic dynamics at its lower boundary, which may importantly affect thermospheric dynamics. The model in its latest version is being constrained by densities observed by the INMS instrument.

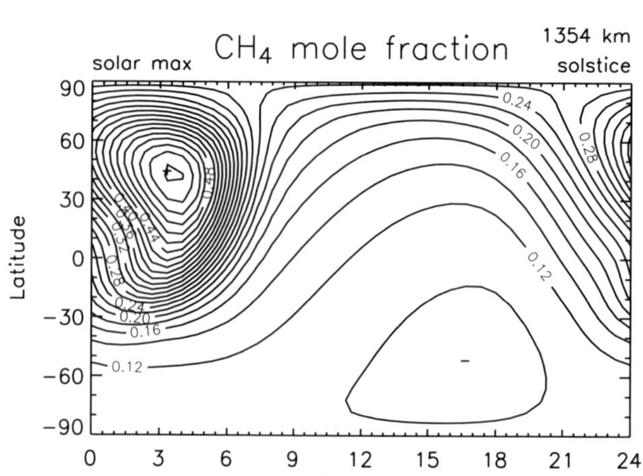

Fig. 12 CH_4 mole fractions near the 1×10^{-3} nbar pressure level for solstice at solar maximum, as calculated by the Titan TGCM. The average height of the pressure level is 1354 km

5 Modeling Frontiers and Problems

5.1 Lower to Upper Atmosphere Coupling

Properly addressing the coupling of the lower and upper atmospheres of planetary environments is a difficult modeling task. "Whole atmosphere" models are ultimately required to capture the physical processes (e.g., thermal, chemical, dynamical) throughout the entire atmosphere from the ground to the exobase. However, diffusion processes are much different above and below the homopause, requiring a method to be employed to bridge the transition between the homosphere and heterosphere regions. In addition, timescales for chemical and radiative processes vary greatly throughout the atmosphere, typically requiring small time-steps within finite-difference codes. Numerical stability (while utilizing longer time-steps) can be achieved in a number of ways; e.g., by employing implicit solvers and various numerical filters. Finally, exercising of multi-dimensional codes on multi-processor computers can also reduce the wall clock time for global simulations.

5.1.1 Separate but Coupled Model Frameworks vs. Whole Atmosphere Model Frameworks

Two approaches have been employed to date to capture the physics of the entire atmosphere (ground to exobase): (a) coupling of separate lower and upper atmosphere codes; and (b) single framework "whole atmosphere" model codes. Each approach has advantages and disadvantages. The coupling of separate codes permits the unique physical processes (and timescales) of the lower and upper atmospheres to be addressed separately within codes which can be optimized for this purpose. Molecular diffusion is one example for the upper atmosphere, for which an implicit (vertical) formulation permits a longer model time-step to be used. However, linking two separate models across an interface is not "seamless". By this we refer to the lack of an exact match of thermal and dynamical processes (e.g., solar heating, IR cooling, diffusion, numerical filtering) across this interface. Furthermore, both upward and downward coupling (i.e., constituent fluxes) is not easily activated across separate models. Whole atmosphere models obviate the need for an "artificial" boundary between 2-separate codes, while at the same time providing a continuous application of processes throughout the ground to exobase model domain. Small time-steps may be needed to accommodate disparate processes and their timescales throughout the model domain. Finally, whole atmosphere model simulations can be visualized from "top to bottom" with a single post-processor.

Examples of both modeling approaches are presented: (Sect. 5.1.2) the whole atmosphere model approach for Earth (NCAR WACCM), (Sect. 5.1.3) the coupled separate model approach for Mars (NASA MGCM and NCAR MTGCM), and (Sect. 5.1.4) the upward extended LMD-MGCM. The coupled model approach for Mars is a precursor to new Mars whole atmosphere models that are presently being developed and validated (see Table 1; see Sects. 4.2.3 and 5.1.4).

5.1.2 NCAR WACCM (Earth)

The WACCM model version 3 (WACCM3) is a state-of-the-art climate model developed at NCAR that extends from the Earth's surface to the lower thermosphere. This model is an outgrowth of three independent models developed separately across three divisions at NCAR. It combines the major features of these three independently developed models of the atmosphere, the Middle Atmosphere Community Climate Model (MACCM) (Boville 1995), the chemical model MOZART (Brasseur et al. 1998) and the TIME-GCM (Roble 2000). This model is one of the few high-top general circulation models that include the Hamburg Model of the Neutral and Ionized Atmosphere (HAMMONIA) (Schmidt et al. 2006) and the extended Canadian Middle Atmosphere Model (CMAM) (Fomichev et al. 2002). These models have been used to study problems such as the solar influence on Earth's climate, constituent transport and trends in the middle atmosphere, the influence of the stratosphere on the tropospheric climate and the connection between climate change and polar mesospheric clouds. WACCM3 extends between the surface and the lower thermosphere near 140 km. But work is now progressing to move the upper boundary to 500–700 km by incorporating the aeronomy of the thermosphere and ionosphere from the TIME-GCM into an upward extended WACCM.

A number of studies are underway with this new model but one, Sassi et al. (2004), showed a coupling between El-Nino/Lanina ocean influences on the stratosphere/mesosphere region and another (Richter et al. 2008) showed the importance of gravity wave forcing on the basic structure of the upper atmosphere. Details of the model can be found on the web site http://www.cgd.ucar.edu/research/models/waccm.html.

5.1.3 NCAR Coupled MGCM-MTGCM (Mars)

The coupled NASA Ames MGCM and the NCAR MTGCM models constitute a numerical framework of 2-independent multi-dimensional codes linked across an interface at 1.32-microbars (~60–80 km) in the Mars atmosphere (see Sect. 4.2.5). This coupled configuration permits both thermal and large scale dynamical processes to be linked across the lower and upper atmospheres of Mars (e.g., Bougher et al. 2004, 2006b). The 2-model treatment is designed to be a testbed for addressing coupling processes in advance of the development and validation of a comprehensive Mars "whole atmosphere" model framework (e.g. Sect. 4.2.3).

The coupled MGCM-MTGCM system itself has been used successfully to address/interpret an assortment of spacecraft observations, including Mars Global Surveyor, 2001 Mars Odyssey, and Mars Reconnaissance Orbiter thermosphere and/or ionosphere data sets (Bougher et al. 1999b, 2000, 2004, 2006b). For example, the recently discovered winter polar warming features of the Mars lower thermosphere (~100–130 km) are found to vary greatly over the Mars seasons (e.g., Keating et al. 2003; Bougher et al. 2006b). Figure 13 illustrates coupled MGCM-MTGCM simulations for $Ls = 90$ (aphelion) and 270 (perihelion) conditions, demonstrating that the basic features of the Martian thermospheric winter polar warming are controlled by seasonal changes in the solar plus tidal forcing, the corresponding variations in the strength of the inter-hemispheric Hadley circulation, and the resulting changes in the magnitude of the adiabatic heating near the winter poles. Calculations of polar warming show that perihelion adiabatic heating can be highly variable from one Mars year to the next, and more than twice as strong as that for aphelion conditions (Bougher et al. 2006b; Bell et al. 2007). Finally, without the deep inter-hemispheric Hadley circulation made possible using these coupled lower and upper atmosphere simulations, winter polar warming features in the Mars thermosphere would not be reproduced at all (Bell et al. 2007).

Several studies are underway utilizing this coupled MGCM-MTGCM framework. For example, the role of interannual variations in horizontal and vertical dust distributions in affecting the thermospheric temperature and wind distributions is being investigated. Factors influencing the seasonal variation in the Mars mesopause heights and minimum temperatures are also being determined (McDunn et al. 2007, 2008).

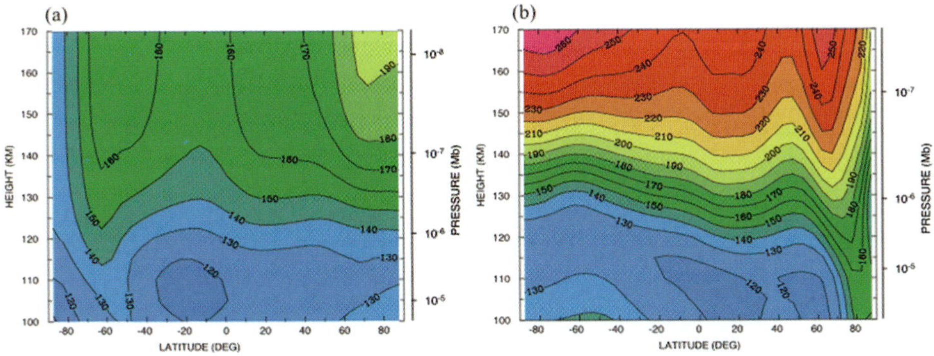

Fig. 13 MGCM-MTGCM zonal averaged temperature slices as a function of height and latitude: (**a**) $Ls = 90$ and (**b**) $Ls = 270$. Contour intervals are 10 K. Color shading is coordinated between these plots. From Bougher et al. (2006b)

5.1.4 LMD-MGCM (Mars)

The Mars LMD-GCM is based on the Terrestrial climate GCM of the Laboratoire de Meteorologie Dynamique (Sadourny and Laval 1984). It has been adapted by Hourdin et al. (1993) and Forget et al. (1999), who developed the first model covering the Martian atmosphere up to 80 km. The transport equations for the dynamics are directly taken from the LMD terrestrial GCM. They are based on a finite-difference formulation of the classical "primitive equations" of meteorology which are a simplified version of the general equations of fluids based on three main approximations: (1) the atmosphere is assumed to be a perfect gas; (2) it is supposed to remain vertically in hydrostatic equilibrium; and (3) the vertical dimension of the atmosphere is supposed to be much smaller than the radius of the planet (thin-layer approximation). The vertical discretization is based on σ-coordinates, where $\sigma = p/p_s$ is the pressure p at a given grid point normalized by its local value p_s at the surface of Mars (normalized pressure coordinates). The grid is chosen to have good coverage of the atmospheric boundary layer.

The energetics solved in the model includes the effects of suspended dust and CO_2 in the infrared in the Mars atmosphere. For the carbon dioxide, the thermal infrared and the near infrared absorption are solved using a NLTE approximation based on a parameterization of the heating and the cooling. Concerning the effect of dust, the radiative transfer of the solar radiation, including absorption and scattering, is modelled through a multi-stream approach. Then, the scattering of the thermal infrared outside the CO_2 15 μm band is modelled using a two-stream radiative transfer model. The CO_2 condensation-sublimation cycle is related to the thermal balance and is likely to a be source or sink of energy, mainly through the latent heat release associated with the change of state. This process is realistically included in the model and the different phases of carbon dioxide are managed consistently (in particular energy and mass conservation).

As mentioned above, the necessary discretization of time and space in the numerical model implies that mechanisms cannot be consistently modelled, because they occur at scales much below the lower limits of the model. They have to be added in an ad hoc way. The relevant processes that have been included near the surface of the planet at sub-grid scales are turbulent diffusion in the planetary boundary layer, convection, orography, and low-level drag.

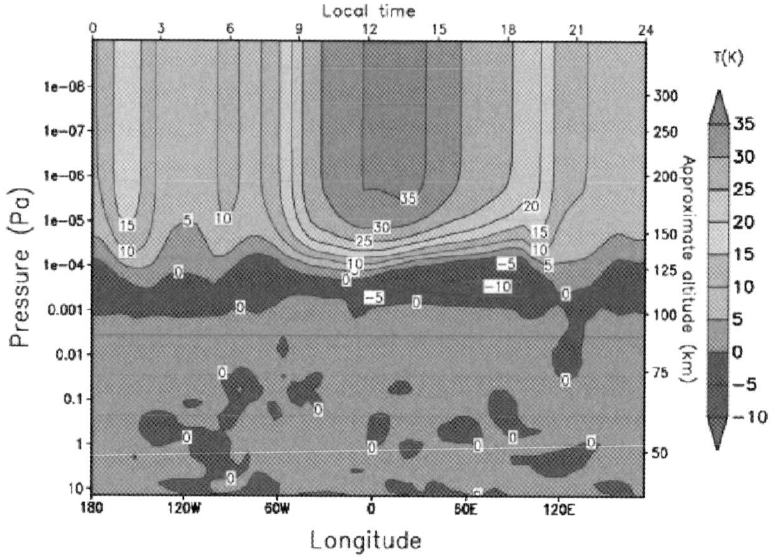

Fig. 14 Difference of temperatures between an LMD-GCM simulation without photochemistry and a simulation with photochemistry

The LMD-GCM extension to the thermosphere and the exosphere is a collaborative work with the University of Oxford and the Instituto de Astrofísicade Andalucía. The model has been successively extended from the ground up to a height of approximately 120 km (Angelats i Coll et al. 2003), 240 km (González-Galindo et al. 2004) and finally up to the exosphere (González-Galindo et al. 2007). The extension to the upper altitudes has been done in such a way that the processes that are important either in the mesosphere or in the thermosphere are taken into account.

In rarefied regions (thermospheres), each component of the neutral gas has an individual behaviour, which can be modelled using the Enskog approach, because this region is still collision dominated. We must distinguish horizontal from vertical dynamics, since the temporal and spatial scales are different. Temperature can be assumed to be the same for all the species, so a single equation for the temperature, which accounts for the thermal conductivity of the mean gas and the UV heating can be used. We do the same for the horizontal velocity, using a mean molecular viscosity. However, vertical diffusion is important in the upper atmosphere, and a multi-species Chapman-Enskog approach is used for multi-component diffusion.

At upper altitudes, it is demonstrated that photochemistry becomes a very important feature of the atmospheric dynamics. A 1-D photochemical model using a complex photochemistry scheme, which includes the 12 major constituents of the C, O and H families, and accounts for 27 reactions between them has been developed and included in the LMD-GCM model. Figure 14 shows the impact that the photochemistry may have on the energy balance in the thermosphere. Temperature differences up to 35 K can be obtained above 200 km if some chemical processes are neglected.

The infrared processes related to CO_2, already mentioned above, are relevant processes for the mesosphere, and strongly NLTE in that region. Some approximation and parameterization has been done to include them, without increasing dramatically the computation time. The main heating source of the Martian thermosphere is the UV heating. In order to be able to reproduce the thermal structure of the Martian upper atmosphere, which is critical for the hydrostatic equilibrium, as well as for dynamics, a parameterization has been included based on a full 1-D UV heating model. This full model includes the absorption by CO_2, O_2, atomic oxygen, H_2, H_2O, H_2O_2 and O_2 in the UV–visible range.

The GCM model has been used to build a database for Martian atmosphere (Lewis et al. 1999), which has become the ESA reference model for the atmosphere of Mars.

5.2 Thermosphere/Ionosphere to Exosphere Coupling

5.2.1 Previous Exosphere Modeling Approaches

As discussed in detail by Johnson et al. (2008), the hot coronae of atomic H, O and C have either been observed or postulated to exist at both Venus and Mars. The major reasons for interest in these hot populations are their importance in the long-term evolution of each atmosphere as well as the role they play in the general solar-wind interaction of each planet.

The history of Venus and Mars exospheric modeling approaches is contained in several early studies (e.g. Cravens et al. 1980; Nagy et al. 1981, 1990, 2001; Nagy and Cravens 1988; Kim et al. 1998; Hodges and Tinsley 1981, 1986; Hodges 2000). Both 1D models (based upon 2-stream calculations and Liouville's equation) plus multi-dimensional Monte Carlo particle trajectory models have been employed in these exospheric simulations. Recent 1D spherical Monte Carlo models have been constructed by Cipriani et al. (2007), extending the original work of Hodges (2000) to include hot O, C, CO_2 and CO. They also have examined more carefully the relative sources of dissociative recombination and atmospheric sputtering (Luhmann and Kozyra 1991) by O^+ pickup ions. Very recently Chaufray et al. (2007) published a study running a 3D Monte Carlo model for exospheric species, where they use the 1D ion profiles from the model of Krasnopolsky (2002) and as well as night-side ionosphere profiles obtained from Viking 1 measurements obtained from Zhang et al. (1990). They extrapolated these profiles around Mars for various SZAs and explored the 3D aspects of the related solar wind interaction processes.

5.2.2 Michigan Mars Exosphere Codes

In this section, the new approach of Valeille et al. (2007a, 2007b, 2008) is described which probes the effects of the 3D structure of the thermosphere on the exosphere. They use the results of a general circulation model, but perform 2D axisymmetric Direct Simulation Monte Carlo calculations for the SZA variation of the hot oxygen exosphere. The next major advance will be for the use of a 3D thermosphere-ionosphere model and a 3D exosphere model.

In order to merge the many local processes into a global picture, a model is needed, which can include these important processes on small spatial scales and which can diversify the different energy scales. The Michigan multi-species kinetic model is based on a technique called the Direct Simulation Monte Carlo (DSMC) method (Bird 1963, 1994). DSMC was developed to simulate the transition regime, where the mean free path of particles is too large for continuum hydrodynamics to be applicable. Individual particles are simulated as they move around within a grid, colliding with other particles and with solid objects (if any). Macroscopic properties, such as density, velocity and temperature are computed by appropriately averaging particle masses, locations, velocities, and internal energies. Momentum and energy exchanges with surfaces allow for chemical reactions and sputtering effects. Mass production can also be introduced as a boundary condition. DSMC is based on the "rarefied-gas" assumption that over a short time 'step' the molecular motion and

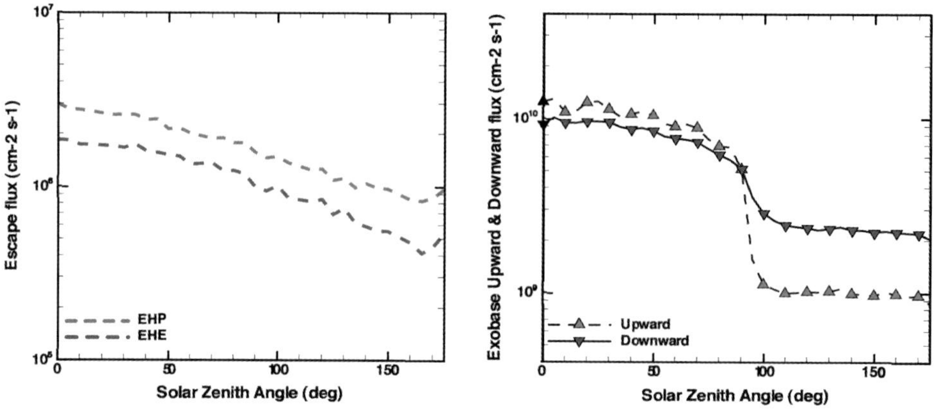

Fig. 15 Escape and upward and downward exobase fluxes of hot oxygen atoms for different solar conditions. Plotted on the *left* is the escape flux as a function of solar zenith angle for our new 2D axisymmetric Direct Simulation Monte Carlo model of the exosphere of Mars. In addition to the escape fluxes we are also calculating the return flux to the thermosphere and compare with what is normally assumed in thermosphere models. The separate upward and downward fluxes at the exobase are given in the plot on the *right*. *Left panel curves*: EHP (Equinox, High Solar Activity, Polar cut)—*top curve*; EHE (Equinox, High Solar Activity, Equatorial)—*bottom curve*

the intermolecular collisions are uncoupled and therefore can be calculated independently. Molecules are moved over the distances appropriate for this time step, followed by the calculation of a representative set of collisions. The time step is small compared to the mean collision time, and the results are independent of its actual value.

The Michigan DSMC code was developed with the dusty-gas comet coma as the first science application (Tenishev et al. 2008) improving in a number of fundamental and technical ways over the original comet DSMC of Combi (1996). It was also developed as a general-purpose gas kinetic solver and tested against a number of standard aerodynamical gas kinetic problems (Tenishev and Combi 2003a, 2003b) as well as comets. The code has been applied both to Mars' exosphere (Valeille et al. 2007a, 2007b, 2008) and plumes of Enceladus (Tenishev et al. 2007, 2008).

Valeille et al. (2007a, 2007b, 2008) have applied DSMC to the hot atomic oxygen corona of Mars, using the 3D MTGCM of Bougher et al. (2006b) (see Sect. 4.2.5) to understand the overall day-night structure of the exosphere as well as to explore day-night, local time, meridian, latitudinal, seasonal and solar activity variations. The goals are to understand the range and limits on the escape rate, exosphere distribution, as well as the return fluxes of exospheric O to the thermosphere.

A set of 2D axisymmetric DSMC models have been run with coordinates of radius and SZA for a variety of upper thermosphere conditions taken from the 3D MTGCM: for quiet and active sun, for equinox conditions, for solstice conditions (both at aphelion and perihelion), and for thermosphere distributions around the equator and along the polar noon-midnight meridian. The escape rates and exobase upward and downward fluxes as a function of SZA for an equinox geometry at solar high (active) and solar low (quiet) conditions are shown in Fig. 15. In addition, Fig. 16 shows the O density distribution in the corona for the solar maximum and solar minimum models for the equinox geometry. Finally, Fig. 17 shows the O atom escape rates for solar maximum and minimum conditions for the new results by Valeille et al. (2007a, 2007b, 2008) compared with published values from the literature.

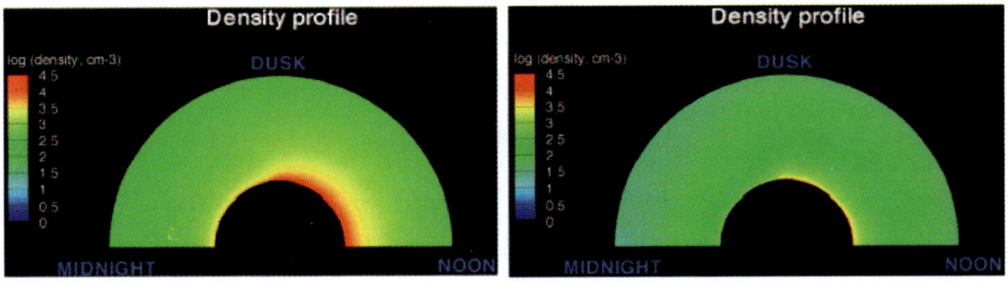

Fig. 16 Hot oxygen density for different solar conditions. Shown is a false *color contour plot* of the hot oxygen atom exospheric density for (**a**) solar active and (**b**) solar quiet conditions of the MTGCM model. Using the same *color-table* highlights the differences of the exosphere for active and quiet solar conditions

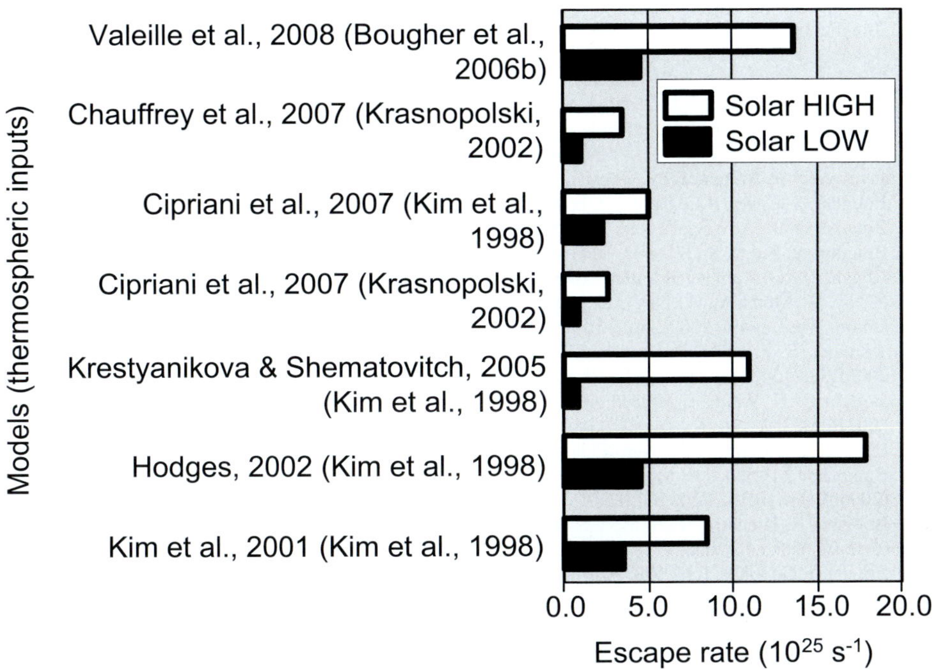

Fig. 17 Comparison of O atom escape rates from various models (from Valeille et al. 2008). Shown are the solar minimum and solar maximum escape rates from several models with the thermosphere conditions used, given in parentheses. Kim et al. (1998) is corrected by a factor of 6.5 according to Nagy et al. (2001). The Valeille et al. (2008) escape rate, averaged over the solar cycle and Mars seasons, is $\sim 1.0 \times 10^{26}$ s^{-1}. Detailed Mars orbiter photochemical escape measurements are needed to validate these models

The general structure of the exosphere and escape fluxes is similar from time to time, but the actual values of the escape rates in the extreme model cases can range over nearly a factor of 50 when comparing conditions from solar minimum to maximum, and from Mars perihelion and aphelion solstices with the equinoxes that are between the two extremes. These results clearly imply that the next step in modeling Mars' exosphere needs to be done with a 3D exosphere model using a realistic 3D thermosphere for its base assumption.

References

N. Achilleos, S. Miller, J. Tennyson, A. Aylward, I. Mueller-Wodarg, D. Rees, J. Geophys. Res. **103**, 20089–20112 (1998). doi:10.1029/98JE00947

N. Achilleos, S. Miller, R. Prangé, G. Millward, M.K. Dougherty, N.J. Phys. **3** (2001). doi:10.1088/1367-2630/3/1/303

M. Angelats i Coll, F. Forget, M.A. López-Valverde, P.L. Read, S.R. Lewis, J. Geophys. Res. **109**, 1011 (2003). doi:10.1029/2003JE002163

J.M. Bell, The dynamics of the upper atmospheres of Mars and Titan. Ph.D Thesis, University of Michigan (2008)

J.M. Bell, S.W. Bougher, V. De LaHaye, J.H. Waite Jr., A. Ridley, Eos Trans. Am. Geophys. Union **87**(Jt. Assem. Suppl.) (2006), abstract U52A-06

J.M. Bell, S.W. Bougher, J.R. Murphy, J. Geophys. Res. **112**, E12002 (2007). doi:10/1029/2006JE002856

J.-L. Bertaux et al., Science **307**, 566–569 (2005)

G.A. Bird, Phys. Fluids **6**, 1518–1519 (1963)

G.A. Bird, *Molecular Gas Dynamics and the Direct Simulations of Gas Flows* (Clarendon, Oxford, 1994)

V. Boqueho, P.-L. Blelly, J. Geophys. Res. **110**, A01313 (2005). doi:10.1029/2004JA010414

S.W. Bougher et al., Icarus **68**, 284–312 (1986)

S.W. Bougher, R.E. Dickinson, E.C. Ridley, R.G. Roble, Icarus **73**, 545–573 (1988)

S.W. Bougher, J.C. Gerard, A.I.F. Stewart, C.G. Fesen, J. Geophys. Res. **95**, 6271–6284 (1990)

S.W. Bougher, W.J. Borucki, J. Geophys. Res. **99**, 3759–3776 (1994)

S.W. Bougher, D.M. Hunten, R.G. Roble, J. Geophys. Res. **99**, 14609–14622 (1994)

S.W. Bougher, M.J. Alexander, H.G. Mayr, in *Venus II*, ed. by S.W. Bougher, D.M. Hunten, R.J. Philips (University of Arizona Press, Tucson, 1997), pp. 259–292

S.W. Bougher, S. Engel, R.G. Roble, B. Foster, J. Geophys. Res. **104**, 16591–16611 (1999a)

S.W. Bougher et al., Adv. Space Res. **23**(11), 1887–1897 (1999b)

S.W. Bougher, S. Engel, R.G. Roble, B. Foster, J. Geophys. Res. **105**, 17,669–17,689 (2000)

S.W. Bougher, R.G. Roble, T.J. Fuller-Rowell, in *Atmospheres in the Solar System: Comparative Aeronomy*, ed. by M. Mendillo, A. Nagy, H. Waite. Geophysical Monograph, vol. 130 (American Geophysical Union, Washington, 2002), pp. 261–288

S.W. Bougher, S. Engel, D.P. Hinson, J.R. Murphy, J. Geophys. Res. **109**, E03010 (2004). doi:10.1029/2003JE002154

S.W. Bougher, J.H. Waite, T. Majeed, G.R. Gladstone, J. Geophys. Res. **110**, E04008 (2005). doi:10.1029/2003JE002230

S.W. Bougher, S. Rafkin, P. Drossart, Planet. Space Sci. **54**, 1371–1380 (2006a)

S.W. Bougher, J.M. Bell, J.R. Murphy, M.A. López-Valverde, P.G. Withers, Geophys. Res. Lett. **33**, L02203 (2006b). doi:10.1029/2005GL024059

S.W. Bougher, A. Brecht, C. Parkinson, S. Rafkin, B. Foster, 37th COSPAR Scientific Assembly 2008, C32-0003-08 Abstract (2008)

B.A. Boville, J. Geophys. Res. **100**, 9017–9039 (1995)

G.P. Brasseur et al., J. Geophys. Res. **103**, 28265–28290 (1998)

A. Brecht, S.W. Bougher, S. Rafkin, B. Foster, Eos Trans. Am. Geophys. Union **88**(52), (Fall Meeting Suppl.) (2007), abstract P33B-1299

J.Y. Chaufray et al., J. Geophys. Res. **112**, E09009 (2007). doi:10.1029/2007JE002915

R.H. Chen, A.F. Nagy, J. Geophys. Res. **83**, 1133–1140 (1978)

R.H. Chen, T.E. Cravens, A.F. Nagy, J. Geophys. Res. **83**, 3871–3876 (1978)

F. Cipriani, F. Leblanc, J.J. Berthelier, J. Geophys. Res. **112**, D07001 (2007). doi:10.1029/2006JE002818

M.R. Combi, Icarus **123**, 207–226 (1996)

T.E. Cravens, T.I. Gombosi, A.F. Nagy, Nature **283**, 178–180 (1980)

T.E. Cravens, A.J. Kliore, J.U. Kozyra, A.F. Nagy, J. Geophys. Res. **86**, 11,323–11,329 (1981)

V. DeLa Haye, Coronal formation and heating efficiencies in Titan's upper atmosphere: Construction of a coupled ion, neutral and thermal structure model to interpret the first INMS Cassini data. Ph.D. Thesis, University of Michigan, Ann Arbor (2005)

Y. Deng, A.D. Richmond, A.J. Ridley, H. Liu, Geophys. Res. Lett. **35**, L01104 (2008). doi:10.1029/2007GL032182

A.L. Dobbin, A.D. Aylward, M.J. Harris, J. Geophys. Res. **111**, A07314 (2006). doi:10.1029/2005JA011543

A. Eviatar, A.D. Barbosa, J. Geophys. Res. **89**, 7393–7398 (1984)

P.R. Field et al., J. Atmos. Sol. Terr. Phys. **60**, 523–543 (1998)

M.O. Fillingim et al., Geophys. Res. Lett. **34**, L12101 (2007). doi:10.1029/2007GL029986

V.I. Fomichev et al., J. Geophys. Res. **107** (2002). doi:1029/2001JD000479

J.M. Forbes, F.G. Lemoine, S.L. Bruinsma, M.D. Smith, X. Zhang, Geophys. Res. Lett. **35**, L01201 (2008). doi:10.1029/2007GL031904
F. Forget et al., J. Geophys. Res. **104**, 24155–24176 (1999). doi:10.1029/1999JE001025
J.L. Fox, Icarus **51**, 248–260 (1982)
J.L. Fox, Planet. Space Sci. **36**, 37–46 (1988)
J.L. Fox, in *Venus and Mars: Atmospheres, Ionospheres and Solar Wind Interaction*, ed. by J.G. Luhmann, M. Tatrallyay, R. Pepin. Geophysical Monograph, vol. 66 (AGU Press, Washington, 1992), pp. 191–222
J.L. Fox, J. Geophys. Res. **109**, A11310 (2004). doi:10.1029/2004JA010380
J.L. Fox, A. Dalgarno, J. Geophys. Res. **84**, 7315–7333 (1979)
J.L. Fox, A. Dalgarno, J. Geophys. Res. **86**, 629–639 (1981)
J.L. Fox, K.Y. Sung, J. Geophys. Res. **106**(A10), 21305–21336 (2001)
J.L. Fox, K.E. Yeager, J. Geophys. Res. **111** (2006). doi:10.1029/2006JA011697
J.L. Fox, P. Zhou, S.W. Bougher, Adv. Space Res. **17**, (11)203–(11)218 (1995)
T.J. Fuller-Rowell, J. Geophys. Res. **103**, 3951–3956 (1998)
T.J. Fuller-Rowell, D. Rees, J. Atmos. Sci. **37**, 2545–2567 (1980)
T.J. Fuller-Rowell, D. Rees, Planet. Space Sci. **10**, 1209–1222 (1983)
T.J. Fuller-Rowell, D. Rees, S. Quegan, R.J. Moffett, G.J. Bailey, J. Geophys. Res. **92**, 7744–7748 (1987)
T.J. Fuller-Rowell et al., in *STEP Handbook on Ionospheric Models*, ed. by R.W. Schunk (Utah State University, Logan, 1996)
T.J. Fuller-Rowell, G.H. Millward, A.D. Richmond, M.V. Codrescu, J. Atmos, Sol.-Terr. Phys. **64**, 1383–1391 (2002)
F. González-Galindo, M.A. López-Valverde, M. Angelats i Coll, F. Forget, J. Geophys. Res. **110**, 9005 (2004). doi:10.1029/2004JE002312
F. González-Galindo, F. Forget, M.A. López-Valverde, M. Angelats i Coll, S.W. Bougher, LPI Contributions **1353**, 3099 (2007)
R.M. Haberle et al., J. Geophys. Res. **104**(E4), 8957–8974 (1999)
W.B. Hanson, S. Sanatani, D.R. Zuccaro, J. Geophys. Res. **82**, 4351–4367 (1977)
M.J. Harris, A new coupled middle atmosphere and thermosphere general circulation model: Studies of dynamic, energetic and photochemical coupling in the middle and upper atmosphere. Ph.D Thesis, University College London (2001)
R.R. Hodges, J. Geophys. Res. **105**, 6971–6981 (2000)
R.R. Hodges, Geophys. Res. Lett. **29**(3), 1038 (2002). doi:10.1029/2001GL013852
R.R. Hodges, B.A. Tinsley, J. Geophys. Res. **86**, 7649–7656 (1981)
R.R. Hodges, B.A. Tinsley, J. Geophys. Res. **91**, 13649–13659 (1986)
F. Hourdin, P. Le Van, F. Forget, O. Talagrand, Meteorological variability and the annual pressure cycle on Mars. J. Atmos. Sci. **50**, 3625–3640 (1993)
W.B. Hubbard et al., Astron. Astrophys. **269**, 541–563 (1993)
R.E. Johnson et al., Space Sci. Rev. (2008, this issue)
W.T. Kasprzak et al., in *Venus II*, ed. by S.W. Bougher, D.M. Hunten, R.J. Philips (University of Arizona Press, Tucson, 1997), pp. 225–257
G.M. Keating, N.C. Hsu, Geophys. Res. Letts. **20**, 2751–2754 (1993)
G.M. Keating, J.Y. Nicholson III, L.R. Lake, J. Geophys. Res. **85**, 7941–7956 (1980)
G.M. Keating et al., Science **279**, 1672–1676 (1998)
G.M. Keating et al., Brief review on the results obtained with the MGS and Mars Odyssey 2001 Accelerometer Experiments. International Workshop: Mars Atmosphere Modeling and Observations, Inst. de Astrofis. de Andalucia, Granada, Spain, paper (2003)
D. Kella, L. Vejby-Christenson, P.J. Johnson, H.B. Pedersen, L.H. Andersen, Science **276**, 1530–1533 (1997)
J.T. Khiel et al., J. Clim. **11**, 1131–1149 (1998)
J. Kim, A.F. Nagy, T.E. Cravens, A.J. Kliore, J. Geophys. Res. **94**, 11,997–12,002 (1989)
J. Kim, A.F. Nagy, J.L. Fox, T.E. Cravens, J. Geophys. Res. **103**(12), 29,339–29,342 (1998)
Y.H. Kim, S. Son, Y. Yi, J. Kim, J. Kor. Ast. Soc. **34**, 25–29 (2001)
A.J. Kliore, G.S. Levy, D.L. Cain, G. Fjeldbo, S.I. Rasool, Science **205**, 99–102 (1967)
V.A. Krasnopolsky, *Photochemistry of the Atmospheres of Mars and Venus* (Springer, Berlin, 1982)
V.A. Krasnopolsky, J. Geophys. Res. **107**(E12), 5128 (2002). doi:10.1029/2001JE001809
M.A. Krestyanikova, V.I. Shematovitch, Sol. Syst. Res. **39**, 22–32 (2005)
S. Kumar, D.M. Hunten, J. Geophys. Res. **79**, 2529–2532 (1974)
S.R. Lewis et al., J. Geophys. Res. **104**, 24,177–24,194 (1999)
H.-L. Liu, R.G. Roble, J. Geophys. Res. **107**(D23), 4695 (2002). doi:10.1029/2001JD001533
H.-L. Liu, R.G. Roble, Geophys. Res. Lett. **32**, L13804 (2005). doi:10.1029/2005GL022939
K. Lodders, B. Fegley Jr., *The Planetary Scientists Companion* (Oxford University Press, New York, 1998)

M.A. López-Valverde, D.P. Edwards, M. López-Puertas, C. Roldán, J. Geophys. Res. **103**(E7), 16,799–16,812 (1998)
J.G. Luhmann, J.U. Kozyra, J. Geophys. Res. **96**, 5457–5467 (1991)
Y. Ma, A.F. Nagy, I.V. Sokolov, K.C. Hansen, J. Geophys. Res. **109**, A07211 (2004). doi:10.1029/2003JA010367
T. Majeed, J.H. Waite Jr., S.W. Bougher, G.R. Gladstone, J. Geophys. Res. **110**, E12007 (2005). doi:10.1029/2004JE002351
T. Majeed, J.H. Waite Jr., S.W. Bougher, G.R. Gladstone, J. Geophys. Res. (2008, submitted)
C.R. Martinis, J.K. Wilson, M.J. Mendillo, J. Geophys. Res. **108**(A10), 1383 (2003). doi:10.1029/2003/JA009973
E. Mazarico, M.T. Zuber, F.G. Lemoine, D.E. Smith, Eos Trans. Am. Geophys. Union **88**(Fall Meeting Suppl.) (2007), abstract P32A-04
T.L. McDunn et al., The 7th International Conference on Mars, Pasadena, CA (2007), abstract XX
T.L. McDunn et al., AGU Chapman conference on the solar wind interaction with Mars. San Diego (2008), abstract A-07
M.B. McElroy, Astrophys. J. **150**, 1125–1138 (1967)
M.B. McElroy, J. Geophys. Res. **73**, 1513–1521 (1968a)
M.B. McElroy, J. Atmos. Sci. **25**, 574–577 (1968b)
M.B. McElroy, J. Geophys. Res. **74**, 29–42 (1969)
M.B. McElroy et al., Science **194**, 1295–1298 (1976)
M.B. McElroy, T.Y. Kong, Y.L. Yung, A.O. Nier, Science **194**, 1295–1298 (1976)
G.H. Millward, R.J. Moffett, S. Quegan, T.J. Fuller-Rowell, in *STEP Handbook on Ionospheric Models*, ed. by R.W. Schunk (Utah State University, Logan, 1996a)
G.H. Millward, H. Rishbeth, R.J. Moffett, S. Quegan, T.J. Fuller-Rowell, J. Geophys. Res. **101**, 5149–5156 (1996b)
G.H. Millward et al., J. Geophys. Res. **106**(A11), 24,733–24,744 (2001)
G.H. Millward, S. Miller, A.D. Aylward, I.C.F. Mueller-Wodarg, N. Achilleos, in *Comparative Atmospheres in the Solar System*, ed. by M. Mendillo, A. Nagy, J.H. Waite (American Geophysical Union, Washington, 2002a), pp. 289–298
G.H. Millward, S. Miller, T. Stallard, A.D. Aylward, Icarus **160**, 95–107 (2002b)
G.H. Millward, S. Miller, T. Stallard, N. Achilleos, A.D. Aylward, Icarus **173**, 200–211 (2005)
L. Moore, M. Mendillo, Geophys. Res. Lett. **34**, L12202 (2007). doi:10.1029/2007GL029381
L. Moore et al., Geophys. Res. Lett. **33**, L22202 (2006). doi:10.1029/2006GL027375
L.E. Moore, M. Mendillo, I.C.F. Mueller-Wodarg, D.L. Murr, Icarus **172**, 503–52 (2004)
I.C.F. Mueller-Wodarg, R.V. Yelle, Geophys. Res. Lett. **29** (2002). doi:10.1029/2002GL016100
I.C.F. Mueller-Wodarg, R.V. Yelle, M. Mendillo, L.A. Young, A.D. Aylward, J. Geophys. Res. **105**, 20833–20856 (2000)
I.C.F. Mueller-Wodarg, R.V. Yelle, M. Mendillo, A.D. Aylward, J. Geophys. Res. **108**(A12), 1453 (2003). doi:10.1029/2003JA010054
I.C.F. Mueller-Wodarg, M. Mendillo, R.V. Yelle, A.D. Aylward, Icarus **180**, 147–160 (2006)
I.C.F. Mueller-Wodarg et al., Space Sci. Rev. (2008, this issue)
A.F. Nagy, T.E. Cravens, Geophys. Res. Lett. **15**, 433–435 (1988)
A.F. Nagy, J. Kim, T.E. Cravens, Ann. Geophys. **8**, 251–256 (1990)
A.F. Nagy, T.E. Cravens, S.G. Smith, H.A. Taylor, H.C. Brinton, J. Geophys. Res. **85**, 7795–7801 (1980)
A.F. Nagy, T.E. Cravens, J.H. Yee, A.I.F. Stewart, Geophys. Res. Lett. **8**, 629–632 (1981)
A.F. Nagy, M.W. Liemohn, J.L. Fox, J. Kim, J. Geophys. Res. **106**(10), 21,565–21,568 (2001)
A.F. Nagy et al., J. Geophys. Res. **111**, A06310 (2006). doi:10.1029/2005JA011519
A.O. Nier, M.B. McElroy, Science **194**, 1298–1300 (1976)
L. Qian, S.C. Solomon, R.G. Roble, T.J. Kane, Geophys. Res. Lett. **33**, L23705 (2006). doi:10.1029/2006GL027185
L. Qian, S.C. Solomon, R.G. Roble, T.J. Kane, Geophys. Res. Lett. **35**, L07811 (2008). doi:10.1029/2007GL033156
S. Quegan et al., J. Atmos. Terr. Phys. **44**, 619–640 (1982)
S. Rafkin, A. Stern, S. Bougher, A. Brecht, Venus Express Science team meeting. Abstract, Thuile, Italy, March 18–24 (2007)
A.D. Richmond, E.C. Ridley, R.G. Roble, Geophys. Res. Lett. **19**, 601–604 (1992)
J.H. Richter, F. Sassi, R.R. Garcia, K. Matthes, C.A. Fischer, *J. Geophys. Res.* (2008, in press)
A.J. Ridley, Y. Deng, G. Toth, J. Atmos. Sol. Terr. Phys. **68**, 839 (2006)
H. Rishbeth, I.C.F. Mueller-Wodarg, Ann. Geophys. **17**, 794–805 (1999)
H. Rishbeth et al., Ann. Geophysicae **18**, 945–956 (2000a)
H. Rishbeth, R.V. Yelle, M. Mendillo, Planet. Space. Sci. **48**, 51–58 (2000b)

R.G. Roble, in *The Upper Mesosphere Lower Thermosphere: A Review of Experiment and Theory*, ed. by R.M. Johnson, T.L. Killeen. Geophysical Monograph, vol. 87 (American Geophysical Union, 1995), p. 1
R.G. Roble, in *Atmospheric Science Across the Stratopause*. Geophysical Monograph, vol. 123 (American Geophysical Union, 2000), pp. 53–67
R.G. Roble, R.E. Dickinson, Geophys. Res. Lett. **16**, 1441–1444 (1989)
R.G. Roble, E.C. Ridley, Geophys. Res. Lett. **21**, 417–420 (1994)
R.G. Roble, E.C. Ridely, R.E. Dickinson, J. Geophys. Res. **92**, 8745–8758 (1987)
C. Roldan, M.A. Lopez-Valverde, M. Lopez-Puertas, D.P. Edwards, Icarus **147**, 11–25 (2000)
R. Sadourny, K. Laval, in *New Perspectives in Climate Modeling*, ed. by A. Berger, C. Nicolis (Elsevier, Amsterdam, 1984), pp. 173–197
F. Sassi, D. Kinneson, B.A. Boville, R.R. Garcia, R.G. Roble, J. Geophys. Res. **109**, D17108 (2004). doi:10.1029/2003JD004434
H.G. Schmidt et al., J. Climate **19** (2006). doi:10.1175/JCLI3829.1
G. Schubert et al., in *Exploring Venus as a Terrestrial Planet*. Geophysical Monograph, vol. 176 (American Geophysical Union, 2007), pp. 101–120
R.W. Schunk, A.F. Nagy, *Ionospheres: Physics, Plasma Physics and Chemistry*. Cambridge Atmospheric and Space Science Series (Cambridge University Press, Cambridge, 2000)
R. Shimazaki, M. Shimizu, Rep. Ionos. Space Res. Jpn. **24**, 80–98 (1970)
H. Shinagawa, T.E. Cravens, J. Geophys. Res. **93**, 11,263–11,277 (1988)
H. Shinagawa, T.E. Cravens, J. Geophys. Res. **94**, 6506–6516 (1989)
C.G.A. Smith, S. Miller, A.D. Aylward, Ann. Geophys. **23**, 1943–1947 (2005)
R.W. Stewart, J. Atmos. Sci. **25**, 578–582 (1968)
R.W. Stewart, J. Atmos. Sci. **28**, 1069–1073 (1971)
D.F. Strobel, in *Atmospheres in the Solar System: Comparative Aeronomy*, ed. by M. Mendillo, A. Nagy, H. Waite. Geophysical Monograph, vol. 130 (American Geophysical Union, Washington, 2002), pp. 7–22
V.T. Tenishev, M.R. Combi, AIAA Paper 2003-3776 (2003a)
V.T. Tenishev, M.R. Combi, RGD, 23rd Int. Symp. 663, 696 (2003b)
V.T. Tenishev, M. Combi, H. Waite, Eos Trans. Am. Geophys. Union **88**(23) (Jt. Assem. Suppl.) (2007), abstract P43A-02
V.T. Tenishev, M.R. Combi, B. Davidsson, Astrophys. J. (2008, in press)
R.H. Tolson et al., J. Spacecr. Rockets **44**(6), 1172–1179 (2007)
A. Valeille, V. Tenishev, S.W. Bougher, M.R. Combi, A.F. Nagy, Eos Trans. Am. Geophys. Union **88**(23) (Jt. Assem. Suppl.) (2007a), abstract SA31B-02
A. Valeille, V. Tenishev, S.W. Bougher, M.R. Combi, A.F. Nagy, B. A. A. S. **39**, #24.01 (2007b)
A. Valeille, M.R. Combi, V.T. Tenishev, S.W. Bougher, A.F. Nagy, Icarus (2008, submitted)
U. von Zahn et al., J. Geophys. Res. **85**, 7829–7840 (1980)
J.H. Waite et al., Science **308**, 982–986 (2005)
O. Witasse et al., Space Sci. Rev. (2008, this issue)
P.G. Withers, Geophys. Res. Letts. **33**, L02201 (2006). doi:10.1029/2005GL024447
R.V. Yelle, S. Miller, in *Jupiter: The Planet, Satellites, and Magnetosphere*, ed. by F. Bagenal, T.E. Dowling, W.B. McKinnon (Cambridge Univ. Press, New York, 2004), pp. 185–218
M.H.G. Zhang, J.G. Luhmann, A.J. Kliore, J. Kim, J. Geophys. Res. **95**, 14829–14839 (1990)
S. Zhang, S.W. Bougher, M.J. Alexander, J. Geophys. Res. **101**, 23195–23205 (1996)

Modeling and Simulating Flowing Plasmas and Related Phenomena

S.A. Ledvina · Y.-J. Ma · E. Kallio

Originally published in the journal Space Science Reviews, Volume 139, Nos 1–4.
DOI: 10.1007/s11214-008-9384-6

Abstract Simulation has become a valuable tool that compliments more traditional methods used to understand solar system plasmas and their interactions with planets, moons and comets. The three popular simulation approaches to studying these interactions are presented. Each approach provides valuable insight to these interactions. To date no one approach is capable of simulating the whole interaction region from the collisionless to the collisional regimes. All three approaches are therefore needed. Each approach has several implicit physical assumptions as well as several numerical assumptions depending on the scheme used. The magnetohydrodynamic (MHD), test-particle/Monte-Carlo and hybrid models used in simulating flowing plasmas are described. Special consideration is given to the implicit assumptions underlying each model. Some of the more common numerical methods used to implement each model, the implications of these numerical methods and the resulting limitations of each simulation approach are also discussed.

Keywords Plasma · Magnetohydrodynamics · Test-particle · Hybrid · Simulations · Numerical methods

1 Introduction

The interaction of solar system plasmas with planets, satellites and comets is a very complex and challenging problem. The interaction depends on the properties of both the incident plasma and the body. The incident plasma properties (i.e. the plasma species, density, incident speed, pressure and magnetic field) are widely varying. Additionally the properties of

S.A. Ledvina (✉)
Space Sciences Lab, University of California, Berkeley, CA 94720, USA
e-mail: ledvina@ssl.berkeley.edu

Y.-J. Ma
IGPP, UCLA, 6877 Slichter Hall, Los Angeles, CA 90095, USA

E. Kallio
Finnish Meteorological Inst., Space Research Unit, Erik Palmenin aukio 1, Helsinki SF-00101, Finland

the body (i.e. does it have an atmosphere or an intrinsic magnetic field) effect the interaction by processes such as mass-loading, ion-neutral interactions, direct interaction with the intrinsic field. The interactions can be classified into three types: 1) interactions with magnetic bodies, 2) interactions with non-magnetized bodies without atmospheres, and 3) interactions with non-magnetized bodies with atmospheres. An excellent review of plasma interactions with exospheres, ionospheres and atmospheres of various celestial bodies can be found in Ma et al. (2008). Some of the computer simulation methods that are used to understand these interactions are described here.

There are three approaches used to study solar system plasmas and their interaction with planets, moons and comets. These approaches are: 1) experimental/observational, 2) theoretical and 3) modeling/simulations. Each approach has its advantages and disadvantages. The experimental/observational approach includes collecting remote and in situ data and laboratory based experiments. This is capable of being the most realistic way of understanding the plasma interaction in question. However, there are some real drawbacks with this approach such as: equipment and operating costs, measurement difficulties, data interpretation and lack of spatial and temporal coverage. The theoretical approach is clean; it provides general information in analytical form. However, it is restricted to simple geometry, physics, chemistry and usually linear problems. The third approach, modeling/simulations is the subject of this paper. No one approach is capable of fully describing all aspects of the problem. They are complimentary and all are needed to understand the complexity of the plasma interactions of interest. The goal is to understand the interaction.

Before proceeding there are a few issues to get straight. The terms model and simulation are often used interchangeably in the literature. This is not strictly correct. A model is defined as a representation of a physical process intended to enhance our ability to understand, predict or control its behavior. These can consist of mathematical equations that describe a physical process or an empirical model of data. Some examples would include models of ionization processes, ion-neutral chemistry, collisions and plasma flow. Simulation is the exercise or use of a model. Typically a simulation is a computer code that executes one or more models to understand how the models interact to form a physical system. An example would be to combine a model of plasma flow with a model of ionization processes and collisions to study planetary exospheres.

There are many modeling approaches used to study solar system plasma interactions. The three most common are magnetohydrodyamic (MHD), test-particle/Monte-Carlo and hybrid simulations. Each approach can be used to study ion motions but only the test-particle/Monte-Carlo method can be applied to electron motion. MHD simulations treat the plasma as a charge neutral fluid (usually a single fluid). Information about the kinetic nature of the ions is lost in this approach. The test-particle/Monte-Carlo approach traces the ion or electron motion through a background magnetic and electric field. The background fields can be from an analytic solution or taken from MHD or hybrid simulations. This approach includes some of the kinetic aspects of the plasma and can easily treat multiple species. However there is no feedback between the charged particles and the fields or between the individual particles. This can lead to significant differences in the results when compared to more self-consistent hybrid simulations. Hybrid simulations treat the ions as kinetic particles and the electrons as a charge neutralizing massless (typically) fluid. The ion motion and the fields are solved self-consistently. Since the electrons are treated as a fluid, electron kinetic effects are absent.

Each modeling approach has its implicit assumptions, region of applicability, advantages and disadvantages. In addition several possible numerical methods can be used for each approach, each with their own assumptions, advantages and disadvantages. It is the goal of this paper to provide the reader with a feel for each simulation method, their implicit assumptions and the issues associated with the choice of algorithm. One cannot cover every aspect of each modeling approach or every possible numerical scheme in this work. In fact to do so would probably fill a small library. The interested reader is encouraged to follow up these ideas using the cited references as a starting point.

Traditionally cgs units were used in both space and plasma physics. Today there is a mixture of cgs and mks used in space physics depending on the context. Now it is common for MHD to be formulated using mks in the literature, while hybrid and kinetic models are still more commonly formulated in cgs. This is a hold over from the fusion community where the hybrid methods were first developed. As a result, mks units will be used in the section on MHD while cgs units will be used in the rest of this paper.

1.1 Basic Plasma Physics Underlying the Models and Simulations

Before proceeding to the discussion of the numerical approaches let's review some of the assumptions made in modeling flowing plasmas. Plasma physics is the study of low density ionized gases. The number of ions should be enough so that the long range Coulomb force is a factor in determining the statistical properties of the plasma, but low enough that the force due to near neighbor ions is much less than the long range Coulomb force exerted by many distant ions. The motion of an individual ion is governed by the equation of motion:

$$m\frac{d\boldsymbol{v}}{dt} = q\left(\boldsymbol{E} + \frac{\boldsymbol{v}\times\boldsymbol{B}}{c}\right) \tag{1.1}$$

where m is the mass of the ion, $\boldsymbol{v}$ is the ions velocity, q is the ions charge and $\boldsymbol{E}$ and $\boldsymbol{B}$ are the electric and magnetic fields the ion is moving through. The position of the individual ion, $\boldsymbol{x}$, is given by:

$$\frac{d\boldsymbol{x}}{dt} = \boldsymbol{v}. \tag{1.2}$$

The fields are affected by the motion of the ions through Maxwell's equations.

$$\nabla\cdot\boldsymbol{B} = 0 \tag{1.3}$$

$$\nabla\cdot\boldsymbol{D} = 4\pi\rho_c \tag{1.4}$$

$$\nabla\times\boldsymbol{H} = \frac{4\pi}{c}\boldsymbol{J} + \frac{1}{c}\frac{\partial\boldsymbol{D}}{\partial t} \tag{1.5}$$

$$\nabla\times\boldsymbol{E} + \frac{1}{c}\frac{\partial\boldsymbol{B}}{\partial t} = 0 \tag{1.6}$$

The interplay between the ion motions and the fields leads to many non-linear processes such as instabilities and waves that are at the core plasma physics.

Solving (1.1)–(1.2) together with Maxwell's equations (1.3)–(1.6) for every electron and ion in a plasma is an intractable task. Since it is the collective behavior or macroscopic properties that one is after many assumptions and simplifications can be made. The choice of assumptions and simplifications will lead to the approach used to model the plasma.

The hybrid and MHD equations are presented but not derived here. The interested reader can find them derived in a variety of plasma physics text (cf. Krall and Trivelpiece 1973; Nicholson 1983; Cravens 1997; Gombosi 1999; Schunk and Nagy 2000; Lipatov 2002). The assumptions used in deriving the model equations and their ramifications are discussed below.

1.1.1 The Hybrid Model

The hybrid approach is typically applied to collisionless plasmas when the electron mass can be ignored. There are finite mass hybrid schemes which are not discussed here (cf. Lipatov 2002 for further details). Hybrid schemes have been around for many years and the interested reader should see the review by Winske et al. (2003), Lipatov (2002) and Brecht and Thomas (1988) and the references therein.

The hybrid scheme solves the following ion momentum and position equations for each particle:

$$\frac{d\boldsymbol{v}_i}{dt} = \frac{q_i}{m_i}\left(\boldsymbol{E} + \frac{\boldsymbol{v}_i \times \boldsymbol{B}}{c} - \eta \boldsymbol{J}\right) \tag{1.7}$$

$$\frac{d\boldsymbol{x}_i}{dt} = \boldsymbol{v}_i \tag{1.8}$$

where $\boldsymbol{J}$ is the total current density and η is the resistivity (discussed more below). The electric field is given by:

$$\boldsymbol{E} = \frac{1}{4\pi n_i e}(\nabla \times \boldsymbol{B}) \times \boldsymbol{B} - \frac{1}{n_i e c}\boldsymbol{J}_i \times \boldsymbol{B} - \frac{1}{n_i e}\nabla(n_e T_e) + \eta \boldsymbol{J}. \tag{1.9}$$

Ampere's law becomes:

$$\nabla \times \boldsymbol{B} = 4\pi/c\,(\boldsymbol{J}_i + \boldsymbol{J}_e) \tag{1.10}$$

where $\boldsymbol{J_i}$ and $\boldsymbol{J_e}$ are the ion and electron current densities. Faraday's law (1.6) is also used to get the magnetic field. In addition the electron temperature has been solved in some simulations (cf. Brecht and Ledvina 2006; Brecht and Thomas 1988 and references therein) using:

$$\frac{\partial T_e}{\partial t} + \boldsymbol{u}_e \cdot \nabla T_e + \frac{3}{2}T_e \nabla \cdot \boldsymbol{u_e} = \frac{2}{3n_e}\eta \boldsymbol{J}^2. \tag{1.11}$$

Here T_e is the electron temperature and $\boldsymbol{u_e}$ is the electron velocity. There is no thermal conduction term in this equation but one could be added for a given problem.

1.1.2 The Ideal Magnetohydrodynamic (MHD) Equations

In the MHD approach the plasma is described by a set of fluid equations that describe the conservation of mass, momentum and energy and the evolution of the magnetic field. In conservative form the MHD equations are:

Continuity:
$$\frac{\partial \rho}{\partial t} + \nabla \cdot \rho \boldsymbol{u} = 0 \tag{1.12}$$

Momentum:
$$\frac{\partial \rho \boldsymbol{u}}{\partial t} + \nabla \cdot (\rho \boldsymbol{u}\boldsymbol{u}) = \frac{\boldsymbol{J} \times \boldsymbol{B}}{c} - \nabla p \tag{1.13}$$

Energy/pressure:
$$\frac{\partial e}{\partial t} + \nabla \cdot (e\boldsymbol{u}) = -p\nabla \cdot \boldsymbol{u} \tag{1.14}$$

Induction:
$$\frac{\partial \boldsymbol{B}}{\partial t} = \nabla \times (\boldsymbol{u} \times \boldsymbol{B}) + \frac{c^2}{4\pi\sigma}\nabla^2 \boldsymbol{B}. \tag{1.15}$$

Here ρ is the fluid mass density, $\boldsymbol{u}$ is the flow velocity, p is the plasma thermal pressure, e is the internal energy density, c is the speed of light, and σ is the plasma conductivity. The thermal pressure is related to the internal energy density by: $p = (\gamma - 1)e$.

1.1.3 The Implicit Assumptions

There are some common assumptions used in both the hybrid and MHD models as well as some model dependent ones.

1. Quasi-neutrality, $n_e = n_i$

Thus the displacement current is ignored in Ampere's law (1.5). This assumption is valid on scales larger than the Debye length λ_D. The assumption breaks down when the grid resolution is finer than the Debye length. This also implies that $\nabla \cdot \boldsymbol{J} = 0$, and removes most electrostatic instabilities.

2. The Darwin approximation

This approximation splits the electric field into a longitudinal part $\boldsymbol{E}_L$ and a solenoidal part $\boldsymbol{E}_T$. Then $\nabla \times \boldsymbol{E}_L = 0$ and $\nabla \cdot \boldsymbol{E}_T = 0$ and $\partial \boldsymbol{E}_T/\partial t$ is neglected in Ampere's law (1.5). This allows the light waves to be ignored. It also removes relativistic phenomena.

3. The electron mass, $m_e = 0$ since $m_e/m_i \ll 1$

This combined with the assumption of quasi-neutrality means that the mass density of the plasma is just the ion number density times the species mass, $\rho_m = n_i m_i$. The electron plasma frequency $(4\pi n_e e^2/m_e)^{1/2}$ and electron gyrofrequency $(eB/m_e c)$ now have zero's in their denominators so they are removed from the calculation. High frequency modes are no longer present, such as the electron whistler. By using these last two assumptions there is no longer a physical mechanism to describe the system behavior at small scales. The Debye length and the magnetic skin depth are not viable scales with this assumption. The viable scale is the ion skin depth c/ω_{pi}. There is now a limit on the smallest cell size that can reliably be used. The minimum cell size should be at least an order of magnitude larger than the electron skin depth c/ω_{pe}. When the incident plasma species are protons the limit becomes 1/4 of the proton inertial length c/ω_{pi}. This assumption has removed the electron processes that are needed to dissipate gradients in the electron densities and pressures that can develop at this scale size leading to unphysical fields. This shows up where ∇p_e plays a role. If smaller scale sizes are needed then the mass of the electron must be included. The interested reader should see Lipatov (2002) for further details.

4. The gas/plasma components are not far from thermodynamic equilibrium, i.e. at every spatial location the distribution function is a Maxwellian

This assumption is used when deriving the fluid equations from statistical mechanics (cf. Gombosi 1994). It is needed in order to get a closed set of transport equations. It is assumed that there are enough collisions in the gas for this assumption to be valid. This limits the use of the fluid approach, for instance rotating plasmas aren't described well by a Maxwellian.

5. $\boldsymbol{J} \times \boldsymbol{B}$ and $d\boldsymbol{J}/dt$ are neglected in Ohm's law

Thus only phenomena of very low frequency and very large spatial scales (compared to λ_D) are valid since:

$$\frac{1}{|\boldsymbol{J}|}\frac{1}{\omega_p}\frac{d\boldsymbol{J}}{dt} \approx \frac{\omega}{\omega_p} \ll 1$$

and

$$\frac{\boldsymbol{J}\times\boldsymbol{B}}{c}\approx\frac{B^2}{L}\ll ne\frac{\boldsymbol{u}\times\boldsymbol{B}}{c}.$$

There are times when the $\boldsymbol{J}\times\boldsymbol{B}$ is kept as part of the Hall term. The Hall term allows ambipolar fields to exist as well as ambipolar flow. When this is done the magnetic field is no longer tied to the plasma flow. The symmetry of the MHD equations is now broken so magnetized flow around a body will not produce symmetric structures or flow patters. The Hall term effects ion and electron motion on the ion inertial length scales: $L < c/\omega_{pi}$. Including the Hall term adds two new wave modes into the system: whistler waves and Hall drift waves. These limit the time step one can take in a simulation. See the review by Hubba (2003) for further details.

6. Isotropic pressure $\nabla\cdot p=\nabla p$

This assumption is valid when the plasma is collisional, with frequent inter particle interactions. It may also be valid in regions where wave activity mimics particle collisions. This assumption greatly simplifies the overhead needed in describing the plasma so it is often applied even when the plasma is nearly collisionless. According to Krall and Trivelpiece (1973) this assumption agrees well with a wide range of experiments, despite the lack of a clear basis for this assumption. However, in a collisionless magnetized plasma it is not clear what time and space scales are required to justify this approximation. This approximation is not valid for simulations that have gyrating plasmas or non-Maxwellian distribution functions.

7. The generalized Ohm's law reduces to: $\boldsymbol{J}=\sigma(\boldsymbol{E}+\frac{\boldsymbol{u}\times\boldsymbol{B}}{c})$

This is a result of assumptions (3–5) and is valid on the following scales. Let L be the length scale for spatial variations of the plasma parameters and $\boldsymbol{U}_0$ be the characteristic plasma velocity. Then the MHD approach and this form of the generalized Ohm's law are valid under the following conditions:

$$\text{if}\quad \frac{L\omega_{pe}^2}{\omega_{ce}}\frac{\boldsymbol{U_0}}{c^2}\gg 1,\quad\text{then}\quad \frac{\boldsymbol{J}\times\boldsymbol{B}}{nec}\ \text{can be neglected,}$$

$$\text{if}\quad \frac{L^2\omega_{pe}^2}{c^2}\gg 1,\quad\text{then}\quad \frac{m_e}{ne^2}\frac{\partial\boldsymbol{J}}{\partial t}\ \text{can be neglected,}$$

$$\text{if}\quad \frac{L\boldsymbol{U_0}\omega_{ce}}{\kappa T_e/m_e}\gg 1,\quad\text{then}\quad \nabla p_e\ \text{can be neglected.}$$

The generalized Ohm's law is further simplified if the conductivity of the plasma is very large. This is a valid assumption if the magnetic Reynolds number, $R_M\equiv\frac{4\pi\sigma LU_0}{c^2}\gg 1$ then $\frac{4\pi\boldsymbol{J}}{\sigma}$ can be neglected. The generalized Ohm's law then reduces to:

$$\boldsymbol{E}+\frac{\boldsymbol{U}\times\boldsymbol{B}}{c}=0. \tag{1.16}$$

8. Finite gyroradii effects can be ignored

Retaining gyro-radius effects will lead to off diagonal terms in the pressure tensor. This would invalidate the isotropic pressure assumption. The condition under which this is a valid assumption is:

$$\frac{L\omega_{ci}}{\sqrt{\kappa T_i/m_i}} \gg 1$$

where ω_{ci}, T_i and m_i denote the ion cyclotron frequency, temperature and mass. These conditions all imply that the plasma properties vary only on very long spatial and low frequency scales.

If the time scales are not slow a new set of more complicated equations is needed to describe the plasma. The gradient terms such as ∇p, $\nabla \boldsymbol{B}$, etc. are small compared to the field terms such as p and $\boldsymbol{B}$. For a plasma in a magnetic field MHD implies that in addition to $L \gg \lambda_D$, the ordering:

$$\left(\frac{r_L}{L}\right)^2 \ll \frac{1}{T\omega_{ci}} \ll 1, \quad \text{MHD ordering}$$

where r_L is the ion gyroradius and T is the time scale of interest. So for MHD to be valid the length scale of interest must be much larger than both the ion gyroradius and the Debye length. Additionally the time scale of interest must be much larger than the ion cyclotron period. When the above MHD ordering is valid gyroradius effects are negligible. MHD theory has been modified to incorporate some gyroradius effects. The modified MHD theory is known as finite Larmor radius MHD (cf. Roberts and Taylor 1962). It is valid for the following ordering:

$$\frac{1}{T\omega_{ci}} \ll \left(\frac{r_L}{L}\right)^2 \ll 1. \quad \text{FLR ordering}$$

Finite Larmor radius MHD differs from ideal MHD in two respects. The first is that the electric field used in the generalized Ohm's law retains the Hall term and includes a term for the gradient in the electron pressure. The second is that the ion pressure is no longer a scalar but has been modified so that it is a non-diagonal tensor. To date FLR MHD has not been applied to global planetary simulations, though aspects of it have been applied in some MHD formulations.

The implicit assumptions that apply in the hybrid and various MHD formulations are listed in Table 1. It is clear that the ion part of the hybrid formulation has the fewest implicit assumptions. The lack of implicit assumptions allows the hybrid model to be applied over a large range of parameter space and makes it well suited for collisionless plasmas. On the other hand the assumptions implicit to MHD makes it much better suited for collisional plasmas and plasma-neutral mixtures than the hybrid model.

Some of the derived solar wind plasma parameters (normalized to the body radius) near the planets and a few of their moons are shown in Table 2. Also shown in Table 2 is what implicit assumptions are valid over the scale size of 0.1 body radii. For each object the Debye length is much smaller than the scale size so quasi-neutrality is a valid assumption. Taking the electron mass to be zero is valid everywhere except Pluto. The electron skin depth is just under half the scale size so electron kinetic effects are important there. The normalized collisional mean free path (λ_{mfp}/R) is much larger than the scale size in every case. Based on this there is no justification to assume the plasma is in thermal dynamic equilibrium, that it can be described by a Maxwellian or that the pressure is isotropic. However, there are many plasma instabilities that will drive the plasma towards thermal dynamic equilibrium. This assumption must be checked against data for validation in each case. The Hall term is important when the scale size approaches the ion skin depth. The isotropic pressure assumption can't be valid if the gyroradii are significant, otherwise it will have to be checked against

Table 1 The assumptions implicit to each model

Model	Implicit assumptions							
	1	2	3	4	5	6	7	8
Hybrid (ion)	X	X						
Hybrid (e^-)	X	X	X	X		X		X
MHD	X		X	X	X	X	X	X
Hall MHD	X		X	X		X		X
Hall Multi-fluid MHD	X		X	X		X		X

Table 2 Derived plasma parameters and the validity of the implicit model assumption using a scale size of 0.1 R

Body	Plasma parameters						Implicit assumptions							
	R (km)	r_L/R	λ_D/R	λ_{mfp}/R	$c/\omega_{pi}/R$	$c/\omega_{pe}/R$	1	2	3	4	5	6	7	8
Mercury	2400	0.041	1.4×10^{-6}	1.8×10^4	0.011	2.5×10^{-4}	X	X	X	?	X	?	X	X
Venus	6052	0.063	9.3×10^{-7}	2.2×10^4	0.0097	2.2×10^{-4}	X	X	X	?	X	?	X	X
Earth	6378	0.10	1.6×10^{-6}	6.4×10^4	0.014	3.1×10^{-4}	X	X	X	?	X		X	
Mars	3395	0.43	6.4×10^{-6}	5.5×10^5	0.047	1.1×10^{-3}	X	X	X	?	X		X	
Jupiter	71492	0.02	6.8×10^{-7}	1.4×10^5	0.0050	1.2×10^{-4}	X	X	X	?	X	?	X	X
Io	1815	0.0039	4.7×10^{-7}	2.0×10^3	0.010	4.9×10^{-5}	X	X	X	?	X	?	X	X
Europa	1569	0.076	5.4×10^{-6}	3.6×10^5	0.12	5.7×10^{-4}	X	X	X	?		?		X
Saturn	60268	0.14	1.5×10^{-6}	5.4×10^5	0.012	2.7×10^{-4}	X	X	X	?	X		X	
Titan H^+	2575	0.097	7.4×10^{-5}	1.6×10^7	0.28	3.8×10^{-3}	X	X	X	?				
Titan O^+	2575	1.6	7.4×10^{-5}	1.4×10^9	0.79	3.8×10^{-3}	X	X	X	?				
Enceladus	250	0.062	5.2×10^{-6}	7.1×10^3	0.48	2.6×10^{-3}	X	X	X	?		?		X
Uranus	25559	0.74	2.8×10^{-6}	8.6×10^5	0.04	9.3×10^{-4}	X	X	X	?	X		X	
Neptune	24764	0.95	5.4×10^{-6}	3.0×10^6	0.10	2.4×10^{-3}	X	X	X	?				
Pluto	1150	20	7.4×10^{-5}	2.6×10^7	2.0	4.6×10^{-2}		X		?				

data. The reduction of Ohm's law on these scale sizes depends on the Hall term, the other terms in assumption (8) are negligible with the exception of the ∇p_e term near Titan. What Table 2 shows is that MHD is a very good approximation at Mercury, Venus, Jupiter and Io for cell sizes of 0.1 R. It is also a good approximation for Earth and Saturn if a larger scale size is used. This may also be the case at Uranus and Neptune depending on the size of their magnetospheres. The Hall term is significant at Neptune, Europa, Titan and Enceldadus. It is not entirely negligible at Mars or Uranus. For these bodies at this scale size, Hall MHD or hybrid models would be better. At Pluto the Hall term is also significant but so is the electron skin depth. The massless electron assumption breaks down there and none of the models discussed here are valid. Again these derived parameters are based on the solar wind and not the magnetospheric plasma found at the magnetized planets. The magnetic fields at these planets have been ignored in this exercise.

1.2 Basic Plasma Physics Summary

The physical assumptions that are implicit in the common models used to simulate plasma interactions have been reviewed. These assumptions limit the applicability of a given model to certain regions. The hybrid model is ideal for collisionless plasmas on scales where the electron kinetic effects are negligible. While MHD approximation is ideal for collisional plasmas and on scale sizes where ion kinetic effects are negligible. It is important to remember these assumptions to ensure that each model is applied correctly so that the results can be taken with confidence.

The next few sections will examine how each modeling approach is solved numerically, what tradeoffs are made in the choice of numerical method and look at some applications. The MHD approach is discussed first because historically this was the first of the modeling approaches used to study plasma interaction with celestial objects. This is followed by the test-particle/Monte-Carlo approach. Many consider this to be an intermediate step between MHD and hybrid approaches. Some of the key numerical schemes used in this approach are also important in hybrid simulations. Finally the hybrid approach is discussed.

2 Magnetohydrodynamic (MHD) Models

The MHD model is the extension of fluid dynamics to electrically conducting fluids such as plasmas, with the inclusion of the effects of electromagnetic forces. The corresponding MHD equations describe the evolution of macroscopic quantities such as density, bulk velocity, magnetic field and pressure of plasma flows. MHD models are especially useful when the exact motion of a single particle is of no interest. Various forms of MHD models have been extensively used in space physics to describe many different kinds of plasma phenomena, such as magnetic reconnection and solar wind interaction with different celestial objects (Otto 2001; Lcboeuf et al. 1978; Brecht et al. 1981; Fedder and Lyon 1987; Shinagawa and Cravens 1988, 1989; Cravens 1989; Keller et al. 1994; Gombosi et al. 1996, 1998; Hansen et al. 2000; Kabin et al. 2000; Ledvina and Cravens 1998; Ledvina et al. 2004a; Bauske et al. 1998; Tanaka and Murawski 1997; Ma et al. 2004a, 2004b; Jia et al. 2007).

2.1 The MHD Equations

The MHD equations consist of the macroscopic transport equations and the magnetic induction equation. The transport equations can be obtained by multiplying the Boltzmann equation with an appropriate function of velocity and then integrating over the velocity space. The induction equation is a combination of Maxwell's equations and the generalized Ohm's law. According to different assumptions made in the derivation, MHD models have various forms. There is the ideal MHD model which is single species and single fluid. There is multi-species MHD where each species is represented by a separate continuity equation but all of the species have a single velocity and temperature. Multi-fluid MHD treats each individual ion species as a separate fluid. Hall MHD retains the Hall term in the generalized Ohm's law. Finally resistive MHD adds a resistive term to the induction equation along with a heating term in the pressure equation. The discussion below starts with the simplest, commonly used, ideal MHD model.

2.1.1 Ideal MHD Equations Plus Sources

The MHD model treats the plasma as a single, quasi-neutral, magnetized fluid and it solves the following set of MHD equations in non-conservative form (cf. Gombosi 1999; Schunk and Nagy 2000), which contains
Continuity equation:

$$\frac{\partial \rho}{\partial t} + \nabla \cdot (\rho \boldsymbol{u}) = m_i \frac{\delta n}{\delta t} \tag{2.1}$$

Momentum equation:

$$\begin{aligned} \rho \frac{\partial (\boldsymbol{u})}{\partial t} + \rho(\boldsymbol{u} \cdot \nabla)\boldsymbol{u} + \nabla \cdot \left(p\boldsymbol{I} + \frac{B^2}{2\mu_0}\boldsymbol{I} - \frac{1}{\mu_0}\boldsymbol{B}\boldsymbol{B} \right) &= \rho \boldsymbol{G} + \frac{\delta M}{\delta t} \\ &= \rho \boldsymbol{G} - \sum_n n_i m_i \nu_{in} (\boldsymbol{u} - \boldsymbol{u}_n) \end{aligned} \tag{2.2}$$

Pressure equation:

$$\begin{aligned} \frac{1}{\gamma - 1}\frac{\partial p}{\partial t} + \frac{1}{\gamma - 1}(\boldsymbol{u} \cdot \nabla p) + \frac{\gamma}{\gamma - 1} p(\nabla \cdot \boldsymbol{u}) &= \frac{\delta E}{\delta t} \\ &= \sum_n \frac{n_i m_i \nu_{in}}{m_i + m_n} [m_n (\boldsymbol{u} - \boldsymbol{u}_n)^2 - 3k(T_i - T_n)] \end{aligned} \tag{2.3}$$

Magnetic induction equation:

$$\frac{\partial \boldsymbol{B}}{\partial t} = \nabla \times \boldsymbol{u} \times \boldsymbol{B} \tag{2.4}$$

where ρ is mass density, $\boldsymbol{u}$ is plasma velocity, p is pressure and $\boldsymbol{B}$ is magnetic field vector. The MHD equations are composed of a continuity equation, momentum equations (for velocity vector), a pressure equation and an induction equation (for magnetic vector). The source terms on the right hand side of (2.1)–(2.3) are not present in the ideal MHD model. These terms are discussed in detail below. These equations together form a complete set of partial differential equations which fully determine the fluid and field quantities. Strictly speaking, this model is only applicable when:

1) The gas components are not far from local thermodynamic equilibrium;
2) The plasma has a Maxwellian distribution function;
3) Heat flow is not important;
4) Charge neutrality assumption is valid;
5) The high-frequency component of the electric field can be neglected.

When the gas system is only partially ionized, the collisions between ions and neutral particles could be important. Generally, there are two kinds of collisions: elastic collisions and inelastic collisions. Inelastic collisions result in charge exchange reactions. Collisions between ions and neutrals do not change the number density of the plasma (charge exchange may change the mass density of the plasma), but generally try to diminish the velocity and temperature differences between ions and neutrals. The inelastic ion-neutral collisional effect can be included in the model by adding the source terms in the right hand sides of (2.1)–(2.3) (Schunk and Nagy 2000).

In the source terms m_i and m_n are the mass of the ion and neutral species, respectively; n_i is the number density of the ions, ν_{in} is the collision frequency between ion and neutral particles. $\boldsymbol{u}_n$ is the bulk velocity of neutrals. γ is the specific heat ratio, it is assumed that the particles have no internal degrees of freedom therefore, $\gamma = 5/3$, T_i and T_n are the temperatures of the ions and neutrals. In the MHD model, only plasma pressure (temperature) is calculated. When the ratio of ion and electron temperatures is unknown, one usually assume $T_i = T_e = T_p/2$. The neutral particles are usually cold and steady compared with the plasma. The main effect of ion-neutral collisions, as shown in (2.2), is to slow down the plasma. The energy effect is more complicated, including a cooling term due to temperature difference and heating caused by the velocity difference.

How well a MHD simulation reproduces the Rankine-Hugoniot jump conditions across a shock is one of the key tests of any MHD code. However, one must keep in mind that shock physics are not present in MHD simulations. Shocks show up as discontinuities in the solution of the MHD equations that cannot be resolved. Several numerical approaches have been developed for capturing shocks in MHD simulations. These approaches are proxies to the missing shock physics that generally give the correct jump conditions and location of the shock.

2.1.2 Hall and Resistive MHD

The magnetic induction equation is an important component of the MHD models. According to the form of induction equations, MHD model can be categorized as ideal, resistive and Hall MHD models.

The magnetic induction equation, which includes the Hall effect and resistivity, can be expressed as:

$$\frac{\partial \boldsymbol{B}}{\partial t} = \nabla \times \left(\boldsymbol{u} \times \boldsymbol{B} - \frac{\boldsymbol{J}}{ne} \times \boldsymbol{B} - \eta \boldsymbol{J} \right) \tag{2.5}$$

where n is total ion number density and e is electron charge. All the other variables have their conventional meanings. A relationship for the current density $\boldsymbol{J}$, such Ampere's law (1.10) is needed for (2.5). Compared with the ideal form of the induction equation (2.4), the right hand side of (2.5) has two extra terms: the Hall and diffusion terms, besides the convection term. The resistive form is necessary to describe the effect of magnetic diffusion due to collisions. All numerical codes produce numerical resistivity, generally enough to enable magnetic reconnection, in some circumstances this may not be sufficient. A notable example is the dynamical evolution of substorms (Raeder 2001). The resistivity dissipates the electromagnetic energy in the system. This energy dissipation can heat the plasma. Hence, a resistive heating term is typically added to the pressure equation.

The inclusion of the Hall term allows the ions and electrons to move at different velocities. The magnetic field lines are still frozen to the electrons, but when there is a significant current, the "frozen-in" condition between ions and magnetic field lines is broken. Strictly speaking, the Hall MHD model is still limited by its fluid assumption, but it captures more essential physics than ideal or resistive MHD. The Hall effect becomes important when the ion skin depth is comparable to the gradient scale size, which is true for Titan. The gyroradii of heavy ion species (such as O^+ or CH_4^+) in the outer magnetosphere were found to be ~5000 km (Hartle et al. 2006). The resulting ion skin depth for these ions is about 2000 km. So the use of the Hall MHD model is more appropriate at Titan. The Hall term also introduces whistler and Hall drift waves into the simulation. These additional waves further restrict the time step that can be used in the simulation. The time step now goes with $\Delta \boldsymbol{x}^2$ instead of with just $\Delta \boldsymbol{x}$. So if the grid spacing is reduced by a factor of 2 the time step must be reduced by a factor of 4.

2.1.3 Multi-species MHD Model

The ideal MHD model is a single species model. When the plasma is composed of different kinds of plasma composition, multi-species MHD models are usually needed to describe more accurately the mass loading effect. The mass loading process is tightly related to the pick-up ions, which is important for the solar wind interaction with weakly magnetized planets, such as Mars and Venus. Pickup ions of planetary origin are mainly created outside the exobase through three different kinds of mechanisms. The neutral atmospheric constituents can be ionized by solar radiation, charge exchange reactions and impact ionization by solar wind electrons. The newly created ions are then picked up by the IMF and the convection electric field. As a consequence of the momentum transfer by electromagnetic fields from the solar wind to the pickup ions, mass loading effectively slows down the solar wind around the planets.

In the framework of the multi-species model, the mass densities of several ion species are tracked, while only one momentum and one energy equation are solved, since all the ions are assumed to have the same bulk velocity and temperature. The set of multi-species MHD equations can be written as:

$$\frac{\partial \rho_i}{\partial t} + \nabla \cdot (\rho_i \boldsymbol{u}) = S_i - L_i \tag{2.6}$$

$$\frac{\partial (\rho \boldsymbol{u})}{\partial t} + \nabla \cdot \left(\rho \boldsymbol{u}\boldsymbol{u} + p\boldsymbol{I} + \frac{B^2}{2\mu_0}\boldsymbol{I} - \frac{1}{\mu_0}\boldsymbol{B}\boldsymbol{B} \right) = \rho \boldsymbol{G} - \sum_{i=ions} \rho_i \sum_{t=neutrals} \nu_{it}\boldsymbol{u} - \sum_{i=ions} L_i \boldsymbol{u} \tag{2.7}$$

$$\frac{\partial \boldsymbol{B}}{\partial t} + \nabla \cdot (\boldsymbol{u}\boldsymbol{B} - \boldsymbol{B}\boldsymbol{u}) = \frac{1}{\sigma_0 \mu_0} \nabla^2 \boldsymbol{B} \tag{2.8}$$

$$\begin{aligned} &\frac{1}{\gamma - 1}\frac{\partial p}{\partial t} + \frac{1}{\gamma - 1}(\boldsymbol{u} \cdot \nabla) p + \frac{\gamma}{\gamma - 1} p (\nabla \cdot \boldsymbol{u}) + \nabla \cdot \boldsymbol{h} \\ &= \sum_{i=ions} \sum_{t=neutrals} \frac{\rho_i \nu_{it}}{m_i + m_t} [m_t (\boldsymbol{u}_n - \boldsymbol{u})^2 - 3k(T_i - T_n)] + \frac{1}{2} \sum_{i=ions} S_i (\boldsymbol{u}_n - \boldsymbol{u})^2 \\ &+ \frac{k}{\gamma - 1} \sum_{i=ions} \left(\frac{S_i T_n - L_i T_i}{m_i} - \frac{\rho_i}{m_i} \alpha_{R,i} n_e T_e \right) + \frac{k}{\gamma - 1} \sum_{i'} n_{i'} (\nu_{ph,i'} + \nu_{imp,i'}) T_n \end{aligned} \tag{2.9}$$

$$\rho = \sum_{i=ions} \rho_i \tag{2.10}$$

$$S_i = m_i n_{i'} \left(\nu_{ph,i'} + \nu_{imp,i'} + \sum_{s=ions} k_{si'} n_s \right) \tag{2.11}$$

$$L_i = m_i n_i \left(\alpha_{R,i} n_e + \sum_{t=neutrals} k_{it} n_t \right) \tag{2.12}$$

where S_i and L_i are the mass production and loss rates of the ith ion species, respectively; $n_{i'}$ is the number density of the neutral parents of the ith ion species; n_e is the electron number density. $\nu_{ph,i'}$, $\nu_{imp,i'}$, $k_{si'}$ and $\alpha_{R,i}$ are the photonionization, impact ionization, charge exchange reaction and recombination reaction rates, respectively. The multi-species MHD model includes the effects of both elastic and inelastic collisions in the equations, as contributions to the source terms. In the multi-species model, the ion mass densities are also controlled by chemical reactions. Thus mass loading effects are adequately treated in the model (Szego et al. 2000).

The multi-species MHD models are important for the study of plasma interaction with weakly or un-magnetized solar system bodies, such as Mars, Venus and Titan (Cravens et al. 1998; Tanaka 1998; Liu et al. 2001; Ma et al. 2004a, 2004b, 2006), where the main component of the incident plasma is different than the major ionospheric ion species.

2.1.4 Multi-fluid MHD Model

A more general approach is the multi-fluid MHD method. The major initial motivation for constructing an MHD model with separate ion momentum equations was observations made in the different plasma environments of comets, Venus and Mars. In addition, the active experiments in space in which barium and lithium were released into the solar wind creating artificial comets (Bryant 1985) provided further stimulus. These observations indicated that the solar wind protons and the heavy ions of the obstacle have their own separate dynamics. The first of such a model was a one dimensional MHD model, with two momentum equations (Sauer et al. 1990; Baumgartel and Sauer 1992). While the classical one-fluid MHD models failed to reproduce a number of important aspects of the observed signatures of the proton flow, the main features of the interaction can be described by an MHD model in which protons and heavy ions develop their own interconnected dynamics (Sauer et al. 1990). The continuity and momentum equations used in the model are given below for the protons (the same equations are also used for the heavy ions; thus interchanging the subscripts p and h, lead to the heavy ion equations):

$$\frac{\partial n_p}{\partial t} + \nabla \cdot \{n_p \boldsymbol{u}_p\} = 0 \tag{2.13}$$

$$\frac{\partial \{n_p \boldsymbol{u}_p\}}{\partial t} + \nabla \cdot \{n_p \boldsymbol{u}_p \boldsymbol{u}_p\} = \frac{1}{m_p}\frac{n_p}{n_e}\left\{ e n_h [\boldsymbol{u}_p - \boldsymbol{u}_h] \times \boldsymbol{B} - \nabla \left[\left(p_e + \frac{B^2}{2\mu_0} \right) I - \frac{BB}{\mu_0} \right] \right\} \tag{2.14}$$

$$\frac{\partial p_e}{\partial t} + \nabla \cdot \{\boldsymbol{u}_e p_e\} + \{\gamma - 1\} p_e \{\nabla \cdot \boldsymbol{u}_e\} = 0 \tag{2.15}$$

$$\frac{\partial \boldsymbol{B}}{\partial t} - \nabla \times \left\{ \frac{1}{n_e} \left[n_p \boldsymbol{u}_p + n_h \boldsymbol{u}_h - \frac{1}{\mu_0} \nabla \times \boldsymbol{B} \right] \times \boldsymbol{B} \right\} = 0 \tag{2.16}$$

where n_p, n_h and n_e are the proton, heavy ion and electron number densities, n_c is the charge density, $\boldsymbol{B}$ is the magnetic field, p_e is the electron pressure and $\boldsymbol{u}$ denotes the relevant velocities. It should be noted that this set of (2.13)–(2.16) does not contain a pressure equation for the ions nor an energy equation. Thus ions cannot be heated nor can they expand thermally other than via electron pressure. This limits the cases where this set of equations adequately describes the system.

Another version of the multi-fluid MHD approach adds kinetic terms to the fluid equations in order to include some ion cyclotron and gyroradii effects to global simulations. The multi-fluid equations in this version are (cf. Winglee 2004; Harnett et al. 2005):

$$\frac{\partial \rho_i}{\partial t} + \nabla \cdot (\rho_i \boldsymbol{u}_i) = 0 \tag{2.17}$$

$$\rho_i \frac{d\boldsymbol{u}_i}{dt} = en_i \left[(\boldsymbol{u}_i \times \boldsymbol{B}) - \sum_i \frac{n_i}{n_e} \boldsymbol{u}_i \times \boldsymbol{B} \right] + \boldsymbol{J} \times \boldsymbol{B} - \nabla(p_i + p_e) + \rho_i \boldsymbol{g}(\boldsymbol{r}) \tag{2.18}$$

$$\frac{\partial p_i}{\partial t} = -\gamma \nabla \cdot (p_i \boldsymbol{u}_i) + (\gamma - 1) \boldsymbol{u}_i \cdot \nabla p_i \tag{2.19}$$

$$\frac{\partial p_e}{\partial t} = -\gamma \nabla \cdot (p_e \boldsymbol{u}_{de}) + (\gamma - 1) \boldsymbol{u}_{de} \cdot \nabla p_e \tag{2.20}$$

$$\frac{\partial \boldsymbol{B}}{\partial \boldsymbol{t}} + \nabla \times \boldsymbol{E} = 0 \tag{2.21}$$

$$\boldsymbol{J} = \nabla \times \boldsymbol{B} \tag{2.22}$$

$$n_e = \sum_i n_i \tag{2.23}$$

$$\boldsymbol{u}_{de} = \sum_i \frac{n_i}{n_e} \boldsymbol{u}_i - \frac{\boldsymbol{J}}{en_e} \tag{2.24}$$

$$\boldsymbol{E} = -\sum \frac{n_i}{n_e} \boldsymbol{u}_i \times \boldsymbol{B} + \frac{\boldsymbol{J} \times \boldsymbol{B}}{en_e} - \frac{1}{en_e} \nabla p_e + \eta \boldsymbol{J} \tag{2.25}$$

where ρ_i is the mass density, n_i is the number density, e is the charge, $\boldsymbol{u}_i$ is the bulk fluid velocity, and p_i is the scalar pressure, each for species i. The electron number density is n_e, p_e is the electron pressure, e is the charge of a electron and $\boldsymbol{u}_{de}$ is the electron bulk velocity. The gravitational acceleration is $\boldsymbol{g}(\boldsymbol{r})$. The current density is $\boldsymbol{J}$, the magnetic and electric fields are $\boldsymbol{B}$ and $\boldsymbol{E}$. The ratio of specific heats is γ and is equal to $5/3$. The resistivity is given by η. The du_i/dt term in (2.18) should be the convective derivative, $\boldsymbol{D}/\boldsymbol{D}t = \partial/\partial t + \boldsymbol{u}_i \cdot \nabla$. The first two terms on the right hand side of (2.18) represent the difference between the acceleration of a given fluid from the acceleration of the center of density, due to the magnetic field. These terms are not present in single fluid versions of MHD since there is only the bulk motion. The remaining terms on the right hand side of (2.18) are identical to the momentum equation in Hall MHD with contributions from the gradient in the electron pressure and gravity.

This set of equations, "incorporates the full spectrum of waves up to the lower hybrid portion of the whistler mode" according to Winglee et al. (2008). Furthermore according to Harnett et al. (2005) equations (2.17)–(2.25) are equivalent to those used in hybrid simulations (1.7)–(1.10) except in the fluid limit. To the best of our knowledge these statements have not been shown in the literature. Nor has it been shown that a simulation based on these equations can reproduce real gyromotion or the wave spectrum. It is known that hybrid codes do not incorporate the full spectrum of waves up to the lower hybrid portion of the whistler mode due to their implicit assumptions.

There is a fundamental difference between this set of equations and those used in hybrid simulations. Equation (2.18) has a scalar pressure term, ∇p_i, in hybrid simulations the pressure is a tensor. Since the pressure is a scalar this approach assumes that the plasma is close to thermodynamic equilibrium and is Maxwellian. Clearly this set of equations cannot include all of the ion gyroradii effects that are present in hybrid simulations. The scalar pressure assumption can have consequences in certain situation. This led to the inclusion of the non-isotropic pressure in FLR MHD by Roberts and Taylor (1962). No information about the plasma distribution function can be learned from this approach, wave-particle interactions are not treated and many plasma instabilities have been assumed away.

The multi-fluid MHD equations include the continuity, momentum and energy equations for each species (note this does not include the model of Sauer et al. 1990, 1994, 1998). The number of equations increases significantly for the multi-fluid MHD model. To get the mass density, velocity and temperature for each fluid, one has to solve very complex equations, which is difficult to do numerically. There are other issues that need special consideration such as what to do when one of the fluids forms a shock? How do the other fluids respond to that shock? If one expects a multi-fluid code to show gyration, the differing gyro radii means that one species might not feel the shock formation if its gyroradii is many cells while a shock is formed in one cell. If it does feel the shock, how does one do the heating for this species? However, this model is needed, when each different species are not tightly coupled with each other and the different components can have different fluid speeds and temperatures. Multi-fluid Hall MHD with a non-isotropic pressure is the simulation approach of the future. A better understanding of the implicit physics in this approach is needed. To the best of our knowledge a careful study showing the importance of each term in the equations and the limitations to the fluid assumption has yet to be done.

2.2 Numerical Solution

MHD equations are non-linear partial differential equations. Analytical solutions of the MHD equations are available only for a few very simple cases. To solve realistic space plasma problems one has to use numerical methods.

2.2.1 Conservative vs. Primitive Form

The Equations listed in the above sections are written in primitive form. The primitive variable formulation leads to numerical schemes that do not strictly conserve momentum and energy, even in the hydrodynamic case. Such schemes do not guarantee correct shock speeds and correct jump conditions at discontinuities (Lu et al. 1989; Raeder 2003). However, a well crafted "primitive" MHD code can meet the jump conditions (cf. Lyon et al. 2004). Furthermore, the convective derivative is difficult to treat numerically. Although the use of the primitive variable formulation leads to algorithms with low memory requirements, sometimes, conservative form of the equations are desirable. As an example, the ideal MHD equations ((2.1)–(2.4) neglecting the source terms) can be rewritten in the following conservative form:

$$\frac{\partial \rho}{\partial t} + \nabla \cdot (\rho \boldsymbol{u}) = 0 \tag{2.26}$$

$$\frac{\partial (\rho \boldsymbol{u})}{\partial t} + \nabla \cdot \left(\rho \boldsymbol{u}\boldsymbol{u} + p\boldsymbol{I} + \frac{B^2}{2\mu_0}\boldsymbol{I} - \frac{1}{\mu_0}\boldsymbol{B}\boldsymbol{B} \right) = \rho \boldsymbol{G} \tag{2.27}$$

$$\frac{\partial \varepsilon}{\partial t} + \nabla \cdot \left(\boldsymbol{u}\left[\varepsilon + p + \frac{1}{2\mu_0} B^2 \right] - \frac{1}{\mu_0}(\boldsymbol{B} \cdot \boldsymbol{u})\boldsymbol{B} \right) = \rho \boldsymbol{u} \cdot \boldsymbol{G} \tag{2.28}$$

$$\frac{\partial \boldsymbol{B}}{\partial t} + \nabla \cdot (\boldsymbol{u}\boldsymbol{B} - \boldsymbol{B}\boldsymbol{u}) = 0 \tag{2.29}$$

where ε is the total energy, defined as:

$$\varepsilon = \frac{1}{2}\rho u^2 + \frac{1}{\gamma - 1}p + \frac{1}{2\mu_0}B^2, \tag{2.30}$$

the first term on the right hand side is the kinetic energy due to the bulk flow, the second term is the thermal energy and the last term is the magnetic energy.

In general there are several benefits to solving the MHD equations in the conservative form. It is possible to develop numerical schemes for this form of the MHD equations which conserve total energy and which obtain the correct jump conditions at discontinuities and shocks. These two properties are desirable in a numerical scheme because they assure that the numerical solution will obey the basic laws of physics represented by the analytic MHD equations.

Remember however, the MHD equations themselves are only an approximation to the actual plasma processes. Shock physics is not reproduced by the MHD equations. Natural shocks often have overshoots and jumps much higher than 4. These features are not reproduced in MHD theory.

There are of course disadvantages to solving the MHD equations in conservative form. When solving the conservative MHD equations, it is important to know the pressure. Rearranging (2.30) one has to compute:

$$p = (\gamma - 1)\left(\varepsilon - \frac{1}{2}\rho u^2 - \frac{1}{2\mu_0}B^2\right) \tag{2.31}$$

analytically this is not a problem. However, if either the kinetic or magnetic energy terms are small, or the terms are near balance numerical round-off errors can lead to unstable pressures. In fact the pressures may become negative. This problem occurs at Saturn, and especially Jupiter, where near the body the magnetic field dominates the pressure.

Accurately solving for the pressure is essential when trying to do temperature dependent ion-neutral chemistry (such as to represent an ionosphere). The temperature is derived from the pressure (usually via the ideal gas law). If the pressure is unstable (or negative) the derived temperature will be unstable (or negative) and the chemistry will be over or under driven. Thus MHD simulations of flowing plasmas interacting with ionospheres will often use the "primitive" form of the MHD equations.

Based on these properties, some simulations chose the combination of the two approaches, solving the conservative MHD equations throughout the most of the computational domain, while solving the primitive MHD equations near the central body (Hansen et al. 2005). This combination is a compromise that in some sense gives the best of both worlds: the correct jump conditions at shocks and discontinuities and positive pressures in the interior region dominated by the intrinsic magnetic field or where chemistry is important.

2.2.2 *Scheme: Finite-Volume Approach*

Finite-difference, finite-element, and finite-volume are the three major numerical approaches to solving partial differential equations (Hirsch 1989). There is much debate over the most accurate approach to use. Often simulations use elements from each approach. The finite difference method is the most straightforward way to solve these equations. It calculates values at each grid point by using a Taylor expansion to approximate the differential equations. This method is relatively cheap in computation time, easy to program and is easily expanded to incorporate additional physical processes. However, discontinuities must be smeared out by adding artificial viscosity. In case of the finite element method, the solutions are approximated by either eliminating the differential equation completely (steady state problems), or rendering the PDE into an equivalent ordinary differential equation, which is then solved using standard techniques such as finite differences, etc. The finite-volume method is a widely used approach, which solves the integral form of the governing equations. In this approach, the physical domain is divided into small volumes, and the dependent variables are evaluated as the volume-averaged value at each of the small volumes. The finite-volume method does not assume smoothness or continuity of the solution; instead, it automatically leads to a conservative discretization. This method is the most robust, but can be computationally expensive.

Consider the model equation:

$$\frac{\partial \boldsymbol{W}}{\partial t} + \nabla \cdot \boldsymbol{F} = \boldsymbol{Q}. \tag{2.32}$$

This represents the plasma part of the conservative MHD equations (2.17)–(2.22). Using the finite-volume approach, the governing equations are integrated over a cell, i, in the grid, giving

$$\int_{cell\,i} \frac{\partial \boldsymbol{W}}{\partial t} dV + \int_{cell\,i} \nabla \cdot \boldsymbol{F}\, dV = \int_{cell\,i} \boldsymbol{Q}\, dV. \tag{2.33}$$

The volume integral of a divergence term is converted to surface integrals using the divergence theorem.

$$\frac{d\boldsymbol{W}_i}{dt} + \frac{1}{V_i} \oint_{cell\,i} \boldsymbol{F} \cdot \hat{\boldsymbol{n}}\, dS = \boldsymbol{Q}_i \tag{2.34}$$

where $\boldsymbol{W}_i$ and $\boldsymbol{Q}_i$ are the cell-averaged conserved state and source vectors, respectively. V_i is the cell volume, and $\hat{\boldsymbol{n}}$ is a unit normal vector, pointing outward from the boundary of the cell. The surface integrals are evaluated as the sum of fluxes at all the surfaces of each finite volume. Using a simple midpoint rule to evaluate the integral yields

$$\frac{d\boldsymbol{W}_i}{dt} + \frac{1}{V_i} \sum_{faces} \boldsymbol{F} \cdot \hat{\boldsymbol{n}}\, dS = \boldsymbol{Q}_i \tag{2.35}$$

the $\boldsymbol{F} \cdot \hat{\boldsymbol{n}}$ terms are evaluated at the midpoints of the cell faces. The algorithms used to calculate the flux at cell interfaces are discussed in the next section. The flux entering a given volume is identical to that leaving the adjacent volume, therefore the mass; momentum and energy are automatically conserved. Another advantage of the finite volume method is that it is easily formulated to allow for unstructured meshes. For example, in a logarithmic spherical (curvilinear) coordinates are used only to define the grid mesh positions, all the physical vectors $\boldsymbol{F}$, $\boldsymbol{u}$ and $\boldsymbol{B}$ can still be taken in an arbitrarily chosen Cartesian frame of reference, and thus the solver does not need to be changed.

2.2.3 Grid

The grid system is a space structure on which the numerical solution is built. There are several ways to discretize a volume of space in order to compute a numerical solution. The two kinds of typical grid structures used are the static or adaptive grids. Both grid types can be non-uniform. A static grid is easy to apply and simple to program. The simplest example is just a uniform Cartesian grid. It provides the lowest programming overhead,

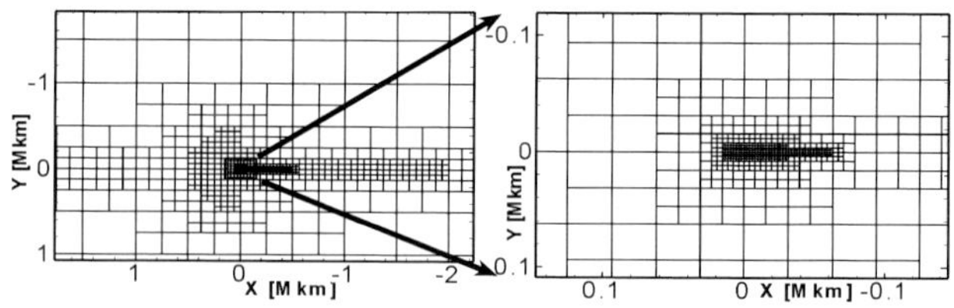

Fig. 1 The multi-scale block system designed with respect to the physical gradient conditions. Each block shown here contains $8 \times 8 \times 8$ cells (not shown). The plot to the *right* is the enlarged view of the rectangular region in the center of *left* plot from Jia et al. (2007)

lowest computing overhead and lowest memory overhead of any grid structure. Grids that can be indexed as though they are Cartesian, such as spherical coordinates, provide the same benefits with a small increase in overhead due to the metric. However, the grid boundary conditions may be more complex. Stretching the mesh so that the cell sizes are non-uniform is a way to increase the resolution in regions of interest while decreasing it in others. They may be better adapted to the solution, with the same advantages as the uniform grid. The main drawback to static grids is that they are not adaptive to the solution. Consequently computational resources may be wasted where they are not needed (in regions where the solutions are smooth) while other regions are under resolved, for example, sharp gradients and shocks (Raeder 2003).

On the other hand, the adaptive grid structures have the potential for the most accurate solution for a given number of cells. This property is particularly important for problems in which there are disparate spatial scales (cf. Gombosi et al. 1996). As an example, in a cometary interaction process, the ionization length scale and the radius of the comet differ by several orders of magnitude. Here an adaptive mesh is a virtual necessity. A recent numerical study of cometary tail disconnection events by Jia et al. (2007) used a grid containing 16 levels of resolution (see Fig. 1). The plate on the left is about one fifth of the calculation domain on each dimension, while the plate on the right is an enlarged view of the black box in the center of the left plate. The grid resolution ranges from a few kilometers close to the nucleus to 10^5 kilometers in regions far from the nucleus in the solar wind. The grid used in the simulation is a block adaptive system, which makes it simple to refine in the interesting region where more resolution is needed and to coarsen the grid in the region of less interest. A big advantage of the block-based data structure is the ease of parallelization (Powell et al. 1999).

Another example is the use of spherical grid structure in Ma et al. (2004a, 2004b, 2006). This grid provides much better altitude resolution, especially in the ionospheric regions. As shown in Fig. 2, the grid is uniformly spaced, throughout each block, with respect to the natural logarithm of the radial distance, r, and the other two spherical coordinates θ and φ.

Adaptive grids are a powerful tool. However, they have a cost. Some computational overhead in needed to handle the changes in the grid structure during the simulation. The selection of the grid refinement criteria is not universal. The refinement criteria can be problem dependent. There are issues associated with propagating the solution from one grid refinement level to another. When going from a course level to a finer level information must be interpolated to the smaller cells. If not done carefully the interpolation will result in regions where the $\nabla \cdot \boldsymbol{B} = 0$ constraint is violated. In addition the numerical resistivity and viscosity are both functions of cell size. Hence they are different at each refinement level. Cell sizes with a jump of 2 have significant changes in these quantities. This can lead to differing propagation speeds of a wave moving along such an interface. Waves propagating into the interface may also experience some reflection off of the interface. Changes in the numerical viscosity when refining the grid can also generate artificial turbulence in the so-

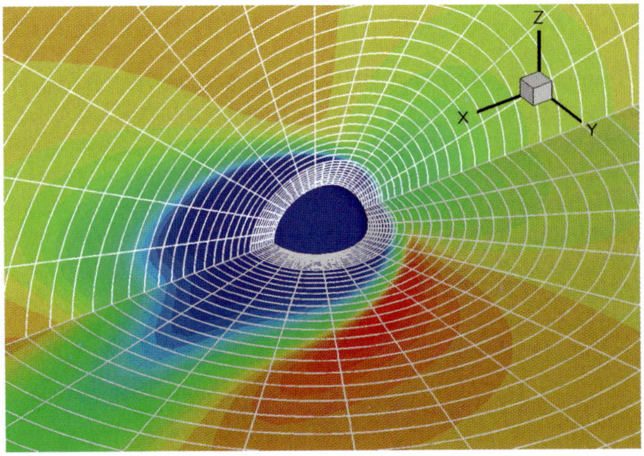

Fig. 2 Sketch of the spherical grid system used in the Titan simulation of Ma et al. (2004a, 2004b, 2006)

lution (D. Odstrcil, private communication). These issues are not problems on non-uniform meshes has long as the cell sizes change gradually.

2.2.4 *Time Stepping*

Solving the MHD equations is an initial value problem, because the ideal MHD equations are hyperbolic with respect to time. The unknown quantities are first assigned to some initial values and then they are advanced to the next time step using a time-stepping scheme. The process is repeated using the newly calculated solutions as the new initial values.

Explicit and implicit are two basic types of time stepping schemes. The explicit approach is straightforward: the solutions at the next time step only require the information about the current solutions. They are simple to implement in a code and computationally inexpensive. However, the maximum stable time step is limited by the CFL requirement (Sod 1985). On the other hand, an implicit method is much more stable and allows larger time steps than an explicit one. While in the implicit approach, the solutions at the next time step depend on the solutions at the same time step. Thus the update of every time step needs the solution of a set of linear equations, and consequently an implicit approach is significantly more expensive per time step. A point-implicit treatment of source terms become necessary, when the source terms are stiff such as the right hand side terms in the multi-species MHD equations (Powell et al. 1999). Implicit time stepping, with the time step larger then the CFL condition, is a good choice when the only thing that matters is the final steady state solution. However, they can not be used to get the correct time evolution of the system if the time step violates the CFL condition.

Generally, explicit methods are easy to program and require minimal computational resources so they run faster per time step. However, they are subject to more stringent stability criteria, limiting the size of the time step that can be used. Implicit methods are much more difficult to program and require a larger amount of computational resources per time step than explicit methods. However they are more stable and can be run with larger time steps than explicit methods, allowing the CFL condition to be circumvented. Circumventing the CFL condition comes at the cost of losing the information about the time evolution of the system.

If a steady state solution is desired, local time stepping, i.e. different cells being updated using different time increments, can be used to accelerate the convergence of the scheme to the steady state solution.

2.2.5 Divergence of B Control

An important difference between the numerical solution of the MHD equations and that of the gas dynamic equations is the constraint that $\nabla \cdot \boldsymbol{B} = 0$. If $\nabla \cdot \boldsymbol{B}$ is not zero, non-physical magnetic forces can arise along magnetic field lines. Additionally the presence of a finite $\nabla \cdot \boldsymbol{B}$ implies that magnetic helicity is no longer a conserved quantity (cf. Lyon et al. 2004 and references therein). Both numerical round-off errors and use of upwind differences can lead to difficulty in fulfilling the $\nabla \cdot B = 0$ condition automatically, especially when applying one-dimensional schemes to multidimensional MHD problem. Enforcing this constraint numerically, particularly in shock-capturing codes, can be done in a number of ways, but each way has its particular strengths and weaknesses. A brief overview of some of the methods is given below. Each of the schemes discussed below is explained more fully in the references cited, and Tóth (2000), has published a numerical comparison of many of the approaches for a suite of test cases.

a. Solve Faraday's Law The most straight forward approach to maintaining the $\nabla \cdot \boldsymbol{B} = 0$ constraint is to directly solve Faraday's law $\partial \boldsymbol{B}/\partial t = -\nabla \times \boldsymbol{E}$. If done in a leap frog fashion it can be shown that the divergence of this equation is zero to all orders in for any orthogonal grid system. The grid can be set up with any advection equations because the flow velocity and the magnetic field will produce $\boldsymbol{E}$ wherever it is needed. The resulting electric field components can then be center differenced to get the components of $\boldsymbol{B}$. This method has been used for uniform and non-uniform meshes as far back as Hain (1977) and Brecht et al. (1981) and is currently used in hybrid codes.

b. Constrained Transport The constrained-transport approach of Evans and Hawley (1988), preserves the $\nabla \cdot \boldsymbol{B} = 0$ constraint to within machine round-off errors. Faraday's law is rewritten using Stoke's theorem so that the magnetic flux through the surface of a grid cell is equal to the line integral of the electric field around the edge of the cell. Thus $\nabla \cdot \boldsymbol{B} = 0$ is conserved in the integral sense; the magnetic flux entering the cell is the same as that leaving. If the initial magnetic field has zero divergence, then at every time step it will be maintained to the accuracy of machine round off error as long as the boundary conditions are compatible with the constraints.

Recently, several approaches have been developed that have combined a Riemann-solver-based scheme with constrained transport approach. Dai and Woodward (1998) and Balsara and Spicer (1999) modified the constrained-transport approach by coupling a Riemann-solver-based scheme for the conservative form of the MHD equations. In their formulations, this required two representations of the magnetic field: a cell-centered one for the Godunov scheme and a face-centered one to enforce the $\nabla \cdot \boldsymbol{B} = 0$ condition. Tóth (2000) subsequently showed that these formulations could be recast in terms of a single cell-centered representation for the magnetic field, through a modification to the flux function used. Advantages of the conservative constrained-transport schemes include the fact that they are strictly conservative and that they meet the $\nabla \cdot \boldsymbol{B} = 0$ constraint to machine accuracy, on a particular stencil. Their primary disadvantage is the difficulty in extending them to general grids. Tóth and Roe (2002) made some progress on this front; they developed divergence-preserving prolongation and restriction operators, allowing the use of conservative constrained-transport schemes on refined meshes. However, they also showed that the conservative constrained transport techniques lose their $\nabla \cdot \boldsymbol{B}$-preserving properties if different cells are advanced at different physical time rates. This rules out the use of local time-stepping. Thus, while for unsteady calculations the cost of the conservative constrained transport approach is comparable to the eight-wave scheme, for steady-state calculations (where one would typically use local time-stepping), the cost can be prohibitive.

c. Divergence-Cleaning Scheme A typical way to solve this problem is the projection method (Ramshaw 1983; Voigt 1989; Tanaka 1993). In the projection method, an additional equation is added for the elimination of artificial magnetic monopoles. The magnetic field $\boldsymbol{B}$ is replaced every few time steps by a new field $\boldsymbol{B}_N$, given as

$$\boldsymbol{B}_N = \boldsymbol{B} + \nabla\phi, \tag{2.36}$$

$$\nabla^2\phi = -\nabla \cdot \boldsymbol{B}. \tag{2.37}$$

The resulting projected magnetic field is divergence-free on a particular numerical stencil, to the level of error of the solution of the Poisson equation. While it is not immediately obvious that the use of the projection scheme in conjunction with the fully conservative form of the MHD equations gives the correct solutions, Tóth (2000) has proven this to be the case. The projection scheme has several advantages, including the ability to use standard software libraries for the Poisson solution, its relatively straightforward extension to general unstructured grids, and its robustness. It does, however, require solution of an elliptic equation at each projection step; this can be expensive, particularly on distributed-memory machines.

d. Powell Scheme An alternative scheme was proposed by Powell (1994), to deal with the problem of the spurious numerical generation of $\nabla \cdot B$. Known as the Powell or 8-wave scheme, the terms including $\nabla \cdot B$, which are typically dropped due to the absence of magnetic monopoles, are kept in the derivation. The MHD equations, having been transformed into the divergence form, have a source vector, which is proportional to $\nabla \cdot B$. This form of MHD equations, although only quasi-conservative, is both symmetrizable and Galilean invariant (Powell et al. 1999). The resulting Riemann solver satisfies the constraint of $\nabla \cdot B = 0$ to truncation-error levels, even for long integration times. Moreover, the addition of the terms proportional to $\nabla \cdot B = 0$ improves results for multidimensional MHD calculations compared to several methods, and reduces errors in the calculated parallel magnetic force (Tóth and Odstrcil 1996). In this approach any magnetic monopoles that are generated do not accumulate at a fixed grid point but rather propagate along with the flow. For many problems this is not a issue however, it may lead to a buildup of the monopoles in stagnation regions which could affect the results. Tóth (2000) has shown that this approach can produce incorrect jump conditions at strong shocks and consequently incorrect results away from the discontinuity. Often this approach is combined with a divergence cleaning step every few time steps to remove the monopoles.

2.2.6 *Solving for Flows with Embedded Steady Fields*

For problems in which a strong intrinsic magnetic field is present, accuracy can be gained by solving for the deviation of the magnetic field from this intrinsic value (Groth et al. 1999). For example, in the interaction of the solar wind with a magnetized planet such as Earth, the planetary magnetic field, a strong dipole, dominates the magnetic-field pattern near the earth. Solving for the perturbation from the dipole field is inherently more accurate than solving for the full field and then subtracting off the dipole field to calculate the perturbation. This approach, first employed by Tanaka (1995), is derived below for the scheme applied to planets with a strong intrinsic magnetic field, which has been used in MHD simulations of Mars when including a crustal magnetic field (Ma et al. 2004a, 2004b).

Given an 'intrinsic' magnetic field, $\boldsymbol{B}_0$, that satisfies

$$\frac{\partial \boldsymbol{B}_0}{\partial t} = 0 \tag{2.38}$$

$$\nabla \cdot \boldsymbol{B}_0 = 0 \tag{2.39}$$

$$\nabla \times \boldsymbol{B}_0 = 0. \tag{2.40}$$

The full magnetic field $\boldsymbol{B}$ may be written as the sum of the intrinsic field and a deviation $\boldsymbol{B}_1$, i.e.,

$$\boldsymbol{B} = \boldsymbol{B}_0 + \boldsymbol{B}_1. \tag{2.41}$$

Nothing in the following analysis assumes that $\boldsymbol{B}_1$ is small in relation to $\boldsymbol{B}_0$. Thus this method can be used in Mars simulations when including a crustal magnetic field, even if the crustal field $\boldsymbol{B}_0$ is small is some region.

2.3 Applications

One major advantage of MHD models is that the lower amount of CPU work needed per time step compared to kinetic models enables one to use a higher spatial resolution. Although the MHD equations are often under scrutiny when applied to space plasmas, experience has proven that they are adequate in many situations where the spatial scale of interest is larger than the ion gyroradius and the ion inertial scales, and the temporal scale is longer than the ion gyroperiod (Raeder 2003). In assessing the validity of the MHD equations one must consider that they are conservation equations. Specifically, MHD describes the conservation of mass, momentum, energy, and magnetic flux.

The fluid model describes the plasma at any location with three parameters: density, velocity and temperature. The concept of temperature only makes sense when the plasma components are not far from local thermodynamic equilibrium. When the ion gyro-radius is large, ion thermal velocity distribution could be far from a Maxwellian distribution. Under such circumstances, the scalar pressure cannot be used; a full pressure tensor is needed to describe the pressure force that acts on the plasma.

As discussed before, the gyroradii near Titan of the heavy ion species (mass 16) are about 1.5 R_T, which is larger than Titan. Strictly speaking, fluid modeling is not applicable in such a case. However it is also important to note that the ion gyroradius is not a constant near the interaction region, and it decreases quite significantly in the area close to Titan due to the pile-up of the magnetic field and the decrease of the ion temperature as a result of mass loading and ion-neutral collision processes (Ma et al. 2007; Ledvina et al. 2000; Cravens et al. 1998). Figure 3 shows the variation of the gyroradii of heavy ions (mass 16) in the equatorial plane for the case of the Cassini T9 flyby. The plasma temperature from MHD simulation results is used to estimate the heavy ion temperature in the calculation of ion gyroradii. The inner boundary (725 km altitude, $\sim$1.28 R_T) of the model is also shown in the figure, with the grey and dark color showing the sunlit and night side, respectively.

The blue region (region A) shows the region where the gyroradii of heavy ions are at least an order of magnitude smaller than Titan's radius. In this region, $R_T > 10\ R_g$, thus the MHD assumptions are valid. Region A is not symmetric about the flow direction and it is also affected by the direction of the solar EUV. The altitude of this region ranges from 1500 km in the upstream side to about 3500 km, and peaks in the dayside. Both Cassini Ta and Tb flybys passed this region, with closest altitude less than 1200 km, and the MHD model results of Backes et al. (2005), Ma et al. (2006) and Neubauer et al. (2006) for the two flybys agreed with the observations quite well.

Region B (light blue) shows where the gyroradii of heavy ions are less than half of Titan's radius. In this region, ions and electrons are not tightly coupled and kinetic effects become

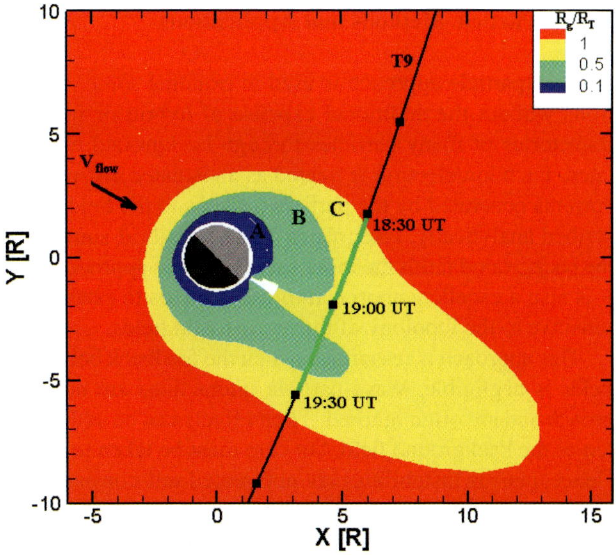

Fig. 3 Contour plot of (R_g/R_T), the ratio of the gyroradii of heavy ions (mass 16) and Titan radius in the equatorial (X–Y) plane. Regions A, B and C correspond to region with gyroradii less than 0.1, 0.5 and 1 respectively. The *green color along the trajectory* of T9 shows the main interaction region for this flyby. The inner boundary (725 km altitude, ~1.28 RT) of the model is also shown in the figure; with the *grey* and *dark color* showing the sunlit and night side, respectively

important. Hall MHD does a much better job at describing the system. Most of the interaction regions of T9 as indicated by the green color along the trajectory, are in this region or very close. This is the reason that Hall MHD simulation results show good agreement with the observations. The better match of Hall MHD simulations with the observations along the trajectory than the multi-species MHD simulations confirms that kinetic effects are important in this region.

Region C (with yellow color) and beyond (red colored area) are the regions with gyroradii are larger than 0.5 R_T. In this region, the kinetic effects become significantly important. However, most of the outer region is unperturbed with no pressure gradient force and the main interaction region is within the area, where the gyroradii is smaller than 1 R_T. Thus a fluid model can still give a reasonable first order estimation of the global interaction structure. In region B and C, some kinetic effects (such as Hall currents) could be significant and there might be noticeable velocity/temperature differences between the different ion species, which are neglected in the single fluid model. In this region hybrid/kinetic models are more appropriate, while multi-fluid models with anisotropic pressure taken into account should also do a fairly good job.

Also there are two white colored regions in the figure. Those regions are cut off because they are either below the ionospheric peak region or inside the current sheet of the tail. In those areas, the magnitude of the magnetic field is quite weak while both the ion and neutral densities are relatively high. Thus collisions are quite important in these regions and the fluid assumption is safe. One also needs to keep in mind that the boundaries of those regions are not fixed, but tightly related with upstream condition and to Titan's relative location in the Saturnian system. The hybrid simulations also show similar trends of the decreasing of ion gyroradii in the interaction region near Titan (R. Modolo, private communication).

In summary, MHD simulations are very powerful tools to understanding plasma physics in space. However, one also needs to remember the implicit assumptions made when using MHD simulations so that they can be used correctly and their limitations appreciated.

3 Test-Particle Methods and Applications

The test-particle approach is used to examine some of the kinetic aspects of ions and electrons without the additional expense of hybrid or fully electromagnetic simulations. The trajectories of many particles through background electric and magnetic fields are calculated. Each ion/electron is treated as an isolated test-particle. There is no feedback from the currents generated by the particle motions to the fields. These background fields can be described analytically, taken from MHD or hybrid simulations. The better the description of the background fields the more successful this approach is when applied to certain processes. It is also a excellent approach to use if one is to establish if a certain particle population is sensitive to the topology of the background fields.

This approach is reasonable when the feedback between the particles and the background fields is negligible. Wave-particle interactions are not treated self-consistently in this approach and are often ignored. Their effects can be added by including a perturbation field on top of the background field, but care must be taken to accurately describe the perturbations. Hence, plasma instabilities are not treated self-consistently in this approach. However if the particles are not sensitive to time variations in the background fields and the fields are not dependent on the set of particles of interest this approach is successful at simulating many kinetic effects. Additional processes can be added to this approach that would further add considerable overhead to more self-consistent simulation such as interactions with neutrals that generate energetic neutral atoms (ENAs). This can therefore be thought of as a "value added" approach to extend the usefulness of previous simulation results.

3.1 The Equations of Motion

The basic equations used in test-particle methods are just the equations of motion given in Sect. 1.1, rewritten here as:

$$m\frac{d\boldsymbol{v}}{dt} = q\left(\boldsymbol{E} + \frac{\boldsymbol{v} \times \boldsymbol{B}}{c}\right) \tag{3.1}$$

$$\frac{d\boldsymbol{x}}{dt} = \boldsymbol{v}. \tag{3.2}$$

Recall m is the mass of the ion, $\boldsymbol{v}$ is the particles velocity, q is the particles charge and $\boldsymbol{E}$ and $\boldsymbol{B}$ are the electric and magnetic fields the particle is moving through, $\boldsymbol{x}$ is the particles position, t is time and c is the speed of light. Other forces such as gravity could be added into (3.1). However, for most problems of interest the Lorentz force is dominant and the other forces are negligible. Additionally collisional interactions of the particles with neutrals can be included in this approach as a separate process after each time step. Here the focus is only on the ion motion through the background fields. The fields are usually assumed to be a function of position and not a function of time. When the fields are obtained from MHD or hybrid simulations they need to be interpolated from the simulation grid to the particle location.

3.2 Integration Schemes

The equations are solved typically for several million particles (for spatial coverage and representing the distribution function) often for a large number of time steps. Due to the shear numbers an efficient integration scheme is highly desirable. The scheme should also be accurate over the entire range of time steps. There are several schemes that can be used to integrate the equations of motion. A few of the more common ones are reviewed here. More details of these schemes and others can be found in several sources (cf. Hockney and Eastwood 1988; Birdsall and Langdon 1985, 2004; Lipatov 2002). When choosing a scheme it is important to consider its convergence, accuracy, stability and efficiency. By convergence the scheme should converge to the exact solution of the equation of motion in the limits that Δt and Δx tend to zero. It should also be time reversible. That is if the velocity is reversed and time is run backwards the particle should traverse that same path. Accuracy means the truncation errors associated with the derivatives. Stability is concerned with how the errors of the scheme change over time. If they grow in time the scheme is considered unstable. Efficiency is important because of the number of particles used and the time step requirements of the scheme. In general lower order schemes are easier to program, require less resources per time step and are more stable. However, they require much smaller time steps to achieve the same accuracy as higher order schemes.

It needs to be mentioned that when using the fields resulting from other simulations that are located on a grid, there is an inherent limitation on the choice of time step. The time step should be small enough, that a particle will not go across a grid zone in a single time step.

3.2.1 Euler's method

Euler's method (cf. MacNeice 1996) is also known as upwind differencing and is first order. Applying Euler's method to the equations of motion gives:

$$\frac{\boldsymbol{v}^{n+1} - \boldsymbol{v}^{n}}{\Delta t} = \frac{\boldsymbol{F}^{n}}{m} \tag{3.3}$$

$$\frac{\boldsymbol{x}^{n+1} - \boldsymbol{x}^{n}}{\Delta t} = \boldsymbol{v}^{n} \tag{3.4}$$

where the superscripts denote the time level of the solution. Here $\boldsymbol{F}$ denotes the net force acting on the particle, the Lorentz force. These equations are solved for the $n+1$ time level. Euler's scheme is first order and it reduces to the correct differential equations as the time step goes to zero. However, it is unconditionally unstable and is not time reversible. It is simple to implement but is generally not a good scheme to use because of its low order and its lack of stability.

3.2.2 Explicit Leap Frog

A common second order scheme used is the leap frog scheme (cf. Birdsall and Langdon 2004; Lipatov 2002; and MacNeice 1996). This scheme is second order accurate in time for a constant time step. The discreatized equations are:

$$\frac{\boldsymbol{v}^{n+1/2} - \boldsymbol{v}^{n-1/2}}{\Delta t} = \frac{\boldsymbol{F}^{n}}{m} \tag{3.5}$$

$$\frac{\boldsymbol{x}^{n+1} - \boldsymbol{x}^{n}}{\Delta t} = \boldsymbol{v}^{n+1/2}. \tag{3.6}$$

Note that the times that the position and velocity of the particle are known are offset by half of a time step. We can center the Lorentz force by averaging $\boldsymbol{v}_i^{n+1/2}$ and $\boldsymbol{v}_i^{n-1/2}$, hence (3.5) becomes:

$$\frac{\boldsymbol{v}^{n+1/2} - \boldsymbol{v}^{n-1/2}}{\Delta t} = \frac{q}{m}\left(\boldsymbol{E} + \frac{\boldsymbol{v}^{n+1/2} + \boldsymbol{v}^{n-1/2}}{2c} \times \boldsymbol{B}\right). \tag{3.7}$$

This equation can be solved for $\boldsymbol{v}^{n+1/2}$ by taking the dot and cross products with $\boldsymbol{B}$ and substituting back into (3.7). Dropping the terms of order larger than 2 gives:

$$\boldsymbol{v}^{n+1/2} = \boldsymbol{v}^{n-1/2}\left(1 - \frac{1}{2}\Omega^2\Delta t^2\right) + \frac{q\Delta t}{m}\left(\boldsymbol{E} + \frac{\boldsymbol{v}^{n-1/2}\times\boldsymbol{B}}{c}\right)$$
$$+ \frac{q^2\Delta t^2}{2m^2c}\boldsymbol{E}\times\boldsymbol{B} + \frac{q^2\Delta t^2}{2m^2c^2}\left(\boldsymbol{v}^{n-1/2}\cdot\boldsymbol{B}\right)\boldsymbol{B} \tag{3.8}$$

where $\Omega = q|\boldsymbol{B}^n|/mc$. Given $\boldsymbol{v}^{n-1/2}$ and $\boldsymbol{x}^n$, (3.8) can be solved for $\boldsymbol{v}^{n+1/2}$ which is then substituted into (3.6) to get $\boldsymbol{x}^{n+1}$. The explicit leap frog scheme is second order accurate and it is time reversible.

Detailed analysis of the stability and convergence of the leap frog scheme for full particle motion is very complicated because of its non-linear nature. Analysis of this scheme for a simple harmonic oscillator (think cyclotron motion) provides valuable insight (cf. Birdsall and Langdon 2004; Lipatov 2002; MacNeice 1996). The equation of motion for a harmonic oscillator is given by:

$$\frac{d^2x}{dt} = -\omega^2 t.$$

When the leap frog scheme is applied to this problem it can be shown that it is stable for $\omega\Delta t \le 2$ and has no amplitude errors and second order phase errors. So choosing Δt such that $\omega\Delta t = 0.3$ gives reasonable accuracy provided the integration is not run beyond about 100 time steps (MacNeice 1996). It was found that increasing the time step size increases the error as the cube of the step size.

3.2.3 Boris's Scheme

The Boris scheme (Boris 1970) operator splits the particle motion into a set of equations with a more simple structure. The electric and magnetic forces are completely separated. It is second order accurate and time centered, hence time reversible. It conserves energy very well, is easily generalized for relativistic particles and is widely used in particle-in-cell (pic) simulations of plasmas. The motion of the particles is split into steps with intermediate values of the velocity being found at the end of each step. The method starts out solving for the motion of the particle due to the electric field, then the motion due to the magnetic field and finally the motion due to the electric field.

$$\begin{aligned}
\boldsymbol{v}_1 &= \boldsymbol{v}^{n-1/2} + \frac{q\Delta t}{2m}\boldsymbol{E}^n \\
\boldsymbol{v}_2 &= \boldsymbol{v}_1 + \frac{q\Delta t}{2m}(\boldsymbol{v}_1\times\boldsymbol{B}^n) \\
\boldsymbol{v}_3 &= \boldsymbol{v}_1 + \frac{2\frac{q\Delta t}{2m}}{1+(\frac{q\Delta t}{2m}\boldsymbol{B}^n)^2}(\boldsymbol{v}_2\times\boldsymbol{B}^n) \\
\boldsymbol{v}^{n+1/2} &= \boldsymbol{v}_3 + \frac{q\Delta t}{2m}\boldsymbol{E}^n.
\end{aligned} \tag{3.9}$$

Again in this scheme one uses (3.4) to find the particles position once $\boldsymbol{v}^{n+1/2}$ is known. According to Lipatov (2002) this scheme gives velocities lying on a circle of radius $|\boldsymbol{v}|$ in velocity space and on a circle of radius R' in coordinate space. The finite time step causes the frequency to be higher than the correct frequency Ω and the radius R' to differ from the

Larmour radius $R = |\boldsymbol{v}|/\Omega$. The error in the rotation angle of the particle as a result of this scheme is less than 1% for a time step such that $\Omega \Delta t < 0.35$. The value of R' is given by:

$$R' = \frac{|\boldsymbol{v}|}{\Omega} \sec\left(\frac{\omega \Delta t}{2}\right),$$

where ω is the angular velocity of the particle. For a time step such that $\omega \Delta t < 0.35$, the error in R' is less than 1.5%.

3.2.4 Higher Order Schemes

There are several higher order schemes that can be used to solve the equations of motion (cf. Birdsall and Langdon 2004; Lipatov 2002). Multistep algorithms such as Runge-Kutta schemes or Adams-Bashford schemes are sometimes used. These have the advantage that they can be extended to much higher orders than the schemes discussed above. The difficulty with these schemes is that they often require velocity information at intermediate times within Δt. Since this information is only available at t^n the missing information is often interpolated from previous values. They are often not time reversible. The higher order nature of these schemes limits their regions of stability. For instance the 4th order Runge-Kutta scheme is not stable for the particle equations of motion. They are also computationally much more expensive than lower order schemes, requiring greater intermediate time levels per Δt and hence more floating point operations. More intermediate values are usually stored in these schemes than lower order schemes increasing the memory overhead needed.

3.2.5 Integration Summary

The choice of a particle integration scheme is a trade off between accuracy and efficiency. On the one hand there are high order schemes that allow the use of a larger time step. On the other, there are low order schemes with a smaller time step. High order schemes are hampered by 1) the need for values (velocities, positions, etc...) at several intermediate time levels, 2) a more restrictive stability limits on the time step, and 3) though they can take larger time steps compared to low order schemes the time step is often limited by the natural frequency of the particles and the grid size the fields may be represented on. Low order schemes (1st order) are 1) generally not accurate enough; 2) have even greater stability issues. The best compromise between accuracy, stability and efficiency is considered to be second order schemes. Of the second order schemes the explicit leap frog and Boris schemes are very popular. Of these two many hybrid and full electromagnetic simulations prefer the Boris scheme because of its accuracy and energy conservation properties over many tens of thousands of time steps.

A recent paper by Mackay et al. (2006) claims that symplectic methods (Methods based on this approach conserve phase space density of Hamiltonian systems, ideally preserving exact constants of the motion.) and interpolating the magnetic vector potential to solve for the particle motion is the only way to accurately integrate test-particle motion in fields from a MHD solution. Future work is needed to determine if their scheme gives better results than the schemes outlined here.

3.3 Injecting and Loading Particles

3.3.1 Injecting Particles

One of the most useful applications of test-particle methods is examining the results of an ambient ion population interacting with a body. In order to extract the maximum benefit from this application it is necessary to inject particles into the simulation with a given velocity distribution function. Several velocity distributions are possible in space plasmas. One of the most common is the drifting Maxwellian. One method of injecting a drifting Maxwellian is discussed in this section. Examples of injecting other ideal distribution functions can be found in Lipatov (2002). A method that discusses the loading of experimental velocity distributions can be found in Schriver et al. (2006).

The Maxwellian distribution in $\boldsymbol{v}$ has the form of $\exp[-(\boldsymbol{v} - \boldsymbol{v}_{drift})^2/2v_t^2]$ where $\boldsymbol{v}_{drift}$ is the bulk flow speed of the plasma and v_t is the thermal speed. What is needed is a way to create a set of particles with the desired Maxwellian velocity distribution. The velocity distribution is mapped to a set of numbers between 0 and 1 such that each number corresponds to unique velocity.

The first thing to realize is that the Maxwellian distribution function can be split into a distribution along the direction of the bulk flow and a distribution that is perpendicular to the bulk flow $f(\boldsymbol{v}) = \exp[-(\boldsymbol{v}_x - \boldsymbol{v}_{drift})^2/2v_t^2]\exp[-\boldsymbol{v}_\perp^2/2v_t^2]$, were v_x is the velocity component in the direction of the bulk flow. The perpendicular part of the injection is outlined first.

The cumulative distribution function for the perpendicular speed ($v_\perp = |\boldsymbol{v}_\perp|$) is:

$$R(0 \to 1) = F(v_\perp) = \int_0^v \exp\left(\frac{-(v_\perp)^2}{2v_t^2}\right) \boldsymbol{dv}_\perp \Big/ \int_0^\infty \exp\left(\frac{-(v_\perp)^2}{2v_t^2}\right) \boldsymbol{dv}_\perp \tag{3.10}$$

The idea is to generate a random number R between 0 and 1 and then invert the distribution function to find $v_\perp$. This is a two-dimensional isotropic thermal distribution involving v_y, v_z with a speed $v_\perp = (v_y^2 + v_z^2)$ and the angle between v_y and v_z, $\theta = \arctan(v_y/v_z)$; $\boldsymbol{dv}_\perp$ is $2\pi v\, dv_\perp$. The integrals can be done explicitly. The inversion for the speed v in terms of R gives:

$$v_s = v_t\sqrt{-2\ln R}. \tag{3.11}$$

Another set of uniform random numbers, R_θ is chosen over the range of 0 to 2π for the angle θ. With v_s and θ one has $v_y = v_s \sin(\theta)$ and $v_z = v_s \cos(\theta)$.

The cumulative distribution function for the speed v along the drift direction is:

$$R(0 \to 1) = F(v) = \int_0^v \exp\left(\frac{-(v_x - v_{drift})^2}{2v_t^2}\right) \boldsymbol{dv} \Big/ \int_0^\infty \exp\left(\frac{-(v_x - v_{drift})^2}{2v_t^2}\right) \boldsymbol{dv}. \tag{3.12}$$

A direct inversion of (3.12) along the drift direction is not straight forward. A simple approach is to create a look up table for v. Several values of v are selected and (3.12) is solved numerically for the probability. Birdsall and Langdon (2004) point out that most of the particles have velocities in the range out to $3v_t$ (99% in $2v_t$) so there is seldom a need to use velocities beyond 3–4v_t to generate the table. When using the table to find v_x a random number is generated representing the probability and then the corresponding v_x is interpolated from the table.

It is tempting to just find v_x using the same procedure that was done for $v_\perp$ and then just add the drift velocity. This would not capture the full velocity range of the distribution function. The larger the drift speed the greater the misrepresentation.

Each injected particle can be weighted so that it represents a much larger number of particles. These representative macro-particles can then be used to calculate ion fluxes for direct comparisons with observations or they can be used to calculate the global distribution and energy deposition of ions into an exosphere. There is no way even with today's high performance computers that the real number of ions/electrons can be simulated. This is the reason to weight the particles. Computationally it is much more efficient to calculate the trajectory of a single macroparticle that represents 10^{15} ions/electrons, than it is to calculate the trajectories of 10^{15} ions or electrons. It is important that enough macroparticles are used to represent the variability of the physical particle distribution.

Before this can be done it is necessary to calculate the rate that particles with the given velocity distribution should enter the simulation region. The number of particles that cross a plane per unit time is just the area of the plane times the particle flux.

$$\frac{\Delta N}{\Delta t} = (area)\,(flux) = (area) \int \boldsymbol{v} f\,(\boldsymbol{x}, \boldsymbol{v}, t)\, d^3\boldsymbol{v}. \tag{3.13}$$

Using the assumed velocity distribution the number of ions/electrons that would enter each plane in the simulation domain per unit time can then be calculated. It is then just a matter of deciding how many macroparticles one is going to use. The weight of each macroparticle is just the number of ions/electrons that would cross that plane of the simulation domain divided by the number of macroparticles to be used.

3.3.2 Loading Particles

Using test-particles to study the pickup ion process near a planet or moon is a natural application. Loading ions into test-particle simulations is straight forward. The particle is just added inside the computational domain. The real trick is to weight the newly created ions properly. A simple approach is to surround the planet or moon with a spherical grid. The newly created macro-particles are loaded into the simulation using this grid. Each macroparticle carries its own unique weight. This weight represents the number density of each macroparticle.

The total number of ions created per unit time in a given cell is found using the background neutral densities and the relevant physical processes such as ionization and ion-neutral chemistry. Doing this on a cell by cell basis allows local effects such as photoionization to be accounted for. Once the total number of ions per time in each cell is know, it is then just a matter of deciding how many macroparticles to use per cell and weighting them accordingly. Of course the larger the number of cells and macroparticles used the better the representation of the pickup process. Each macroparticle is then loaded at a random location in each cell and the particle integration can begin.

It is worth noting that a lot of research was performed in the 1960's, 1970's and early 1980's within the fusion community to address how to inject and load ion/electrons to create a "quiet" start for particle codes. Further, this was done for a variety of distribution function. The research of the day was using particle codes fully electromagnetic, Darwin and even hybrid, to study the stability of waves to differing distribution functions. References to some of this research can be found in Birdsall and Langdon (1985, 2004). Many papers on the topic were published in Physics of Fluids which is where most of the plasma fusion papers were published.

3.4 Applications

Test-particle methods have been used to study aspects of several plasma interactions. Examples of their use can be found for comets (cf. Cravens 1986; Kimmel et al. 1987; Luhmann et al. 1988; McKenzie et al. 1994), Mars (cf. Cipriani et al. 2007; Gunell et al. 2006; Cravens et al. 2002; Kallio et al. 1997; Kallio and Koskinnen 1999), Venus (cf. Luhmann

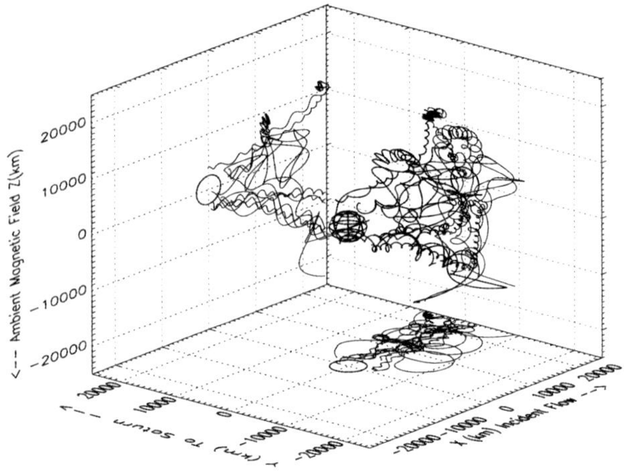

Fig. 4 Ten sample trajectories of 10 keV protons moving in Titan's induced magnetosphere

et al. 2006; Lammer et al. 2006), Pluto (cf. Kecskemety and Cravens 1993) and Titan (cf. Tseng et al. 2008; Ledvina et al. 2005, 2004b, 2000; Luhmann 1996). They have also been used by Brecht et al. (2001) to examine how the Jovian radiation belts were altered by comet Shoemaker-Levy 9.

The application of test-particle methods to aspects of these interactions is vast. Test-particles have been used to study the non-linear nature of the ion trajectories. They have been used to study ion distribution functions, examine ion deposition into planetary atmospheres and ion-neutral interactions in those atmospheres. They have even been used to simulate instrument observations and to explain those observations.

Figure 4 shows 10 sample trajectories of 10 keV protons moving in the tail region behind Titan. The motion of the ions is very complex. However, it shows that 10 keV protons are sensitive to the topology of Titan's induced magnetosphere, even though the nominal gyroradii of these ions is larger than the size of Titan. This figure illustrates the potential of test-particle methods to understand in-situ plasma observations.

Test-particles have been used extensively to understand the distributions of pickup ions. The process is usually forward modeled, meaning that many pickup ions are created and sampled by a instrument in the simulation. A more efficient approach would be to backward model the instrument response. That is the observed ion distributions may be placed into a simulation at the location of the observations. The ion trajectories could then be followed backwards in time to their source (this is why a time reversible method is important).

Careful applications of test-particle methods can be very successful at describing, explaining and predicting many aspects of flowing plasmas and electrons. Their results have not been rigorously tested against more self-consistent hybrid simulations. However, there is one test case where they compare very favorably. The ion flux into Titan's exobase for Voyager 1 plasma conditions was calculated by Ledvina et al. (2005) using test-particle/Monte-Carlo methods. They found that the incident ion flux was dependent on the ambient ion distribution function. If the heavy ambient ion species had a Maxwellian distribution then 1.7×10^{24} ions/s entered Titan's atmosphere. Recent self-consistent hybrid simulations by Sillanpää et al. (2007) calculate a flux of 1.3×10^{24} ions/s entering Titan's atmosphere using a ambient Maxwellian distribution. These results are in very good agreement. Hence, the test-particle/Monte-Carlo approach is reasonable in this case.

Test-particle methods have their limitations and are dependent on the accuracy of the background fields. When applied carefully they can provide a wealth of information about a system. In some cases they can reproduce many of the same features found in hybrid simulations with a savings in computational expense.

4 Quasi-neutral Hybrid Model

Historically, quasi-neutral hybrid simulations have been used to study various objects and plasma phenomena, especially kinetic effects that take place near the bow shock (see, for example Winske and Omidi 1996, and references therein). The increase of available computer power has made it possible to apply global two dimensional (2D) and three dimensional (3D) hybrid simulations to study how flowing plasma interacts with various solar system objects, such as: Mercury (Kallio and Janhunen 2002, 2003; Trávnìček et al. 2007), Venus (Brecht and Ferrante 1991; Shimazu 1999; Terada et al. 2002; Kallio et al. 2006), the Moon (Kallio 2005; Trávníček et al. 2005), Mars (Brecht and Ferrante 1991; Kallio and Janhunen 2001; Bößwetter et al. 2004; Modolo et al. 2005; Brecht and Ledvina 2006), Saturn's moon Titan (Brecht et al. 2000; Kallio et al. 2004; Ledvina et al. 2004a, 2004b; Simon et al. 2006a; Modolo et al. 2007), asteroids (Omidi et al. 2002; Simon et al. 2006b) and comets (Bagdonat and Motschmann 2002).

In a quasi-neutral hybrid (QNH) model (1) positively charged particles are modeled as ions, (2) electrons form a charge neutralizing massless (typically) fluid, and (3) the macroscopic plasma parameters determine the evolution of the magnetic field. Thus hybrid simulations self-consistently solve for the ion motion and the fields. They have all of the kinetic processes needed to self-consistently treat shocks as real phenomena and not just as a proxy. Since the electrons are treated as a fluid, electron kinetic effects are absent. The goal of this section is to briefly describe basic hybrid assumptions and to point out some issues about the frequently used hybrid algorithms.

4.1 Basic equations

The equations solved in the hybrid scheme are given in Sect. 1.1.1 (1.7)–(1.11), there are rewritten here for the readers convenience. The following ion momentum and position equations for each ion species:

$$\frac{d\boldsymbol{v}_i}{dt} = \frac{q_i}{m_i}\left(\boldsymbol{E} + \frac{\boldsymbol{v}_i \times \boldsymbol{B}}{c} - \eta \boldsymbol{J}\right) \tag{4.1}$$

$$\frac{d\boldsymbol{x}_i}{dt} = \boldsymbol{v}_i \tag{4.2}$$

where $\boldsymbol{x}_i$, $\boldsymbol{v}_i$, m_i and q_i are the position, velocity, mass and charge of each ion, $\boldsymbol{J}$ is the total current density, η is resistivity. The total current density is the sum of the ion and electron current densities, $\boldsymbol{J} = \boldsymbol{J}_i + \boldsymbol{J}_e$. The electron momentum equation is:

$$\boldsymbol{E} = \frac{1}{4\pi n_e e}(\nabla \times \boldsymbol{B}) \times \boldsymbol{B} - \frac{1}{n_i e c}\boldsymbol{J}_i \times \boldsymbol{B} - \frac{1}{n_e e}\nabla(n_e T_e) + \eta \boldsymbol{J} \tag{4.3}$$

where $\boldsymbol{E}$ and $\boldsymbol{B}$ are the electric and magnetic fields, n_e is the electron density, e is the electron charge, c is the speed of light, and T_e is the electron temperature. The $\nabla(n_e T_e)$ is often recast as the gradient of the electron pressure or ∇p_e. The total current density is found by Ampere's law:

$$\nabla \times \boldsymbol{B} = \frac{4\pi}{c}\boldsymbol{J}. \tag{4.4}$$

Faraday's law is also used to advance $\boldsymbol{B}$ in time:

$$\nabla \times \boldsymbol{E} + \frac{1}{c}\frac{\partial \boldsymbol{B}}{\partial t} = 0. \tag{4.5}$$

The electric field is found directly from (4.3) so there is no need for an equation to solve for the time advance of $\boldsymbol{E}$. The electric field contains contributions from the electron pressure gradient, resistive effects and the Hall currents.

The scheme correctly simulates electromagnetic modes well below the electron cyclotron frequency, $\omega \ll \omega_{\text{ce}}$. The time step is determined by the ion cyclotron frequency. This comes at the price of the loss of electron particle effects and charge separation. Some small scale electrostatic effects can be included through the resistivity terms. The resistivity terms can also be used to stabilize the numerical scheme used to solve the equations by adding it as a small amount of artificial resistivity.

Typical hybrid simulations often ignore the ∇p_e term in (4.3) or they treat the electron temperature T_e as a fixed quantity. There are situations however, were the ∇p_e term is important. For example the atmospheric loss rates from Mars were found to be sensitive to this term by Brecht and Ledvina (2006). In these situations it is desirable to also evolve the electron temperature as done by Brecht and Ledvina (2006) using:

$$\frac{\partial T_e}{\partial t} + \boldsymbol{u}_e \cdot \nabla T_e + \frac{3}{2} T_e \nabla \cdot \boldsymbol{u}_e = \frac{2}{3n_e}\eta J^2. \tag{4.6}$$

Here $\boldsymbol{u}_e$ is the electron velocity. There is no thermal conduction term in this version of the equation but one can be included if needed. The electron velocity is found using (4.4) to get the total current density and then subtracting off the ion current density ($\boldsymbol{J}_i = \Sigma_i q_i n_i \boldsymbol{v}_i$) to get the electron current density. The electron current density is then divided by the electron charge and the electron density to get the fluid velocity for (4.6). This equation for T_e has been extensively tested and found to work very well through and behind collisionless shock regions and compares well with data from planets such as Uranus and Mars (S. Brecht, private communication). The next section discusses some of the numerical schemes used in the hybrid approach.

4.2 Numerical Implementation

In many respects the numerical implementation of the hybrid approach is simple and straightforward, even more so than the implementation of the MHD approach. There are several possible numerical implementations of the hybrid approach, two commonly used to study global plasma interactions with solar system bodies are described here. Other hybrid methods can be found in the review by Winske et al. (2003).

Each approach has some common characteristics. The particles are Lagrangian, they are free to move anywhere in the simulation domain. Other quantities are represented on a computational grid, making the field solving part of the code Eulerian. Thus, the hybrid code combines a Lagrangian and Eulerian approach to addressing the kinetic plasma interactions. The magnetic field is located on the grid (for example the cell centers) such that the curl of E and the curl of B are performed in a centered fashion. Thus Faraday's law (4.5) can be used to advance $\boldsymbol{B}$ in time while maintaining $\nabla \cdot \boldsymbol{B} = 0$ (Yee 1966). The electric fields, the plasma, electron and current densities and the electron temperature are all staggered from the locations of B (such as at the vertices of the cell). As in other particle-in-cell (PIC) codes, the fields are interpolated to the particle positions, to obtain the particle accelerations,

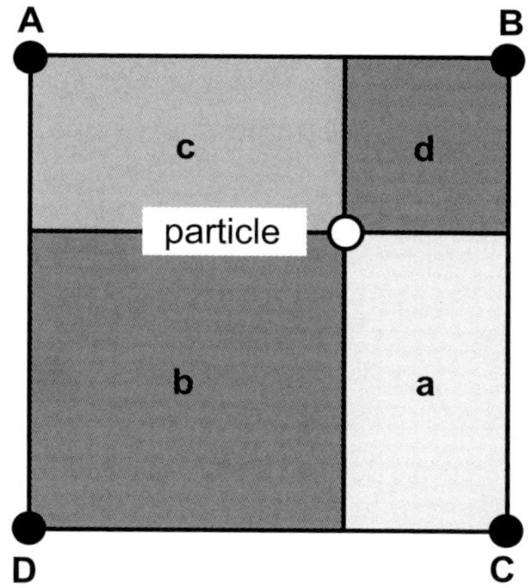

Fig. 5 Charge assignment for area weighting in 2-dimensions. Areas are assigned to grid points; i.e. $(area\ a)/(total\ area) \times (particle\ charge)$ *to grid point* A, etc.

after the particles are moved the densities and currents are redeposited back on the grid. Regardless of the numerical scheme used in the hybrid approach the particle push is critical. Common particle pushers used in hybrid simulations include explicit leap-frog, the method of Buneman (1967), and the method of Boris (cf. Birdsall and Langdon 1985, 2004). The explicit leap-frog and the method of Boris were discussed in Sect. 3.2.

There are many possible ways one could weight the particles to the grid and the fields to the particles. High order splines have been employed. These however are very computationally expensive. This expense really adds up when using several millions of particles in a typical simulation. Weighting the particles to the nearest grid point (zero-order) is quick and cheap. However, it can be shown to give very poor results. Typically the best balance is linear weighting (first-order) or area weighting (2D, volume weighting in 3D) due to its geometric interpretation (see Fig. 5). The particle location divides the cells area into sub-areas. The ratio of the sub-area to cell area is then used to weight the particle to the surrounding grid points.

4.2.1 The Predictor-Corrector Scheme

Historically one of the first numerical implementations of the hybrid approach has been a predictor-corrector. The basic idea is to 1) make a prediction for the fields at time $n+1$, 2) advance the particles in the predicted fields in order to compute the ion source terms at time $n+3/2$, 3) use the currents and densities to compute the fields at time $n+3/2$, 4) use the average of the electric field at $n+1/2$ and the predicted field at $n+3/2$ to get electric field at $n+1$. The predictor-corrector procedure outlined in Harned (1982) is as follows. The quantities $\boldsymbol{J}_i^{n+1/2}$, $\boldsymbol{v}_i^{n+1/2}$, $n_i^{n+1/2}$, $\boldsymbol{B}^n$ and $\boldsymbol{E}^n$ are known; the magnetic field is advanced to $n+1/2$ by:

$$\boldsymbol{B}^{n+1/2} = \boldsymbol{B}^n - (c\Delta t/2)\nabla \times \boldsymbol{E}^n. \tag{4.7}$$

a prediction is made for $\boldsymbol{E}$ and $\boldsymbol{B}$ at time $n+1$ by

$$\boldsymbol{E}_{pred}^{n+1} = -\boldsymbol{E}^n + 2\boldsymbol{E}(\boldsymbol{J}_i, n_i, \boldsymbol{B}, p_e)^{n+1/2}, \tag{4.8}$$

$$\boldsymbol{B}_{pred}^{n+1} = \boldsymbol{B}^{n+1/2} - (c\Delta t/2)\nabla \times \boldsymbol{E}_{pred}^{n+1}. \tag{4.9}$$

The predicted fields are now used to do a predicted particle move. The predicted particles are deposited on the grid to get $n^{n+3/2}_{i,pred}$ and $\boldsymbol{J}^{n+3/2}_{i,pred}$, then $\boldsymbol{B}^{n+3/2}_{pred}$ is predicted by:

$$\boldsymbol{B}^{n+3/2}_{pred} = \boldsymbol{B}^{n+1}_{pred} - (c\Delta t/2)\nabla \times \boldsymbol{E}^{n+1}_{pred}. \tag{4.10}$$

The new electric and magnetic fields are obtained from:

$$\boldsymbol{E}^{n+1} = \frac{1}{2}\boldsymbol{E}(\boldsymbol{J}_i, n_i, \boldsymbol{B}, p_e)^{n+1/2} + \frac{1}{2}\boldsymbol{E}(\boldsymbol{J}_i, n_i, \boldsymbol{B}, p_e)^{n+3/2}_{pred}, \tag{4.11}$$

$$\boldsymbol{B}^{n+1} = \boldsymbol{B}^{n+1/2} - (c\Delta t/2)\nabla \times \boldsymbol{E}^{n+1}. \tag{4.12}$$

The corrected particle positions can now be advanced to time $n + 3/2$ using the new fields. The algorithm is second order accurate in space and time. The corrector iteration prevents the appearance of large amplitude odd–even oscillations. The method gives very good energy conservation and is rather robust. However there can be a significant amount of short wavelength whistler noise generated by this technique (Winske et al. 2003), which can be reduced by filtering the electric fields and the densities. This method may be thought of as slow by some because the particles are pushed twice. However, the energy conservation properties of this approach means that often far fewer particles are needed in the simulation to get the desired results when compared to other methods.

4.2.2 The Current Advance Method and Cyclic Leapfrog Scheme

Another numerical method that has gained in popularity recently is the Current Advance Method and Cyclic Leapfrog (CAM-CL) method of Matthews (1994). The CAM-CL is distinguished by four main features: 1) Only a single computational pass through the particles is needed per time step. This is achieved without the need to extrapolate the electric field in time. The particles are advanced by a leapfrog procedure which requires the electric field to be a half time step ahead of the particle velocities. 2) CAM advances the ion current density a half time step to avoid the pre-push of the velocities. 3) A free streaming ion current density is collected (velocities are collected at positions a half time step ahead). 4) CL is a leapfrog scheme for advancing the magnetic field. It is an adaptation of the modified midpoint method described by Press et al. (1993).

The algorithm of the CAM-CL scheme as implemented in Bagdonat and Motschmann (2002) is briefly described here. A detailed description of the procedure along with the results of several numerical tests can be found in Matthews (1994). Given a magnetic field $\boldsymbol{B}^n$ and a set of particles with positions $\boldsymbol{x}_i$ at time step n, with velocities $\boldsymbol{v}_i^{n+1/2}$ at time step $n + 1/2$ the CAM-CL cycle is as follows:

1. Deposit the charge densities ρ_c^n at each grid point from the particles. Calculate the ion currents $\boldsymbol{J}^{n+1/2}_{ion}$ from the particle velocities $\boldsymbol{v}_i^{n+1/2}$.
2. Push the particles to their new positions $\boldsymbol{x}_i^{n+1}$ using (4.1) and (4.2) via a leapfrog scheme with $\boldsymbol{v}_i^{n+1/2}$.
3. Deposit the charge densities $\rho_c^{n+1/2}$ from the new particle positions and calculate the charge densities at $n + 1/2$, $\rho_c^{n+1/2} = 1/2(\rho_c^n + \rho_c^{n+1/2})$.
4. Update the magnetic field to $\boldsymbol{B}^{n+1}$ using $\boldsymbol{J}^{n+1/2}_{ion}$ and $\rho_c^{n+1/2}$ using Faraday's law (4.5) together with the electric field given by (4.3). This can be done with smaller time steps using a cyclic leapfrog method described in Matthews (1994).
5. Extrapolate the ion current densities to $\boldsymbol{J}^{n+1}_{ion}$ using the current advance method (CAM) described in Matthews (1994).

6. Update the electric field to $\boldsymbol{E}^{n+1}$ using (4.3). Update the velocities to $\boldsymbol{v}_i^{n+3/2}$ by interpolating the fields to the particles and using (4.1) and (4.2).

There are other implementations of the hybrid approach that will not be described here. The interested reader can find a description of these approaches in Winske et al. (2003).

4.2.3 Comparing the Schemes

Comparisons of the predictor-corrector and the CAM-CL schemes can be found in Krauss-Varban (2005) and Karimabadi et al. (2004). When both schemes were run at the same time step they compared well on some tests. They found that the CAM-CL method is stable at longer time steps, even when the CFL condition on the fastest waves was marginally violated. Further tests also gave temperatures that were slightly too low. This implies that the CAM-CL method is more diffusive than the predictor-corrector. When anti-parallel magnetic field configurations were embedded in streaming plasmas the CAM-CL method was found to be far more diffusive than the predictor-corrector (Krauss-Varban 2005). The predictor-corrector uses more CPU time due to its second particle push. However, it requires far fewer particles to conserve energy to the same degree as the CAM-CL scheme. According to Matthews (1994), CAM-CL shows a 4% energy gain using 32 particles per cell after $t = 100\omega_{ci}$ for a two-dimensional simulation of a quiet plasma. Using fewer particles resulted in a larger energy gain. By contrast Brecht and Ledvina (2006) reported that the three-dimensional predictor-corrector version of the scheme developed by Harned (1982) and discussed above, conserves energy to within 1% after $t = 112\omega_{ci}$ using only 4 particles per cell. Since far more particles are needed in the CAM-CL scheme to achieve the same degree of energy conservation as the predictor-corrector scheme the actual difference in CPU time needed by each scheme may not be that great. Karimabadi et al. (2004) concludes that the CAM-CL method is not very suitable to applications in moving plasma. Both Karimabadi et al. (2004) and Krauss-Varban (2005) concluded that under circumstances where high accuracy and the best conservation properties are required, the predictor-corrector method greatly outshines the CAM-CL method.

4.3 Applications and Limitations

There are several things that should be considered when applying hybrid simulations to plasma interactions with planets and satellites with atmospheres. Some of the issues are discussed in this section.

One crucial issue in a hybrid model, as is the case in a every 3D modeling approach, is the inner obstacle boundary conditions. In some of the current 3D models the whole modeled region is in the non-collisional regime and no collision terms are included (e.g. Kallio and Janhunen 2001; Modolo et al. 2005). The ionosphere is represented by a conducting sphere which absorbs the ions that hit the surface. Ionospheric outflow is represented by injecting ions into the simulation at the surface of the sphere. This approach causes the simulation results to depend strongly on the choice of the injected ionospheric ion outflow. Other simulations represent the object as a large ionized gaseous body (Shimazu 1999) or being formed with heavy non-moving ions (Bößwetter et al. 2004). A couple of simulations have tried to model the ionosphere with ionospheric chemistry (cf. Terada et al. 2002; Brecht and Ledvina 2006). This approach has several challenges that must be faced, but is the most realistic. Ion-neutral collisions are easily treated in a MHD simulation. They are much more difficult to treat in hybrid simulations. Some hybrid simulations have simply modeled them as a neutral drag force (Bößwetter et al. 2004). However, a drag force fails to capture the effects that ion-neutral collisions have on the fields, it also does not scatter the ion trajectories. Brecht and Ledvina (2006) have included ion-neutral collisional effects by adding the Pederson and Hall conductivity tensors to the calculation of the electric field. Their approach does add the effects on the fields and some ion scattering but only in a averaged sense. Monte-Carlo collision models have been used in 1D and 2D hybrid simulations (cf. Puhl-Quinn and Cravens 1995; Terada et al. 2002). This approach is perhaps the most complete treatment of ion-neutral collisions. However, it quickly becomes computationally untenable in three-dimensions even for large parallel computers. Adding a accurate model of ion-neutral collisions that does not overwhelm 3D hybrid simulations remains an outstanding issue.

The ion motion equations (4.1)–(4.2) describe how an ion is moving under the effect of different forces. However, they do not show how ions are formed and that an ion can be destroyed, in other word, that there are ion loss and source processes. In hybrid simulations ion loss and source processes can be modeled in a very elementary manner by adding a new particle (ion) into the simulation, changing the weight of a particle or removing a particle. The main ion production processes included in hybrid simulations are typically (1) photoionization, (2) electron impact ionization (3) charge exchange process and (4) chemical reactions. The same processes which act as a source for one ion species can act as a loss process for another. For example charge exchange can act as a loss process for a solar wind proton but a source process for a ionospheric ion species. Additional ion loss processes include radiative and/or dissociative recombination.

In the hybrid model ions are treated in a self-consistent manner with respect to the fields. The simulations self-consistently give results for both the fields and the ions (see Fig. 6). Hence, various interaction processes between the ions and the fields are naturally included. The approach automatically includes both Hall and ion gyroradius effects and. can be easily expanded to multispecies plasmas. The ions are not frozen to the magnetic field as they are in ideal MHD simulations, they can also counter-stream. One critical issue in multifluid MHD is what to do if one of the fluids shocks in the simulation. What should happen

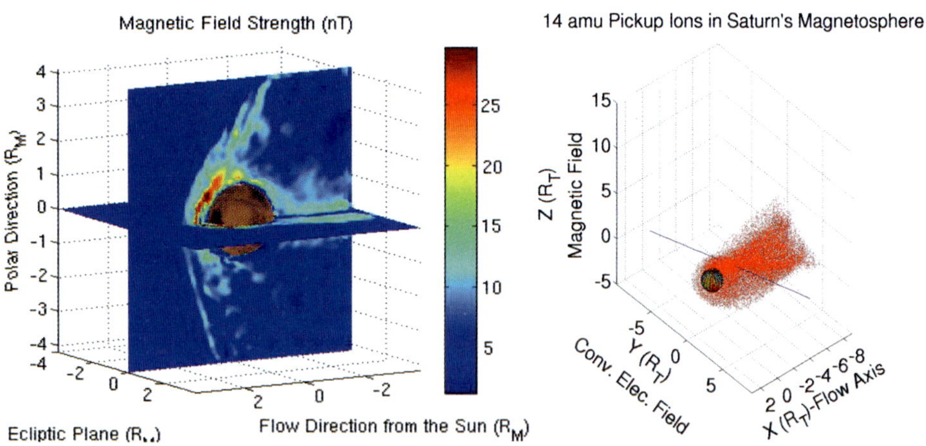

Fig. 6 Hybrid simulations are capable of providing both the fields and full particle information. The *left panel* is the magnetic field strength from a hybrid simulation of Mars. The incident magnetic field lies in the ecliptic plane. The asymmetry in the magnetic topology about the ecliptic plane is a result of finite ion gyroradii effects. The *right panel* shows the 14 amu pickup ion positions from a hybrid simulation of Titan interacting with Saturn's magnetosphere at the time of the Voyager encounter (from Ledvina et al. 2004b)

to the other fluids? This is not an issue with hybrid simulations. If one species shocks in a hybrid simulation it will generate waves. These waves will interact with ions from the others species. If the cyclotron frequencies of the other species matches the frequency of the waves generated by the shocked species then they may also shock. This wave-particle feedback is naturally present in hybrid simulations but must be treated in a ad-hoc fashion in MHD no matter the formulations.

No presumptions about correlations between ions or between different ion species are implicit to hybrid simulations. The ion distribution can be fully three dimensional, for example, a ring distribution or a shell distribution. It does not have to be Maxwellian. Thus the effective ion pressure is a vector and not a isotropic scalar as in MHD simulations. Diffusion of mass, momentum and energy from one place or space to another place are also automatically taken into account, because ions can move from place to place carrying momentum and energy with them. Different ion species can also have different bulk velocities and temperatures. These properties make the hybrid model approach a powerful tool to analyze situations when the ion velocity distribution function and/or finite gyroradius plays an important role or when plasma includes several ions species with different mass per charge ratio.

However, there are physical and practical issues which cause limitations of the hybrid approach. The approach does not include electron kinetics. The hybrid model can handle spatial scales down to about an order of magnitude larger than the electron skin depth and time scales on the order of the ion gyrofrequency $(q_i|\boldsymbol{B}|/m_i c)$. Below this scale size gradients in the electron pressure are not represented because electron diffusion and electron inertial would modify the gradients and the electric fields. Such modifications can't be represented with the current set of hybrid assumptions. Thus electric fields calculated on grids near the electron skin depth will be too large and the results unphysical. For simulations where the ions are protons (such as the solar wind) the resulting scale limit is $1/4$ the ion inertial length (c/ω_{pi}). If electron kinetic scales are of a crucial importance in the physical process, studies using a fully kinetic code for both ions and electrons are required.

Setting up a hybrid simulation is a trade off between the physics of interest, the electron skin depth, the number of particles and the size of the machine. Cell sizes for global simulations have ranged from a few hundred of kilometers at Mars and Titan to 10–20 km at Europa and Enceladus. These cell sizes were chosen as the best compromise between the computer resources available and the physics of interest. It is true that less expensive MHD simulations have allowed for higher resolution to be used at these bodies. However, modern parallel computers will enable hybrid simulations to close the resolution gap. Hybrid simulations of other phenomena have been performed with 100's of millions up to billions of particles with cell numbers reaching 100's of millions (S. Brecht private communication). The same simulation techniques can be applied to simulations of non-magnetized bodies. The real limit on the resolution is not necessarily the size of the machine but the missing electron physics in the hybrid framework.

Hybrid simulations are more computationally expensive than MHD, including interaction processes between ion, electrons and neutrals would make that even more so. For this reason hybrid simulations have been traditionally limited to the collisionless regions. The next frontier is to extend hybrid simulations into the collisional regions of the ionospheres. It has been shown that hybrid simulations that contain ion-ion, ion-neutral, ion-electron and electron-neutral collisions and ion chemistry can reproduce a planetary ionosphere self-consistently (Terada et al. 2002; Brecht and Ledvina 2006). Many challenges remain in extending 3D hybrid simulations into this region. However, it is necessary task in order properly address many features of plasma interactions with ionospheres.

5 Verification, Validation and Pre-simulation

Each of the simulation approaches mentioned above are complex. There are many points in the process where mistakes can be made. Verification, validation and quality management are critical to a simulation's success. It is hard to determine if a simulation result is right or wrong. A referee cannot detect many things that can be wrong with a simulation paper. There can be defects hidden in the code, programming errors, applying a model incorrectly; the spatial or temporal resolutions might be too course. A simulation is only a model of a physical system. Does the model accurately reflect the physical system of interest? Verification and validation examines the errors in the code and the simulation results. The more complex the code, the more models included in it, the harder it is to verify and validate. The accuracy of the code will depend on the validity of each of its the component models, the completeness of the set of all models, the solution method, the interaction between the models, the quality of the input data, the grid and temporal resolution and the users ability to set up the problem and interpret the results. Because of the complexity one must first verify and validate each component and then do the same for progressively larger sets of components until the entire integrated code has been verified and validated for the problem of interest.

5.1 Verification

> Verification tests that a code or simulation accurately represents the conceptual model or intended design of the code—that we're "solving the equations right." The process involves identification and quantification of error; the main strategy for finite-volume, finite-difference, and finite-element methods is a systematic study of the effect of mesh and time-step refinement on simulation accuracy. Verification requires comparing the results of simulations to a correct answer of the model's equations, which might be an analytic solution or a "highly accurate" benchmarked solution (Calder et al. 2004).

Common verification techniques include:

1. Comparing code results to a test problem with an exact answer. Does the code reproduce the correct dispersion relation or plasma instability of the analytical solution?
2. Comparing calculated with expected results for a problem manufactured to test the code. For example can the code propagate a given density distribution, magnetic field or electric field around the grid without smearing them out? The field solver portion of the code is different than the plasma portion of the code although they are nonlinearly coupled in most MHD and hybrid codes. Is the resulting particle trajectory smooth with the correct gyro-radius? Is the particle trajectory time-reversible? The results from Euler's and the explicit leap-frog methods suggest limits to the accuracy of the particle trajectory.
3. Monitoring conserved quantities and parameters, preservation of symmetry properties and other predictable outcomes. Does the code conserve mass, momentum and energy? Is the flow symmetric about an object without the magnetic field?
4. Benchmarking or comparing results with those from other codes on similar problems. It is important to use the same boundary conditions and models when doing this. Do they agree?
5. Establishing that the convergence rate of the code errors with changing grid spacing and time step. Are they consistent with expectations?

A code should be verified with as many of these techniques as possible. Verification must happen before proceeding with anything else. If the model or the scheme is new then the verification tests and results should be documented and made available to the community. This is the best way to establish the reliability of the code and the trust in it by the community.

5.2 Validation

> Validation tests that a code or simulation meaningfully describes nature. The process involves investigating the applicability of conceptual models—that we're "solving the right equations" for a given problem. The test compares simulation results to experimental or observational data, so validation's scope is therefore much larger than verification's, requiring understanding and quantifying error or uncertainty in the experimental results as well as in models and simulation results (Calder et al. 2004).

Validating global simulations of space plasmas is challenging. Typically various types of experimental data are used for the validation simulations:

1. Controlled experiments designed to investigate a physical process.
2. Experiments designed to certify the performance of a component.
3. Experiments specifically designed to validate code calculations.
4. Passive observations of physical events.

The first three of these are difficult to do. There has been some efforts to validate codes against experiments that have already been performed. Examples include the AMPTE release and laser target experiments. However, simulations are usually validated against in-situ observations made by various spacecraft. Successful prediction before the in-situ observations are seen by the simulator is a better test than reproduction after the fact, since agreement is often achieved by tuning a simulation to what's already known.

Consider building a multi-fluid MHD simulation for Mars or Venus. It is the logical next step to take to the MHD approach. However, multi-fluid means many more degrees of freedom. This requires extensive testing and design decisions at each step in building the simulation. What form of the MHD equations will be solved? What physics does this form of the MHD equations include? What shock scheme will be used? How does the shock scheme work? Does it give the proper values for the jump conditions across the shock? If one fluid forms a shock, how do the other fluids respond? How do fluids with different gyroradii respond to the shock or various plasma waves? How accurate is the transport portion of the simulation. Will it transport a density distribution or electromagnetic pulse around the grid with minimal diffusion? Does the code give the proper gyromotion (radius and period) for a test fluid element? How many gyroperiods can the code accurately follow? How are regions where the density of one or more species are zero handled? If the Hall term is included, does the code recover the correct dispersion relations for whistler and ion cyclotron waves? What is the parameter space that the code is valid for? These are just some of the questions that have to be answered and documented in the verification and validation process.

Verification and validation are essential to establishing the reliability of a simulation and building community trust in the results. Failure to document the verification and validation process of a new model or computational scheme is like reporting data collected by an un-calibrated instrument, namely worthless. An excellent discussion about the need for verification and validation is given by Post and Votta (2005). The moral of the story is to test, test, test and when you think you are done, test some more. Even the best of scientific codes is estimated to contain 7 errors for every 1000 lines of code (Hatton and Roberts 1994).

5.3 Pre-simulation

With the physical assumption, models, their strengths, their limitation, verification and validation in mind there are a few things to consider before even starting a simulation. The first thing to do is decide what the problem is that needs a solution? What are the physical processes that need to be understood? What physics and chemistry is involved in those processes. What physical processes are not important and can be ignored? What are the boundary conditions that are going to be used and what are the implications of those boundary conditions? It has been said that "One should never start a calculation before you know the answer" (J. Wheeler, via S. Brecht, private communication).

Having the answers to these questions before beginning is important. It aids in the selection of the right physical model to use (kinetic, MHD, etc...) and the best computational/numerical methods to implement. It keeps the simulations simple and not over burdened by including unnecessary physics. The boundary conditions drive the simulation so they need special consideration. The wrong choice of boundary conditions can give wonderful results that don't describe the physical interaction and are therefore wrong. This stresses the importance of John Wheeler's comment. It is a sanity check. If the simulation results do not agree with the expected results it is important to know why. Often, it is because something is overlooked in setting up the simulation, such as a typo in the code or a poor choice of boundary conditions. Less often but more exciting is when the anticipated answer before the calculation is wrong, leading to a new understanding of the key physical processes in the problem. In summary before starting a simulation one should do the following:

1. Identify the problem of interest.
2. Identify the key physical processes involved.
3. Select proper boundary conditions.
4. Using ones understanding of the problem estimate the expected answer.
5. Select the proper physical model to use.
6. Select the best computational/numerical method.

With answers to the above, one is now ready to start working on the simulation.

That being done the goal of simulation is not to match the observations! The goal is to understand the problem of interest. Simulations are complex with many variables that can be fine tuned to match a desired outcome. This doesn't necessarily lead to a better understanding of the problem. Fine tuning the variables to match data could lead to a better understanding of the key physical processes and that is what is important. Matching the data is fine if the simulation is used to explain the data. One has to be careful, given all the various adjustable model parameters an apparent fit to the data, does not necessarily imply that the simulation correctly describes the physics being studied. Also one should keep in mind that the data are not perfect. There may be errors in its collection, processing and interpretation. It may be time dependent because of varying upstream condition. There are also many assumptions that are made in analyzing the data. Matching the data can lend confidence in the simulation results but it is not sufficient to proclaim the simulation is successful. It is important to ask what has been learned.

6 Selecting a Scheme: The Solar Wind Interaction with Pluto

As an example of the issues discussed in the last section consider the solar wind interacting with Pluto. The interaction of the solar wind with Pluto has been studied with hybrid simulations by Delamere and Bagenal (2004) using a 2200 km cell size. Higher resolution (358 km) multi-fluid simulations have been performed by Harnett et al. (2005). Ionospheric loss rates from Pluto have been investigated by Kecskemety and Cravens (1993) using test-particles. Recall the initial steps one should consider before starting a simulation.

1. Identify the problem of interest

What is the topology of Pluto's induced magnetosphere and what is the atmospheric loss rate from Pluto to the solar wind? The atmospheric loss rate has been studied using test-particles. However, this approach can not address the magnetic topology of the induced magnetosphere. This part of the problem requires a self-consistent treatment of mass loading and the magnetic field.

The solar wind conditions at Pluto are as follows. The magnetic field strength is 0.2 nT, the proton density is 0.01 cm^{-3}, the solar wind speed is 450 km/s, the proton temperature is 1.3 eV (Bagenal et al. 1997). The electron temperature can be taken to be 0.65 eV ($T_e = 1/2\,T_P$). For the purpose of this exercise the ionosphere will be assumed to be spherically symmetric about Pluto. It consists of N_2^+, with a scale height of 800 km and a peak ion density of 200 cm^{-3} (Ip et al. 2000). Pluto has a radius (r_P) of 1150 km.

2. Identify the key physical processes involved

What are the important physical processes? Are kinetic effects important? The gyroradii of the solar wind protons and the ionospheric N_2^+ are 23,000 km (about 20 r_P) and 658,000 km (about 550 r_P) respectively. The Debye lengths for the protons and the electrons are 85 m and 60 m respectively. The ion skin depth is 2280 km, about 2 r_P or 1 $r_P + 1.4$ times the ionospheric scale height.

3. Select proper boundary conditions

Part of selecting the boundary conditions is selecting the simulation domain size. Since the solar wind proton gyroradii is 20 r_P the inflow boundary should be put at least 2–3 times this distance upstream of Pluto. The further upstream the inflow boundary is placed the less likely reflections from the interaction are to interfere with the inflowing plasma and fields. The downstream and side boundaries should also be far enough away from the interaction so that they do not interfere with the simulation. Some simulations will use periodic boundary conditions on the side boundaries. This can be potentially dangerous. Ions and fields leaving one boundary will enter the opposite boundary with the wrong properties propagating into the simulation domain. It is better to treat these as outflow boundaries. The location of the downstream boundary will depend on the interest in the tail structures. Near Pluto the ion density is determined by the ionosphere. Pluto itself can be treated as a conducting sphere.

4. Using ones understanding of the problem estimate the expected answer

Given the expected comet like nature of Pluto one might expect Pluto's induced magnetosphere to resemble that of a comet's. Differences might arise do to differences in the ionospheric properties or the solar wind condition. Asymmetric structures should be present due to the large pickup ion gyroradii.

5. Select the proper physical model to use

These gyroradii are much larger than the radius of Pluto so kinetic effects are important. This suggests that ideal MHD would not be the best choice, non-ideal MHD, including the Hall term and/or multi-fluid MHD are better approaches. Of course since the gyroradii are large compared to the interaction region, the isotropic pressure assumption is not valid for Pluto. A hybrid simulation so far seems like it is the best choice of the approaches examined here. The Debye lengths of the protons and electrons are much smaller than 1 r_P, so that requirement is met. The H^+ ion skin depth is a red flag however. Below about 1/4 of this scale size electron kinetic effects are important. The massless electron fluid assumption implicit to non-ideal/multi-fluid MHD and hybrid approaches breaks down at on this scale. This limits the cell size that can be used to about 570 km or $1/2\ r_P$. Using cell sizes smaller than this will yield unreliable results. However, using cell sizes of $1/2\ r_P$ does not resolve Pluto and its atmosphere very well. This is a major drawback to the hybrid simulations of Pluto by Delamere and Bagenal (2004).

Since it is highly desirable to resolve Pluto and its ionosphere smaller cell sizes are needed. The kinetic effects of the electrons can not be ignored. None of the simulation approaches described here work in this parameter space. What is needed is a simulation approach that includes the electron kinetic effects. A fully kinetic electromagnetic particle code is in order. There is not enough space here to discuss this approach. The interested reader can find more details in Birdsall and Langdon (2004). This little exercise highlights the importance of knowing the limitations of the modeling approaches so that the proper approach can be used for the given problem.

7 Conclusions

Simulation is a valuable tool for understanding flowing plasma interactions. The companion paper by Ma et al. (2008) discusses some of the insights simulations have provided towards our understanding of plasma interactions with planets, moons and comets. Simulations do not replace the more traditional experimental or theoretical approaches, but compliment them. It is possible to examine the whole interaction region with simulations and thus get an idea of the global perspective at any given time. It is also possible to selectively add or remove physical processes to study their importance to the interaction and how the processes are interconnected.

The MHD, test-particle and hybrid methods have been reviewed here. All of these approaches can be used to study ion motions. Only the test-particle method can be used to study electron motion. None of these methods is capable of simulating the transition of flowing plasma from the collisionless to the collisional regimes. Each of these approaches has several implicit assumptions made in their formulation. Both MHD and hybrid methods assume quasi-neutrality and neglect the mass of the electron. This implies that the scale sizes used in both methods must be larger than the Debye length and no smaller than ten times the electron skin depth. Assumptions implicit to the hybrid approach make this approach well suited for the study of collisionless plasmas. Additional assumptions made in formulating the MHD approach make this model very well suited for collisional plasmas. Further do to these simplifications adding additional models such as ion-neutral chemistry and collisions to MHD is straightforward. The next frontier in these simulation methodologies is to push each methodology to bridge the collisional/collisionless transition.

There are several numerical issues and assumptions made when implementing each modeling approach into a simulation. These include for example, what type of grid to use, how best to represent each approach on that grid, how to insure $\nabla \cdot \boldsymbol{B} = 0$, how to weight particles and what numerical scheme to use. The choices made and schemes used are a trade-off between accuracy and efficiency. Once the choice of models and schemes is made into a simulation, it must be verified that the simulation accurately represents the conceptual models. The simulation must also be validated to insure that it meaningfully describes nature. The moral of this process is to test, test, test and test again.

There are several steps that should be performed before even starting a simulation. These include 1) identifying the problem of interest, 2) identifying the key physical processes, 3) selecting the proper boundary conditions, 4) use the understanding of the problem to predict the answer, 5) select the proper physical model to use and 6) select the best computational/numerical method. Performing these tasks insures that effort will not be wasted with unnecessary processes, keeping the simulation as simple as possible and insuring that the proper modeling approach is chosen. They also provide a sanity check against the final results. The goal of simulation is not to match the data but to understand the plasma interaction. It is important to ask what was learned from the simulation before it can be proclaimed a success. When this is done the full potential of simulations to understanding plasma interactions can be realized.

Acknowledgements S.A. Ledvina would like to thank Stephen H. Brecht for many useful discussions. The authors would like to thank T.E. Cravens and A.F. Nagy for their helpful comments on the manuscript. SAL greatfully acknowledge support from NASA grant NNG05GA04G and the Cassini Ion Neutral Mass Spectrometer Investigation. Y.-J. Ma was supported by NASA/JPL contract 1279285, and NASA grant NNG06GF31G.

References

H. Backes et al., Titan's magnetic field signature during the first Cassini encounter. Science **308**, 992 (2005)

T. Bagdonat, U. Motschmann, From a weak to a strong comet—3D global hybrid simulation studies. Earth, Moon Planets **90**, 305–321 (2002)

F. Bagenal, T.E. Cravens, J.G. Luhmann, R.L. McNutt, A.F. Cheng, Pluto's interaction with the solar wind, in *Pluto and Charon*, ed. by S.A. Stern, D.J. Tholen (University of Arizona Press, Tucson, 1997), p. 523

D.S. Balsara, D.S. Spicer, A staggered mesh algorithm using high order Godunov fluxes to ensure solenoidal magnetic fields in magnetohydrodynamic simulations. J. Comput. Phys. **149**, 270 (1999)

K. Baumgartel, K. Sauer, Interaction of a magnetized plasma stream with an immobile ion cloud. Ann. Geophys. **10**, 763 (1992)

R. Bauske, A.F. Nagy, T.I. Gombosi, D.L. DeZeeuw, K.G. Powell, J.G. Luhmann, A three-dimensional MHD study of solar wind mass loading processes at Venus: Effects of photoionization, electron impact ionization and charge exchange. J. Geophys. Res. **103**, 23,625 (1998)

C.K. Birdsall, A.B. Langdon, *Plasma Physics via Computer Simulations* (McGraw-Hill, USA, 1985)

C.K. Birdsall, A.B. Langdon, *Plasma Physics via Computer Simulation* (Taylor & Francis, London, 2004)

A. Bößwetter, T. Bagdonat, U. Motschmann, K. Sauer, Plasma boundaries at Mars: A 3-D simulation study. Ann. Geophys. **22**, 4363–4379 (2004)

J.P. Boris, Relativistic plasma simulation-optimization of a hybrid code, in *Proc. Fourth Conf. Num. Sim. Plasmas*, Naval Res. Lab., Wash., D.C., 2–3 November 1970, pp. 3–67

S.H. Brecht, J.F. Ferrante, Global hybrid simulation of unmagnetized planets: Comparison of Venus and Mars. J. Geophys. Res. **96**, 11,209–11,220 (1991)

S.H. Brecht, S.A. Ledvina, The solar wind interaction with the Martian ionosphere/atmosphere. Space Sci. Rev. **126**, 15–38 (2006). doi:10.1007/s11214-006-9084-z

S.H. Brecht, V.A. Thomas, Multidimensional simulations using hybrid particle codes. Comput. Phys. Commun. **48**, 135–143 (1988)

S.H. Brecht, J. Lyon, J.A. Fedder, K. Hain, A simulation study of east-west IMF effects on the magnetosphere. Geophys. Res. Lett. **8**(4), 397–400 (1981)

S.H. Brecht, J.G. Luhmann, D.J. Larson, Simulation of the Saturnian magnetospheric interaction with Titan. J. Geophys. Res. **105**, 13,119–13,130 (2000)

S.H. Brecht, I. de Pater, D.J. Larson, M.E. Pesses, Modification of the jovian radiation belts by Shoemaker-Levy 9: An explanation of the data. Icarus **151**(1), 25–38 (2001)

D.A. Bryant, Ion release experiments in the solar wind. Plasma Phys. Control. Fusion **27**, 1369 (1985)

O. Buneman, Time-reversible difference procedures. J. Comput. Phys. **1**, 517 (1967)

A. Calder et al., Validating astrophysical simulation codes. Comput. Sci. Eng. **6**, 10 (2004)

F. Cipriani, F. Leblanc, J.J. Berthelier, Martian corona: Nonthermal sources of hot heavy species. J. Geophys. Res. **112**, E07001 (2007). doi:10.1029/2006JE002818

T.E. Cravens, Ion distribution functions in the vicinity of Comet Giacobini-Zinner. Geophys. Res. Lett. **13**, 276 (1986)

T.E. Cravens, A magnetohydrodynamical model of the inner coma of comet Halley. J. Geophys. Res. **94**, 15025 (1989)

T.E. Cravens, *Physics of Solar System Plasmas* (Cambridge University Press, Cambridge, 1997)

T.E. Cravens, C.J. Lindgren, S.A. Ledvina, A two-dimensional multifluid MHD model of Titan's plasma environment. Planet. Space Sci. **46**, 1193 (1998)

T.E. Cravens, A. Hoppe, S.A. Ledvina, S. McKenna-Lawlor, Pickup ions near Mars associated with escaping oxygen atoms. J. Geophys. Res. **107**(A8), 1170 (2002). doi:10.1029/2001JA000125

W. Dai, P.R. Woodward, On the divergence-free condition and conservation laws in numerical simulations for supersonic magnetohydrodynamic flows. Astrophys. J. **494**, 317 (1998)

P.A. Delamere, F. Bagenal, Pluto's kinetic interaction with the solar wind. Geophys. Res. Lett. **31**, L04807 (2004). doi:1029/2003GL018122

C.R. Evans, J.F. Hawley, Simulation of magnetohydrodynamic flows—a constrained transport method. Astrophys. J. **332**, 659 (1988)

J.A. Fedder, J.G. Lyon, The solar wind-magnetosphere-ionosphere current-voltage relationship. Geophys. Res. Lett. **14**, 880–883 (1987)

T.I. Gombosi, *Gaskinetic Theory* (Cambridge University Press, Cambridge, 1994)

T.I. Gombosi, *Physics of the Space Environment* (Cambridge University Press, Cambridge, 1999)

T.I. Gombosi, D.L. De Zeeuw, R.M. Haberli, K.G. Powell, Three-dimensional multiscale MHD model of cometary plasma environments. J. Geophys. Res. **101**, 15233 (1996)

T.I. Gombosi, D.L. De Zeeuw, C.P.T. Groth, K.G. Powell, P. Song, The length of the magnetotail for northward IMF: Results of 3D MHD simulations. Phys. Space Plasmas **15**, 121–128 (1998)

C.P.T. Groth, D.L. DeZeeuw, T.I. Gombosi, K.G. Powell, A parallel adaptive 3D MHD scheme for modeling coronal and solar wind plasma flows. Space Sci. Rev. **87**, 193 (1999)

H. Gunell, M. Holmström, S. Barabash, E. Kallio, P. Janhunen, A.F. Nagy, Y. Ma, Planetary ENA imaging: Effects of different interaction models for Mars. Planet. Space Sci. **54**, 117–131 (2006)

K. Hain, Behavior of ionized plasma in the high latitude topside ionosphere. NRL Memorandum Report 3713 (1977), p. 1

K.C. Hansen, T.I. Gombosi, D.L. De Zeeuw, C.P.T. Groth, K.G. Powell, A 3D global MHD simulation of Saturn's magnetosphere. Adv. Space Res. **26**(10), 1681–1690 (2000)

K.C. Hansen et al., Global MHD simulations of Saturn's magnetosphere at the time of Cassini approach. Geophys. Res. Lett. **32**, L20S06 (2005). doi:10.1029/2005GL022835

D.S. Harned, Quasineutral hybrid simulation of macroscopic plasma phenomena. J. Comput. Phys. **47**, 452–462 (1982)

E.M. Harnett, R.M. Winglee, P.A. Delemere, Three-dimensional multi-fluid simulations of Pluto's magnetosphere: A comparison to 3D hybrid simulations. Geophys. Res. Lett. **32**, L19104 (2005). doi:10.1029/2005GL023178

R.E. Hartle et al., Preliminary interpretation of titan plasma interaction as observed by the Cassini plasma spectrometer: Comparisons with Voyager 1. Geophys. Res. Lett. **33**, L08201 (2006). doi:10.1029/2005GL024817

L. Hatton, A. Roberts, IEEE Trans. Softw. Eng. **20**, 785 (1994)

C. Hirsch, *Numerical Computation of Internal and External Flows,* vol. 1 (Wiley, New York, 1989)

R.W. Hockney, J.W. Eastwood, *Computer Simulation Using Particles* (Adam Hilger, Bristor ans Philadelphia, 1988). ISBN 0-85274-392-0

J. Hubba, Hall magnetohydrodynamics – A tutorial, Space Plasma Simulation. Edited by J. Büchner, C. Dum, M. Scholer. Lect. Notes Phys. **615**, 166–192 (2003)

W.-H. Ip et al., Pluto's ionospheric models and solar wind interaction. Adv. Space Res. **26**, 1559–1563 (2000)

Y.-D. Jia, M.R. Combi, K.C. Hansen, T.I. Gombosi, A global model of cometary tail disconnection events triggered by solar wind magnetic variations. J. Geophys. Res. **112**, A05223 (2007). doi:10.1029/2006JA012175

K. Kabin, T.I. Gombosi, D.L. De Zeeuw, K.G. Powell, Interaction of Mercury with the solar wind. Icarus **143**, 397–406 (2000)

E. Kallio, J.G. Luhmann, S. Barabash, Charge exchange near Mars: the solar wind absorption and energetic neutral atom production. J. Geophys. Res. **102**, 22183–22197 (1997)

E. Kallio, H. Koskinnen, A test particle simulation of the motion of oxygen ions and solar wind protons near Mars. J. Geophys. Res. **104**, 557–579 (1999)

E. Kallio, P. Janhunen, Atmospheric effects of proton precipitation in the Martian atmosphere and its connection to the Mars-solar wind interaction. J. Geophys. Res. **106**, 5617–5634 (2001)

E. Kallio, P. Janhunen, Ion escape from Mars in a quasineutral hybrid model. J. Geophys. Res. **107**, A3 (2002)

E. Kallio, P. Janhunen, Solar wind and magnetospheric ion impact on Mercury's surface. Geophys. Res. Lett. **30**, 17,1877 (2003). doi:10.1029/2003GL017842

E. Kallio, I. Sillanpää, P. Janhunen, Titan in subsonic and supersonic flow. Geophys. Res. Lett. **31**, L15703 (2004). doi:10.1029/2004GL020344

E. Kallio, Formation of the lunar wake in quasi-neutral hybrid model. Geophys. Res. Lett. **32**, L06107 (2005). doi:10.1029/2004GL021989

E. Kallio, R. Järvinen, P. Janhunen, Venus-Solar wind interaction: Asymmetries and the escape of O^+ ions. Planet. Space Sci. **54**, 1472–1481 (2006)

H. Karimabadi, D. Krauss-Varban, J.D. Huba, H.X. Vu, On magnetic reconnection regimes and associated three-dimensional asymmetries: Hybrid, Hall-less hybrid, and Hall-MHD simulations. J. Geophys. Res. **109**, A09205 (2004). doi:10.1029/2004JA010478

K. Kecskemety, T.E. Cravens, Pickup ions at Pluto. Geophys. Res. Lett. **20**, 543 (1993)

C.D. Kimmel, J.G. Luhmann, J.L. Phillips, J.A. Fedder, Characteristics of cometary pickup-up ions in a global model of Giacobini-Zinner. J. Geophys. Res. **92**, 8536 (1987)

N.A. Krall, A.W. Trivelpiece, *Principles of Plasma Physics* (McGraw-Hill, New York, 1973)

D. Krauss-Varban, From theoretical foundation to invaluable research tool: Modern hybrid simulations, in *Proceedings of the 7th International Symposium for Space Simulations (ISSI 7)*, Kyoto University, 2005, pp. 15–18

C.N. Keller, T.E. Cravens, L. Gan, One-dimensional multispecies magnetohydrodynamic models of the ramside ionosphere of Titan. J. Geophys. Res. **99**, 6511 (1994)

H. Lammer, H.I.M. Lichtenegger, H.K. Biernat, N.V. Erkaev, I.L. Arshukova, C. Kolb, H. Gunell, A. Lukyanov, M. Holmstrom, S. Barabash, T.L. Zhang, W. Baumjohann, Planet. Space Sci. **54**, 1445–1456 (2006)

J.N. Lcboeuf, T. Tajima, C.F. Kennel, J.M. Dauhon, Glohal simulations of the time-dependent niagnetosphcrc. Geophys. Res. Lett. **5**, 609 (1978)

S.A. Ledvina, T.E. Cravens, A three-dimensional MHD model of plasma flow around Titan. Planet. Space Sci. **46**, 1175 (1998)

S.A. Ledvina, T.E. Cravens, A. Salman, K. Kecskeméty, Ion trajectories in Saturn's magnetosphere near Titan. Adv. Space Res. **26**, 1691 (2000)

S.A. Ledvina, J.G. Luhmann, S.H. Brecht, T.E. Cranvens, Titan's induced magnetosphere. Adv. Space Res. **33**, 2092 (2004a)

S.A. Ledvina, S.H. Brecht, J.G. Luhmann, J. Geophys. Res. **31** (2004b). CiteID L17S10

S.A. Ledvina, T.E. Cravens, K. Kecskemety, Ion distributions in Saturn's magnetosphere near Titan, J. Geophys. Res. **110** (2005). CiteID A06211

A. Lipatov, *The Hybrid Multiscale Simulation Technology* (Springer, Berlin, 2002)

Y. Liu, A.F. Nagy, T.I. Gombosi, D.L. DeZeeuw, K.G. Powell, The solar wind interaction with Mars: Results of three-dimensional three-species MHD studies. Adv. Space Res. **27**, 1837 (2001)

G. Lu, P.H. Reiff, M.R. Hairston, R.A. Heelis, J.L. Karty, Distribution of convection potential around the polar cap boundary as a function of the interplanetary magnetic field. J. Geophys. Res. **94**, 13447 (1989)

J.G. Luhmann, S.A. Ledvina, J.G. Lyon, C.T. Russell, Planet. Space Sci. **54**, 1457–1471 (2006)

J.G. Luhmann, Titan's ion exospheric wake: A natural ion mass spectrometer? J. Geophys. Res. **101**, 29,387 (1996)

J.G. Luhmann, J.A. Fedder, D. Winske, A test particle model of pickup ions at comet Halley. J. Geophys. Res. **93**, 7532 (1988)

J.G. Lyon, J.A. Fedder, C.M. Mobarry, The Lyon-Fedder-Mobarry (LFM) global MHD magnetospheric simulation code. J. Atmos. Sol.-Terr. Phys. **66**, 1333–1350 (2004)

Y.J. Ma, A.F. Nagy, T.E. Cravens, I.V. Sokolov, J. Clark, K.C. Hansen, 3-D Global MHD model Prediction of the first close flyby of Titan by Cassini. Geophys. Res. Let. **31**, L22803 (2004a)

Y.J. Ma, A.F. Nagy, I.V. Sokolov, K.C. Hansen, Three-dimensional, multispecies, high spatial resolution MHD studies of the solar wind interaction with Mars. J. Geophys. Res. **109**, A07211 (2004b). doi:10.1029/2003JA010367

Y. Ma, A.F. Nagy, T.E. Cravens, I.V. Sokolov, K.C. Hansen, J.-E. Wahlund, F.J. Crary, A.J. Coates, M.K. Dougherty, Comparisons between MHD model calculations and observations of Cassini flybys of Titan. J. Geophys. Res. **111**, A05207 (2006). doi:10.1029/2005JA011481

Y. Ma et al., 3D global multi-species Hall-MHD simulation of the Cassini T9 flyby. Geophys. Res. Lett. (2007). doi:10.1029/2007GL031627

Y. Ma, A.F. Nagy, K. Altwegg, T. Breus, M.R. Combi, T.E. Cravens, E. Kallio, S.A. Ledvina, J.G. Luhmann, S. Miller, A.J. Ridley, D.F. Strobel, Plasma flow and related phenomena in planetary aeronomy. Space Sci. Rev. (2008). doi:10.1007/s11214-008-9389-1

P. MacNeice, Particle-mesh techniques (1996). http://citeseer.ist.psu.edu/496901.html

F. Mackay, R. Marchand, K. Kabin, Divergence-free magnetic field interpolation and charged particle trajectory integration. J. Geophys. Res. **111**, A06208 (2006). doi:10.1029/2005JA011382

A.P. Matthews, Current advance method and cyclic leapfrog for 2D multispecies hybrid plasma simulations. J. Comput. Phys. **112**, 102–116 (1994)

M.L. McKenzie, T.E. Cravens, G. Ye, Theoretical calculations of ion acceleration in the vicinity of comet Giacobini-Zinner. J. Geophys. Res. **99**, 6585 (1994)

R. Modolo, G.M. Chanteur, E. Dubinin, A.P. Matthews, Influence of the solar EUV flux on the Martian plasma environment. Ann. Geophys. **23**, 433–444 (2005)

R. Modolo, G.M. Chanteur, J.-E. Wahlund, P. Canu, W.S. Kurth, D. Gurnett, A.P. Matthews, C. Bertucci, A. Law, M.K. Doughert, The plasma environment in the wake of Titan from hybrid simulation—a case study. Geophys. Res. Lett. (2007, submitted)

Neubauer et al., Titan's near magnetotail from magnetic field and electron plasma observations and modeling: Cassini flybys TA, TB, and T3. J. Geophys. Res. **111**, 10220 (2006)

D.R. Nicholson, *Introduction to Plasma Theory* (Wiley, New York, 1983)

N. Omidi, X. Blanco-Cano, C.T. Russell, H. Karimabadi, M. Acuna, Hybrid simulations of solar wind interaction with magnetized asteroids: General characteristics. J. Geophys. Res. **107**, A12 (2002)

A. Otto, Geospace Environment Modeling (GEM) magnetic reconnection challenge: MHD and Hall MHD—constant and current dependent resistivity models. J. Geophys. Res. **106**(A3), 3751–3758 (2001)

D.E. Post, L.G. Votta, Computational science demands a new paradigm. Phys. Today **58**, 35 (2005)

K.G. Powell, An approximate Riemann solver for magnetohydrodynamics (that works in more than one dimension). ICASE Report, No. 94-24, Langley, VA, 1994

K.G. Powell, P.L. Roe, T.J. Linde, T.I. Gombosi, D.L. DeZeeuw, A solution-adaptive upwind scheme for ideal magnetohydrodynamics. J. Comput. Phys. **154**, 284 (1999)

W.H. Press, S.A. Teukolsky, W.T. Vetterling, B.P. Flannery, *Numerical Recipes in Fortran: The Art of Scientific Computing*, 2nd edn. (Cambridge Univ. Press, Cambridge, 1993)

P. Puhl-Quinn, T.E. Cravens, One-dimensional hybrid simulations of the diamagnetic cavity boundary region of comet Halley. J. Geophys. Res. **100**, 21631 (1995)

J. Raeder, Global geospace modeling: Tutorial and review. Space plasma simulation. Edited by J. Büchner, C. Dum, M. Scholer. Lect. Notes Phys. **84** (2001)

J. Raeder, Global Magnetohydrodynamics – A tutorial, space plasma simulation. Edited by J. Büchner, C. Dum, M. Scholer. Lect. Notes Phys. **615**, 212 (2003)

J. Ramshaw, A method for enforcing the solenoidal condition on magnetic fields in numerical calculations. J. Comput. Phys. **52**, 592 (1983)

K.V. Roberts, J.B. Taylor, Magnetohydrodynamic equations for finite Larmor radius. Phys. Rev. Lett. **8**, 197 (1962)

K. Sauer, K. Baumgartel, I. Axnas, N. Brenning, A fluid simulation of the AMPTE solar wind lithium release. Adv. Space Res. **10**, 95 (1990)

K. Sauer, A. Bogdanov, K. Baumgartel, Evidence of an ion composition boundary (protonopause) in bi-ion fluid simulations of solar wind mass loading. Geophys. Res. Lett. **21**, 2255 (1994)

K. Sauer, E. Dubinin, K. Baumgartel, Nonlinear MHD waves and discontinuities in the Martian magnetosheath: Observations and 2D bi-ion MHD simulations. Earth Planets Space **50**, 793 (1998)

D. Schriver, J.-N. Leboeuf, M. Ashour-Abdalla, M. El-Alaoui, J.-M. Bosqued, V. Sotnikov, Loading experimental velocity distributions into particle-in-cell simulation of space and fusion plasmas. J. Plasma Phys. **72**, 949 (2006)

R.W. Schunk, A.F. Nagy, *Ionospheres: Physics, Plasma Physics, and Chemistry* (Cambridge Univ. Press, New York, 2000)

I. Sillanpää, E. Kallio, R. Jarvinen, P. Janhunen, Oxygen ions at Titan's exobase in a Voyager 1–type interaction from a hybrid simulation. J. Geophys. Res. **112**, A12205 (2007). doi:10.1029/2007JA012348

H. Shimazu, Three-dimensional hybrid simulation of magnetized plasma flow around an obstacle. Earth Planets Space **51**, 383–393 (1999)

H. Shinagawa, T.E. Cravens, A one-dimensional multispecies magnetohydrodynamic model of the dayside ionosphere of Venus. J. Geophys. Res. **93**, 11263 (1988)

H. Shinagawa, T.E. Cravens, A one-dimensional multispecies magnetohydrodynamic model of the dayside ionosphere of Mars. J. Geophys. Res. **94**, 6506 (1989)

S. Simon, A. Bößwetter, U. Motschmann, K.H. Glassmeier, Plasma environment of Titan: A 3-D hybrid simulation study. Ann. Geophys. **24**, 1113–1135 (2006a)

S. Simon, T. Bagdonat, U. Motschmann, K.-H. Glassmeier, Plasma environment of magnetized asteroids: A 3D hybrid simulation study. Ann. Geophys. **24**, 407–414 (2006b)

G. Sod, *Numerical Methods in Fluid Dynamics* (Cambridge University Press, Cambridge, 1985)

K. Szego et al., Physics of mass-loaded plasmas. Space Science Rev. **94**, 429–671 (2000)

T. Tanaka, Configurations of the solar wind flow and magnetic field around the planets with no magnetic field: Calculation by a new MHD simulation scheme. J. Geophys. Res. **98**, 17251 (1993)

T. Tanaka, Generation mechanisms for magnetosphere-ionosphere current systems deduced from a three-dimensional MHD simulation of the solar wind-magnetosphere-ionosphere coupling process. J. Geophys. Res. **100**, 12057 (1995)

T. Tanaka, Effects of decreasing ionospheric pressure on the solar wind interaction with non-magnetized planets. Earth Planets Space **50**, 259 (1998)

T. Tanaka, K. Murawski, Three-dimensional MHD simulation of the solar wind interaction with the ionosphere of Venus: Results of two-component reacting plasma simulation. J. Geophys. Res. **102**, 19,805 (1997)

N. Terada, S. Machida, H. Shinagawa, Global hybrid simulation of the Kelvin–Helmholtz instability at the Venus ionopause. J. Geophys. Res. **107**(A12), 1471 (2002). doi:10.1029/2001JA009224

G. Tóth, The div $B = 0$ constraint in shock-capturing magnetohydrodynamics codes. J. Comput. Phys. **161**, 605–652 (2000)

G. Tóth, D. Odstrcil, Comparison of some flux corrected transport and total variation diminishing numerical schemes for hydrodynamic and magnetohydrodynamic problems. J. Comput. Phys. **128**, 82 (1996)

G. Tóth, P.L. Roe, Divergence and curl-preserving prolongation and restriction formulas. J. Comput. Phys. **180**, 736 (2002)

P. Trávníček, P. Hellinger, D. Schriver, S.D. Bale, Structure of the lunar wake: Two-dimensional global hybrid simulations. Geophys. Res. Lett. **32**, L06102 (2005). doi:10.1029/2004GL022243

P. Trávnìček, P. Hellinger, D. Schriver, Structure of Mercury's magnetosphere for different pressure of the solar wind: Three dimensional hybrid simulations. Geophys. Res. Lett. **34**, L05104 (2007). doi:10.1029/2006GL028518

W.-L. Tseng, W.-H. Ip, A. Kopp, Exospheric heating by pickup ions at Titan. Adv. Space Res. **42**, 54–60 (2008)

M.S. Voigt, Time-dependent MHD simulations for cometary plasma. Astron. Astrophys. **210**, 433 (1989)

R.M. Winglee, Ion cyclotron and heavy ion effects on reconnection in a global magnetotail. J. Geophys. Res. **109**, A09206 (2004). doi:10.1029/2004JA010385

R.M. Winglee, E. Harnett, A. Stickle, J. Porter, Multiscale/multifluid simulations of flux ropes at the magnetopause within a global magnetospheric model. J. Geophys. Res. **113**, A02209 (2008). doi:10.1029/2007JA012653

D. Winske, N. Omidi, A nonspecialist's guide to kinetic simulations of space plasmas. J. Geophys. Res. **101**(A8), 17,287–17,303 (1996)

D. Winske, L. Yin, N. Omidi, H. Karimabadi, K. Quest, Hybrid simulation codes: Past, present and future – A tutorial. Space Plasma Simulation. Edited by J. Büchner, C. Dum, M. Scholer. Lect. Notes Phys. **615**, 136–165 (2003)

K.S. Yee, Numerical solution of initial boundary value problems involving Maxwell's equations in isotropic media. IEEE Trans. Ant. Propagat **14**, 302 (1966)

Neutral Atmospheres

**I.C.F. Mueller-Wodarg · D.F. Strobel · J.I. Moses ·
J.H. Waite · J. Crovisier · R.V. Yelle · S.W. Bougher ·
R.G. Roble**

Originally published in the journal Space Science Reviews, Volume 139, Nos 1–4.
DOI: 10.1007/s11214-008-9404-6

Abstract This paper summarizes the understanding of aeronomy of neutral atmospheres in the solar system, discussing most planets as well as Saturn's moon Titan and comets. The thermal structure and energy balance is compared, highlighting the principal reasons

I.C.F. Mueller-Wodarg (✉)
Blackett Laboratory, Imperial College, Prince Consort Road, London SW7 2BW, UK
e-mail: i.mueller-wodarg@imperial.ac.uk

D.F. Strobel
Johns Hopkins University, 3400 North Charles Street, Baltimore, MD 21218, USA
e-mail: strobel@jhu.edu

J.I. Moses
Lunar and Planetary Institute, 3600 Bay Area Boulevard, Houston, TX 77058, USA
e-mail: moses@lpi.usra.edu

J.H. Waite
Southwest Research Institute, 6220 Culebra, San Antonio, TX 78228, USA
e-mail: hwaite@swri.edu

J. Crovisier
LESIA, Observatoire de Paris, 92195 Meudon, France
e-mail: Jacques.Crovisier@obspm.fr

R.V. Yelle
Lunar and Planetary Laboratory, University of Arizona, Tucson, AZ 85721, USA
e-mail: yelle@lpl.arizona.edu

S.W. Bougher
Department of Atmospheric, Oceanic and Space Sciences, College of Engineering, University of Michigan, 2455 Hayward Street, Ann Arbor, MI 48109, USA
e-mail: bougher@umich.edu

R.G. Roble
High Altitude Observatory, National Center for Atmospheric Research, P.O. Box 3000, Boulder, CO 80307, USA
e-mail: roble@hao.ucar.edu

for discrepancies amongst the atmospheres, a combination of atmospheric composition, heliocentric distance and other external energy sources not common to all. The composition of atmospheres is discussed in terms of vertical structure, chemistry and evolution. The final section compares dynamics in the upper atmospheres of most planets and highlights the importance of vertical dynamical coupling as well as magnetospheric forcing in auroral regions, where present. It is shown that a first order understanding of neutral atmospheres has emerged over the past decades, thanks to the combined effects of spacecraft and Earth-based observations as well as advances in theoretical modeling capabilities. Key gaps in our understanding are highlighted which ultimately call for a more comprehensive programme of observation and laboratory measurements.

Keywords Atmospheres · Planets · Moons · Comets · Dynamics · Composition · Thermal structure · Observations · Theoretical modeling

1 Introduction

All planets, many moons and some of the smaller objects in the solar system are surrounded by gas envelopes which may in the widest definition be referred to as atmospheres. However, important distinctions can be made between these atmospheres. The first important distinction is that between permanent and transient atmospheres; the former are atmospheres bound gravitationally by the solid body, giving them long lifetimes on the scale of the age of the solid body. Permanent atmospheres can be formed either from accretion of primordial material or outgassing processes and volcanism. Planets Venus, Earth, Mars, Pluto and all Gas Giants including moons Titan and Triton host atmospheres that are referred to as permanent. In contrast, transient atmospheres are those not sufficiently bound gravitationally, where constituents escape the gravitational field of the solid body. In these cases the presence of an atmosphere relies upon an active gas source, such as outgassing from the interior, volcanism or sputtering processes. Typical examples of transient atmospheres are comets, Jupiter's moon Io and most other moons in the solar system, planet Mercury and larger asteroids.

In this chapter permanent atmospheres of planets and moons as well as transient atmospheres of comets are discussed. The focus will remain on neutral gases only, which form the principal mass and volume of most atmospheres. Small fractions of the neutral atmosphere are ionized by solar radiation or charged particle impact processes, forming ionospheres. These may affect neutral gas properties as well via dynamical or chemical or energetic coupling, and some of the principal ion-neutral coupling processes are also reviewed in this chapter.

Table 1 compares main properties of permanent atmospheres in the solar system, such as heliocentric distance, rotation period of the solid body, principal gas composition and exospheric temperatures. It is evident from this overview that one can in terms of composition classify the atmospheres into mainly three classes, those dominated by N_2 (Earth, Titan, Triton, Pluto), CO_2 (Venus, Mars) and hydrogen/helium (Jupiter, Saturn, Uranus, Neptune). Of the terrestrial planets, Venus has the most substantial atmosphere with a surface pressure of around 90 bars and temperature of 733 K. Differences between the terrestrial planet atmospheres are considerable, despite their relative proximity in the solar system, while in the outer solar system most planets are Gas Giants with essentially similar atmospheres. The temperature reached asymptotically in the thermosphere, or exosphere temperature (see also Sect. 2), should to a first approximation be largely influenced by solar EUV radiation

Table 1 Main properties of non-transient atmospheres in the solar system

Object	Mean heliocentric distance	Duration of mean solar day	Principal gases	Surface or 1 bar level P, T	Exobase height	Dayside exosphere temperature
Venus	0.72 AU	116.75 days[a]	CO_2, N_2, SO_2, O	92 bar, 735 K	160 km	230–325 K
Earth	1.00 AU	1 day	N_2, O_2, O	1 bar, 185–331 K	450 km	600–2000 K[c]
Mars	1.52 AU	1.03 days	CO_2, N_2, O	6 mbar, 140–270 K	180 km	180–325 K
Jupiter	5.20 AU	$9^h 55^m 27^s$	H_2, He, H, CH_4	1 bar, 165 K	1600 km[b]	940 K[c]
Saturn	9.54 AU	$10^h 39^m 22^s$	H_2, He, H, CH_4	1 bar, 160 K	2500 km[b]	420 K[c]
Titan	9.54 AU	15.95 days	N_2, CH_4, H	1.4 bar, 94 K	1430 km	150 K
Uranus	19.19 AU	$17^h 14^m 24^s$	H_2, He, H, CH_4	1 bar, 76 K	4700 km[b]	670 K[c]
Neptune	30.07 AU	$16^h 6^m 36^s$	H_2, He, H, CH_4	1 bar, 73 K	2200 km[b]	470 K[c]
Triton	30.07 AU	5.88 days[a]	N_2, CO, CH_4, H_2	14 μbar, 38 K	930 km	100 K
Pluto	39.48 AU	6.39 days[a]	N_2, CO, CH_4, H_2	3-90 μbar, 35–50 K	1800 km	100 K

[a] Retrograde rotation

[b] Above the 1 bar level

[c] In the non-auroral regions

which is absorbed in that region of the atmosphere, and yet Table 1 shows no consistent trend of these temperatures with distance from the Sun. Venus, the planet closest to the Sun, has the lowest exosphere temperature of all planets, while the Gas Giants Jupiter and Saturn have exospheric temperatures larger than those of Mars, Venus and often Earth, despite their much larger distances from the Sun. Understanding these first order differences between atmospheres in the solar system will be one of the aims of this chapter. Comparative studies between different planets give us a deep understanding of the main physics in planetary atmospheres since applying the same basic physics to different boundary conditions (distance from Sun, mass of solid body, composition) produces very different worlds. The paper will discuss the thermal structures (Sect. 2), composition and chemistry (Sect. 3) as well as dynamics (Sect. 4) and highlight open questions to be addressed by future studies (Sect. 5). Although discussions concentrate on our solar system, many of the basic physics are believed to apply similarly to extrasolar planets and understanding our own solar system forms the basis for understanding other solar systems.

2 Thermal Structure and Energy Balance

2.1 Planets and Moons

The thermal structure of a planetary atmosphere is normally described with nomenclature originally introduced to describe the globally averaged temperature profile of the Earth's atmosphere, using as principal criterion the vertical thermal gradient. The Earth's lower atmosphere is known as the troposphere (see Fig. 1). It is characterized by a relatively constant negative temperature gradient, ~ -6.5 K km^{-1}. This lapse rate is determined both by an intermediate value of the moist adiabatic lapse rate (which can be as low as -3 K km^{-1} up to the dry adiabat of -9.8 K km^{-1}) and the large scale dynamics of the troposphere. At the equator the top of the troposphere, known as the tropopause is approximately 17 km above

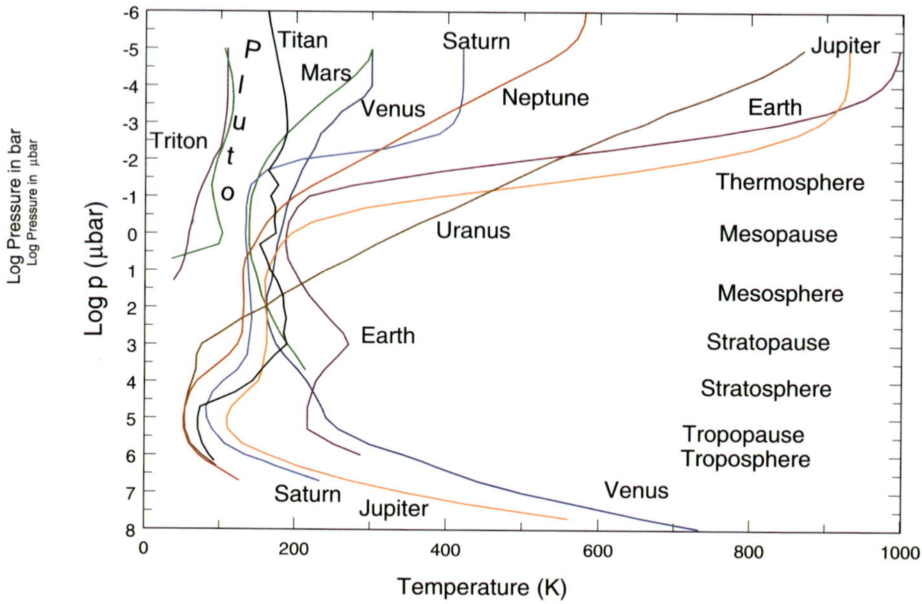

Fig. 1 Temperatures versus pressure for Venus, Earth, Mars, Jupiter, Saturn, Titan, Uranus, Neptune, Triton and Pluto

the surface with a temperature of ~190 K and it slopes downward towards the poles to an altitude of only ~8 km but with a higher temperature ~210–230 K, depending on season. By comparison the surface temperature varies from 300 K at the equator to 250 K at the poles.

Photochemistry of molecular oxygen leads to the formation of ozone, whose photochemistry driven by absorption of solar ultraviolet radiation results in atmospheric heating and the formation of the stratosphere, which has a positive temperature gradient of ~ 2 K km^{-1}. The upper boundary of the stratosphere (called the stratopause) is at 50 km and the 1 mbar level, where the temperature reaches a relative maximum of ~290 K at the summer pole and ~250 K at the winter pole with a global average of ~270 K. Above the stratosphere is the region known as the mesosphere, which is characterized by a negative temperature gradient of ~ -3 K km^{-1}, as a consequence of the ozone heating rate decreasing more rapidly with altitude than the CO_2 infrared cooling rate and consistent with the observed ratio of ozone density to CO_2 density declining with height. Collectively, the stratosphere and mesosphere are known as the middle atmosphere.

The upper boundary of the mesosphere is known as the mesopause and is typically at 85–90 km and approximately the 1 μbar level. The globally averaged mesopause temperature is ~185 K, but over the summer pole it drops to ~130 K and can be a high as 220 K over the winter pole due to a large scale meridional circulation, which transports heat to the winter pole with adiabatic cooling over the summer pole and adiabatic heating over the winter pole. Calculations have shown that the thermal structure and dynamics of the mesosphere cannot be reproduced when assuming a radiative equilibrium case (Geller 1983) and a wave drag term is needed. Physically, this represents the momentum deposited by dissipating or breaking gravity waves, tides and planetary waves in the atmosphere, and for simplicity it is often approximated by a linear Rayleigh friction term (Schoeberl and Strobel 1978). More comprehensive models use gravity wave parameterization schemes to simulate the momentum deposition, as described further in Sect. 4. Note that this affects both the dynamics and the thermal structure via advection of energy and adiabatic heating and cooling.

The region above the mesopause is the thermosphere which is characterized by a very rapid temperature increase of ~10–20 K km^{-1} and asymptotically attains a constant (isothermal) temperature at high altitudes, commonly called the exospheric temperature and denoted by T_∞. This temperature can be as low as 600 K during solar minimum activity and as high as 2000 K during solar maximum conditions. In addition, there is a diurnal variation in exospheric temperature of ~30%, which has a phase-shifted maximum from the subsolar point and minimum from the antisolar point of ~2–3 hr caused by thermospheric winds. The basic physics of the thermosphere is intense heating by absorption of short wavelength solar ultraviolet radiation (<170 nm) in the dissociation and/or ionization of molecules and atoms and the downward transport of thermal energy by heat conduction due to the absence of infrared active molecules, such as CO_2, capable of radiating thermal energy away locally. The mesopause is the region where radiative cooling of downward conducted energy becomes most effective, giving it the lowest temperature. The mesopause can therefore essentially be regarded as the heat sink for the thermosphere. Thermospheric heating is supplemented by auroral energy deposition in the polar regions (primarily Joule heating and partly particle precipitation heating) which may exceed solar heating rates.

Even though the heating and cooling processes in the atmospheres of other planets and satellites may have a different relative balance, vertical temperature profiles on all planets show similar features (Fig. 1) and this nomenclature originally used for the Earth's atmosphere is adopted for their atmospheres and temperature profiles as well. The Earth's twin planet Venus has a massive CO_2 atmosphere with a surface temperature of 733 K at low latitudes and slightly warmer (737 K) at higher latitudes with essentially constant surface pressure of 92.1 bar (Table 1). It has a very extended troposphere with a lapse rate of −8.98 K km^{-1} up to about 35 km and 6 bar, which is close to the adiabatic lapse rate, but stable. Above 35 km where haze layers are present on Venus, the temperature notably departs from this lapse rate, indicating heating and this trend continues into the cloud region between 50–70 km (1–0.03 bar), with one exception of a convective layer within the clouds at ~55 km (0.5 bar). At 58 km, where the pressure is 0.3 bar and temperature is ~270 K, the atmosphere becomes substantially more stable in analogy with the Earth's atmosphere, this level is denoted as the tropopause and the region above as the stratosphere, even though the temperature continues to decrease up to 95 km to ~162–167 K and pressure ~0.1 mbar. On the dayside the temperature ramps up to an asymptotic value of ~230–325 K (the range from solar minimum to solar maximum activity), whereas on the nightside it decreases to ~100 K. Given the slow rotation of Venus, the dayside-nightside temperature contrast is a subsolar bulge to antisolar hole. On the dayside the region between 95–160 km is known as the dayside thermosphere, whereas on the nightside in effect the stratosphere extends up to the exobase at 160 km, where the pressure is ~20 pbar, but the "nightside thermosphere" is called the cryosphere. The dayside model atmosphere in Fig. 1 is based on measurements by the Pioneer Venus atmospheric probes (Seiff 1983).

The other terrestrial CO_2 atmosphere is the thin atmosphere of Mars with surface pressure of 6 mbar and temperature ~214 K, but which varies from 140–270 K. On Mars the troposphere extends to about 50 km and 10 μbar, where occasionally CO_2-ice clouds are present and the temperature is ~145 K. Above this level, as for Venus, the change in atmospheric stability justifies calling this region the stratosphere and the temperature gradually decreases to about 140 K depending on latitude at 100 km and 40 nbar. This level is the mesopause, as the temperature increases rapidly above it to an asymptotic value of

180 K during low solar activity to as high as 325 K during solar maximum and when Mars is closest to the Sun. With an eccentricity of 0.093, the solar flux varies by 45% over the orbit.

The key to understanding the differences in exospheric temperatures on Venus and Mars appears to lie in the role of radiative cooling. As shown by Bougher et al. (1994), 15 μm emissions from CO_2 largely originate from vibrationally excited CO_2 and cause significant cooling in regions of non-local thermodynamics equilibrium (LTE). Collisions of O atoms with CO_2 molecules cause the main vibrational excitation of CO_2 and hence enhance this radiative cooling process, implying that larger values of O/CO_2 ratios favor radiative cooling. Typical values for Venus are $O/CO_2 \approx 7$–16%, while on Mars they have, in the absence of direct O measurements (Sect. 5.6), been calculated to lie between 2–4%. Despite the uncertainty of this ratio on Mars due to the lack of direct O measurements there, the values indicate a significantly lower ratio on Mars than Venus, making CO_2 15 μm cooling more effective on Venus than Mars. This finding is supported by the relatively weak temperature changes with solar activity on Venus compared with Mars and also Earth (Forbes et al. 2006). On Venus, CO_2 cooling acts as thermostat to solar energy deposition. At higher solar activity levels, dissociation of O_2 and CO_2 becomes more effective, and hence the abundances of O, leading to more effective vibrational excitation of CO_2 and ultimately more effective cooling, offsetting much of the temperature change which would result from the larger solar flux. Theoretical calculations of the cooling rate rely on knowledge of the CO_2–O relaxation rate, which to-date is poorly constrained, with values from laboratory measurements ranging from 1.2 ± 0.2 to $6 \pm 3 \times 10^{-12}$ $cm^3 s^{-1}$ (Pollock et al. 1993; Sharma and Wintersteiner 1990; Shved et al. 1991; Castle et al. 2006), as discussed also by Huestis et al. (2008).

The giant planets are mostly H_2–He atmospheres with deep tropospheres that have slightly superadiabatic lapse rates consistent with the outward convection of internal heat slowly working its way out from gravitational contraction during formation and condensational heating associated with phase changes. Convection extends up to the cloud tops where the transition to radiative equilibrium occurs. The stratospheres are in radiative equilibrium with solar heating in near-IR CH_4 bands at 3.3, 2.3, 1.6 μm augmented by solar heating of haze particles in the lower stratosphere. The radio occultation experiments performed with the Voyager spacecrafts provide a comparative data set for the pressure regions from 1–6 bars to 1–0.35 mbar (cf. Table 1 in Lindal 1992). All thermospheres are substantially warmer than would be expected on the basis of solar UV and EUV heating, an issue as yet unresolved (see Sect. 5). The Voyager Ultraviolet Spectrometer (UVS) solar and stellar occultation data are still the primary data sets for the thermal structure of the upper atmospheres of Saturn, Uranus, and Neptune. For the case of Saturn, the Cassini Ultraviolet Imaging Spectrograph (UVIS) is carrying out occultation measurements which will further constrain the thermal structure of Saturn's middle and upper atmosphere.

The best vertical profiles of density, pressure, and temperature for any giant planet were obtained from the Galileo Probe's descent through Jupiter's atmosphere down to ~20 bar (Seiff et al. 1997). The transition from convection to radiative equilibrium occurs at approximately 0.7 bar on Jupiter ($T \approx 136$ K), which is well below the tropopause temperature minimum of 110 K at 0.1 bar. Unlike the Earth's atmosphere, the tropopause on Jupiter is purely determined by radiative processes rather than the transition from tropospheric convection to radiative equilibrium in the stratosphere. The latter is heated to ~160 K by near-IR solar radiation absorbed by CH_4 and is fairly isothermal up to 5 μbar. Here the thermosphere begins and the temperature ramps up to about 940 K due to an unknown heating mechanisms.

Wave heating is widely suspected to be important, but the waves detected in the Galileo Probe data were not of sufficient amplitude and did not penetrate high enough to account for the observed temperature (Sect. 5). Significant amounts of Joule heating are expected to be deposited in the polar regions of Jupiter and the other Gas Giants. The main problem however turns out to be the meridional redistribution of this energy, which is largely inhibited by the fast rotation of the planets and hence primarily zonal flow in the atmosphere. A recent study by Majeed et al. (2005) proposed that dynamical heating induced by the low-latitude convergence of the high-latitude-driven thermospheric circulation would deposit thermal energy in the equatorial thermosphere of Jupiter via adiabatic heating rather than direct meridional energy transport by winds. In a separate study by Smith et al. (2007) the opposite was found, whereby the equatorial thermosphere is cooled as a result of energy and momentum deposition in the auroral zones due to poleward lower thermosphere winds which close the circulation cell initiated by the equatorward flow at higher altitudes and transports energy poleward. While their study was for Saturn, the situation would largely be the same for Jupiter.

For Saturn, so far there are no publications of radio or solar occultation data taken by the Cassini spacecraft to compare with the equivalent Voyager data. Thus the data in Fig. 1 are from Voyager data sets. The basic structure mimics the structure of Jupiter's atmosphere from 5 bar up to 5 μbar with a 20–30 K offset to lower temperatures. However the Saturnian thermosphere is considerably colder than Jupiter's for reasons not entirely understood. The ice giants Uranus and Neptune have comparable temperatures in their tropospheres and lower stratospheres up to 10 mbar. Above 10 mbar, Neptune's stratosphere increases much faster than on Uranus, probably associated with the higher abundance of CH_4 on Neptune, as the Voyager UVS detected the Raman-scattered Lyman-α line at 127 nm in the Uranian airglow, which places severe limits on the stratospheric CH_4 abundance (Yelle et al. 1987). The thermal structure of Neptune's thermosphere in Fig. 1 is based on unpublished analyses by the Voyager UVS team, whereas the thermal structure of the Uranian thermosphere is based on the analysis of Stevens et al. (1993).

Titan's thermal structure has been measured most extensively by the Cassini-Huygens Mission from multiple data sets. The Voyager Mission radio and solar occultations have now been complemented by the equivalent UVIS measurements by the Cassini spacecraft, although the data have yet to be published. The most remarkable data is from the Huygens Atmospheric Science Investigation (HASI) by the Huygens Probe which yielded the density, pressure, and temperature structure from 1270 km down to the surface (Fulchignoni et al. 2005). The remote sensing data acquired by the Cassini Composite Infrared Spectrometer (CIRS) instrument provides a global distribution of temperature over a pressure range of 10 mbar to 1 μbar. At the Probe entry latitude CIRS data is not in agreement with the Huygen ASI data as to the location and temperature of the stratopause at equatorial latitudes (Achterberg et al. 2008). Titan's stratosphere results from intense solar heating of its optically thick photochemical haze. Cooling of the upper atmosphere is dominated at high altitudes by rotational line cooling by HCN (Yelle 1991). The topside thermosphere has a negative temperature gradient associated with adiabatic cooling from Titan's hydrodynamically expanding atmosphere (Strobel 2008a).

The other N_2 atmospheres in the solar system with observational data are Triton and Pluto. For Triton the thermal structure has been assembled from the stellar occultation measurements reported in Elliot et al. (2000) from the surface up to about 100 km and then above from Voyager UVS solar occultation data and analysis (cf. Strobel and Summers 1995). For Pluto, the lower atmosphere is based on stellar occultations (cf. Elliot et al. 2007) augmented at higher altitudes by theoretical modeling reported in Strobel et al. (1996) and

Strobel (2008b). The higher temperatures in the lower atmosphere of Pluto are due to the thermostat capability of CH_4, whereby the near-IR 3.3 μm band absorbs solar radiation and the 7.6 μm band radiates heat away and maintains an equilibrium temperature ~100 K (Yelle and Lunine 1989). The former is augmented by heating in the 2.3 μm band and the latter by local thermal equilibrium (LTE) CO rotational line cooling. Thermal heat conduction delivers heat to the cold surface generating a steep temperature gradient for pressures greater that 2 μbar. Elliot et al. (2000) found that CH_4 heating played a non-negligible role in Triton's lower atmosphere, but the radiative equilibrium temperature is only ~50 K and not as well understood as that of Pluto. At higher altitudes, solar EUV and UV heating of N_2 and CH_4, respectively, and thermal heat conduction are important for Pluto's thermosphere, whereas in Triton's thermosphere magnetospheric electron precipitation accounts for about 2/3 the total power input.

2.2 Comets

The thermodynamics of cometary atmospheres is quite different from that of the permanent atmospheres of planets due to their expansion and conversion between a collisional regime in the inner coma to a supersonic free molecular flow in the outer coma. As a result of the large size of the atmosphere compared to the nucleus, plane-parallel models cannot apply. A spherical geometry with a density proportional to $1/r^2$ (r being the distance from the centre of mass) may be adopted in a first approach. Although the individual processes involved in the thermodynamics of the coma are well understood, their coupling with chemistry, molecular excitation and radiative trapping render their modelling difficult. Indeed, the completion of a comprehensive model including all relevant processes for a realistic description of the cometary environment is a formidable project which is far from being achieved at the present time. A recent review of this topic was given by Combi et al. (2005).

Besides rare in situ observations, information on the expansion velocity of cometary atmospheres comes from the analysis of spectral line shapes, especially of rotational lines which can be observed at high spectral resolution at radio wavelengths. Information on the temperature comes, for the collisional region of the inner coma where LTE prevails, from the simultaneous observation of several rotational lines, or from the rotational structure of vibrational or electronic bands. However, the velocities and temperatures which are retrieved by remote sensing are averaged over the instrumental field of view and along the line of sight: we do not have direct access to their profiles as a function of distance from the nucleus. Typical velocities are in the range 0.5–2.0 km s^{-1}, and temperatures in the range 10–150 K. Velocities and temperatures are both observed to increase when the heliocentric distance r_h decreases and the total gas production of the comet increases (Biver et al. 2002; Tseng et al. 2007).

Heating in a cometary atmosphere occurs from photolysis of molecules. When a molecule is photodissociated, part of the excess energy is transferred to the fragments. Heating subsequently occurs through thermalization by collisions of the fast fragments. The process is dominated by water photolysis (Crovisier 1989), with a heating rate of 3.3×10^{-23} W per H_2O molecule at 1 AU from the Sun. The contribution of other species is much smaller, in part because of their small abundances and partly because their photolytic rates are smaller. Heating by other exothermic chemical reactions is also negligible. Heating is effective only if the fragments can be thermalized. Water photodissociates mainly into H and OH. Because of momentum conservation, most (84%) of the translational energy is transmitted to fast hydrogen atoms. Thermalization occurs principally through collisions with water molecules, but because of the large difference in mass of H and H_2O, only a small fraction of the energy is exchanged at each collision. Many collisions are thus necessary for total thermalization (more than 20 collisions are needed for a 90% thermalization). The thermalization efficiency may be evaluated by Monte Carlo simulations.

Cooling occurs through emissions of molecular lines. Due to the low temperature of the coma, only rotational lines can be excited by collisions and contribute to cooling. Here again water plays the major role, and emissions of water rotational lines are the most efficient mechanism. Water radiative cooling can be accurately computed in the optically thin case for LTE conditions from H_2O spectroscopic data. Typical cooling rates are 10^{-23} W per H_2O molecule at 20 K, 4×10^{-22} W at 100 K. The problem is that, for high densities, optical depth effects are critical for water lines and seriously limit the efficiency of radiative cooling (the fundamental 1_{10}–1_{01} rotational line of water at 555 GHz is thick within 7000 km from the nucleus even for a moderately active comet with $Q[H_2O] = 10^{28}$ s^{-1}). A full theoretical treatment of the problem would require knowledge of cross-sections for collisions which are still largely unknown for water and self-consistent evaluation of water excitation, optical depth effects and thermodynamics of the coma atmosphere (Bockelée-Morvan and Crovisier 1987).

Initial conditions are determined by sublimation at the nucleus surface. Exposed water ice, when freely sublimating under solar insolation at $r_h \approx 1$ AU, is expected to equilibrate at a temperature near 180 K. Indeed, analysis of the Deep Impact mission results revealed the presence of exposed water ice on the nucleus of comet 9P/Tempel 1, but with a very small surface which can in no way explain the sublimation rate of water observed in this comet (Sunshine et al. 2007). If sublimation occurs below the nucleus surface, it is likely that the initial conditions are affected by percolation of water vapour through the surface porous mantle. The initial velocity at the surface is about 0.3 km s^{-1}, rapidly increasing to become supersonic at 0.5–0.8 km s^{-1}.

For a small comet such as 67P/Churyumov-Gerasimenko (the target of the Rosetta mission), which has a diameter of 4 km and a gas production rate of 10^{28} s^{-1} at 1 AU from the Sun, the gas density close to the nucleus surface is $\sim 7 \times 10^{11}$ cm^{-3}, and the pressure is $\sim 1.7 \times 10^{-10}$ bar. Cooling and heating are slow processes, so that close to nucleus, the coma undergoes an almost adiabatic expansion. The result is that the temperature first drops rapidly as the coma is expanding. Indeed, temperatures as low as ~10 K have been observed in some cases (small comets; comets far from the Sun).

In the inner, collisional coma, purely hydrodynamical treatment of the atmosphere is possible. The thermal and velocity profiles result from the balance of the heating and cooling processes. Heating (favoured by collisions) prevails over cooling (inhibited by optical trapping) and the temperature increases with the distance to the nucleus, as shown in Fig. 2. Indeed, temperatures as large as 100–150 K have been observed for comet Hale-Bopp. Outside the collisional region, the outflow regime evolves to a free molecular flow. The large transitional region—which is out of equilibrium—is difficult to model. Monte Carlo simulations are usually employed (Hodges 1990). The size of the collisional coma depends on the gas production rate of the comet, being typically ~200 km for a small comet such as 67P/Churyumov-Gerasimenko with $Q[H_2O] = 10^{28}$ s^{-1} and ~200 000 km for a giant comet such as Hale-Bopp with $Q[H_2O] = 10^{31}$ s^{-1}.

Current models reproduce well, at least qualitatively, the observed expansion velocity and temperature, and their evolution with heliocentric distance and gas production rate. Comprehensive models have to consider interaction with dust and, on a larger scale, interaction with the solar wind (Ma et al. 2008).

Water sublimation is inhibited far from the Sun (at typically heliocentric distances $r_h > 4$ AU), where cometary activity is rather dominated by CO sublimation. The transition

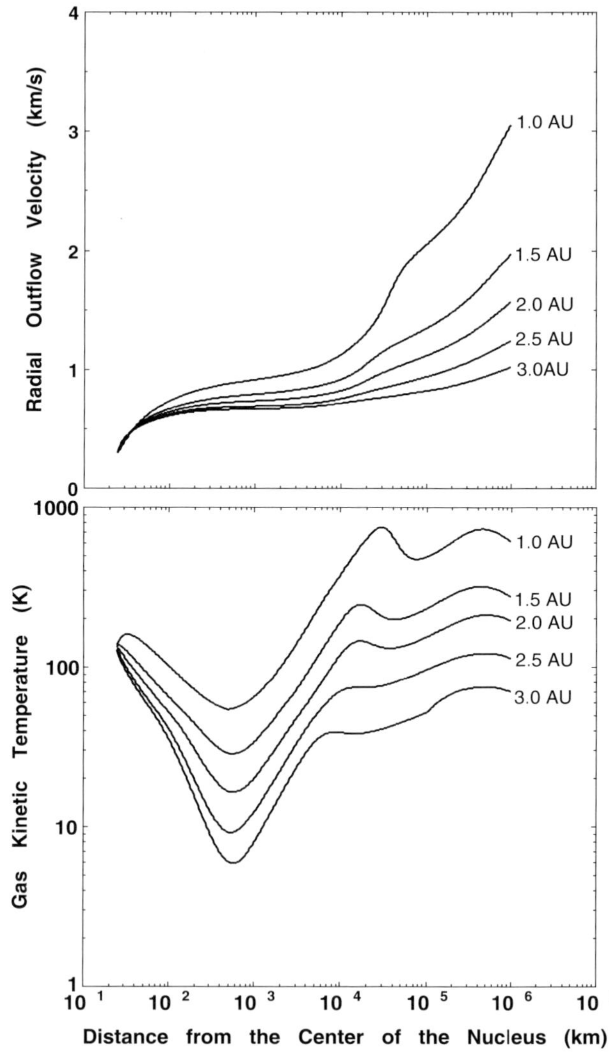

Fig. 2 Kinetic temperature and expansion velocity of a cometary atmosphere as a function of distance to nucleus, for various heliocentric distances, from a 1-D spherical hybrid kinetic/dusty-gas hydrodynamical model (from Combi et al. 1999). The model is applied here to comet C/1995 O1 (Hale-Bopp), with water production rates varying from $4 \times 10^{29}\ s^{-1}$ at 3 AU to $8 \times 10^{30}\ s^{-1}$ at 1 AU. Other examples may be seen in Combi et al. (2005)

between CO-driven and H_2O-driven activity was observed in comet Hale-Bopp for which the gas production was monitored up to $r_h = 14$ AU (Biver et al. 2002). CO is the only parent molecule observed in 27P/Schwassmann-Wachmann 1 (Senay and Jewitt 1994), a permanently active comet in a nearly circular orbit at 5.8 AU. Modelling of a CO-dominated comet is discussed by Ip (1983).

Real comets do not have spherical symmetry, as is obvious from most images of comets. This is because cometary nuclei are not spherical and not homogeneous, and because cometary activity and aeronomic processes are driven by the Sun. Axisymmetric, or even 3-D models, are needed. The modelling of the interface between the nucleus and the coma is complex, as it involves the nucleus topography, its rotation and thermal inertia, as well as its long-term thermal history (Crifo et al. 2005).

3 Composition and Chemistry

3.1 Vertical Structure

Table 1 lists the main gases found in planetary atmospheres in our solar system. As previously mentioned, one may hence subdivide the planets and moons into families of those with N_2 atmospheres, CO_2 atmospheres and H_2/He atmospheres. One may vertically subdivide each atmosphere into two main regions, the homosphere and heterosphere. In the homosphere gases are mixed by turbulent diffusion and their abundances controlled by chemistry, while in the heterosphere the mean free path between gas molecules becomes sufficiently large for molecular diffusion to dominate the vertical gas distribution. In the heterosphere gases tend to adopt a vertical distribution according to their individual scale heights, thus separating from one another. Heavy gases in the heterosphere accumulate at lower altitudes, while the lighter gases become dominant at higher altitudes. The region separating the heterosphere and homosphere is referred to as the homopause. Mathematically speaking the homopause is the altitude/pressure where the eddy and molecular diffusion coefficients are equal, however this criterium has to be treated with caution since other dynamical effects not necessarily captured by the eddy diffusion coefficient (such as global scale dynamics and atmospheric escape) further influence the vertical distribution of gases and thereby the homopause location. The homopause may rather be regarded as a transition region (as opposed to a specific altitude/pressure) in which the vertical distribution of a given gas changes from being described by the overall atmospheric scale height to being described by its individual scale height. Since the mixing ratios of heavy constituents begin to decrease significantly near the homopause, the chemistry in the heterosphere becomes considerably simpler than at lower altitudes.

The homopause usually lies within the lower thermosphere, which energetically is a natural consequence. Heavy molecules and those efficient in radiatively cooling the atmosphere for reasons outlined above tend to decrease in mixing ratio significantly near the homopause or below, so above the homopause the atmospheric temperatures tend to rapidly increase with altitude, a distinct feature of the lower thermosphere region. On Earth the average homopause is typically located near 105 km altitude (100 nbar) and principal gases above that region are O, N_2 and O_2. On Mars the homopause lies near 125 km (1 nbar), on Venus near 135 km (10 nbar), on both planets principal gases above the homopause being O, CO_2 and N_2. The CH_4 homopause on Jupiter is located near the 10 μbar and on Saturn near 1 μbar level, above which the dominant gases are H, H_2 and He. Recent observations of Titan's upper atmosphere by the Cassini Ion Neutral Mass Spectrometer (INMS) have placed the CH_4 homopause there near 850 km (1.6 nbar) (Yelle et al. 2008), above which the dominant gases are CH_4, N_2 and H_2.

While molecular diffusion is dominant above the homopause, thermospheric dynamics may play a key role in redistributing gases. As first proposed for Earth by Duncan (1969) and later calculated amongst other by Rishbeth and Mueller-Wodarg (1999), vertical winds in particular affect the distribution of gases, enhancing the abundances of heavier constituents in regions of upwelling and lighter constituents in regions of downwelling, a process also sometimes referred to as "wind-induced diffusion". The effects of this are particularly strong in the Earth's auroral regions where local Joule and particle precipitation heating generate regions of strong upwelling and downwelling, reducing and increasing the O/N_2 ratios locally (Rishbeth and Mueller-Wodarg 1999). This principle is also responsible for many of the upper atmospheric changes observed during geomagnetic storms, via changes in the O/N_2 ratio, which in turn, via recombination processes, controls ionospheric electron densities. On Venus the nightside hydrogen bulge first observed by the Pioneer Venus spacecraft

(Brinton et al. 1980) is believed to be caused by the downwelling as well. The vertical wind redistribution of constituents is believed to act equally on Jupiter (Bougher et al. 2005), Saturn (Mueller-Wodarg et al. 2006a) and Titan (Mueller-Wodarg et al. 2003), although for these atmospheres observations yet have to unambiguously confirm the process happening.

3.2 Composition and Evolution

The atmospheric composition of a planet or satellite originates from two primary sources, outgassing of volatiles trapped in the icy grains and/or planetesimals that formed the primary body, and material accumulated via accretion or impact of small bodies during or after the differentiation of the primary body. Such a case is quantitatively developed for the evolution of Earth's atmosphere by Dauphus (2003), who invokes fractionated nebular gases and accreted cometary volatiles. The fractionation stage resulted in a high Xe/Kr ratio, with xenon being more isotopically fractionated than krypton. Comet-delivered volatiles having low Xe/Kr ratios and solar isotopic composition. The resulting atmosphere had a near solar Xe/Kr ratio, almost unfractionated krypton delivered by comets, and fractionated xenon inherited from the fractionation episode. Similar arguments can be made for Mars based on the noble gas abundances and their isotopic ratios as measured by Viking and from ground based analysis of SNC meteorites (Owen et al. 1992; Bogard et al. 2001). Venus on the other hand shows a distinct difference: the abundances of neon and argon per gram of planet are excessively high (Owen et al. 1992). This must be considered in the context that the Venusian noble gas inventory is less sampled given the lack of a SNC analog. However, for the most part the noble gas inventories of the terrestrial planets (Venus, Earth, and Mars) indicate a similar origin and evolution.

The gas giants have two distinct evolutionary differences. Although they started from similar planetesimal fragments, once they reached a critical size they began to gravitationally accrete protosolar nebular gases (H, He, and Ne) in copius quantities. Furthermore, their distance from the protosun resulted in bombardment by a continuous stream of icy grains whose formation temperatures were below 50 K (Atreya et al. 1999; Owen 2007). The primary evidence for a significant contribution of heavy elements in the form of icy grains comes from the super solar ratios (4 ± 2) of argon, krypton, and xenon as well as carbon and sulfur bearing volatiles measured by the Galileo probe in its descent through the Jovian atmosphere, thus the name Solar Composition Icy Planetesimals (SCIPs) (Owen 2007).

Owen (2007) argues that the abundances of nitrogen and carbon in planetary atmospheres show a similar trend throughout the solar system (15–20 times solar) largely due to the fact that the carbon is easily converted into refractory materials via solar extreme ultraviolet/energetic particle induced chemistry in the interstellar medium, whereas the primary carrier of nitrogen is the highly volatile N_2 molecule requiring cold temperatures for retention on silicate/mineral surfaces. This is most directly illustrated at Venus where the carbon and nitrogen volatile inventory appears to still be present in the atmosphere today. Earth's nitrogen-dominated atmosphere is largely attributed to the creation and burial of carbonates a process mediated by the liquid water in the oceans (apparently absent at Venus). This C/N ratio also implies that for Mars with a CO_2 dominated atmosphere there cannot be large deposits of carbonates or the atmosphere would be predominantly nitrogen—the exception would be if significant burial of nitrogen bearing minerals occurs as well. The domination of N_2 in Titan's atmosphere again suggest the loss of copius quantities of carbon via methane escape (Strobel 2008a) and/or the conversion of carbon to heavy refractory compounds that are sequestered on Titan's surface (Lorenz et al. 2008), based on radar observations of surficial features (methane seas and organic dunes). However, both nitrogen in the form of a water-ammonia ocean and methane bound in clathrates may be significant sources of carbon and nitrogen yet accounted for. This is backed up by the amount of ^{40}Ar in the atmosphere (mixing ratio $4.3 \pm 0.1 \times 10^{-5}$) equivalent to $\sim 2\%$ outgassing of the interior (Niemann et al. 2005; Waite et al. 2005).

Water is the primary form in which oxygen (and substantial amounts of hydrogen) is found throughout the solar system, although the degree that it exists as vapor is highly dependent on the planet and the temperature of the body. Venus' high D/H ratio in both water and molecular hydrogen gas found in the atmosphere as well as its proximity to the sun are primary arguments for the escape of much of Venus' water to space over geological time (Donahue 1999). Earth's water is largely sequestered in its oceans. Mars' small size and weak gravitational binding of its atmosphere have been used to argue for significant loss of H, C, N, and O to space over geological time (Owen et al. 1992). However, the discovery of increasing quantities of sub-surface water at high latitudes leaves the question open as to how much water has been lost and how much retained in the interior and the polar caps.

The abundance of water in the outer solar system is a question of significant importance. The Galileo entry probe did not measure a super solar abundance of water as it did for carbon and sulfur. However, at the highest pressures measured the water mixing ratio was still rising. This as well as similar altitude dependent variations for sulfur and carbon bearing volatiles has led theorists to postulate that the probe landed in a 5 micron hotspot with a significant downdraft that led to an anomalous altitude distribution of volatiles and that future measurements must sample below this level to determine the water content of Jupiter (Atreya et al. 1999). This is a key factor for determining whether the SCIP delivery of heavy elements (water should be 4 ± 2 times solar) is correct or whether the heavy elements were delivered in the form of clathrates (water should be ~ 10 times solar).

In addition to N_2, CO_2 (inner solar system), CH_4 (outer solar system) and H_2O (see Table 1), each of the planets and satellites have their own unique sources of minor volatile compounds that provide important clues to the chemical evolution of the atmosphere. Venus' clouds contain a host of acidic compounds (HCl, HF, H_2SO_4) as well as OCS, SO, SO_2, and CO all presumably derived from late volcanic outgassing in an environment where there is no ocean to mitigate the chemistry (see Table 1 of de Bergh et al. 2006). Viking found both O_2 and CO in the atmosphere of Mars (Owen et al. 1977)—photochemical products of CO_2 photolysis. However, more recently CH_4 has been tentatively detected at the tens of parts per billion level (Formisano et al. 2004; Krasnopolsky et al. 2004; Krasnopolsky 2007b) suggesting either an internal geological process such as serpentinization or perhaps evidence for life in the interior, but the detection is as yet still controversial. Earth has a host of minor hydrocarbon volatiles that are byproducts of biology and many of which are human-induced pollutants largely due to our every increasing appetite for energy.

Jupiter's and Saturn's atmospheres contain the condensable volatiles NH_3, H_2S, as well as H_2O and their by products (i.e., NH_4SH). These gas giant atmospheres also contain a plethora of minor hydrocarbons derived from CH_4 including CH_3, C_2H_2, C_2H_4, C_2H_6, C_3H_4, C_3H_8, C_4H_2, and C_6H_6 (see Table 1 of Atreya et al. 2002). The induced hydrocarbon complexity is especially prevalent at high-latitudes where energetic particle precipitation from the powerful planetary aurorae leads to the formation of polycyclic aromatic hydrocarbons (PAHs) (Wong et al. 2000). This hydrocarbon complexity is also exemplified in the atmosphere of Titan where C_2H_2, C_2H_4, C_2H_6, C_3H_4, C_3H_6, C_4H_2, C_6H_6, and C_2N_2 have been observed (see Table 1 of Waite et al. 2005 and Table 1 of Waite et al. 2007). In fact the relative lack of hydrogen (for recombination with the photochemical initiator CH_3 to reform CH_4) as compared to Jupiter and Saturn, combined with the existence of N_2 and sufficient sources of free energy from solar extreme ultraviolet and energetic particle precipitation provide a unique environment at 1000 km above Titan's surface for the initiation of ion neutral chemistry that reaches complexities of over 10,000 atomic mass units (Waite et al. 2007) and contains a host of nitrile compounds (HCN, HC_3N, C_2H_3CN, CH_2NH, CH_3CN, C_2H_5CN, CH_3C_3N, HC_5N, C_5H_5N, CH_3C_5N, and C_6H_7N) (Vuitton et al. 2007) as well as the hydrocarbons cited above. This chemistry appears to have similar character and complexity to processes in interstellar clouds apart from the substitution of predominantly nitrogen bearing compounds for the oxygenated hydrocarbons of interstellar clouds. The large negative ions formed via this process are postulated as the precursor of the organic haze found throughout Titan's atmosphere.

In summary, the noble gases and their isotopic ratios are the best way to compare the origin and evolution of the volatiles that make up an atmosphere. The predominant form of carbon is CO_2 for the inner planets and CH_4 at the outer planets. Oxygen and hydrogen in the form of water is found at all the planets and satellites. However, the uncertainty of the water content of the outer solar system planets is a major open issue in our understanding of planetary formation. Nitrogen is primarily in the form of N_2 in the inner solar system and NH_3 in the outer solar system with the exception of Titan, which has a thick N_2 atmosphere. Minor species are involved in cloud and haze formation and can tell us much about the chemical evolution of the atmosphere.

3.3 Atmospheric Chemistry

The middle and upper atmospheres of planets in our solar system are cold enough that thermochemical equilibrium tends to play a negligible role in controlling atmospheric composition. Instead, aeronomy in these regions is dominated by disequilibrium photochemistry and chemical kinetics. Atmospheric chemistry is initiated when atoms or molecules absorb solar ultraviolet photons and/or when they interact with charged particles from the solar wind or planetary magnetospheres; that chemistry can in turn affect many properties of the atmosphere. For example, catalytic chemical cycles can allow trace photochemical constituents to profoundly affect the composition and properties of the bulk atmosphere, such as with odd-hydrogen chemistry maintaining carbon dioxide on Mars or chlorine chemistry influencing the composition of the middle atmospheres of Venus and the Earth. Trace photochemical constituents can also profoundly affect radiative balance and energy transport in an atmosphere, which in turn can affect atmospheric thermal structure and dynamics, such as with hydrocarbon photochemical products controlling radiative transport on the giant planets and Titan. Understanding the quantitative details of the chemical production and loss of minor species in a planetary atmosphere is therefore important for fully understanding the behavior of planetary atmospheres.

3.3.1 Terrestrial Planets

The chemistry of the middle and upper atmosphere of the Earth is too broad a subject to be effectively covered in this short review. Seinfeld and Pandis (2006), Brasseur and Solomon (2005), Wayne (2000), Finlayson-Pitts and Pitts (1999), and Brasseur et al. (1999) provide good general reviews of the subject.

Photochemistry on Mars is not completely understood. Many of the basic processes have been worked out, but current photochemical models have difficulty simultaneously reproducing the observed abundances of O_2, CO, H_2O_2, O_3, H_2O and H_2, indicating shortcomings with the models (e.g., Nair et al. 1994; Atreya and Gu 1994; Krasnopolsky 1995). Global-average one-dimensional (1-D) models have now been augmented by time-variable 1-D models (e.g., Clancy and Nair 1996; García Muñoz et al. 2005; Krasnopolsky 2006a;

Zhu and Yee 2007) and by multi-dimensional photochemistry-transport models (e.g., Moreau et al. 1991; Lefèvre et al. 2004; Moudden and McConnell 2007; Moudden 2007), which can better explain some of the seasonal, diurnal, and spatial variability of atmospheric constituents, but some puzzles still remain.

The Martian atmosphere is composed predominantly of CO_2, which can be photolyzed by solar ultraviolet photons to produce CO and O. The recombination of CO and O is spin forbidden and very slow. The oxygen atoms will instead combine with other oxygen atoms to form O_2. Large amounts of CO and O_2 in a 2:1 ratio would be expected to form in a pure CO_2 atmosphere. The Martian atmosphere is not pure CO_2, however, and McElroy and Donahue (1972) and Parkinson and Hunten (1972) first pointed out the fundamental importance of catalytic cycles involving the so-called "odd hydrogen" HO_x species (e.g., H, OH, HO_2), which together with H_2O_2 help recycle the CO_2 and reduce the CO and O_2 abundances (and the CO/O_2 ratio) on Mars. Water photolysis, along with the reaction of H_2O with $O(^1D)$ from CO_2, O_2, and O_3 photolysis, provides the source of the HO_x species. One important HO_x catalytic cycle is

$$\begin{aligned} \mathrm{H + O_2 + M} &\longrightarrow \mathrm{HO_2 + M} \\ \mathrm{O + HO_2} &\longrightarrow \mathrm{O_2 + OH} \\ \mathrm{CO + OH} &\longrightarrow \mathrm{CO_2 + H} \\ \hline \mathrm{Net: CO + O} &\longrightarrow \mathrm{CO_2}, \end{aligned}$$

but others involving O_3, H_2O_2, and perhaps odd-nitrogen (NO_x: NO, NO_2) species operate as well (e.g., Krasnopolsky 1986; Yung and DeMore 1999). The HO_x catalytic cycles are so effective at removing CO that photochemical modelers (e.g., Nair et al. 1994; Atreya and Gu 1994; Krasnopolsky 2006a) have resorted to invoking heterogeneous reactions on aerosol surfaces or changes in measured reaction rate coefficients as a means of reducing HO_x abundances in order to reproduce the observed CO abundance on Mars. Note that no measurements of O densities have to-date been published for Mars.

HO_x chemistry also strongly affects the abundance of many other atmospheric species, including O_3, H_2O_2, and H_2. The photochemistry of ozone on Mars is discussed by Lefèvre et al. (2004 and references therein). Ozone is produced through three-body recombination of O with O_2 and is lost through reactions with HO_x species. Because HO_x species are derived from water photochemistry, an anticorrelation between O_3 and H_2O abundances is expected (e.g., Clancy and Nair 1996), although the actual O_3–H_2O relationship becomes more complicated than this simple picture when full seasonal, meridional, and vertical variations are taken into account (e.g., Lefèvre et al. 2004, 2007; Krasnopolsky 2006a). Dayglow observations from the $a^1\Delta_g \rightarrow X^3\Sigma_g^-$ electronic transition of O_2 have also been used as a proxy to track ozone abundances, and $O_2(a^1\Delta_g)$ chemistry is usually included in photochemical models (e.g., García Muñoz et al. 2005; Krasnopolsky 2006a). The chemistry of H_2O_2, which is observed to vary with location and season on Mars (e.g., Encrenaz et al. 2004), has been extensively studied. Hydrogen peroxide is produced mainly through reaction of HO_2 with HO_2 and is lost through photolysis. Because H_2O_2 is derived from HO_x chemistry, the abundances of H_2O and H_2O_2 are expected to be correlated to some extent (e.g., Krasnopolsky 2006a; Moudden 2007), although again complications with this correlation are present in observations and models. HO_x species and O_3, $O_2(a^1\Delta_g)$, and H_2O_2 are all short-lived enough that diurnal variations in abundances are expected.

Molecular hydrogen, on the other hand, is long-lived on Mars. Water dissociation provides a source of atomic H, and reaction of H with HO_2 produces H_2. The long-lived H_2 molecules can then diffuse into the upper atmosphere where ion reactions convert H_2 back into H, and the H can escape. Hydrogen chemistry on Mars is a classic example of chemical coupling between the upper and lower atmospheres, and the bulk Martian atmospheric composition cannot be understood without examining the behavior of the entire atmosphere, from the surface to the exobase. Hydrogen escape over time could potentially affect the oxidation state of the Martian atmosphere; however, nonthermal escape mechanisms for oxygen also exist (as well as potential surface sinks of reactive oxygen), and the Martian atmosphere may be self-regulating in the sense that a steady-state loss of two hydrogen atoms for every one oxygen atom is maintained (see Yung and DeMore 1999 for further details), although this has to-date not yet been proven to be the case on Mars.

The recent tentative detection of CH_4 on Mars (Formisano et al. 2004; Krasnopolsky et al. 2004; Krasnopolsky 2007b) has sparked much interest due to potential astrobiological implications and due to a puzzling observed variability with time and location, but much uncertainty still surrounds this detection. Methane is expected to have a relatively long (~300–700 year) photochemical lifetime on Mars (Wong et al. 2003; Krasnopolsky et al. 2004)—short-term temporal and spatial variability is therefore unexpected, and a significant source must exist to replace loss by photolysis and reaction with OH, O, and $O(^1D)$. Although it is possible that some hitherto unidentified photochemical production mechanism may exist in the Martian atmosphere (Bar-Nun and Dimitrov 2006) or that the methane is derived from meteoroid impacts, it is more likely that the methane derives from crustal or interior sources (e.g., fluid-rock interactions, volcanism, methane clathrate hydrates, methanogenic bacteria; e.g., Max and Clifford 2000; Krasnopolsky et al. 2004; Oze and Sharma 2005; Lyons et al. 2005; Krasnopolsky 2006b; Atreya et al. 2007). The observed temporal variability suggests that local surface sources and sinks could be operating.

Venus, with its massive CO_2 atmosphere, shares some similarity in atmospheric chemistry with Mars, but the catalytic removal of CO and O_2 to recycle the CO_2 atmosphere is apparently considerably more efficient on Venus than on Mars. Early photochemical models (e.g., Winick and Stewart 1980; Yung and DeMore 1982; Krasnopolsky and Parshev 1983) overpredicted the O_2 abundance and underpredicted the efficiency of CO_2 recycling. Aside from the HO_x catalytic cycles described for Mars, chlorine catalytic cycles, such as the following, likely operate to stabilize CO_2 on Venus:

$$\begin{aligned}
\mathrm{Cl + CO + M} &\longrightarrow \mathrm{ClCO + M} \\
\mathrm{ClCO + O_2 + M} &\longrightarrow \mathrm{ClC(O)O_2 + M} \\
\mathrm{ClC(O)O_2 + O} &\longrightarrow \mathrm{CO_2 + O_2 + Cl} \\
\hline
\mathrm{Net: CO + O} &\longrightarrow \mathrm{CO_2},
\end{aligned}$$

where the chlorine atoms are derived from HCl. The chlorine cycles are expected to have much more influence than the HO_x cycles on Venus. The above cycle, first proposed by Yung and DeMore (1982), is used in more recent photochemical models, along with oxidation of CO on cloud particles, to help explain the efficiency of CO_2 recycling on Venus (e.g., Pernice et al. 2004; Mills et al. 2006; Mills and Allen 2007), but significant uncertainties remain in the details of the CO_2 cycle on Venus. Slanger et al. (2006) have proposed that reaction between excited $O_2(c^1\Sigma_u^-, v=0)$ and CO may help explain both the CO_2 stability and an observed variability of 557.7 nm oxygen green line emission and 1.27-μm $O_2(c - X)$ emission on Venus.

Thermochemical equilibrium and high-temperature kinetic processes prevail in the lowest regions below the cloud decks on Venus (e.g., Krasnopolsky 2007a). Near ~60 km (i.e., the upper cloud region), SO_2 is oxidized through photochemical processes to form H_2SO_4, which condenses and falls to lower altitudes, where it thermally decomposes to produce SO_2 and H_2O. This sulfur oxidation cycle may be influenced by coupled CO and SO_2 oxidation, whose net result is $CO + SO_2 + O_2 + H_2O \longrightarrow CO_2 + H_2SO_4$ (e.g., Yung and DeMore 1982, 1999). An up-to-date discussion of the unknown cloud constituent that absorbs at blue wavelengths is provided by Krasnopolsky (2006c). The chemistry of sulfur in the Venus atmosphere is rich and complex, and other interesting sulfur cycles besides SO_2 oxidation may exist, including coupled chlorine and sulfur chemistry (e.g., DeMore et al. 1985; Mills 1998; Mills and Allen 2007) and elemental sulfur ("polysulfur") cycles (e.g., Krasnopolsky and Pollack 1994; Mills 1998; Yung and DeMore 1999). Nitrogen chemistry on Venus has also been examined, and NO_x chemistry has the potential to affect the major photochemical products (e.g., Yung and DeMore 1999). Atmospheric dynamics likely affects photochemical processes on Venus, but multi-dimensional coupled chemical-dynamical models of the lower and middle atmosphere of Venus have yet to be developed.

The neutral upper atmosphere of Venus is affected by ionospheric chemistry, solar-cycle variations, dynamics, and escape. Nonthermal escape processes such as charge exchange, sputtering, solar-wind pick up, detached plasma clouds, and collisional ejection (via hot-atom production during photochemical reactions) dominate over thermal escape (see Lammer et al. 2006). The review by Fox (2004) discusses recent advances related to the aeronomy of the Venusian thermosphere (see also the older reviews of Fox and Bougher 1991; Fox and Sung 2001).

Although our understanding of atmospheric chemistry on Venus has advanced greatly in the past three decades, numerous outstanding questions remain (e.g., Mills and Allen 2007; Krasnopolsky 2006c). Venus Express data will hopefully provide new key measurements on the three-dimensional distribution of atmospheric constituents, maps of airglow emission, temperature retrievals, and wind derivations that will improve our understanding of chemical and physical processes in Venus' atmosphere. First measurements by the Spectroscopy for Investigation of Characteristics of the Atmosphere of Venus (SPICAV) instrument of minor mesospheric constituents and an unexpected warm layer near 100 km on the nightside have been reported by Bertaux et al. (2007). Mesospheric emissions in the infrared by CO_2 (non-LTE at 4.3 μm on the dayside) and O_2 (1.27 μm on the nightside) have been observed by the Venus Express Visible and Infrared Thermal Imaging Spectrometer (VIRTIS) (Drossart et al. 2007). These data, combined with much-needed laboratory experiments, theoretical models, and continued Earth-based observations will provide vital clues for deciphering the remaining questions regarding the intricate chemical coupling mechanisms in the Venus atmosphere.

3.3.2 Giant Planets

The hydrogen-dominated giant planets Jupiter, Saturn, Uranus, and Neptune share many similarities in upper-atmospheric chemistry, although some notable differences exist. Temperatures are low enough on the giant planets that major equilibrium volatiles like H_2O, NH_3, and H_2S will condense in the tropospheres of all the giant planets, and even CH_4 will condense on Uranus and Neptune. Methane, the most volatile of all these major hydrides, can survive past the tropopause cold trap and be transported to the upper stratosphere or mesosphere, where it will interact with solar and stellar ultraviolet radiation and be photodissociated. Methane photochemistry therefore dominates middle-atmospheric chemistry on the giant planets, and the major photochemical products are complex hydrocarbons. Molecular diffusion of the relatively heavy methane molecules in the lighter background hydrogen gas eventually limits the vertical extent to which the CH_4 can be carried on the giant planets and provides a natural boundary between the middle atmosphere and thermosphere. The long-lived hydrocarbon photochemical products eventually diffuse down into the deep troposphere, where they are thermochemically converted back to methane, completing the methane cycle. Uranus' weak internal heat source (unlike that of the other giant planets) seems to suppress vertical motions, allowing molecular diffusion to take over at relatively low altitudes and preventing the buildup of the large column abundances of complex hydrocarbons that are observed on the other giant planets.

Our current understanding of hydrocarbon photochemistry on the giant planets is reviewed by Strobel (2005), Moses et al. (2004), and Yung and DeMore (1999). Further details of the photochemistry on each giant planet can be found in models presented by Romani and Atreya (1988), Summers and Strobel (1989), Romani et al. (1993), Gladstone et al. (1996), Romani (1996), Bishop et al. (1998), Moses et al. 2000a, 2005, Lebonnois (2005), and Moses and Greathouse (2005). Methane is photolyzed predominantly by Lyman α photons at 121.6 nm, and the major photolysis products are CH_3, $CH_2(a^1A_1)$, and CH. The methane photolysis branching ratios at 121.6 nm are not completely known. The CH_3 radicals will either recombine with themselves to form C_2H_6, or combine with H to reform methane. $CH_2(a^1A_1)$ radicals will react with H_2 to either produce CH_3 or to quench to the ground state and eventually react with H to form CH. The CH radicals can insert into methane to form C_2H_4, into ethane to form C_3H_6, or into H_2 to recycle CH_2.

The major hydrocarbon photochemical products on the giant planets are ethane (C_2H_6), acetylene (C_2H_2), ethylene (C_2H_4), propane (C_3H_8), methylacetylene (C_3H_4), diacetylene (C_4H_2), benzene (C_6H_6), and methyl radicals (CH_3), all of which have been observed on one or more of the giant planets. The primary mechanisms for producing complex hydrocarbons are radical-radical combination reactions (e.g., $2CH_3 + M \longrightarrow C_2H_6 + M$), CH insertion reactions (e.g., $CH + CH_4 \longrightarrow C_2H_4 + H$), and C_2H insertion reactions (e.g., $C_2H + C_2H_2 \longrightarrow C_4H_2 + H$). Photolysis, cracking by atomic hydrogen (e.g., $H + C_2H_5 \longrightarrow 2CH_3$), and disproportionation reactions (e.g., $CH_3 + C_3H_2 \longrightarrow C_2H_2 + C_2H_3$) are the primary mechanisms for destroying carbon-carbon bonds. Atomic hydrogen is a major product of this hydrocarbon photochemistry, and reactions of H with hydrocarbons are of critical importance in defining the relative abundances of the major constituents. Ethane and acetylene are major coolants in giant-planet stratospheres, and photochemistry therefore strongly influences the atmospheric thermal structure (e.g., Bézard and Gautier 1985; Yelle et al. 2001). Although hydrocarbon photochemistry on the giant planets is qualitatively understood, some quantitative details are lacking, especially for the photochemistry of C_3H_x hydrocarbons and benzene. Laboratory measurements and theoretical calculations are needed to fill in uncertain model parameters and uncertain spectroscopic parameters for abundance derivations (see also Huestis et al. 2008).

The recent detection of oxygen compounds that are unambiguously in the stratospheres of the giant planets (aside from the Comet Shoemaker-Levy 9 impact debris on Jupiter) indicates that external material from meteoritic dust, ring/satellite debris, and/or cometary impacts frequently enters giant-planet atmospheres (e.g., Feuchtgruber et al. 1997, 1999; Bergin et al. 2000; Moses et al. 2000b; Bézard et al. 2002; Lellouch et al. 2002, 2005, 2006; Kunde et al. 2004; Flasar et al. 2005a; Burgdorf et al. 2006; Hesman et al. 2007). The relative importance of each of these sources is not well understood and may differ from planet to planet. The influx of oxygen species can have a minor effect on hydrocarbon abundances (e.g., Moses et al. 2000b, 2005). Water derived from external sources will condense in the stratospheres of all the giant planets and contribute to the stratospheric haze load. Above the condensation region, water is lost primarily by photolysis to form $OH + H$, but the OH reacts with H_2, CH_4, and C_2H_6 to efficiently recycle H_2O. Some of the water can be permanently converted to other oxygen compounds (mainly CO) through three-body addition of OH with C_2H_2 and C_2H_4, which forms complexes that eventually produce CO. These reactions of OH with unsaturated hydrocarbons most likely dominate over $O + CH_3$ or $OH + CH_3$ in the photochemical production of CO on the giant planets (Moses et al. 2000b, 2005). CO is quite stable in giant-planet stratospheres and will be lost through diffusion into the troposphere. Carbon dioxide can form through the reaction of $OH + CO \rightarrow CO_2 + H$. Interestingly, the large CO abundance on Neptune and the vertical distribution of that CO suggest that Neptune may have experienced a large cometary impact on the order of 200 years ago (Lellouch et al. 2005; see also Hesman et al. 2007). Between this event, and the 1994 Shoemaker-Levy 9 impacts with Jupiter, the fate of cometary debris on the giant planets is of great interest not only for understanding current atmospheric chemistry on the giant planets but for understanding impact rates in the solar system.

Observers have now mapped the meridional distribution of several stratospheric constituents on Jupiter and Saturn (e.g., Kunde et al. 2004; Flasar et al. 2005a; Greathouse et al. 2005, Prangé et al. 2006, Nixon et al. 2007; Howett et al. 2007), and the observations indicate the need for coupled photochemistry-transport models (see Moses and Greathouse 2005; Liang et al. 2005; Lellouch et al. 2006; Moses et al. 2007). Another focus of current research is the extent to which auroral chemistry affects the stratospheric composition locally and perhaps even globally on Jupiter and Saturn (e.g., Wong et al. 2000, 2003; Friedson et al. 2002). Outstanding problems include the underlying reasons for the differences in the hydrocarbon abundances on the giant planets, the underlying physical and chemical mechanisms controlling the meridional distribution of hydrocarbons on the giant planets, the details of the intricate coupling between chemistry, temperatures, and dynamics on the giant planets, and the reasons for the observed changes in composition over time.

3.3.3 Titan, Pluto, Triton, and Io

Atmospheric chemistry on Titan is vigorous and complex and understood only in its basics. As on the giant planets, hydrocarbon photochemistry plays a large role in the middle and upper atmosphere, but Titan is composed predominantly of N_2 rather than H_2, and the resulting photochemistry differs considerably. On the giant planets, the dominant nitrogen-bearing constituent—in this case NH_3—is physically separated from the methane photolysis region due to tropospheric condensation. Coupled carbon-nitrogen chemistry is therefore suppressed on the giant planets (although Neptune, which may contain a large amount of N_2, is an exception, see Lellouch et al. 1994). That is not the case on Titan. On Titan, as on the Earth, N_2 can be dissociated at high altitudes. Although 127 nm photons have sufficient energy to break the strong N_2 bond, N_2 photoabsorption only becomes significant at wavelengths below 100 nm, and photolysis occurs indirectly through excitation into predissociating electronic states, which provides only a minor source of N atoms. Instead of direct photolysis, the N_2 bond is broken predominantly through dissociative ionization from impacting magnetospheric electrons or solar EUV photons. The ultimate products of this process are $N(^2D)$, N, and N^+. The N atoms can react with CH_3 to produce H_2CN, and eventually HCN and other nitriles (e.g., through dissociation of HCN to CN, followed by reaction of CN with unsaturated hydrocarbons like C_2H_2 and C_2H_4 to form HC_3N and

C_2H_3CN, respectively). The N^+ ions also can react with methane and follow subsequent pathways that can lead to nitrile production. The excited $N(^2D)$ atoms can react with CH_4 to produce NH, which predominantly ends up recycling the N_2. $N_2{}^+$ ions can also react with methane to produce $CH_3{}^+$ or $CH_2{}^+$ ions, followed by ion-neutral reactions to produce complex hydrocarbon ions, which can recombine with electrons to produce complex neutral hydrocarbon molecules.

The suggestion that ion chemistry has a significant impact on Titan's neutral atmospheric chemistry (e.g., Banaszkiewicz et al. 2000; Molina-Cuberos et al. 2002; Wilson and Atreya 2004) and aerosol formation has been dramatically demonstrated from measurements acquired with the Ion-Neutral Mass Spectrometer (INMS) aboard the Cassini spacecraft (e.g., Waite et al. 2005, 2007; Cravens et al. 2006; Vuitton et al. 2006, 2007) and is further confirmed from laboratory experiments (e.g., Imanaka and Smith 2007). The production of nitriles, polycyclic aromatic hydrocarbons (PAHs), and "tholins" in particular may be augmented by ion chemistry.

Hydrocarbon photochemistry in Titan's middle atmosphere is very similar to that described above for the giant planets, with some minor differences resulting from the fact that H_2 is much less abundant on Titan, and hydrogen escape affects abundances (see Yung et al. 1984; Toublanc et al. 1995; Lara et al. 1996; Yung and DeMore 1999; Wilson et al. 2003; Wilson and Atreya 2004, and Lebonnois 2005 for detailed descriptions of Titan photochemistry). Unlike the giant planets, however, there is no obvious mechanism for converting the complex hydrocarbons back to methane. Methane will be irreversibly destroyed by photochemistry on a time scale of $\sim$10–100 million years in Titan's atmosphere, and the complex hydrocarbon products will rain down or diffuse onto the satellite surface. The fact that methane is still present in Titan's atmosphere suggests either that the human species observe Titan at a point of time where much of its methane is still present in the atmosphere or, more likely, that a significant surface or interior source mechanism must exist. Similar to what has been suggested for methane production on Mars, several hydrogeochemical sources are possible (e.g., Atreya et al. 2006). As on the giant planets, the presence of H_2O in Titan's upper atmosphere suggests an external source of oxygen (Coustenis et al. 1998). The photochemistry of oxygen species in Titan's atmosphere is discussed by Wong et al. (2002) and Wilson and Atreya (2004). Coupled photochemistry-dynamical models have recently been developed for Titan to help explain observed meridional and seasonal variations in species abundances (e.g., Lebonnois et al. 2001; Luz et al. 2003).

Many outstanding problems regarding neutral atmospheric chemistry on Titan remain, including details of the complex ion-neutral coupling, the pathways that convert gas-phase species to "tholins" and other aerosols, the dominant physical and chemical mechanisms controlling the latitude, altitude, and time variation of hydrocarbons and nitriles, the evolutionary history of atmospheric chemistry on Titan, the source of methane, the origin and fate of oxygen species, and the mechanisms and rates or escape of atmospheric constituents.

Pluto and Triton have vapor-pressure-controlled atmospheres composed largely of N_2, CO, CH_4, and Ar that are buffered by surface ices. The atmospheric photochemistry on Pluto and Triton is expected to be similar and share some general characteristics with hydrocarbon, nitrogen, and oxygen photochemistry on Titan. Methane is expected to produce hydrocarbons like C_2H_2 and C_2H_6, which would condense to form hazes. Coupled N_2–CH_4 photochemistry can produce nitriles, which would also condense. The presence of abundant CO adds some interesting atomic carbon chemistry that differs from the situation on Titan. Ion chemistry is again intimately linked with neutral chemistry. Atmospheric escape is prevalent. For details concerning atmospheric chemistry on Pluto and Triton, see Strobel and Summers (1995), Krasnopolsky and Cruikshank (1995, 1999), Summers et al. (1997), Lara et al. (1997), Yung and DeMore (1999).

Io's atmosphere is also buffered by surface frosts, but in this case by SO_2. Carbon and hydrogen appear to be virtually absent from the system, so Io's atmospheric chemistry is unique in the solar system. Volcanism is the ultimate source of the SO_2, and active volcanic outgassing may locally affect atmospheric chemistry. The SO_2 atmosphere as a whole appears to be dominated by frost sublimation (e.g., Jessup et al. 2004; Saur and Strobel 2004; Strobel and Wolven 2001), except at high latitudes and near the terminators (e.g., at high solar zenith angles) or on the night side, where local volcanic regions can be important. Temporal variability is also evident in observations. Sulfur dioxide is dissociated to form SO and O (dominant branch) or $S + O_2$, and subsequent chemistry produces an atmosphere in which SO_2, SO, O, S, and O_2 dominate the neutral chemistry. Observations of S_2 at ultraviolet wavelengths (e.g., Spencer et al. 2000; Jessup et al. 2007), NaCl at millimeter wavelengths (Lellouch et al. 2003), and SO emission at 1.7 μm (de Pater et al. 2002, 2007; Laver et al. 2007) indicate that the background atmosphere can be perturbed by active volcanoes. S_2 is short-lived and is readily photolyzed to form S or can react with O to form SO. Other products of S_2–SO_2 photochemistry include S_x species (e.g., S_3, S_4, S_8) and S_2O. Alkali chemistry leads to NaCl, Na, Cl, K, KCl, and perhaps species such as $NaSO_2$.

Details of SO_2 photochemistry on Io are discussed by Summers and Strobel (1996), and the influence of volcanic species like S_2 and NaCl are discussed by Moses et al. (2002a, 2002b). These 1-D hydrostatic-equilibrium models are useful for elucidating the dominant chemical production and loss mechanisms on Io but do not provide realistic descriptions the actual atmosphere, in which the SO_2 gas will flow rapidly away from the subsolar point or volcanic sources. Multi-dimensional coupled chemistry-transport models have been developed by Wong and Johnson (1996), Wong and Smyth (2000), Smyth and Wong (2004) (see also Zhang et al. 2003, 2004; Saur and Strobel 2004). Thermochemistry in volcanic gases has been studied by Zolotov and Fegley (1998a, 1998b, 1999, 2000), Fegley and Zolotov (2000), Schaefer and Fegley (2004, 2005a, 2005b, 2005c). The latter papers show that volcanic sources can introduce more exotic species into Io's atmosphere, albeit at minor abundances.

3.3.4 Comets

Except for some limited in situ investigations (e.g., comet Halley with the mass spectrometers of VEGA and Giotto), the chemical composition of comets is assessed by remote sensing. In this investigation, radio and infrared spectroscopy are the main techniques used, since visible and ultraviolet spectroscopy are only sensitive to secondary products radicals, ions and atoms with the notable exception of CO and its UV bands. The current census of these cometary molecules is reviewed by Bockelée-Morvan et al. (2005) and production rates of volatiles shown in Fig. 3. The major cometary species are probably all known now, but many minor constituents are still to be identified.

The overwhelming chemical process in cometary atmospheres is photolysis, leading to photodegradation and photoionization of the cometary molecules (Huebner et al. 1992; Crovisier 1994). Hence the paradigm of parent molecules, directly sublimated from nucleus ices, and their degradation product, the daughter molecules (in fact a misnomer, since they are indeed radicals, atoms or molecular ions). Two-body chemical reactions between neutrals have very low rates because of the low temperature and of the energy threshold of such reactions. This is not the case for ion-neutral reactions, but they have little influence on the chemical composition because of the low ionization state of the inner coma (Rogers et al. 2005). Thus it is unlikely that the minor constituents observed in the coma (Fig. 3) result

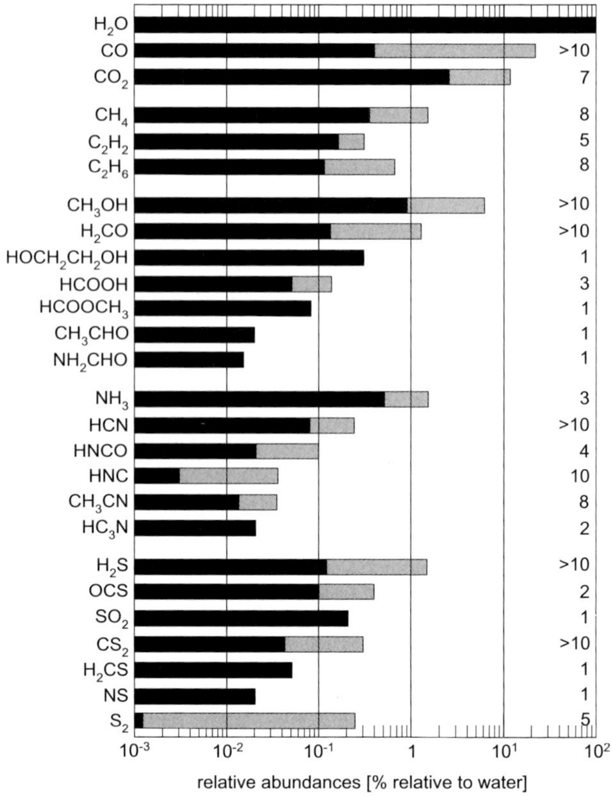

Fig. 3 Production rates of cometary volatile molecules relative to water, in percent. These rates are believed to trace the relative abundances in cometary ices. The *grey portions of the bars* show the range of the comet-to-comet variations. The number of comets for which data are available is listed on the *right*. (Adapted and updated from Bockelée-Morvan et al. 2005)

from chemical reactions. They rather come from nucleus ices, or from the degradation of organic grains, the so-called distributed sources.

Photodestruction rates, normalised at 1 AU from the Sun, are typically $\beta = 10^{-6}$ to 10^{-4} s^{-1}. They are 1.3×10^{-5} s^{-1} for water, 2×10^{-6} s^{-1} for a long-lived species such as CO_2, and around 2×10^{-4} s^{-1} for short-lived species such as NH_3, H_2S or SO_2 (Huebner et al. 1992; Crovisier, 1989, 1994). These rates are not yet known with sufficient accuracy for all species, especially for radicals, for which quantitative absorption spectra are difficult to obtain.

Imaging the lines of cometary molecules with radio interferometers, or long-slit spectroscopy in the visible or the infrared, give direct measurements of the molecular space distributions. In this way, one can put stringent constraints to the photodestruction rates, test parent molecule–daughter molecule filiations, and investigate distributed sources. For example, in a study of sulfur-bearing molecules in comet Hale-Bopp with the Institut de Radio Astronomie Millimétrique (IRAM) interferometer, Boissier et al. (2007) obtained interesting constraints to the lifetimes of the CS and SO radicals.

The origin of some radicals is not yet fully understood: this is the case for NS and S_2 which are unlikely to come from nucleus ices or from a parent molecule. In this situation, coma chemistry has been invoked (Canaves et al. 2007). Another puzzling case is that of hydrogen isocyanide HNC (e.g., Lis et al. 2008). Chemical models are unable to account for a HCN⟶HNC conversion. The HNC/HCN abundance ratio is found to increase when the distance to the Sun decreases. This would favour the production of HNC from the thermal degradation of organic grains. Among the major cometary molecules, several bona fide parent molecules are known to come from a significant extended source in addition to the native, nuclear source. This is the case for H_2CO, OCS, CO.

Comets are not all alike. Large variations in the relative molecular abundances are observed from comet to comet. From a dynamical point of view, one distinguishes Jupiter-family comets coming from the Kuiper belt and nearly-ecliptic comets coming from the Oort cloud. If these two main dynamical classes correspond to different formation scenarios, one would expect different chemical compositions (Crovisier 2007).

The diversity of the chemical composition of comets has been studied for daughter molecules with narrow-band spectro-photometry in the visible (about 100 comets; e.g., A'Hearn et al. 1995), for parent molecules from radio spectroscopy (about 30 comets; Biver et al. 2002) and infrared spectroscopy (about 15 comets; e.g., Mumma et al. 2003). The diversity of relative abundances of parent molecules is indicated in Fig. 3. The scatter is peculiarly large for species such as CO, CH_3OH, H_2CO, H_2S. However, the two main fragments of the split comet 73P/Schwassmann-Wachmann 3 were found to have similar compositions, suggesting that the building blocks of this comet had a uniform composition (Villanueva et al. 2006; Dello Russo et al. 2007).

From their taxonomic study based upon daughter molecules, A'Hearn et al. (1995) proposed two classes of comets according to their C_2/CN ratio: typical comets and C_2-depleted comets. The taxonomy studies based upon parent molecules are still in their infancy. There is no obvious link between cometary composition and dynamical origin, except that the C_2-depleted class of comets is found to be linked to Jupiter-family comets (A'Hearn et al. 1995). It is not known how this translates into terms of parent molecules, because the parents of the C_2 radical are still poorly identified.

4 Dynamics and Vertical Coupling

4.1 Global Dynamics and Waves on Earth

Atmospheric dynamics can in terms of time be separated into several components, the "mean circulation" which varies on seasonal time scales, the "tidal component" which varies on time scales of a day (rotation) or fractions thereof and the "turbulent component" which describes variations on time scales of small fractions of a rotation (hours to minutes or less). Additionally, one may subdivide dynamics into their geographical extent, namely global (planetary-scale), intermediate (extending over several degrees in latitude/local time) and local (on scales of km or less). This chapter will concentrate on the global mean circulation in planetary atmospheres, describing a summary of the main characteristics of atmospheric dynamics and the importance of atmospheric waves.

Often atmospheric winds are associated with solar heating, but in addition to the thermal drivers of winds there are other momentum sources, including turbulence, wave drag and in the ionosphere/thermosphere region of magnetized bodies ion drag. While particular momentum sources may play a role in specific regions only, the atmosphere is ultimately an integrated and coupled system, whereby conservation of mass and other coupling processes (wave drag, viscosity amongst other) require us to look at an atmosphere as a global entity. Mostly, dynamical coupling occurs upward, so the stratosphere/mesosphere region will affect the thermosphere more than vice-versa. However, this may not be true in terms of composition, where downward flux of gases formed at higher altitudes often plays an important

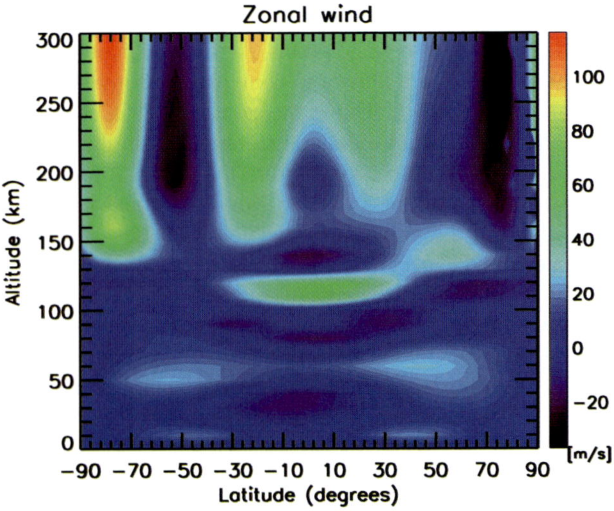

Fig. 4 Diurnally and longitudinally averaged zonal winds (positive eastward) in the Earth's atmosphere, as given by the Horizontal Wind Model (HWM) of Hedin et al. (1996) for equinox conditions at low geomagnetic activity (ap = 6) and moderate solar activity (F10.7 = 100)

role. A striking example of this is the atmosphere of Titan, where ionospheric chemistry is believed to profoundly affect the composition of the stratosphere and below.

Extensive wind observations have been made on Earth over the past decades using satellites such as the Atmospheric Explorer (AE), Dynamics Explorer (DE), Upper Atmosphere Research Satellite (UARS) and the Thermosphere Ionosphere Mesosphere Energetics and Dynamics (TIMED) satellite as well as rockets, balloons and ground based telescopes. Some of these measurements have been summarized in an empirical model, the Horizontal Wind Model (HWM) which represents a spectral fit to the data (Hedin et al. 1996). Figure 4 shows diurnally and longitudinally averaged zonal winds in the Earth's atmosphere between the surface and 300 km altitude, as given by the HWM for equinox conditions at low geomagnetic and moderate solar activity conditions.

Wind profiles such as these are successfully reproduced also by General Circulation Models (GCMs), including the Coupled Mesosphere and Thermosphere (CMAT) model (Harris 2000; Harris et al. 2002). This model is one example of codes developed for Earth that reach from the stratosphere to the thermosphere, covering most of the vertical extent of the atmosphere. Another example of such a model is the Thermosphere Ionosphere Mesosphere Electrodynamic General Circulation Model (TIME-GCM) of Roble and Ridley (1994). Development of such models over the past decade was motivated by the recognition that vertical coupling in the atmosphere was important. The need is now recognized to study the atmosphere in its entirety from the lower to the upper atmosphere, rather than examining regions such as the thermosphere in isolation (see also Bougher et al. 2008).

The success of physical models such as CMAT in reproducing observed winds in the stratosphere, mesosphere and thermosphere allows us to investigate the momentum balance in the Earth's atmosphere. In the stratosphere and mesosphere pressure gradients balance Coriolis forces, resulting in geostrophic flow parallel to isobaric surfaces. The eastward jets in Fig. 4 at mid-latitudes near 50 km are driven by stratospheric ozone heating which generates poleward winds (not shown) which become eastward due to Coriolis forces. In the thermosphere the momentum balance is found to be between pressure gradients, Coriolis forces, vertical viscosity and ion drag, giving a much more complex situation that requires numerical models for accurate calculation due to the non-linearity of the problem. Wind flow in the low and mid latitude thermosphere is primarily from the sub-solar to anti-solar hemisphere perpendicular to isobars. At high latitudes ion drag leads to the wind velocities considerably differing from the simple sub-solar to anti-solar flow. The strong zonal winds in the polar thermosphere (above 200 km) in Fig. 4 are driven primarily by ion drag due to the high latitude convective electric fields (see also Sect. 4.3). In the low and mid latitude thermosphere the wind flow is primarily diurnal above around 200 km altitude and semidiurnal in the lower thermosphere.

One important process in the upper mesosphere (60–90 km) region that is not captured by GCMs driven by solar heating alone is the dissipation or breaking of atmospheric gravity waves which deposits momentum in that region. The wave drag gives rise to the closure and reversal of the zonal jets shown in Fig. 4, a feature observed in the atmosphere which cannot be reproduced by models without including the gravity wave drag term in the momentum equation (Geller 1983).

Internal gravity waves propagate vertically and may be driven either thermally, by surface topography, convection or shear instabilities in the background wind flow. Most families of gravity waves have horizontal phase speeds, c, of up to tens of meters per second with respect to the surface, but their vertical propagation is determined rather by the phase speed with respect to the mean flow, $\overline{u}$, in the atmosphere. As long as $\overline{u} - c \neq 0$ the waves do not interact with the background atmosphere. In the absence of any dissipation of the waves their amplitudes grow vertically as $\rho^{-1/2}$, with ρ being the mean atmospheric density, so amplitudes increase substantially with altitude. When $\overline{u} - c \approx 0$ the mean flow absorbs the wave and thereby acts as a barrier, preventing further vertical propagation. This and other dissipation mechanisms such as eddy and molecular diffusion or viscosity prevent further wave amplitude growth, filtering the wave spectrum and leading to deposition of horizontal momentum in the region of breaking or dissipation. Waves therefore act as an effective means of vertical coupling in the atmosphere, whereby horizontal momentum is transported vertically. The consequences of gravity wave dissipation and breaking are three-fold, they affect the zonal and meridional wind flow, lead to potentially significant turbulent mixing which affects the composition and thirdly they affect the thermal structure via primarily adiabatic heating and cooling. A comprehensive review of wave coupling on the terrestrial planets is given by Forbes (2002).

The gravity waves transporting most of the momentum in the Earth's atmosphere have horizontal wavelengths from 10's to 100's of km and are hence not resolved by global models. Gravity wave momentum deposition therefore needs to me included in these models in parameterized form. The gravity wave parameterization scheme used in CMAT is a hybrid Matsuno-Lindzen scheme (Meyer 1999), one of several available in the literature. The Lindzen scheme (Lindzen 1981) assumes that amplitudes of gravity waves propagating vertically grow exponentially with height until a critical altitude is reached where wave perturbations create super-adiabatic lapse rates and the wave begins to break. The scheme gives an expression for eddy diffusion that translates into an expression of horizontal momentum that prevents the further growth of the wave above the critical level. Many alternative gravity wave drag parameterizations have been developed since, most recently by Medvedev and Klaassen (2000), and are used in General Circulation Models to better represent the middle atmosphere dynamics, although none of the schemes work for all scenarios. The TIME-GCM also includes gravity wave schemes which are adapted to the particular modeled scenarios to best match observations (Zhou et al. 1997). Gravity wave parameterizations can be regarded almost as a free parameter which is adjusted for resulting dynamics to best fit observations.

4.2 Dynamics on Other Planets and Moons

No direct measurements have been made to-date of mesosphere/thermosphere winds in any planetary atmosphere except Earth. The techniques used to directly measure winds by remote sensing rely either on Doppler-shift determinations of emission spectra or time-tracking of distinct features such as clouds. Another technique used on Venus and most recently on Titan is that of tracking atmospheric probes as they descend through the atmosphere. None of these techniques however work in the upper cloud-free regions of atmospheres due to the low densities which hardly affect the motion of atmospheric probes there and make detection of Doppler shifts in atmospheric emissions challenging.

The planetary atmosphere currently best known apart from Earth is Venus, largely due to the Pioneer Venus mission which was in orbit from 1978 to 1992 and included 4 atmospheric probes which were dropped into the atmosphere early in the mission. One striking feature of Venus' atmosphere between the surface and the upper cloud top (around 70 km) is its super-rotation (Schubert 1983; Gierasch et al. 1997) characterized by retrograde zonal winds reaching average velocities of $\sim$100 m/s over the equator. Several theories have been proposed to explain this, in particular meridional transport of momentum from mid-latitudes as a result of eddy mixing (Gierasch 1975) and momentum pumping by thermal tides (Fels and Lindzen 1974). Dynamics near the cloud top have been described as being the combination of a mean zonal super-rotating flow and two time-dependent features with each an amplitude of around 10 m/s, the solar tide and a 4-day traveling wave (Gierasch et al. 1997). Theoretical calculations of dynamics in Venus' thermosphere have predicted the presence of a strong sub-solar to anti-solar (SS-AS) flow driven by the large day–night temperature difference in the thermosphere (see Sect. 2) as well as a super-imposed time-varying retrograde super-rotating flow (Bougher et al. 1997). This circulation pattern has been indirectly confirmed by ground-based and spacecraft observations of nightside emissions at 1.27 μm. These emissions result from recombination of atomic oxygen atoms generated on the dayside from CO_2 photolysis and are associated with regions of downwelling in the atmosphere. Other emissions used as tracers include UV emissions in the NO γ and δ bands which result from the recombination of N and O. The reviews by Lellouch et al. (1997), Bougher et al. (1997, 2006a) and Schubert et al. (2007) give a comprehensive summary of emissions in Venus' atmosphere and their role in deriving wind speeds at different altitudes.

Waves, like on Earth, are thought to play a key role in the atmosphere of Venus. An early example of observed solar thermal tides was obtained from measurements by the Orbiter IR Radiometer (OIR) on board Pioneer Venus (Schofield and Taylor 1983) which identified semidiurnal temperature oscillations peaking near 95 km altitude that are replaced at higher altitudes by diurnal changes due to thermospheric heating. In situ thermospheric gas density observations by the Pioneer Venus Orbiter Neutral Mass Spectrometer (PV-ONMS) revealed the presence of waves of horizontal scales ranging from 100–600 km (Kasprzak et al. 1988) which have been interpreted as signatures of gravity waves propagating upward from the upper mesosphere region (Mayr et al. 1988) and depositing momentum into the background atmosphere at thermospheric altitudes, possibly supporting a super-rotating flow (Alexander 1992). To-date, theoretical calculations of the Venus thermosphere cannot reproduce the large day-night temperature gradient (Sect. 2) without imposing an empirical drag term which slows down the sub-solar to anti-solar flow (Bougher et al. 2006a). Without this term, dynamics via adiabatic heating on the nightside reduce the day-night temperature difference to considerably smaller than observed values. It is possible that this drag is also a result of dissipating or breaking gravity waves, but further observations are needed to confirm this. Gravity wave drag parameterizations developed for Earth (see Sect. 4.1) are increasingly

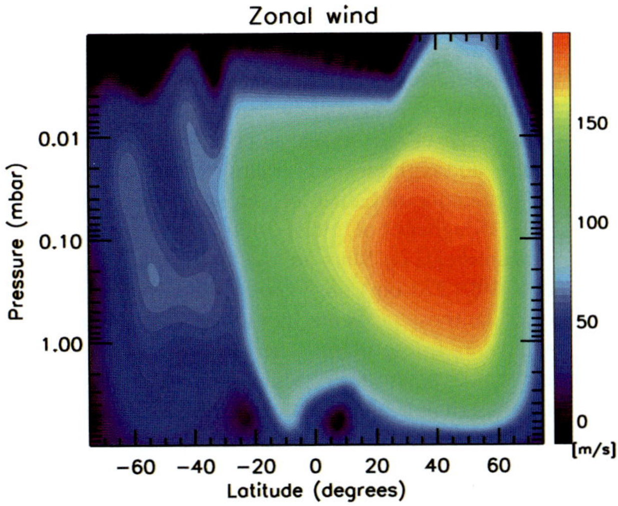

Fig. 5 Zonal winds (positive eastward) in Titan's stratosphere, as inferred from temperature observations by the Cassini Composite Infrared Spectrometer (CIRS) by Achterberg et al. (2008)

adapted to other planets like Venus, but remain "trial-and-error" approaches due to lack of sufficient observations to characterize the gravity wave spectra there (Bougher et al. 1997). The important role that gravity waves play for dynamics on Venus therefore pose a serious challenge to theoretical modelers. Observations by the ongoing Venus Express mission may help better constrain the dynamics on Venus and improve our understanding of the key momentum sources.

The other slowly rotating body with a substantial atmosphere is Titan. There, like on Venus, super-rotation has been detected in the stratosphere (Hubbard et al. 1993; Flasar et al. 2005b; Sicardy et al. 2006) with prograde wind velocities of around 100 m/s. Theories about the origin have been proposed which are essentially similar to those for Venus, and numerical simulations of Titan's stratospheric circulation have in part succeeded in reproducing the super-rotation without the need to artificially introduce further momentum sources (Hourdin et al. 1995). Measurements of the Doppler shift of the radio signals from the Huygens Probe gave in situ measurements of zonal wind on Titan near 10°S latitude between the surface (1.5 bar) and 140 km altitude (3 mbar) (Bird et al. 2005). Recent observations by the Cassini Composite Infrared Spectrometer (CIRS) have helped construct temperature maps of Titan's stratosphere up to around 500 km altitude, allowing the derivation of zonal wind profiles using the thermal wind equation (Flasar et al. 2005b; Achterberg et al. 2008). Figure 5 shows that a strong eastward jet dominates the stratosphere on Titan. The current southern hemisphere summer conditions generate a hemispherically asymmetric profile. No direct measurements are available for higher altitudes on Titan, but calculations by Mueller-Wodarg et al. (2000, 2003, 2008) and Bell (2008) have shown that thermally driven winds could reach velocities in the order of 50–100 m/s in the thermosphere. One important factor in this was vertical coupling to the lower altitudes. Cassini observations in Titan's mesosphere (500–1000 km) region are too scarce to constrain dynamics, so little is known of winds at 1000 km, the bottom boundary of the region systematically examined by the Ion Neutral Mass Spectrometer (INMS). Further observations from Cassini during the extended mission will help further constrain dynamics on Titan and assess the role of waves identified in the Cassini and Huygens data (Fulchignoni et al. 2005; Mueller-Wodarg et al. 2006b).

No in situ wind measurements are presently available for the Mars upper atmosphere. Few ground-based observations were made to-date, focusing mostly on the middle atmosphere (~50–100 km) (Lellouch et al. 1991). Due to their scarcity and coarse resolution, these measurements did not allow characterization of short term and seasonal trends in Martian winds. Most of our current knowledge of dynamics on Mars is derived from density, temperature, and nightglow observations from spacecraft and the inferred winds that result from the application of General Circulation Models to interpret these observations (Keating et al. 1998; Bougher et al. 1999, 2000, 2006b; Withers et al. 2003). One example of nightglow measurements on Mars are the hydrogen Lyman α (121.6 nm) and NO γ and δ band emissions (190–270 nm) observed in the winter polar night (Bertaux et al. 2005). These emissions, like those on Venus, are thought to be a result of dayside production of N and O atoms which are transported to the nightside by horizontal winds and to lower altitudes by subsiding winds, where they radiatively recombine to produce the NO nightglow. Density measurements such as those obtained from the Mars Global Surveyor Accelerometer readings (Keating et al. 1998) have shown the presence of features which have been interpreted as nonmigrating tides, one of many families of waves found to be present in the Mars upper atmosphere (Forbes and Hagan 2000; Forbes et al. 2002; 2004; Forbes 2002). While dynamics in the upper atmosphere are thought to be driven primarily by solar EUV heating, studies have suggested that vertical dynamical coupling nevertheless plays an equally significant role on Mars as it does on Venus and Earth, most prominently in the form of strong wave signatures in the thermosphere that are not only formed in situ at those altitudes, but also include effects due to topography (Forbes et al. 2002). Bougher et al. (2006b) and Bell et al. (2007) additionally proposed the existence of a deep inter-hemispheric circulation in order to enable thermospheric winter polar warming to operate. While most models available to-date still rely on implementing this coupling to lower altitudes as a lower boundary condition (Bougher et al. 2008), codes are now appearing for Mars with a vertical range from the ground to the thermosphere (González-Galindo et al. 2005), having the advantage of more self-consistently including processes of vertical dynamical coupling.

Winds on the Gas Giants are dominated by the fast rotation periods of the solid planets (Table 1) which gives rise to primarily geostrophic circulation due to the large Coriolis accelerations. The presence of strong zonal jets in the troposphere Saturn is well known (Flasar et al. 2004, 2005a), with some evidence of their presence reaching into the stratosphere and mesosphere as well (Hubbard et al. 1997). Saturn's zonal winds are still not well understood. Wind speeds derived from cloud tracking indicated values of 500 m/s near the equator (Ingersoll et al. 1984; Barnet et al. 1992) which would imply a considerable excess of axial angular momentum present in the atmosphere relative to Saturn's interior. Cloud tracking observations made more recently (1996–2002) with the Hubble Space Telescope (HST) suggest smaller equatorial winds of up to 275 m/s (Sánchez-Lavega et al. 2003), values largely confirmed by the latest Cassini-CIRS observations (Flasar et al. 2005a). The discrepancy between earlier and more recent observations is currently not well understood, and it is unclear whether the difference represents a real change in the atmosphere or an observational uncertainty. No observational evidence is available for winds in Saturn's mesosphere and thermosphere, but calculations by Mueller-Wodarg et al. (2006a) demonstrated the likely dominance of zonal flow there as well. Little is known also about stratospheric and (especially) tropospheric winds on Jupiter. In situ measurement were obtained with a Doppler wind profiler on the Galileo Probe (Atkinson et al. 1996), but the probe sampled an anomalous, cloud-free location (Orton et al. 1996), so the winds it measured may not be representative for Jupiter. As for Saturn, all other wind measurements on Jupiter have been indirect and based on simplifying assumptions such as a non-scattering atmosphere (in the temperature retrievals) and the gradient wind balance (Flasar et al. 2004). General circulation models for the upper atmospheres of Jupiter and Saturn (Achilleos et al. 1998; Bougher et al. 2005; Mueller-Wodarg et al. 2006a) have helped predict what dynamics in the thermosphere and ionosphere could be, but many aspects of these calculations still await validation by observations. One particular uncertainty is the source of unexpectedly high temperatures in the thermospheres of all Gas Giants (see Sect. 2), which will affect not only the thermal structure but also the global dynamics. Upward propagating waves, one possible candidate energy source (Yelle and Miller 2004), may also deposit significant amounts of momentum in the upper atmospheres. One additional source of momentum in the thermospheres of Earth, Jupiter and Saturn is ion drag, discussed in the following. In the auroral regions additional thermal energy sources are Joule heating and particle precipitation, which via temperature and pressure gradients further alter the dynamics.

4.3 Auroral Dynamics

It is well known that both solar EUV radiation and auroral energy and momentum sources have a significant effect on the Earth's thermospheric and ionospheric structure and dynamics. A schematic of the auroral particle and joule heat inputs and current systems is shown in Fig. 6.

Upper atmosphere general circulation models using these forcings (Roble et al. 1988; Fuller-Rowell et al. 1996) have been reasonably successful in simulating the thermosphere and ionosphere responses for a wide range of auroral activity. Figure 7 is a schematic of the zonally averaged mass flow stream function illustrating the response of the circulation to increasing levels of auroral activity. The aurora increases from geomagnetic quiet conditions with an energy input of 10^{10} W to geomagnetic moderate conditions with an energy input of 10^{11} W to a geomagnetic storm with and energy input of 10^{12} W. During equinox when

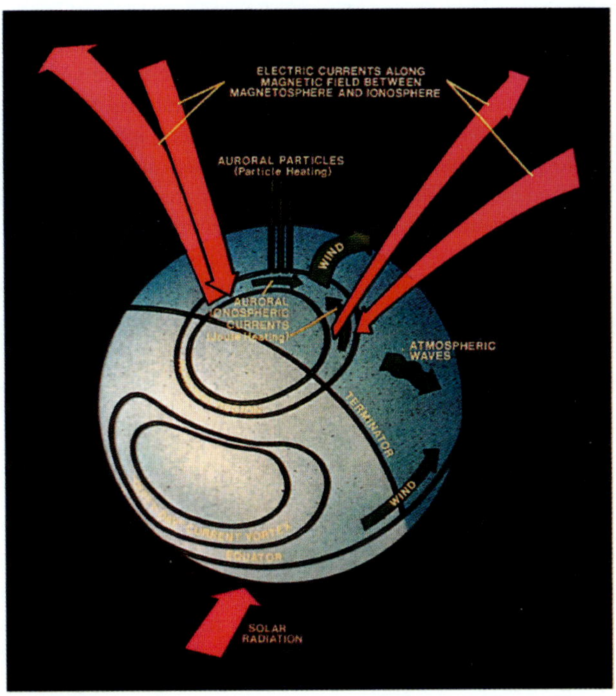

Fig. 6 Schematic of the auroral particle and joule heating inputs and current systems in the Earth's polar upper atmosphere (Roble 1996)

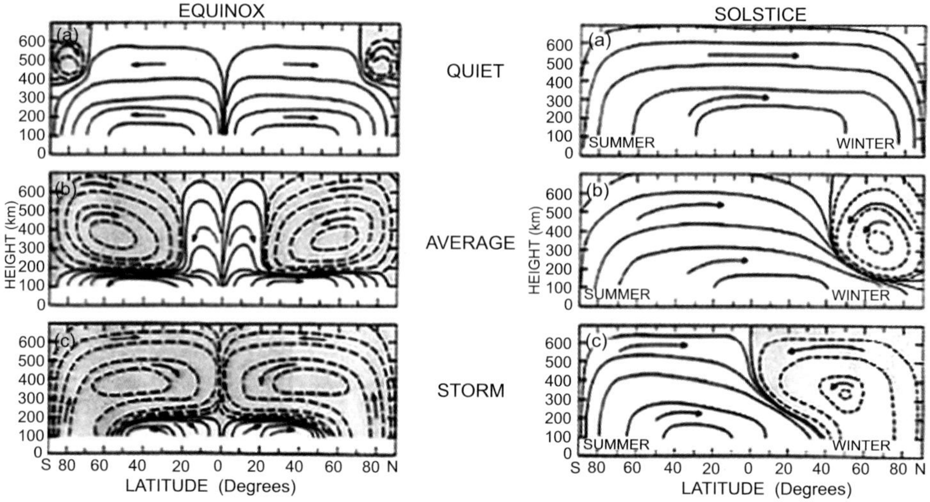

Fig. 7 Zonally averaged mass flow stream function illustrating the circulation in the Earth's upper atmosphere for equinox (*left*) and southern hemisphere summer (*right*) conditions at quiet, average and strong levels of geomagnetic activity (Roble et al. 1983)

the sun is directly over the equator, the circulation is from the equator toward the poles in both hemispheres. Auroral energy input occurs at high magnetic conjugate latitudes in both hemispheres and the increased heating drives a circulation cell from high latitudes toward the equator. As geomagnetic activity increases the auroral equatorward circulation intensifies, forcing the reversal of the solar directed poleward flow and the equatorward directed auroral flow to move equatorward and to lower altitudes as shown in the left panels.

For solstice there is a similar situation. The December solstice solar circulation is from the summer pole in the southern hemisphere to the winter pole in the northern hemisphere. In the summer hemisphere the joule and auroral heating reinforces the summer-to-winter circulation whereas in the winter hemisphere, joule and auroral heating is sufficiently strong to produce a small equatorward flow at high latitudes during geomagnetic quiet conditions. For moderate conditions the circulations are enhanced and for geomagnetic storms both the solar driven summer hemisphere and winter reverse cells are both enhanced. With geomagnetic activity being highly variable, the pattern varies between the quiet, moderate and storm conditions. These schematics represent the zonal mean circulation or mean flow. During impulsive events the auroral heat and momentum source can launch large scale thermospheric waves that can travel equatorward at speeds of up to 700 m/s at F-region altitudes near 300 km and waves in the lower thermosphere near 150 km of 300–400 m/s. Thus, the thermospheric response is a complex mixture of the overturning of the mean circulation with various impulsive waves superimposed.

Figure 8 (Roble et al. 1983) is a schematic showing the response of the high latitude thermosphere to different levels of auroral forcing. The top left panel shows a TGCM calculation of perturbation temperature (from a global mean) for a case where the circulation is due to solar heating alone, without auroral particle precipitation or joule heating. Here, the flow is from the dayside to the nightside, essentially perpendicular to isobars, with some influence of Coriolis forces. The top right panel shows the perturbation temperature and circulation for the case of geomagnetic quiet conditions with only a 20 kV cross polar cap potential drop driving ion convection and the lower panel is for moderate steady state forc-

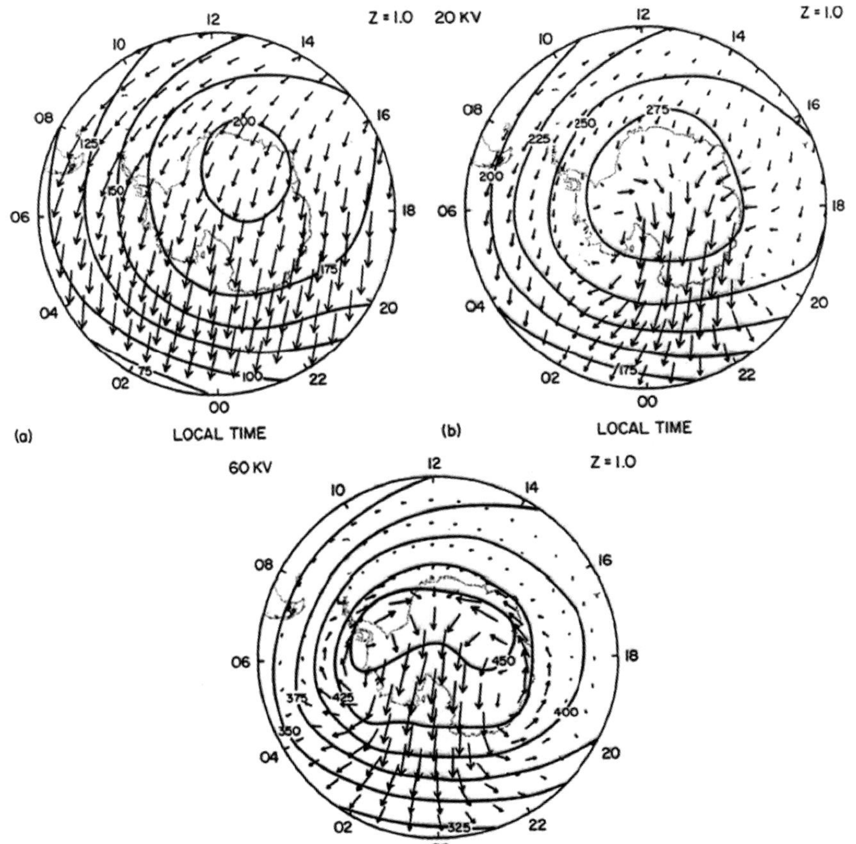

Fig. 8 Schematic showing the response of the high latitude thermosphere to different levels of auroral forcing. Contour lines are exospheric temperatures, arrows are neutral winds. The *upper left panel* shows the solar-driven situation only, the *upper right panel* shows a situation with quiet geomagnetic conditions (cross-cap potential 20 kV) and the *bottom panel* is for moderate geomagnetic conditions (cross-cap potential 60 kV)

ing of 60 kV cross polar cap. The interaction of the solar wind with the earth's geomagnetic field produces a dawn-to-dusk potential drop and the corresponding ExB drift generates a two cell pattern in the ion drift that transfers momentum to the neutral gas which tends to follow the ion convection pattern. The principle can be compared to a giant "egg beater" operating over the high latitude thermosphere. In addition to the ion drag momentum source, auroral Joule and particle heating occur along the convecting ion boundaries in an oval in the lower thermosphere. Thus, the aurora heats the thermosphere 100–500 K and generates winds approaching 200–600 m/s at 300 km. Again, the pattern is highly variable because of the variability in the aurora.

With Jupiter and Saturn having internal magnetic fields as well of sufficient strength to produce gyrofrequencies smaller than ion-neutral collision frequencies in the ionospheres near the peak density heights, similar dynamical coupling is expected to occur there as well. No direct neutral wind observations exist in the auroral regions of Jupiter or Saturn, but ground-based observations of Doppler shifts in the H_3^+ $\nu 2$ $Q(1, 0^-)$ line at 3.953 μm have detected westward ion velocities in the auroral regions of Jupiter reaching 1.5 km/s (Stallard et al. 2001). Similar observations for Saturn detected H_3^+ westward velocities implying the

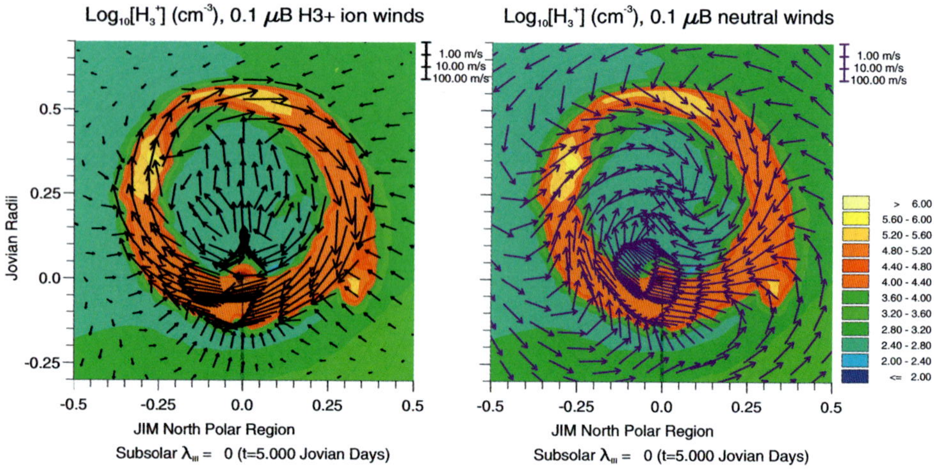

Fig. 9 Global model profiles of ionization and velocity in the co-rotating frame for the northern polar region of Jupiter. *Left panel*: Colour contours of H_3^+ density over a 0.1 μbar pressure surface (location of the auroral ion peak). *Arrows* represent the horizontal ion velocities, following the mixed linear/log velocity *scale bar* shown. The meridian (vertically downward in plot) indicates the zero longitude and subsolar point. *Right panel*: As for the *left panel*, but with *arrows* now representing the neutral horizontal wind velocity. From Achilleos et al. (2001)

auroral region was sub-corotating by up to a factor of 0.43 (Stallard et al. 2007). Simulations with the General Circulation Model of Jupiter by Achilleos et al. (1998, 2001) have shown that ion velocities driven by convection electric fields in the auroral regions can through ion-neutral collisions drive neutral wind velocities of around half the ion speeds (Fig. 9). To-date, accurate models of the convection electric field on Jupiter are not available, so most of these simulations rely on simplified electric field models adapted from Earth conditions which do not reflect the high time variability of the real auroral regions. It is likely that the factor of momentum transfer from the ions to the neutrals is nevertheless accurate within an order of magnitude, implying significant influence of the ion velocities on thermospheric winds, as on Earth. Similar results were found by calculations of Bougher et al. (2005) using their JTGCM code of Jupiter's thermosphere and ionosphere, for which ion velocities (up to 1.5 to 3.0 km/sec) drove neutral velocities up to 1.0 to 1.2 km/sec.

5 Key Outstanding Problems and Future Needs

As the result of highly sophisticated and successful space missions, Earth based observations and theoretical models, the past decades have seen major advances in our understanding of the atmospheres of Earth and other planets or moons, in particular Venus and Titan, followed by Jupiter, Mars and Saturn. Little is known yet about Uranus and Neptune as well as Triton and Pluto. The New Horizons mission will help considerably in advancing our knowledge about Pluto's atmosphere after its scheduled arrival in 2015, providing the first ever close-up view of Pluto and Charon, which have never been visited by any spacecraft. Several future missions to the outer solar system are currently in planning or under study, including the Juno mission to Jupiter which is due for launch in 2011 and either a Jupiter/Europa mission or Titan/Enceladus mission. The next planned mission to Venus is Planet-C, lead by the Japanese Space Agency (Jaxa) and is due for launch in summer of 2010. Future planned missions to Mars include Nasa's Mars Science Laboratory (launch in Fall 2009), NASA's Mars Scout mission (launch in or after 2013) and ESA's Aurora Flagship mission currently named Exomars which is due for launch in 2011 or later.

Despite the significant advances over the past decades, a large number of scientific questions remain open about the aeronomy of many planets in our solar system, including our own. Our knowledge has now advanced enough about most atmospheres that we begin to obtain a first understanding of aeronomic processes under different conditions and can ask increasingly sophisticated questions. The following outlines only a few of these scientific questions.

5.1 Energy Crisis on Giant Planets

As outlined in Sect. 2 we currently do not understand the thermal balance on Gas Giants. With solar EUV heating and magnetospheric energy sources being insufficient to explain observed thermospheric temperatures outside the auroral regions, the most likely candidate sources for the energy are the interiors of the planets. Energy can be transferred to the outer regions of the atmospheres via propagating waves, but the only in situ observations available to-date for Jupiter by the Galileo probe (Seiff et al. 1997) did not resolve this issue. The observed gravity waves could not unambiguously be shown to be able to transport the required amount of energy into Jupiter's thermosphere (Yelle and Miller 2004). A more substantial programme of observation is needed to resolve this issue, one of the key aeronomy problems to-date. Atmosphere probes such as the Galileo probe may considerably help in answering this question.

5.2 Dynamics in Planetary Atmospheres

Despite the plentiful availability of atmospheric wind profiles obtained through remote sensing for most planets, aeronomy studies suffer from a severe shortage of direct wind measurements. Only on Earth have upper atmosphere winds been sampled directly, with some direct measurements in the lower atmospheres available for Venus and Titan. Mostly, winds are inferred from thermal profiles by applying either simple approximations (thermal wind equation, geostrophic approximation) to derive winds from horizontal or vertical thermal gradients or "tuning" General Circulation Models to reproduce the thermal structure and obtain the necessary global dynamic structure from those. Another method commonly used is observation of doppler shifts in atmospheric gas emissions, but those are restricted to regions of significant (detectable) atmospheric gas emissions, which is usually below the thermosphere only. As a result of limited vertical coverage of wind measurements we still have an incomplete understanding of the global circulation system on most planets. For all planets we are in need for more direct and more comprehensive measurements of winds throughout the atmosphere, either by tracking balloons or more systematically observing emissions from the upper atmospheres.

5.3 Stratospheric Constituents on the Giant Planets

Systematic measurements are needed of the latitude, altitude, and time variation of stratospheric constituents on the giant planets. This information is necessary to help constrain heating rates and circulation. The stratospheric energy balance is largely determined from absorption of solar radiation and the re-emission of this energy and radiative transfer it undergoes. Understanding these problems requires a detailed inventory of stratospheric constituents and, importantly, their horizontal and vertical distribution. Not only is their distribution important for understanding the heating rates, but some of the more inert gases can also be seen as tracers for atmospheric dynamics. Understanding their latitudinal and seasonal distribution will considerably help us understand the dynamics and chemistry of Gas Giants.

5.4 Atmospheric Waves

As shown in Sect. 4 atmospheric waves on all scales play a key role in vertical coupling within an atmosphere. Studies of the Earth's atmosphere have shown that we cannot understand the global dynamics without considering the effects of gravity wave drag. In the thermosphere the diurnal tide leads to removal of O via an enhancement in the effective recombination rate due to vertical (downward) transport by the tide. The problem of gravity waves on Earth is still not fully understood and often taken as a degree of freedom. "Tuning" the gravity wave drag parameterization often allows theoreticians to fit their calculations to observations, but this approach ultimately remains unsatisfying. On other planets the wave spectra are even less known and will probably never be fully understood, but as systematic programme of observation of larger-scale atmospheric waves would be of considerable benefit. This can best be achieved by orbiting spacecraft which dedicate part of their observations to monitoring of the atmosphere and its periodic fluctuations.

5.5 Heating Efficiencies and CO_2(v2)–O Relaxation Rates

The thermal balance on planets Venus, Earth and Mars critically relies on IR cooling by the CO_2 molecule, mostly at 15 μm, which is considerably affected by vibrational excitation of CO_2 by O (Sect. 2). The CO_2(v2)–O relaxation rate however is still poorly known, allowing considerable degrees of freedom in our understanding of the thermal balance on these planets. Systematically carrying out experiments such as recent simultaneous observation of the responses of Venus, Earth and Mars to changes in solar flux (Forbes et al. 2006) may help resolve this problem, but additionally laboratory measurements need to be carried out to constrain this rate. Laboratory experiments are also needed to measure the heating efficiencies of CO_2. As discussed by Fox (1998), theoretical models rely on lower heating efficiencies than can be justified from molecular calculations in order to reproduce observed thermospheric temperatures. This shortcoming highlights another uncertainty in our understanding of the thermal balance on terrestrial planets. Our current knowledge of the CO_2(v2)–O relaxation rates is discussed by Huestis et al. (2008).

5.6 Atomic Oxygen on Mars

No direct measurements exist to-date of atomic oxygen on Mars. This constituent is not only an important tracer of dynamics, but plays a key role in the thermal balance of the thermosphere and photochemistry of the ionosphere. This therefore constitutes a major unknown for our understanding of aeronomy on Mars.

Acknowledgements IMW is funded through a University Research Fellowship by the Royal Society London. JM gratefully acknowledges support from the NASA Planetary Atmospheres program grant NNX08AF05G. SWB acknowledges NASA grant NNX07A084G that supported research and writing for this chapter.

References

M.F. A'Hearn, R.L. Millis, D.G. Schleicher, D.J. Osip, P.V. Birch, The ensemble properties of comets: Results from narrowband photometry of 85 comets, 1976–1992. Icarus **118**, 223–270 (1995)

N. Achilleos, S. Miller, J. Tennyson, A.D. Aylward, I.C.F. Mueller-Wodarg, D. Rees, JIM: A time-dependent, three-dimensional model of Jupiter's thermosphere and ionosphere. J. Geophys. Res. **103**, 20089–20112 (1998)

N. Achilleos, S. Miller, R. Prangé, G. Millward, M.K. Dougherty, A dynamical model of Jupiter's auroral electrojet. New J. Phys. **3**, 3.1–3.20 (2001)

R.K. Achterberg, B.J. Conrath, P.J. Gierasch, F.M. Flasar, C.A. Nixon, Titan's middle-atmospheric temperatures and dynamics observed by the Cassini composite infrared spectrometer. Icarus **194**, 262–277 (2008)

M.J. Alexander, A mechanism for the Venus thermospheric superrotation. Geophys. Res. Lett. **19**, 2207–2210 (1992)

D.M. Atkinson, J.B. Pollack, A. Seiff, Galileo Doppler measurements of the deep zonal winds at Jupiter. Science **272**, 842–843 (1996)

S.K. Atreya et al., A comparison of the atmospheres of Jupiter and Saturn: deep atmospheric composition, cloud structure, vertical mixing, and origin. Planet. Space Sci. **47**, 1243–1262 (1999)

S.K. Atreya, Z.G. Gu, Stability of the Martian atmosphere: Is heterogeneous catalysis essential? J. Geophys. Res. **99**, 13133–13145 (1994)

S.K. Atreya, P.R. Mahaffy, H.B. Niemann, M.H. Wong, T.C. Owen, Composition and origin of the atmosphere of Jupiteran update, and implications for the extrasolar giant planets. Planet. Space Sci. **51**, 105–112 (2002). doi:10.1016/S0032-0633(02)00144-7

S.K. Atreya, E.Y. Adama, H.B. Niemann, J.E. Demick-Montelara, T.C. Owen, M. Fulchignoni, F. Ferri, E.H. Wilson, Planet. Space Sci. **54**, 1177–1187 (2006)

S.K. Atreya, P.R. Mahaffy, A.-S. Wong, Methane and related trace species on Mars: Origin, loss, implications for life, and habitibility. Planet. Space Sci. **55**, 358–369 (2007)

M. Banaszkiewicz, L.M. Lara, R. Rodrigo, J.J. López-Moreno, G.J. Molina-Cuberos, A coupled model of Titan's atmosphere and ionosphere. Icarus **147**, 386–404 (2000)

A. Bar-Nun, V. Dimitrov, Methane on Mars: A product of H_2O photolysis in the presence of CO. Icarus **181**, 320–322 (2006)

C.D. Barnet, J.A. Westphal, R.F. Beebe, L.F. Huber, Icarus **100**, 499 (1992)

J.M. Bell, S.W. Bougher, J.R. Murphy, Vertical dust mixing and the interannual variations in the Mars thermosphere. J. Geophys. Res. **112**, E12002 (2007). doi:10.1029/2006JE002856

J.M. Bell, The dynamics of the upper atmospheres of Mars and Titan. Ph.D. Thesis, U. of Michigan (2008)

E.A. Bergin et al., Submillimeter Wave Astronomy Satellite observations of Jupiter and Saturn: Detection of 557 GHz water emission from the upper atmosphere. Astrophys. J. **539**, L147–L150 (2000)

J.L. Bertaux et al., Nightglow in the upper atmosphere of Mars and implications for atmospheric transport. Science **307**, 566–569 (2005). doi:10.1126/science.1106957

J.-L. Bertaux et al., A warm layer in Venus' cryosphere and high-altitude measurements of HF, HCl, H_2O and HDO. Nature **450** (2007). doi:10.1038/nature05974

B. Bézard, D. Gautier, A seasonal climate model of the atmospheres of the giant planets at the Voyager encounter time. I. Saturn's stratosphere. Icarus **61**, 296–310 (1985)

B. Bézard, E. Lellouch, D. Strobel, J.-P. Maillard, P. Drossart, Carbon monoxide on Jupiter: Evidence for both internal and external sources. Icarus **159**, 95–111 (2002)

M.K. Bird et al., The vertical profile of winds on Titan. Nature **438**, 800–802 (2005). doi:10.1038/nature04060

J. Bishop, P.N. Romani, S.K. Atreya, Voyager 2 ultraviolet spectrometer solar occultations at Neptune: Constraints on the abundance of methane in the stratosphere. J. Geophys. Res. **97**, 11681–11694 (1998)

N. Biver, D. Bockelée-Morvan, J. Crovisier et al., Chemical composition diversity among 24 comets observed at radio wavelengths. Earth Moon Planets **90**, 323–333 (2002)

D. Bockelée-Morvan, J. Crovisier, The role of water in the thermal balance of the coma. ESA SP **278**, 235–240 (1987)

D. Bockelée-Morvan, J. Crovisier, M.J. Mumma, H.A. Weaver, The composition of cometary volatiles, in *Comets II*, ed. by M.C. Festou, H.U. Keller, H.A. Weaver (Univ. Arizona Press, Tucson, 2005), pp. 391–423

D.D. Bogard et al., Martian volatiles: Isotopic composition, origin, and evolution. Chronol. Evol. Mars **96**, 425–458 (2001)

J. Boissier, D. Bockelée-Morvan, N. Biver et al., Interferometric imaging of the sulfur-bearing molecules H_2S, SO and CS in comet C/1995 O1 (Hale-Bopp). Astron. Astrophys. **475**, 1131–1144 (2007)

S.W. Bougher, D.M. Hunten, R.G. Roble, CO_2 cooling in terrestrial planet thermospheres. J. Geophys. Res. **99**, 14609–14622 (1994)

S.W. Bougher, M.J. Alexander, H.G. Mayr, Upper atmosphere dynamics: global circulation and gravity waves, in *Venus II*, ed. by S.W. Bougher, D.M. Hunten, R.J. Phillips (University of Arizona Press, Tucson, 1997), pp. 259–291

S.W. Bougher, S. Engel, R.G. Roble, B. Foster, Comparative terrestrial planet thermospheres: 2. Solar cycle variation of global structure and winds at equinox. J. Geophys. Res. **104**, 16591–16611 (1999)

S.W. Bougher, S. Engel, R.G. Roble, B. Foster, Comparative terrestrial planet thermospheres 3. Solar cycle variation of global structure and winds at solstices. J. Geophys. Res. **105**, 17,669–17,689 (2000)

S.W. Bougher, J.H. Waite Jr., T. Majeed, G.R. Gladstone, Jupiter Thermospheric General Circulation Model (JTGCM): Global structure and dynamics driven by auroral and Joule heating. J. Geophys. Res. **110**, E04008 (2005). doi:10.1029/2003JE002230

S.W. Bougher, S. Rafkin, P. Drossart, Dynamics of the Venus upper atmosphere: Outstanding problems and new constraints expected from Venus Express. Planet. Space Sci. **54**, 1371–1380 (2006a). doi:10.1016/j.pss.2006.04.023

S.W. Bougher, J.M. Bell, J.R. Murphy, M.A. López-Valverde, P.G. Withers, Polar warming in the Mars thermosphere: Seasonal variations owing to changing insolation and dust distributions. Geophys. Res. Lett. **33**, L02203 (2006b). doi:10.1029/2005GL024059

S.W. Bougher, P.-L. Blelly, M. Combi, J. Fox, I.C.F. Mueller-Wodarg, A. Ridley, R.G. Roble, Neutral upper atmosphere and ionosphere modeling. Space Sci. Rev. (2008, this issue)

G.P. Brasseur, S. Solomon, *Aeronomy of the Middle Atmosphere* (Springer, Dordrecht, 2005)

G.P. Brasseur, J.J. Orlando, G.S. Tyndall (eds.), *Atmospheric Chemistry and Global Change* (Oxford University Press, New York, 1999)

H.C. Brinton, H.A. Taylor, H.B. Niemann, H.G. Mayr, A.F. Nagy, T.E. Cravens, D.F. Strobel, Venus night time hydrogen bulge. Geophys. Res. Lett. **7**, 865–868 (1980)

M. Burgdorf, G. Orton, J. van Cleve, V. Meadows, J. Houck, Detection of new hydrocarbons in Uranus' atmosphere by infrared spectroscopy. Icarus **184**, 634–637 (2006)

M.V. Canaves, A.A. de Almeida, D.C. Boice, G.C. Snzovo, On the chemistry of CS and NS in cometary comae. Adv. Space Res. **39**, 451–457 (2007)

K.J. Castle, K.M. Kleissas, J.M. Rhinehart, E.S. Hwang, J.A. Dodd, Vibrational relaxation of CO2(?2) by atomic oxygen. J. Geophys. Res. **111**, A09303 (2006). doi:10.1029/2006JA011736

R.T. Clancy, H. Nair, Annual (perihelion-aphelion) cycles in the photochemical behavior of the global Mars atmosphere. J. Geophys. Res. **101**, 12785–12790 (1996)

T.E. Cravens et al., Composition of Titan's ionosphere. Geophys. Res. Lett. **33**, L07105 (2006). doi:10.1029/2005GL025575

J.F. Crifo, M. Fulle, N.I. Kmle, K. Szego, Lessons from physical models, in *Comets II*, ed. by M.C. Festou, H.U. Keller, H.A. Weaver (Univ. Arizona Press, Tucson, 2005), pp. 471–503

M.R. Combi, W.M. Harris, W.H. Smyth, Gas dynamics and kinetics in the cometary coma: Theory and observations, in *Comets II*, ed. by M.C. Festou, H.U. Keller, H.A. Weaver (Univ. Arizona Press, Tucson, 2005), pp. 523–552

M.R. Combi, K. Kabin, D. De Zeeuw, T.I. Gombosi, K.G. Powell, Dust-gas interrelations in Comets: Observations and theory. Earth Moon Planets **79**, 275–306 (1999)

A. Coustenis, A. Salama, E. Lellouch, T. Encrenaz, G.L. Bjoraker, R.E. Samuelson, T. de Graauw, H. Feuchtgruber, M.F. Kessler, Evidence for water vapor in Titan's atmosphere. Astron. Astrophys. **336**, L85–L89 (1998)

J. Crovisier, The photodissociation of water in cometary atmospheres. Astron. Astrophys. **213**, 459–464 (1989)

J. Crovisier, Photodestruction rates for cometary parent molecules. J. Geophys Res. Planets **99**, 3777–3781 (1994)

J. Crovisier, Cometary diversity and cometary families, in *Proceedings of the XVIIIemes Rencontres de Blois: Planetary Science: Challenges and Discoveries* (2007). astro-ph/0703785

N. Dauphus, The dual origin of the terrestrial atmosphere. Icarus **165**, 326–339 (2003)

C. de Bergh et al., The composition of the atmosphere of Venus below 100 km altitude: An overview. Planet. Space Sci. **54**, 1389–1397 (2006)

N. Dello Russo, R.J. Vervack Jr., H.A. Weaver et al., Compositional homogeneity in the fragmented comet 73P/Schwassmann-Wachmann 3. Nature **448**, 172–175 (2007)

W.B. DeMore, M.T. Leu, R.H. Smith, Y.L. Yung, Laboratory studies on the reactions between chlorine, sulfur dioxide, and oxygen: Implications for the Venus stratosphere. Icarus **63**, 247 (1985)

I. de Pater, H. Roe, J.R. Graham, D.F. Strobel, P. Bernath, Detection of forbidden SO $a^1\Delta$–$X^3\Sigma^-$ rovibronic transition on Io at 1.7 μm. Icarus **156**, 296–301 (2002)

I. de Pater, C. Laver, F. Marchis, H.G. Roe, B.A. Macintosh, Spatially resolved observations of the forbidden SO a$\Delta \rightarrow X\Sigma$ rovibronic transition on Io during an eclipse and a volcanic eruption at Ra Patera. Icarus **191**, 172–182 (2007)
T.M. Donahue, New analysis of hydrogen and deuterium escape from Venus. Icarus **141**, 226–235 (1999)
P. Drossart et al., A dynamic upper atmosphere of Venus as revealed by VIRTIS on Venus Express. Nature **450** (2007). doi:10.1038/nature06140
R.A. Duncan, F-region seasonal and magnetic storm behaviour. J. Atmos. Terr. Phys. **31**, 59–70 (1969)
J.L. Elliot, D.F. Strobel, X. Zhu, J.A. Stansberry, L.H. Wasserman, O.G. Franz, The thermal structure of Triton's middle atmosphere. Icarus **143**, 425–428 (2000)
J.L. Elliot, M.J. Person, A.A.S. Gulbis et al., Changes in Pluto's atmosphere: 1988–2006. Astron. J. **134**, 1–13 (2007)
T. Encrenaz, B. Bézard, T.K. Greathouse, M.J. Richter, J.H. Lacy, S.K. Atreya, A.S. Wong, S. Lebonnois, F. Lefèvre, F. Forget, Hydrogen peroxide on Mars: evidence for spatial and seasonal variations. Icarus **170**, 424–429 (2004)
B. Fegley Jr., M.Yu. Zolotov, Chemistry of sodium, potassium, and chlorine in volcanic gases on Io. Icarus **148**, 193–210 (2000)
S.B. Fels, R.S. Lindzen, The interaction of thermally excited gravity waves with mean flows. Geophys. Fluid Dyn. **6**, 149–191 (1974)
H. Feuchtgruber, E. Lellouch, T. de Graauw, B. Bézard, T. Encrenaz, M. Griffin, External supply of oxygen to the atmospheres of the giant planets. Nature **389**, 159–162 (1997)
H. Feuchtgruber, E. Lellouch, T. Encrenaz, B. Bézard, A. Coustenis, P. Drossart, A. Salama, T. de Graauw, G.R. Davis, Oxygen in the stratospheres of the giant planets and Titan. In The Universe as Seen by ISO, ed. by P. Cox, M.F. Kessler. ESA SP **427**, 133–136 (1999)
B.J. Finlayson-Pitts, J.N. Pitts Jr., *Chemistry of the Upper and Lower Atmosphere* (Academic Press, San Diego, 1999)
F.M. Flasar et al., An intense stratospheric jet on Jupiter. Nature **427**, 132–135 (2004)
F.M. Flasar et al., Temperatures, winds, and composition in the Saturn system. Science **307**, 1247–1251 (2005a)
F.M. Flasar et al., Titan's atmospheric temperatures, winds, and composition. Science **308**, 975–978 (2005b). doi:10.1126/science.1111150
J.M. Forbes, Wave coupling in terrestrial planetary atmospheres, in *Atmospheres in the Solar System: Comparative Aeronomy*, ed. by M. Mendillo, A. Nagy, J.H. Waite. Geophysical Monograph, vol. 130 (American Geophysical Union, Washington, 2002), pp. 171–190
J.M. Forbes, M.E. Hagan, Diurnal Kelvin wave in the atmosphere of Mars: towards an understanding of stationary density structures observed by the MGS accelerometer. Geophys. Res. Lett. **27**, 3563–3566 (2000)
J.M. Forbes, A.F.C. Bridger, S.W. Bougher, M.E. Hagan, J.L. Hollingsworth, G.M. Keating, J. Murphy, Non-migrating tides in the thermosphere of Mars. J. Geophys. Res. **107** (2002). doi:10.1029/2001JE001582
J.M. Forbes, X. Zhang, M. Angelats i Coll, G.M. Keating, Nonmigrating tides in the thermosphere of Mars: a quasi-empirical description. Adv. Space Res. **34**, 1690–1695 (2004)
J.M. Forbes, S. Bruinsma, F.G. Lemoine, Solar rotation effects on the thermospheres of Mars and Earth. Science **312**, 1366–1368 (2006). doi:10.1126/science.1126389
V. Formisano, S.K. Atreya, T. Encrenaz, N. Ignatiev, M. Giuranna, Detection of methane in the atmosphere of Mars. Science **306**, 1758–1761 (2004)
J.L. Fox, Heating efficiencies in the thermosphere of Venus reconsidered. Planet. Space Sci. **36**, 37–46 (1998)
J.L. Fox, Advances in the aeronomy of Venus and Mars. Adv. Space Res. **33**, 132–139 (2004)
J.L. Fox, S.W. Bougher, Structure, luminosity, and dynamics of the Venus thermosphere. Space Sci. Rev. **55**, 357–489 (1991)
J.L. Fox, K.Y. Sung, Solar activity variations of the Venus thermosphere/ionosphere. J. Geophys. Res. **106**, 21305–21335 (2001)
A.J. Friedson, A.-S. Wong, Y.L. Yung, Models for polar haze formation in Jupiter's stratosphere. Icarus **158**, 389–400 (2002)
M. Fulchignoni, F. Ferri, F. Angrilliet et al., Titan's Physical Characteristics Measured by the Huygens Atmospheric Structure Instrument (HASI). Nature **438**, 785–791 (2005). doi:10.1038/nature04314
T.J. Fuller-Rowell, D. Rees, S. Quegan, R.J. Moffett, M.V. Codrescu, G.H. Millward, A coupled thermosphere-ionosphere model (CTIM), in *STEP Handbook of Ionospheric Models*, ed. by R.W. Schunk (Utah State University, Logan, 1996), pp. 217–238
A. García Muñoz, J.C. McConnell, I.C. McDade, S.M.L. Melo, Airglow on Mars: Some model expectations for the OH Meinel bands and the O_2 IR atmospheric band. Icarus **176**, 75–95 (2005)
M.A. Geller, Dynamics of the Middle Atmosphere. Space Sci. Rev. **34**, 359–375 (1983)

P. Gierasch, Meridional circulation and the maintenance of the Venus atmospheric rotation. J. Atmos. Sci. **32**, 1038–1044 (1975)

P.J. Gierasch, R.M. Goody, R.E. Young, D. Crisp, C. Edwards, R. Kahn, D. Rider, A. Del Genio, R. Greeley, A. Hou, C.B. Leovy, D. McCleese, M. Newman, The general circulation of the Venus atmosphere: an assessment, in *Venus II*, ed. by S.W. Bougher, D.M. Hunten, R.J. Phillips (University of Arizona Press, Tucson, 1997), pp. 459–500

G.R. Gladstone, M. Allen, Y.L. Yung, Hydrocarbon photochemistry in the upper atmosphere of Jupiter. Icarus **119**, 1–52 (1996)

F. González-Galindo, M.A. López-Valverde, M. Angelats i Coll, F. Forget, Extension of a Martian general circulation model to thermospheric altitudes: UV heating and photochemical models. J. Geophys. Res. **110** (2005). doi:10.1029/2004JE002312

T.K. Greathouse, J.H. Lacy, B. Bézard, J.I. Moses, C.A. Griffith, M.J. Richter, Meridional variations of temperature, C_2H_2 and C_2H_6 abundances in Saturn's stratosphere at southern summer solstice. Icarus **177**, 18–31 (2005)

M.J. Harris, A new coupled terrestrial mesosphere-thermosphere general circulation model: Studies of dynamic, energetic, and photochemical coupling in the middle and upper atmosphere. PhD Thesis, University of London, UK (2000)

M.J. Harris, N.F. Arnold, A.D. Aylward, A study into the effect of the diurnal tide on the structure of the background mesosphere and thermosphere using the new coupled middle atmosphere and thermosphere (CMAT) general circulation model. Ann. Geophys. **20**, 225–235 (2002)

A.E. Hedin, E.L. Fleming, A.H. Manson, F.J. Schmidlin, S.K. Avery, R.R. Clark, S.J. Franke, G.J.A. Fraser, T.A. Tsuda, F.A. Vial, R.A. Vincent, Empirical wind model for the upper, middle and lower atmosphere. J. Atmos. Terr. Phys. **58**, 1421–1447 (1996)

B.E. Hesman, G.R. Davis, H.E. Matthews, G.S. Orton, The abundance profile of CO in Neptune's atmosphere. Icarus **186**, 342–353 (2007)

R.R. Hodges, Monte Carlo simulation of nonadiabatic expansion in cometary atmospheres–Halley. Icarus **83**, 410–433 (1990)

F. Hourdin, O. Talagrand, R. Sadourny, R. Courtin, D. Gautier, C.P. Mc Kay, Numerical simulation of the general circulation of the atmosphere of Titan. Icarus **117**, 358–374 (1995)

C.J.A. Howett, P.G.J. Irwin, N.A. Teanby, A. Simon-Miller, S.B. Calcutt, L.N. Fletcher, R. de Kok, Meridional variations in stratospheric acetylene and ethane in the Southern hemisphere of the saturnian atmosphere as determined from Cassini/CIRS measurements. Icarus **190**, 556–572 (2007)

W.B. Hubbard et al., The occultation of 28 SGR by Titan. Astron. Astrophys. **269**, 541–563 (1993)

W.B. Hubbard et al., Structure of Saturn's mesosphere from the 28 Sgr occultations. Icarus **130**, 404–425 (1997)

W.F. Huebner, J.J. Keady, S.P. Lyon, Astrophys. Space Sci. **195**, 1–294 (1992)

D.L. Huestis, S.W. Bougher, J.L. Fox, M. Galand, R.E. Johnson, J.I. Moses, J.C. Pickering, Cross sections and reaction rates for comparative planetary aeronomy. Space Sci. Rev. (2008, this issue)

H. Imanaka, M.A. Smith, Role of photoionization in the formation of complex organic molecules in Titan's upper atmosphere. Geophys. Res. Lett. **34**, L02204 (2007). doi:10.1029/2006GL028317

A.P. Ingersoll, R.F. Beebe, B.J. Conrath, G.E. Hunt, in *Saturn*, ed. by T. Gehrels, M.S. Matthews (Univ. of Arizona Press, Tucson, 1984), pp. 195–238

W.-H. Ip, On photochemical heating of cometary comae – The cases of H_2O and CO-rich comets. Astrophys. J. **264**, 726–732 (1983)

K.L. Jessup, J.R. Spencer, G.E. Ballester, R. Howell, F. Roessler, M. Vigel, R.V. Yelle, The atmospheric signature of Io's Prometheus plume and anti-jovian hemisphere: Evidence for a sublimation atmosphere. Icarus **169**, 197–215 (2004)

K.L. Jessup, J. Spencer, R.V. Yelle, Sulfur volcanism on Io. Icarus **192**, 24–40 (2007)

W.T. Kasprzak, A.E. Hedin, H.G. Mayr, H.B. Niemann, Wavelike perturbations observed in the neutral thermosphere of Venus. J. Geophys. Res. **93**, 11237–11245 (1988)

G.M. Keating et al., The structure of the upper atmosphere of Mars: in situ accelerometer measurements from Mars Global Surveyor. Science **279**, 1672–1675 (1998)

V.A. Krasnopolsky, *Photochemistry of the Atmospheres of Mars and Venus* (Springer, Berlin, 1986)

V.A. Krasnopolsky, Uniqueness of a solution of a steady-state photochemical problem: Applications to Mars. J. Geophys. Res. **100**, 3263–3276 (1995)

V.A. Krasnopolsky, Photochemistry in the martian atmosphere: Seasonal, latitudinal, and diurnal variations. Icarus **185**, 153–170 (2006a)

V.A. Krasnopolsky, Some problems related to the origin of methane on Mars. Icarus **180**, 359–367 (2006b)

V.A. Krasnopolsky, Chemical composition of Venus atmosphere and clouds: Some unsolved problems. Planet. Space Sci. **54**, 1352–1359 (2006c)

V.A. Krasnopolsky, Chemical kinetic model for the lower atmosphere of Venus. Icarus **191**, 25–37 (2007a)

V.A. Krasnopolsky, Long-term spectroscopic observations of Mars using IRTF/CSHELL: Mapping of O_2 dayglow, CO, and search for CH_4. Icarus **190**, 93–102 (2007b)
V.A. Krasnopolsky, D.P. Cruikshank, Photochemistry of Triton's atmosphere and ionosphere. J. Geophys. Res. **100**, 21271–21286 (1995)
V.A. Krasnopolsky, D.P. Cruikshank, Photochemistry of Pluto's atmosphere and ionosphere near perihelion. J. Geophys. Res. **104**, 21979–21996 (1999)
V.A. Krasnopolsky, V.A. Parshev, Photochemistry of the Venus atmosphere, in *Venus*, ed. by D.M. Hunten, L. Colin, T.M. Donahue, V.I. Moroz (University of Arizona Press, Tucson, 1983), pp. 431–458
V.A. Krasnopolsky, J.B. Pollack, H_2O-H_2SO_4 system in Venus' clouds and OCS, CO, and H_2SO_4 profiles in Venus' troposphere. Icarus **109**, 58–78 (1994)
V.A. Krasnopolsky, J.P. Maillard, T.C. Owen, Detection of methane in the martian atmosphere: evidence for life? Icarus **172**, 537–547 (2004)
V. Kunde et al., Jupiter's atmospheric composition from the Cassini thermal infrared spectroscopy experiment. Science **305**, 1582–1587 (2004)
H. Lammer et al., Loss of hydrogen and oxygen from the upper atmosphere of Venus. Planet. Space Sci. **54**, 1445–1456 (2006)
L.M. Lara, E. Lellouch, J.J. López-Moreno, R. Rodrigo, Vertical distribution of Titan's atmospheric neutral constituents. J. Geophys. Res. **113**, 2–26 (1996)
L.M. Lara, W.-H. Ip, R. Rodrigo, Photochemical models of Pluto's atmosphere. Icarus **130**, 16–35 (1997)
C. Laver, I. de Pater, H. Roe, D.F. Strobel, Temporal behavior of the SO 1.707 μm ro-vibronic emission band in Io's atmosphere. Icarus **189**, 401–408 (2007)
S. Lebonnois, D. Toublanc, F. Hourdin, P. Rannou, Seasonal variations of Titan's atmospheric composition. Icarus **152**, 384–406 (2001)
S. Lebonnois, Benzene and aerosol production in Titan and Jupiter's atmospheres: a sensitivity study. Planet. Space Sci. **53**, 486–497 (2005)
F. Lefèvre, S. Lebonnois, F. Montmessin, F. Forget, Threee-dimensional modeling of ozone on Mars. J. Geophys. Res. **109**, E07004 (2004). doi:10.1029/2004JE002268
F. Lefèvre, J.-L. Bertaux, S. Perrier, S. Lebonnois, O. Korablev, A. Fedorova, F. Montmessin, F. Forget, The Martian ozone layer as seen by SPICAM/Mars-Express. Paper presented at the Seventh International Conference on Mars, 9–13 July 2007, Pasadena, California (2007), p. 3137
E. Lellouch, J. Rosenquist, J.J. Goldstein, S.W. Bougher, G. Paubert, First absolute wind measurements in the middle atmosphere of Mars. Astrophys. J. **383**, 401–406 (1991)
E. Lellouch, T. Clancy, D. Crisp, A. Kliore, D. Titov, S.W. Bougher, Monitoring of mesospheric structure and dynamics, in *Venus II* (University of Arizona Press, Tucson, 1997), pp. 295–324
E. Lellouch, P.N. Romani, J. Rosenqvist, The vertical distribution and origin of HCN in Neptune's atmosphere. Icarus **108**, 112–136 (1994)
E. Lellouch, B. Bézard, J.I. Moses, G.R. Davis, P. Drossart, H. Feuchtgruber, E.A. Bergin, R. Moreno, T. Encrenaz, The origin of water vapor and carbon dioxide in Jupiter's stratosphere. Icarus **159**, 112–131 (2002)
E. Lellouch, G. Paubert, J.I. Moses, N.M. Schneider, D.F. Strobel, Volcanically emitted sodium chloride as a source for Io's neutral clouds and plasma torus. Nature **421**, 45–47 (2003)
E. Lellouch, R. Moreno, G. Paubert, A dual origin for Neptune's carbon monoxide. Astron. Astrophys. **430**, L37–L40 (2005)
E. Lellouch, B. Bézard, D.F. Strobel, G.L. Bjoraker, F.M. Flasar, P.N. Romani, On the HCN and CO_2 abundance and distribution in Jupiter's stratosphere. Icarus **184**, 478–497 (2006)
M.-C. Liang, R.-L. Shia, A.Y.-T. Lee, M. Allen, A.J. Friedson, Y.L. Yung, Meridional transport in the stratosphere of Jupiter. Astrophys. J. **635**, L177–L180 (2005)
G.F. Lindal, The atmosphere of Neptune: An analysis of the radio occultation data acquired with Voyager 2. Astron. J. **103**, 967–972 (1992)
R.S. Lindzen, Turbulence and stress owing to gravity wave and tidal breakdown. J. Geophys. Res. **86**, 9707–9714 (1981)
D. Lis, D. Bockelée-Morvan, N. Biver et al., Hydrogen isocyanide in comet 73P/Schwassmann-Wachmann (fragment B). Astrophys. J. **675**, 931–936 (2008)
R.D. Lorenz, K.L. Mitchell, R.L. Kirk, A.G. Hayes, O. Aharonson, H.A. Zebker, P. Paillou, J. Radebaugh, J.I. Lunine, M.A. Janssen, S.D. Wall, R.M. Lopes, B. Stiles, S. Ostro, G. Mitri, E.R. Stofan, Titan's inventory of organic surface materials. Geophys. Res. Lett. **35**, L02206 (2008). doi:10.1029/2007GL032118
D. Luz, F. Hourdin, P. Rannou, S. Lebonnois, Latitudinal transport by barotropic waves in Titan's stratosphere. II. Results from a coupled dynamics-microphysics-photochemistry GCM. Icarus **166**, 343–358 (2003)
J.R. Lyons, C. Manning, F. Nimmo, Formation of methane on Mars by fluid-rock interaction in the crust. Geophys. Res. Lett. **32**, L13201 (2005). doi:10.1029/2004GL022161

Y.-J. Ma, K. Altwegg, T. Breus, M.R. Combi, T.E. Cravens, E. Kallio, S.A. Ledvina, J.G. Luhmann, S. Miller, A.F. Nagy, A.J. Ridley, D.F. Strobel, Plasma flow and related phenomena in planetary aeronomy. Space Sci. Rev. (2008). doi:10.1007/s11214-008-9389-1

T. Majeed, J.H. Waite Jr., S.W. Bougher, G.R. Gladstone, Processes of equatorial thermal structure at Jupiter: An analysis of the Galileo temperature profile with a three-dimensional model. J. Geophys. Res. **110**, E12007 (2005). doi:10.1029/2004JE002351

H.G. Mayr, I. Harris, W.T. Kasprzak, M. Dube, F. Varosi, Gravity waves in the upper atmosphere of Venus. J. Geophys. Res. **93**, 11247–11262 (1988)

M.D. Max, S.M. Clifford, The state, potential distribution, and biological implications of methane in the Martian crust. J. Geophys. Res. **105**, 4165–4172 (2000)

M.B. McElroy, T.M. Donahue, Stability of the Martian atmosphere. J. Atmos. Sci. **28**, 879–884 (1972)

A.S. Medvedev, G.P. Klaassen, Parameterization of gravity wave momentum deposition based on nonlinear wave interactions: basic formulation and sensitivity tests. J. Atmos. Sol.-Terr. Phys. **62**, 1015–1033 (2000)

C.K. Meyer, Gravity wave interactions with mesospheric planetary waves: A mechanism for penetration into the thermosphere-ionosphere system. J. Geophys. Res. **104**, 28181–28196 (1999)

F.P. Mills, I. Observations and photochemical modeling of the Venus middle atmosphere. II. Thermal infrared spectroscopy of Europa and Callisto. Ph.D. Thesis, California Institute of Technology, Pasadena, CA (1998)

F.P. Mills, M. Allen, A review of selected issues concerning the chemistry in Venus' middle atmosphere. Planet. Space Sci. **55**, 1729–1740 (2007)

F.P. Mills, M. Sundaram, T.G. Slanger, M. Allen, Y.L. Yung, Oxygen chemistry in the Venus middle atmosphere, in *Advances in Geosciences*, vol. 3, ed. by W.-H. Ip, A. Bhardwaj (World Scientific Publishing Co., Singapore, 2006), pp. 109–117

G.J. Molina-Cuberos, K. Schwingenschuh, J.J. López-Moreno, R. Rodrigo, L.M. Lara, V. Anicich, Nitriles produced by ion chemistry in the lower ionosphere of Titan. J. Geophys. Res. **107**, 5099 (2002). doi:10.1029/2000JE001480

D. Moreau, L.W. Esposito, G. Brasseur, The chemical composition of the dust-free Martian atmosphere: Preliminary results of a two-dimensional model. J. Geophys. Res. **96**, 7933–7945 (1991)

J.I. Moses, T.K. Greathouse, Latitudinal and seasonal models of stratospheric photochemistry on Saturn: Comparison with infrared data from ITRF/TEXES. J. Geophys. Res. **110**, E09007 (2005). doi:10.1029/2005JE002450

J.I. Moses, B. Bézard, E. Lellouch, G.R. Gladstone, H. Feuchtgruber, M. Allen, Photochemistry of Saturn's atmosphere. I. Hydrocarbon chemistry and comparisons with ISO observations. Icarus **143**, 244–298 (2000a)

J.I. Moses, E. Lellouch, B. Bézard, G.R. Gladstone, H. Feuchtgruber, M. Allen, Photochemistry of Saturn's atmosphere. II. Effects of an influx of external material. Icarus **145**, 166–202 (2000b)

J.I. Moses, M.Y. Zolotov, B. Fegley Jr., Photochemistry of a volcanically driven atmosphere on Io: Sulfur and oxygen species from a Pele-type eruption. Icarus **156**, 76–106 (2002a)

J.I. Moses, M.Y. Zolotov, B. Fegley Jr., Alkali and chlorine photochemistry in a volcanically driven atmosphere on Io. Icarus **156**, 107–135 (2002b)

J.I. Moses, T. Fouchet, R.V. Yelle, A.J. Friedson, G.S. Orton, B. Bézard, P. Drossart, G.R. Gladstone, T. Kostiuk, T.A. Livengood, The stratosphere of Jupiter, in *Jupiter: Planet, Satellites and Magnetosphere*, ed. by F. Bagenal, W. McKinnon, T. Dowling (Cambridge Univ. Press, New York, 2004), pp. 129–157

J.I. Moses, T. Fouchet, B. Bézard, G.R. Gladstone, E. Lellouch, H. Feuchtgruber, Photochemistry and diffusion in Jupiter's stratosphere: Constraints from ISO observations and comparisons with other giant planets. J. Geophys. Res. **110**, E08001 (2005). doi:10.1029/2005JE002411

J.I. Moses, M.-C. Liang, Y.L. Yung, R.-L. Shia, Meridional distribution of hydrocarbons on Saturn: Implications for stratospheric transport. Paper presented at the Planetary Atmospheres Workshop, 6–7 Nov. 2007, Greenbelt, MD (2007), pp. 85–86

Y. Moudden, Simulated seasonal variations of hydrogen peroxide in the atmosphere of Mars. Planet. Space Sci. **55**, 2137–2143 (2007)

Y. Moudden, J.C. McConnell, Three-dimensional on-line chemical modeling in a Mars general circulation model. Icarus **188**, 18–34 (2007)

I.C.F. Mueller-Wodarg, R.V. Yelle, M. Mendillo, L.A. Young, A.D. Aylward, The thermosphere of Titan simulated by a global 3-dimensional time-dependent model. J. Geophys. Res. **105**, 20833–20856 (2000)

I.C.F. Mueller-Wodarg, R.V. Yelle, M. Mendillo, A.D. Aylward, On the global distribution of neutral gases in Titan's upper atmosphere and its effect on the thermal structure. J. Geophys. Res. **108** (2003). doi:10.1029/2003JA010054

I.C.F. Mueller-Wodarg, M. Mendillo, R.V. Yelle, A.D. Aylward, A global circulation model of Saturn's thermosphere. Icarus **180**, 147–160 (2006a)

I.C.F. Mueller-Wodarg, R.V. Yelle, N. Borggren, J.H. Waite, Waves and horizontal structures in Titans thermosphere. J. Geophys. Res. **111**, A12315 (2006b). doi:10.1029/2006JA011961

I.C.F. Mueller-Wodarg, R.V. Yelle, J. Cui, J.H. Waite, Horizontal structures and dynamics of Titan's thermosphere. J. Geophys. Res. (2008, in press)

M.L. Mumma, M.A. DiSanti, N. Dello Russo et al., Remote infrared observations of parent volatiles in comets: A window on the early solar system. Adv. Space Res. **31**, 2563–2575 (2003)

H. Nair, M. Allen, A.D. Anbar, Y.L. Yung, R.T. Clancy, A photochemical model of the Martian atmosphere. Icarus **111**, 124–150 (1994)

H.B. Niemann, S.K. Atreya, S.J. Bauer, G.R. Carignan, J.E. Demick, R.L. Frost, D. Gautier, J.A. Haberman, D.N. Harpold, D.M. Hunten, G. Israel, J.I. Lunine, W.T. Kasprzak, T.C. Owen, M. Paulkovich, F. Raulin, E. Raaen, S.H. Way, The abundances of constituents of Titan's atmosphere from the GCMS instrument on the Huygens probe. Nature **438**, 779–784 (2005). doi:10.1038/nature04122

C.A. Nixon, R.K. Achterberg, B.J. Conrath, P.G.J. Irwin, N.A. Teanby, T. Fouchet, P.D. Parrish, P.N. Romani, M. Abbas, A. LeClair, D. Strobel, A.A. Simon-Miller, D.J. Jennings, F.M. Flasar, V.G. Kunde, Meridional variations of C_2H_2 and C_2H_6 in Jupiter's atmosphere from Cassini CIRS infrared spectra. Icarus **188**, 47–71 (2007)

G. Orton et al., Earth-based observations of the Galileo probe entry site. Science **272**, 839–840 (1996)

T. Owen et al., The compostion of the atmosphere at the surface of Mars. J. Geophys. Res. **82**, 4635–4639 (1977)

T. Owen, A. Bar-Nun, I. Kleinfeld, Possible cometary origin of heavy noble gases in the atmospheres of Venus, Earth, and Mars. Nature **358**, 43–46 (1992). doi:10.1038/358043a0

T. Owen, Planetary atmospheres. Space Sci. Rev. **130**, 97–104 (2007). doi:10.1007/s11214-007-9233-z

C. Oze, M. Sharma, Have olivine, will gas: Serpentization and the abiogenic production of methane on Mars. Geophys. Res. Lett. **32**, L10203 (2005). doi:10.1029/2005GL022691

T.M. Parkinson, D.M. Hunten, Spectroscopy and aeronomy of O_2 on Mars. J. Atmos. Sci. **29**, 1380–1390 (1972)

H. Pernice, P. Garcia, H. Willner, J.S. Francisco, F.P. Mills, M. Allen, Y.L. Yung, Laboratory evidence for a key intermediate in the Venus atmosphere: Peroxychloroformyl radical. Proc. Natl. Acad. Sci. **101**, 14007–14010 (2004)

D.S. Pollock, G.B.I. Scott, L.F. Phillips, Rate constant for quenching of CO2(010) by atomic oxygen. Geophys. Res. Lett. **20**, 727–729 (1993)

R. Prangé, T. Fouchet, R. Courtin, J.E.P. Connerney, J.C. McConnell, Latitudinal variation of Saturn photochemistry deduced from spatially-resolved ultraviolet spectra. Icarus **180**, 379–392 (2006)

H. Rishbeth, I.C.F. Mueller-Wodarg, Vertical circulation and thermospheric composition: a modelling study. Ann. Geophys. **17**, 794–805 (1999)

R.G. Roble, in *On Solar Induced Variability in the Earth's Upper Atmosphere and Ionosphere*, ed. by K.S. Balasubramaniam, S.L. Keil, R.N. Smartt (Astron. Soc. of the Pacific, 1996), p. 609

R.G. Roble, R.E. Dickinson, E.C. Ridley, B.A. Emery, P.B. Hays, T.L. Killeen, N.W. Spencer, The high latitude circulation and temperature structure of the thermosphere near solstice. Planet. Space Science **31**, 1479–1499 (1983). doi:10.1016/0032-0633(83)90021-1

R.G. Roble, E.C. Ridley, A.D. Richmond, D.E. Dickinson, A coupled Thermosphere-Ionosphere General Circulation Model. Geophys. Res. Lett. **15**, 1325–1328 (1988)

R.G. Roble, E.C. Ridley, A thermosphere-ionosphere-mesosphere-electrodynamics general circulation model (TIME-GCM): Equinox solar cycle minimum simulations (30500 km). Geophys. Res. Lett. **21**, 417–420 (1994)

S.D. Rogers, S.B. Charnley, W.F. Huebner, D.C. Boice, Physical processes and chemical reactions in cometary comae, in *Comets II*, ed. by M.C. Festou, H.U. Keller, H.A. Weaver (Univ. Arizona Press, Tucson, 2005), pp. 505–522

P.N. Romani, Recent rate constant and product measurements of the reactions $C_2H_3 + H_2$ and C_2H_3+H—Importance for photochemical modeling of hydrocarbons on Jupiter. Icarus **122**, 233–241 (1996)

P.N. Romani, S.K. Atreya, Methane photochemistry and haze production on Neptune. Icarus **74**, 424–445 (1988)

P.N. Romani, J. Bishop, B. Bézard, S. Atreya, Methane photochemistry on Neptune: Ethane and acetylene mixing ratios and haze production. Icarus **106**, 442–463 (1993)

A. Sánchez-Lavega, S. Pérez-Hoyos, J.F. Rojas, R. Hues, R.G. French, Nature **423**, 623 (2003)

J. Saur, D.F. Strobel, Relative contributions of sublimation and volcanoes to Io's atmosphere from its plasma interaction during solar eclipse. Icarus **171**, 411–420 (2004)

L. Schaefer, B. Fegley, A thermodynamic model of high temperature lava vaporization on Io. Icarus **169**, 216–241 (2004)

L. Schaefer, B. Fegley, Predicted abundances of carbon compounds in volcanic gases on Io. Astrophys. J. **618**, 1079–1085 (2005a)

L. Schaefer, B. Fegley, Alkali and halogen chemistry in volcanic gases on Io. Icarus **173**, 454–468 (2005b)

L. Schaefer, B. Fegley, Silicon tetrafluoride on Io. Icarus **179**, 252–258 (2005c)

M.R. Schoeberl, D.F. Strobel, The zonally averaged circulation of the middle atmosphere. J. Atmos. Sci. **35**, 577–591 (1978)

J.T. Schofield, F.W. Taylor, Measurements of the mean solar-fixed temperature and cloud structure in the middle atmosphere of Venus. Q.J.R. Meteorol. Soc. **109**, 57–80 (1983)

G. Schubert, General circulation and dynamical state of the Venus atmosphere, in *Venus*, ed. by D.M. Hunten et al. (Univ. of Arizona Press, Tucson, 1983), pp. 681–765

G. Schubert et al., Venus atmosphere dynamics: A continuing enigma, in *Exploring Venus as a Terrestrial Planet*. Geophysical Monograph, vol. 176 (American Geophysical Union, 2007), pp. 101–120

A. Seiff, Models of Venuss atmospheric structure, in *Venus*, ed. by D.M. Hunten, L. Colin, V.I. Moroz (Univ. Arizona Press, Tucson, 1983), pp. 1045–1048

A. Seiff et al., Thermal structure of Jupiter's upper atmosphere derived from the Galileo probe. Science **276**, 102–104 (1997)

J.H. Seinfeld, S.N. Pandis, *Atmospheric Chemistry and Physics: From Air Pollution to Climate Change* (Wiley, Hoboken, 2006)

M.C. Senay, D. Jewitt, Coma formation driven by carbon-monoxide release from comet Schwassmann-Wachmann 1. Nature **371**, 229–231 (1994)

R.D. Sharma, P.P. Wintersteiner, Role of carbon dioxide in cooling planetary thermospheres. Geophys. Res. Lett. **17**, 2201–2204 (1990)

G.M. Shved, L.E. Khvorostovskaia, I.Iu. Potekhin, A.I. Dem'ianikov, A.A. Kutepov, The measurement of the rate constant of CO2/01 super-1 0/ quenching by atomic oxygen and the importance of the rate constant magnitude for the thermal regime and radiation of the lower thermosphere. Akad. Nauk SSSR, Izv. Fiz. Atmos. Okeana **27**, 431–437 (1991) (ISSN 0002-3515)

B. Sicardy et al., The two Titan stellar occultations of 14 November 2003. J. Geophys. Res. **111**, E11S91 (2006). doi:10.1029/2005JE002624

T. Slanger, D.L. Huestis, P.C. Cosby, N.J. Chanover, T.A. Bida, The Venus nightglow: Ground-based observations and chemical mechanisms. Icarus **182**, 1–9 (2006)

C.G.A. Smith, A.D. Aylward, G.H. Millward, S. Miller, L.E. Moore, An unexpected cooling effect in Saturn's upper atmosphere. Nature **445**, 399–401 (2007). doi:10.1038/nature05518

W.H. Smyth, M.C. Wong, Impact of electron chemistry on the structure and composition of Io's atmosphere. Icarus **171**, 171–182 (2004)

J.R. Spencer, K.L. Jessup, M.A. McGrath, G.E. Ballester, R. Yelle, Discovery of gaseous S_2 in Io's Pele plume. Science **288**, 1208–1210 (2000)

T. Stallard, S. Miller, G. Millward, R.D. Joseph, On the dynamics of the jovian ionosphere and thermosphere. I. The measurement of ion winds. Icarus **154**, 475–491 (2001)

T. Stallard, S. Miller, H. Melin, M. Lystrup, K. Dougherty, N. Achilleos, Saturn's auroral/polar H_3^+ infrared emission—I. General morphology and ion velocity structure. Icarus **189**, 1–13 (2007). doi:10.1016/j.icarus.2006.12.027

M.H. Stevens, D.F. Strobel, F. Herbert, An analysis of the Voyager 2 ultraviolet spectrometer occultation data at Uranus: Inferring heat sources and model atmospheres. Icarus **100**, 45–63 (1993)

D.F. Strobel, Photochemistry in outer solar system atmospheres. Space Sci. Rev. **116**, 155–170 (2005)

D.F. Strobel, Titan's hydrodynamically escaping atmosphere. Icarus **193**, 588–594 (2008a). doi:10.1016/j.icarus.2007.08.014

D.F. Strobel, N_2 Escape rates from Pluto's atmosphere. Icarus **193**, 612–619 (2008b). doi:10.1016/j.icarus.2007.08.021

D.F. Strobel, M.E. Summers, Triton's upper atmosphere and ionosphere, in *Neptune and Triton*, ed. by D.P. Cruikshank (Univ. Arizona Press, Tucson, 1995), pp. 1107–1148

D.F. Strobel, B.C. Wolven, The atmosphere of Io: Abundances and sources of sulfur dioxide and atomic hydrogen. Astrophys. Space Sci. **277**, 271–287 (2001)

D.F. Strobel, X. Zhu, M.E. Summers, On the vertical thermal structure of Pluto's atmosphere. Icarus **120**, 266–289 (1996)

M.E. Summers, D.F. Strobel, Photochemistry of the atmosphere of Uranus. Astrophys. J. **346**, 495–508 (1989)

M.E. Summers, D.F. Strobel, Photochemistry and vertical transport in Io's atmosphere and ionosphere. Icarus **120**, 290–316 (1996)

J.M. Sunshine, M.F. A'Hearn, O. Groussin et al., Exposed water ice deposits on the surface of comet 9P/Tempel 1. Science **311**, 1453–1455 (2007)

M.E. Summers, D.F. Strobel, G.R. Gladstone, Chemical models of Pluto's atmosphere, in *Pluto*, ed. by S.A. Stern, D.J. Tholen (Univ. Arizona Press, Tucson, 1997), pp. 391–434

D. Toublanc, J.P. Parisot, J. Brillet, D. Gautier, F. Raulin, C.P. McKay, Photochemical modeling of Titan's atmosphere. Icarus **113**, 2–26 (1995)

W.-L. Tseng, D. Bockelée-Morvan, J. Crovisier, P. Colom, W.-H. Ip, Cometary water expansion velocity from OH line shapes. Astron. Astrophys. **467**, 729–735 (2007)

G.L. Villanueva, B.P. Bonev, M.J. Mumma et al., The volatile composition of the split ecliptic comet 73P/Schwassmann-Wachmann 3: A comparison of fragments C and B. Astrophys. J. **650**, L87–L90 (2006)

V. Vuitton, R.V. Yelle, V.G. Anicich, The nitrogen chemistry of Titan's upper atmosphere revealed. Astrophys. J. **647**, L175–L178 (2006)

V. Vuitton, R.V. Yelle, M.J. McEwan, Ion chemistry and N-containing molecules in Titan's upper atmosphere. Icarus **191**, 722–742 (2007)

J.H. Waite et al., Ion Neutral Mass Spectrometer results from the first flyby of Titan. Science **308**, 982–986 (2005)

J.H. Waite, D.T. Young, T.E. Cravens, A.J. Coates, F.J. Crary, B. Magee, J. Westlake, The process of tholin formation in Titan's upper atmosphere. Science **316**, 870–875 (2007)

R.P. Wayne, *Chemistry of Atmospheres* (Oxford University Press, Oxford, 2000)

E.H. Wilson, S.K. Atreya, Current state of modeling the photochemistry of Titan's mutually dependent atmosphere and ionosphere. J. Geophys. Res. **109**, E06002 (2004). doi:10.1029/2003JE002181

E.H. Wilson, S.K. Atreya, A. Coustenis, Mechanisms for the formation of benzene in the atmosphere of Titan. J. Geophys. Res. **108**, 5014 (2003). doi:10.1029/2002JE001896

J.R. Winick, A.I. Stewart, Photochemistry of SO_2 in Venus' upper cloud layers. J. Geophys. Res. **85**, 7849–7860 (1980)

P. Withers, S.W. Bougher, G.M. Keating, The effects of topographically-controlled thermal tides in the martian upper atmosphere as seen by the MGS accelerometer. Icarus **164**, 14–32 (2003)

M.C. Wong, R.E. Johnson, A three-dimensional azimuthally symmetric model atmosphere for Io. 1. Photochemistry and the accumulation of a nightside atmosphere. J. Geophys. Res. **101**, 23243–23254 (1996)

M.C. Wong, W.H. Smyth, Model calculations for Io's atmosphere at eastern and western elongations. Icarus **146**, 60–74 (2000)

A.-S. Wong, A.Y.T. Lee, Y.L. Yung, J.M. Ajello, Jupiter: Aerosol chemistry in the polar atmosphere. Astrophys. J. **534**, L215–L217 (2000)

A.-S. Wong, C.G. Morgan, Y.L. Yung, Evolution of CO on Titan. Icarus **155**, 382–392 (2002)

A.-S. Wong, Y.L. Yung, A.J. Friedson, Benzene and haze formation in the polar atmosphere of Jupiter. Geophys. Res. Lett. **30**, 1447 (2003). doi:10.1029/2002GL016661

R.V. Yelle, Non-LTE models of Titan's upper atmosphere. Astrophys. J. **383**, 380–400 (1991)

R.V. Yelle, J.I. Lunine, Evidence for a molecule heavier than methane in the atmosphere of Pluto. Nature **339**, 288–290 (1989)

R.V. Yelle, L.R. Doose, M.G. Tomasko, D.F. Strobel, Analysis of Raman scattered Ly-Alpha emissions from the atmosphere of Uranus. Geophys. Res. Lett. **14**, 483 (1987)

R.V. Yelle, C.A. Griffith, L.A. Young, Structure of the Jovian stratosphere at the Galileo probe entry site. Icarus **152**, 331–346 (2001)

R.V. Yelle, S. Miller, Jupiter's thermosphere and ionosphere, in *Jupiter, The Planet, Satellites and Magnetosphere*, ed. by F. Bagenal, T.E. Dowling, W.B. McKinnon (Cambridge University Press, Cambridge, 2004)

R.V. Yelle, J. Cui, I.C.F. Mueller-Wodarg, Eddy diffusion and methane escape from Titan's atmosphere. J. Geophys. Res. (2008, in press)

Y.L. Yung, W.B. DeMore, Photochemistry of the stratosphere of Venus: Implications for atmospheric evolution. Icarus **51**, 199–247 (1982)

Y.L. Yung, W.B. DeMore, *Photochemistry of Planetary Atmospheres* (Oxford University Press, New York, 1999)

Y.L. Yung, M. Allen, J.P. Pinto, Photochemistry of the atmosphere of Titan: Comparison between model and observations. Astrophys. J. Suppl. Ser. **55**, 465–506 (1984)

J. Zhang, D.B. Goldstein, P.L. Varghese, N.E. Gimelshein, S.F. Gimelshein, D.A. Levin, Simulation of gas dynamics and radiation in volcanic plumes on Io. Icarus **163**, 182–197 (2003)

J. Zhang, D.B. Goldstein, P.L. Varghese, L. Trafton, C. Moore, K. Miki, Numerical modeling of ionian volcanis plumes with entrained particulates. Icarus **172**, 479–502 (2004)

Q.H. Zhou, M.P. Sulzer, C.A. Tepley, C.G. Fesen, R.G. Roble, M.C. Kelley, Neutral winds and temperature in the tropical mesosphere and lower thermosphere during January 1993: Observation and comparison with TIME-GCM results. J. Geophys. Res. **102**, 11,507–11,519 (1997)

X. Zhu, J.-H. Yee, Wave-photochemistry coupling and its effect on water vapor, ozone and airglow variations in the atmosphere of Mars. Icarus **189**, 136–150 (2007)

M.Yu. Zolotov, B. Fegley Jr., Volcanic origin of disulfur monoxide (S_2O) on Io. Icarus **133**, 293–297 (1998a)

M.Yu. Zolotov, B. Fegley Jr., Volcanic production of sulfur monoxide (SO) on Io. Icarus **132**, 431–434 (1998b)

M.Yu. Zolotov, B. Fegley Jr., Oxidation state of volcanic gases and the interior of Io. Icarus **141**, 40–52 (1999)

M.Yu. Zolotov, B. Fegley Jr., Eruption conditions of Pele volcano on Io inferred from chemistry of its volcanic plume. Geophys. Res. Lett. **27**, 2789–2792 (2000)

Solar System Ionospheres

O. Witasse · T. Cravens · M. Mendillo · J. Moses · A. Kliore · A.F. Nagy · T. Breus

Originally published in the journal Space Science Reviews, Volume 139, Nos 1–4.
DOI: 10.1007/s11214-008-9395-3

Abstract This article reviews our understanding of the ionospheres in the solar system. It provides some basic information on the sources and sinks of the ionospheric plasma, its dynamics, the energetics and the coupling to the neutral atmosphere. Ionospheres in the solar system are reviewed and comparative ionospheric topics are discussed.

Keywords Ionospheres · Planetology

O. Witasse (✉)
European Space Agency, P.O. Box 299, 2200 AG Noordwijk, The Netherlands
e-mail: owitasse@rssd.esa.int

T. Cravens
University of Kansas, Lawrence, KS 66045, USA
e-mail: cravens@ku.edu

M. Mendillo
Boston University, One Sherborn Street, Boston, MA 02215, USA
e-mail: mendillo@bu.edu

J. Moses
Lunar and Planetary Institute, Houston, TX 77058, USA
e-mail: moses@lpi.usra.edu

A. Kliore
Jet Propulsion Laboratory, Pasadena, CA 91109, USA
e-mail: akliore@jpl.nasa.gov

A.F. Nagy
Department of Atmospheric, Oceanic and Space Sciences, University of Michigan, Ann Arbour, MI 48109, USA
e-mail: anagy@umich.edu

T. Breus
Space Research Institute, 84/32 Profsoyouznaya, 117810 Moscow, Russia
e-mail: breus36@mail.ru

1 Introduction

The purpose of this chapter is to provide an overview of our understanding of the ionospheres in the solar system. This is certainly not the first review of this kind, and the reader will find useful information in previous works (e.g. Nagy and Cravens 2002; Schunk and Nagy 2000). The next section provides some generalities on the physics of the ionospheres. Section 3 reviews the individual bodies. Section 4 discusses some selected comparative ionospheric topics.

2 Current Understanding of Planetary Ionospheres

2.1 Sources and Sinks

An ionosphere is created by ionization of the neutral atmosphere of a planet or moon. The main source of ionization, especially in the inner solar system, is the solar extreme ultraviolet and X-ray photon flux. The consequence of the ionization process is the creation of ions and photo-electrons, which can lead to additional ionization of the neutrals. Other sources include energetic magnetospheric or solar wind electrons and ions, and even meteoroids. The reader is referred to the Galand and Fox chapter (this issue) for additional information of energy deposition into the upper atmosphere.

Once ions are produced by photo-ionization, there are several ways whereby ions are transformed into other ions, or lost due to recombination with electrons. An ion-transformation example via ion-neutral chemistry is:

$$O^+ + N_2 \rightarrow NO^+ + N.$$

This is the main mechanism of formation of NO^+ ion in the Earth's ionosphere, in the so-called E-region. In the ionospheres of Venus and Mars, a similar example of ion transformation is:

$$CO_2^+ + O \rightarrow O_2^+ + CO.$$

Plasma loss via dissociative recombination of a molecular ion is an effective process for removing ions and electrons:

$$O_2^+ + e^- \rightarrow O + O.$$

Composition and dynamics can be used to separate an ionosphere into various regions. When the production of ionization is balanced by the sinks, the region is chemically controlled. These are the D, E and F1 regions of the Earth, in the altitude range 60–300 km. At Venus and Mars, this region is at altitudes below about 180 km, while at Saturn this altitude occurs below ~2200 km. Above these altitudes, the ionosphere is strongly influenced by transport (see next section). In the Bougher and Blelly chapter (this issue) overall modelling of ionospheric sources and sinks is explained in detail.

2.2 Dynamics

Dynamics plays an important role in determining the character of an ionosphere, particularly at higher altitudes where chemical lifetimes are long. Plasma flow from one region to another, either horizontally or vertically, can be described either kinetically or using fluid methods (for a general discussion read Kelley 1989; Cravens 1997; Schunk and Nagy 2000). The dynamics of individual plasma species (i.e., specific ion species and electrons) must often be considered, but this short review focuses on the plasma as a whole (i.e., single-fluid theory). The force (or momentum) balance controls the dynamics. The single-fluid equation of motion (or momentum equation) can be written as:

$$\rho\left(\frac{\mathbf{Du}}{Dt}+\mathbf{u}\cdot\nabla\mathbf{u}\right)=\mathbf{J}\times\mathbf{B}-\nabla(p_\mathrm{e}+p_\mathrm{i})-\rho\mathbf{g}-\rho\nu_\mathrm{in}(\mathbf{u}-\mathbf{u}_\mathrm{n}) \tag{1}$$

where the mass density, ρ and bulk flow velocity, $\mathbf{u}$, are functions of position and time. The left-hand side is the mass density times the convective derivative of the fluid velocity. p_e and p_i are electron and ion thermal pressures, $\mathbf{g}$ is the acceleration due to gravity, $\mathbf{J}$ is the current density, $\mathbf{B}$ is the magnetic field, ν_in is the ion-neutral collision frequency, and $\mathbf{u}_\mathrm{n}$ is the neutral flow velocity. This equation states that the acceleration per unit volume equals the net force per unit volume on the parcel.

The relative importance of different force terms depends on the planet in question. The second term on the right-hand side (RHS) is the pressure gradient term. The plasma pressures are given by $p_\mathrm{e}=n_\mathrm{e}k_\mathrm{B}T_\mathrm{e}$ and $p_\mathrm{i}=n_\mathrm{i}k_\mathrm{B}\ T_\mathrm{i}$, where k_B is Boltzmann's constant, n_e and n_i are the electron and ion densities, respectively, with $n_\mathrm{e}=n_\mathrm{i}$ from quasi-neutrality. T_e and T_i are the electron and ion temperatures, respectively. Clearly, the pressure gradient force term depends on the overall plasma density (and hence on the sources and sinks discussed earlier) as well as on the plasma temperatures (part of the energetics, to be discussed next). The other force terms on the RHS include the force on the plasma due to magnetic fields ($\mathbf{J}\times\mathbf{B}$), the gravitational force, and the frictional/collisional force due to the interaction with the neutral atmosphere.

The acceleration term (i.e., LHS of the equation) can be neglected for flow speeds much less than the sound speed. In this case, a diffusion equation for the flow velocity can be found by isolating the ion-neutral collision term. As shown by Schunk and Nagy (2000) this diffusion approach can also be used for individual ion species. An important special case is hydrostatic equilibrium, along $\mathbf{B}$, for which $\mathbf{u}\approx 0$ is assumed in the momentum equation. Only the pressure gradient and gravitational terms are left and one finds that the plasma pressure varies exponentially with the plasma scale-height, $H=k_\mathrm{B}\ (T_\mathrm{e}+T_\mathrm{i})/(mg)$, where m is the average ion mass. This so-called diffusive equilibrium approximation is often used to describe the density distribution in topside planetary ionospheres.

Magnetized plasmas experience the Lorentz force—the force per unit volume is given by $\mathbf{J}\times\mathbf{B}$. For planets with substantial intrinsic magnetic fields (e.g., Earth, Jupiter, Saturn) it is advantageous to separately keep track of the current density, $\mathbf{J}$, using a generalized Ohm's law with a conductivity tensor organized by the magnetic field direction (cf., Kelley 1989). However, the magnetohydrodynamic (MHD) approach is usually taken for objects lacking significant intrinsic magnetic fields such as Venus, Titan, Io, and comets (cf., Cravens 1997; Luhmann and Cravens 1991; Cravens et al. 1997). Magnetic fields for such bodies are generated by currents induced in the ionosphere by the interaction with external flowing plasma (e.g., the solar wind or magnetosphere). Using Ampere's law, the $\mathbf{J}\times\mathbf{B}$ term can be split into a magnetic pressure gradient term and a magnetic curvature term:

$$\mathbf{J}\times\mathbf{B}=\frac{1}{\mu_0}(\nabla\times\mathbf{B})\times\mathbf{B}=-\nabla\frac{B^2}{2\mu_0}+\frac{1}{\mu_0}(\mathbf{B}\cdot\nabla)^{\mathbf{B}} \tag{2}$$

where the magnetic pressure is identified as $p_\mathrm{B}=B^2/2\mu_0$.

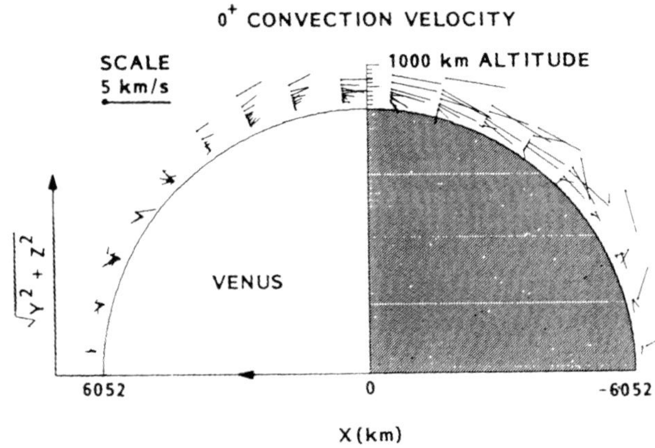

Fig. 1 Ion flow velocities in the ionosphere of Venus measured by the Pioneer Venus Orbiter retarding potential analyzer experiment. The Sun is to the left. From Knudsen et al. (1982)

The ionosphere of Venus provides two relatively simple examples of ionospheric force balance (i.e., dynamics). The ionosphere remains free of large-scale magnetic fields during low solar wind dynamic pressure conditions and a sharp boundary (called the ionopause) separates the cold dense ionospheric plasma from the magnetized shocked solar wind plasma (Brace and Kliore 1991; Luhmann and Cravens 1991; Cravens et al. 1997). The momentum equation can then be approximated near the boundary with the expression $p_B + p_e + p_i$ = constant. Data from experiments onboard the Pioneer Venus Orbiter demonstrated that ionospheric thermal pressure below the ionopause balances the magnetic pressure in the magnetic barrier region above the ionopause (Luhmann and Cravens 1991, and references therein).

Consider a second example of ionospheric dynamics at Venus. Fast day-to-night flows were observed in the unmagnetized ionosphere and it has been demonstrated that they are driven by large day to night thermal pressure gradient forces (Knudsen et al. 1981; Elphic et al. 1984; Nagy et al. 1991). The thermal pressure gradient is associated with density gradients from higher dayside ionospheric densities produced by solar radiation (Sect. 2.1). Note that Venus has a very low rotation rate (retrograde with a 243 Earth day period; effective night is about 58 Earth days) so that solar radiation cannot contribute to a nightside ionosphere. The pressure gradient force is balanced by ion-neutral collisions at lower altitudes and by inertial effects at higher altitudes. Figure 1 shows O^+ flow velocity vectors measured by the Pioneer Venus Retarding Potential Analyzer experiment (Knudsen et al. 1982). The day to night ionospheric flow speeds reach several kilometers per second at higher altitudes. The day-to-night flow is the main source of the nightside ionosphere of Venus during low solar wind dynamic pressure conditions (Cravens et al. 1983; Spenner et al. 1981; Dobe et al. 1995). Chemically long-lived O^+ and O_2^+ ions produced on the dayside flow to the night where a fraction of them diffuse/flow downward and help to maintain the nightside ionosphere (Ma et al. 2004).

A terrestrial example of ionospheric dynamics is provided by the high-speed anti-sunward ionospheric flows observed in the polar ionosphere. These flows are driven by the dawn-to-dusk convection electric field, which "maps" along the Earth's geomagnetic field from the magnetosphere down to the ionosphere (Heelis et al. 1980; Kelley 1989). Ionospheric dynamics at Earth, Jupiter or Saturn, is organized by the large intrinsic magnetic fields.

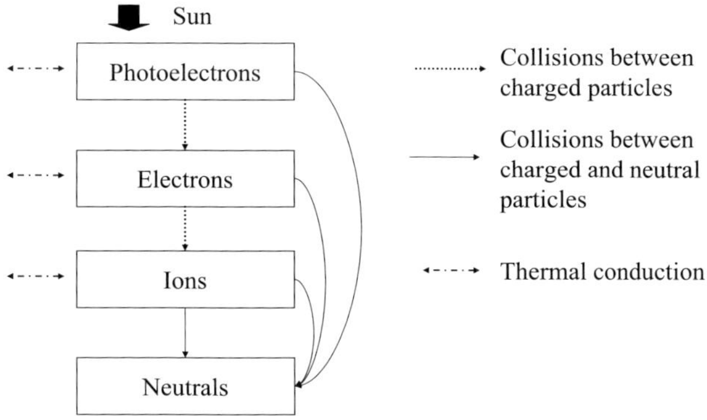

Fig. 2 Illustration of energy exchanges between neutrals, electrons and ions in a planetary ionosphere

2.3 Energetics

The plasma temperatures are the results of the energy exchanges between the key players of the upper atmospheres, namely, the neutrals, ions and electrons. The full story is rather complex (Schunk and Nagy 1978; Lilensten and Blelly 1999), and only a short summary will be given in this section. The photo-ionization process leads to the creation of energetic electrons called photo-electrons that lose their energy in the medium via collisions. Electrons with the highest energy having collisions with the neutral components produce ions, secondary electrons, and neutrals in excited states. Suprathermal electrons of low energy (below 10 eV) lose their energy by collisions with the ambient thermal electron population. The thermal electrons lose energy by inelastic collisions with the neutrals and by elastic collisions with the ions, while the ions transfer their energy to the neutrals. Within a given population, energy is transferred via thermal conduction. These exchanges are illustrated in Fig. 2.

As a result of these interactions, the thermal profile exhibits the following behaviour: at lower altitudes, in the terrestrial D and E ionospheric regions, the three populations of neutrals, ions and electrons have the same temperatures, since the dominating process is inelastic collisions with the neutrals. At higher altitudes, the heating by the suprathermal population is more efficient and the electron temperature departs from the neutral temperature, while the ions are still being thermalized via collisions with the neutrals. Above a certain altitude (400–500 km for the Earth), the ion population becomes warmer than the neutrals due to the ion-electron collisions that are more frequent that the ion-neutral collisions.

The situation is even more complex in the presence of an induced and horizontal magnetic field, like at Mars and Venus. Such a field prevents the energy from being transferred by conduction in the direction perpendicular to the magnetic field. At Venus, both electron and ion temperatures were observed to be higher than the neutral temperatures above 140 km. These high temperatures could not be explained by simply invoking electron heating and the classical thermal conduction as is the case at the Earth. It has been suggested that either a somewhat *ad hoc* energy input at the top of the ionosphere is present, or the thermal conductivities are reduced as a result of magnetic fluctuations. It is quite likely that both of the mechanisms play a role, but whether one or the other is dominant remains an open question. Clearly, the processes controlling the energetics of the ionosphere of Venus and of Mars need to be clarified.

Figure 3 displays computed thermal structures for Titan.

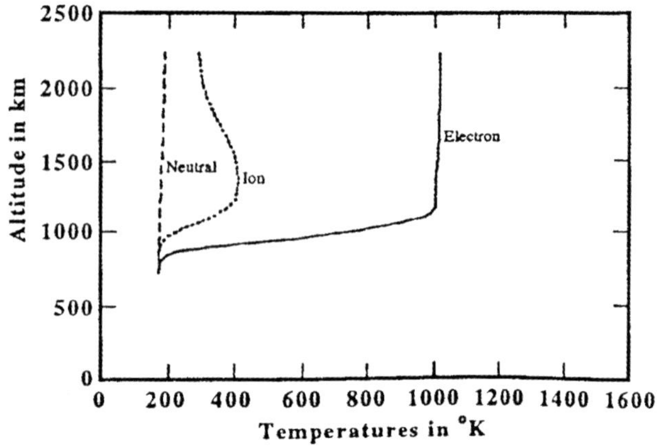

Fig. 3 Computed temperature profiles on Titan (Roboz and Nagy 1994)

2.4 Ionospheric Coupling to Neutral Atmosphere

In addition to the neutral atmosphere's central role in ionospheric production and loss via ionization and ion-neutral chemistry described above (Sect. 2.1), there are several aspects of plasma energetics and dynamics due to neutrals as well. As described above, the electron, ion and neutral temperatures satisfy the relation $T_e \geq T_i \geq T_n$, and thus to solve for plasma diffusion driven by spatial gradients in plasma temperatures, it is first required to obtain the starting point of correct neutral temperatures at each planet. Electrical conductivities depend on collision rates between electrons, ions and neutrals, and thus ionospheric electrodynamics also depends on proper representation of plasma coupling to the neutral constituents. Neutral winds can cause bulk plasma motions of ion-electron pairs, frictional heating and thus the correct specification of winds is also extremely important for the correct portrayal of ionospheric dynamics. This is particularly important on planets with strong intrinsic magnetic fields that result in a horizontal neutral wind (U_n) causing vertical plasma motion, $V_p = U_n \sin I \cos I$, where I is the inclination angle of the local magnetic field. Finally, in ionospheric layers embedded in very dense neutral atmospheres, neutral winds and tides exert differential effects upon ions and electrons, resulting in so-called dynamo currents and polarization driven electrodynamics. These processes are treated in detail in Rishbeth and Garriott (1969) and Schunk and Nagy (2000).

Two dramatic cases of a planet's neutral atmosphere coupling to its ionosphere come from recent spacecraft observations at Earth and Mars. At Earth, the well known Equatorial Ionization Anomaly (EIA) refers to the bi-hemispheric zones of maximum ionospheric electron densities at low latitudes to either side of the geomagnetic equator. These belts of enhanced F-layer plasma produce airglow called *inter-tropical arcs* that can be imaged in visible light from the ground and in UV from space—the optical signatures of the EIA. Figure 4 gives an example of such a global dataset (Immel et al. 2006). The electrodynamics that establishes the EIA (the so-called fountain effect) is clearly not uniform in longitude. The "wave-4" pattern shown is interpreted as a modulation of the fountain effect mechanism by similar longitude structures in the upward propagating tides, as described by the Global Scale Wind Model of Hagan et al. (2001). Thus, while longitude structure in the inter-tropical arcs had been known as far back as OGO measurements of 6300 Å emission in 1967 (Chandra et al. 1973), it is the recent appreciation of upward neutral coupling via electrodynamics that offers a comprehensive framework for its understanding.

The radio occultation experiment of the Mars Global Surveyor (MGS) spacecraft revealed a similar type of neutral-ionospheric coupling, namely, a longitudinal pattern in the

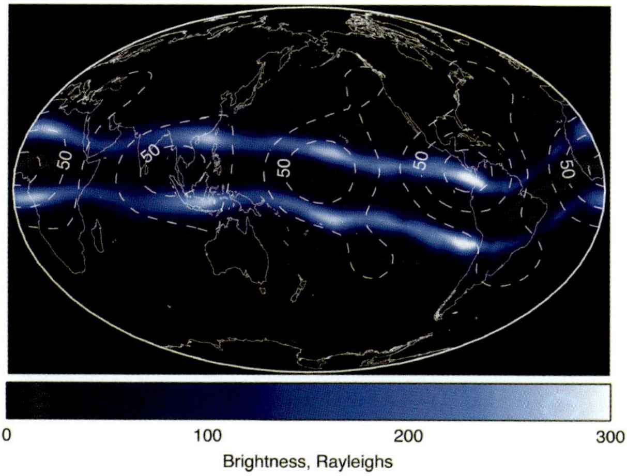

Fig. 4 Longitudinal (wave-4) structure in the inter-tropical arcs associated with the Earth's equatorial ionization anomaly, as obtained using data from the FUV detector on the IMAGE spacecraft. The pattern shown is a reconstruction from 30 days of observations at a local time of 21:00, taken in the northern hemisphere in 2002, and reproduced as a mirror image for the southern hemisphere. Overlaid with white *dashed contours* are amplitudes of the diurnal temperature variation at 115 km due to upward-propagating atmospheric tides from the global scale wind model (after Immel et al. 2006)

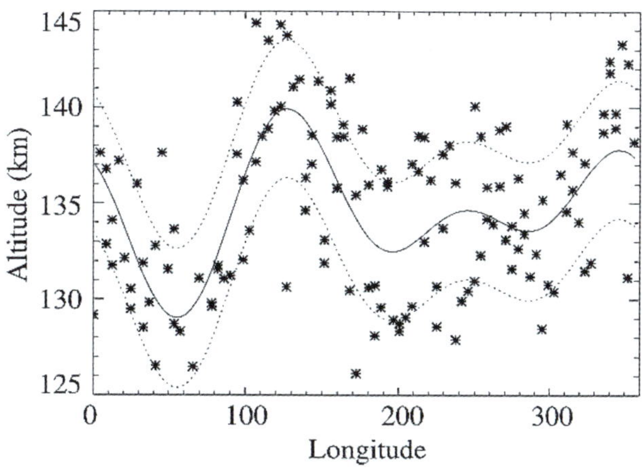

Fig. 5 Longitudinal pattern (wave-3) in the heights of the primary peak in the ionospheric electron density profiles at Mars taken by the MGS radio occultation experiment in December 2000. The *solid line* gives a least squares fit, with $\pm\sigma$ uncertainties given by the *dashed lines* (after Bougher et al. 2004)

height of the ionosphere's peak electron density (h_{max}). As shown in Fig. 5, the h_{max} values scaled from many MGS electron density profiles exhibit a wave-3 pattern in longitude (Bougher et al. 2004; Seth et al. 2006). Since a photochemical ionospheric layer is produced at the height of unit optical depth in a neutral atmosphere, such modulations in the $N_e(h)$ profiles can serve as a method to study the tidal modes most prominent in Mars' upper atmosphere (see Bougher et al. 2004, and references therein).

At the giant planets, coupling between neutral winds, ionospheric conductivities, and the mapping of electrodynamics along high latitude magnetic field lines into the magnetosphere has been suggested as the most dramatic case of upward atmospheric coupling in the Solar System. Smith (2006) showed that the variability from pure co-rotation of magnetospheric processes at Jupiter and Saturn might be attributed to outward interactions driven by coupled thermosphere-ionosphere-magnetosphere systems.

3 Review of the Individual Bodies

3.1 Venus

Most of our knowledge of the ionosphere at Venus comes from the 14-year mission of Pioneer Venus Orbiter (PVO). On the dayside, the main electron density peak is located at about 135–140 km altitude, with a typical value of 5×10^5 cm^{-3}. The scheme of the chemical reactions involving neutrals, ions and electrons is relatively well known. Although the major neutral species is CO_2 the major ion is O_2^+, because of rapid charge exchange processes. CO_2^+ is a minor ion at all altitudes. O^+ becomes the major ion above about 200 km. The other minor ions are C^+, N^+, CO^+, N_2^+, H^+ and the doubly-charged ion O^{++}. The variability of the plasma composition as a function of altitude, solar zenith angle and solar activity has been relatively well established. Figure 6, reproduced from Nagy et al. (1980), shows the calculated mean ion densities at Venus, which are in reasonably good agreement with observations.

Recent radio-occultation data from the Venus Express spacecraft have revealed the ionospheric structure in the altitude range 100–500 km and show large variations in the topside ionosphere on the dayside, a clear layer due to meteoritic activity, and a very variable nighttime ionosphere (Paetzold et al. 2007). The authors report that the majority of electron density profiles display a bulge in the topside ionosphere between 160 and 180 km that is not documented (according to the authors) in theoretical models of ion and electron production. An ionospheric ledge near this altitude was observed by the radio occultation experiments on the Mariner 5 and 10 spacecraft and later explained as being a MHD effect associated with the solar wind interaction (Brace and Kliore 1991).

Some still unresolved issues have been reviewed in Witasse and Nagy (2006), and include the chemistry of the O^+ ions in their excited state, the presence (or not) of molecular doubly-charged ions CO_2^{++}, the energetics, the ion dynamics around the terminator, interpretation of nighttime features such as holes, clouds and channels, plasma temperatures, and the characterization of the lower ionosphere.

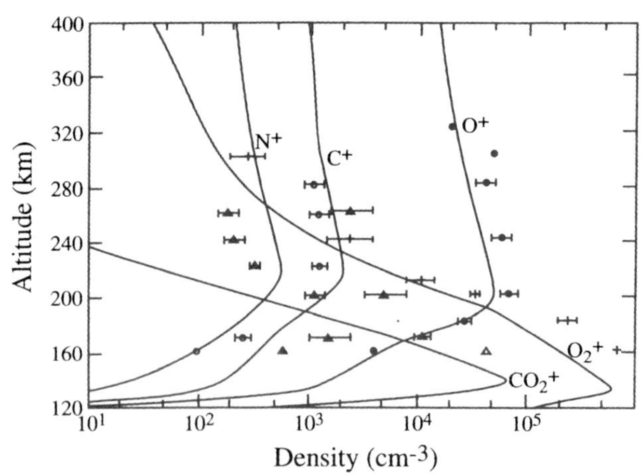

Fig. 6 Measured and calculated daytime ion densities at Venus (Nagy et al. 1980)

3.2 Earth

In contrast to the other planets, where discovery-mode studies are still underway, the data rich field of terrestrial aeronomy concentrates on highly focused topics that deal with a very mature level of understanding of key ionospheric processes. Traditionally, these areas are grouped by latitude region: equatorial-and-low latitudes, mid-latitudes, and auroral-polar latitudes. A theme that has emerged recently is that coupling between these regions can occur far more frequently and have far greater impact that previously considered. For example, the penetration of magnetospheric electric fields to middle and low latitudes is now an accepted fact, in contrast to its status as a highly unlikely mechanism of importance in studies of ionospheric storms in the 1980s (Mendillo 2006). Today, the largest ionospheric storm effects monitored by the worldwide network of Global Positioning System (GPS) observations of total electron content (TEC) are routinely attributed to electrodynamical effects (Foster and Rideout 2005). Similarly, the poleward intrusion of equatorial plasma instability processes into the midlatitudes is an active area of research (Mendillo et al. 2005). Finally, in the domain of temporal coupling, considerable advances have been made in the last decade in studies of the short time scales associated with auroral and meteoritic effects. These range, for example, from high time resolution observations and modeling of filamentary structures in the aurora (Lanchester et al. 1997; Semeter and Blixt 2006) to the micro-scale evolution of plasma structures within meteoritic layers (Dyrud et al. 2005). Remarkably, auroral flickering has been observed at frequencies greater than 100 Hz, and discrete arc widths are now known to be as small as $\sim$30 meters (McHarg et al. 1998; Sakanoi and Fukunishi 2004).

The investigation of small scale ionospheric irregularities is a topic capable of being studied only at Earth, again due to the unique availability of multiple diagnostic techniques (radio, optical, *in-situ* satellite data). The plasma instabilities that create such irregularities were all discovered decades ago using line-of-site radar observations. In recent years, the increased availability of 2-dimensional optical images from both groundbased (visible) and spacebased (ultraviolet) detectors have been able to characterize the broad spatial envelopes that contain ionospheric irregularities (Kelley et al. 2003). The airglow signatures of equatorial spread-F (ESF) irregularities have long been known to appear as dark, irregular structures aligned approximately in the north-south direction, seen upon a background of bright airglow of the intertropical arcs (Weber et al. 1978; Martinis et al. 2006). Examples of groundbased, high-resolution images in 6300 Å airglow depletions are shown in Fig. 7 (top). Such patterns arise from the gravitationally driven Rayleigh-Taylor fluxtube interchange instability processes, and their day-to-day occurrence patterns remain one of the major unexplained phenomena in all of space physics. A second type of ionospheric irregularity appears optically as alternating bands of bright and dark airglow (Mendillo et al. 1997), as shown in the bottom panel of Fig. 7. These structures are typically aligned at large angles to a north-south meridian, with drift patterns in longitude (equatorward and westward) that are different from typical ESF patterns (eastward). A modified version of the Perkins instability is the leading candidate to explain the 'airglow bands" phenomena (Kelley et al. 2002; Shiokawa et al. 2005). An unanticipated effect was the recent discovery of geomagnetic fluxtube-aligned ESF airglow depletions, originating at the equator and extending to lower midlatitudes, on the same night as airglow bands generated at midlatitudes and propagating equatorward (Martinis 2007). How distinct electrodynamical processes can occur simultaneously in the same ionospheric location is an area of investigation in need of further study—observationally, theoretically and via modeling.

The Earth is still the only planet for which geomagnetic storm effects are known to cause ionospheric perturbations. While aurora have been observed repeatedly at Jupiter (Clarke et

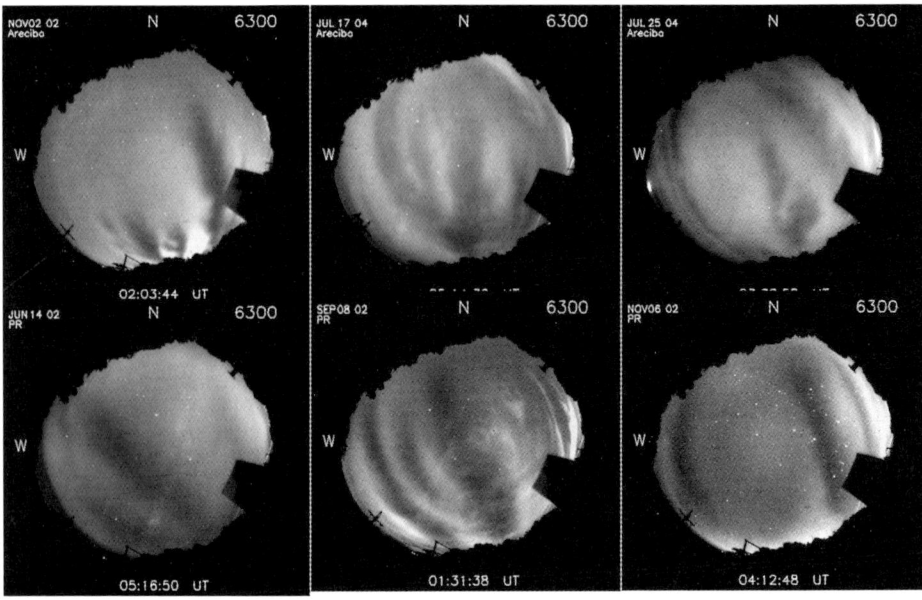

Fig. 7 (*Top*) Examples of 6300 Å airglow depletions from the Arecibo Observatory. These depletions are related to Rayleigh-Taylor fluxtube interchange instability processes and show a typical eastward motion. (*Bottom*) Examples of alternating bands of bright and dark airglow, whose origin could be related to a modified version of the Perkins instability. These structures are typically aligned at large angles with respect to the north-south magnetic meridian; their motion is equatorward and westward (courtesy of C. Martinis, Boston University)

al. 2004); and Saturn (Grodent et al. 2005), and recently at Mars (Bertaux et al. 2005), unambiguous ionospheric disturbances associated with those patterns have yet to be documented. At Mars, characteristics of the ionosphere in the vicinity of crustal magnetic fields have been shown to be different from those at non-magnetized locations (Krymskii et al. 2003; Withers et al. 2005), yet there has been no observational evidence that such patterns are due to auroral processes, as opposed to local plasma instability effects. Thus, the study of solar-planetary relationships continues to look mainly at the Earth for studies of processes initiated by changes in the solar wind at a planet (Prölss 2004).

A sudden input of magnetospheric energy sources to the terrestrial thermosphere-ionosphere system can occur via energetic particle precipitation and the dissipation of strong currents at auroral latitudes (Schunk and Nagy 2008). Modeling studies have shown that the initial response to such heating is the launch of a pulse within the thermosphere, a type of soliton-wave propagating equatorward called a traveling atmospheric disturbance (TAD), summarized in Prölss (2004). Recent observations and modeling have advanced our understanding of such processes. The key effect is that the pulse of equatorward winds initially moves plasma upwards (due to the geomagnetic field inclination angle at midlatitudes, as described in Sect. 2.4) to regions of reduced chemical loss. As the waves from both polar regions cross the equator and pass in opposite hemispheres, the vertical motions are reversed. The full consequences of such effects are not yet understood in detail, though recent modeling advances have started to approach closure between observations and models (Shiokawa et al. 2007).

3.3 Mars

The observational studies of the Martian ionosphere have not followed a steady path. The Viking 1 and 2 landers provided in 1976 the only two *in-situ* profiles of the ionosphere measured so far. A very long gap occurred until the arrival of the Mars Global Surveyor spacecraft in 1998 equipped with a magnetometer, a suprathermal electron detector and a radio-occultation experiment, and in 2003, Mars Express was inserted into orbit, carrying radio-occultation and radar experiments.

The main characteristics of the ionosphere are known. A number of radio occultation measurements provided the general shape of the electron density profiles, between about 90 and 350–400 km. The maximum electron density is located around 125–140 km, with a typical value of about 1 to 2×10^5 cm^{-3}. The existence and the location of an ionopause are still a matter of debate (Schunk and Nagy 2008). Some radio occultation profiles gave a possible indication of an ionopause located between 200 and 300 km (Kliore 1992; Paetzold, private communication). Information about the ion composition essentially comes from the two profiles measured by the Retarding Potential Analyzer instrument on board the Viking landers (Hanson et al. 1977). Like on Venus, the major ion is O_2^+, while CO_2^+ and O^+ have also been detected, with peak densities around 10^4 cm^{-3} at 140 km and 8×10^2 cm^{-3} at 230 km, respectively. The ion temperature is 150–200 K at 120 km and almost 2500 K at 300 km (Hanson et al. 1977), while the thermal electron temperature reaches 3500 to 4000 K at 300 km (Hanson and Mantas 1988). According to these authors, two other electron populations with higher temperatures have been detected. The first one is the photoelectron component, while the other one originates in the solar wind. More recently, Mitchell (2001), using the electron reflectometer onboard Mars Global Surveyor (Albee et al. 1998), showed that the energetic electron population present in the Martian ionosphere has two origins, one being the solar wind and the other the photoionization of the neutral atmosphere. Their results clearly show a sharp decrease in the electron flux above 1 keV when crossing an altitude around 380 km. This is associated with an effective magnetic separation of ionospheric and solar wind plasmas and is interpreted as a possible sign of an ionopause.

Recent updates in the knowledge of the ionosphere are highlighted in Sect. 4 and concern the existence of a lower ionosphere layer and its variability, the result of meteoric impact.

The MARSIS radar onboard Mars Express has been operating since June 2005 and brings a brand new set of information on the ionosphere, through three channels: first, the radar can work in the so-called 'active ionosphere sounding' mode, and thus provide the vertical electron density profile as a function of altitude, above the main ionospheric peak. Interesting double echos have been reported (Gurnett et al. 2008), and are correlated with regions controlled by the crustal magnetic field. In addition, the radar also obtains electron density and magnetic field results from local resonances. Secondly, the radar, when in subsurface mode (the main objective of the radar being the study of the subsurface of Mars), provide data from which the total electron content (TEC) can be retrieved (Safaeinili et al. 2007). A very interesting behaviour of the TEC has been found, especially on the nightside, where sudden jumps are clearly visible and are probably linked to the crustal magnetic field. Some TEC variations are clearly correlated with observations of aurora by the ultraviolet spectrometer SPICAM aboard Mars Express. Thirdly, and also to be treated in Sect. 4, the surface radar experiences some periods of total blackout. The explanation is that the radar signal is attenuated when there are enough free electrons around 80 km altitude, which tells us about the state of the ionosphere and its variability at these altitudes. The external factors responsible for this variability include the penetration of energetic solar particles and, possibly, solar flares, dust storms and meteor showers.

3.4 Giant Planets and Their Moons

All the giant planets have ionospheres, and some of their moons do as well. Recent progress in our understanding of these ionospheres has been reviewed by Nagy and Cravens (2002), Majeed et al. (2004), Mueller-Wodarg (2004), Yelle and Miller (2004), Miller et al. (2005) and Schunk and Nagy (2008). The main focus in this subsection is on Saturn and Titan as representative of current research topics.

3.4.1 Saturn

Much of what is known about the ionospheric structure on Saturn comes from the radio occultation experiments from the Pioneer, Voyager, and Cassini-Huygens missions (Kliore et al. 1980a, 1980b; Tyler et al. 1981, 1982; Lindal et al. 1985; Nagy et al. 2006). Peak electron densities are found to be as high as $\sim 10^4$ cm^{-3} on Saturn, but the electron density profiles are highly variable with location and time. Other important information on Saturn's ionosphere comes from near-infrared observations of H_3^+ (e.g., Geballe et al. 1993; Stallard et al. 1999, 2004, 2007; Melin et al. 2007; see also Miller et al. 2000, 2006) and X-ray and ultraviolet observations of the aurorae and airglow (e.g., Clarke et al. 1982, 2005; Sandel et al. 1982; Sandel and Ajello 1983; Jaffel et al. 1995; Parkinson et al. 1998; Bhardwaj and Gladstone 2000; Ness et al. 2004a, 2004b; Esposito et al. 2005; Grodent et al. 2005; Gerard et al. 2005; Bhardwaj et al. 2005, 2007; Cravens et al. 2006). The H_3^+ observations provide useful information about ion column densities, temperatures, and winds, as well as provide insight into the energy balance of the ionosphere and thermosphere. X-ray and ultraviolet observations can provide information concerning ionospheric temperatures, the properties of the ionizing radiation, and the structure, source, and characteristics of the aurora. Our understanding of Saturn's ionospheric chemistry and structure is also aided by observations of the neutral upper atmosphere in which the ionosphere is embedded (e.g., Festou and Atreya 1982; Smith et al. 1983; Hubbard et al. 1997; Cooray et al. 1998; Moses and Bass 2000; Vervack 2004; Moses and Vervack 2006; Shemansky 2006).

For Saturn, new observational data have been supplied by the Cassini radio occultations; Nagy et al. (2006) present results from twelve of these near-equatorial occultations recorded in the May-September 2005 time period. The most striking characteristic of the derived electron-density profiles is the huge amount of variability between profiles. Although all the profiles have electron density maxima in the 1000–10^4 cm^{-3} range, none exhibit the broad, smooth, Chapman-like structure that would be expected if the ionosphere were controlled by photochemical processes; moreover, the height of the peak electron density varies from occultation to occultation. Layered structures are common and are not confined to lower altitudes, as was suggested from the sparser Voyager and Pioneer data (e.g., Kliore et al. 1980a, 1980b; Lindal et al. 1985). The Cassini data also exhibit peak electron densities that are significantly smaller on average than those derived from the Voyager and Pioneer occultations. Nagy et al. (2006) point out that some, but not all, of this difference in peak electron density can be explained by the fact that the Pioneer and Voyager encounters occurred near solar-cycle maximum, whereas the Cassini data were acquired closer to solar minimum. Despite the overall high observed structural variability, some general trends in the Cassini data can be noted (Nagy et al. 2006): (1) the magnitude of the peak electron density tends to be smaller in occultations acquired near the dawn terminator as compared with the dusk terminator, (2) the altitude of the peak tends to be higher at dawn than at dusk, (3) dusk profiles tend to have smaller topside scale heights than dawn profiles, and (4) the observed dawn-to-dusk variation is smaller than the roughly two-orders-of-magnitude day-to-night differences in peak electron density implied by the Voyager 2 observations of Saturn Electrostatic Discharges (SED) (Kaiser et al. 1984). As is discussed by Nagy et al. (2006), the first two general trends suggest that a molecular ion such as H_3^+ or H_3O^+, which can decay relatively quickly during the short Saturnian night, and is an important component of the middle and lower ionosphere of Saturn during the day.

Theoretical models of Saturn's ionosphere have advanced significantly in the past decade, although our general view of H^+ dominating at high altitudes, H_3^+ or H^+ at intermediate altitudes, and hydrocarbon or metal ions at low altitudes has remained intact since Voyager-era and earlier modeling efforts (e.g., Waite 1981; Atreya and Waite 1981; Connerney and Waite 1984; Majeed and McConnell 1991, and references therein). In our current view of ion chemistry on Saturn, H_2^+ is the main ion initially produced on Saturn; H_2^+ forms predominantly from photoionization or electron impact of H_2. The fast reaction of H_2^+ with H_2 leads to the production of H_3^+. The H_3^+ ions are lost through dissociative recombination or, in the lower ionosphere, by charge-exchange reactions with methane and other hydrocarbons to form hydrocarbon ions. In the topside ionosphere, H_2^+ is also lost through charge-exchange reactions with H atoms to form H^+. Dissociative ionization and direct photoionization of H are key formation mechanisms for H^+. Radiative recombination of H^+ is slow, and H^+ is expected to be the dominant ion in the topside ionosphere of Saturn. Charge exchange of H^+ with vibrationally excited H_2 is potentially a major loss process for H^+ on Saturn (e.g., McElroy 1973; Cravens 1987; Majeed et al. 1991), as is charge-exchange of H^+ with inflowing H_2O molecules from ring or micrometeoric sources (e.g., Shimizu 1980; Connerney and Waite 1984). Note, however, that Huestis (2005) has questioned the effectiveness of the charge-exchange reaction of H^+ with vibrationally excited H_2, given that a likely consequence of the reaction is vibrational relaxation of the H_2 rather than charge exchange. Dissociative recombination is the major loss process for hydrocarbon ions in the lower ionosphere, unless the influx rate of metals due to micrometeoroids is large enough to allow charge exchange of metal atoms with hydrocarbon ions to dominate (Moses and Bass 2000).

Recent theoretical investigations of Saturn's ionospheric chemistry and structure include systematic studies of the expected diurnal variation and the effects of ring-derived water (Majeed and McConnell 1996; Moses and Bass 2000; Moore et al. 2004), the expected variation with latitude (Moore et al. 2004; Mendillo et al. 2005), the details of hydrocarbon-, and metal-ion chemistry (Moses and Bass 2000), the effects of ad hoc plasma drifts (Majeed and McConnell 1996; Moses et al. 2000) and gravity waves (Moses and Bass 2000), the effects of ring shadowing (Moore et al. 2004; Mendillo et al. 2005), and the effects of a transient, enhanced water source (Moore and Mendillo 2007). Moore et al. (2004) provide a significant advance in modeling efforts to date by providing full time-dependent and latitude-dependent calculations and by using a three-dimensional thermospheric global circulation model (GCM; see Mueller-Wodarg et al. 2006) to specify the time-dependent neutral background densities. However, the neutral atmosphere and ionosphere are not fully coupled in this model or in models that focus on the thermospheric circulation and energy transport (e.g., Smith et al. 2007). Such fully coupled models will be a major focus of future efforts. Given the abundance of good review papers currently available regarding ionospheric modeling of Saturn (e.g., Nagy and Cravens 2002; Majeed et al. 2004; Miller et al. 2005), this article will discuss only the most recent theoretical results.

Moore et al. (2004) find that H_3^+ can become the dominant ion in the afternoon on Saturn at altitudes near the peak, especially in regions where the daily average flux of ultraviolet radiation is lowest (e.g., in winter, at middle and high latitudes, at latitudes that experience ring shadowing for some portion of the day, and at solar minimum); the H^+/H_3^+ ratio therefore depends on latitude, local time, and solar flux. Ring shadowing significantly reduces the electron densities in the winter hemisphere by as much as a factor of three. As with previous photochemical models, Moore et al. (2004) find electron densities at dawn and dusk that are consistent with those observed by Nagy et al. (2006), but the height of the peak is generally lower than is observed, and the overall modeled day-to-night variation is not high enough to explain the SED measurements (e.g., Kaiser et al. 1984). Mendillo et al. (2005) further explore whether radio frequency "windows" caused by a reduction in the electron density during ring shadowing can explain the SED measurements. They find that ring shadowing can lead to peak electron densities that are considerably reduced from typical daytime values and suggest that ring shadowing might explain the Voyager nighttime SED measurements; however, the high peak electron densities inferred from the daytime SED measurements (i.e., $\sim 1-3 \times 10^5$ cm^{-3}), in combination with the measured dawn and dusk peak electron densities of $< 10^4$ cm^{-3}, are still difficult to explain with photochemical models. Mendillo et al. (2005) go on to suggest that overall Voyager-Cassini differences in SED observations may be related to the differences in ring-shadowing geometry in the Cassini era vs the Voyager era and to the localized nature of lightning-bearing storms on Saturn.

Moore et al. (2006a) studied the sensitivity of the electron density profile to the assumed influx of external H_2O (from either ring sources or Enceladus plume-derived water) and determined that a constant influx rate of $0.5-1 \times 10^7$ H_2O molecules cm^{-2} s^{-1} provides the best fit to the averaged Cassini dawn and dusk equatorial occultation profiles presented by Nagy et al. (2006). Note that this water influx rate is higher by a factor of $\sim$3–10 than the global-average water influx rate determined by the observations of neutral H_2O and CO_2 in Saturn's stratosphere (e.g., Feuchtgruber et al. 1997; Moses et al. 2000), suggesting that the equatorial regions are receiving a much greater H_2O influx than other regions of the planet if the influx rate inferred by Moore et al. (2006a) is correct. Alternatively, the current water influx rate at Saturn could be greater than the global-average rate for the past few hundred years: the ISO observations sample the average influx rate over a several-hundred-year time scale because of the long stratospheric diffusion time constants, whereas water would diffuse through the ionosphere in less than a couple days. Moore and Mendillo (2007) examined the effects of enhanced, but transient water influx rates, both from an arbitrary point source release within the ionosphere, and from a short-lived enhanced influx from the top of the atmosphere. They find that a point-source release of a few times 10^{27} H_2O molecules in the ionospheric peak region can lead to strong depletions in electron density in the peak region, similar to what is sometimes observed in the occultation profiles. An enhanced influx of a few times 10^8 H_2O molecules cm^{-2} s^{-1} occurring at the top of the atmosphere for tens of minutes leads to a similar transient depletion in electron density. Although there is no known source of water that would resemble a point-source release at ionospheric peak altitudes (ablation of the appropriate-sized icy bodies would occur lower in the atmosphere; see Moses et al. 2000), the point that a variable influx of water could have interesting effects on the ionosphere is still valid.

However, making quantitative conclusions about the chemistry and structure of Saturn's ionosphere from theoretical models that include only photochemistry and diffusion should be regarded with caution. The amazing variability of the occultation profiles demonstrates that dynamical processes, particle precipitation and perhaps electrodynamical processes play a major role in influencing Saturn's ionosphere. Majeed and McConnell (1996) and Moses and Bass (2000) demonstrated with simple examples how plasma drifts (derived from neutral winds or electric fields) or wind shears from gravity waves can dramatically affect plasma structure and chemistry. Matcheva et al. (2001) examine a more realistic model of the interaction of upward propagating gravity waves on the H^+ density structure on Jupiter, using gravity-wave properties constrained by the Galileo probe atmospheric structure instrument (Young et al. 1997). Matcheva et al. (2001) find that gravity waves can have a profound effect on the localized structure of the ionosphere (with significant layering and alterations of scale heights) and important and long-lasting effects on the horizontally averaged structure (with a wave-induced downward electron flux at high altitudes leading to an overall reduction on the electron densities). Such processes could be important on Saturn, as well.

3.4.2 Titan

An ionosphere was first detected on Titan by the radio occultation experiment during the Voyager 1 encounter in 1980 (Bird 1997) and this remained the only experimental data until Cassini made the first in situ measurements in Titan's ionosphere in 2005 (Wahlund et al. 2005). Nevertheless Titan's ionosphere was studied theoretically in the intervening years. It was recognized early on that both solar photons and magnetospheric electrons could act as ionization sources (Strobel 1985; Atreya 1986; Bauer 1987) and due the presence of methane in the neutral atmosphere it was also recognized that the ion chemistry was probably quite complex (Strobel 1985; Atreya 1986). Ionospheric layers near 600 km due to meteor ablation were also suggested (Molina-Cuberos et al. 2001) and also an ionospheric layer in the lower atmosphere due to cosmic ray ionization (e.g., Capone et al. 1976). Plasma and field data in Saturn's outer magnetosphere and in Titan's magnetotail was also acquired during the Voyager's passage 2.7 Titan radii downstream (Hartle 1982) and this data also prompted many modeling studies of the plasma dynamics associated with the magnetospheric interaction with Titan's upper atmosphere and ionosphere (Ip 1990; Gan et al. 1992; Keller et al. 1994; Keller and Cravends 1994). More recent dynamical studies are described in the companion chapter, and thus only the purely ionospheric issues will be considered here.

The chemistry of Titan's ionosphere was extensively studied in the decade or so prior to the arrival of Cassini at the Saturn system (Keller et al. 1998; Fox and Yelle 1997; Wilson and Atreya 2004; Anicich and Huntress 1986; Anicich and McEwan 1997; Cravens 2004; Galand et al. 1999). The main ion species produced in the ionosphere (N_2^+, N^+, CH_4^+, CH_3^+, ...) result from photoionization or electron impact ionization of the major neutral species, molecular nitrogen and methane. The ion chemistry associated with minor neutrals (i.e., acetylene, ethylene, ethane, hydrogen cyanide, benzene...) then acts to create a large number of complex hydrocarbon and nitrile ion species. For example, two of the most abundant ion species predicted by pre-Cassini models ($C_2H_5^+$ and $HCNH^+$) are created via the following sequence of reactions (Anicich and Huntress 1986; Anicich and McEwan 1997):

$$N_2^+ + CH_4 \rightarrow CH_3^+ + N_2 + H,$$

$$CH_3^+ + CH_4 \rightarrow C_2H_5^+ + H_2,$$

$$C_2H_5^+ + HCN \rightarrow HCNH^+ + C_2H_4.$$

Many other higher-mass hydrocarbon or nitrile species are formed due to further ion-neutral reactions of the "major" species. The electron density, which equals the total ion density, then depends on the dissociative recombination of many "terminal" ion species.

The first in situ measurements of Titan's main ionospheric layer were made during the first (the "Ta" pass) Cassini Orbiter encounter on October 26, 2004 (Mahaffy 2005). The neutral upper atmosphere composition and structure was measured (Waite et al. 2005), ionospheric magnetic fields were detected (Backes et al. 2005), and ionospheric electron densities and temperatures were measured (Wahlund et al. 2005). The Ta pass started out on the dayside and ended up on the nightside, with closest approach (at an altitude of

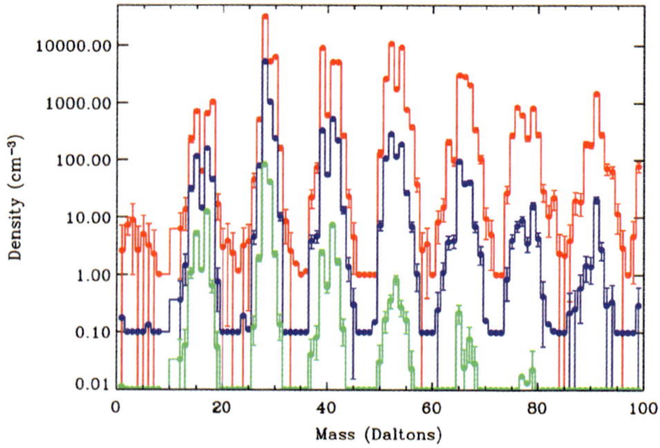

Fig. 8 Ion mass spectra from Cassini-INMS (Cravens et al. 2006). The most abundant species is $HCNH^+$. The number of detected ion species is very high and are grouped into "families" separated by a mass number of 12 (carbon). Other important species are $C_2H_5^+$, CH_5^+, and $C_3H_5^+$, as well as heavier ion species such as $C_6H_7^+$

1174 km) being right at the terminator (i.e., a solar zenith angle of 91°). Clearly, a substantial ionosphere is present with a higher density ($\approx$3000 cm^{-3}) on the inbound dayside pass. The electron temperature in the ionosphere is about 1000 K (or $kT_e \approx 0.1$ eV). Ionospheric models subsequently demonstrated that solar radiation could account for most of the observed ionosphere even a good distance beyond the terminator due to the extended "sphericity" of Titan's ionosphere (Cravens et al. 2006).

Titan's ionospheric composition was first measured during the T5 encounter (April 2005) by the Cassini Ion and Neutral Mass Spectrometer (INMS), which was entirely on the nightside (Cravens et al. 2006). The total ion density measured by INMS agrees quite well with the electron density measured by RPWS, with both instruments showing a broad ledge of ionospheric plasma ($n_e \approx 1000$ cm^{-3}) between closest approach (altitude of 1027 km) and an altitude of $\approx$1450 km. Figure 8 shows some ion mass spectra from INMS for 3 altitude ranges. The data demonstrates that the most abundant species is the nitrile species $HCNH^+$ (mass number, $m = 28$), as predicted by pre-Cassini models (Keller et al. 1998; Fox and Yelle 1997; Wilson and Atreya 2004). The number of detected ion species is very high and are grouped into "families" separated by a mass number of 12 (carbon). Other important predicted species evident in Fig. 8 include $C_2H_5^+$, CH_5^+, and $C_3H_5^+$, as well as heavier ion species such as $C_6H_7^+$. However, a number of important species evident in Fig. 8 were not predicted to exist in any significant abundance by pre-Cassini models (e.g., Keller et al. 1998; Fox and Yelle 1997; Wilson and Atreya 2004), including species at mass numbers 18, 30, 54, and 66.

Vuitton et al. (2006, 2007) explained most of the "new" ion species observed as being due to reaction of major ion species with several neutral nitrile species whose abundances are low enough such that they were not measured by the INMS in its neutral mode. For example, Vuitton et al. (2006) (and Cravens et al. 2006) suggested that mass 18 was NH_4^+ ions formed by the reaction of major species with ammonia. Vuitton et al. (2006) suggested that mass 30 was $CH_2NH_2^+$ produced by a reaction of $HCNH^+$ with CH_2NH and Vuitton et al. (2007) explained mass 66 as $C_2H_3CNH^+$ produced by reaction of $HCNH^+$ with C_2H_3CN. The ion chemistry for heavier species has significant implications for aerosol formation lower in the atmosphere (e.g., Waite et al. 2005). Coates et al. (2007) reveal the existence of negative ions. These ions, with densities up to $\sim$100 cm^{-3}, are in mass groups of 10–30, 30–50, 50–80, 80–110, 110–200 and 200+ amu/charge. Due to their unexpectedly high densities at $\sim$950 km altitude, these negative ions must play a key role in the ion chemistry and they

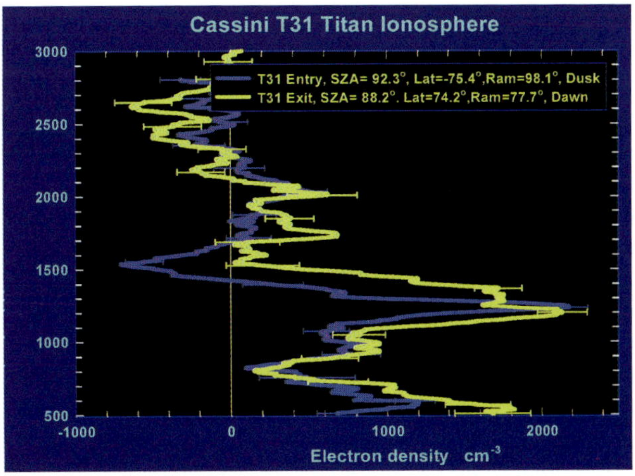

Fig. 9 Titan electron vertical profiles (Kliore et al. 2007). Electron densities peak at about 1200 km altitude. Interesting ionospheric structures are found below 1000 km height

may be important in the formation of organic-rich aerosols (tholins) eventually falling to the surface

The most likely source of the nightside ionosphere observed by the RPWS and INMS instruments during the T5 encounter is ionization by electrons transported down magnetic field lines from the magnetosphere (Cravens et al. 2006; Agren et al. 2007). As discussed earlier such "auroral" ionization at Titan was not unexpected (Atreya 1986; Gan et al. 1992), although recent modeling work (using as inputs electron fluxes measured by the Cassini CAPS instrument) shows that how such electron fluxes interact with Titan's atmosphere is not quantitatively understood. It does appear though that the electron energies (few hundred to a few keV) observed are sufficient to reach the altitudes at which the ionosphere is observed (Agren et al. 2007).

The Cassini Radio Science (RSS) Experiment has measured several electron density profiles in the dawn-dusk regions (Kliore et al. 2007). One of these profiles is shown in Fig. 9 and the peak region near an altitude of $\approx$1200 km is what one would expect from solar radiation (Bird 1997; Wahlund et al. 2005; Cravens et al. 2006). But the RSS profile for altitudes below 1000 km is particularly interesting; note from Fig. 9 that substantial electron densities are found down to altitudes as low as 500 km. An intermittent (about half the time) ionospheric layer with substantial electron densities ($n_e \approx 200-2000$ cm^{-3}) between 500 km and 900 km was observed (Kliore et al. 2007). Creation of such a low altitude layer by precipitating magnetospheric electrons would require substantial fluxes of electrons with energies well in excess of 1 keV (cf. Gan et al. 1992; Agren et al. 2007), and this would be very sensitive to magnetic field topology. Cravens et al. (2007) recently suggested that the precipitation of energetic ions found in Saturn's magnetosphere can produce such a low-altitude ionization layer. Energetic protons and oxygen ions have been observed in Saturn's outer magnetosphere (Hartle 1982; Krimigis et al. 2005) and Cravens et al. suggested that the ion fluxes at higher energies (30 keV to 4 MeV) measured by Cassini are sufficiently high to explain the RSS observations, although many details remain to be worked out quantitatively.

The CAPS ELS instrument observed high abundances (in excess of 100 cm^{-3}) of negative ions (Coates et al. 2007) in the lower ionosphere (960–1200 km), including some very heavy (10^4 amu or higher) aerosol-like species. No clear explanation exists at this time for these negative ions, although Cravens et al. (2007) suggested that creation of negative ions by charge exchange between precipitating ions and neutrals could contribute. Dissociative attachment by superthermal electrons is another possible source of negative ions.

The measured electron temperature ($T_e \approx 1000$ K) in Titan's ionosphere is several times greater than the neutral temperature ($T_n \approx 140-150$ K, Waite et al. 2005) even near 1000 km. From our knowledge of the energetics of other ionospheres, including the Earth's, and from pre-Cassini modeling (Gan et al. 1992; Roboz and Nagy 1994; Cravens 2004) this was not entirely surprising at least qualitatively. Superthermal electrons, either magnetospheric or photoelectrons associated with photoionization of neutrals by solar radiation, heat the colder thermal electron population via Coulomb collisions. However, quantitative modeling using Cassini inputs (Galand et al. 2006) demonstrated that for terminator Ta type conditions, the electron temperature should only be about half the observed temperature unless superthermal electron fluxes from the dayside were transported along draped magnetic field lines into the region observed by Cassini.

3.4.3 Jupiter and Its Satellites, Uranus and Neptune

The first radio occultation measurements of the ionosphere of Jupiter were made with the Pioneer 10 and 11 spacecraft in the early 1970s (Fjeldbo et al. 1975, 1976; Kliore et al. 1977). They showed an extensive ionosphere with multiple narrow layers below the main peak. Pioneer 10 also provided the first evidence of an atmosphere on Io by measuring its ionosphere (Kliore et al. 1975). In the late 1970s the Voyager 1 and 2 flybys provided four more ionospheric profiles (Eshleman et al. 1979a, 1979b). Finally, in the 1990s, the Galileo orbiter provided numerous measurements of the Jovian ionosphere at various latitudes (Hinson et al. 1998). They revealed great variability in the maximum density and the peak altitude from one measurement to the next, with little dependence on latitude, except for a marked depletion in the North auroral zone.

Galileo also provided multiple occultations of the Galilean satellites of Jupiter. On Io, the structure of the ionosphere was found to be controlled by its location relative to the magnetospheric ram direction, with plasma being carried off into the wake direction by Jupiter's corotating magnetosphere (Hinson et al. 1997). Ionospheric plasma was detected at Europa (Kliore et al. 1977), and a well-formed ionosphere was detected on Callisto on two occasions (Kliore et al. 2002). On Ganymede, only one weak ionospheric profile was detected out of eight measurements, and this may be due to shielding of Jupiter's corotating plasma by Ganymede' own magnetic field (Kliore 1998).

The ionospheres of Uranus (Tyler et al. 1986; Lindal et al. 1987) and Neptune (Tyler et al. 1989) were observed by Voyager 2 during its final two flybys of its remarkable tour of the outer solar system. On both planets, multiple sharp layers were observed below the main peak, which seems to be a prevalent feature of outer planet ionospheres. A well-developed ionosphere was also observed on Neptune's satellite Triton (Tyler et al. 1989).

4 Comparative Aspects of Planetary Ionospheres

4.1 Ionospheres at Low Altitudes

It is natural that the first aspect of comparative ionospheres dealt with studies and models of the peak electron densities observed at the planets. The current status of such work is summarized in Nagy and Cravens (2002). The radio occultation technique used to observe planetary ionospheres also covers altitudes below the peak, and these have revealed both persistent and sporadic lower layers. The terminology for such layers is often as diverse (and contentious) as the physics involved. At Earth, the progression from low to high altitudes of the D, E, F1 and F2 layers is linked to the photon wavelengths most responsible for their formation (hard X-rays for the D-layer, soft X-rays for the E-layer, and EUV for the F-layers). At Mars, two layers are found rather consistently, and thus it was tempting to some (e.g., Fox and Yeager 2006) to refer to them are F1 and E layers since they are produced by the same wavelengths as those at Earth. Others (Mendillo et al. 2003) have argued that the main martian layer is composed overwhelmingly of molecular ions (O_2^+) that do not diffuse and thus is different from the terrestrial F1 layer where atomic ions (O^+) can be prominent and their diffusion can be important (Banks and Kockarts 1973). It was then suggested that planet-specific names might be more appropriate, e.g., the M1 and M2 layers for the secondary and primary electron density layers at Mars (Rishbeth and Mendillo 2004), and S1 and S2 for lower and main layers at Saturn (Moore et al. 2006a). The relevance to the topic at hand is that layers below these regular features occur sporadically at the planets, and thus it might not make sense to call them D-layers in analogy to Earth since the D-layer is considered a regular feature. Thus, this article will treat such variable layers by names associated with their sources.

To create a temporary layer below the diurnally-occurring layers of a planetary ionosphere, an ionizing source that is very energetic has to be operating. The possibilities include a burst of very hard X-rays from a solar flare, the subsequent sudden arrival of very energetic charged particles from the flare or magnetosphere, an influx of galactic cosmic rays, or the impingement of meteoritic particles. For the terrestrial case, the D-layer ($h \sim 95$ km and below) can be attributed to all of these causes, both routinely and sporadically (Rishbeth and Garriott 1969). The energy required for photons is typically 0.1 Mev ($\sim$1 A), about 1 Mev for electrons and protons, with higher values for meteoritic particles due to their higher mass. Thus, the study of sporadic ionospheric layers is, equivalently, an investigation of some of the more energetic sources found in the solar system. When the same process can be studied at several venues, validations of and constraints for that process become more robust.

Meteoritic Layers In their excellent review of meteoric material as a source of planetary ionospheric layers, Grebowsky et al. (2002) gave a concise history of the field, together with a comprehensive summary of observations and models. There is little that can be done here to improve upon that summary just five years later, except for some new observations since that time. Grebowsky et al. (2002) pointed out that past radio occultation experiments at the giant planets often showed evidence of highly structured layers well below the main ionospheric peak(s). Since then, a new set of $N_e(h)$ profiles were obtained at Saturn (Nagy et al. 2006) and indeed that trend continues. The model predictions of Moses and Bass (2000) described the production of meteoritic-induced plasmas in the hydrocarbon rich altitude regime below Saturn's S1 and S2 layers of H^+ and H_3^+ (Moore et al. 2006a). Ion-neutral-plasma chemistry between the meteoritic ions and the mix of ambient plasma creates a remarkable number of different ions that collectively match the concentration of electrons (and any negative ions produced). Detailed comparisons of such scenarios with the new Cassini data at low altitudes are yet to be made.

The situation was somewhat different at Mars. As pointed out by Grebowsky et al. (2002), low altitude layers did not appear in early radio occultation results from Mars, presumably because of observational limitations on extracting such effects from the low signal-to-noise conditions at those altitudes. That changed with the arrival of the Mars Global Surveyor (MGS) and Mars Express (MEX) in 1998 and 2004, respectively. In an early set of MEX radio occultation profiles, Paetzold et al. (2005) presented strong evidence of meteor-induced

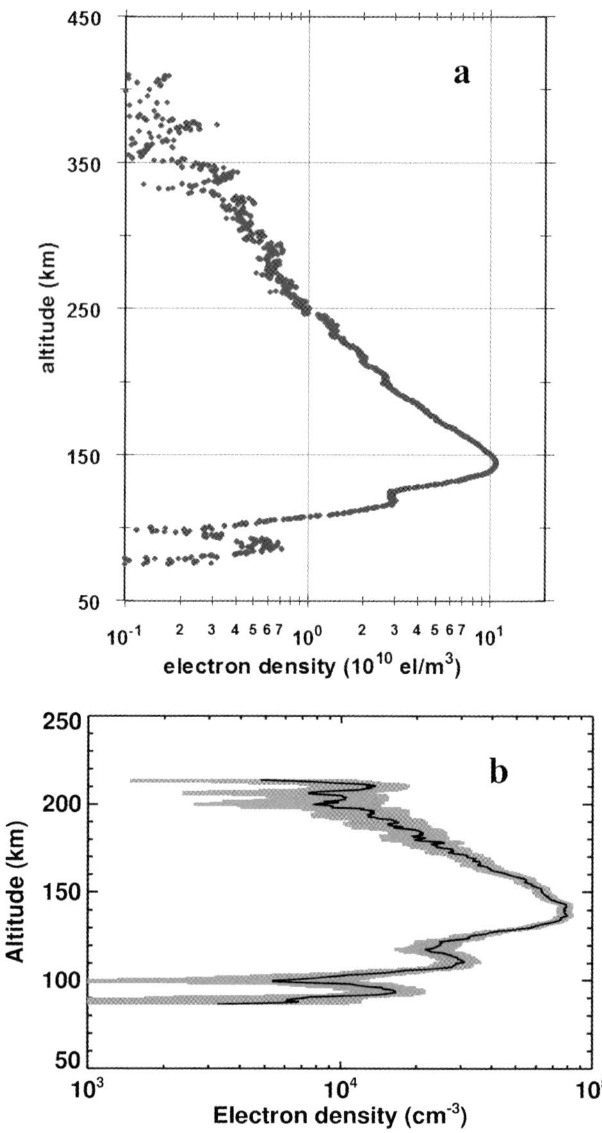

Fig. 10 (**a**) Example of a low altitude ionospheric layer attributed to meteoric input, as observed by the Mars Express radio occultation experiment (Paetzold et al. 2005).
(**b**) Example of a low altitude ionospheric layer attributed to meteoric input, as observed by the Mars Global Surveyor radio occultation experiment (Withers et al. 2006)

layers at ~80–90 kilometers, as shown in Fig. 10a. A subsequent examination of MGS profiles by Withers et al. (2006) revealed equally convincing detections in the same altitude regime (panel b of Fig. 10). The occurrence rate of detections of a meteoritic layer is considerably higher in the MEX data sets than those of MGS, an effect still in need of explanation. By combining both data sets, Withers et al. (2008) offer the possibility of recurrent meteor streams providing a predictable source function for such layers.

The radio occultation method is sensitive to electron densities and thus no observational evidence exists for the types of ions present in meteoritic layers to test the predictions of Mg^+ dominated plasmas by Pesnell and Grebowsky (2000) and Molina-Cuberos et al. (2003). Upcoming aeronomy missions to Mars will not fly at such low altitudes, and thus

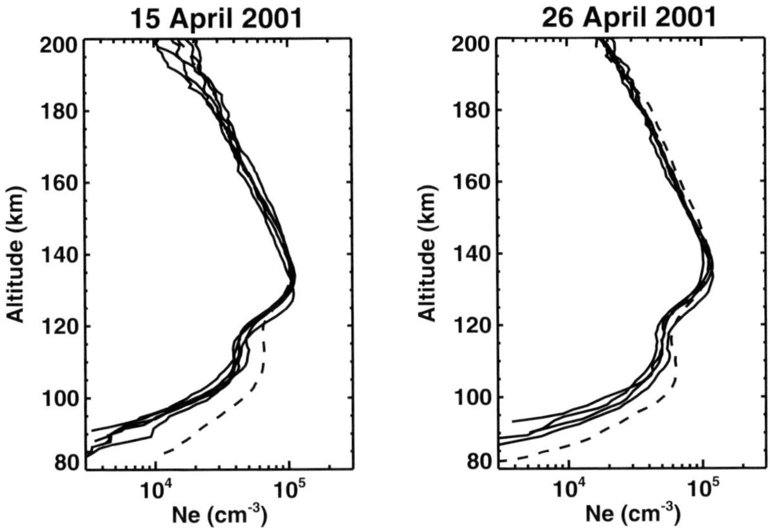

Fig. 11 Two examples of solar flare induced enhancements to the ionosphere at Mars (*dashed curves*), in comparison to non-flare profiles on the same day (*solid curves*), as observed by the radio occultation experiment on Mars Global Surveyor, (**a**) 15 April 2001, a day of a strong X14-class flare, and (**b**) 26 April 2001, a day of a moderate M8-class flare. After Mendillo et al. (2006)

ion composition studies await future landers with suitable spectrometers operating during descent. Alternately, spacebased LIDARs and optical imagers might be used, employing techniques developed for ground-based observations on Earth to detect structures in Na, K, Fe, Ca and Ca^+ emissions (Zhou et al. 2005; Smith et al. 2005).

Flare-induced Layers Solar flares change the solar irradiance impinging upon a planet in two fundamental ways: the flux at all wavelengths increases, but not in the same proportion. The most energetic photons are the ones most enhanced, and these lead to enhanced ionospheric layers at low altitudes. The most comprehensive summary of solar flare effects upon the terrestrial ionosphere appears in Mitra (1974). For planets other than Earth, solar flare effects have been detected, to date, only in the ionosphere of Mars, and these are shown in Fig. 11 (Mendillo et al. 2006). For the 15 April 2001 event, the MGS profile was obtained $\sim$20 minutes after the flare's peak fluxes reached Mars. By that time the EUV component of the flare had returned to ambient levels, while the soft X-ray component persisted, thereby accounting for the enhancements confined to regions at and below the M1 layer. For the 26 April 2001 event, a flare of far less flux, the MGS measurement was just 90 seconds after the flare's peak emission arrival time at Mars. With no significant increases in the M2-layer, the soft X-rays-only created a similar enhancement to the lower M1-layer, as on 15 April 2001. The hard X-rays measured by the GOES spacecraft at 1 AU (used to obtain the timing above) would have created a third layer near 60 km altitude, one below the MGS detection threshold. Energetic X-ray induced layers may yet be found in MEX datasets. At Earth, such hard X-rays create considerable ionization enhancements at low altitudes, causing the so-called "D-layer absorption" of radio signals. This effect was prominent for the 15 April 2001 flare, making routine ionosonde observations impossible for extended periods (Mendillo et al. 2006). Comparative studies of the same flare emission striking multiple planets can be used to study and perhaps constrain uncertainties in secondary ionization processes (Fox and

Yeager 2006), as well as electrodynamical process associated with flare-induced changes in planetary magnetic fields, as discuss in Mendillo and Withers (2008).

Energetic Particle Induced Layers A third type of low altitude ionospheric enhancement occurs when energetic particles create low altitude layers which on the Earth occur in concert with the onset of aurora. Prompt effects can follow a large solar outburst when highly energetic protons arrive at polar latitudes within 10 s of minutes to hours causing sudden ionospheric disturbances (SIDs) and polar cap absorption (PCA) events. Just as with solar flare photons, the Mev protons spread throughout that side of the solar system from which the solar disturbance occurred. The first detection of radio absorption events due to such fluxes at another planet came from the MARSIS radar on MEX (Nielsen et al. 2007; Morgan et al. 2006; Esplay et al. 2007). Thus, radio communication and navigation systems on Mars will have to deal with the same types of ionospheric effects that can lead to disruptions that occur on Earth (Mitra 1974; Mendillo et al. 2004). Finally, there is a newly discovered significant, intermittent ionospheric layer at Titan in the altitude range of 500–600 km, believed to be the results of either energetic particle precipitation or meteoric impact.

4.2 Ionospheric Variability

Studies of ionospheric disturbances caused by solar flares and geomagnetic storms serve as test cases for our understanding of a fully coupled system. If the system response functions can be modeled properly during average conditions (e.g., monthly mean, solar minimum vs. solar maximum), and during times of maximum stress (flares and storms), then the remaining requirement to achieve full understanding is to account for day-to-day variability. This is one of the dominant themes in current terrestrial aeronomy, ranging from long term basic morphology patterns of the F-layer (Forbes et al. 2000; Fuller-Rowell et al. 2000; Rishbeth and Mendillo 2001) and the E-layer (Moore et al. 2006b), to highly focused 30-day observational experiments (Zhang et al. 2005), to the variability of plasma instability onset conditions for equatorial spread-F (Mendillo et al. 1992, 2001).

At the present time, plasma instability patterns have been studied to a significant degree only at Earth. However, solar-induced variability patterns have been documented among the planets of the inner solar system. For example, Forbes et al. (2006) were the first to examine thermospheric responses to solar forcings at Mars, Earth and Venus. For ionospheric variability at these planets, a longer history of results is available due to the past and current missions with radio occultation experiments.

To illustrate the new types of comparative investigations made possible by $\sim$5600 electron density profiles obtained by MGS at Mars, Fig. 12 uses same-day ionospheric measurements on Earth and Mars during March 1999 to examine the photochemical equilibrium condition expected for the terrestrial E- and martian M2-layers. From standard Chapman theory, the peak density varies as the square root of the solar flux and this is shown; yet, there is considerable residuals from that correlation at both planets showing that dynamical effects are important as well. Follow-up studies of day-to-day solar forcing of both the M1 and M2 ionospheric layers at Mars and the E and F1 layers at Earth were described in Rishbeth and Mendillo (2004). Month-to-month periods of variability using MGS data were then examined for peak density by Withers and Mendillo (2005), and in the topside ionosphere by Breus et al. (2004). Studies over solar cycle time periods were conducted prior to MGS at Mars (Bauer and Hantsch 1989; Hantsch and Bauer 1990), and at Venus and Mars by Kliore and Mullen (1989) and Kliore (1992). All of these studies essentially agreed with basic Chapman theory for photochemical layers, except for

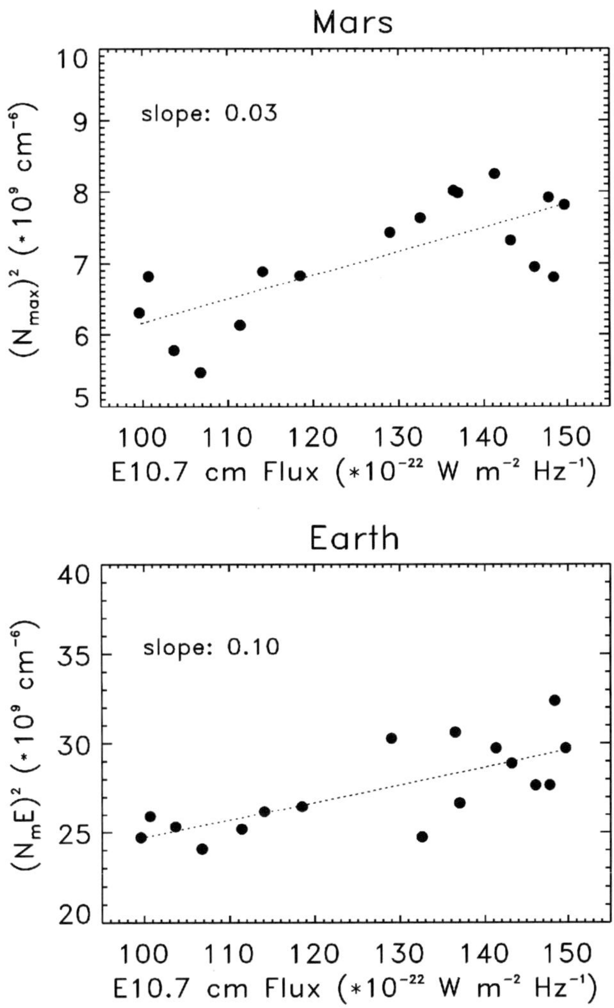

Fig. 12 A comparison of same-day ionospheric variability patterns caused by changes in solar flux at Mars and Earth. The N_{max} densities at Mars come from MGS observations in March 1999, and the terrestrial NmE values are daytime averages from two midlatitude ionosonde stations. The E10.7 index is a proxy for solar irradiance at EUV and soft X-rays. From Mendillo et al. (2003)

the ionosphere at Mars above regions of crustal magnetic fields (Krymskii et al. 2003; Withers and Mendillo 2005). The sources of variability from other processes (dynamical, auroral, instabilities) must be substantial in these regions, but to date no significant progress has been made on such non-photochemical effects.

Returning to Earth, where again the most data exists and general circulation models (GCMs) are most advanced, it is now possible to investigate the sources of day-to-day variability patterns for every day of a year. A new series of GCMs have been created that couple models of the lower atmosphere to that of an upper atmosphere (Roble 2000; Mendillo et al. 2002). With daily values of solar irradiance, time dependent input due to geomagnetic activity, and daily parameterizations of waves and tides from below, the Thermosphere-Ionosphere-Mesosphere-Electrodynamical General Circulation Model (TIME-GCM) was run for the full year of 2002 (Roble 2007). Figure 13 presents a sample comparison using hourly values of observed peak electron density (N_{max}) from the midlatitude site at Wallops Island (VA) in the left panel, with model output given in the right panel. Clearly, the variable inputs to a self-consistent first-principles model produce results that

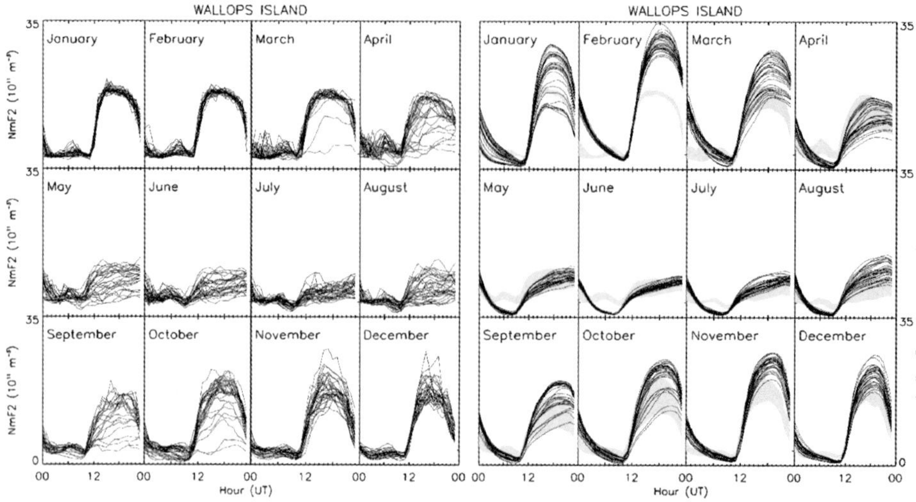

Fig. 13 A comparison of ionospheric variability, observed and modeled. (*Left*) Daily values of peak electron density as observed by the ionosonde at Wallops Island (VA), USA, for every day of 2002. (*Right*) Model output from the TIMEGCM1.2 for the same year using daily input of solar irradiance and 3-hour geomagnetic activity as variable input sources from above, plus daily input from below from the NCEP (National Centers for Environmental Predictions). The *solid curves* give hourly output from the model grid point closest to Wallops Island. The *shading* represents $\pm 1\sigma$ about the monthly mean of the data in the *left panel*. From Roble (2007)

capture many aspects of average diurnal and seasonal morphology patterns. The seasonal anomaly (N_{max} higher in winter than in summer) is over-portrayed in the model, summer solstice variability is under-portrayed, and nighttime behavior still requires work. Yet, such models are clearly approaching observed trends in day-to-day variability patterns is very impressive. By conducting a full series of 'computer experiments' with such models, it will be possible to characterize the individual contributions of specific processes to the general problem of a variable ionosphere—for Earth. Much remains to be done for the planets.

5 Conclusion

This chapter outlined our current knowledge of key topics associated with the ionospheres of the solar system. At the time of writing, new data sets are arriving from several planets. These are leading to substantial progress being made on our understanding of the ionospheres of Titan, Saturn, Venus and Mars. As in other fields of research, comparative aeronomy helps to shed new light on the thin air of the planetary atmospheres.

Acknowledgements MM acknowledges discussions with P. Withers, C. Martinis, R. Roble, H. Rishbeth and J. Wroten, and manuscript preparation assistance from C. Narvaez. Work at Boston University was conducted with support from the NSF, NASA and ONR. The authors thank the committee of the aeronomy workshop held at ISSI (Bern, June 2007) for its excellent organisation.

References

K. Agren et al., On magnetospheric electron impact ionisation and dynamics in Titan's ram-side and polar ionosphere—a Cassini case study. Ann. Geophys. **25**(11), 2359–2369 (2007)

A.L. Albee et al., Mars global surveyor mission: overview and status. Science **279**(5357), 1671 (1998)
V.G. Anicich, W.T. Huntress, A survey of bimolecular ion-molecule reactions for use in modeling the chemistry of planetary atmospheres, cometary comae, and interstellar clouds. Astrophys. J. Suppl. Ser. **62**, 553–672 (1986)
V.G. Anicich, M.J. McEwan, Ion-molecule chemistry in Titan's ionosphere. Planet. Space Sci. **45**, 897–921 (1997)
S.K. Atreya, *Atmospheres and Ionospheres of the Outer Planets and Their Satellites* (Springer, Berlin, 1986), 224 pp. Also in Phys. Chem. Space **15**. 90 figs. (partly in color)
S.K. Atreya, J.H. Waite Jr., Saturn ionosphere: theoretical interpretation. Nature **292**, 682–683 (1981). doi:10.1038/292682a0
H. Backes et al., Titan's magnetic field signature during the first Cassini encounter. Science **308**(5724), 992–995 (2005)
P.M. Banks, G. Kockarts, *Aeronomy* (Academic Press, San Diego, 1973)
S.J. Bauer, Titan's ionosphere and atmospheric evolution. Adv. Space Res. **7**(5), 65–69 (1987)
S.J. Bauer, M.H. Hantsch, Solar cycle variations of the upper atmosphere temperature of Mars. Geophys. Res. Lett. **16**, 373–376 (1989). doi:10.1029/GL016i005p00373
J.-L. Bertaux, F. Leblanc, W. Witasse, E. Quemerais, J. Lilensten, A.S. Stern et al., Discovery of aurora on Mars. Nature **435**, 790–794 (2005). doi:10.1038/nature03603 Medline
A. Bhardwaj, 13 co-authors, X-rays from solar-system objects. Planet. Space Sci. **55**, 1135–1189 (2007). doi:10.1016/j.pss.2006.11.009
A. Bhardwaj, G.R. Gladstone, Auroras on Saturn, Uranus, and Neptune. Adv. Space Res. **26**, 1551–1558 (2000). doi:10.1016/S0273-1177(00)00096-X
A. Bhardwaj, R.F. Elsner, J.H. Waite Jr., G.R. Gladstone, T.E. Cravens, P.G. Ford, Chandra observation of an X-ray flare at Saturn: Evidence of direct solar control on Saturn's disk X-ray emissions. Astrophys. J. **624**, L121–L124 (2005). doi:10.1086/430521
M.K. Bird, Detection of Titan's ionosphere from Voyager 1 radio occultation observations. Icarus **130**(2), 426–436 (1997)
S.W. Bougher, S. Engel, D.P. Hinson, J.R. Murphy, MGS Radio Science electron density profiles: interannual variability and implications for the martian neutral atmosphere. J. Geophys. Res. **109**, E03010 (2004). doi:10.1029/2003JE002154
L.H. Brace, A.J. Kliore, The structure of the Venus ionosphere, in *Venus Aeronomy*, ed. by C.T. Russell. Space Science Reviews, vol. 55 (1991), pp. 81–163
T.K. Breus, A.M. Krymskii, D.H. Crider, N.F. Ness, D. Hinson, K.K. Barashyan, Effect of the solar radiation in the topside atmosphere/ionosphere of Mars: Mars Global Surveyor observations. J. Geophys. Res. **109** (2004). doi:10.1029/2003JE002154
L.A. Capone et al., The lower ionosphere of Titan. Icarus **28**, 367–378 (1976)
S. Chandra, E.I. Reed, B.E. Troy Jr., J.E. Blamont, Equatorial airglow and the Ionospheric geomagnetic anomaly. J. Geophys. Res. **78**, 4630–4640 (1973). doi:10.1029/JA078i022p04630
J.T. Clarke, 12 co-authors, Morphological differences between Saturn's ultraviolet aurorae and those of Earth and Jupiter. Nature **433**, 717–719 (2005). doi:10.1038/nature03331 Medline
J.T. Clarke, H.W. Moos, P.D. Feldman, The far-ultraviolet spectra and geometric albedos of Jupiter and Saturn. Astrophys. J. **255**, 806–818 (1982). doi:10.1086/159879
J.T. Clarke, D. Grodent, S.W.H. Cowley, E.J. Bunce, P. Zarka, J.E.P. Connerney et al., Jupiter's Aurora, in *Jupiter: The Planet, Satellites and Magnetosphere*, ed. by F. Bagenal, T.E. Dowling, W.B. McKinnon (Camb. Univ. Press., Cambridge, 2004)
A. Coates, F.J. Crary, G.R. Lewis, D.T. Young, J.H. Waite, E.C. Sittler, Discovery of heavy negative ions in Titan's ionosphere. Geophys. Res. Lett. **34**(22), L22103 (2007)
J.E.P. Connerney, J.H. Waite Jr., New model of Saturn's ionosphere with an influx of water from the rings. Nature **312**, 136–138 (1984). doi:10.1038/312136a0
A.R. Cooray, J.L. Elliot, A.S. Bosh, L.A. Young, M.A. Shure, Stellar occultation observations of Saturn's north-polar temperature structure. Icarus **132**, 298–310 (1998). doi:10.1006/icar.1998.5901
T.E. Cravens, Vibrationally excited molecular hydrogen in the upper atmosphere of Jupiter. J. Geophys. Res. **92**, 11083–11100 (1987). doi:10.1029/JA092iA10p11083
T.E. Cravens, *Physics of Solar System Plasmas* (Cambridge Univ. Press, Cambridge, 1997)
T.E. Cravens, The ionosphere of Titan: an updated theoretical model. Adv. Space Res. **33**(2), 212–215 (2004)
T.E. Cravens, S.L. Crawford, A.F. Nagy, T.I. Gombosi, A two-dimensional model of the ionosphere of Venus. J. Geophys. Res. **88**, 5595 (1983). doi:10.1029/JA088iA07p05595
T.E. Cravens, H. Shinagawa, J.G. Luhmann, Magnetohydrodynamic processes: Magnetic fields in the ionosphere of Venus, in *Venus II*, ed. by S.W. Bougher, D.M. Hunten, R.J. Phillips (Univ. of Arizona Press, Tucson, 1997), pp. 61–95

T.E. Cravens, J. Clark, A. Bhardwaj, R. Elsner, J.H. Waite, A.N. Maurellis, G.R. Gladstone, G. Branuardi-Raymont, X-ray emission from the outer planets: Albedo for scattering and fluorescence of solar X-rays. J. Geophys. Res. **111**(A&), A07308 (2006). doi:10.1029/2005JA011413
T.E. Cravens et al., Energetic Ion Precipitation at Titan. American Geophysical Union, Fall Meeting 2007, abstract #P43A–1011 (2007)
Z. Dobe et al., A theoretical study concerning the solar cycle dependence of the nightside ionosphere of Venus. J. Geophys. Res. **100**, 14507 (1995). doi:10.1029/95JA00331
L.P. Dyrud, L. Ray, M. Oppenheim, S. Close, K. Denney, Modelling high-power large-aperture radar meteor trains. J. Atmos. Sol. Terr. Phys. **67**, 1171–1177 (2005). doi:10.1016/j.jastp.2005.06.016
R.C. Elphic et al., Nightward ion flow in the Venus ionosphere: Implications of momentum balance. Geophys. Res. Lett. **11**, 1007 (1984). doi:10.1029/GL011i010p01007
V.R. Eshleman et al., Radio science with Voyager at Jupiter—initial Voyager 2 results and a Voyager 1 measure of the Io torus. Science **206**(23), 959–962 (1979a)
V.R. Eshleman et al., Radio science with Voyager 1 at Jupiter—preliminary profiles of the atmosphere and ionosphere. Science **204**(1), 976–978 (1979b)
J.R. Esplay et al., Absorption of MARSIS radar signals: Solar energetic particles and the daytime ionosphere. Geophys. Res. Lett. **34**(9), L09101 (2007)
L. Esposito, 15 co-authors, Ultraviolet imaging spectroscopy shows an active Saturnian system. Science **307**, 1251–1255 (2005). doi:10.1126/science.1105606 Medline
M.C. Festou, S.K. Atreya, Voyager ultraviolet stellar occultation measurements of the composition and thermal profiles of the Saturnian upper atmosphere. Geophys. Res. Lett. **9**, 1147–1150 (1982). doi:10.1029/GL009i010p01147
H. Feuchtgruber, E. Lellouch, T. de Graauw, B. Bezard, T. Encrenaz, M. Griffin, External supply of oxygen to the atmospheres of the giant planets. Nature **389**, 159–162 (1997). doi:10.1038/38236 Medline
G. Fjeldbo et al., The Pioneer 10 radio occultation measurements of the ionosphere of Jupiter. Astron. Astrophys. **39**(1), 91–96 (1975)
G. Fjeldbo et al., The Pioneer 11 radio occultation measurements of the Jovian ionosphere. Jupiter 238–246 (1976)
J.M. Forbes, S.E. Palo, X. Zhang, Variability of the ionosphere. J. Atmos. Sol. Terr. Phys. **62**, 685–693 (2000). doi:10.1016/S1364-6826(00)00029-8
J.M. Forbes, S. Bruinsma, F.G. Lemoine, Solar rotation effects on the thermospheres of Mars and Earth. Science **312**, 1366–1368 (2006). doi:10.1126/science.1126389 Medline
J.C. Foster, W. Rideout, Midlatitude TEC enhancements during the October 2003 superstorm. Geophys. Res. Lett. **44**, L12S04 (2005). doi:10.1029/2004GL021719
J.L. Fox, K.E. Yeager, Morphology of the near-terminator Martian ionosphere: A comparison of models and data. J. Geophys. Res. **111**, A10309 (2006). doi:10.1029/2006JA011697
J.L. Fox, R.V. Yelle, Hydrocarbon ions in the ionosphere of Titan. Geophys. Res. Lett. **24**, 2179 (1997)
T.J. Fuller-Rowell, M. Codrescu, P. Wilkinson, Quantitative modeling of the ionospheric response to geomagnetic activity. Ann. Geophys. **18**, 766–781 (2000). doi:10.1007/s00585-000-0766-7
M. Galand et al., The ionosphere of Titan: ideal diurnal and nocturnal cases. Icarus **140**, 92–105 (1999)
M. Galand et al., Electron temperature of Titan's sunlit ionosphere. Geophys. Res. Lett. **33**(21), L21101 (2006)
L. Gan et al., Electrons in the ionosphere of Titan. J. Geophys. Res. **97**(A8), 12,137–12,151 (1992)
T.R. Geballe, M.-F. Jagod, T. Oka, Detection of H_3^+ infrared emission lines in Saturn. Astrophys. J. **408**, L109–L112 (1993). doi:10.1086/186843
J.-C. Gerard, E.J. Bunce, D. Grodent, S.W.H. Cowley, J.T. Clarke, S.V. Badman, Signature of Saturn's auroral cusp: Simultaneous Hubble Space Telescope FUV observations and upstream solar wind monitoring. J. Geophys. Res. **110**(A11), A11201 (2005). doi:10.1029/2005JA011094
J.M. Grebowsky, J.I. Moses, W.D. Pesnell, Meteoric material—An important component of planetary atmospheres, in *Atmospheres in the Solar System: Comparative Aeronomy*, ed. by M. Mendillo, A. Nagy, J.H. Waite (Am. Geophys. Union, Washington, 2002), pp. 235–244
D. Grodent, J.-C. Gerard, S.W.H. Cowley, E.J. Bunce, J.T. Clarke, Variable morphology of Saturn's southern ultraviolet aurora. J. Geophys. Res. **110**, A07215 (2005). doi:10.1029/2004JA010983
D.A. Gurnett et al., An overview of radar soundings of the martian ionosphere from the Mars Express spacecraft. Adv. Space Res. **41**(9), 1335–1346 (2008)
M.E. Hagan, R.G. Roble, J. Hackney, Migrating thermospheric tides. J. Geophys. Res. **106**, 12,739–12,752 (2001). doi:10.1029/2000JA000344
W.B. Hanson, G.P. Mantas, Viking electron temperature measurements—evidence for a magnetic field in the Martian ionosphere. J. Geophys. Res. **93**(1), 7538–7544 (1988)
W.B. Hanson et al., The Martian ionosphere as observed by the Viking retarding potential analyzers. J. Geophys. Res. **82**(30), 4351–4363 (1977)

M.H. Hantsch, S.J. Bauer, Solar control of the Mars ionosphere. Planet. Space Sci. **38**, 539–542 (1990). doi:10.1016/0032-0633(90)90146-H

R.E. Hartle, Titan's ion exosphere observed from Voyager 1. J. Geophys. Res. **87**(1), 1383–1394 (1982)

R.A. Heelis, J.D. Winningham, W.B. Hanson, J.L. Burch, The relationships between high-latitude convection reversals and the energetic particle morphology observed by Atmosphere Explorer. J. Geophys. Res. **85**, 3315 (1980). doi:10.1029/JA085iA07p03315

D.P. Hinson et al., Jupiter's ionosphere: Results from the first Galileo radio occultation experiment. Geophys. Res. Lett. **24**, 2107 (1997)

D.P. Hinson et al., Galileo radio occultation measurements of Io's ionosphere and plasma wake. J. Geophys. Res. **103**(A12), 29343–29358 (1998)

W.B. Hubbard et al., Structure of Saturn's mesosphere from the 28 Sgr occultations. Icarus **130**, 404–425 (1997). doi:10.1006/icar.1997.5839

D. Huestis, $H^+ + H_2$ ion-molecule reactions in the ionospheres of the outer planets. Bull. Am. Astron. Soc. **37**, 757 (2005)

T.J. Immel, E. Sagawa, S.L. England, S.B. Henderson, M.E. Hagan, S.B. Mende et al., Control of equatorial ionospheric morphology by atmospheric tides. Geophys. Res. Lett. **33**, L15108 (2006). doi:10.1029/2006GL026161

W.-H. Ip, Titan's upper ionosphere. Astrophys. J. Part 1 **362**(10), 354–363 (1990)

L. Jaffel, R. Prange, B.R. Sandel, R.V. Yelle, C. Emerich, D. Feng et al., New analysis of the Voyager UVS H Lyman alpha emission of Saturn. Icarus **113**, 91–102 (1995). doi:10.1006/icar.1995.1007

M. Kaiser, M.D. Desch, J.E.P. Connerney, Saturn's ionosphere: Inferred electron densities. J. Geophys. Res. **89**, 2371–2376 (1984). doi:10.1029/JA089iA04p02371

C.N. Keller, T.E. Cravends, One-dimensional multispecies hydrodynamic models of the wakeside ionosphere of Titan. J. Geophys. Res. A **99**(4), 6527–6536 (1994)

C.N. Keller et al., One-dimensional multispecies magnetohydrodynamic models of the ramside ionosphere of Titan. J. Geophys. Res. A **99**(4), 6511–6525 (1994)

C.N. Keller et al., Model of Titans ionosphere with detailed hydrocarbon ion chemistry. Planet. Space Sci. **46**(9–10), 1157–1174 (1998)

M.C. Kelley, *The Earth's Ionosphere, Plasma Physics, and Electrodynamics* (Academic Press, San Diego, 1989)

M. Kelley, J. Makela, A. Saito, The mid-latitude F region at the mesoscale: Some progress at last. J. Atmos. Sol. Terr. Phys. **64**, 1525–1529 (2002). doi:10.1016/S1364-6826(02)00090-1

M. Kelley, J. Makela, L. Paxton, F. Kamalabadi, J. Comberiate, H. Kil, The first coordinated ground- and space-based optical observations of equatorial plasma bubbles. Geophys. Res. Lett. **30**, 1766 (2003). doi:10.1029/2003GL017301

A.J. Kliore, Radio occultation observations of the ionospheres of Mars and Venus, in *Venus and Mars: Atmospheres, Ionospheres and Solar Wind Interactions*, ed. by J.G. Luhman, M. Tatrallyay, R.O. Pepin. Geophys. Monograph Series, vol. 66 (AGU, Washington, 1992), pp. 265–276

A.J. Kliore, The ionosphere of Io and the plasma environments of Europa, Ganymede, and Callisto. Bull. Am. Astron. Soc. **30**(4), 1450–1451 (1998)

A.J. Kliore, L.F. Mullen, The long-term behavior of the main peak of the dayside Ionosphere of Venus during solar cycle 21 and its implications on the effect of the solar cycle upon the electron temperature in the main peak region. J. Geophys. Res. **94**, 13,339–13,351 (1989). doi:10.1029/JA094iA10p13339

A.J. Kliore, I.R. Patel, G.F. Lindal, D.N. Sweetnam, H.B. Hotz, T. McDonough, Vertical structure of the ionosphere and upper neutral atmosphere of Saturn from the Pioneer radio occultation. Science **207**, 446–449 (1980a). doi:10.1126/science.207.4429.446 Medline

A.J. Kliore, I.R. Patel, G.F. Lindal, D.N. Sweetnam, H.B. Hotz, J.H. Waite et al., Structure of the ionosphere of Saturn from Pioneer 11 Saturn radio occultation. J. Geophys. Res. **85**, 5857–5870 (1980b). doi:10.1029/JA085iA11p05857

A.J. Kliore et al., The atmosphere of Io from Pioneer 10 radio occultation measurements. Icarus **24**, 407–410 (1975)

A.J. Kliore et al., Pioneer 10 and 11 radio occultations by Jupiter, in *Space Research XVII; Proceedings of the Open Meetings of Working Groups on Physical Sciences, June 8–19, 1976 and Symposium on Minor Constituents and Excited Species*, Philadelphia, PA, June 9, 10, 1976 (Pergamon Press, Elmsford, 1977), pp. 703–710

A.J. Kliore et al., The ionosphere of Europa from Galileo radio occultations. Science **277**, 355–358 (1997)

A.J. Kliore et al., Ionosphere of Callisto from Galileo radio occultation observations. J. Geophys. Res. (Space Phys.) **107**(A11), 1407 (2002). doi:10.1029/2002JA009365

A.J. Kliore et al., The Structure of the Titan Ionosphere. American Astronomical Society, DPS meeting #39, #56.15 (2007)

W.C. Knudsen et al., Anti-solar acceleration of ionospheric plasma across the Venus terminator. Geophys. Res. Lett. **8**, 241 (1981). doi:10.1029/GL008i003p00241

W.C. Knudsen, P.M. Banks, K.L. Miller, A model of plasma motion and planetary magnetic fields for Venus. Geophys. Res. Lett. **9**, 765 (1982). doi:10.1029/GL009i007p00765

S.M. Krimigis et al., Dynamics of Saturn's magnetosphere from MIMI during Cassini's orbital insertion. Science **307**(5713), 1270–1273 (2005)

A.M. Krymskii, T.K. Breus, N.F. Ness, D.P. Hanson, D.I. Bojkov, Effect of crustal magnetic fields on the near terminator ionosphere of Mars: Comparison of in situ magnetic field measurements with the data of radio science experiments on board Mars Global Surveyor. J. Geophys. Res. **108**, 1431 (2003). doi:10.1029/2002JA009662

B.S. Lanchester, M.H. Rees, D. Lummerzheim, A. Otto, H.U. Frey, K.U. Kaila, Large fluxes of auroral electrons in filaments of 100 m width. J. Geophys. Res. **102**, 1741–9748 (1997). doi:10.1029/97JA00231

J. Lilensten, P.-L. Blelly, Du Soleil a la Terre. Collection Grenoble France (1999)

G.F. Lindal, D.N. Sweetnam, V.R. Eshleman, The atmosphere of Saturn: An analysis of the Voyager radio occultation measurements. Astron. J. **90**, 1136–1146 (1985). doi:10.1086/113820

G.F. Lindal et al., The atmosphere of Uranus—results of radio occultation measurements with Voyager 2. J. Geophys. Res. **92**(30), 14987–15001 (1987)

J.G. Luhmann, T.E. Cravens, in *Venus Aeronomy*, ed. by C.T. Russell. Space Science Reviews, vol. 55 (1991), pp. 201–274

Y. Ma et al., Three-dimensional, multispecies, high spatial resolution MHD studies of the solar wind interaction with Mars. J. Geophys. Res. **109**, A07211 (2004). doi:10.1029/2003JA010367

P.R. Mahaffy, Intensive Titan exploration begins. Science **308**(5724), 969–970 (2005)

T. Majeed, J.C. McConnell, The upper ionospheres of Jupiter and Saturn. Planet. Space Sci. **39**, 1715–1732 (1991). doi:10.1016/0032-0633(91)90031-5

T. Majeed, J.C. McConnell, Voyager electron density measurements on Saturn: Analysis with a time dependent ionospheric model. J. Geophys. Res. **101**, 7589–7598 (1996). doi:10.1029/96JE00115

T. Majeed, J.C. McConnell, R.V. Yelle, Vibrationally excited H_2 in the outer planets thermosphere: Fluorescence in the Lyman and Werner bands. Planet. Space Sci. **39**, 1591–1606 (1991). doi:10.1016/0032-0633(91)90085-O

T. Majeed, J.H. Waite Jr., S.W. Bougher, R.V. Yelle, G.R. Gladstone, J.C. McConnell et al., The ionospheres-thermospheres of the giant planets. Adv. Space Res. **33**, 197–211 (2004). doi:10.1016/j.asr.2003.05.009

C. Martinis, Private communication, 2007

C. Martinis, J. Baumgardner, S.M. Smith, M. Colerico, M. Mendillo, Imaging science at El Leoncito. Ann. Geophys. **24**, 1375–1385 (2006)

K.I. Matcheva, D.F. Strobel, F.M. Flasar, Interaction of gravity waves with ionospheric plasma: Implications for Jupiter's ionosphere. Icarus **152**, 347–365 (2001). doi:10.1006/icar.2001.6631

M.B. McElroy, The ionospheres of the major planets. Space Sci. Rev. **14**, 460–473 (1973). doi:10.1007/BF00214756

M.G. McHarg, D.L. Hampton, H.C. Stenbaek-Nielsen, Fast photometry of flickering in discrete auroral arcs. Geophys. Res. Lett. **25**, 2637–2640 (1998). doi:10.1029/98GL01972

H. Melin, S. Miller, T. Stallard, L.M. Trafton, T.R. Geballe, Variability in the H_3^+ emission of Saturn: Consequences for ionisation rates and temperatures. Icarus **186**, 234–241 (2007). doi:10.1016/j.icarus.2006.08.014

M. Mendillo, Storms in the ionosphere: Patterns and processes for total electron content. Rev. Geophys. **47**, RG4001 (2006). doi:10.1029/2005RG000193

M. Mendillo, P. Withers, Solar flare effects upon the ionospheres of Earth and Mars, in *Proceedings of the Symposium on Radio Sounding and Plasma Physics* (Am. Inst. Phys., New York, 2008, in press)

M. Mendillo, J. Baumgardner, X.-Q. Pi, P.J. Sulton, R.T. Tsunoda, Onset conditions for equatorial spread F. J. Geophys. Res. **97**, 13,865–13,876 (1992). doi:10.1029/92JA00647

M. Mendillo, J. Baumgardner, J. Nottingham, D. Aarons, J. Reinisch, B. Scali et al., Investigations of thermospheric-ionospheric dynamics with 6300-Å images from the Arecibo observatory. J. Geophys. Res. **102**, 7331–7344 (1997). doi:10.1029/96JA02786

M. Mendillo, J. Meriwether, M. Biondi, Testing the thermospheric neutral wind suppression mechanism for day-to-day variability of equatorial spread F. J. Geophys. Res. **106**, 3655–3663 (2001). doi:10.1029/2000JA000148

M. Mendillo, H. Rishbeth, R.G. Roble, J. Wroten, Modelling F2-layer seasonal trends and day-to-day variability driven by coupling with the lower atmosphere. J. Atmos. Sol. Terr. Phys. **64**, 1911–1931 (2002). doi:10.1016/S1364-6826(02)00193-1

M. Mendillo, S. Smith, J. Wroten, H. Rishbeth, D. Hinson, Simultaneous ionospheric variability on Earth and Mars. J. Geophys. Res. **108**, 1432 (2003). doi:10.1029/2003JA009961

M. Mendillo, X. Pi, S. Smith, C. Martinis, J. Wilson, D. Hinson, Ionospheric effects upon a satellite navigation system on Mars. Radio Sci. **39**, RS2028 (2004). doi:10.1029/2003RS002933

M. Mendillo, L. Moore, J. Clarke, I. Mueller-Wodarg, W.S. Kurth, M.L. Kaiser, Effects of ring-shadowing on the detection of electrostatic discharges at Saturn. Geophys. Res. Lett. **32**, L05107 (2005). doi:10.1029/2004GL021934

M. Mendillo, P. Withers, D. Hinson, H. Rishbeth, B. Reinisch, Effects of solar flares on the ionosphere of Mars. Science **311**, 1135–1138 (2006). doi:10.1126/science.1122099 Medline

S. Miller, N. Achilleos, G.E. Ballester, T.R. Geballe, R.D. Joseph, R. Prange et al., The role of H_3^+ in planetary atmospheres. Philos. Trans. R. Soc. Lond. A **358**, 2485–2502 (2000). doi:10.1098/rsta.2000.0662

S. Miller, A. Aylward, G. Millward, Giant planet ionospheres: The importance of ion-neutral coupling. Space Sci. Rev. **116**, 319–343 (2005). doi:10.1007/s11214-005-1960-4

S. Miller, T. Stallard, C. Smith et al., H_3^+: The driver of giant-planet atmospheres. Philos. Trans. R. Soc. Lond. A **364**, 3121–3137 (2006). doi:10.1098/rsta.2006.1877

D.L. Mitchell, Probing Mars' crustal magnetic field and ionosphere with the MGS Electron Reflectometer. J. Geophys. Res. **106**(E10), 23419–23428 (2001)

A.P. Mitra, *Ionospheric Effects of Solar Flares* (Reidel, Dordrecht, 1974)

G.J. Molina-Cuberos et al., Ionospheric layer induced by meteoric ionization in Titan's atmosphere. Planet. Space Sci. **49**(2), 143–153 (2001)

G. Molina-Cuberos et al., Planet. Space Sci. (2003)

L. Moore, M. Mendillo, Are plasma depletions in Saturn's ionosphere a signature of time-dependent water input? Geophys. Res. Lett. **34**, L12202 (2007). doi:10.1029/2007GL029381

L.E. Moore, M. Mendillo, I.C.F. Müller-Wodarg, D.L. Murr, Modeling of global variations and ring shadowing in Saturn's ionosphere. Icarus **172**, 503–520 (2004)

L. Moore, A.F. Nagy, A.J. Kliore, I. Muller-Wodarg, J.D. Richardson, M. Mendillo, Cassini radio occultation of Saturn's ionosphere: Model comparisons Using a constant water flux. Geophys. Res. Lett. **33**, L22202 (2006a). doi:10.1029/2006GL027375

L. Moore, M. Mendillo, C. Martinis, S. Bailey, Day-to-day variability of the E layer. J. Geophys. Res. **111**, A06307 (2006b). doi:10.1029/2005JA011448

D. Morgan et al., Solar control of radar wave absorption by the Martian ionosphere. Geophys. Res. Lett. **33**(13), L13202 (2006)

J.I. Moses, S.F. Bass, The effects of external material on the chemistry and structure of Saturn's ionosphere. J. Geophys. Res. **105**, 7013–7052 (2000). doi:10.1029/1999JE001172

J.I. Moses, R.J. Vervack Jr., The structure of the upper atmosphere of Saturn. Lunar Planet. Sci. Conf. **37**, 1803 (2006)

J.I. Moses, B. Bézard, E. Lellouch, G.R. Gladstone, H. Feuchtgruber, M. Allen, Photochemistry of Saturn's atmosphere. II. Effects of an influx of external oxygen. Icarus **145**, 166–202 (2000). doi:10.1006/icar.1999.6320

I. Mueller-Wodarg, Planetary upper atmospheres, in *Advances in Astronomy (J.M.T. Thompson)* (Imperial College Press, London, 2004), pp. 331–353

I.C.F. Mueller-Wodarg, M. Mendillo, R.V. Yelle, A.D. Aylward, A global circulation model of Saturn's thermosphere. Icarus **180**, 147–160 (2006). doi:10.1016/j.icarus.2005.09.002

A.F. Nagy, T.E. Cravens, Solar system ionospheres, in *Atmospheres in the Solar System: Comparative Aeronomy*, ed. by M. Mendillo, A. Nagy, J.H. Waite. Geophysical Monograph, vol. 130 (Am. Geophys. Union, Washington, 2002)

A.F. Nagy et al., Model calculations of the dayside ionosphere of Venus: Ionic composition. J. Geophys. Res. **85**, 7795 (1980). doi:10.1029/JA085iA13p07795

A.F. Nagy et al., A two dimensional shock capturing, hydrodynamic model of the Venus ionosphere. Geophys. Res. Lett. **18**, 801 (1991). doi:10.1029/91GL00362

A.F. Nagy, A.J. Kliore, E. Marouf, R. French, M. Flasar, N.J. Rappaport et al., First results from the ionospheric radio occultations of Saturn by the Cassini spacecraft. J. Geophys. Res. **111**, A06310 (2006). doi:10.1029/2005JA011519

J.-U. Ness, J.H.M.M. Schmitt, J. Robrade, Detection of Saturnian X-ray emission with XMM-Newton. Astron. Astrophys. **414**, L49–L52 (2004a). doi:10.1051/0004-6361:20031761

J.-U. Ness, J.H.M.M. Schmitt, S.J. Wolk, K. Dennerl, V. Burwitz, X-ray emission from Saturn. Astron. Astrophys. **418**, 337–345 (2004b). doi:10.1051/0004-6361:20035736

E. Nielsen, D.D. Morgan, D.L. Kirchner, J. Plaut, G. Picardi, Absorption and reflection of radio waves in the Martian ionosphere. Planet. Space Sci. **55**, 864–870 (2007). doi:10.1016/j.pss.2006.10.005

M. Paetzold, S. Tellmann, B. Häusler, D. Hinson, R. Schaa, G.L. Tyler, A sporadic third layer in the ionosphere of mars. Science **310**, 837–839 (2005). doi:10.1126/science.1117755 Medline

M. Paetzold et al., The structure of Venus' middle atmosphere and ionosphere. Nature **450**, 657–660 (2007). doi:10.1038/nature06239

C.D. Parkinson, E. Griffioen, J.C. McConnell, G.R. Gladstone, B.R. Sandel, He 584 A dayglow at Saturn: A reassessment. Icarus **133**, 210–220 (1998). doi:10.1006/icar.1998.5926
W.D. Pesnell, J. Grebowsky, Meteoric magnesium ions in the Martian Atmosphere. J. Geophys. Res. **105**, 1695–1708 (2000). doi:10.1029/1999JE001115
G.W. Prölss, *Physics of the Earth's Space Environment* (Springer, Berlin, 2004)
H. Rishbeth, O.K. Garriott, *Introduction to Ionospheric Physics* (Academic Press, San Diego, 1969)
H. Rishbeth, M. Mendillo, Patterns of F2-layer variability. J. Atmos. Sol. Terr. Phys. **63**, 1661–1680 (2001). doi:10.1016/S1364-6826(01)00036-0
H. Rishbeth, M. Mendillo, Ionospheric layers at Earth and Mars. Planet. Space Sci. **52**, 849–852 (2004). doi:10.1016/j.pss.2004.02.007
R. Roble, *On the Feasibility of Developing a Global Atmospheric Model Extending from the Ground to the Exosphere*. Geophysical Monograph, vol. 123 (Am. Geophys. Union, Washington, 2000), pp. 53–67
R. Roble, Private communication, 2007
A. Roboz, A.F. Nagy, The energetics of Titan's ionosphere. J. Geophys. Res. **99**(A2), 2087–2093 (1994)
A. Safaeinili et al., Estimation of the total electron content of the Martian ionosphere using radar sounder surface echoes. Geophys. Res. Lett. **34**(23), L23204 (2007)
K. Sakanoi, H. Fukunishi, Temporal and spatial structures of flickering aurora derived from high-speed imaging photometer observations at Syowa Station in the Antarctic. J. Geophys. Res. **109**, A01221 (2004). doi:10.1029/2003JA010081
B.R. Sandel, 12 co-authors, Extreme ultraviolet observations from the Voyager 2 encounter with Saturn. Science **215**, 548–553 (1982). doi:10.1126/science.215.4532.548 Medline
D.E. Sandel, J.M. Ajello, The Saturn spectrum in the EUV: Electron excited hydrogen. J. Geophys. Res. **88**, 459–464 (1983). doi:10.1029/JA088iA01p00459
R.W. Schunk, A.F. Nagy, Electron temperatures in the F region of the ionosphere: theory and observation. Rev. Geophys. Space Phys. **16**, 355–399 (1978). doi:10.1029/RG016i003p00355
R.W. Schunk, A.F. Nagy, *Ionospheres: Physics, Plasma Physics and Chemistry*, 1st edn. (Cambridge University Press, Cambridge, 2000)
R.W. Schunk, A.F. Nagy, *Ionospheres: Physics, Plasma Physics and Chemistry*, 2nd edn. (Cambridge University Press, Cambridge, 2008)
J. Semeter, E.M. Blixt, Evidence for Alfven wave dispersion identified in high-resolution auroral imagery. Geophys. Res. Lett. **33**, L13106 (2006). doi:10.1029/2006GL026274
S.P. Seth, V. Brahmananda Rao, C.M. Espirito Santo, S.A. Haider, V.R. Choksi, Zonal variations of peak ionization rates in upper atmosphere of Mars at high latitude using Mars global surveyor accelerometer data. J. Geophys. Res. **111**, A09308 (2006). doi:10.1029/2006JA011753
D. Shemansky, Atmospheric structure of Saturn and Titan: Results from the Cassini UVIS experiment, in *36th COSPAR meeting*, 16–23 July 2006, #2756
M. Shimizu, Strong interaction between the ring system and the ionosphere of Saturn. Moon Planets **22**, 521–522 (1980). doi:10.1007/BF00897291
K. Shiokawa, Y. Otsuka, T. Tsugawa et al., Geomagnetic conjugate observations of nighttime medium-scale and large-scale traveling ionospheric disturbances: FRONT3 campaign. J. Geophys. Res. **110**, A05303 (2005). doi:10.1029/2004JA010845
K. Shiokawa, G. Liu, Y. Otsuka, T. Ogawa, M. Yamamoto, N. Nishitani et al., Ground observations and AMIE-TIEGCM modeling of a storm-time traveling ionospheric disturbance. J. Geophys. Res. **112**, A05308 (2007). doi:10.1029/2006JA011772
C.G.A. Smith, Periodic modulation of gas giant magnetospheres by neutral upper atmosphere. Ann. Geophys. **24**, 2709–2717 (2006)
G.R. Smith, D.E. Shemansky, J.B. Holberg, A.L. Broadfoot, B.R. Sandel, J.C. McConnell, Saturn's upper atmosphere from the Voyager 2 EUV solar and stellar occultations. J. Geophys. Res. **88**, 8667–8678 (1983). doi:10.1029/JA088iA11p08667
S.M. Smith, J. Friedman, S. Raizada, C. Tepley, J. Baumgardner, M. Mendillo, Evidence of mesospheric bore formation from a breaking gravity wave event: Simultaneous imaging and lidar measurements. J. Atmos. Sol. Terr. Phys. **67**, 345–365 (2005). doi:10.1016/j.jastp.2004.11.008
C.G.A. Smith, A.D. Aylward, G.H. Millward, S. Miller, L.E. Moore, An unexpected cooling effect in Saturn's upper atmosphere. Nature **445**, 399–401 (2007). doi:10.1038/nature05518 Medline
K. Spenner et al., On the maintenance of the Venus nightside ionosphere: Electron precipitation and plasma transport. J. Geophys. Res. **86**, 9170–9178 (1981). doi:10.1029/JA086iA11p09170
T. Stallard, S. Miller, G.E. Ballester, D. Rego, R.D. Joseph, L.M. Trafton, The H_3^+ latitudinal profile of Saturn. Astrophys. J. **521**, L149–L152 (1999). doi:10.1086/312189
T. Stallard, S. Miller, L.M. Trafton, T.R. Geballe, R.D. Joseph, Ion winds in Saturn auroral/polar region. Icarus **167**, 204–211 (2004). doi:10.1016/j.icarus.2003.09.006

T. Stallard, S. Miller, H. Melin, M. Lystrup, M. Dougherty, N. Achilleos, Saturn's auroral/polar H_3^+ infrared emission. I. General morphology and ion velocity structure. Icarus **189**, 1–13 (2007). doi:10.1016/j.icarus.2006.12.027

D.F. Strobel, Photochemistry of Titan. ESA Spec. Publ., ESA SP-241, pp. 145–148 (1985)

G.L. Tyler, V.R. Eshleman, J.D. Anderson, G.S. Levy, G.F. Lindal, G.E. Wood et al., Radio science investigations of the Saturn system with Voyager 1: Preliminary results. Science **212**, 201–206 (1981). doi:10.1126/science.212.4491.201 Medline

G.L. Tyler, V.R. Eshleman, J.D. Anderson, G.S. Levy, G.F. Lindal, G.E. Wood et al., Radio science with Voyager 2 at Saturn: Atmosphere and ionosphere and the masses of Mimas. Tethys, and Iapetus. Science **215**, 553–558 (1982). doi:10.1126/science.215.4532.553 Medline

G.L. Tyler et al., Voyager 2 radio science observations of the Uranian system atmosphere, rings, and satellites. Science **233**(4), 79–84 (1986)

G.L. Tyler et al., Voyager radio science observations of Neptune and Triton. Science **246**(15), 1466–1473 (1989)

R.J. Vervack, An updated view of Saturn's upper atmosphere from a reanalysis of the Voyager 1 and 2 UVS occultations. American Geophys. Union, Fall Meeting, #P51B-1436 (2004)

V. Vuitton et al., Experimental and theoretical study of hydrocarbon photochemistry applied to Titan stratosphere. Icarus **185**(1), 287–300 (2006)

V. Vuitton et al., Ion chemistry and N-containing molecules in Titan's upper atmosphere. Icarus **191**(2), 722–742 (2007)

J.-E. Wahlund et al., Cassini measurements of cold plasma in the ionosphere of Titan. Science **308**(5724), 986–989 (2005)

J.H. Waite Jr., The ionosphere of Saturn. Ph.D. thesis, Univ. Michigan, Ann Arbor, 1981

H. Waite et al., Ion neutral mass spectrometer results from the first flyby of Titan. Science **308**(5724), 982–986 (2005)

E.J. Weber, J. Buchau, R. Eather, S.B. Mende, North-south aligned equatorial airglow depletions. J. Geophys. Res. **83**, 712–716 (1978). doi:10.1029/JA083iA02p00712

E.H. Wilson, S.K. Atreya, Current state of modeling the photochemistry of Titan's mutually dependent atmosphere and ionosphere. J. Geophys. Res. **109**(E6), E06002 (2004)

O. Witasse, F. Nagy, Outstanding aeronomy problems at Venus. Planet. Space Sci. (2006). doi:10.1016/j.pss.2006.04.028

P. Withers, M. Mendillo, Response of peak electron densities in the martian ionosphere to day-to-day changes in solar flux due to solar rotation. Planet. Space Sci. **53**, 1401–1418 (2005). doi:10.1016/j.pss.2005.07.010

P. Withers, M. Mendillo, H. Rishbeth, D.P. Hinson, J. Arkani-Hamed, Ionospheric characteristics above Martian crustal magnetic anomalies. Geophys. Res. Lett. **32**, L16204 (2005). doi:10.1029/2005GL023483

P. Withers, M. Mendillo, D. Hinson, Space weather effects on the Mars Ionosphere due to solar flares and meteors, in *European Planetary Science Congress*, EPSC2006-A-00190, 2006

P. Withers, M. Mendillo, M. Pätzold, S. Tellmann, A.A. Christou, J. Vaubaillon, Comparison of ionospheric observations and dynamical predictions of Meteor showers at Mars. Astron. Astrophys. (2008, submitted)

R.V. Yelle, S. Miller, Jupiter's thermosphere and ionosphere, in *Jupiter: The Planet, Satellites and Magnetosphere*, ed. by F. Bagenal, T.E. Dowling, W.B. McKinnon (Cambridge University Press, Cambridge, 2004)

L.A. Young, R.V. Yelle, R.E. Young, A. Seiff, D.B. Kirk, Gravity waves in Jupiter's thermosphere. Science **267**, 108–111 (1997). doi:10.1126/science.276.5309.108

S.-R. Zhang, 10 co-authors, October 2002 30-day incoherent scatter radar experiments at Millstone Hill and Svalbard and simultaneous GUVI/TIMED observations. Geophys. Res. Lett. **32**, L01108 (2005). doi:10.1029/2004GL020732

Q. Zhou, J. Friedman, S. Raizada, C. Tepley, Y.T. Morton, Morphology of nighttime ion, potassium and sodium layers in the meteor zone above Arecibo. J. Atmos. Sol. Terr. Phys. **67**, 1245–1257 (2005). doi:10.1016/j.jastp.2005.06.013

Photoemission Phenomena in the Solar System

T.G. Slanger · T.E. Cravens · J. Crovisier · S. Miller ·
D.F. Strobel

Originally published in the journal Space Science Reviews, Volume 139, Nos 1–4.
DOI: 10.1007/s11214-008-9387-3

Abstract Much of what we know about the atmospheres of the planets and other bodies in the solar system comes from detection of photons over a wide wavelength range, from X-rays to radio waves. In this chapter, we present current information in various categories—measurements of the airglows of the terrestrial planets, the dayglows of the outer planets and satellites, aurora throughout the solar system, observations of cometary spectra, and the emission of X-rays from a variety of planetary bodies.

Keywords Nightglow · Dayglow · Mars · Earth · Venus · Jupiter · Saturn · Uranus · Neptune · Titan · Comets · X-rays · Aurora

1 Introduction

Each of the bodies in our solar system is unique, but there are substantial common attributes. The variety of phenomena that we find will presumably be reflected in any solar system that will ultimately be investigated. In this chapter we deal with photoemission over a wide range of energies, and data collection by different methods—ground-based, satellite-based, and from planetary orbiters and fly-bys. The subject matter will often be closely related to

T.G. Slanger (✉)
Molecular Physics Laboratory, SRI International, Menlo Park, CA 94025, USA
e-mail: tom.slanger@sri.com

T.E. Cravens
Dept. of Physics and Astronomy, University of Kansas, Lawrence, KS 66045, USA

J. Crovisier
LESIA, Observatoire de Paris, 92195 Meudon, France

S. Miller
Dept. of Physics and Astronomy, University College London, London WC1E 6BT, UK

D.F. Strobel
Dept. of Earth and Planetary Sciences, Johns Hopkins University, Baltimore, MD 21218, USA

that appearing in other chapters of this collection, particularly Chap. 1 (Fox et al. 2008, this issue, Energy Deposition), Chap. 2 (Huestis et al. 2008, this issue, Reaction Rates of Relevance to Aeronomic Studies), and Chap. 5 (Witasse et al. 2008, this issue, Ionospheres).

1.1 Remote Sensing of Atmospheres

One of the basic ways through which atmospheres are studied is through the observation of emission and absorption of radiation, from X-rays to radio waves. Ground-based observations consist of a subset in this broad wavelength range, while space-based observations show the entire range, although trade-offs are necessary in comparing how measurements are made.

Although there is far more information on the terrestrial atmosphere than that of other solar system objects, over the last thirty years a great deal has been learned about other atmospheres. All the planets with atmospheres have been visited with fly-bys at least once, and only Uranus and Neptune have not had orbital missions. We are therefore in a position to discuss planetary and satellite atmospheres in a comparative manner, where the terrestrial-like inner planets—Venus, Earth, Mars—comprise one group, the outer planets with predominantly hydrogen/helium atmospheres are another, while the outer planet satellites and comets exhibit an interesting variety of differences. Titan stands alone, with an atmosphere having a high density of both nitrogen and methane.

1.2 Excited States

Optical emission measurements involve excited state photochemistry and photophysics. To observe emission, the atom or molecule must be in an electronically or vibrationally/rotationally excited state, with photoemission occurring rapidly or slowly, depending on spectroscopic and collisional details.

If the excited state has an optically allowed pathway to a lower level, emission is prompt and not subject to collisions. Examples are the electronically excited states of NO, with sub-microsecond radiative lifetimes. For the cases—very common in aeronomy—where the emitters are metastable, collisional deactivation can play an important role. Prime examples of this sort of behavior are $O(^1D)$ and $O_2(a^1\Delta_g, v)$, where the former has a radiative lifetime of 110 s (Tachiev and Fischer 2002), and is rapidly quenched by all significant atmospheric colliders—O_2, N_2 (Sander et al. 2003) and $O(^3P)$ (Closser et al. 2005). The $O_2(a, v)$ molecule exhibits both properties, in that the strongly emitting $v = 0$ level is almost impervious to quenching (Sander et al. 2003), in spite of having a radiative lifetime of $\sim$4000 s (Newman et al. 1999), whereas vibrationally-excited levels of $O_2(a)$ are removed so rapidly by O_2 collisions (Slanger and Copeland 2003) that they are never observed in the atmosphere, in spite of being copiously generated. Table 1 gives examples of the various ways that excited atoms and molecules are produced in the atmosphere.

1.3 Tools for Interpretation of Emissions

The appropriate tools for the study of optical emissions depend on the desired goals, although in all cases it is essential to know what to expect. For example, the 557.7 nm oxygen green line in the terrestrial nightglow is a feature that always dominates other emissions in the vicinity. Thus, a simple two-filter photometer, on- and off-resonance with the line, is often adequate for a measurement. In contrast, the O-atom 777.2–777.5 nm triplet is contaminated by an OH line pair (Slanger et al. 2000), and a filtered photometer will lead to misleading results unless the oxygen lines are very strong.

Table 1 Sources of excitation

1.	Atom recombination
	$[O + O + M \rightarrow O_2^*]$
	$[N + O \rightarrow NO^*]$
2.	Radiative recombination
	$[O^+ + e \rightarrow O^*]$
3.	Dissociative recombination
	$[O_2^+ + e \rightarrow O^* + O]$
4.	Dissociative photoexcitation
	$[CO_2 + hv \rightarrow CO^* + O^*]$
	$[O_3 + hv \rightarrow O^* + O_2^*]$
5.	Dissociative electron excitation
	$[CO_2 + e \rightarrow O^* + CO]$
6.	Direct electron or heavy particle excitation
	$[e_{fast} + O \rightarrow O^*]$
7.	Reactive excitation
	$[H + O_3 \rightarrow OH(v) + O_2]$
8.	Excitation exchange
	$[O(^1D) + O_2 \rightarrow O(^3P) + O_2(b)]$
9.	Resonance fluorescence
	Electronic $[O + hv \rightarrow O^*]$
	Vibrational $[NO(v = 0) + hv \rightarrow NO(v = 1)]$

In general, it is safest to disperse the radiation to eliminate ambiguities, whether by a spectrometer or an interferometer. A Fabry-Perot system has the advantage of very high resolution, but operates over a narrow spectral range. A spectrometer generally has the opposite characteristics. Thus to measure temperature and winds the interferometer has inherent advantages, whereas to look at the relationships between features in an aurora, it is important to have the large bandwidth of a spectrometer system.

The available instrumentation for the study of the terrestrial atmosphere has recently been expanded by the recognition that the sky spectra generated at large ground-based telescopes give a superb view of the nightglow. Sky spectra represent the terrestrial background, which must be subtracted from the astronomer's target spectrum in order to exclude the atmospheric features. At the Keck I telescope on Mauna Kea and the VLT in Chile, the echelle spectrographs produce high-quality spectra with broad wavelength coverage—from the ozone cut-off near 310 nm to the infrared at 1.05 μ—spectral resolution of 40,000–50,000, and the ability to detect lines with intensities less than 100 mR (Cosby and Slanger 2007). Thus, the best qualities of the various systems are brought together in such an instrument (Cosby et al. 2006; Cosby and Slanger 2007).

1.4 Platforms—Ground-based, Terrestrial/Planetary Orbiters

An issue that is always present in aeronomical discussions is that of ground-based *vs* space-based instrumentation. Space-based system provide a global view, emission altitudes can be determined, and spectral features are not blocked by the atmosphere—ground-based instruments can be larger and more sophisticated, and are much more accessible to modification. It is obvious that both types of measurements are valuable.

The same considerations apply to ground-based observations *vs* measurements from planetary orbiters. When a spacecraft is sent to another planet, the instrumentation choice is fixed, and typically the technology used has already been superseded, whereas ground-based measurements use hardware that is more up-to-date. On the other hand, *in situ* measurements give global coverage and are not limited by atmospheric transmission, at least from a perspective above the atmosphere. Terrestrial orbiters (cf. the Hubble Space Telescope) also give an unimpeded view of the planet, but there are still restrictions. For instance, the Mars nightglow can only be investigated from Mars orbit, because in the view from the terrestrial vicinity, at most a 10% dark crescent is all that is accessible.

2 Outer Planet Airglows

2.1 Intensities of Outer Planet Airglows—Background and Overview

The atmospheres of the gas giant planets Jupiter and Saturn and of the icy giant planets Uranus and Neptune are mostly H_2 and He and their airglow is dominated by H_2 electronic bands, the He 58.4 nm line, the H Lyman line series, and H_3^+ near-IR bands. Each atmosphere has a thermosphere significantly hotter than would be predicted by solar EUV and UV heating and, potentially, indicating that other energy sources may be relevant to understanding their airglow emission.

For Jupiter, the Pioneer 10 two-channel ultraviolet photometer inferred H Lyman-α 121.6 and He 58.4 nm lines under the assumption that they were the strongest emitters in each channel during its Jupiter flyby in 1973. The former was an anomalously low 0.4 kR and the latter was inferred to be 5.1 R. Carlson and Judge (1976) discussed the difficulty in interpreting these lines when the [He]/[H_2] ratio, temperature, and vertical mixing were all uncertain, and the analysis focused more on the magnitude of the vertical eddy diffusion coefficient at the homopause than other properties of the atmosphere. The Voyager Ultraviolet Spectrometers (UVS) made measurements during its Jupiter flybys in 1979 with a spectral resolution of ~3 nm and found strong emission from H_2 electronic bands (~2 kR, mostly from the Lyman and Werner bands) in addition to a very strong disk averaged H Lyman-α brightness of ~14 kR, but only an upper limit for the He 58.4 nm line of 0.1 R (Broadfoot et al. 1979). Earlier Lyman-α measurements from rocket experiments (Rottman et al. 1973) and with the Copernicus satellite (Atreya et al. 1977) had intermediate intensities of 1.2–4.4 kR, but the International Ultraviolet Explorer (IUE) measurement at the time of Voyager confirmed the higher intensity (~13 kR) (Clarke et al. 1980b). The differences in the Lyman-α brightness values were almost certainly due to calibration problems with instruments on the early missions. Both Clarke et al. (1980b) and Sandel et al. (1980) found that the Lyman-α brightness had a strong longitudinal asymmetry in its variation ("the Lyman-α bulge") with minimum values at longitudes 200–300°W (~14.4 kR) and a peak value at ~110°W of 19.6 kR.

For Saturn, the two Voyager UVSs measured H Lyman-α as ~3.3 kR (V1) and 3.0 kR (V2), while for He 58.4 nm, the respective values were 2.2 ± 0.3 and 4.2 ± 0.5 R (Sandel et al. 1982). At 90–110 nm, the V2 H_2 band intensities were measured as 210 R and the total intensity was inferred to be 650 R (cf. Strobel et al. 1991a). The average Lyman-α disk brightness from 29 IUE observations was 1.1 ± 0.36 kR (McGrath and Clarke 1992). This discrepancy between UVS and IUE is perplexing in light of their agreement on the Jovian Lyman-α brightness.

For Uranus, V2 UVS measured H Lyman-α as 1.6 ± 0.25 kR and its Raman-scattered line at 128 nm as 45 ± 15 R. From the latter Yelle et al. (1987b) inferred ~300–600 R of the measured H Lyman-α was due to H_2 Rayleigh scattering. H Lyman-β was inferred as ~3 R, but only an upper limit of 0.05 R could be given for He 58.4 nm (Strobel et al. 1991a). At 90–110 nm, the V2 H_2 band intensities were measured as ~50 R and the total intensity was inferred to be 150 R (cf. Strobel et al. 1991a). The detection of Raman-scattered Lyman-α radiation implied severe depletion of hydrocarbons in the middle atmosphere, which would otherwise absorb the 128 nm line. Clarke et al. (1986) reported IUE measurements of H Lyman-α with considerable variability over the four-year period 1982–1986 (a factor of ~3) and on one occasion a factor of 2 variation in a 24 h period. However, it must be remembered that IUE lacked spatial resolution to distinguish airglow from aurora and required substantial correction for absorption of Uranian Lyman-α by interplanetary hydrogen between Uranus and the Earth.

For Neptune, V2 UVS measured H Lyman-α as 340 ± 20 R and He 58.4 nm as 0.5 ± 0.2 R. At 80–110 nm, the V2 H_2 band intensities were measured as 19 ± 3 R and the total intensity was inferred to be ~60 R.

2.2 H_2 Dayglow Mechanisms—Jupiter and Saturn Results

It should be kept in mind that calibration in the EUV/FUV has been a long-term problem for space-borne spectrometers. In addition, the low spectral resolution of the Voyager UVSs and "no resolution" of the Pioneer 10 photometer made interpretation of the data difficult. A lively debate emerged about the excitation mechanism(s) that generated the surprisingly large H_2 EUV/FUV dayglow intensities observed by Voyager. The three principal mechanisms were electron excitation (Shemansky 1985), dynamo-plasma acceleration (Clarke et al. 1987) and solar fluorescence (Yelle et al. 1987a; Yelle 1988; cf. Strobel et al. 1991a). The phenomenon was called "electroglow" (Broadfoot et al. 1986), yet the measured intensities exhibited a dependence on the incident solar EUV and UV fluxes at each planet. All proposed mechanisms also included photoelectron production of emission. Broadfoot et al. (1986) emphasized excitation by low energy electrons as a necessary component of the phenomenon. However, the power requirements to energize these electrons exceeded substantially what the Sun could supply in the UV.

The solar-induced H_2 fluorescence creates a spectrum distinctly different from electron impact on H_2, because most of the fluorescing photons originate from the $B^1\Sigma_u^+$ state in the $v' = 6$ vibrational level and $J = 1$ rotational level (Feldman and Fastie 1973). The poor 3 nm Voyager resolution data and the low signal-to-noise IUE Jovian dayglow data at 1 nm were inadequate to resolve the controversy, although the latter slightly favored electron impact (McGrath et al. 1989). Better spectral resolution and higher signal to noise were required, which was supplied by measurements at 0.3 nm with the Hopkins Ultraviolet Telescope (HUT) in December 1990 (Feldman et al. 1993). The HUT data shown in Fig. 1 yielded H_2 electronic band intensities of 275 R at 90–110 nm and a total at all wavelengths of 2.3 kR. By comparison, the Voyager intensity at 90–110 nm was 640 R and the inferred total intensity was 2 kR. The difference in the measured quantities immediately raises the calibration issue for the UVS spectrometers (cf. Strobel et al. 1991a). HUT also measured intensities of the H Lyman series—Ly-α: 15.1 ± 2 kR, Ly-β: 30 ± 7 R, Ly-γ: 16 ± 5 R.

Liu and Dalgarno (1996) demonstrated by detailed calculations of solar fluorescence and photoelectron excitation that the HUT 0.3 nm resolution spectra of Jupiter's equatorial dayglow could be explained by these mechanisms, as hypothesized by Yelle (1988), without any significant residual intensity and that no additional mechanisms or energy sources were required. They benefited from recent advances in knowledge of the solar EUV spectrum (cf. Tobiska 1993). The strongest fluorescence is due to the solar Lyman-β line at

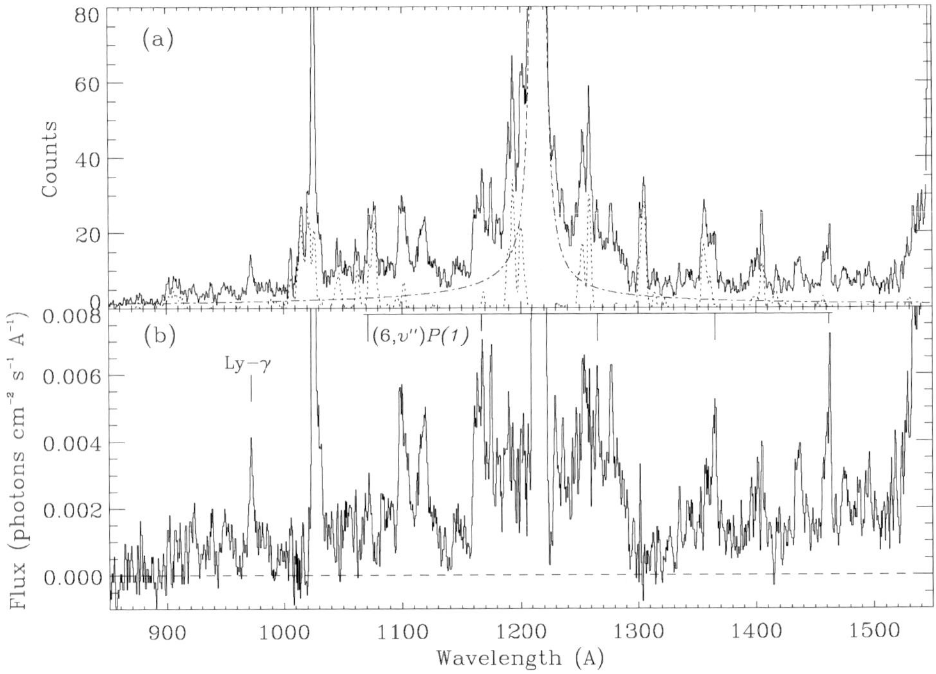

Fig. 1 Composite spectrum of Jupiter's equatorial region from HUT observations 1990 December 7 and 10 with integration time of 712 s and smooth over three 0.51 A bins. Note this spectrum contains contributions from the terrestrial geocorona, grating-scattered Lyman-alpha (*dot-dashed line*) and the Io plasma torus (*dashed line*). Subtracting the latter two components yields the spectrum in panel (**b**), which is in physical units, whereas panel (**a**) is in counts. The solar HI Lyman-beta induced fluorescence line are indicated in (**b**). After Feldman et al. (1993)

102.572 nm, ~14% of the total, which is coincident with the P(1) line of the Lyman 6-0 band at 102.593 nm. Liu and Dalgarno (1996) found that the Jovian H_2 dayglow illustrated in Fig. 1 is characterized by an atmospheric temperature of ~530 K and H_2 column density of $\sim 10^{20}$ cm^{-2}. Their paper effectively ended the lively debate on dayglow excitation mechanisms for Jupiter (but not necessarily for Saturn and Uranus) with a definitive verdict for solar fluorescence as the major dayglow contributor. They did not rule out minor contributions from other excitation sources, in particular, low-energy electron precipitation, which may be required to account for the H_3^+ dayglow of Jupiter at mid-to-low latitudes (see Sect. 2.4). There are still open questions on the quantitative significance of these other excitation sources.

For Saturn, Cassini UVIS data have not been published with an equivalent analysis of its H_2 dayglow to determine whether the solar fluorescence and photoelectron excitation can explain the measured intensities. The colder temperatures in Saturn's thermosphere should lead to enhanced solar Lyman-β fluorescence, because it is driven by the coincidence with the $P(1)$ line of the Lyman 6-0 band. Likewise, analyses for H_2 dayglow on Uranus and Neptune need updates beyond what was reported in Strobel et al. (1991a).

2.3 Dayglows of Atomic H and He Lines

The Lyman-α dayglow on the giant planets should, in principle, be quite straightforward to explain. Atomic hydrogen above the CH_4 absorbing region can efficiently resonantly scatter the strong solar Lyman-α line, which is broad with a line width of $\sim$0.1 nm, characteristic of line formation in a region of the solar atmosphere at $\sim 10^{4-5}$ K. The intrinsic planetary line width is determined by thermospheric temperatures, which range from 300 to 1000 K. The H column density above the absorbing CH_4 region coupled with the planetary line width determines what fraction of the solar line may be resonantly scattered. While the scattering optical depth at line center can be very large—up to 10^5—it does not extend out to the wings of the solar line, due to the mismatch of line widths indicative of their respective line formation temperatures. For Jupiter, the thermospheric temperature is $\sim$900 K and the Lyman-α dayglow intensities are large, $\sim$12–20 kR during solar maximum conditions—with low values outside and high values inside the bulge regions. Jupiter's Lyman-α is bright, because the planetary line width is considerably broader than the Doppler width characteristic of 900 K hydrogen atoms.

Observations consistently give a flat center-to-limb variation outside the bulge region, which implies optically thin emission (e.g. Jaffel et al. 2007 and references therein). However, the Jupiter line is optically thick at line center with scattering optical depths $\sim 10^5$ and total emission is dominated by line center photons to yield a conservatively scattering atmosphere with a center-to-limb cosine-like variation. But near and at the limbs, the optically thin wings of the Jupiter line are more important and the overall line emission decreases more rapidly. This combination of very optically thick at line center and very thin in the wings yields the observed center-to-limb behavior. Ben Jaffel et al. (2007) show the importance of measuring spectrally resolved line profiles at the limb to understand Lyman-α emission.

Gladstone et al. (2004) found that Galileo UVS observations of Jupiter's corona could be explained with a thermal H column density of $\sim 10^{17}$ cm^{-2}, augmented by a column density of 10^{14} cm^{-2} hot ($T \sim 25{,}000$ K) H atoms with a scale height of 1000 km. Note that all values pertain to regions above the homopause/absorbing CH_4 layer. These values are not unique and other combinations are possible. In the bulge region the planetary Lyman-α line profile is broader than in the non-bulge region (Clarke et al. 1991) and simply increasing the H column density in the bulge region does not yield the correct line profile (Jaffel et al. 2007).

For Saturn, Jaffel et al. (1995) concluded that an H column density of $\sim 9 \times 10^{16}$ cm^{-2} could explain, by resonance scattering of solar and interplanetary Lyman-α radiation, the Voyager observed intensity of 3.3 kR and its center-to-limb variation. However, this H column density above the absorbing CH_4 region is a factor of three larger than can be produced by any photochemical model, even with input parameters and assumptions biased to favor large column densities (Moses et al. 2000). Hence the latter authors argue for a calibration problem with Voyager UVS data and a strong preference for the validity of the IUE intensities, as their nominal model yields a Lyman-α brightness of $\sim$0.94 kR and their largest calculated value was 1.5 kR.

For Uranus, Strobel et al. (1991a) reviewed previous work and gave estimates for (1) resonance scattering of solar Lyman-α with an intensity of $\sim$300 R, (2) ISM Lyman-α scattering of $\sim$400 R, and (3) a contribution of $\sim$300–600 R from H_2 Rayleigh-scattered solar Lyman-α (Yelle et al. 1989), which result in a total brightness of 1–1.3 kR, quite close to the measured value of 1.6 ± 0.25 kR, but indicative that scattering alone cannot account fully for the observed emission from Uranus. Since the Voyager flyby of Neptune, no analysis has been performed for its Lyman-α dayglow.

The interpretation of the He I 58.4 nm line requires accurate knowledge of the $[He]/[H_2]$ ratio, temperature, and vertical mixing at and above the homopause, all uncertain to various degrees. Fundamentally, planetary He absorbs solar He I 58.4 nm radiation and re-emits/scatters with a probability equal to 0.9989. The amount of scattering (i.e. brightness) depends on the atmospheric temperature for planetary line width, and the He column density above the unit optical depth due to absorbing H_2 and H. Only on Jupiter is the $[He]/[H_2]$ ratio known accurately deep in the atmosphere. Extreme Ultraviolet Explorer (EUVE) satellite observations during the impact of Comet Shoemaker-Levy 9 in 1996 yielded up to a factor of 10 enhancement in the brightness of He 58.4 nm ~2–4 h after the impact of large fragments. The most probable explanation was the enhancement of $[He]/[H_2]$ in the upper atmosphere by lofting the more He-enriched lower atmosphere upward (Gladstone et al. 1996).

Parkinson et al. (2006) have done a preliminary analysis of the Cassini Ultraviolet Imaging Spectrometer (UVIS) data and found that Jupiter's disk brightness contains both a dayglow contribution (equivalent to a disk-averaged brightness of ~6 R) and an auroral contribution (a few R). Without spatial resolution, these components cannot be separated. Parkinson et al. (2006) interpreted these observations in terms of a vertical eddy mixing coefficient $K_{zz} \sim 2 \times 10^6$ cm^2 s^{-1}, associated with the dayglow and $K_{zz} \sim (8\text{–}40) \times 10^6$ cm^2 s^{-1} associated with the auroral regions. However, a dynamically active planet with strong auroral heating cannot be adequately described dynamically by only two vertical eddy mixing coefficients.

Parkinson et al. (1998) performed the most recent analysis of the Saturnian He 58.4 nm line brightness. With a range of $[He]/[H_2]$ mixing ratios from a minimum from the Voyager IRIS measurements (~0.03) to a maximum from the Conrath and Gautier (2000) reanalysis of IRIS data (~0.14), they inferred a range for $K_{zz} \sim 10^{7-9}$ cm^2 s^{-1} for various assumptions about the solar 58.4 nm flux as a function of solar cycle and upper atmospheric temperatures. Hopefully, analysis of the Cassini UVIS solar and stellar occultation and the He 58.4 dayglow will lead to a more definitive understanding of the distribution of He at and above the homopause.

2.4 Thermal H_3^+ Emission

The H_3^+ ion plays the fundamental role of the thermospheric thermostat for the giant planets in a manner similar to NO in the Earth's thermosphere, especially for Jupiter. The near-IR H_3^+ emissions are the principal means of remotely sensing portions of their ionospheres. H_3^+ emissions were first discovered in the auroral zones of Jupiter by Drossart et al. (1989). However, H_3^+ emission is also planet-wide (i.e. dayglow) on Jupiter (Miller et al. 1997), perhaps on Saturn (Stallard et al. 1999), and apparently also on Uranus (Lam et al. 1997). To date, it has not been detected on Neptune. The H_3^+ υ_2 band, between 3.4–4.1 microns, is mostly thermal emission and hence is a measure of the H_3^+ column density and the temperature of the region of the atmosphere where H_3^+ is an abundant ion. As it is near-IR emission, elevated temperatures and ionospheric densities are required for detection in the dayglow. It is no surprise then that it has not been detected on Neptune, which is furthest from the Sun with the weakest aurora. Jupiter has the hottest thermosphere (~900 K) and the largest ionospheric densities in the non-auroral regions. Saturn's thermosphere is colder (~420 K) with fewer H_3^+ ions, whereas Uranus has a hotter thermosphere (~880 K) than Saturn to compensate for presumably a lower H_3^+ density (temperatures quoted are asymptotic values). Temperatures deduced from H_3^+ are ~900–1100 K for Jupiter and ~700 K for Uranus. From the Wien displacement law, peak blackbody radiation at 4 microns occurs at 750 K and at 1000 K for 3 micron radiation. Thus H_3^+ emission in the υ_2 band occurs at the Planck function maximum for the thermospheres of Jupiter and Uranus.

The H_3^+ dayglow thermal emission from Jupiter requires some explaining. H_3^+ is produced as a result of solar EUV ionization of molecular hydrogen, and this certainly contributes to the emission. But Rego et al. (2000) found that even if all of the H_3^+ found at the (magnetic) equator is produced by solar EUV, this mechanism cannot account for all of the emission observed between latitudes corresponding to the Jovian auroral oval (see Sect. 4.2) and latitudes up to 30° equatorward of the oval. These authors consistently found that a "mid-to-low" (MTL) latitude component was required, which could be as much as twice as intense as the EUV-produced H_3^+ emission.

Rego et al. (2000) considered whether this additional MTL H_3^+ could be the result of aurorally-produced ions drifting equatorward. But they rejected this mechanism on the grounds that meridional winds were not sufficiently rapid to carry H_3^+ to low latitudes before dissociative recombination could occur.

2.5 Nitrogen at Titan and Triton

The nitrogen atmospheres of Titan and Triton were remotely sensed by Voyagers 1 and 2, respectively, at a nominal 3 nm spectral resolution. At the time of these flybys, the understanding of the nitrogen dayglow, even for the Earth's thermosphere, had major gaps, which influenced the interpretation of the Voyager data. The analysis was further complicated by the initial inability to accurately determine spacecraft pointing for bright limb scans. Strobel and Shemansky (1982) identified the N_2 Carroll-Yoshino (CY) $c_4'^1\Sigma_u^+ - X^1\Sigma_g^+$ 0-0 and 0-1 Rydberg band emissions near 95.8 and 98.1 nm as significant emitters in the EUV based on their absence/very low brightness in the Earth's dayglow and on laboratory data where electron impact excitation of optically thin amounts of N_2 generates these bands as the strongest emitters. The other key spectral features were the N^+ multiplet at 108.5 nm and the LBH bands in the UV. At the time of the Voyager flybys, all of these spectral features were thought to be due principally to electron impact excitation. The CY 0-0 band is optically thick at the altitude of peak photoelectron excitation and could only be detected from space if it were produced near the exobase.

A major breakthrough was the laboratory measurements of Samson et al. (1991), as discussed by Meier et al. (1991), which demonstrated that many multiplets of N and N^+ are produced by the photon excitation of N_2 into the repulsive $N_2^+ H^2\Sigma_g^+$ state that dissociates into excited fragments that radiate above 90 nm. Another major advance was the development of a multiple scattering model by Stevens et al. (1994) to treat the radiative transfer within the N_2 CY bands and with the inclusion of predissociation, branching, escape to space, and extinction. Stevens (2001) applied this model to Titan and demonstrated that the N_2 CY 0-0 band at 95.8 nm was misidentified and that Voyager UVS actually detected two NI multiplets at 95.32 and 96.45 nm generated from the photodissociative ionization via the $N_2^+ H^2\Sigma_g^+$ state. Confirmation of the Stevens (2001) analysis is found in the Cassini UVIS data (Fig. 2), which reveal that the N_2 CY 0-0 band at 95.8 nm is absent and the NI multiplets at 95.32 ($^4S^0$-^4P, $^4S^0$-^4D) and 96.45 ($^4S^0$-^4P) nm are present (Ajello et al. 2007). The FUV shown in Fig. 3 contains principally the LBH band system and also the NI multiplets at 123.4 and 149.3 nm. Ajello et al. (2007) found no nitrogen emission features in the nightglow.

For Triton, only Voyager 2 UVS observations are available and need to be updated based on the Cassini UVIS Titan data and the Stevens (2001) theoretical model. Strobel et al. (1991b) attributed most nitrogen emissions (N_2, N, N^+) to be due principally to magnetospheric electron impact excitation. Scaling the Cassini UVIS Titan data, which can be explained by solar radiation only, to 30 AU and solar maximum conditions at the time of the Voyager 2 flyby cannot explain the UVS intensities (cf. Strobel et al. 1991b; Ajello et al. 2007).

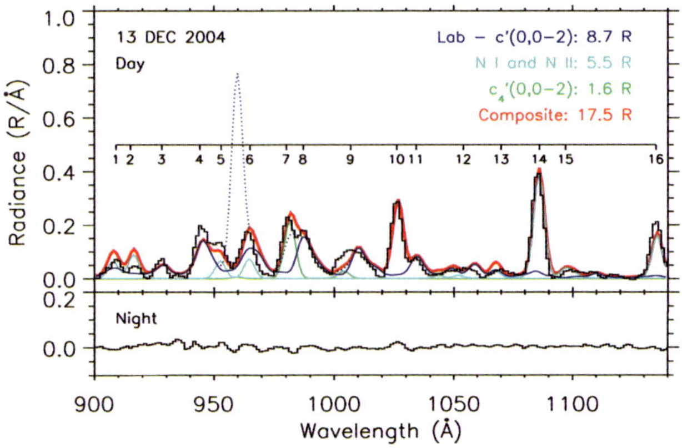

Fig. 2 (*Top Panel*) Model fit to UVIS dayglow spectrum from 13 December 2004 (*red*). Two N I multiplets at 95.32 and 96.45 nm (Stevens 2001) account for most of what was originally identified as $c_4'(0, 0)$ (Broadfoot et al. 1981). The optically thin model $c_4'(0 - v)$ bands that were removed are shown in *dotted-light blue* and the loss in area under this $c_4'(0, 0)$ band represents multiple-scattering predissociation of N2 to create fast N-atoms. (*Lower Panel*) Nightglow with a single H Ly-α intensity of ~0.1 R (~20 times weaker than in the dayglow). After Ajello et al. (2007)

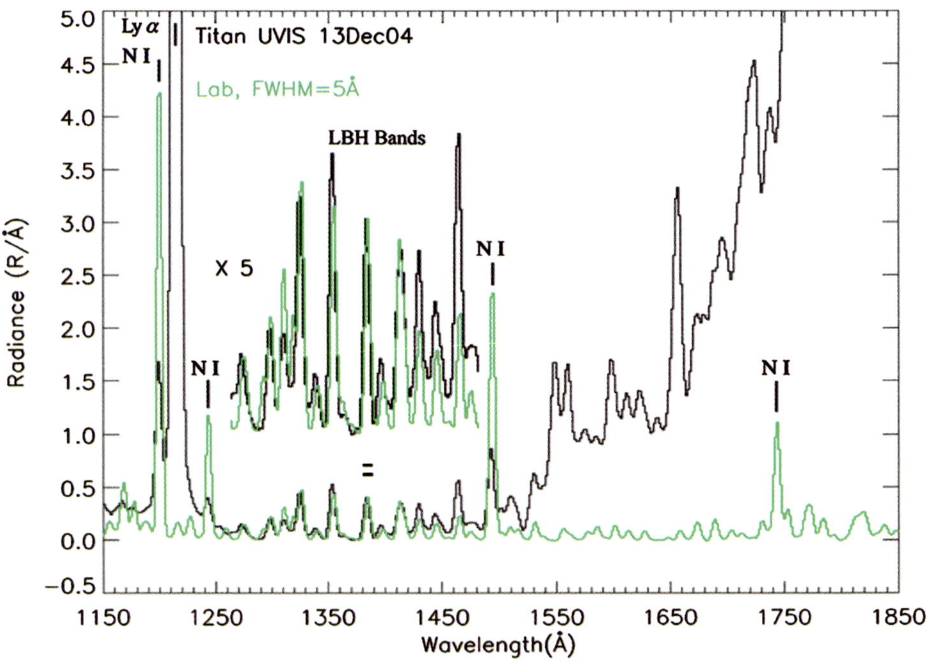

Fig. 3 The FUV limb spectrum of 13 December 2004 with the same geometry and timing as the EUV observation in Fig. 2, includes extensive Rayleigh scattering longward of 150 nm and some CI multiplets from atmosphere of Titan. A laboratory spectrum of electron impact induced fluorescence at 30 eV is over-plotted with the FUV spectrum with the two spectra normalized at LBH (2,0) band, 138.3 nm. The principal features and disk averaged intensities are: LBH bands (a 1.g PI-X 1Sg+) at ~60 R, N I 120.0 nm at ~15 R, H I Ly α at ~250 R, N I 124.3 nm at ~4 R, and N I 149.3 nm at ~90 R. After Ajello et al. (2007)

3 Inner Planet Observations

3.1 Comparative Aeronomy of Earth/Venus/Mars

The atmospheres of the three terrestrial planets are distinctly different—the Mars and Venus atmospheres are similar in being composed of CO_2 with a few percent N_2, but the surface densities differ by four orders of magnitude. The terrestrial atmosphere is O_2/N_2, with only about 0.03% CO_2. However, in terms of the airglows there are substantial similarities, the primary reason being that the principal energy absorbers—O_2 and CO_2—both give oxygen atoms upon dissociation. Thus, in the upper atmospheres of all three planets the oxygen atoms are important species, and influence the nature of the airglows. Furthermore, despite the large differences in CO_2 mole fraction, even the CO_2 in the terrestrial atmosphere plays a critical role in the overall energy balance, a fact of great current concern in discussions of global climate change.

Table 1 shows dayglow and nightglow sources of excitation relevant in some measure to the terrestrial planets. Process (1)—atom recombination—is of particular importance in the nightglows, although the manifestation of O-atom recombination is distinctly different in the earth's upper atmosphere and that of the CO_2 planets. The initial recombination products are likely to be the same, but the effects of collisions with CO_2 of the nascent O_2^* molecules results in a different distribution of populations among the so-called Herzberg states of O_2, where in the Venus and Mars atmospheres there is much less vibrational excitation in the emitting O_2^* electronic states. As a result, in the terrestrial atmosphere, Herzberg state emission is seen over the 250–450 nm range, while at Venus (and presumably Mars) the emission lies in the visible spectral region (Broadfoot and Bellaire 1999; Krasnopolsky and Parshev 1983).

The nightglow features most common to all three terrestrial planets are those from the *C-X* delta and *A-X* gamma bands of NO. The NO ($C^2\Pi$, $A^2\Sigma^+$) states are formed by two-body recombination of N+O, and radiate immediately upon formation. Thus, the collisional environment is unimportant, and one sees the same distribution of emission at all three planets (Eastes et al. 1992; Huestis and Slanger 1993; Stewart et al. 1980; Bertaux et al. 2005a). In all cases, the N and O atoms are formed on the dayside of the planets by photon and electron processes. For the earth and Mars, these atoms are brought to the nightside principally by the rotation of the planets, but for Venus rotation is very slow, and it is high altitude winds that sweep the atoms to the nightside before they are lost by recombination on the sunlit side.

Figure 4 shows the NO nightglow emissions from Mars Express (courtesy of F. Leblanc) and from a terrestrial spectrum (Eastes et al. 1992). They are identical in structure for the NO *C-X* and *A-X* bands, but quite different in the longer wavelength region, where the nadir-viewed terrestrial spectrum shows the emergence of the O_2 Herzberg bands. The Mars and terrestrial NO nightglow intensities are comparable, while that at Venus is far more intense.

The $O(^1S)$ state, emitting in two lines—$^1S \rightarrow ^3P$ at 297.2 nm and $^1S \rightarrow ^1D$ at 557.7 nm—is one of the most studied features in the terrestrial atmosphere, and the 557.7 nm green line has been known as a nightglow radiator for almost 150 years (Ångström 1869). The exact means by which it is produced in the atmosphere is still not completely understood, but is believed to involve electronic energy transfer from an electronically excited O_2 molecule to $O(^3P)$.

Because both of these species are found in the Venus atmosphere, it seems likely that $O(^1S)$ emission should be observed. However, the first Venus measurements made from a planetary orbiter (Krasnopolsky et al. 1976) showed no evidence for the green line, and thus

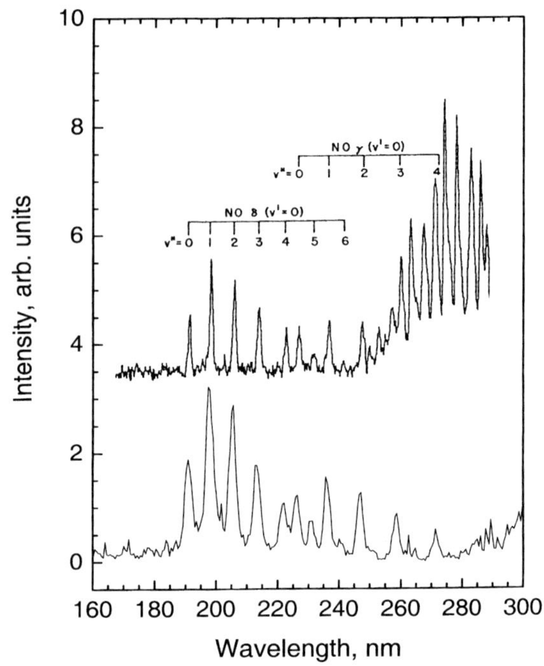

Fig. 4 The terrestrial nightglow (*upper*) (Eastes et al. 1992) compared to the Mars nightglow (*lower*) (Bertaux et al. 2005a). The 180–260 nm NO features are identical in appearance, but the O_2 Herzberg band terrestrial features at $\lambda > 250$ nm are not reproduced at Mars

it became accepted that in some manner collisions with CO_2 impede $O(^1S)$ observation, although it was well-known from Mariner dayglow observations that $O(^1S)$ is a strong feature in the UV (Barth et al. 1971), and we know that the green line emission is ten times more intense (Slanger et al. 2006a). However, in 1999, a ground-based Venus nightglow study demonstrated that the green line can be as intense as it is in the terrestrial atmosphere (Crisp 2001; Slanger et al. 2001), although at times (as in the Venera study) it is indiscernible (Slanger et al. 2006b).

Figure 5 demonstrates this behavior, contrasting two ground-based measurements of the green line in the Venus nightglow. The upper spectrum was taken with Keck I/HIRES in 1999, and shows the strong Venus green line, red-shifted with respect to the terrestrial line. The lower spectrum was taken in 2002 at APO with the ARCES echelle spectrograph at somewhat lower resolution, and although the terrestrial line intensity is comparable in the two spectra, there is only a hint of the Venusian line. This temporal variability urgently requires an explanation.

Yet another nightglow feature which is common to all three atmospheres is the $O_2(a\text{-}X)$ Infrared Atmospheric 0-0 band, at 1.27 μ (Crisp et al. 1996; McDade et al. 1987; Slanger and Copeland 2003). In the nightglows, this emission is produced by O-atom recombination, and because the O_2 $(a, v = 0)$ level is unreactive with O_2, N_2, and particularly CO_2 (Sander et al. 2003), emission from the level can be very intense. In the dayglows, this intensity can be enhanced even further by reaction (4b) in Table 1, dissociative excitation of O_3. The fact that in the case of both the Earth (Mlynczak et al. 2001) and Mars (Noxon et al. 1976) the dayglow intensities are very high is diagnostic of the presence of ozone, and ozone is mapped over the sunlit hemispheres by this emission (Krasnopolsky 2003; Novak et al. 2002). For Venus, the 1.27 μ emission from O-atom recombination is extremely variable (Crisp et al. 1996). Early measurements had shown that dayglow and nightglow intensities were comparable (Connes et al. 1979), an indication that there is very little ozone present,

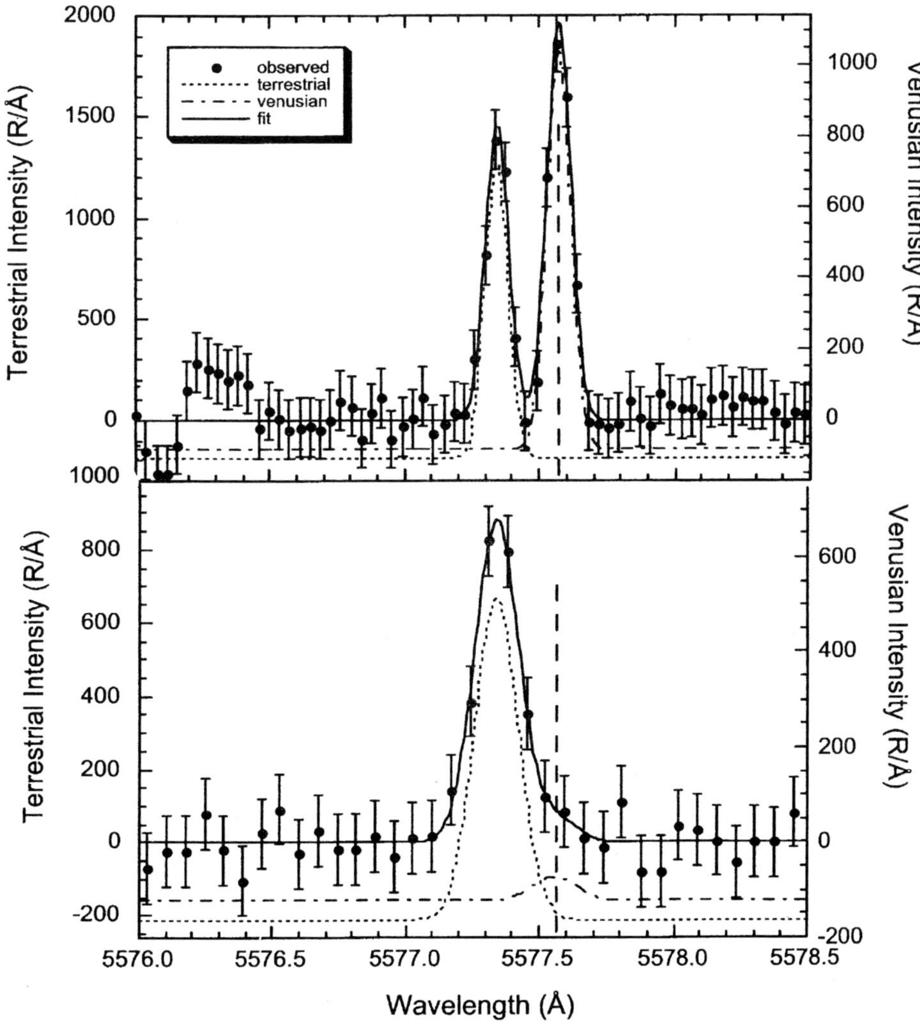

Fig. 5 Examples of the appearance of the Venus *green line* in ground-based observations. The *upper* spectrum is the discovery spectrum of the line, as measured with Keck I/HIRES (Slanger et al. 2001). The *lower* spectrum shows a much weaker manifestation, measured at APO/ARCES (Slanger et al. 2006b)

but with the great temporal and spatial changes that are now seen, the picture is not entirely clear.

3.2 Results from Sky Spectra

As explained above, sky spectra represent the terrestrial background, which astronomers must subtract from their target spectrum in order to separate them from the atmospheric features. In Figs. 3–7 are presented examples of the spectra and the data obtained from their analysis.

Figure 6 shows a newly-discovered O_2 band system, the $c^1\Sigma_u^- - b^1\Sigma_g^+$ transition, as seen both at Keck I and the VLT (Slanger et al. 2003). The emitting $O_2(c)$ state is the one seen

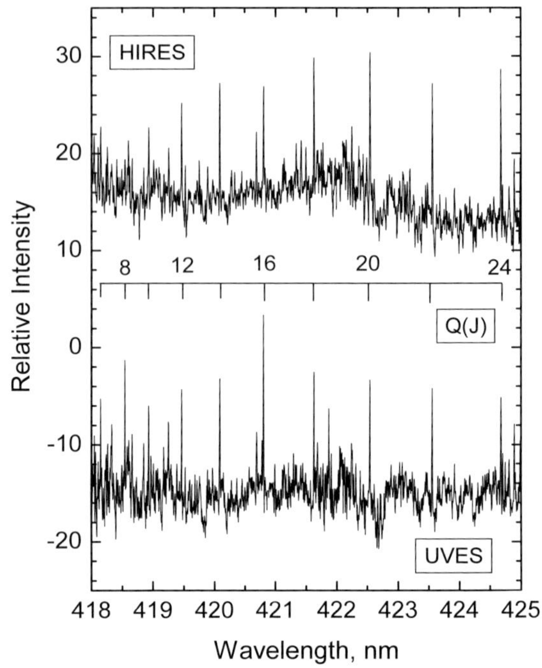

Fig. 6 The 9-1 band of the $O_2(c\text{-}b)$ system as seen by Keck I/HIRES and UVES/VLT. The rotational lines in the Q-branch are indicated (Slanger et al. 2003)

in the visible region at Venus, in the *c*-*X* bands (Krasnopolsky et al. 1976). In the terrestrial atmosphere, it is advantageous to use the new system to study the behavior of $O_2(c)$, because the *c*-*X* system is less distinct.

Figure 7 shows measurements of the sodium D_2/D_1 ratio in the terrestrial nightglow (Slanger et al. 2005). It had been assumed that on statistical grounds the ratio was fixed at 2.0 (Sipler and Biondi 1978), but the sky spectra show that this is far from the case. The error bars on the points are extremely small, the lines being intense features in these observations. The mechanism causing this variable effect is under study in the laboratory, and may be related to the $[O]/[O_2]$ ratio within the sodium layer.

Figure 8 shows a portion of the UV spectra from the VLT, taken during the period when a strong geomagnetic storm was interacting with the magnetosphere. The Chamberlain 5-3 band is part of the normal nightglow, but the $O^+(^2D\text{-}^4S)$ lines are unique, and virtually never seen except from very high altitudes (Sivjee 1991), even in dayglow and aurora. The implication is that the emission comes from far higher in the atmosphere (600–1000 km) than the usual ionospheric features, because $O^+(^2D)$ has a radiative lifetime of hours and is very rapidly quenched by collisions with O-atoms and N_2 (Strickland et al. 1999; D.L. Huestis, Chap. 2, this issue). It is interesting to note that the O^+ doublet can be seen in the Io torus (Morgan and Pilcher 1982), a demonstration that the requirement for its observation is low density.

Table 2 presents a guide to typical intensities of the major emissions found among the terrestrial planets. It is impossible to be rigorous in such a presentation, because the intensities are very variable, depending on solar cycle, altitude, latitude, season, solar zenith angle, time since the atmosphere was last illuminated (for nightglow observations), and other parameters. The table should be considered only as an indicator of significant emissions, with order of magnitude estimates of intensities coming from a variety of literature sources.

Fig. 7 The $Na(D_2/D_1)$ nightglow intensity ratio as a function of time of year. Data are from Keck I/HIRES and Keck II/ESI, over several years (Slanger et al. 2005)

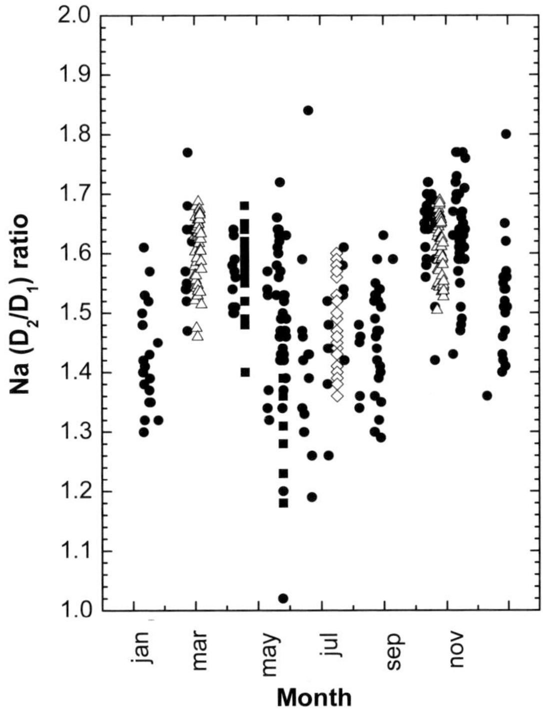

Fig. 8 (**a**) Appearance of the $O^+(^2D\text{-}^4S)$ lines during a geomagnetic storm; (**b**) Simulation of the $O_2(A'\text{-}a)$ Chamberlain 5-3 band, normally the strongest nightglow feature in this region

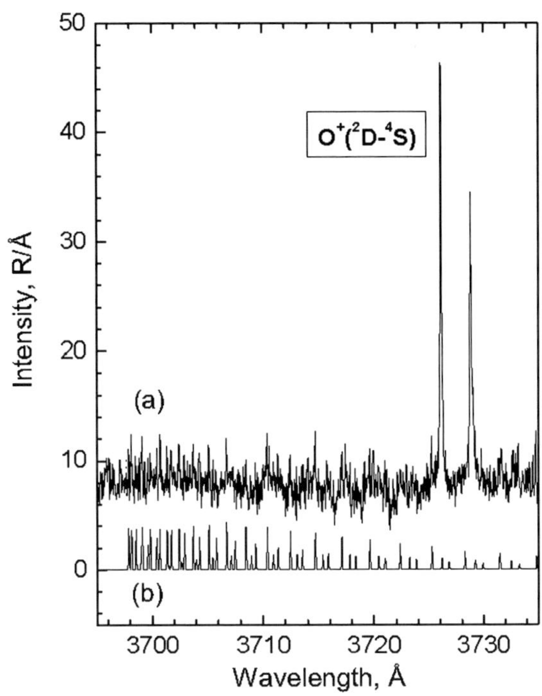

Table 2 Some airglow emissions of the terrestrial planets

Emitter	Planet	Wavelength, nm	Typical Int., night	Typical Int., day
$O(^1D)$	E	630.0/636.4	0–50 R	50 kR
$O(^1S)$	E	557.7/297.2	300/30 R	(4–13)/(0.4/1.3) kR
	V		(0–200)/(0–20) R	800/80 kR (slant)
	M			200/20 kR (slant)
$O^+(^2D)$	E	372.6/372.9	~0	~0
$O^+(^2P)$	E	247.0/(731.9–733)	<1,0–2 R	5 R
$N(^2D)$	E	519.8/520.0	10 R	~60 R
$N(^2P)$	E	346.6	2 R	~40 R
$O_2(a)$	E	0–0, 1270	50 kR	
	M			1–30 MR
	V		0–1 MR	1–2 MR
$O_2(b)$	E	0–0, 762.1	5 kR	100 kR (complex)
$O_2(c)$	E	UV	*c-X* ~50 R	~same
			c-b ~30 R	
	V	Visible	*c-X*, 3–6 kR	~same
$O_2(A')$	E	320–480	150 R	~150 R
$O_2(A)$	E	260–440	600 R	~600 R
$OH(X)$	E	500–4000	500 kR	
$N_2(a)$	E	140–180	–	4 kR
$CO(a)$	M	200–250	–	1 MR (slant)
	V		–	4 MR (slant)
$NO(C,A)$	E	190–260	~100 R	4 kR
	M		85 R	–
	V		3 kR	–

Similarly, Table 3 represents an attempt to determine the radiating efficiency of the various metastable species. Again, this is not well defined because a metastable particle emits over a range of altitudes (densities and temperatures), and thus the efficiency is higher at higher altitudes. The important thing is to distinguish between atoms and molecules with high efficiencies (>0.3) and those with low efficiency (<0.1), the latter being shown in the table in bold. We also now know that it is not enough to consider a molecule in a particular electronic state. There are often large differences in radiating efficiency for different vibrational levels. For example, in studies of relaxation of the $O_2(b^1\Sigma_g^+, v = 1\text{–}3)$ levels at 200 K (Slanger and Copeland 2003), it is found that removal by O_2 becomes much slower with increasing v, the difference between $v = 1$ and $v = 3$ being more than two orders of magnitude. There are similar effects for the $N_2(A^3\Sigma_u^+)$ state (Dreyer et al. 1974).

3.3 Space-based Measurements

Measurements from space supply the ingredients missing from ground-based measurements—the altitude of the emission and the UV emission region masked at the ground by ozone. This comes at the expense of having a moving field of view for satellite observations, and a relatively short observation time for rocket flights. Also, the complexity and weight of ground-based instrumentation may preclude missions to space.

Table 3 Determination of the radiating fraction (F) for typical emissions

Emitter	Planet	Alt. (km)	Radiat. life. (s)	Primary Quench.	k(300 K) $cm^3\ s^{-1}$	$[M]$ cm^{-3}	$\tau\Sigma k[M]$	F
$O(^1D)$	E	250	110	$O(^3P)$	2E–11	1E9	2.2	0.31
$O(^1S)$	E	95	0.8	O_2	6E–14	6E12	0.32	0.76
	V	115	0.8	CO_2	5E–15	1E14	0.4	0.71
	M	90	0.8	CO_2	5E–15	5E13	0.2	0.83
$O^+(^2D)$	E	200	$\sim 10^4$	O/N_2	1E–10	3E9	3E5	**3E–6**
$O^+(^2P)$	E	200	5	N_2	2E–10	3E9	3	0.25
$N(^2D)$	E	200	$\sim 10^5$	O_2	7E–12	2E8	140	**0.007**
$N(^2P)$	E	200	15	O	2E–11	4E9	1.2	0.45
$O_2(a, v = 0)$	E	95	4000	O_2	2E–18	6E12	0.05	0.95
	M	90		CO?				~ 1
	V	115		CO?		1E14		~ 1
$O_2(b, v = 0)$	E	95	12	N_2	2E–15	3E13	0.72	0.58
$O_2(c, v = 0)$	E	95	~ 3	O_2	3E–14	6E12	0.5	0.67
	V	115		CO_2	<6E–14	1E14	<20	>0.05
$O_2(A', \text{high-}v)$	E	95	~ 1	N_2	3E–11	3E13	500	**0.002**
$O_2(A, \text{high-}v)$	E	95	0.6	N_2	3E–11	3E13	540	**0.002**
$OH(X, v = 3)$	E	87	5 ms	O_2	3E–13	3E13	0.045	0.96
$N_2(A, v = 0)$	E	150	~ 1	O_2	4E–12	3E9	0.01	0.99
$CO(a)$	M	120	5 ms	CO_2	2E–11	2E11	0.02	0.98
	V	~ 140	5 ms	CO_2	2E–11	1E11	0.01	~ 1

Nevertheless, the impact of airglow studies in space has been enormous, and there is a long list of successful flight programs in which airglow has been studied (Meier 1991; Broadfoot and Bellaire 1999; Budzien et al. 1994; Eastes et al. 1992; Cleary et al. 1995; Hecht et al. 2000; Minschwaner et al. 2004; Stephan et al. 2004), including Atmospheric Explorer (AE), the Upper Atmospheric Research Satellite (UARS), MSX, ARGOS, and the TIMED satellite. A prime example of long-term satellite monitoring of the atmosphere is the WINDII experiment on UARS, in which a Michelson interferometer has been used to look at a variety of dayglow and nightglow emissions and to study winds and temperatures over the 80–300 km range. From such data, it has been shown that the $O(^1S)$ dayglow is double-peaked, with maxima near 100 and 150 km (Zhang and Shepherd 2005).

3.4 Early Observations of Venus and Mars

The differentiation of the Venus/Mars airglows from the terrestrial airglow is most in evidence from the CO and CO_2^+ emissions. These are seen in the dayglow originating from processes (4) and (5) in Table 1. The Mariner spacecraft showed that the 190–400 nm region is dominated by emission from the CO(a-X) Cameron bands, the $O(^1S\text{-}^3P)$ line at 297.2 nm, and the CO_2^+ A-X and B-X systems (Barth et al. 1971). Pioneer Venus made dayglow measurements in the same spectral region (LeCompte et al. 1989). The Mars Express SPICAM spectrometer is providing similar information, although the CO_2^+ A-X bands are not recorded because of the 300 nm cut-off of the instrument (Leblanc et al. 2006a). The SPICAV instrument on Venus Express can be expected to show the same spectra, with higher intensity.

It is unfortunate that most planetary probes pay little attention to the visible spectral region, given that it can be a rich source of information. Discovery of the oxygen green line at Venus in a ground-based nightglow observation (Slanger et al. 2001) is an example where planetary orbiters could have contributed to an improved understanding of the atmosphere at a much earlier phase of exploration. There is evidence that process (4) in Table 1 generates not only the $CO(a^3\Pi)$ state, the source of the Cameron bands, but also higher CO states that cascade down to the $CO(a)$ state in fully-allowed transitions (Judge and Lee 1973; Lee and Judge 1973). Emission from these higher CO states ($a'^3\Sigma^+$, $d^3\Delta$, $e^3\Sigma^-$) is to be found in the visible spectral region, and might cumulatively be as intense as the Cameron band emission, but there is currently no spacecraft with the capability of making the measurements.

There have been no reports of the oxygen red line in the airglows of either Venus or Mars. Again, this is because no observations have been made. In the Venus nightglow, there was no red line observed from the ground-based measurements in which the green line was detected (Slanger et al. 2001), but the green line is believed to be mesospheric and due to O-atom recombination—the red line would not be seen from such a source. However, lack of detection implies that a nighttime ionospheric source, cf. dissociative recombination of O_2^+, is weak (Fox 1990). It would of course be extremely valuable to be able to measure the dayglow red line, because it would be a measure of reactions 3–6 in Table 1.

3.5 Mars/Venus Express Studies

Mars Express has been in orbit since December 2003, while Venus Express started transmitting scientific data in June 2006. For the purpose of investigating the range of dayglow and nightglow emissions, the relevant instruments are the SPICAM spectrograph on Mars Express and the SPICAV spectrograph on Venus Express. These have essentially the same design, and in both cases the spectral range covered is ~110–300 nm. The 300–400 nm region in the Mars dayglow was observed by the Mariner spacecraft, and contains principally the $CO_2^+(A\text{-}X)$ system, but the visible spectral region remains unexplored by *in situ* observations since the days of Venera 9/10 (Krasnopolsky et al. 1976), and continues to be to a large extent the province of ground-based measurements.

Venus Express data are not yet available, but there are now many papers published from the Mars Express observations (Leblanc et al. 2006a, 2006b, 2007; Bertaux et al. 2005a; Fedorova et al. 2006), and we can expect substantial similarities when the Venus data appear. As Mariner showed, the most characteristic emissions in the dayglow of the CO_2 atmosphere of Mars are the $CO(a\text{-}X)$ Cameron band system and the $CO_2^+(B\text{-}X)$ 0-0 band. Figure 9 shows the appearance of the 180–300 nm region in the dayglow (Leblanc et al. 2006a), and in addition to these two features, one sees the trans-auroral line of OI at 297.2 nm. Recent work has established that the oxygen green line at 557.7 nm, arising from the same $O(^1S)$ upper level, has ten times the intensity of the trans-auroral line (Slanger et al. 2006a), so measurements of the latter provide estimates for the former, which in turn can be seen from terrestrial ground-based measurements, at least in the nightglow (Slanger et al. 2006b).

One of the unexpected results of the SPICAM measurements is that the Cameron bands can be seen at night, when the orbiter is viewing regions where the crustal magnetic field is known to be high (Leblanc et al. 2006b). Thus, over these sites there are Martian aurorae (Bertaux et al. 2005b). Analysis of the spectra indicates that the auroral intensity is on the order of 15 times weaker than the dayglow intensity of the Cameron bands. The latter peaks near the $a\text{-}X$ 0-0 band at 206 nm, at an intensity of about 700 R/nm (at peak altitude), and the relative intensities of the Cameron band system and the $CO_2^+(B\text{-}X)$ band are similar in the two environments. These aurorae are further discussed in the next section.

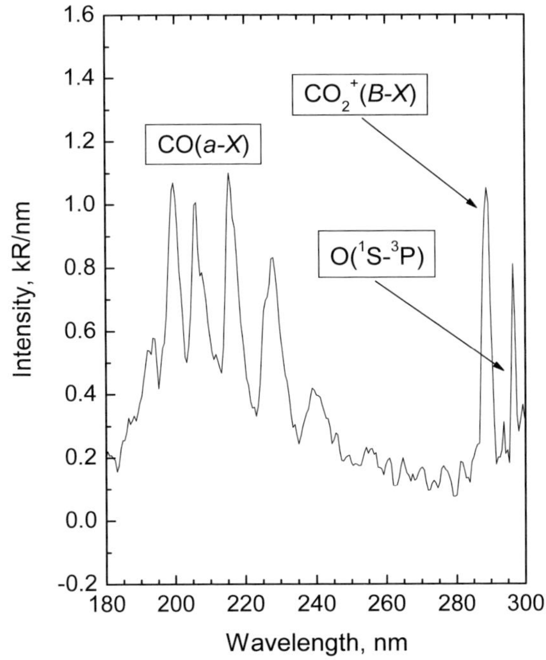

Fig. 9 Features in the 180–300 nm region of Mars as seen by SPICAM on Mars Express. The tangent ray height was 130 km

Because of the low intensity of the aurora these spectra were taken with wide slits and low resolution, resulting in no separation between the CO_2^+ band and the neighboring OI 297.2 nm line. With narrower slits, the dayglow spectra completely resolve these features.

4 Aurora in the Solar System

4.1 Auroral Mechanisms

Auroral emissions from planetary atmospheres are ubiquitous in the Solar System. In their excellent review published in 2002, Galand and Chakrabarti (2002) listed 10 solar system bodies for which aurorae had been detected: these included six of the (then) nine planets, the exceptions being Mercury, Mars and Pluto, three of the four Galilean moons of Jupiter, the exception being Callisto, and comets.

The definition Galand and Chakrabarti used for the term aurora was "any optical manifestation of the interaction of extra-atmospheric energetic electrons, ions and neutrals with an atmosphere". The term "optical" in this definition is misleading, since it may imply "visible to the human eye", although these authors explicitly state that it applies to emissions in wavelength ranges from the far infrared to gamma radiation. In this chapter we use a similar definition to Galand and Chakrabarti, but without the possible misapprehension as to wavelength: aurorae are atmospheric photoemissions caused by the impact of (energetic) primary particles from the near-space environment, either directly or through the agency of the secondary particles produced by the impacts (see Chap. 1, Fox et al.). It should also be noted that—as well as the wavelength ranges identified by Galand and Chakrabarti—radio wave emissions are often associated with aurorae, although they are not usually produced by particle impacts themselves. Radio emission is therefore not reviewed here and readers are referred to reviews such as those by Zarka (1992, 1998) and Zarka and Kurth (2005). According to the definition used here, bodies without (permanent) atmospheres do not exhibit aurorae. So Mercury could not be added to the list, at present, since it is not supposed to have an atmosphere, although our understanding of this innermost planet of the Solar System is likely to be transformed by the Messenger and Bepi-Colombo missions.

The most observed auroral phenomena are those of our own Earth, and terrestrial aurorae span the entire wavelength range, although γ-ray emission is associated only with the most energetic events. Earth has both an atmosphere—a *sine qua non* for aurorae—and a magnetic field. So the particles that produce terrestrial aurorae are mediated by and accelerated to high energies by the planet's extensive magnetosphere. Typically, the aurorae are produced by impacting electrons with energies in the keV to 100 s of keV range, and ions with energies in the MeV range. The literature on Earth's auroral emission is far too extensive to review here (see Carlson and Egeland 1995; Galand and Chakrabarti 2002; Mauk et al. 2002; Paschmann et al. 2003). There are emissions due to the atmospheric species H, N, N_2, N_2^+, O and O_2. Well studied are the visible emissions due to the forbidden lines of atomic oxygen, the green line at 557.7 nm and the red lines at 630.0 and 636.4 nm. These latter lines have a relatively long lifetime of ~120 s, which makes them particularly useful for studying thermospheric winds (e.g. Rees et al. 1990; Davies et al. 1995; Aruliah et al. 1996).

Different auroral emissions not only reflect emission from different species—providing information about atmospheric composition and chemistry—but they also occur at different altitudes. This is because much auroral emission originates in the upper atmosphere—the thermosphere and the accompanying ionosphere—which is a region of the atmosphere where different species settle out according to their own scale height, $H_A = kT/m_A g$, with m_A the mass of atmospheric species A. Thus auroral emission from lighter atomic species—N, O, for example—tends to predominate at the top of the atmosphere (with H at the very top), while molecular emissions—from N_2 and O_2—come from lower altitudes. Since high energy particles are required to reach deep into the atmosphere, emissions from different atmospheric species give information about the energy of the impacting particles. For example, the intensities of the red O lines, which are produced by lower energy electron impact, compared to those of the high energy blue lines of N_2^+ (391.4 nm and 427.8 nm) can yield important information about the relative energy spectrum of particle precipitating at different locations (see Chap. 1, Fox et al.). This is an important point that can be generalised to the auroral emissions of other planets.

In their review of magnetosphere-ionosphere coupling, Mauk et al. (2002) point to the multi-scaled nature of the Earth's aurorae, phenomena that are still poorly understood. This multi-scaled nature refers to both space and time, with features as small as 100 m to global phenomena like the main auroral oval observed by polar orbiting satellites. Again, multi-spatial and multi-temporal scales appear to be universal phenomena for the aurorae of those planets—such as Jupiter and Saturn—that have been extensively studied (e.g. Clarke et al. 2004; Miller et al. 2006).

Earth's aurorae are highly sensitive to conditions in the solar wind, although aurorae are visible on roughly 50% of all nights, whatever the level of solar activity. In the sunward direction, the Earth's magnetosphere extends ~10 Earth radii ($R_\oplus$) to the boundary, and it presents a cone-like obstacle to the solar wind. Thus, the Earth's magnetosphere presents a cross section of ~100 $R_\oplus^2$—over 4 billion square kilometers—to the solar wind. The main zones or ovals are centred roughly at a magnetic co-latitude of ~23° in both hemispheres and are ~10° co-latitude in width. The poleward edge of this locus probably maps the boundary between open and closed magnetic field lines, although Mauk et al. (2002) caution against over-simplistic interpretations of the processes that give rise to the high energy precipitating particles that power the observed emissions.

An overview of the auroral emissions of the giant planets has been provided by Bhardwaj and Gladstone (2000), covering wavelengths from the infrared to the X-ray. Following this, a good review of the similarities and differences between the terrestrial and Jovian aurorae is given by Waite and Lummerzheim (2002). UV emission from Jupiter's auroral/polar regions has been known since the time of Voyager 1 (Broadfoot et al. 1979), and was studied for many years using the International Ultraviolet Explorer (Clarke et al. 1980b; Livengood et al. 1992). With the advent of the Hubble Space Telescope and the availability of high-resolution imagers and spectrometers on ground-based telescopes, various authors have demonstrated that the morphology of Jupiter's auroral emission is extremely complex (Gérard et al. 1994). The main auroral oval rotates with the planet in the longitude frame known as System III, which rotates with the planet's magnetic field. Connerney et al. (1981, 1998) have shown that this is best represented by an offset dipole, tilted at $\sim 10°$ from the rotational axis, plus several additional multi-polar terms. Additionally, the current sheet has an effect, stretching out field lines in the equatorial plane. Since the precipitating particles that are responsible for the aurora follow the field lines into the atmosphere, the main oval is offset with respect to the rotational axis. In the northern hemisphere, this results in the oval being centered on latitude $\sim 80°$, longitude $\sim 180°$; offsets in the southern hemisphere are less pronounced. There are emissions both poleward and equatorward of the main oval.

4.2 Jupiter Aurora

Unlike Earth, the mechanism responsible for producing the main auroral oval on Jupiter arises from the breakdown in co-rotation of the planet's equatorial plasma sheet at distances around 15 to 30 Jovian radii (R_J; $1 R_J \sim 71{,}400$ km). The plasma for this sheet is supplied mainly from Io's volcanic activity: neutral atoms of sulfur and oxygen are ionized and then spun into co-rotation by the action of the planet's magnetic field; plasma instabilities cause a centrifugal migration that extends the sheet outwards in the equatorial plane. Co-rotation breakdown was first described by Hill (1979) at the time of the Voyager encounters, but it was not explicitly recognized as the mechanism producing the main Jovian auroral oval emissions until 20 years later (Cowley and Bunce 2001; Hill 2001). There may also be a role for the solar wind in influencing overall auroral brightness (Baron et al. 1996; Cowley and Bunce 2001; Southwood and Kivelson 2001). Whether this is a positive correlation as Baron et al. (1996) propose, or an anti-correlation as Cowley and Bunce (2001) and Southwood and Kivelson (2001) suggest, has yet to be determined by observations. In addition to its main oval, Jupiter also has bright and very variable poleward auroral emission, some of which maps to the open-closed field line boundary (Pallier and Prangé 2004). The plasma flows that result from these processes are described in this issue in the chapter by Ma et al. Auroral emission is also observed at the footprints of the Galilean moons, accompanied by trails (Connerney et al. 1993; Clarke et al. 1996; Prangé et al. 1996).

The most studied auroral emission from Jupiter is in the UV, where H Lyman-alpha (121.6 nm) and the H_2 Lyman and Werner bands, spanning the approximate range from ~ 85 to 170 nm, were first observed by the Voyager spacecraft (Broadfoot et al. 1979; Sandel et al. 1979), and then by the Earth-orbiting International Ultraviolet Explorer (Clarke et al. 1980b). Various auroral "hot spots" in the infrared were also soon identified from studies of hydrocarbon emission (Caldwell et al. 1989; Kostiuk et al. 1983; Drossart et al. 1986). H_2 emission in the infrared, around 2 microns, was also detected in the 1980s (Trafton et al. 1989a, 1989b; Drossart et al. 1989), although this has not been seriously exploited as a source of information about the Jovian atmosphere.

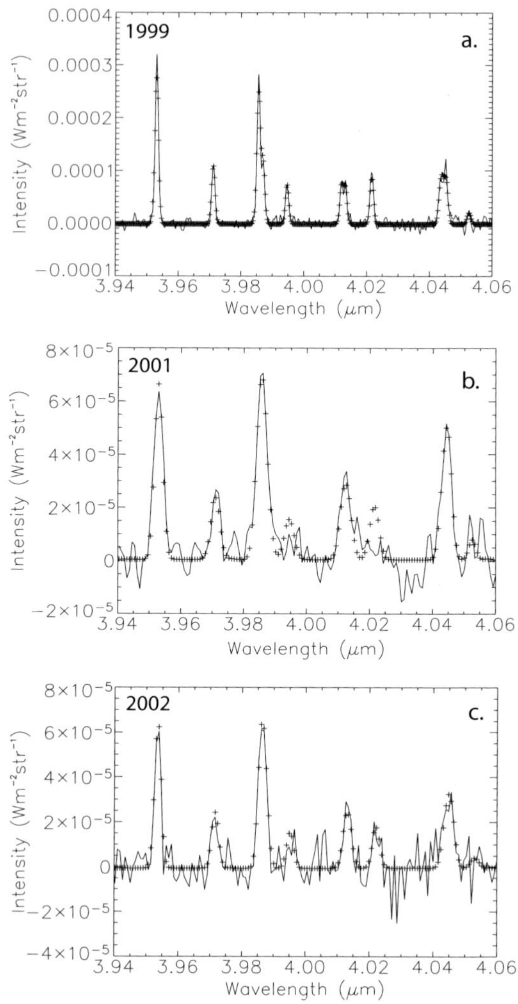

Fig. 10 Series of near infrared spectra of Uranus taken from the observatories on Mauna Kea, Hawaii. The spectrum is almost entirely fitted by lines from the H_3^+ ν_2 manifold, Q branch (Q(1) to Q(5)). (**a**) Spectrum (*line*) and fit (*crosses*) for 15 Sept 1999. (**b**) Spectrum (*line*) and fit (*crosses*) for 16 jun 2001. (**c**) Spectrum (*line*) and fit (*crosses*) for 18 jul 2002. Note that the intensity in the 1999 spectrum is approximately five times that for 2001 and 2002, showing the temporal variability of the planet's H_3^+ emission

4.3 H_3^+ Emission

A major breakthrough in infrared auroral studies of Jupiter came with the detection of emission from the overtone ($v_2 = 2 \rightarrow 0$) band of the H_3^+ molecular ion around 2 microns by Drossart et al. (1989) and—independently—by Trafton et al. (1989a, 1989b), although this latter group did not initially identify the emitter responsible for producing their previously unobserved spectral features. Jupiter H_3^+ auroral emission at 3.42 microns is shown in Fig. 10 (Connerney and Satoh 2000).

The strength of the emission—it was as intense as the $S(1)$ emission line of H_2 at 2.12 μm—was something of a surprise, since the ion was modeled to have a column abundance many orders of magnitude less than H_2. However, H_2 emission is from quadrupole transitions, with Einstein A_{if} coefficients of typically $\sim 10^{-6}$ s^{-1}. Calculations carried out in the 1980s showed that H_3^+ transitions—due to the dipole moment induced by its vibrations—typically had A_{if} values of 10^1–10^2 s^{-1} (Miller and Tennyson 1987; Majewski et al. 1989). H_3^+ emission was then detected from Uranus (Trafton et al. 1993)

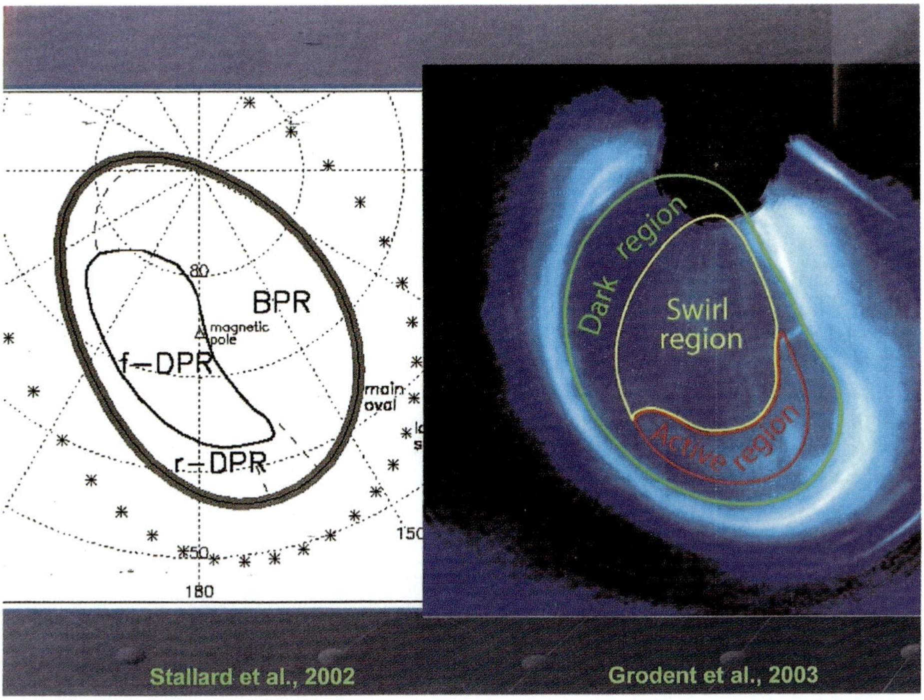

Fig. 11 Comparison of the auroral morphologies of Jupiter's northern polar region observed in the infrared (*left*) and ultraviolet (*right*) spectral regions. The region identified by f-DPR (fixed-Dark Polar Region) by Stallard and the Dark Region by Grodent correspond to the Jovian polar cap. The main oval in both wavelengths corresponds to the magnetic footprint of the region in the equatorial plasma sheet where co-rotation breakdown occurs. Figure courtesy of Denis Grodent

and Saturn (Geballe et al. 1993) in the same year, making use of transitions from the fundamental ($v_2 = 1 \rightarrow 0$) band around 3–4 μm (see Fig. 10). This band, in particular, has been used to probe conditions in planetary upper atmospheres ever since (see reviews by Miller et al. 2000; 2006). Hot band ($v_2 = 2 \rightarrow 1$) (Stallard et al. 2002) and ($v_2 = 3 \rightarrow 2$) (Raynaud et al. 2004) lines have also been detected. H_3^+ is produced in the auroral regions by the electron impact ionization of H_2, followed by the rapid reaction:

$$H_2^+ + H_2 \rightarrow H_3^+ + H \tag{1}$$

and thus is a key tracer of energy inputs into the upper atmosphere (see Connerney and Satoh 2000).

Prangé (1991) first proposed that the UV and the hydrocarbon IR aurorae on Jupiter should have a common origin. Clarke et al. (2004) show that comparison of images obtained in the UV and H_3^+ IR shows much morphological similarity (see Fig. 11). However, these authors also point out that H_3^+ emission originates mainly in the 100—1 nanobar pressure region, ~100–400 km above the Jovian homopause (around 2 microbars), while the UV emission comes largely from below the homopause. Thus the UV emission is more sensitive to higher energy electrons—≥100 keV—that penetrate deeper into the atmosphere, while the H_3^+ emission is produced by electrons with individual energies of some tens of keV. This may account for the differences in morphology noted by Stallard et al. (2001), who observed that the darker polar regions associated with the polar cap were comparatively brighter in H_3^+ than in the UV. One further point to note is that the UV emission is prompt, while H_3^+ emission is "quasi-thermalised" (Miller et al. 1990), which means it can be used to probe ion densities and atmospheric temperatures. Recently, it has proved possible to derive the altitude distribution of the auroral H_3^+ emission (Lystrup et al. 2008). In both the UV and H_3^+ total emission from the Jovian auroral/polar regions is between 10^{12} and 10^{13} Watts (1–10 TeraWatts, TW). For comparison, this figure is greater than the $\sim$1 TW total solar EUV emission absorbed by the upper atmosphere planet-wide. Jupiter's auroral output is thus about 100 times greater than Earth's. Above the homopause, H_3^+ cooling can offset the energy input due to auroral particle precipitation (Miller et al. 1994), but not that resulting from major heating events (Melin et al. 2006).

Auroral emission in the visible, from H_α, was observed by the Galileo spacecraft, tracing a narrow oval on the nightside of the planet (Ingersoll et al. 1998) that seems to coincide with the UV/IR oval. X-ray emission has also been detected from the Jovian poles (Waite et al. 1997; Bhardwaj et al. 2007a, 2007b), which appears to be poleward of the main oval. This has been linked with the Jovian cusp region (Pallier and Prangé 2004) and dayside pulsed reconnection events (Bunce et al. 2004). Overall, the emission intensities in the visible and X-ray regions are several orders of magnitude lower than for the UV and H_3^+ (see Bhardwaj and Gladstone 2000). Radio emission in wavelengths from decametric to kilometric is also associated with the acceleration of particles responsible for auroral emission, although this falls outside our definition of auroral emission *per se*.

4.4 Saturn Aurora

Like Jupiter, Saturn also has auroral emission in the UV (Broadfoot et al. 1981; Sandel et al. 1982) and IR (Geballe et al. 1993), but with much lower emission levels; globally, UV emission from the planet is $\sim 10^{10}$ W (10 GigaWatts, GW), but is sometimes as low as 10^8 W (Trauger et al. 1998). Similar variability in the H_3^+ emission has been noted, with values from $\sim$4 GW to over ten times that amount (Melin et al. 2007). According to Bhardwaj et al. (2007b), there is no convincing evidence of Saturnian auroral X-ray emission. Saturn's magnetic axis and its rotational axis nearly coincide, which makes the auroral emission more symmetric around the poles. The morphology of Saturn's aurora is represented—on average—by an oval centered on the rotational pole, usually at co-latitudes of 10° to 15° (Trauger et al. 1998; Cowley et al. 2004). Brightenings in Saturn's overall auroral emission, on timescales of a number of Saturnian revolutions, have been linked to enhancements of the solar wind magnetic field and plasma density (Crary et al. 2005). The morphology of Saturn's auroral emission shows similarities and differences with both Earth and Jupiter (Clarke et al. 2005). Emission morphology is now known to be as variable as its intensity, with brightenings on both dawn (Grodent et al. 2005) and dusk (Stallard et al. 2008) sectors, as well as features that partially or fully co-rotate with the planet (Stallard et al. 2004, 2007; Cowley et al. 2005; Grodent et al. 2005). Compression of Saturn's magnetosphere by the solar wind has been linked to poleward shifts of the auroral oval, on the dawn side, followed by the development of a spiral structure of auroral arcs (Cowley et al. 2005).

There is considerable debate still over the mechanism causing the main auroral oval emission: Cowley et al. (2004) originally proposed that the main oval corresponds to the footprint between open and closed field lines, making it more "Earth-like" than "Jupiter-like"; they have followed this interpretation in their subsequent analyses of HST images. On the other hand, Hill (2005) has argued that the oval corresponds to the magnetic mapping of co-rotation breakdown in the Saturnian equatorial plasma sheet. Sittler et al. (2006) suggest that the oval maps to the outer boundary of the plasma sheet, where plasma instabilities cause enhanced particle precipitation. Evidence from the radio emissions of Saturn (Kurth et al. 2005) and from H_3^+ spectra (Stallard et al. 2008) lend weight to the Cowley et al. (2004) proposal. Making use of the framework provided by their model, Cowley and coworkers have identified emission from the Saturnian cusp (Bunce et al. 2005).

4.5 Aurora at Uranus and Neptune

The aurorae of the ice giant planets, Uranus and Neptune, are less studied than the gas giants. The magnetic field structure around both planets is somewhat complicated, especially for Uranus (Ness et al. 1991). For a start, the planet rotates with its axis nearly in the plane of the ecliptic, rather than perpendicular to it, as is the case for the other solar system planets. The planet's own magnetic field may be approximated to a dipole that is tilted at 58.6° to the rotational axis, greater than for any other planet, and then offset from the planetary centre by about 0.3 planetary radii. In Connerney et al. (1987) Q3 model, there are also large quadrupole and octopole magnetic field terms. This puts the north magnetic pole at about +14° latitude, and the south at ∼−45°. Voyager 2 detected UV auroral emission from H and H_2 around both magnetic poles (Broadfoot et al. 1986; Herbert and Sandel 1994), but attempts to identify UV *auroral* emission, as distinct from a planet-wide glow, using HST, have proved unsuccessful. In the infrared, Uranus has emission from both H_2 and H_3^+ (see Fig. 11). Both S-branch lines and Q-branch lines from the $v = 1 \rightarrow 0$ H_2 band have been characterized, along with H_3^+ emission in both the fundamental and overtone bands (Trafton et al. 1999). Attempts to locate infrared aurorae unambiguously have not proved successful so far, and it is generally thought that the auroral emission of Uranus is probably not more than ∼20% of the total infrared emission planet-wide (Lam et al. 1997). Generally, attempts to identify the uranian auroral regions post-Voyager have been hindered by the fact that the planet's rotational period—derived from radio emission accompanying the auroral activity when the spacecraft flew by Desch et al. (1986)—is not known sufficiently precisely for observers to know where the magnetic poles are, and thus where to concentrate their search.

Although Neptune's rotational axis is only inclined at 28° to the perpendicular of the plane of the ecliptic, its magnetospheric configuration is also complicated by the large tilt to the dipole component of its magnetic field −47°. Voyager 2 detected weak UV auroral emission (with an intensity of ∼30 microWatts m^{-2}) on the nightside of the planet (Broadfoot et al. 1989). Attempts to find infrared aurorae have so far been unsuccessful.

4.6 Aurora at Mars and Venus

Of the non-magnetised planets, Venus was the first for which faint, possibly auroral emissions were observed: the UV spectrometer onboard the Pioneer Venus Orbiter picked up the signal of atomic oxygen at 130.4 nm on the nightside of the planet (Phillips et al. 1986). Supra-thermal and "soft" electrons have been proposed as the excitation source for these emissions (Fox and Stewart 1991), whose intensity was between a few and 10 μW m^{-2}. The draping of solar wind magnetic field lines passing the obstacle of Venus is thought to be the mechanism by which those electrons are channeled into the Venusian atmosphere. More recently, enhanced emission in the Cameron bands of CO (185–240 nm) has been detected by the Mars Express orbiter (Bertaux et al. 2005a, 2005b) and linked to areas of strong crustal field magnetism in the southern hemisphere around 178° east longitude and −50° latitude. Figure 12 shows that the Cameron band emission from these regions and the dayglow are indistinguishable, except that the dayglow is fifteen times more intense. Leblanc et al. (2006b)

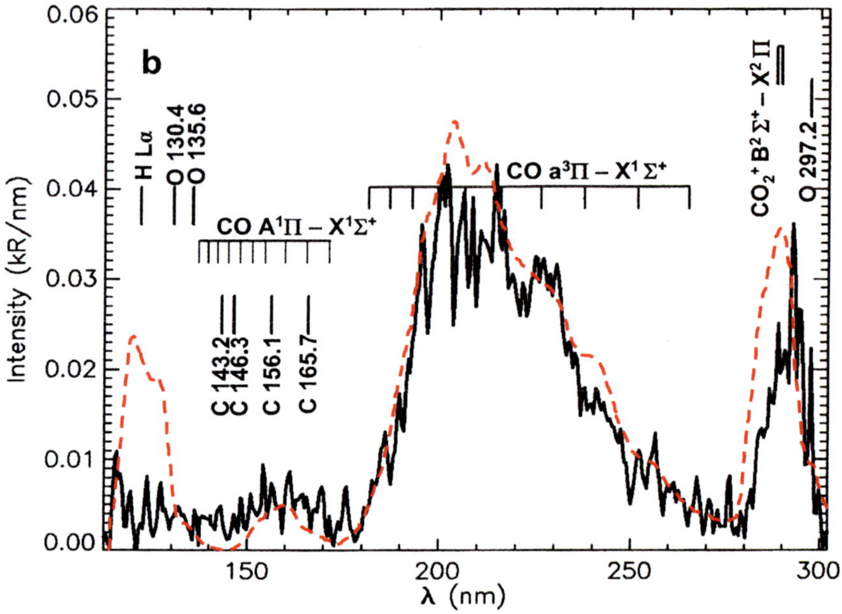

Fig. 12 Comparison of the Mars ultraviolet aurora and the dayglow, adapted from Leblanc et al. (2006b). The dayglow is shown as a *dotted line*, and has 15 times the intensity of the aurora. The nightglow has been subtracted from the auroral spectrum, which is seen over magnetic hotspots of the planet. The dayglow is seen at higher resolution in Fig. 9

have proposed that the observed emission could be produced by precipitating electrons with individual energies peaking around 10 eV.

The existence of aurorae on these non-magnetized bodies, as well as the previously known emission from Io, Europa, and Ganymede, has raised the question as to whether Titan might have auroral features. So far none have been observed. However, nightside electron density observations and electron flux measurements in the magnetosphere of Saturn near Titan indicate that electrons in the range of 10 eV to several keV must be present at Titan, which will lead to soft auroral emissions (Cravens et al. 2008).

One final area for speculation is whether auroral emission might be observed from some of the closely orbiting giant extrasolar planets, such as HD 209458b and HD 189733b, for which there is now some information about their atmospheric composition (Vidal-Madjar et al. 2003, 2004; Ballester et al. 2007; Tinetti et al. 2007). Searches so far have been negative (Shkolnik et al. 2006), but search conditions have not been ideal, and it is just possible that the largest telescopes with the most sensitive instruments may pick up telltale auroral emission.

5 Atomic and Molecular Emissions of Comets

5.1 Atomic and Molecular Emission Processes in Cometary Atmospheres

Comets have two components: their nucleus, which is their hard, permanent core, and their atmosphere and tails, which are the transient manifestation of their activity. The motor of cometary activity is the Sun which heats the nucleus, sublimating its ices, and which drives the cometary tails, by radiation pressure and interaction with the solar wind. Two basic phenomena occur in cometary atmospheres, also driven by the Sun: photolysis, which is the prevailing chemical mechanism, and fluorescence, which is the prevailing photoemission mechanism.

We discuss here the photoemission processes in cometary atmospheres. Chemical processes, the chemical composition of comets, the formation and interaction of cometary ionospheres with the solar wind, are discussed in other chapters of this issue.

5.2 Historical Perspective

The first reported spectroscopic observation of a comet was made visually in 1864 by Giovanni Donati in Florence, on a comet (C/1864 N1) just discovered by Wilhelm Tempel in Marseilles (Donati 1864). Donati reported the presence of bright emission bands, which coincided with the absorption bands observed in the spectra of stars and of the Sun. This was the discovery of fluorescence emission by cometary molecules. These bands were identified as the Swan bands (now known to be due to the C_2 radical) four years latter by Huggins (1868), in comet C/1868 L1 (Winnecke). Huggins used a clever device which allowed him to observe simultaneously the comet's spectrum and the spectrum of hydrocarbon vapors undergoing an electric discharge, thus enabling a direct comparison between observations at the telescope and measurements in the laboratory.

5.3 Molecular Fluorescence

The fluorescence of cometary molecules, radicals, ions, and atoms is now a well understood phenomenon which is observed from infrared to UV wavelengths (e.g., Bockelée-Morvan et al. 2005; Feldman et al. 2005), as seen in Figs. 13 and 14. The same phenomenon also occurs in the infrared in the upper atmospheres of planets for the vibrational bands of molecules such as CO, CO_2, CH_4, HCN. However, it was only recognized recently because in planetary atmospheres it is superimposed over the strong absorption bands of the same molecules (Billebaud et al. 1991; Crovisier et al. 2006).

The fluorescence rates, or g-factors, (number of photons emitted per second per molecule) are typically 0.01–0.1 s^{-1} for the electronic bands of radicals in the visible (Table 4, Fig. 13). Molecules from cometary ices (except CO) do not fluoresce in the visible or UV, because excitation of their electronic bands lead to photodissociation rather than fluorescence. However, their fundamental bands of vibration in the infrared show fluorescence with typical rates of 10^{-4} s^{-1} (Table 4, Fig. 14). Since the photodestruction rate of such molecules is typically $\beta = 10^{-6}$ to 10^{-4} s^{-1}, these species can only emit a small number of photons by fluorescence during their lifetime.

5.4 Prompt Emission Following Photodissociation

The production of daughter species, following photodissociation, in electronic, vibrational or rotational states, leads to the emission of radiation. This phenomenon is known as *prompt emission*, *chemiluminescence*, or *chemical pumping*. Although it happens only once during the lifetime of the species, this phenomenon may be competitive with other emission mechanisms such as fluorescence, especially for short-lived species or when one observes the inner coma where these species are created. Compared to fluorescence, prompt emission can account for at most a proportion β/g, β being the photodestruction rate of the parent and g the fluorescence rate.

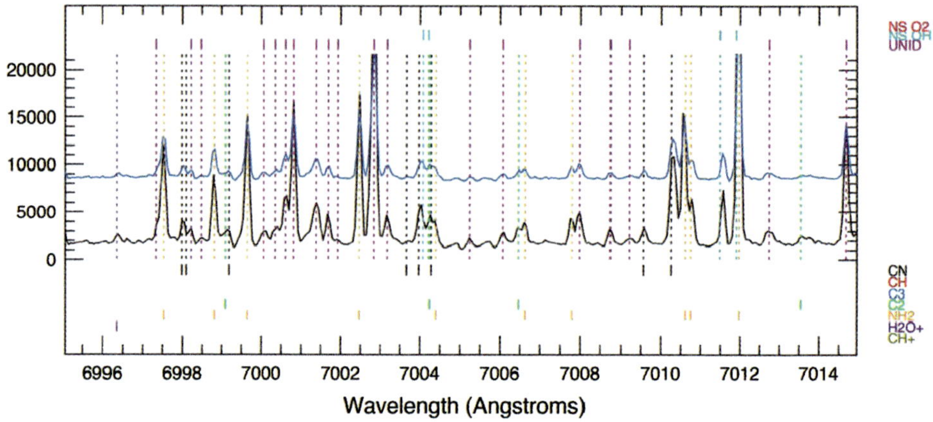

Fig. 13 An excerpt of the high-resolution spectrum atlas of comet 122P/de Vico in the visible. One can see lines of the CN, C_2 and NH_2 radicals, as well as many unidentified lines (Cochran and Cochran 2002)

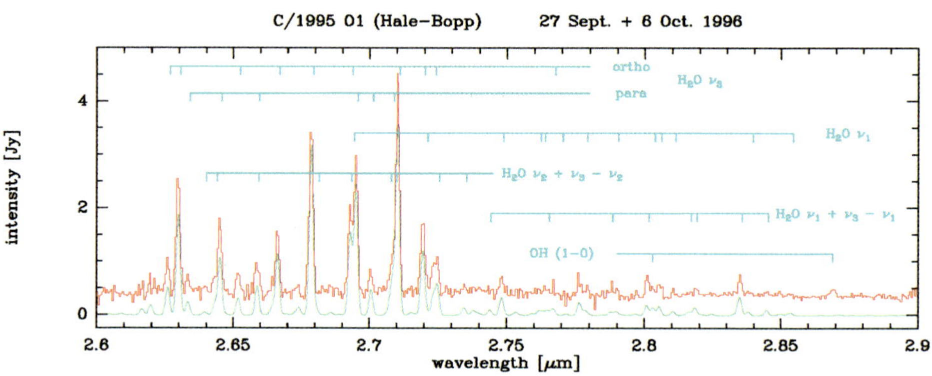

Fig. 14 The region of the υ_3 band of water observed with the *Infrared Space Observatory* in comet C/1995 O1 (*top*). Line assignments are indicated, and a synthetic fluorescence spectrum of water is shown (*bottom*) (Hale-Bopp). Lines from the OH radical are also present (Crovisier et al. 1997)

Table 4 Fluorescence rates for a selection of cometary species at 1 AU from the Sun

Species	Wavelength	Band or line	Rate [s^{-1}]
H	121 nm	Lyman α	1.4×10^{-3}
CO	151 nm	4th Positive	1.2×10^{-6}
OH	315 nm	*A-X*	3.1×10^{-4}
CN	388 nm	*B-X*	9.3×10^{-2}
C_2	516 nm	Swan	3.8×10^{-2}
Na	589 nm	D lines	15
H_2O	2.66 μm	υ_3	2.6×10^{-4}
CH_4	3.31 μm	υ_3	4.0×10^{-4}
CO_2	4.26 μm	υ_3	2.6×10^{-3}
CO	4.67 μm	υ(1–0)	2.6×10^{-4}
H_2O	6.27 μm	υ_2	4.7×10^{-4}
CO_2	15.0 μm	υ_2	8.2×10^{-5}

Such a process has been invoked for emission of OH created from H_2O in the near-UV (Piehler and Kegel 1986; Bertaux 1986; Budzien and Feldman 1991) or in the infrared (Bonev et al. 2004), the forbidden lines (red doublet and green line) of O created from H_2O or from other species (Bierman and Trefftz 1964), and the Cameron bands of CO (near 205 nm) created from CO_2 (Weaver et al. 1994).

5.5 Rotational Lines

The rotational lines of cometary species, which are observed at radio wavelengths, are important for probing the temperature, the kinetics and the chemical composition of cometary atmospheres. Several molecular species are only observed in this spectral domain (e.g., complex molecules such as HCOOH, $HCOOCH_3$, CH_3CHO, NH_2CHO, $(CH_2OH)_2$, Bockelée-Morvan et al. 2000; Crovisier et al. 2004), as seen in Fig. 15. The emission of rotational lines is governed by the balance between radiative excitation, collisions and spontaneous emission. It is thermal in the inner coma, dominated by collisional excitation. Specific modelling is required for the interpretation of these lines, requiring the knowledge of collisional cross-sections. For optically thick strong lines such as those of water, a radiative transfer treatment is necessary.

A special case is that of the Λ-doubling lines of OH at 18 cm wavelength. The cometary OH radicals act as a weak maser, whose inversion is governed by solar UV excitation. Because of coincidence between the exciting OH UV lines and the Fraunhofer lines, the OH maser inversion strongly depends upon the comet heliocentric velocity (Despois et al. 1981).

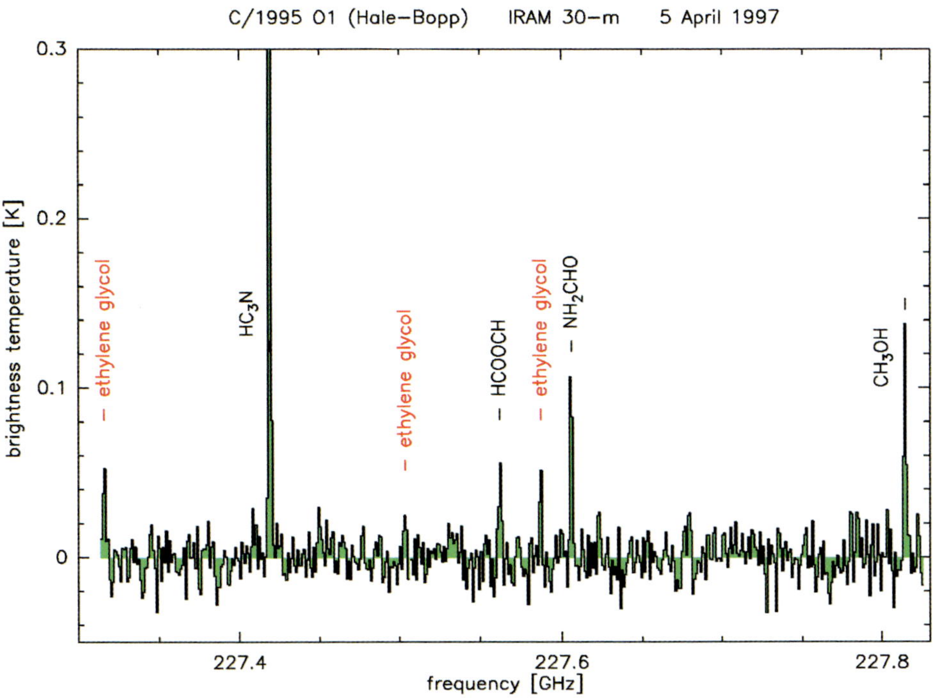

Fig. 15 A radio spectrum of comet C/1995 O1 (Hale-Bopp) observed at the IRAM 30-m radio telescope. Lines of methanol (CH_3OH), methyl formate ($HCOOCH_3$), cyanoacetylene (HC_3N), formamide (NH_2CHO) and ethylene glycol ($(CH_2OH)_2$) are present (Crovisier et al. 2004)

5.6 Anisotropy Issues: Photolysis Self-Shielding; Polarization due to Unidirectional Excitation; Anisotropic Ejection of Radicals

Simple models of cometary atmospheres (e.g., Haser (1957) model which describes the molecular density distribution) simply assume spherical symmetry. Cometary atmospheres, however, are not spherical.

Anisotropy comes first from the very origin of the outgassing, the non-uniform sublimation of nucleus ices. Evidence for localised jets are provided indirectly from the existence of non-gravitational forces (Whipple 1950), from Earth-based imaging, and from *in situ* imaging (Feaga et al. 2007). This is due to the nucleus heterogeneity and irregular shape, and to the unidirectional insolation. Anisotropy is also emphasized by several aeronomical processes occurring in the coma. They are discussed below.

5.7 Photodissociation Self-Shielding

Photodissociation is inhibited when solar radiation is absorbed by an optically thick atmosphere, the main cometary absorber being water. Usually, the effect is negligible, but it may become significant for a highly productive comet, or close to the Sun.

For a Haser parent distribution, along a nucleus-Sun line of sight, the medium becomes optically thick at wavelength λ for a distance from the nucleus, given by:

$$\rho_{\text{thick}} \leq \frac{Q_x \sigma_x(\lambda)}{4\pi\, v_{\text{exp}}} \tag{2}$$

At the wavelength of the Lyman α line, the cross-section for photodissociation of water is $\sigma = 1.5 \times 10^{-21}$ m^2. For a water production rate $Q_{H_2O} = 10^{31}$ s^{-1}, $v_{\text{exp}} = 1$ km s^{-1}, $\rho_{\text{thick}} = 12000$ km. This is to be compared to the unattenuated photodissociation scale length $l_{H_2O} = 80000$ km at $r_h = 1$ AU (case for comet Hale-Bopp near perihelion).

Thus this effect is of little importance for most comets observed from the Earth. It certainly must be considered, however, for in situ studies with focus on the inner coma (Rosetta mission), or for comets coming close to the Sun. In the case of C/2006 P1 (McNaught), for which Q_{H_2O} reached 3×10^{31} s^{-1} at $r_h = 0.25$ AU, $\rho_{\text{thick}} = 3600$ km, to be compared to $l_{H_2O} = 4800$ km.

5.8 Atomic Tails

Active comets show two conspicuous tails: the dust tail resulting from radiation pressure on cometary dust particles, and the ionic tail (mainly composed of H_2O^+ and CO^+ molecular ions) resulting from interaction with the solar wind (see Chaps. by Witasse et al. and Ma et al.). An additional third type of tail, composed of neutral atoms, has been observed in some comets.

Fluorescence is an anisotropic process in comets: the exciting photons come from the direction of the Sun, whereas the reemitted photons are released isotropically. A recoil force results, causing an acceleration opposite to the solar direction:

$$b = \frac{\sum g_i h v_i}{mc} \tag{3}$$

where g_i is the fluorescence rate and v_i is the excitation frequency of the band or line i, and m is the mass of the atom (or molecule). In most cases (and especially for radicals and molecules), this acceleration is negligible. However, it is important for atomic hydrogen, due to Lyman α fluorescence, causing a strong distortion of the hydrogen coma.

It is also important for sodium, due to the large fluorescence rate of its D line. The resulting acceleration is large, causing the sodium atoms to spread over a thin, linear tail. This sodium tail, already noticed in comets C/1910 A1 (Great January Comet) and C/1957 P1 (Mrkos), was *rediscovered* in comet C/1996 O1 (Hale-Bopp) (Cremonese et al. 1997).

In comet C/2006 P1 (McNaught), observations with the STEREO Hemispheric Imager when the comet was only 0.17 AU from the Sun revealed a tail which was inconsistent with the shapes of dust, ionic and sodium tails (Fulle et al. 2007). It was attributed to iron because of its orientation which corresponds to the acceleration expected for fluorescing iron.

The origin (production mechanism) of the cometary sodium and iron atoms is unknown. The sputtering mechanism, which is invoked for the release of sodium from the Moon and Mercury, does not seem to work.

5.9 Fluorescence Polarization

The polarization of sun radiation scattered in the dust coma was first observed by Arago in the Great Comet of 1819, using his visual polarimeter. The fluorescence emission of cometary molecular bands is also polarized, as was first noticed by Öhman (1941) in comet C/1940 R2 (Cunningham).

The theory of the polarization of fluorescence emission was outlined by Feofilov (1959), and the emissions in the visible of several cometary radicals (e.g., CN, C_2, C_3) are observed to be linearly polarized (Le Borgne and Crovisier 1987). Such observations are difficult (molecular polarization has to be separated from dust polarization) and little data are available.

The linear polarization P and intensity I of the molecular bands, as a function of phase angle β, are:

$$P(\beta) = \frac{P_{max}\sin^2(\beta)}{P_{max}\cos^2(\beta)+1} \tag{4}$$

$$I(\beta) = \frac{P_{max}+\cos^2(\beta)}{\frac{1}{2}P_{max}+1} \tag{5}$$

where P_{max} is the maximum polarization, occurring at $\beta = \pi/2$.

P_{max} is not always precisely known. It depends on the rotational distribution, on the presence or absence of hyperfine structure and on possible depolarization by collisions and Larmor precession. The observation of $P(\beta)$ is thus crucial to evaluate polarization theory. The same process is at the origin of a phase effect on the intensity of the emissions. This effect, which could amount to 10% for the red system of CN, or even 28% for the Lyman α line, directly affects the determinations of gas production rates, although ignored in most analyses.

5.10 Anisotropic Ejection of Radicals

Two popular models are used to describe the density distribution of molecules and radicals in cometary atmospheres (e.g., Combi et al. 2005). The Haser model assumes spherical symmetry and a purely radial expansion for both parent and daughter species. The vectorial model assumes spherical symmetry, radial expansion of the parent molecules, and isotropic ejection of the daughter molecules with an additional velocity at random directions.

Here again, the unidirectional nature of the solar radiation is a source of anisotropy. Both theoretical work and laboratory experiments show that in this case, fragments resulting from photodissociation may not be ejected isotropically in the rest frame of the dissociated molecule (Okabe 1978): they show prolate or oblate distributions. Up to now, this aspect of photodissociation has not been considered in cometary models; it could significantly affect the distribution of secondary products. Laboratory experiments have measured this effect for several molecules under specific conditions, but they do not yet allow us to make quantitative evaluations for comets.

6 X-ray Emission in the Solar System

6.1 X-ray Emission Mechanisms

Solar system X-ray astronomy has developed rapidly over the past 10–20 years and has mainly been driven by observations made by orbiting observatories: Einstein Observatory (e.g., Metzger et al. 1983), Roentgen satellite (ROSAT) (e.g., Waite et al. 1994), Chandra X-ray Observatory (CXO) (e.g., Gladstone et al. 2002), XMM-Newton (e.g., Branduardi-Raymont et al. 2004 and Fujimoto et al. 2007). Several reviews of this topic exist (Cravens 2002a, 2002b; Bhardwaj et al. 2007b; Krasnopolsky et al. 2004).

The most powerful X-ray source in the solar system is the solar corona. X-ray emission has also been observed from Venus, Earth, Mars, Saturn, and Jupiter, from the Moon, from a large number of comets, from the Io Plasma Torus, and from interstellar gas in the heliosphere. X-ray emission is typically thought to originate from hot collisional plasmas such as the million-degree gas residing in the solar corona, where X-rays are produced by thermal bremsstrahlung and by excitation of highly stripped ions. However, planetary atmospheres and comets are rather cold and X-ray emission requires external power sources (primarily related to the Sun). For example, X-ray emission from the disks of Venus, Mars, Earth, Moon, and the outer planets results from the scattering and K-shell fluorescence of solar X-rays interacting with atoms and molecules in the atmospheres. For Jupiter and Saturn elastic scattering (i.e., Thomsen scattering) of solar X-ray photons dominates disk emission (Maurellis et al. 2000; Cravens et al. 2006; Bhardwaj et al. 2005a, 2005b). Disk emission for Venus and Mars is thought to be mainly fluorescent X-rays emitted after K-shell electrons have been ejected from carbon and oxygen atoms (in atmospheric CO_2) due to photoionization by solar radiation (Cravens and Maurellis 2001; Dennerl et al. 2002, 2006; Dennerl 2002).

Charged particle collisions also produce X-rays if the energies are sufficiently high (see Chap. 1, Fox et al.). X-rays can be produced during charged particle collisions with atoms or ions if the particles have sufficient energy. Line emission results from excitation of bound states followed by radiative de-excitation. Continuum radiation comes from free-free (i.e., bremsstrahlung radiation) or bound-free transitions (e.g., dielectronic recombination). These collisional processes explain X-ray emission by the solar corona. Auroral X-ray emission at Earth and Jupiter can be produced by energetic electron and ion precipitation from the magnetospheres of these planets. Bremsstrahlung X-ray emission due to energetic auroral electron precipitation has been observed both at Earth (Berger and Seltzer 1972) and at Jupiter (Branduardi-Raymont et al. 2006a, 2007).

X-rays can also be produced as a by-product of the solar wind interaction with neutral gas. A small fraction ($\approx$0.1%) of the solar wind is composed of ions heavier than helium and

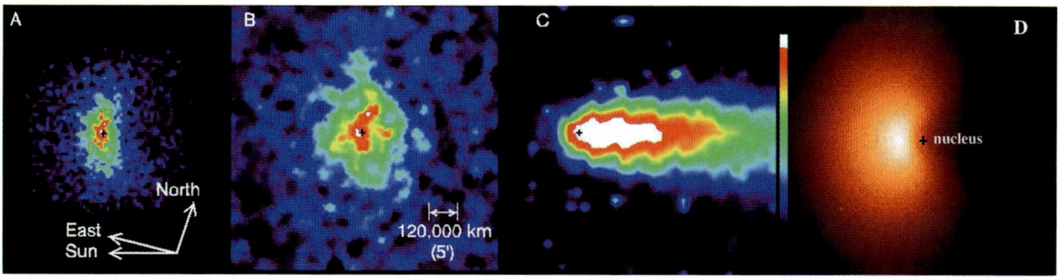

Fig. 16 **A**; Chandra X-ray image of comet c/1999 S4 (LINEAR), **B**: Extreme ultraviolet image of same comet by EUVE satellite, **C**: Optical image of same comet, **D**: X-ray image from MHD simulation of comet Hyakutake (Cravens 2002a, 2002b)

due to their origin in the hot solar corona they are highly-charged (e.g., O^{7+}, C^{6+}, Fe^{12+}) (cf., Bame 1972). At solar wind speeds high charge-state ions readily undergo charge transfer collisions with neutral atoms and molecules (Cravens 1997; Krasnopolsky et al. 2004). Ions produced by such collisions are highly excited and emit extreme ultraviolet (EUV) or soft X-ray photons. This solar wind charge exchange (SWCX) mechanism is thought to account for X-ray emission from comets, the terrestrial geocorona, and from the heliosphere (where the targets are interstellar neutrals entering our solar system). A related mechanism has been suggested for the production of Jovian auroral X-rays (cf., Cravens et al. 2003).

6.2 X-ray Emission from Comets

The cometary nucleus is a several-kilometer size object consisting of ice and dust which generates an extensive atmosphere of water vapor, dust, and water dissociation products as it approaches the Sun. Cold cometary gas was not originally thought to be a promising X-ray source, so that the discovery in 1996 by ROSAT that comet Hyakutake produces about 1 GW of X-ray power was very surprising (Lisse et al. 2006). X-rays have subsequently been detected from almost every active comet observed (Dennerl et al. 1997). Several mechanisms were quickly suggested (cf., Krasnopolsky 1997), one of which was the solar wind charge exchange mechanism described above (Cravens 1997). Solar wind heavy ions can undergo charge transfer reactions with cometary neutrals over a very large volume of space. Many observational, experimental, and theoretical studies of this subject have taken place over the past decade (see reviews by Cravens 2002a, 2002b; Krasnopolsky et al. 2004; Bhardwaj et al. 2007b).

The spatial morphology of cometary X-ray (or EUV) emission is very different from the optical emission, with the peak X-ray intensity located sunward of the nucleus, and the emission extending up to a million km from the nucleus. Figure 16 shows images of cometary X-ray, EUV, and optical emissions, plus a simulated image from a theoretical model (Häberli et al. 1997). The SWCX mechanism predicts that the emission is in the form of many atomic lines from high charge-state ions (e.g., O^{6+} lines at $\approx$560 eV). The definitive proof that SWCX was the dominant cometary X-ray source came when Chandra measured spectra exhibiting high-charge state species line emission (Lisse et al. 2001; Dennerl et al. 2003). Figure 17 shows a cometary X-ray spectrum measured by XMM Newton (Dennerl et al. 2003). Lisse et al. (2007) discuss some recent observational work.

6.3 X-ray Emission from Venus, Mars, Earth, and the Moon

X-ray emission from Venus and Mars due to solar fluorescence and scattering was predicted by Cravens and Maurellis (2001). Chandra observations of Venus (Dennerl et al. 2002) re-

Fig. 17 Spectrum of comet C/2000 WM1 measured by XMM-Newton (Dennerl et al. 2003)

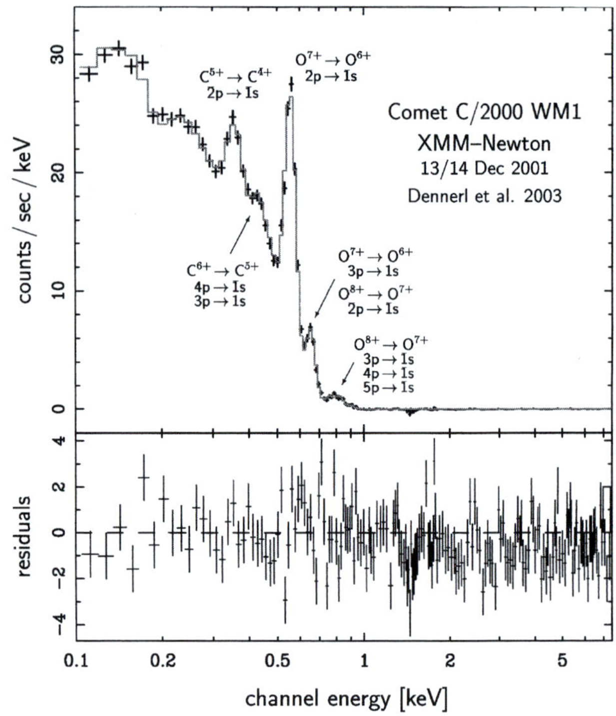

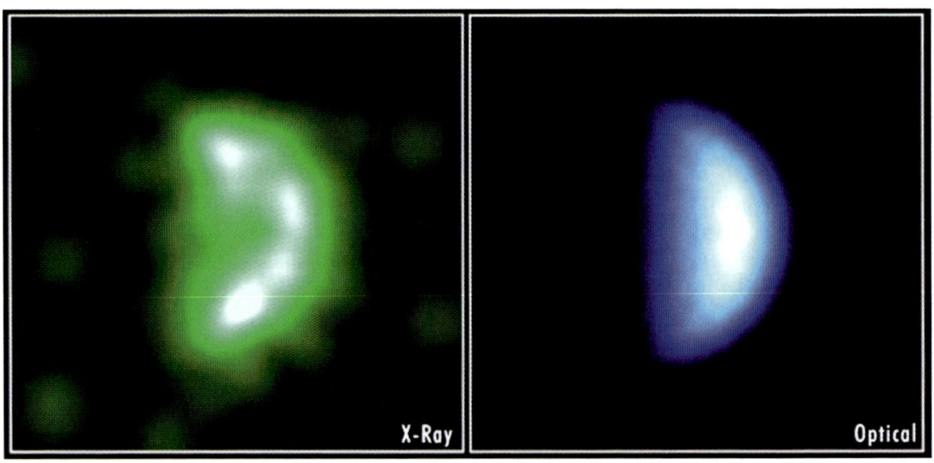

Fig. 18 X-ray image of Venus obtained on January 13, 2001 by the ACIS-I instrument on CXO (Dennerl et al. 2002)

vealed emission with a distinctive disk morphology (Fig. 18) with a spectrum having carbon and oxygen K-shell lines expected from the solar fluorescence mechanism. Observations of Mars (Dennerl 2002) showed the presence of disk emission due to solar fluorescence. But at Mars, a faint, high-noise soft X-ray halo extending several Martian radii was also observed. Previously, Holmstrom et al. (2001) had carried out hybrid simulations of the solar

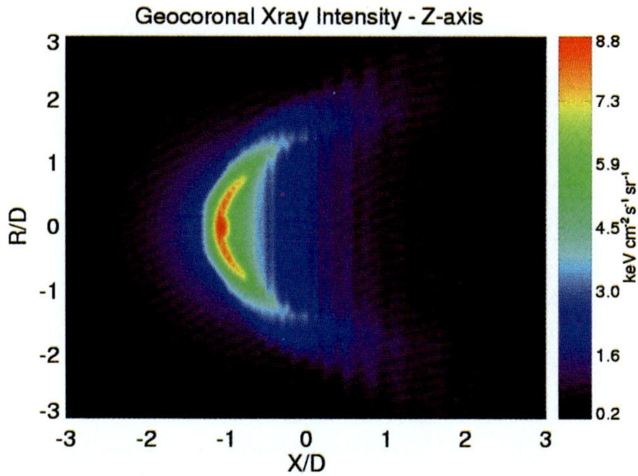

Fig. 19 Simulated soft X-ray image of the geocorona and magnetosheath as viewed from outside the magnetosphere (Robertson and Cravens 2003a, 2003b)

wind interaction with Mars predicting the existence of X-rays due to the solar wind charge exchange mechanism. XMM spectra of Martian soft X-rays have confirmed that the halo is due to the SWCX mechanism (Dennerl et al. 2006).

Auroral X-ray emission has long been seen at Earth and is due to bremmstrahlung associated with the precipitation of energetic electrons in the polar atmosphere (Berger and Seltzer 1972). Soft X-ray emission due to the SWCX mechanism operating in the terrestrial geocorona was also predicted (Cravens 2000a; Cravens et al. 2001; Cox 1998; Robertson and Cravens 2003a, 2003b). Figure 19 shows a model of geocoronal X-ray intensities as it would be seen from outside the magnetosphere (Robertson et al. 2006), the emission arising from outside the magnetopause. Some fraction of the soft X-ray background (SXRB) emission varies with time (Snowden et al. 1995) and it was demonstrated that these variations correlated with the measured solar wind flux (Cravens et al. 2001), confirming that the SWCX mechanism must be responsible. X-rays from the Moon have also been seen (Schmitt et al. 1991) with a higher intensity from the sunlit side but with very measurable emission from the darkside. Dayside emission has been attributed to solar fluorescence from the solid surface. Spectra of the darkside lunar X-ray emission measured by CXO had the characteristic SWCX lines, confirming that geocoronal X-rays are due to the SWCX mechanism (Wargelin et al. 2004) as is true for other observed spectra of geocoronal X-ray emission (Snowden et al. 2004).

6.4 X-ray Emission from the Outer Solar System and the Heliosphere

X-rays from the SWCX mechanism are also produced by the solar wind interaction with interstellar neutrals entering the heliosphere (Cravens 2000; Robertson and Cravens 2003b; Cox 1998). Variations in the solar wind flux result in large variations in X-ray emission from interstellar helium and from geocoronal H, but X-rays produced from interstellar H are rather steady. The observed soft X-ray background (SXRB) contains a mixture of this steady heliospheric emission as well as emission from the hot interstellar medium and the galactic halo. Astrophysicists are attempting to remove the heliospheric component of the SXRB (Cox 1998; Robertson and Cravens 2003b; Lallement 2004; Koutroumpa et al. 2007; Snowden et al. 1995). X-ray emission from other astrospheres due to stellar winds interacting with interstellar gas has been modeled but not yet observed (Wargelin et al. 2005; Medvedev et al. 2006).

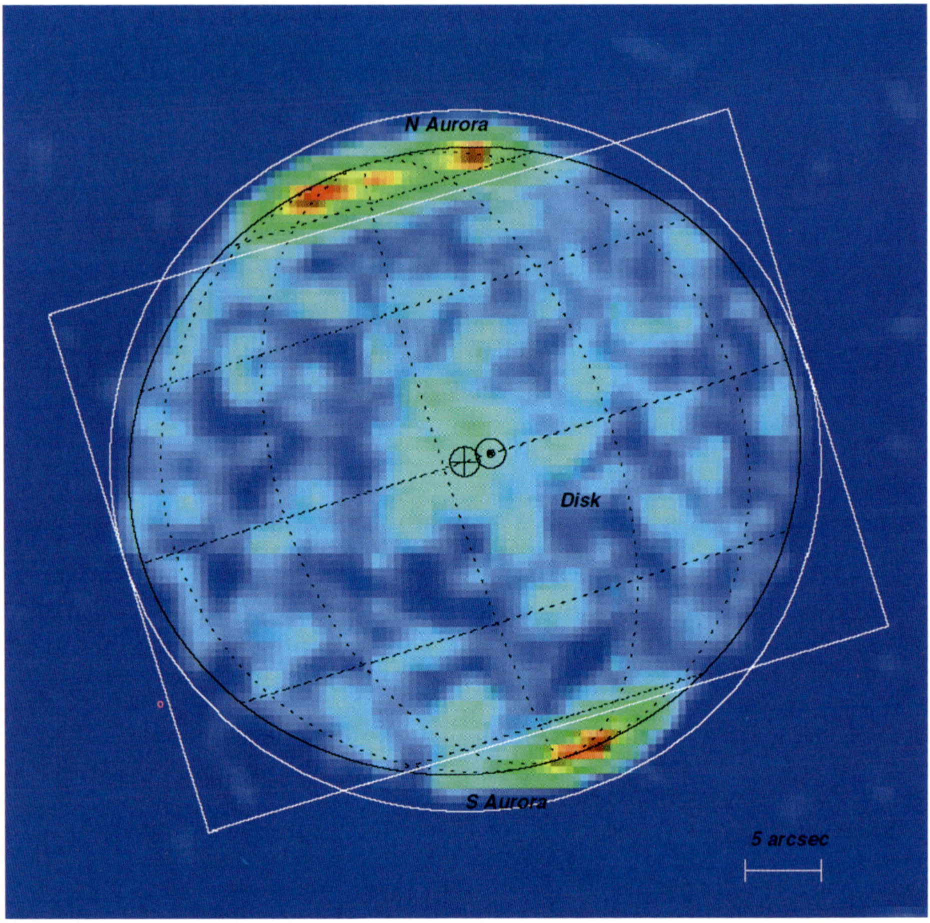

Fig. 20 Soft X-ray image of Jupiter measured by the Chandra X-Ray Observatory (Elsner et al. 2005)

Jupiter has long been known to be a source of X-rays (Metzger et al. 1983; cf., Bhardwaj et al. 2007b). ROSAT images showed that the X-rays originated from both low latitudes (i.e., disk emission) and high latitudes (i.e., auroral), but the spatial resolution was too low to accurately define the spatial morphology of the emission. Both bremsstrahlung associated with electron precipitation (cf. Barbosa 1990; Waite 1991) and energetic ion precipitation (cf., Cravens et al. 1995) were invoked to explain the auroral X-ray emission and even the low-latitude X-ray emission (Waite et al. 1997), but the ROSAT spectral resolution was too low to settle this issue.

High spatial resolution Chandra images of the Jovian soft X-ray emission revealed both uniform disk emission and high-latitude auroral emission (see Fig. 20; Elsner et al. 2005). Spectral measurements of the disk emission from Jupiter and Saturn by CXO and XMM-Newton (e.g., Cravens et al. 2006; Branduardi-Raymont et al. 2006a, 2006b; see review by Bhardwaj et al. 2007b) strongly support the identification of disk emission as being due to scattered solar X-rays. The observed temporal correlation of Saturnian X-ray intensity with a solar flare provided a dramatic confirmation of the scattering hypothesis (Bhardwaj et al. 2005b).

Production of Jovian X-rays by precipitating energetic sulfur and oxygen atoms was originally suggested by Metzger et al. (1983) and the detailed atomic mechanism was developed by Cravens et al. (1995), Kharchenko et al. (1998), and Liu and Schultz (2000). The incident ions, originating in the magnetosphere as S^+, S^{++}, S^{+++}, O^+, and O^{++} are required to have energies of at least several MeV in order for collisions with atmospheric gas to convert them to more highly-charged ions such as O^{7+}. Charge transfer collisions of these highly-charged ions then lead to X-ray production (e.g., $O^{7+} + H_2 \rightarrow O^{6+*} + H_2^+$, followed by the emission of a photon by the excited O^{6+*}), just as in the SWCX mechanism.

These theoretical studies and ROSAT-based observational studies (e.g., Waite et al. 1994, 1997) envisioned the energetic ion populations being associated with the middle magnetosphere, but higher-resolution CXO images (Gladstone et al. 2002) clearly demonstrated that the X-ray emission was located well poleward of the main UV auroral oval. The X-ray intensity also showed a 40-minute periodicity. Cravens et al. (2003) suggested two possible sources of these polar cap X-rays: (1) solar wind ions entering the magnetospheric cusp and/or (2) acceleration of outer magnetospheric sulfur and oxygen ions to several MeV energies by field-aligned potentials. Bunce et al. (2004) then suggested that reconnection processes near the dayside magnetopause could be the ion source. Recently measured spectra of the Jovian X-ray aurora contain O^{7+} and O^{6+} lines, as expected, but the evidence for sulfur lines is not as clear (Elsner et al. 2005; Branduardi-Raymont et al. 2004, 2006a; Bhardwaj et al. 2007b). X-ray emission from the main auroral oval with photon energies in excess of 1 keV (the ion precipitation produces softer X-rays) has just been detected with XMM-Newton, albeit at relatively low power, and has been attributed to bremsstrahlung associated with electron precipitation (Branduardi-Raymont et al. 2007).

7 Summary

There is a broad range of interrelated phenomena associated with observations of the planetary atmospheres. The optical emissions for Earth/Venus/Mars are linked by the fact that O-atoms are a product of photodissociation at all three planets. Cometary and planetary emissions have common features because some of the same atoms and molecules are present. The gas giants and the inner planets have very different atmospheres, but again the satellites of the outer planets have inner planet features because N_2 and sometimes O_2 are present. It seemed that aurora were only associated with planets with a large intrinsic magnetic field, but the observations from Mars Express have shown that Nature often has ways of circumventing our classifications. X-ray emissions at planets and comets were also very surprising, but as more sophisticated measurements are made, this seems to be a general phenomenon.

Observations of planetary airglow and auroral emissions are able to remotely provide diagnostic information on upper atmospheric composition, energy deposition processes, and dynamics that can often not be obtained in other ways and in this way have historically made important contributions to our understanding of planetary aeronomy. In our increasingly detailed search for extra-solar planets, we must keep in mind the full range of phenomena found in our solar system, and expect that solar systems elsewhere will be at least as varied and complex.

References

J.M. Ajello, M.H. Stevens, A.I. Stewart et al., Geophys. Res. Lett. **34**, L24204 (2007). doi:10.1029/2007GL031555

A.J. Ångström, Pogg. Ann. **137**, 161–163 (1869)
A.L. Aruliah, A.D. Farmer, D. Rees, U. Brändström, J. Geophys. Res. **101**, 15701–15712 (1996). doi:10.1029/96JA00360
S. Atreya, Y.L. Yung, T.M. Donahue, E.S. Barker, Astrophys. J. Lett. **218**, L83–87 (1977). doi:10.1086/182581
G.E. Ballester, D.K. Sign, F. Herbert, Nature **445**, 511–514 (2007). doi:10.1038/nature05525
S.J. Bame, in *Solar Wind*, ed. by C.P. Sonnett et al. NASA Publ. SP-308, (1972) p. 529
D.D. Barbosa, J. Geophys. Res. **95**, 14969–14976 (1990). doi:10.1029/JA095iA09p14969
R.L. Baron, T. Owen, J.E.P. Connerney, T. Satoh, J. Harrington, Icarus **120**, 437–442 (1996). doi:10.1006/icar.1996.0063
C.A. Barth, C.W. Hord, J.B. Pearce, K.K. Kelly, G.P. Anderson, A.I. Stewart, J. Geophys. Res. **76**, 2213–2227 (1971). doi:10.1029/JA076i010p02213
L.B. Jaffel, R. Prangé, B.R. Sandel, R.V. Yelle, C. Emerich, D. Feng et al., Icarus **113**, 91–102 (1995). doi:10.1006/icar.1995.1007
L.B. Jaffel, Y.J. Kim, J. Clarke, Icarus **190**, 504–527 (2007). doi:10.1016/j.icarus.2007.03.013
M.J. Berger, S.M. Seltzer, J. Atmos. Terr. Phys. **34**, 85–108 (1972). doi:10.1016/0021-9169(72)90006-2
J.-L. Bertaux, F. Leblanc, S. Perrier et al., Science **307**, 566–569 (2005a). doi:10.1126/science.1106957
J.-L. Bertaux, Astron. Astrophys. **160**, L7–L10 (1986)
J.-L. Bertaux, F. Leblanc, O. Witasse et al., Nature **435**, 790–794 (2005b). doi:10.1038/nature03603
A. Bhardwaj, G.R. Gladstone, Rev. Geophys. **38**, 295–353 (2000). doi:10.1029/1998RG000046
A. Bhardwaj, G. Branduardi-Raymont, R.F. Elsner et al., Geophys. Res. Lett. **32**, L03S08 (2005a). doi:1029/2004GL021497
A. Bhardwaj, R.F. Elsner, J.H. Waite Jr., G.R. Gladstone, T.E. Cravens, P. Ford, Astrophys. J. Lett. **624**, L121–L124 (2005b). doi:10.1086/430521
A. Bhardwaj et al., Planet. Space Sci. **55**, 1126–1134 (2007a). doi:10.1016/j.pss.2006.11.017
A. Bhardwaj et al., Planet. Space Sci. **55**, 1135–1189 (2007b). doi:10.1016/j.pss.2006.11.009
L. Bierman, E. Trefftz, Zeit. Astrophys. **59**, 1–28 (1964)
F. Billebaud, J. Crovisier, E. Lellouch, T. Encrenaz, J.-P. Maillard, Planet. Space Sci. **39**, 213–218 (1991). doi:10.1016/0032-0633(91)90144-Y
D. Bockelée-Morvan, J. Crovisier, M.J. Mumma, H.A. Weaver, in *Comets II*, ed. by M.C. Festou, H.U. Keller, H.A. Weaver (University Arizona Press, Tucson, 2005), pp. 391–423
D. Bockelée-Morvan, D.C. Lis, J.E. Wink et al., Astron. Astrophys. **353**, 1101–1114 (2000)
B.P. Bonev, M.J. Mumma, N. Dello Russo et al., Astrophys. J. **615**, 1048–1053 (2004). doi:10.1086/424587
G. Branduardi-Raymont, R.F. Elsner, G.R. Gladstone et al., Astron. Astrophys. **424**, 331–337 (2004). doi:10.1051/0004-6361:20041149
G. Branduardi-Raymont, A. Bhardwaj, R. Elsner et al., in *Proceedings of the Symposium 'The X-ray Universe'*, El Escorial, Spain, 26–30 September 2005. ESA SP-604 (2006a), pp. 15–20
G. Branduardi-Raymont, A. Bhardwaj, R. Elsner, G. Gladstone, G. Ramsay, P. Rodriguez, R. Soria, J.H. Waite, T.E. Cravens, XMM-Newton observations of X-ray emission from Jupiter. Adv. Geosci. **3**, 203–214 (2006b)
G. Branduardi-Raymont, A. Bhardwaj, R.F. Elsner et al., Astron. Astrophys. **463**, 761 (2007). doi:10.1051/0004-6361:20066406
A.L. Broadfoot et al., Science **204**, 979–982 (1979). doi:10.1126/science.204.4396.979
A.L. Broadfoot et al., Science **212**, 206–211 (1981). doi:10.1126/science.212.4491.206
A.L. Broadfoot et al., Science **233**, 74–79 (1986). doi:10.1126/science.233.4759.74
A.L. Broadfoot et al., Science **246**, 1459–1456 (1989). doi:10.1126/science.246.4936.1459
A.L. Broadfoot, P.J. Bellaire, J. Geophys. Res. **104**, 17127–17138 (1999). doi:10.1029/1999JA900135
S.A. Budzien, P.D. Feldman, R.R. Conway, J. Geophys. Res. **99**, 23275–23287 (1994). doi:10.1029/94JA01543
S.A. Budzien, P.D. Feldman, Icarus **90**, 308–318 (1991). doi:10.1016/0019-1035(91)90109-7
E.J. Bunce, S.W.H. Cowley, T.K. Yeoman, J. Geophys. Res. **109**, A09S13 (2004). doi:10.1029/2003JA010280
E.J. Bunce, S.W.H. Cowley, S.E. Milan, Ann. Geophys. **23**, 1405–1431 (2005)
J. Caldwell, A.T. Tokunaga, F.C. Gillett, Icarus **53**, 133–140 (1989). doi:10.1016/0019-1035(83)90026-X
H.C. Carlson Jr., A. Egeland, in *Introduction to Space Physics*, ed. by M.G. Kivelson, C.T. Russell (Cambridge University Press, Cambridge, 1995), pp. 459–500
R.W. Carlson, D.L. Judge, in *Jupiter*, ed. by T. Gehrels (University of Arizona Press, Tuscon, 1976), pp. 418–440
J.T. Clarke, H.W. Moos, S.K. Atreya, A.L. Lane, Astrophys. J. **241**, L179–L182 (1980a). doi:10.1086/183386
J.T. Clarke et al., Science **274**, 404–409 (1996). doi:10.1126/science.274.5286.404

J.T. Clarke, D. Grodent, S.W.H. Cowley et al., in *Jupiter: The Planet, Satellites and Magnetosphere*, ed. by F. Bagenal, T. Dowling, W. McKinnon (Cambridge University Press, Cambridge, 2004), pp. 639–670
J.T. Clarke et al., Nature **433**, 717–719 (2005). doi:10.1038/nature03331
J.T. Clarke, H.A. Weaver, P.D. Feldman, H.W. Moos, W.G. Fastie, C.B. Opal, Astrophys. J. **240**, 696–701 (1980b). doi:10.1086/158277
J.T. Clarke, S. Durrance, S.K. Atreya et al., J. Geophys. Res. **91**, 8771–8781 (1986). doi:10.1029/JA091iA08p08771
J.T. Clarke, M. Hudson, Y.L. Yung, J. Geophys. Res. **92**, 15,139–15,147 (1987). doi:10.1029/JA092iA13p15139
J.T. Clarke, G.R. Gladstone, L.B. Jaffel, Geophys. Res. Lett. **18**, 1935–1938 (1991). doi:10.1029/91GL02091
D.D. Cleary, S. Gnanalingam, R.P. McCoy, K.F. Dymond, F.G. Eparvier, J. Geophys. Res. **100**, 9729–9739 (1995). doi:10.1029/94JA03145
K.D. Closser, D.A. Pejaković, K.S. Kalogerakis, in *AGU Fall Meeting*, San Francisco, CA, 2005, pp. SA11A-0215
A.L. Cochran, W.D. Cochran, Icarus **157**, 297–308 (2002). doi:10.1006/icar.2002.6850
M.R. Combi, W.M. Harris, W.H. Smyth, in *Comets II* ed. by M.C. Festou, H.U. Keller, H.A. Weaver (University Arizona Press, Tucson, 2005), pp. 523–552
J.E.P. Connerney, M.H. Acuña, N.F. Ness, J. Geophys. Res. **86**, 8370–8384 (1981). doi:10.1029/JA086iA10p08370
J.E.P. Connerney, M.H. Acuña, N.F. Ness, J. Geophys. Res. **92**, 15329–15336 (1987). doi:10.1029/JA092iA13p15329
J.E.P. Connerney, R. Baron, T. Satoh, T. Owen, Science **262**, 1035–1038 (1993). doi:10.1126/science.262.5136.1035
J.E.P. Connerney, M.H. Acuña, N.F. Ness, T. Satoh, J. Geophys. Res. **103**, 11929–11939 (1998). doi:10.1029/97JA03726
J.E.P. Connerney, T. Satoh, Philos. Trans. R. Soc. **358**, 2471–2483 (2000). doi:10.1098/rsta.2000.0661
P. Connes, J.F. Noxon, W.A. Traub, N.P. Carleton, Astrophys. J. **233**, L29–L32 (1979). doi:10.1086/183070
B.J. Conrath, D. Gautier, Icarus **144**, 124–134 (2000). doi:10.1006/icar.1999.6265
P.C. Cosby, B.D. Sharpee, D.L. Huestis, T.G. Slanger, R. Hanuschik, J. Geophys. Res. **111**, A12307 (2006). doi:10.1029/2006JA012023
P.C. Cosby, T.G. Slanger, Can. J. Phys. **85**, 77–99 (2007). doi:10.1139/P06-088
S.W.H. Cowley, E.J. Bunce, Planet. Space Sci. **49**, 1067–1088 (2001). doi:10.1016/S0032-0633(00)00167-7
S.W.H. Cowley, E.J. Bunce, R. Prangé, Ann. Geophys. **22**, 1379–1394 (2004)
S.W.H. Cowley, S.V. Badman, E.J. Bunce et al., J. Geophys. Res. **110**, A02201 (2005). doi:10.1029/2004JA010796
D.P. Cox, in *The Local Bubble and Beyond*, ed. by D. Breitschwerdt, M.J. Freyberg, J. Trümper (Springer, New York, 1998), p. 121
F.J. Crary et al., Nature **433**, 720–722 (2005). doi:10.1038/nature03333
T.E. Cravens, Geophys. Res. Lett. **24**, 105 (1997). doi:10.1029/96GL03780
T.E. Cravens, Astrophys. J. **532**, L153 (2000). doi:10.1086/312574
T.E. Cravens, Adv. Space Res. **26**, 1443–1451 (2000a). doi:10.1016/S0273-1177(00)00100-9
T.E. Cravens, Science **296**, 1042–1046 (2002a). doi:10.1126/science.1070001
T.E. Cravens, in *Atomic Processes in Plasmas: 13th APS Topical Conference on Atomic Processes in Plasmas*, ed. by D.R. Schultz et al. (AIP, Melville, 2002b), p. 173
T.E. Cravens, A.N. Maurellis, Geophys. Res. Lett. **28**, 3043–3046 (2001). doi:10.1029/2001GL013021
T.E. Cravens, E. Howell, J.H. Waite Jr., G.R. Gladstone, J. Geophys. Res. **100**, 17153–17162 (1995). doi:10.1029/95JA00970
T.E. Cravens, I.P. Robertson, S.L. Snowden, J. Geophys. Res. **106**, 24883–24892 (2001). doi:10.1029/2000JA000461
T.E. Cravens, J.H. Waite, T.I. Gombosi, N. Lugaz, G.R. Gladstone, B.H. Mauk et al., J. Geophys. Res. **108**, 1465 (2003). doi:10.1029/2003JA010050
T.E. Cravens, J. Clark, A. Bhardwaj et al., J. Geophys. Res. **111**, A07308 (2006). doi:10.1029/2005JA011413
T.E. Cravens, J.P. Robertson, S.A. Ledvina et al., Geophys. Res. Lett. **35**, L03103 (2008). doi:10.1029/2007GL032451
G. Cremonese, H. Boehnhardt, J. Crovisier et al., Astrophys. J. **490**, L199–L202 (1997). doi:10.1086/311040
D. Crisp, Science **291**, 444–445 (2001). doi:10.1126/science.1057630
D. Crisp, V.S. Meadows, B. Bezard, C.D. Bergh, J.-P. Maillard, F.P. Mills, J. Geophys. Res. **101**, 4577–4593 (1996). doi:10.1029/95JE03136
J. Crovisier, K. Leech, D. Bockelée-Morvan et al., Science **275**, 1904–1907 (1997). doi:10.1126/science.275.5308.1904

J. Crovisier, D. Bockelée-Morvan, N. Biver et al., Astron. Astrophys. **418**, L35–L37 (2004). doi:10.1051/0004-6361:20040116
J. Crovisier, E. Lellouch, C. de Bergh et al., Planet. Space Sci. **54**, 1398–1414 (2006). doi:10.1016/j.pss.2006.04.027
J.J. Davies, A.D. Farmer, A.L. Aruliah, Ann. Geophys. **13**, 541–550 (1995)
K. Dennerl, J. Englhauser, J. Trümper, Science **277**, 1625 (1997). doi:10.1126/science.277.5332.1625
K. Dennerl, V. Burwitz, J. Englhauser, C. Lisse, S. Wolk, Astron. Astrophys. **386**, 319 (2002). doi:10.1051/0004-6361:20020097
K. Dennerl, Astron. Astrophys. **394**, 119–1128 (2002). doi:10.1051/0004-6361:20021116
K. Dennerl, B. Aschenbach, V. Burwitz, J. Englhauser, C.M. Lisse, P.M. Rodriguez-Pascual, Proc. SPIE **4851**, 277–288 (2003). doi:10.1117/12.461137
K. Dennerl et al., Astron. Astrophys. **451**, 709–722 (2006). doi:10.1051/0004-6361:20054253
M.D. Desch, J.E.P. Connerney, M.L. Kaiser, Nature **322**, 42–43 (1986). doi:10.1038/322042a0
D. Despois, E. Gérard, J. Crovisier, I. Kazès, Astron. Astrophys. **99**, 320–340 (1981)
G.B. Donati, Astron. Nachr. **62**, 375–378 (1864)
J.W. Dreyer, D. Perner, C.R. Roy, J. Chem. Phys. **61**, 3164–3169 (1974). doi:10.1063/1.1682472
P. Drossart, B. Bézard, S.K. Atreya et al., Icarus **66**, 610–618 (1986). doi:10.1016/0019-1035(86)90094-1
P. Drossart et al., Nature **340**, 539–541 (1989). doi:10.1038/340539a0
R. Eastes, R.E. Huffman, F.J. LeBlanc, Planet. Space Sci. **40**, 481–493 (1992). doi:10.1016/0032-0633(92)90168-N
R.F. Elsner, N. Lugaz, J.H. Waite Jr., et al. J. Geophys. Res. **110**, A01207 (2005). doi:10.1029/2004JA010717
L.M. Feaga, M.F. A'Hearn, J.M. Sunshine, O. Groussin, T.L. Farnham, Icarus **190**, 345–356 (2007). doi:10.1016/j.icarus.2007.04.009
A. Fedorova, O. Korablev, S. Perrier, J.-L. Bertaux, F. Lefevre, A. Rodin, J. Geophys. Res. **111**, E09S07 (2006). doi:10.1029/2006JE002694
P.D. Feldman, A.L. Cochran, M.R. Combi, in *Comets II*, ed. by M.C. Festou, H.U. Keller, H.A. Weaver (University of Arizona Press, Tucson, 2005), pp. 425–447
P.D. Feldman, M.A. McGrath, H.W. Moos, S.T. Durrance, D.F. Strobel, A.F. Davidsen, Astrophys. J. **406**, 279–284 (1993). doi:10.1086/172439
P.D. Feldman, W.G. Fastie, Astrophys. J. **185**, L101–L104 (1973). doi:10.1086/181330
P.P. Feofilov, *The Physical Basis of Polarized Emission* (State Physico-Mathematical Press, Moscow, 1959) (in Russian)
J.L. Fox, Adv. Space Res. **10**, 31–36 (1990). doi:10.1016/0273-1177(90)90162-S
J.L. Fox, A.I.F. Stewart, J. Geophys. Res. **96**, 9821–9828 (1991). doi:10.1029/91JA00252
R. Fujimoto et al., Publ. Astron. Soc. Jpn. **59**(SP1), 133–140 (2007)
M. Fulle, F. Leblanc, R.A. Harrison et al., Astrophys. J. **661**, L93–L96 (2007). doi:10.1086/518719
M. Galand, S. Chakrabarti, in *Atmospheres in the Solar System: Comparative Aeronomy*, ed. by M. Mendillo, A. Nagy, J.H. Waite (American Geophysical Union, Washington, DC, 2002), pp. 55–76
T.R. Geballe, M.-F. Jagod, T. Oka, Astrophys. J. **408**, L109–L112 (1993). doi:10.1086/186843
J.-C. Gérard, V. Dols, R. Prangé, F. Paresce, Planet. Space Sci. **42**, 905–917 (1994). doi:10.1016/0032-0633(94)90051-5
G.R. Gladstone, D.T. Hall, J.H. Waite Jr., Science **268**, 1595–1597 (1996). doi:10.1126/science.268.5217.1595
G.R. Gladstone et al., Nature **415**, 1000 (2002). doi:10.1038/4151000a
G.R. Gladstone, W.R. Pryor, W.K. Tobiska, A.I.F. Stewart, K.E. Simmons, J.M. Ajello, Planet. Space Sci. **52**, 415–421 (2004). doi:10.1016/j.pss.2003.06.012
P.A. Greet, W.J.R. French, G.B. Burns, P.F.B. Williams, R.P. Lowe, K. Finlayson, Ann. Geophys. **16**, 77–89 (1998). doi:10.1007/s00585-997-0077-3
D. Grodent, J.-C. Gérard, S.W.H. Cowley, E.J. Bunce, J.T. Clarke, J. Geophys. Res. **110**, A07215 (2005)
R.M. Häberli, T.I. Gombosi, D.L. de Zeeuw, M.R. Combi, K.G. Powell, Science **276**, 939–942 (1997). doi:10.1126/science.276.5314.939
L. Haser, Bull. Acad. Sc. Liège **43**, 740–750 (1957)
J.H. Hecht, S. Collins, C. Kruschwitz, M.C. Kelley, R.G. Roble, R.L. Walterscheid, Geophys. Res. Lett. **27**, 453–456 (2000). doi:10.1029/1999GL010853
F. Herbert, B.R. Sandel, J. Geophys. Res. **99**, 4143–4160 (1994). doi:10.1029/93JA02673
T.W. Hill, J. Geophys. Res. **84**, 6554–6558 (1979). doi:10.1029/JA084iA11p06554
T.W. Hill, J. Geophys. Res. **106**, 8101–8108 (2001). doi:10.1029/2000JA000302
T.W. Hill, Rotationally Driven Dynamics in the Magnetospheres of Jupiter and Saturn. Magnetospheres of the Outer Planets, Leicester, 12 August 2005
M. Holmstrom, S. Barabash, E. Kallio, Geophys. Res. Lett. **28**, 1287–1290 (2001). doi:10.1029/2000GL012381

D.L. Huestis, T.G. Slanger, J. Geophys. Res. **98**, 10839–10847 (1993). doi:10.1029/93JE00997
W. Huggins, Philos. Trans. R. Soc. **158**, 529 (1868). doi:10.1098/rstl.1868.0022
A.P. Ingersoll, A.R. Vasavada, B. Little et al., Icarus **135**, 251–264 (1998). doi:10.1006/icar.1998.5971
D.L. Judge, L.C. Lee, J. Chem. Phys. **58**, 104–107 (1973). doi:10.1063/1.1678892
V. Kharchenko, W.H. Liu, A. Dalgarno, J. Geophys. Res. **103**, 26687–26698 (1998). doi:10.1029/98JA02395
T. Kostiuk, M.J. Mumma, F. Espenak et al., Astrophys. J. **265**, 564–569 (1983). doi:10.1086/160699
D. Koutroumpa, F. Acero, R. Lallement, J. Ballet, V. Kharchenko, Astron. Astrophys. **475**, 901–914 (2007). doi:10.1051/0004-6361:20078271
V.A. Krasnopolsky, Icarus **165**, 315–325 (2003). doi:10.1016/S0019-1035(03)00214-8
V.A. Krasnopolsky, A.A. Krysko, V.N. Rogachev, V.A. Parshev, Cosm. Res. **14**, 789–795 (1976)
V.A. Krasnopolsky, V.A. Parshev, in *Venus*, ed. by D.M. Hunten, L. Colin, T.M. Donahue, V.I. Moroz (University of Arizona Press, Tucson, 1983), pp. 431–458
V.A. Krasnopolsky, Icarus **128**, 368–385 (1997). doi:10.1006/icar.1997.5722
V.A. Krasnopolsky, J.B. Greenwood, P.C. Stancil, Space Sci. Rev. **113**, 271–374 (2004)
W.S. Kurth et al., Nature **433**, 722–725 (2005). doi:10.1038/nature03334
R. Lallement, Astron. Astrophys. **418**, 143–150 (2004). doi:10.1051/0004-6361:20040059
H.A. Lam, S. Miller, R.D. Joseph et al., Astrophys. J. **474**, L73–L76 (1997). doi:10.1086/310424
F. Leblanc, J.Y. Chaufray, J.-L. Bertaux, Geophys. Res. Lett. **34**, L02206 (2007). doi:10.1029/2006GL028437
F. Leblanc, J.Y. Chaufray, J. Lilensten, O. Witasse, J.-L. Bertaux, J. Geophys. Res. Planets **111**(E9), E09S11 (2006a). doi:10.1029/2005JE002664
F. Leblanc, O. Witasse, J. Winningham et al., J. Geophys. Res. **111**, A09313 (2006b). doi:10.1029/2006JA011763
J.-F. Le Borgne, J. Crovisier, ESA SP **278**, 171–175 (1987)
M.A. LeCompte, L.J. Paxton, A.I.F. Stewart, J. Geophys. Res. **94**, 208–216 (1989). doi:10.1029/JA094iA01p00208
L.C. Lee, D.L. Judge, Can. J. Phys. **51**, 378–381 (1973)
C.M. Lisse, D.J. Christian, K. Dennerl et al., Science **292**, 1343–1348 (2001). doi:10.1126/science.292.5520.1343
C.M. Lisse, 11 colleagues, Science **274**, 205–209 (2006)
C.M. Lisse et al., Icarus **190**, 391–405 (2007). doi:10.1016/j.icarus.2007.03.004
W. Liu, D.R. Schultz, Astrophys. J. **530**, 500–503 (2000). doi:10.1086/308367
W. Liu, A. Dalgarno, Astrophys. J. **462**, 502–518 (1996). doi:10.1086/177168
T.A. Livengood, H.W. Moos, G.E. Ballester, R. Prangé, Icarus **97**, 26–45 (1992). doi:10.1016/0019-1035(92)90055-C
M.B. Lystrup, S. Miller, N. Dello Russo, R.J. Vervack, T. Stallard, Astrophys. J. (2008, in press)
W.A. Majewski, P.A. Feldman, J.K.G. Watson, S. Miller, J. Tennyson, Astrophys. J. **347**, L51–L54 (1989). doi:10.1086/185605
B.H. Mauk, B.J. Anderson, R.M. Thorne, in *Atmospheres in the Solar System: Comparative Aeronomy*, ed. by M. Mendillo, A. Nagy, J.H. Waite (American Geophysical Union, Washington, DC, 2002), pp. 97–114
A.N. Maurellis, T.E. Cravens, G.R. Gladstone, J.H. Waite Jr., L.W. Acton, Geophys. Res. Lett. **27**, 1339–1342 (2000). doi:10.1029/1999GL010723
I.C. McDade, Planet. Space Sci. **39**, 1049–1057 (1991). doi:10.1016/0032-0633(91)90112-N
I.C. McDade, E.J. Llewellyn, R.G.H. Greer, D.P. Murtagh, Planet. Space Sci. **35**, 1541–1552 (1987). doi:10.1016/0032-0633(87)90079-1
M.A. McGrath, P.D. Feldman, G.E. Ballester, H.W. Moos, Geophys. Res. Lett. **16**, 583–586 (1989). doi:10.1029/GL016i006p00583
M.A. McGrath, J.T. Clarke, J. Geophys. Res. **97**, 13691–13703 (1992). doi:10.1029/92JA00143
M.V. Medvedev, I.P. Robertson, T.E. Cravens, G.P. Zank, V. Florinski, AIP Conf. Proc. **858**, 348 (2006). doi:10.1063/1.2359349
A.E. Metzger, D.A. Gilman, J.L. Luthey et al., J. Geophys. Res. **88**, 7731–7741 (1983). doi:10.1029/JA088iA10p07731
R.R. Meier, Space Sci. Rev. **58**, 1–185 (1991). doi:10.1007/BF01206000
R.R. Meier, J.A.R. Samson, Y. Chung, E.-M. Lee, Z.-X. He, Planet. Space Sci. **39**, 1197–1207 (1991). doi:10.1016/0032-0633(91)90171-6
H. Melin, S. Miller, T. Stallard, C. Smith, D. Grodent, Icarus **181**, 256–265 (2006). doi:10.1016/j.icarus.2005.11.004
H. Melin, S. Miller, T. Stallard, Icarus **186**, 234–241 (2007). doi:10.1016/j.icarus.2006.08.014
S. Miller, N. Achilleos, G.E. Ballester, H. Lam, J. Tennyson, T.R. Geballe et al., Icarus **130**, 57–67 (1997). doi:10.1006/icar.1997.5813
S. Miller, R.D. Joseph, J. Tennyson, Astrophys. J. **360**, L55–L58 (1990). doi:10.1086/185811

S. Miller, J. Tennyson, J. Mol. Spectrosc. **126**, 183–192 (1987). doi:10.1016/0022-2852(87)90089-0
S. Miller, H.A. Lam, J. Tennyson, Can. J. Phys. **72**, 760–771 (1994)
S. Miller et al., Philos. Trans. R. Soc. **358**, 2485–2502 (2000). doi:10.1098/rsta.2000.0662
S. Miller, T. Stallard, C. Smith et al., Philos. Trans. R. Soc. **364**, 3121–3137 (2006). doi:10.1098/rsta.2006.1877
K. Minschwaner, J. Bishop, S.A. Budzien et al., J. Geophys. Res. **109**, A01304 (2004). doi:10.1029/2003JA009941
M.G. Mlynczak, F. Morgan, J.-H. Yee, P. Espy, D. Murtagh, B. Marshall et al., Geophys. Res. Lett. **28**, 999–1002 (2001). doi:10.1029/2000GL012423
J.S. Morgan, C.B. Pilcher, Astrophys. J. **253**, 406–421 (1982). doi:10.1086/159645
J.I. Moses, B. Bézard, E. Lellouch, G.R. Gladstone, H. Feuchtgruber, M. Allen, Icarus **143**, 244–298 (2000). doi:10.1006/icar.1999.6270
N.F. Ness, J.E.P. Connerney, R.P. Lepping, M. Schulz, G.-H. Voigt, in *Uranus*, ed. by J.T. Bergstralh, E.D. Miner, M.S. Matthews (University of Arizona Press, Tucson, 1991), pp. 739–779
S.M. Newman, I.C. Lane, A.J. Orr-Ewing, D.A. Newnham, J. Ballard, J. Chem. Phys. **110**, 10749–10757 (1999)
R.E. Novak, M.J. Mumma, M.A. DiSanti, N.D. Russo, K. Magee-Sauer, Icarus **158**, 14–23 (2002). doi:10.1006/icar.2002.6863
J.F. Noxon, W.A. Traub, N.P. Carleton, P. Connes, Astrophys. J. **207**, 1025–1035 (1976). doi:10.1086/154572
Y. Öhman, Stockholms Obs. Ann. **11**(13), 1–15 (1941)
H. Okabe, *Photochemistry of Small Molecules* (Wiley-Interscience, New York, 1978)
L. Pallier, R. Prangé, Geophys. Res. Lett. **31**, L06701 (2004). doi:10.1029/2003GL018041
C.D. Parkinson, E. Griffioen, J.C. McConnell, G.R. Gladstone, B.R. Sandel, Icarus **133**, 210–220 (1998). doi:10.1006/icar.1998.5926
C.D. Parkinson, A.I.F. Stewart, A.-S. Wong, Y.L. Yung, J.M. Ajello, J. Geophys. Res. **111**, E02002 (2006). doi:10.1029/2005JE002539
G. Paschmann, S. Haaland, R. Treumann (ed.), *Auroral Plasma Physics*. ISSI Space Sciences Series, vol. 15 (Springer, Dordrecht, 2003)
J.L. Phillips, J.G. Luhmann, A.I.F. Stewart, Geophys. Res. Lett. **13**, 1047–1050 (1986). doi:10.1029/GL013i010p01047
G. Piehler, W.H. Kegel, Astrophys. J. **155**, L13–L16 (1986)
R. Prangé, Astron. Astrophys. **251**, L15–L18 (1991)
R. Prangé, D. Rego, D. Southwood, P. Zarka, S. Miller, W. Ip, Nature **379**, 323–325 (1996). doi:10.1038/379323a0
E. Raynaud, E. Lellouch, J.-P. Maillard et al., Icarus **171**, 133–152 (2004). doi:10.1016/j.icarus.2004.04.020
D. Rees, A.L. Aruliah, T.J. Fuller-Rowell, V.B. Wickwar, R.J. Sica, Geophys. Res. Lett. **17**, 1259–1262 (1990). doi:10.1029/GL017i009p01259
D. Rego, S. Miller, R. Prangé, R.D. Joseph, Icarus **147**, 366–385 (2000). doi:10.1006/icar.2000.6444
I.P. Robertson, T.E. Cravens, Geophys. Res. Lett. **30**, 1439 (2003a). doi:10.1029/2002GL016740
I.P. Robertson, T.E. Cravens, J. Geophys. Res. **108**, 8031 (2003b). doi:10.1029/2003JA009873
I.P. Robertson, M.R. Collier, T.E. Cravens, M.-C. Fok, J. Geophys. Res. **111**, A12105 (2006). doi:10.1029/2006JA011672
G.J. Rottman, H.W. Moos, C.S. Freer, Astrophys. J. **184**, L83–L87 (1973). doi:10.1086/181295
B.R. Sandel et al., Science **206**, 962–966 (1979). doi:10.1126/science.206.4421.962
B.R. Sandel et al., Science **215**, 548–553 (1982). doi:10.1126/science.215.4532.548
B.R. Sandel, A.L. Broadfoot, D.F. Strobel, Geophys. Res. Lett. **7**, 5–8 (1980). doi:10.1029/GL007i001p00005
J.A.R. Samson, Y. Chung, E.M. Lee, J. Chem. Phys. **95**, 717–719 (1991). doi:10.1063/1.461424
S.P. Sander, R.R. Friedl, D.M. Golden et al., Chemical kinetics and photochemical data for use in stratospheric modeling. Evaluation Number 14, National Aeronautics and Space Administration, Jet Propulsion Laboratory, 2003
J.H.M.M. Schmitt, S.L. Snowden, B. Aschenbach, G. Hasinger, E. Pfeffermann, P. Predehl et al., Nature **349**, 583–587 (1991). doi:10.1038/349583a0
D.E. Shemansky, J. Geophys. Res. **90**, 2673–2694 (1985). doi:10.1029/JA090iA03p02673
E. Shkolnik, E. Gaidos, N. Moskovitz, Astrophys. J. **132**, 1267–1274 (2006)
D.P. Sipler, M.A. Biondi, Planet. Space Sci. **26**, 65–73 (1978). doi:10.1016/0032-0633(78)90037-5
E.C. Sittler, M.F. Blanc, J.D. Richardson, J. Geophys. Res. **111**, A06208 (2006). doi:10.1029/2005JA011191
G.G. Sivjee, Planet. Space Sci. **39**, 777–784 (1991). doi:10.1016/0032-0633(91)90072-I
T.G. Slanger, R.A. Copeland, Chem. Rev. **103**, 4731–4765 (2003). doi:10.1021/cr0205311
T.G. Slanger, P.C. Cosby, D.L. Huestis, J. Geophys. Res. **108**, 1089 (2003). doi:10.1029/2002JA009677

T.G. Slanger, P.C. Cosby, D.L. Huestis, T.A. Bida, Science **291**, 463–465 (2001). doi:10.1126/science.291.5503.463
T.G. Slanger, P.C. Cosby, D.L. Huestis, D.E. Osterbrock, J. Geophys. Res. **105**, 20557–20564 (2000). doi:10.1029/2000JD900256
T.G. Slanger, P.C. Cosby, D.L. Huestis et al., J. Geophys. Res. **110**, D23302 (2005). doi:10.1029/2005JD006078
T.G. Slanger, P.C. Cosby, B.D. Sharpee, K.R. Minschwaner, D.E. Siskind, J. Geophys. Res. **111**, A12318 (2006a). doi:10.1029/2006JA011972
T.G. Slanger, D.L. Huestis, P.C. Cosby, N.J. Chanover, T.A. Bida, Icarus **182**, 1–9 (2006b). doi:10.1016/j.icarus.2005.12.007
T.G. Slanger, D.L. Huestis, P.C. Cosby, R.R. Meier, J. Geophys. Res. **109**, A10309 (2004). doi:10.1029/2004JA010556
S.L. Snowden, D. McCammon, D.N. Burrows, J.A. Mendenhall, Astrophys. J. **424**, 714–728 (1994). doi:10.1086/173925
S.L. Snowden, M.J. Freyberg, P.P. Plucinsky et al., Astrophys. J. **454**, 643 (1995). Part 1
S.L. Snowden, M.R. Collier, K.D. Kuntz, Astrophys. J. **610**, 1182–1190 (2004). doi:10.1086/421841
D.J. Southwood, M.G. Kivelson, J. Geophys. Res. **106**, 6123–6130 (2001). doi:10.1029/2000JA000236
T. Stallard, S. Miller, G. Millward, R.D. Joseph, Icarus **154**, 475–491 (2001). doi:10.1006/icar.2001.6681
T. Stallard, S. Miller, G. Millward, R.D. Joseph, Icarus **156**, 498–514 (2002). doi:10.1006/icar.2001.6793
T. Stallard, S. Miller, L.M. Trafton, T.R. Geballe, R.D. Joseph, Icarus **167**, 204–211 (2004). doi:10.1016/j.icarus.2003.09.006
T. Stallard, S. Miller, M. Lystrup, N. Achilleos, C. Arridge, M. Dougherty, Astrophys. J. **673**, L203–L206 (2008). doi:10.1086/527545
T. Stallard, C. Smith, S. Miller, H. Melin, M. Lystrup, A. Aylward et al., Icarus **191**, 678–690 (2007). doi:10.1016/j.icarus.2007.05.016
T. Stallard, S. Miller, G.E. Ballester, R.D. Joseph, L.M. Trafton, Astrophys. J. **521**, L149–L152 (1999). doi:10.1086/312189
A.W. Stephan, K.F. Dymond, S.A. Budzien, S.E. Thonnard, J. Geophys. Res. **109**, A09208 (2004). doi:10.1029/2004JA010557
M.H. Stevens, J. Geophys. Res. **106**, 3685–3690 (2001). doi:10.1029/1999JA000329
M.H. Stevens, R.R. Meier, R.R. Conway, D.F. Strobel, J. Geophys. Res. **99**, 417–433 (1994). doi:10.1029/93JA01996
A.I.F. Stewart, C.A. Barth, Science **205**, 59–62 (1979). doi:10.1126/science.205.4401.59
A.I.F. Stewart, J.-C. Gerard, D.W. Rusch, S.W. Bougher, J. Geophys. Res. **85**, 7861–7870 (1980). doi:10.1029/
JA085iA13p07861
D.J. Strickland, J. Bishop, J.S. Evans et al., J. Quant. Spectrosc. Radiat. Transfer **62**, 689–742 (1999). doi:10.1016/S0022-4073(98)00098-3
D.F. Strobel, D.E. Shemansky, J. Geophys. Res. **87**, 1361–1368 (1982). doi:10.1029/JA087iA03p01361
D.F. Strobel, R.V. Yelle, D.E. Shemansky, S.K. Atreya, in *Uranus*, ed. by J. Bergstrahl, M.S. Matthews (University of Arizona Press, Tucson, 1991a), pp. 65–109
D.F. Strobel, R.R. Meier, M.E. Summers, D.J. Strickland, Geophys. Res. Lett. **18**, 689 (1991b). doi:10.1029/91GL00133
D.F. Strobel, M.E. Summers, X. Zhu, Icarus **100**, 512 (1992). doi:10.1016/0019-1035(92)90114-M
G.I. Tachiev, C.F. Fischer, Astron. Astrophys. **385**, 716–723 (2002). doi:10.1051/0004-6361:20011816
M.J. Taylor, W.R. Pendleton, H.-L. Liu et al., Geophys. Res. Lett. **28**, 1899–1902 (2001). doi:10.1029/2000GL012682
G. Tinetti et al., Nature **448**, 169–171 (2007). doi:10.1038/nature06002
W.K. Tobiska, J. Geophys. Res. **98**, 18879–18893 (1993)
L.M. Trafton, J. Carr, D. Lester, P. Harvey, in *Time Variable Phenomena in the Jovian System*, ed. by M.J.S. Belton, R.A. West, J. Rahe (NASA, Washington, DC, 1989a), pp. 229–233
L.M. Trafton, D.F. Lester, K.L. Thompson, Astrophys. J. **343**, L73–L76 (1989b). doi:10.1086/185514
L.M. Trafton, T.R. Geballe, S. Miller, J. Tennyson, G.E. Ballester, Astrophys. J. **405**, 761–766 (1993). doi:10.1086/172404
L.M. Trafton, S. Miller, T.R. Geballe, J. Tennyson, G.E. Ballester, Astrophys. J. **524**, 1059–1083 (1999). doi:10.1086/307838
J.T. Trauger et al., J. Geophys. Res. **103**, 20237–20244 (1998). doi:10.1029/98JE01324
D.N. Turnbull, R.P. Lowe, Planet. Space Sci. **37**, 723–738 (1989). doi:10.1016/0032-0633(89)90042-1
A. Vidal-Madjar, A. Lecavalier des Etangs, J.M. Desert, G.E. Ballester, R. Ferlet, G. Hebrard, M. Mayor, Nature **422**, 143–146 (2003). doi:10.1038/nature01448
A. Vidal-Madjar et al., Astrophys. J. **604**, L69–L72 (2004). doi:10.1086/383347

J.H. Waite Jr., J. Geophys. Res. **96**, 19529–19532 (1991). doi:10.1029/91JA02143
J.H. Waite Jr., F. Bagenal, F. Seward et al., J. Geophys. Res. **99**, 14799–14809 (1994). doi:10.1029/94JA01005
J.H. Waite Jr., G.R. Gladstone, W.S. Lewis et al., Science **276**, 104–108 (1997). doi:10.1126/science.276.5309.104
J.H. Waite, D. Lummerzheim, in *Atmospheres in the Solar System: Comparative Aeronomy*, ed. by M. Mendillo, A. Nagy, J.H. Waite (American Geophysical Union, Washington, DC, 2002), pp. 115–140
B.J. Wargelin, M. Markevitch, M. Juda, V. Kharchenko, R. Edgar, A. Dalgarno, Astrophys. J. **607**, 596–610 (2004). doi:10.1086/383410
B.J. Wargelin, P. Beiersdorfer, P.A. Neill, R.E. Olson, J.H. Scofield, Astrophys. J. **634**, 687–697 (2005). doi:10.1086/496874
H.A. Weaver, P.D. Feldman, J.B. McPhate et al., Astrophys. J. **422**, 374–380 (1994). doi:10.1086/173732
F.L. Whipple, Astrophys. J. **111**, 374–395 (1950)
R.V. Yelle, Geophys. Res. Lett. **15**, 1145–1148 (1988). doi:10.1029/GL015i010p01145
R.V. Yelle, J.C. McConnell, B.R. Sandel, A.L. Broadfoot, J. Geophys. Res. **92**, 15110–15124 (1987a). doi:10.1029/JA092iA13p15110
R.V. Yelle, L.R. Doose, M.G. Tomasko, D.F. Strobel, Geophys. Res. Lett. **14**, 483–486 (1987b). doi:10.1029/GL014i005p00483
R.V. Yelle, J.C. McConnell, D.F. Strobel, L.R. Doose, Icarus **77**, 439–456 (1989). doi:10.1016/0019-1035(89)90098-5
P. Zarka, Adv. Space Res. **12**, 99–115 (1992). doi:10.1016/0273-1177(92)90383-9
P. Zarka, J. Geophys. Res. **103**, 20159–20194 (1998). doi:10.1029/98JE01323
P. Zarka, W.S. Kurth, in *The Outer Planets and their Moons*, ed. by T. Encrenaz, R. Kallenbach, T.C. Owen, C. Sotin. ISSI Space Sciences Series, vol. 19 (Springer, Dordrecht, 2005)
S.P. Zhang, G. Shepherd, J. Geophys. Res. **110**, A03304 (2005). doi:10.1029/2004JA010887

Plasma Flow and Related Phenomena in Planetary Aeronomy

Y.-J. Ma · K. Altwegg · T. Breus · M.R. Combi · T.E. Cravens · E. Kallio · S.A. Ledvina · J.G. Luhmann · S. Miller · A.F. Nagy · A.J. Ridley · D.F. Strobel

Originally published in the journal Space Science Reviews, Volume 139, Nos 1–4.
DOI: 10.1007/s11214-008-9389-1

Abstract Understanding the processes involved in the interaction of solar system bodies with plasma flows is fundamental to the entire field of space physics. The features of the interaction can be very different, depending upon the properties of the incident plasma as well as the nature of the obstacle. The properties of the atmosphere/ionosphere associated with the obstacle are of particular importance into understanding the plasma interaction process, especially for non-magnetized obstacle. This paper discusses in detail the roles of the atmosphere and ionosphere systems of plasma interaction around Venus, Mars, comets

---

Y.-J. Ma (✉)
IGPP, UCLA, 6877 Slichter Hall, Los Angeles, CA 90095, USA
e-mail: yingjuan@igpp.ula.edu

K. Altwegg
Physikalisches Institut, University of Bern, Sidlerstr. 5, 3012 Bern, Switzerland

T. Breus
Space Research Institute RAS, Profsoyznaya Str. 84/32, 117997 Moscow, Russia

M.R. Combi · A.F. Nagy · A.J. Ridley
University of Michigan, Space Res. Bldg., 2455 Hayward St., Ann Arbor, MI 48109, USA

T.E. Cravens
Department of Physics and Astronomy, University of Kansas, Lawrence, KS 66045, USA

E. Kallio
Finnish Meteorological Inst, Space Research Unit, Erik Palmenin aukio 1, Helsinki, SF-00101, Finland

S.A. Ledvina · J.G. Luhmann
Space Sciences Laboratory, University of California, Berkeley, CA 94720, USA

S. Miller
Department of Science and Technology Studies/Physics and Astronomy, University College London, Gower Street, London WC1E 6BT, UK

D.F. Strobel
Johns Hopkins University, 3400 N. Charles St., Baltimore, MD 21218, USA

and some particular satellites. The coupling between magnetosphere and ionosphere is also discussed for Earth and Giant planets.

Keywords Plasma flow · Interaction · Ionosphere

1 Introduction

The nature of the interaction between a fast-moving plasma and a relatively slow obstacle in the solar system has been studied for a long time, yet many of the relevant processes are still far from being fully understood. Such a fast-moving plasma can be the solar wind or the corotational plasma flow inside a planetary magnetosphere. The various obstacles include all of the planets, satellites, comets and asteroids in our solar system. The features of the interaction can be very different, depending upon the properties of the incident plasma and the nature of the obstacle. The characteristics of the atmosphere/ionosphere associated with the obstacle are of particular importance into understanding the plasma interaction process, especially for non-magnetized obstacle. The properties of the ionosphere are discussed in an accompanying paper by Witasse et al. (2008), as is the energy deposition to the ionosphere by Fox et al. (2008). Also the numerical simulations of the interaction are discussed by Ledvina et al. (2008) in this issue/book. This chapter will only focus on the ionospheric responses to the plasma flow.

1.1 Categories of Different Kind of Interactions with Fast Plasma Flows

The solar wind is a stream of fully ionized plasma, composed of high energy particles (mainly protons and electrons), which originates from the solar corona. The density, composition, temperature and flow velocity of the solar wind plasma is highly variable as is the interplanetary magnetic field frozen-in with the plasma. The solar wind plasma becomes both supersonic and super-Alfvenic tens of solar radii from the Sun. There are basically two types of planetary obstacles to the solar wind flow: the planetary magnetosphere and the planetary atmosphere/ionosphere system (here we do not include any discussion of interactions with solar system objects that have neither an atmosphere or a strong intrinsic magnetic field). In either case, the supersonic nature of the solar wind necessitates the formation of a bow shock to deflect the flow. The bow shock is a standing wave in front of an obstacle at which the supersonic plasma flow is slowed, heated, and deflected around the obstacles. The strength of this shock depends on the flow velocity of the solar wind relative to the velocity of compressible waves (denoted by the fast magnetosonic Mach number) in the plasma.

In the case of magnetized planets the obstacle to the solar wind is the planetary magnetic field. The outer limit of the planet's field of influence is known as the "magnetopause". The volume within, which the planet's field dominates, is the "magnetosphere". The size of the magnetosphere is governed by the relative strengths of the magnetic field and the solar wind pressure at the planet. Planets with strong intrinsic magnetic field, such as Earth, Jupiter and Saturn, have large and complex magnetospheres. The magnetospheres of these planets differ both in size and internal energy sources but also in the strength of the solar wind flow past them. Thus, the interaction of each of the magnetospheres with the solar wind differs in some degree from the others. Russell (1991) reviewed the basic physical processes that occur in those magnetospheres, and briefly highlighted our exploration of these planets and the problems remaining. The terrestrial magnetosphere is of course the one that has been most thoroughly studied, but given the recent Gallio and Cassini missions, we now also have significant information on Jupiter and Saturn's magnetospheres. This type of interaction will not be discussed in this chapter any further, as there exists a very extensive literature on this subject, including books, book chapters and review articles (e.g. Cravens 1997a; Dessler 1983; Khurana 2004)

In the case of a planet without a global magnetic field, like Mars and Venus, the standing bow shock location is much closer to the obstacle body, and the shocked solar wind interacts directly with the ionosphere and the upper atmosphere of the planet (e.g. Nagy et al. 2004; Russell et al. 2007). Many of the planetary satellites and asteroids are either exposed directly to the solar wind or to the planet's magnetospheric plasma, because they have small or no intrinsic magnetic fields and possess no significant atmospheres. However, some of the satellites such as Saturn's largest moon Titan, for example, and some comets (when close enough to the Sun), do have a substantial atmosphere and thus an ionosphere, and possess no or only small intrinsic magnetic fields. The interaction of these non-magnetic objects with the planet's magnetospheric plasma or the solar wind is expected to be similar to the solar wind interaction with Mars and Venus.

1.2 The Potential Importance of the Ionosphere/Atmosphere System

At planets and moons that have an atmosphere, the neutral atoms and molecules may be photo-ionized by sunlight (or starlight, in the case of extra-solar planets) or by the impact of energetic particles. In some cases, the degree of ionization may reach 10^{-6} or even 10^{-5}, generating a region in the atmosphere that is known as the ionosphere. The resulting situation is one in which, at altitudes that usually correspond to the thermosphere, ions and electrons coexist with the much more abundant neutral atmosphere.

The ionosphere/atmosphere system is especially important to understand the plasma interaction around non-magnetic solar bodies. Without the shielding of a strong intrinsic magnetic field, the upstream plasma can easily come close to the body, thus interact more directly with the neutral atmosphere/ionosphere. The interaction of the inflowing plasma with the ionosphere/atmosphere system is through electro-magnetic forces and collisions.

This chapter will discuss in details the plasma interaction with the ionosphere/atmosphere system. The chapter is organized as follows. The next section discusses plasma flow interaction with un-magnetized planets: Venus and Mars. Plasma interactions around satellites and comets are discussed in Sects. 3 and 4, respectively. Section 5 provides a brief discussion of the coupling between an ionosphere and magnetosphere for magnetized planets. A few brief comments are given in the last section.

2 Un-magnetized Planets: Venus and Mars

Venus and Mars are non-magnetized or weakly magnetized inner planets with substantial atmospheres and ionospheres. As a consequence, the plasma environments around these two bodies are similar in many ways.

2.1 Venus; General Interaction Features

Many spacecraft flew by, orbited, probed or landed on Venus. However, our present understanding of the solar wind interaction with Venus primarily comes from observations made during the Pioneer-Venus Orbiter (PVO) mission, which lasted 14 years, had a broad and relevant instrument payload and covered more than an entire solar cycle (Russell et al. 2006).

PVO observations of Venus plasma environment and upper atmosphere are reviewed in several comprehensive volumes (J. Geophys Res., **85**, A13, 1980; Space Sci., Rev., **55**, 1991) and two books Venus I and Venus II by the University of Arizona Press. In addition the literature through the 1990s contains the descriptions of many individual in-depth analyses. Two recent reviews by Russell et al. (2006, 2007) have also been published. The reader is encouraged to explore beyond the short list of references provided here. The recently launched Venus Express spacecraft (Titov et al. 2006), equipped with a magnetometer (Zhang et al. 2006) and a comprehensive plasma instrument package with an ion mass analyzer, an electron spectrometer and energetic neutral atom (ENA) images (Barabash et al. 2007a, 2007b) is now also adding new, important and complimentary information on the Venus plasma environment.

The observations by Venera 9, 10 and PVO spacecraft showed that Venus has no intrinsic dipole magnetic field (Dolginov et al. 1978; Yeroshenko 1979; Russell et al. 1980). Figure 1 illustrates the major features of the solar wind interaction with Venus's ionosphere deduced from the solar maximum measurements obtained by PVO. The most basic features are the bow shock that defines the outer boundary of the interaction, followed by the large-scale draping of the interplanetary magnetic field over the ionosphere in the magnetosheath and the related comet-like "induced magnetotail" made up of the innermost draped flux tubes that have been slowed where they pass through the upper atmosphere and ionosphere.

The solar cycle causes changes in the effective obstacle presented by Venus to the solar wind. Figure 2, from Phillips and McComas (1991), illustrates the reduction in the termi-

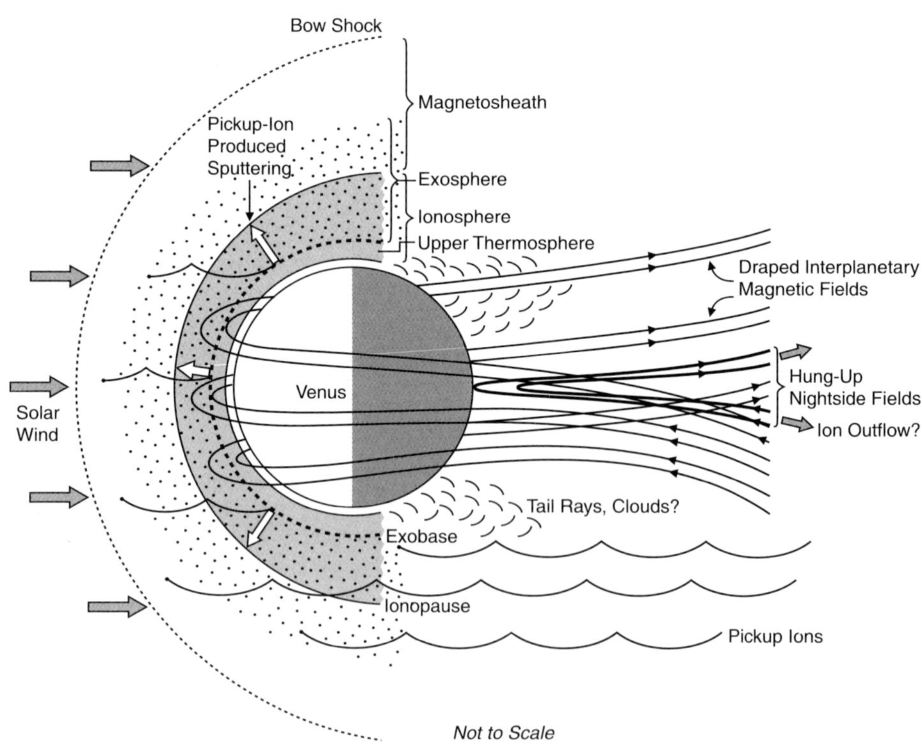

Fig. 1 Schematic diagram of the solar wind interaction with Venus, including the oxygen escape processes (Russell et al. 2007)

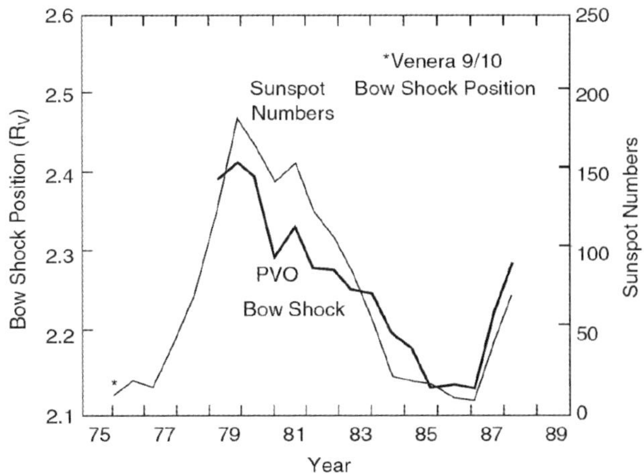

Fig. 2 Location of the bow shock, normalized to the terminator, measured by PVO through 1987 (*heavy trace*). *Light trace* shows the sunspot number averaged over each PVO observing season of roughly 100 Earth days. The asterisk at lower left marks the bow shock location observed by Venera 9 and 10 (Phillips and McComas 1991)

nator location of the bow shock with solar EUV flux, implying that the obstacle has either become smaller or is no longer impenetrable (Zhang et al. 1991a, 1991b). The shape of the shock can be fit quite well with a simple conic function: $r = R_T/(1 + \varepsilon\cos(\chi))$, where r is the radial distance to the shock, χ is the Sun–Venus–satellite angle and R_T is the radius of the terminator shock (Russell 1977). Slavin et al. (1980) found that the best fit to the observed PVO bow shock crossing corresponds to an ε of 0.88 and an R_T of 2.21 R_V. Venus Express has obtained 147 clear crossings from April to August 2006 (Zhang et al. 2007). The best fit to the bow-shock location from a solar-zenith angle of 20° to 120° gives a terminator bow-shock location of 2.14 R_V, and subsolar bow shock at 1.32 R_V, about 1,900 km above the surface of the planet. Statistical studies also showed that the location of the Venus bow shock is not only dependent on the solar cycle and solar EUV flux, but is also tightly related to the upstream solar wind parameters, including the orientation of the interplanetary magnetic field (Russell et al. 1988; Zhang et al. 1991a).

Signatures of the Venus obstacle have also been seen upstream of the bow shock in the high resolution PVO magnetic field data and as well as in the plasma wave data (Crawford et al. 1993). Some of the solar wind protons and electrons incident on the nose of the bow shock are accelerated back into the oncoming solar wind along interplanetary magnetic field lines as they connect to the shock. This phenomenon produces the foreshock, a feature observed as upstream magnetic field and plasma fluctuations, and plasma waves, in front of virtually every planetary bow shock encountered in the solar system. The magnetic field and plasma fluctuations associated with the foreshock can be convected with the solar wind into the magnetosheath where they produce turbulent conditions on magnetosheath streamlines connected to the portion of the shock downstream from the foreshock (Luhmann et al. 1983). These fluctuations may affect the ion pickup process, discussed below, in the magnetosheath.

In the subsolar magnetosheath the magnetic field lines "pile up" and close to the top of the ionosphere form a magnetic barrier. Features of this magnetic pile-up region (MPR) near Venus are rather well established, because the PVO crossed this region hundreds of time (Zhang et al. 1991b). The peak pressure of the IMF measured within the MPR is typically about 83% of that of the upstream solar wind value and is balanced by the ionospheric thermal pressure. The mean distance between the topside boundary of the ionosphere, the ionopause, and the surface where the IMF pressure reaches 50% of the local dynamic pres-

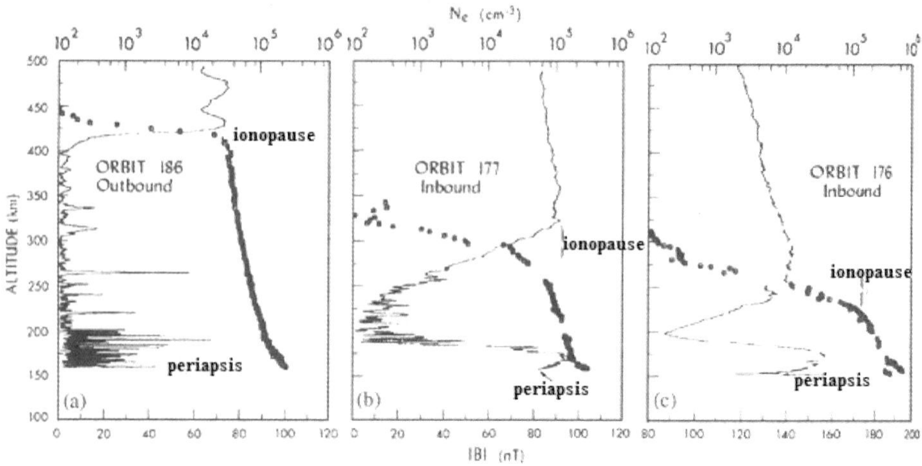

Fig. 3 The altitude profile of the magnetic field strength and the plasma density in the Venusian ionosphere in period of low (orbit 186), moderate (orbit 177) and high (orbit 176) dynamic pressure of the SW (from Elphic et al. 1981)

sure of the solar wind is about 200 km at the subsolar point and 800 km at the terminator. The peak IMF pressure decreases from the subsolar point to the terminator approximately following a $\cos^2\theta$ law (θ is the solar zenith angle). The so called ionopause is the location of the transition from magnetic to ionospheric thermal pressure. The ionopause has been identified by Brace et al. (1980) and Knudsen et al. (1980) as the altitude where the thermal plasma density is approximately equal 100 cm^{-3}. Cloutier et al. (1983) defined the ionopause simply as a drastic change in the plasma density. Phillips et al. (1984) used a definition of the ionopause as the point at which the magnetic and thermal pressure crossed upon entry or exit from the ionosphere.

Figure 3 shows examples of ionopause measured by PVO. The left figure, Orbit 186, represents condition when the ionospheric thermal pressure well exceeds the solar wind dynamic pressure and the transition region is above the collisional region of the ionosphere, resulting in a sharp ionopause. Under these conditions the magnetic barrier magnetic field is essentially excluded from the ionosphere by currents flowing in the ionopause layer (Elphic et al. 1981; Luhmann and Cravens 1991). The two plots on the right hand side of Fig. 3, corresponding to Orbits 177 and 176, show what happens when the solar wind pressure increases and pushes the pressure balance altitude into the collisional region of the ionosphere. This situation can arise either because of very high solar wind pressures, mostly during solar cycle maximum, and/or reduced ionospheric pressure conditions. Under these circumstances the magnetic field penetrates the ionosphere resulting in magnetic field strength in excess of 100 nT. This is accompanied by a "thick" ionopause. These observations have been successfully modeled by Shinagawa and Cravens (1988) using a one-dimensional multi-fluid MHD model as being due to downward convection into the ionosphere due to both magnetic pressure and aeronomically controlled pressure gradients combined with ohmic dissipation near 150 km.

Solar wind dynamic pressure at Venus varies according to a distribution dictated mostly by solar conditions. For the period of PVO observations of the ionosphere, the overall spread and occurrence frequency of values obtained from the Plasma Analyzer is given in Fig. 4 (Phillips et al. 1984). Even at solar maximum, the number of times the solar wind pres-

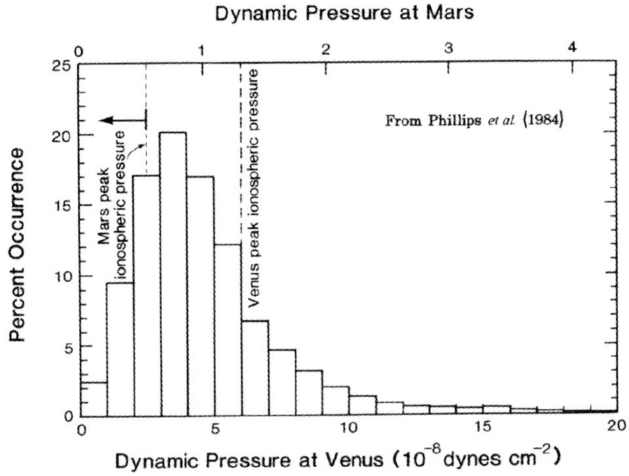

Fig. 4 Histogram of solar wind dynamic pressure at Mars and Venus. Dashed lines are peak effective ionospheric thermal pressure (from Phillips et al. 1984)

sure exceeds the peak thermal pressure of the ionosphere, a measure of its robustness as an obstacle, is small relative to the situation inferred for Mars. The related solar zenith angle dependence of the ionopause altitude determined from both the topside ionosphere density gradients and the altitudes where ionospheric thermal and magnetic barrier magnetic pressures balance is shown in Fig. 5a. As the solar wind pressure increases the height of the ionopause decreases and actually levels of near 300 km, when the pressure exceeds about 4×10^8 dyn cm^{-2}. The observations also showed that the mean ionopause height increases from about 350 km at the subsolar point to about 900 km at the terminator (see Fig. 5b). The observed ionopause altitude behavior follows a roughly cosine solar zenith angle squared dependence because the component of the dynamic pressure normal to the boundary is the one that is important for the pressure balance. It is also much more variable at high solar zenith angles than low, in part because of the cosine squared dependence but also because attached structures such as ionospheric clouds and tail rays are often observed at the ionopause there. In addition Phillips et al. (1988) showed that the ionopause has an elliptical cross section at the terminator, with generally higher altitudes found at the poles of the magnetosheath field draping pattern.

2.2 Atmosphere Escape from Venus

The typical altitude of pressure balance with the Venus ionosphere, and in the absence of balance, the altitude at which the transition from solar wind to the ionospheric plasma occurs, is such that the exosphere is exposed to the solar wind plasma and/or magnetic field. In addition, the ionospheric chemistry is such that the dissociative recombination reaction of O_2^+ results in pairs of suprathermal oxygen atoms populating the exosphere; there is also a significant hydrogen exosphere present (Nagy and Cravens 1998; Kim et al. 1998; Hodges 2000). Any ionization mechanism that operates in these upper regions of the neutral atmosphere, including photoionization, charge exchange with solar wind protons, or electron impact by solar wind electrons, produces a seed population for the ion pickup process. Ion pickup, illustrated by Fig. 6, can lead to either atmosphere escape or to impact of the pickup ion on the atmosphere, with energy deposition and possible sputtering loss of additional atmosphere (Luhmann and Kozyra 1991). The signature large cycloidal trajectories of Venus O^+ pickup ions suggested in Fig. 6 are

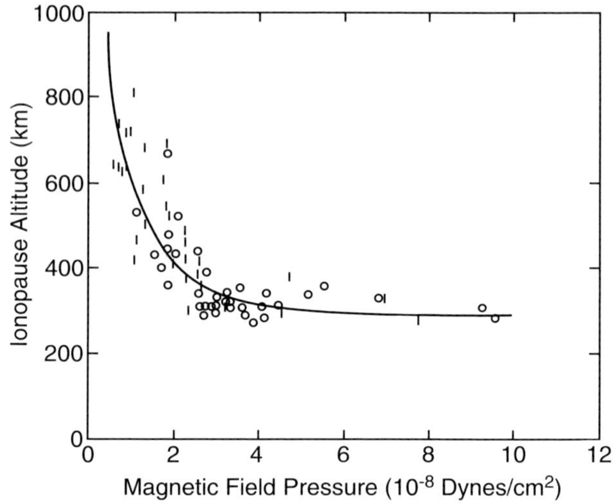

Fig. 5a The response to the ionopause altitude, inbound (I) and outbound (O), to the orbit to orbit changes in magnetic pressure. The ionopause altitude declines with increasing solar wind pressure, but levels above about 4×10^{-8} dyn cm^{-2} (Brace et al. 1980)

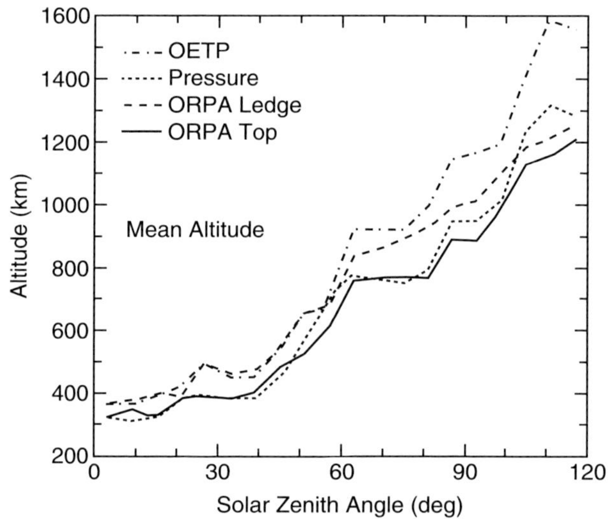

Fig. 5b The solar zenith angle variations of the ionopause height using ionopause definitions based on various PVO measurements. All definitions show similar SZA variation of the ionopause height (Phillips et al. 1988)

most relevant for high altitude pickup of the atomic oxygen coronal atoms. Smaller cycloids and helices are expected for smaller mass hydrogen pickup, and for O^+ picked up in the slowed background plasma flows and stronger, draped magnetic fields in the Venus magnetic barrier region where most of the pickup must occur. The suprathermal oxygen ions observed on PVO at the top of the dayside ionosphere (Kasprzak and Niemann 1982; Grebowsky et al. 1993) and in the low altitude wake (Kasprzak et al. 1991) are likely evidence of this population.

An ongoing debate concerns the importance of this mechanism for atmosphere evolution at Venus, as for Mars (see Russell et al. 2007). At Venus, however, the gravitational field is such that pickup is one of the only ways that oxygen can be lost to space. Venus is notoriously dry compared to Earth, and while it is easy to envision the hydrogen from photodissociated water vapor in the upper atmosphere escaping, oxygen should be retained.

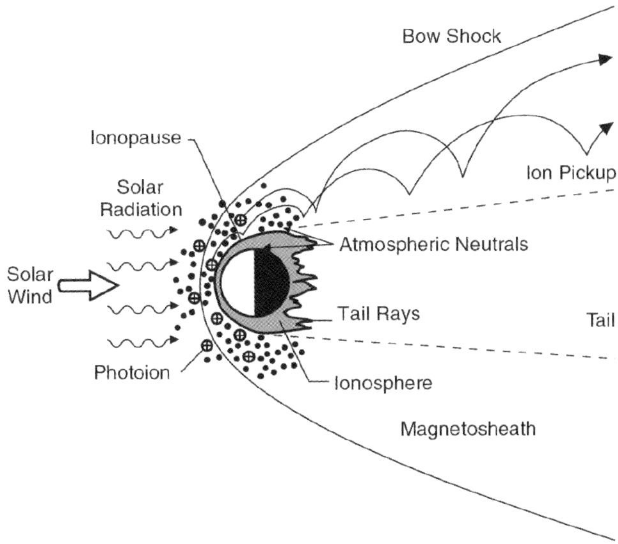

Fig. 6 Schematic illustration of the pickup of heavy planetary ionospheric ions at Venus and Mars. Only those ions produced in the region where the solar wind plasma flows are removed. They are either carried away with the solar wind or "precipitate" into the lower atmosphere, depending on the location where they are produced and the direction of the interplanetary magnetic field (Luhmann and Kozyra 1991)

Over time, this oxygen would have had to be removed from the atmosphere by some process such as oxidation of the surface. But if Venus originally had an Earth-like inventory of water, it would be difficult for the crust to have accommodated this amount. Thus escape to space by ion pickup and associated sputtering is a viable candidate for further examination. This historical aspect, as well as an understanding of the present-day Venus-solar wind interaction consequences, has motivated several special analyses of the PVO data.

The average indirectly inferred escape rates from PVO observations, of the order of 5×10^{24} sec^{-1} (McComas et al. 1986), fall far short of those needed to remove an ocean's worth of oxygen. However, conditions for this process to be effective may have been more favorable in the past. In particular, periods of disturbed solar wind from coronal eruptions and stream interactions have been found to enhance lower energy oxygen ion escape, as observed by the PVO Neutral Mass Spectrometer (Luhmann et al. 2007). Observations of ion escape by the ASPERA-4 experiment on Venus Express (Barabash et al. 2007a, 2007b) are well suited to measure the ion escape rates from Venus. Preliminary observations indicate an escape rate in the order of 10^{25} sec^{-1}.

Several "bulk removal" processes for ion loss have also been suggested based on the PVO observations. These observations include the "clouds", attached and detached regions of ionospheric plasma detected above the nominal ionopause, as shown in the Langmuir Probe time series, and "tail rays" of thermal electrons and ions detected in the near-Venus wake (see Brace et al. 1995). The clouds were analyzed by Ong et al. (1991) and found to be associated with interplanetary field rotations, suggesting they may be transient features such as ridges on the ionopause, associated with changing interplanetary field draping. Hybrid model calculations by Terada et al. (2002) also indicate that Kelvin–Helmholtz instabilities at the ionopause result in filaments/streamers that can move down the tail and escape.

2.3 Mars

The data from the Phobos 2 mission created two, controversial hypotheses: one stated that Mars has a substantial intrinsic magnetic moment ($\sim 1.4 \times 10^{22}$ G cm^3, Dolginov and Zhuzgov 1991) and while the other claimed that the intrinsic magnetic field of Mars is negligible

(Yeroshenko et al. 1990). This controversy was settled by magnetometer (MAG/ER) carried by the Mars Global Surveyor (MGS) which clearly established the presence of intense crustal magnetic anomalies and set an upper bound of 2×10^{17} A m^2 = 2×10^{20} G cm^3 on the intrinsic dipole field at Mars (Acuna et al. 1998, 1999; Connerney et al. 1999; Ness et al. 1999). This corresponds to an equatorial surface field of less than 0.5 nT. The magnetic fields measured by the MGS magnetometer over certain localized regions were found to range up to 1600 nT. Such fields are sufficiently strong to withstand the solar wind pressure up to near 400 km. The crustal sources are distributed mainly over the surface of the southern hemisphere with the largest concentration in the geographic range of 120–210°W and 30–85°S. These anomalies complicate at least locally the plasma environment of Mars.

2.3.1 The Size, Shape and Variability of the Bowshock (BS) at Mars

The location of the bow shock at Mars is more distant and also more variable than for Venus: in the terminator plane, 2.66 R_M versus 2.39 R_V with a standard deviation of 0.49 R_M versus 0.21 R_V during solar maximum, based on Phobos 2 observations (Slavin et al. 1991). The MGS results showed that, during periods of medium solar activity, the terminator distance of the shock is at 2.62 R_M with a standard deviation of 0.33 R_M. Also note that the sunspot number was 140–180 and 30–90 respectively, for the Phobos 2 and MGS missions, so this also suggests that the mean Martian BS position is independent of the solar cycle and EUV flux. This is different from Venus as discussed in the previous section.

Trotignon et al. (2006) recently reanalyzed a total of 700 shock crossings, using the mixed data set of Phobos 2 and MGS. As shown in Fig. 7, the shock locations are highly variable. The best fit corresponds to a subsolar shock location at 1.63 ± 0.1 R_M, these results are consistent with previous studies (Vignes et al. 2000)

Several factors that control the Martian shock locations were examined by Vignes et al. (2002) based on 553 MGS shock crossings. Contrary to many expectations, the high crustal magnetic sources, found in the southern hemisphere, do not seem responsible for the bow shock (BS) variability. Data results show no obvious strong correlation between the location of the highest crustal sources and the variability of the shock position. However, a recent study by Edberg et al. (2008) found that the shock locations are further away from the planet when the crossing was observed over regions with strong crustal fields. The new results are

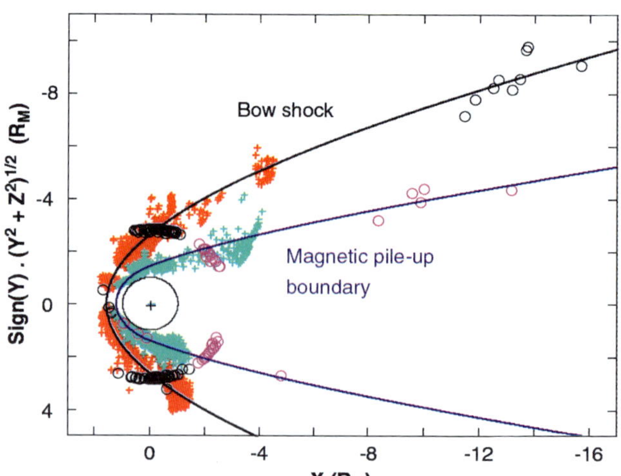

Fig. 7 Martian bow shock (*black line*) and magnetic pileup boundary (*blue line*) models that best fit to the observations made by both the plasma wave system of the Phobos 2 mission (*black rings* for the BS and *purple rings* for the MPB) and the MAG/ER MGS (*red crosses* for the BS and *blue crosses* for the MPB). (Trotignon et al. 2006)

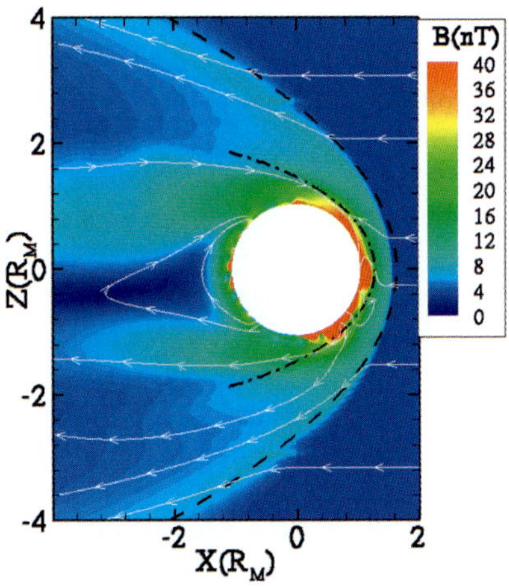

Fig. 8 The calculated magnetic field in the meridianal plane. The *color plots* show the magnitudes; the *white lines* marked with arrows indicate the vector direction of the magnetic field. The *dashed line* represents the mean bow shock and the *dash-dot line* is the mean MPB locations from Vignes et al. (2000) (Ma et al. 2004)

also consistent with the global multi-species MHD model results of Ma et al. (2002), which found that the crustal sources do move the shock boundary outward significantly. Numerous complex 3-D models have been developed during the last decade to model solar wind interaction with Mars. These models are either based on MHD formulations (e.g. Ma et al. 2004; Harnett and Winglee 2006; Ma and Nagy 2007) or hybrid formulations (e.g. Kallio and Janhunen 2002; Bösswetter et al. 2004; Modolo et al. 2005). Both of these approaches have advantages and disadvantages as discussed in an accompanying paper/chapter by Ledvina et al. (2008) and in some respect complement each other. These models have been successful in reproducing the observed bow shock and magnetic pile-up locations rather well. An example of such successful model and observation comparison is shown in Fig. 8, from the work of Ma et al. (2004).

Vignes et al. (2002) also found that when the classical definitions of quasi-parallel ($\theta_{\mathrm{Bn}} \leq 45°$) and quasi-perpendicular ($45° \leq \theta_{\mathrm{Bn}} \leq 90°$) shocks are used, the mean value of the terminator distance for the quasi-parallel shock crossings is about 3% closer than for the quasi-perpendicular shock crossings. This creates a dawn–dusk asymmetry in the bow shock location, because according to the IMF orientation, quasi-parallel shocks dominate the dawn side while quasi-perpendicular shocks dominate the dusk side. This asymmetry is consistent with a previous study on Venus and Mars (Zhang et al. 1991a), which found the quasi-parallel shock is closer to the planet than the quasi-perpendicular shock. Similar to Venus, the Martian shock also appears farther from Mars in the hemisphere of the locally upward interplanetary electric field (Vignes et al. 2002), which is consistent with the idea that mass loading plays a role in controlling the bow shock location,

In summary, the shock position at Mars is observed to be different than that of Venus. Although the shock location around Mars is found to be highly variable, according to the statistical study, the shock location has only a slight dawn–dusk asymmetry depending on the IMF direction, small dependence on the convection electric field direction, no significant effect due to solar cycle conditions, while the effect of crustal sources is still under debate. Although there is no direct solar wind pressure data for the shock crossings, we can expect that the solar wind dynamic pressure will also contribute to the variations of Martian shock location, as is the case for Venus.

2.3.2 Magnetic Pile up Region and Magnetic Pileup Boundary (MPB) Characteristics

The magnetic pileup boundary (MPB) is a sharp, thin, and well-defined plasma boundary located between the bow shock and the upper ionosphere boundary at Mars (Fig. 9). This boundary separates the magnetosheath, a region of low magnetic fields with significant wave activity, from the magnetic pileup region dominated by strong, highly organized magnetic fields, which are a result of the pileup and draping of the interplanetary magnetic field.

The MPB appears in the MAG/ER data as a layer in which the measured magnetic fields increase over a short radial distance (Vignes et al. 2000). Simultaneously, the electron fluxes measured by the electron reflectometer (ER) attenuate in a manner consistent with electron impact ionization of the oxygen and hydrogen exosphere by solar wind electrons (Crider et al. 2000). Coincident with this location, wave activity in the MAG data declines with decreasing altitude. Previously, instruments onboard Phobos-2 detected a boundary consistent with the MPB location when comparing the respective fits (Trotignon et al. 1996; Vignes et al. 2000). The Phobos-2 instruments, ASPERA (Lundin et al. 1989) and TAUS

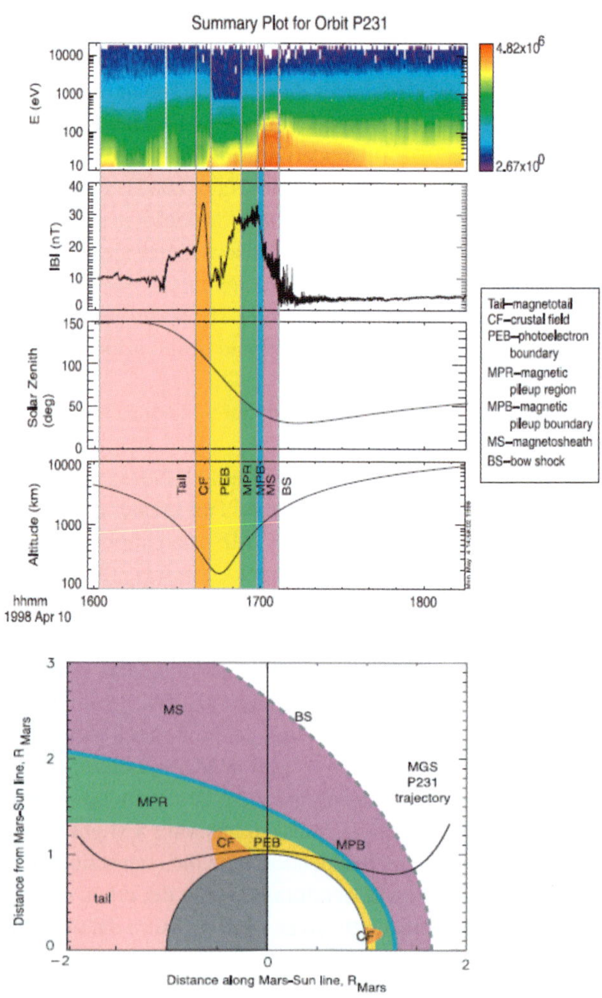

Fig. 9 Magnetic pile-up boundary (MPB) detected at MGS in the Martian magnetosheath (*left*) and cylindrical projection of the trajectory of MGS to show the locations of the different regimes of plasma regions at Mars (Crider et al. 2003)

(Rosenbauer et al. 1989) saw the MPB (then called "planetopause") as a termination in solar wind proton flux in the same location as the magnetometers sensed an increase in magnetic field values (Riedler et al. 1989). These attributes indicate that solar wind protons are interacting through charge exchange with the planetary exosphere. If one puts this together with the complementary MGS observations of electron impact ionization and wave activity then the MPB appears to be the transition to the region in which the planetary exosphere becomes substantial in determining the plasma properties. A complete discussion on the Martian MPB based on MGS and Phobos 2 measurements is given by Nagy et al. (2004).

It is important to note that the MPB is neither a pressure balance boundary nor a discontinuity in the solar wind flow. ER data show the presence of solar wind electrons everywhere above the photoelectron boundary (Mitchell et al. 2000). Therefore, the solar wind plasma and magnetic field do penetrate the MPB. The MPB simply marks the transition from shocked solar wind plasma to plasma that is interacting and/or has interacted with the exosphere.

Recently, new features were discovered which indicated that the MPB is a boundary of the induced magnetosphere. Bertucci et al. (2003) found a strong and sudden enhancement of the magnetic field draping at the MPB of Mars, in contrast with the picture of a gradual draping between the bow shock and the final planetary obstacle. The boundary marks then the entrance into an induced magnetosphere, where draping is strong. In this region, the magnetic field lines frozen into the electron gas follow the denser and cooler plasma. This is a consequence of the mass loading and ionization mechanisms, which contribute to the dominance of heavy ions of exospheric origin (Lundin et al. 1990; Crider et al. 2000). The lower limit of this region is the final planetary obstacle (Luhmann 1986; Mitchell et al. 2001). These results are highly consistent with Phobos 2 observations across the magnetic tail boundary at 2.86 R_M (Yeroshenko et al. 1990).

2.3.3 The Ionopause at Mars

The ionopause is a feature almost always observed in the ionosphere of Venus by radio occultation and in-situ instruments onboard the Pioneer Venus Orbiter (Kliore 1992; Brace and Kliore 1991). The ionopause is marked by sharp decrease in the electron density at the top of the ionosphere as discussed in the previous section. However, no clear ionopause signatures at Mars have been reported so far. A boundary between the Magnetic Pileup Region and the ionosphere below is evident in the supra-thermal electron observations ($>$10 eV) by MGS. This boundary has been called the "photoelectron boundary" (Crider et al. 2003; Nagy et al. 2004). However, it is not clear at all how this transition is related to the classical ionopause. Both the Mars Express radio occultation and radar data seem to give hints of thermal electron density transitions, but so far no definitive results have been published. It has long been known from Viking measurements, that the ionospheric thermal pressure at Mars is usually insufficient to balance the total pressure in the overlying Magnetic Pileup Region (see Fig. 4). Thus, it is expected that the Martian ionosphere is magnetized, much like Venus' ionosphere during times of high solar wind dynamic pressure and/or low solar cycle conditions. It was found at Venus that under such conditions the ionopause is not a sharp feature. The magnetic pressure associated with this ionospheric field will supplement the ionospheric thermal pressure, possibly enough to balance the overlying MPR pressure. Crustal magnetic fields vastly complicate the topology of the ionosphere and thus conditions for the formation of a "classical" ionopause.

2.3.4 Atmosphere Escape from Mars

The gravity at Mars is significantly lower than it is at Venus thus more processes can contribute to the escape flux. Some of the neutral oxygen atoms created from the dissociative recombination of O_2^+ ions have enough energy to escape. This process is considered to be the main contributor to the escape of neutral oxygen atoms, although charge exchange with hot H^+ and O^+ ions can also make a contribution. The estimates of the neutral oxygen escape flux vary widely from about 2×10^{26} to 1×10^{25} sec^{-1} (Kim et al. 1998; Hodges 2002; Chaufray et al. 2007). Ma and Nagy (2007), using a 3-D MHD model estimated that the escape flux of ionospheric and pick up ions varies between about 3×10^{24} to 3×10^{23} sec^{-1} under "normal" solar cycle maximum and minimum conditions. Chaufray et al. (2007) using a comprehensive, combined hybrid and precipitation models find that the pick up ion, ENA and sputtered escape fluxes are in the range of 2×10^{23}–3×10^{24}, 4×10^{22}–4×10^{23} and 2×10^{23}–7×10^{23} sec^{-1} respectively. The Mars Express ASPERA instrument measured the escape flux of ions in excess of about 30 eV; they estimated a total flux of about 3.2×10^{23} sec^{-1} during solar minimum (Barabash et al. 2007a, 2007b).

3 Satellites

As discussed before, satellites with an atmosphere can interact with the corotational plasma flow inside a planetary magnetosphere. As a consequence, the corotational plasma near the moon will be disturbed. This section focuses on plasma flow interactions around several Galliean satellites and Titan, the largest moon of Saturn.

3.1 Galilean Satellites

Each of the Galilean satellites of Jupiter has a tenuous atmosphere by terrestrial standards. In the discussion below, supporting references may be found in the book chapter by McGrath et al. (2004). Io is unique in the solar system with an SO_2 atmosphere that is ultimately derived from continual volcanic eruptions, either directly injected into the atmosphere or indirectly via deposition on its surface and subsequent sublimation. Normally, Io's atmosphere is most abundant at lower latitudes where most of the volcanic activity occurs. In the southern polar region the atmosphere almost becomes an exosphere, and also in the northern polar region when the large volcano Tvastar is inactive. The atmosphere is longitudinally asymmetric with higher densities in the trailing and anti-Jupiter quadrant. Accordingly, it is characterized by spatial and temporal variability. Besides SO_2, the other driving gas for volcanoes is S_2, which has a short chemical lifetime in the atmosphere and hence a minor constituent. Other detected gases in the atmosphere include SO, O, S, NaCl, Na, K, as listed in Table 1. SO and NaCl are thermochemical components of hot volcanic gaseous plumes. Photochemistry combined with the magnetospheric interaction with the atmosphere yields SO, O, S, Na, and K. At the equator the surface pressure is about 1 nbar. The canonical mass loss rate from Io is $\sim 2 \times 10^{30}$ amu sec^{-1} and mostly neutrals, rather than ions. This implies that the residence time or lifetime of the atmosphere is a mere 3 days and must be continually re-supplied on a time scale less than a week.

The magnetospheric interaction with Io's atmosphere produces a substantial ionosphere with electron densities as large as 3×10^5 cm^{-3}. The electron densities are organized with respect to the ram direction rather than the subsolar point (Hinson et al. 1998). The highest densities are found on the flanks and the lowest in upstream and downstream directions

Table 1 General properties of the Galilean satellites' atmospheres, ionospheres, and magnetospheric interaction[a]

Quantity	IO	Europa	Ganymede	Callisto
Atmospheric source	Volcanoes	Sputtering H_2O ice	Sputtering H_2O ice	Sputtering H_2O ice
$P_{surface}$ (nanobar)	1	0.0007	0.0007	0.008–0.3
Column density (cm^{-2})	3×10^{16} $(0.03–10) \times 10^{16}$	5×10^{14} $(3–7) \times 10^{14}$	5×10^{14} $(1–10) \times 10^{14}$	8×10^{14} CO_2 $\sim 3 \times 10^{16}$ O_2
Major gas	SO_2	O_2	O_2	O_2 or CO_2?
Other species	S_2, SO, O, S, NaCl, Na, K	O, H_2O, H_2, H, Na, K	O, H_2O, H_2, H	CO_2, O, H_2O, H_2, H, CO, C
Thermospheric temperature (K)	$\sim$2000	$\sim$2000	$\sim$1000	$\sim$2000
Escape rate (s^{-1})	1.6×10^{28} SO_2	5×10^{26} O_2	?	?
Lifetime (d)	3	4	?	?
Peak electron density (cm^{-3})	$\sim 3 \times 10^5$	$\sim 10^4$	$<10^3$	$\sim 2 \times 10^4$
Σ_P (mho)	$\sim$100 (ec)	$\sim$14	$\leq$100 (pu)	$\sim 10^4$ (ec)
Σ_H (mho)	$\sim$150 (ec)	$\sim$10	$\leq$100 (pu)	$\sim 10^4$ (ec)
Σ_A (mho)	5	1.7	2	1.3
$V_{rel\ plasma}$ ($km\ s^{-1}$)	57	90	180	170
V_{alfven} ($km\ s^{-1}$)	220	460	375	600
M_A	0.25	0.2	0.5	0.3
M_S	1.65	1.75	2.4	2.4
B (nT)	1800, J	420, J	750, G	35, J
E_0 ($V\ m^{-1}$)	0.1	0.04	0.02	0.006
$\alpha = E_i/E_0$	0.1	0.2	–	0.001
I_{iono} (10^6 A)	10	1.4	–	0.15
Joule heating (W)	4×10^{11}	3×10^{10}	?	$\leq 10^9$
$\lvert \Delta B/B_J \rvert$	0.45	0.4	–	0.35
Magnetospheric ion power (W)	2×10^9	1×10^8	$\sim 10^7$	$\sim 10^4$
Solar EUV/UV power input (W)	7×10^9	$<10^8$	$<10^8$	5×10^7, CO_2

ec = elastic collisions, pu = pickup, J = Jupiter's magnetic field, G = Ganymede's magnetic field
[a]Adopted from Strobel (2005)

with a downstream wake or tail consistent with acceleration of plasma up to local corotation speed. There is no evidence that Io possesses either an intrinsic or an induced magnetic field that exceeds the local Jovian magnetic field strength above the surface.

Europa and Ganymede have atmospheres produced by ion sputtering of the water ice surfaces, comprised mostly of O_2, with approximately picobar surface pressures (Johnson 1990). Whereas the sputtering yield of H_2O molecules is about 7 times larger than the O_2 yield, H_2O molecules stick on every collision with the cold surface, whereas the O_2 sticking coefficient is much lower ($\sim$0.001) and hence O_2 emerges as the dominant atmospheric constituent. Other products of ion sputtering such as H, H_2, O, and OH escape the weak gravitational field of these satellites, especially H and H_2. For Europa, Smyth and Marconi (2006) estimate an escape rate of 2×10^{27} H_2 s^{-1}, which creates a gas torus. These O_2 atmospheres have column densities in the range of $(1–10) \times 10^{14}$ cm^{-2}. Europa's O_2 atmosphere escapes by magnetospheric interaction with a rate $\sim 5 \times 10^{26}$ O_2 s^{-1} leading to a short lifetime of ~ 2 days. A big difference between Europa and Ganymede is the extent of their ionospheres and the fact the Ganymede has an intrinsic magnetic field, while Europa has an induced magnetic field presumably from a conducting ocean below the surface moving through the strong Jovian magnetic field. Peak electron densities on Europa reach $\sim 10^4$ cm^{-3} (Kliore et al. 1997), whereas on Ganymede an ionosphere is only marginally detectable with at most densities of $\leq 10^3$ cm^{-3} in the equatorial regions. However for Ganymede, other estimates are available based on interpretation of HST UV auroral observations by Eviatar et al. (2001). It is highly probable that the electron density could reach $\sim 10^5$ cm^{-3}, along the auroral ovals located at the separatrix between Ganymede's intrinsic magnetic field and the Jovian magnetic field. Note that these latitudes were not probed by Galileo radio occultations.

For Callisto, Carlson (1999) provided a direct measurement of the CO_2, using near-IR bright limb emission and inferred a vertical column density of 8×10^{14} cm^{-2}, whereas Kliore et al. (2002) inferred from radio occultation measurements of the electron density that another, larger component must be present in the atmosphere to explain electron densities of $\sim 10^4$ cm^{-3}. They suggested neutral column densities on the order of 3×10^{16} cm^{-2}, most probably O_2 by analogy with Europa and Ganymede. Thus the surface pressure is at most ~ 0.3 nbars. Of particular importance, Kliore et al. (2002) concluded that a necessary condition to have an ionosphere is that the trailing hemisphere be partially solar illuminated.

For Io and Europa, the harsh environment of Jupiter's inner magnetosphere produces atmospheric chemistry driven by magnetospheric electrons as well as by solar radiation. In fact, electron impact dominates atmospheric ionization and UV emission. With the much lower magnetospheric electron densities at the orbits of Ganymede and Callisto, solar photoionization exceeds electron impact ionization.

With observations available for neutral and electron densities specifically measured ionospheric electron density profiles, ($\sim 10^4$–10^5 cm^{-3}) for Io, Europa, and Callisto, it is straightforward, with a reasonable assumption about the main ion's mass, to compute Pedersen (Σ_P) and Hall (Σ_H) conductances and compare them with the Alfven conductance (Σ_A). Typical values are given in Table 1. The ionospheric Pedersen and Hall conductances in all cases substantially exceed the respective Alfven conductances. For satellite electrodynamic interactions, other relevant quantities include the velocity of magnetospheric plasma relative to the satellite, $V_{\text{rel plasma}}$, the Alfven velocity, V_{alfven}, the Alfven Mach number, M_A, the sound speed Mach number, M_S, the magnetic field strength, B, and the corotation electric field, E_0. The strength of the electrodynamic interactions of the Galilean satellites with the Io plasma torus (plasma in Jupiter's inner magnetosphere) can be quantified by the ratio of the ionospheric electric field to the corotation electric field, α, the total ionospheric electric current, I_{iono}, the Joule heating, and the perturbation magnetic field due to ionospheric currents relative to the background field, $|\Delta B/B_J|$, for which typical values may be found in Table 1. As a result of these interactions large electric currents flow through their ionospheres ($\sim$10, 1.4, 0.15 MA for Io, Europa, and Callisto, respectively) accompanied by respective large Joule heating rates ($\sim$40, 3, $\leq 0.1 10^{10}$ W) that are the dominant heating mechanism of their atmospheres, leading to estimates of high temperatures $\sim$1000–2000 K for altitudes 2 scale heights above their surfaces (cf. Table 1).

The consequences of Callisto's large distance from Jupiter, the large electron and neutral densities and only an induced magnetic field, are a very highly conducting ionosphere with

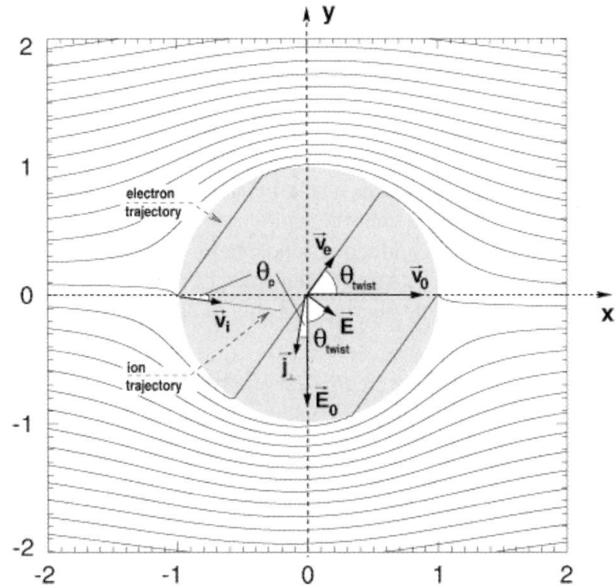

Fig. 10 Isocontours of the electric potential for an illustrative analytic solution with constant conductances $\Sigma_A = 5$, $\Sigma_H = 50$, $\Sigma_P = 25$ S, where the ionospheric field $\vec{E_i} = \vec{E}$, the velocities of the corotation plasma, $\vec{v_0}$, electrons, ions, $\vec{v_i}$, and electrons, $\vec{v_e}$ (from Saur et al. 1999)

conductances on the order of 10^4 mho, that greatly shield the atmosphere from penetration of magnetospheric plasma. One measure of this shielding is the ratio of the ionospheric electric field to the corotation electric field ($\alpha = E_i/E_0$), which is only 0.001 for Callisto in comparison to 0.1 and 0.2 for Io and Europa, respectively.

In order to gain physical insight into magnetospheric interaction with the Galilean satellites, which do not have an intrinsic magnetic field, the analytic Alfven wind model of Saur et al. (1999) is adopted. With the standard magnetospheric Cartesian coordinate system (x in the direction of corotation flow, y directed towards Jupiter and z upward with the magnetic field $\vec{B}$ in the minus z direction), Saur et al. obtained for the ionospheric electric field (cf. Fig. 10)

$$\vec{E}_i = -E_0 \frac{2\Sigma_A}{\Sigma_H^2 + (\Sigma_P + 2\Sigma_A)^2} \begin{pmatrix} -\Sigma_H \\ \Sigma_P + 2\Sigma_A \\ 0 \end{pmatrix}. \tag{1}$$

With the ionospheric conductances much larger than the Alfven conductance, the magnitude of the ionospheric electric field is considerably reduced from the external corotation electric field, $\vec{E_0}$. This is the $\alpha = E_i/E_0$ quantity in Table 1. Note that Hall conductances cause the ionospheric electric field to have a positive component in the x direction and the ionospheric electric field, $\vec{E}_i$, to be rotated counterclockwise from the direction of the corotation electric field, $\vec{E_0}$, by an angle

$$\tan\theta_{\text{twist}} = -\frac{\Sigma_H}{\Sigma_P + 2\Sigma_A} \tag{2}$$

(cf. Fig. 10). The perpendicular electric current through one hemisphere of Io is

$$\vec{J}_\perp = -E_0 \frac{2\Sigma_A}{\Sigma_H^2 + (\Sigma_P + 2\Sigma_A)^2} \begin{pmatrix} 2\Sigma_H\Sigma_A \\ \Sigma_H^2 + \Sigma_P(\Sigma_P + 2\Sigma_A) \\ 0 \end{pmatrix}. \tag{3}$$

The perpendicular electric current is rotated clockwise from the direction of the corotation electric field, $\overrightarrow{E_0}$, because of Hall conductance, by an angle

$$\tan\theta_\mathrm{p} = \frac{2\Sigma_\mathrm{H}\Sigma_\mathrm{A}}{\Sigma_\mathrm{H}^2 + \Sigma_\mathrm{P}(\Sigma_\mathrm{P} + 2\Sigma_\mathrm{A})} \quad (4)$$

(cf. Fig. 10) and generates a small Hall current oppositely directed from the corotation flow. This is also the angle that the incident magnetospheric ions are deflected when entering the highly anisotropic conducting ionosphere. The angle, θ_twist, is much larger than the angle, θ_P, because $\Sigma_\mathrm{P,H} \gg \Sigma_\mathrm{A}$. As a consequence of the current flowing through the highly conducting ionosphere, the Joule heating rate due to dissipation is given by

$$P = 2\pi R_\mathrm{S}^2 E_0^2 \Sigma_\mathrm{P} \frac{(2\Sigma_\mathrm{A})^2}{\Sigma_\mathrm{H}^2 + \Sigma_\mathrm{P}(\Sigma_\mathrm{P} + 2\Sigma_\mathrm{A})}. \quad (5)$$

The magnetospheric plasma interaction with these atmospheres has been remotely sensed by HST/STIS observations. The regions of brightest UV line emissions from $\mathrm{O_I}$ (and $\mathrm{S_I}$ on Io) correspond to the regions of maximum deposition of magnetospheric electron power. On Io, some of this UV auroral emission is produced in equatorial spots organized by Jupiter's magnetospheric field and brightest where the field is closest to the surface, but above the limb. Theoretical models can explain why UV emission is preferentially brighter at the equator than the poles (Saur et al. 2000). On Europa with a thinner atmosphere, the UV emission is primarily limb glow with possibly one bright region on the disk. The latter is not understood. On Ganymede UV emission is mostly confined to the auroral oval regions, whereas on Callisto no UV emission has been detected other than resonance scattering of solar Lyman-α by a hydrogen atom corona. For Callisto the absence of UV emission is due in part to the very low magnetospheric electron densities and to strong shielding by the highly conducting ionosphere.

3.2 Titan

As the largest moon of Saturn, Titan has an extensive atmosphere/ionosphere system (Hartle et al. 1982; Nagy and Cravens 1998; Wahlund et al. 2005; Waite et al. 2005; Cravens et al. 2006) with no appreciable intrinsic magnetic field (Ness et al. 1982; Backes et al. 2005). Titan's orbit is at 20 R_S (R_S is the radius of Saturn) from Saturn and is located inside Saturn's magnetosphere for nominal solar wind conditions. The early knowledge of Titan and its plasma interaction are based on the observations by Voyager 1 spacecraft when it encountered Titan on November 12, 1980 (Ness et al. 1982; Neubauer et al. 1984). The ongoing Cassini mission will have made more than 40 Titan flybys, by the end of its nominal mission. These flybys provide a great opportunity to study in detail Titan's interaction with Saturn's magnetospheric plasma under a wide variety of conditions.

The magnetospheric conditions near Titan play an important role in the evolution of Titan's atmosphere and ionosphere. The corotational plasma proves to be both a source and a loss of ions for Titan. The incident plasma can add a significant amount of H^+ and O^+ into Titan's atmosphere. These ions will in turn deposit their energy in the atmosphere, influencing chemical reactions, ionizing neutral species and heating the atmosphere. In addition the electric and magnetic fields present in the Saturn's outer magnetosphere can pick up ionized species from Titan's ionosphere and thus remove them. In the next section the properties of the plasma near Titan's orbit are outlined. General interaction features are discussed next. The current estimates of the amount of H^+ and O^+ that is deposited into Titan's atmosphere and estimates of the loss rates for various ionospheric species by the pick up process are also presented.

3.2.1 Plasma Properties Near Titan's Orbit

The plasma conditions in Saturn's outer magnetosphere along Titan's orbit are highly variable. The morphology of the magnetosphere is sensitive to the incident solar wind. Observations of Saturn's magnetospheric magnetic field using the Cassini magnetometer suggest that the magnetic equator is distorted northward (for northern hemisphere winter) not only in the magnetotail, but also on the dayside and flanks (Arridge et al. 2008). Thus, Saturn's magnetic equator and equatorial current sheet form a "bowl-shape" geometry. A magnetodisc is prevented from forming near the sub solar region by the solar wind, but forms tailward of 0900 SLT (cf. Arridge et al. 2006 and references there in). The sensitivity of the magnetosphere to the solar wind means that there is a possibility for Titan to find itself outside of Saturn's magnetosphere in the magnetosheath or even possibly the solar wind. A statistical study by Lundberg et al. (2005) found that for Titan located near the sub solar point, 56.7% of its time would be in the magnetosphere, 42.1% in the magnetosheath and 1.3% in the solar wind. It is not known yet how much of Titan's orbit is spent within Saturn's magnetodisc. The plasma conditions in all of these regions are still being investigated by Cassini. The discussion here is confined to plasma conditions during the initial Cassini flybys of Titan.

The Cassini Plasma Spectrometer (CAPS) observations of the Ta and Tb encounters indicated a flow speed of 120–160 km/s for the incident plasma with respect to Titan (Szego et al. 2005). They also found that the plasma is composed of H^+ and an additional heavy species of mass 14–16 amu. CAPS found the plasma conditions during the TA and T5 encounters to be quite variable. The plasma consists of a mixture of hot, low-density flux tubes and cold, high-density ones. These structures had size scales on order of one Saturn radius (60330 km or about 23 R_T). The electron number density across these flux tubes varied by an order of magnitude centered around 10^{-1} cm^{-3}. The mean temperature of the heavy ion component was found by Cassini to be about 1 keV, compared with 2.9 keV temperature derived by Voyager. Further analysis by Hartle et al. (2006a) found that an additional light H_2^+ species was present. They found that the ambient plasma was hot, with energies of $\sim$0.16–6.3 keV for H^+, $\sim$0.1–4 keV for H_2^+ and $\sim$1–4 keV for O^+. The distribution function of the ambient ion species is often considered to be a Maxwellian. However, analysis by Szego et al. (2005) found that the velocity distribution of the ambient O^+ is a shell. It was shown by Ledvina et al. (2005) that the ambient heavy ion component observed by Voyager could be described in terms of either a Maxwellian or a shell distribution.

In addition to the ion observations made by the CAPS instrument the magnetospheric imaging instrument (MIMI) has observed very energetic (up to a few MeV) H^+ near Titan's orbit. The MIMI instrument has found a difference in their observed ion spectra depending on whether or not Titan is in Saturn's magnetodisc. The ion spectra above $\sim$600 keV is fairly constant regardless of Titan's location. Below this energy the flux of H^+ is larger when Titan is in the magnetodisc (see Cravens et al. 2008 for further discussion). They also find evidence for very energetic O^+ ions near Titan.

3.2.2 General Interaction Features

Titan's interaction with the Saturnian magnetospheric plasma flow is similar in many ways to the solar wind interaction with Venus/Mars, but normally with subsonic rather than supersonic flow conditions. The subsonic nature means that a bow shock is not expected to form

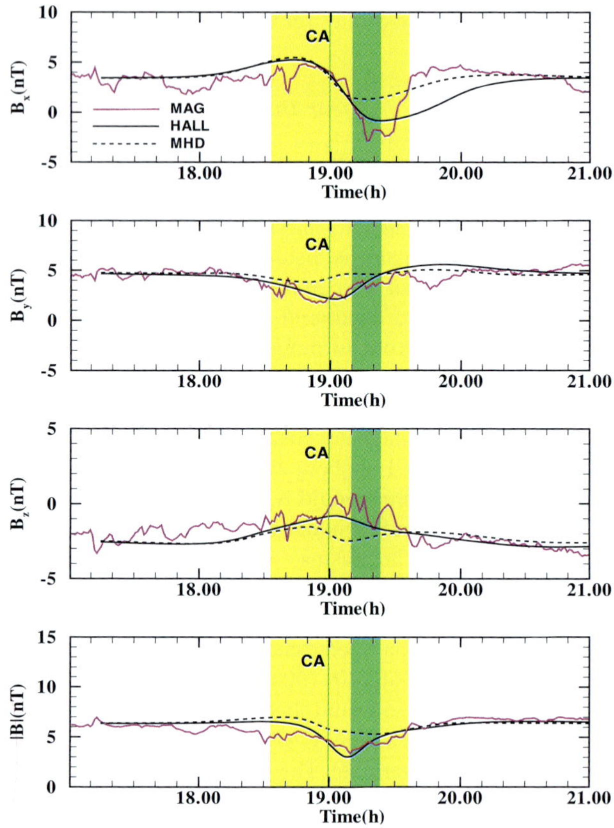

Fig. 11 Model and data comparison of magnetic field along T9 flyby. The green lines indicate the closest approach time. (From Ma et al. 2007a)

in front of Titan. A pair of Alfven wings forms along the magnetic field lines due to perturbations of field-aligned currents (Ness et al. 1982; Luhmann et al. 2004). On the upstream side, Saturn's magnetospheric plasma flow starts to slow down from about 8 R_T (Titan radius) due to mass loading with Titan's extended atmosphere (Hartle et al. 2006a, 2006b), while significant pile-up and strong draping of the magnetic field lines begins around 2 to 3 R_T in front of the satellite. In the tail region, an induced bipolar magnetic tail was observed by Voyager spacecraft and three thin current regions were crossed (Ness et al. 1982). A clear tail structure was seen by the Cassini TA, TB and T3 observations (Backes et al. 2005; Neubauer et al. 2006), as the magnetic field reversed direction suddenly when Cassini passed from one magnetic tail lobe to the other. The wake region is highly dynamic, and both the location and width of the current sheet are closely related to the upstream plasma pressure and magnetic field directions.

The Cassini spacecraft passed through the distant downstream region of Titan at 18:59:30 UT on Dec. 26, 2005, which is referred to as the T9 flyby. This flyby provided important information of the highly dynamic wake region and has been discussed in detail by a series papers in a special issue of *Geophysical Research Letters* (Bertucci et al. 2007; Wei et al. 2007; Coates et al. 2007; Kallio et al. 2007; Simon et al. 2007). Figure 11 shows the comparison between the calculated and observed magnetic field values along the T9 trajectory for plus and minus 2 hours of closest approach (CA) from Ma et al. (2007a). The overall trends of magnetic field vectors, especially the sharp decrease of B_X, are reasonably well repro-

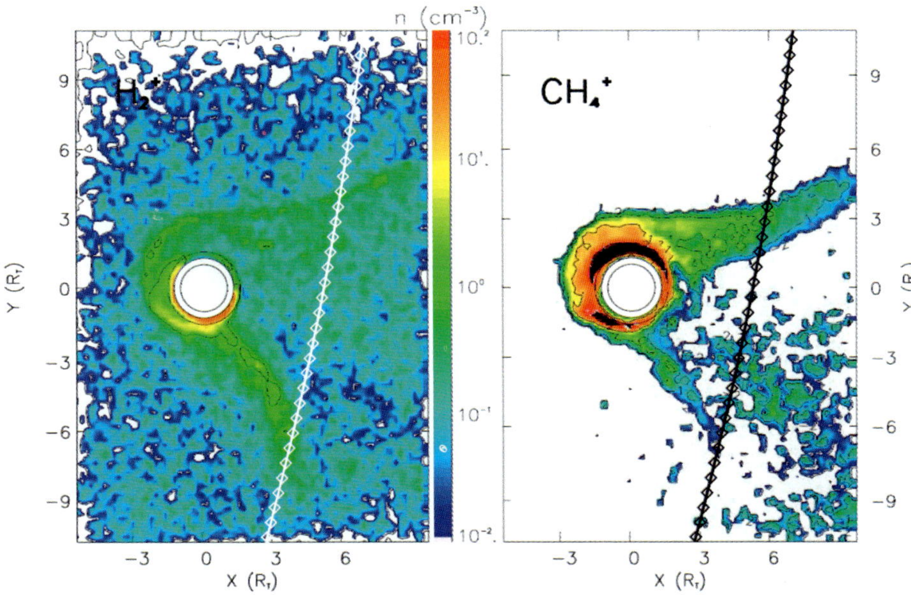

Fig. 12 Density maps of (*left*) H_2^+ and (*right*) CH_4^+ in the *XY* plane of Titan. The trajectory of Cassini is indicated in the simulation coordinate system. (From Modolo et al. 2007a)

duced by the Hall MHD model, indicating the importance of kinetic effects in the interaction process. They also discussed in detail why the Hall MHD model results fit the observations fairly well even when the ion gyroradii of upstream plasma ions are larger than Titan's size. A difference in the plasma composition, between the inbound and outbound segments of the flybys was observed by the CAPS experiment (Szego et al. 2007). Heavy ions of mass 14–18 AMU and mass 30 AMU were only observed on the Saturn-facing edge (positive *Y* direction) of the wake while the light ions with an ion mass of 2 AMU are present on the anti-Saturn edge (negative *Y* direction). As shown in Fig. 12, such features were successfully reproduced by Modolo et al. (2007a).

Although there is significant magnetic enhancement in the upstream region close to Titan, no statistical study of the location of the magnetic pile-up boundary has been performed. Clear ionopause signatures have been observed at Titan only during one Cassini pass to date: T34 flyby (Ma et al. 2007b). During this pass, Titan was located near 18 SLT, so that the upstream region was corresponding to the dayside ionosphere. The spacecraft passed through the upstream equatorial plane. As can be seen from Fig. 13, the magnetic field strength drops sharply near the closest approach, corresponding to what could be considered an ionopause around 1500 km. Since the peak ionospheric thermal pressure in the upstream region varies significantly when Titan moves along its orbit, whether the ionopause exists or not during other Titan orbital location is still an open question.

3.2.3 Ambient Ion Fluxes into Titan's Atmosphere

The gyroradii of the ambient ions can be significant when compared to the size of Titan. For instance the gyroradius of a 5 keV H^+ and O^+ ion is 0.79 R_T and 3.2 R_T respectively. The trajectories of ambient ions near Titan have been simulated by Ledvina et al. (2000,

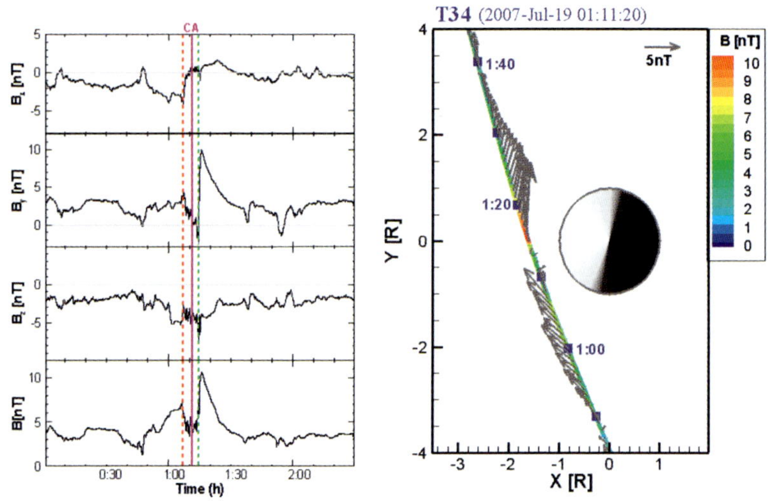

Fig. 13 Cassini Magnetometer observation during T34 flyby. The *dashed lines* indicate the ionopause locations (Ma et al. 2007b)

2005). They combined the electric and magnetic fields from a MHD simulation of Titan's plasma interaction during the Voyager encounter, with a Monte Carlo/test particle approach. Their approach though not self-consistent was able to predict, describe and explain much of the collective ion behavior near Titan. They argue that the larger the ion gyroradius the less sensitive the ambient ion is to Titan's induced magnetic field and the more likely it is to enter Titan's atmosphere. Their simulations predicted that most of the ambient H^+ ions (energies <3 keV) are shielded from Titan's atmosphere by its induced magnetic field. In contrast the heavier ambient ion species were much less sensitive to the induced magnetosphere and easily entered Titan's atmosphere. They further calculated that for conditions present during the Voyager 1 encounter 3.9×10^{22} H^+ ions/s would enter Titan's atmosphere. The ion flux into Titan's atmosphere was found to be dependent on the distribution function for the heavy ambient ion species. If the heavy ambient ion species had a Maxwellian distribution then 1.7×10^{24} ions/s entered Titan's atmosphere. This flux dropped to 2.9×10^{23} ions/s if the distribution function was a shell.

Recent self-consistent hybrid simulations by Sillanpää et al. (2007) calculate a flux of 1.3×10^{24} ions/s of O^+ entering Titan's atmosphere using a ambient Maxwellian distribution. They found an H^+ flux of 1.3–1.4×10^{23} ions/s. This is more than twice that of the Ledvina et al. (2005) result. Sillanpää et al. (2007) attribute this to the symmetric field results in the MHD simulation used by Ledvina et al. (2005). Another possibility is that Ledvina et al. (2005) did not allow for the possibility of H^+ ions moving along the draped field towards Titan. This effect was included in the Sillanpaa et al., simulation. What the difference in these results illustrates is that the H^+ flux into Titan's atmosphere is sensitive to the configuration of Titan's induced magnetosphere.

Hartle et al. (2006a) reported that both the Voyager PLS instrument and the Cassini CAPS instrument observed an O^+ clearing region around Titan. From these observations they estimated that Titan absorbs $\sim 5.6 \times 10^{23}$ ions/s of ambient O^+. They are quick to point out that the interaction region is complex requiring a full 3D kinetic treatment for a more accurate rate. Still their estimate is within a factor of 3 of the estimates given above.

There has been much interest recently in the very high energy H^+ ions observed by MIMI and their atmospheric consequences. The MIMI team (D. Mitchell, private communication) has observed ions (mostly H^+) precipitating into Titan's atmosphere below an altitude of 1000 km. They estimate that these precipitating ions (energies above 10 keV) deliver from 5×10^{-5} to 5×10^{-3} ergs/cm^2/s to Titan's atmosphere below 1000 km. Energetic O^+ is often present, but they have not surveyed its contribution to the energy deposition yet. Recently Cravens et al. (2008) have examined the atmospheric consequences of these energetic ions. They found that these precipitating ions can lead to significant ionization around 600–800 km, which is well below the altitude of the peak ion production from photoionizataion. This ionization process may explain recent ionospheric layers observed by the radio occultation observations at Titan (Kliore et al. 2007).

3.2.4 Pickup Ion Fluxes into Titan's Atmosphere

Another source of ion precipitation into Titan's atmosphere comes from Titan itself. Neutrals in Titan's upper atmosphere and exosphere can become ionized. These ions can be picked up by the surrounding plasma flow and be accelerated back into Titan's atmosphere. This can in turn heat the atmosphere. The motional electric field that picks up the ions is directional so these ions will tend to be accelerated away from Titan in the anti-Saturn facing hemisphere and into Titan on the Saturn facing hemisphere. This will result in asymmetric heating of the atmosphere.

The study of Ledvina et al. (2005) also examined the re-entering pickup ions. They found that 1.4×10^{22}, 5.6×10^{23} and 8.7×10^{23} ion/s for 1, 14 and 28 amu ions re-entered Titan's atmosphere. Most of these ions tended to impact the Saturn facing hemisphere of Titan. However, they found that Titan's induced magnetosphere also allowed a fair amount of ion energy to be deposited in the anti-Saturn hemisphere. In this study a symmetric ionosphere was produced which is a significant model limitation since it is known that photoionization is directional.

Tseng et al. (2008) improved upon the study of Ledvina et al., by removing the symmetry assumptions. They used the field output from a MHD simulation of Titan's plasma interaction at different regions in Saturn's magnetosphere. Thus these fields were not symmetric about Titan but included the effects of a solar zenith angle production source. They also ran a symmetric case for comparisons with Ledvina et al. (2005). They found that their ion fluxes ranged from 2.22–2.31 $\times 10^{22}$ ions/s for H_2^+, 3.33–4.56 $\times 10^{23}$ ions/s for CH_4^+ and 5.70–8.02 $\times 10^{23}$ ions/s for N_2^+. These values are in good agreement with the estimates of Ledvina et al. (2005). The variations are due to the change in the solar zenith angle. Both Tseng et al. (2008) and Ledvina et al. (2005) are also in agreement with the estimated pickup ion energy deposited into Titan's atmosphere.

Michael and Johnson (2005) used a direct simulation Monte Carlo (DSMC) method to study the energy deposition and heating of Titan's upper atmosphere. They found that pickup ions can deposit more energy near the exobase than solar radiation. However sputtering and collisional transport to the lower atmosphere causes this energy to be lost resulting in a temperature increase near the exobase of just a few degrees.

3.2.5 Ion Loss from Titan

Neutral species becoming ionized and getting picked up by the motional electric field is thought to be one of the chief loss mechanisms at Titan. This process was examined by Ledvina et al. (2005). They found that Titan's induced magnetosphere could act to prevent ions from being picked up by the external flow or even scatter some of the ions back into Titan. They calculated that the pick up process would result in a loss of 6.6×10^{23}, 7.4×10^{23} and 6.3×10^{23} ions/s for 1, 14 and 28 amu pick up ion species. Most of these ions would have energies of less than 3 keV near Titan. Their velocity space distributions would be non-gyrotropic rings.

MHD simulations computed with upstream plasma parameters similar to those encountered for Ta and Tb encounters, gave an estimate of the total ion escape fluxes of 5.1×10^{24} ions/s and 2.6×10^{24} ions/s respectively (Ma et al. 2006). Hybrid simulations performed by Sillanpää et al. (2006) indicate a net outflow of methane ions, of the order of 1.1×10^{25} ions/s, with different Titan's orbital positions. Hybrid simulation by Modolo et al. (2007a) performed with the T9 conditions, suggests a total escape flux of 5.6×10^{25} ions/s (with a contribution of 1.3, 2.4 and 1.9×10^{25} ions/s for N_2^+, CH_4^+ and H_2^+ respectively).

Observations of pickup ions by the CAPS instrument are reported in Hartle et al. (2006a). They find that the principal pickup ion species are H^+, H_2^+, N^+, CH_4^+, and N_2^+. Hartle et al. (2006a) state that the pickup ion distribution function should be a ring. However, close to Titan the distribution can be interpreted in terms of narrow energy ion beams. This is because the ion gyroradii are much larger than the scale height of the source neutral profile. To the best of our knowledge CAPS has not yet published estimates of the pickup ion loss rates. However, Eviatar et al. (1982) estimated a net loss of 3×10^{24} ions/s based on the Voyager encounter. The total ion outflow is estimated at 2–7 $\times 10^{25}$ ions/s based on Langmuir Probe observations of the T9 flyby, if one assumes a cylindrical geometry for the plasma wake (Modolo et al. 2007b). Further observations and simulations are needed to better understand the extent of atmospheric loss by the pickup process.

4 Comets

Comets are small solar system bodies surrounded by extensive dusty atmospheres that evolve along their orbit. The plasma environment around comets is affected by the cometary dust and gas coma, as well as the persisting solar wind. Most of our current knowledge of comets and their plasma environment are based on optical observations using telescopes, and direct observations from a few flybys to comets 21P/Giacobini-Zinner, 1P/Halley, 26P/Grigg-Skjellerup and 19P/Borrelly. Nevertheless, the plasma environment around comets provides us a unique laboratory filled with current systems, waves and dynamic boundaries. Most of the experimental evidence on ionospheres of comets comes from the Giotto mission to comet 1P/Halley in 1986 and to comet 26P/Grigg-Skjellerup in 1992. The most surprising cometary ion data were probably the data gathered with the solar wind experiment and the magnetometer on the solar mission Ulysses 3.4 AU away from comet C/1996 B2 (Hyakutake) and 1.6 AU away from C/2006 P1 (McNaught).

4.1 Overview of the Comet Interaction with the Solar Wind Plasma

Neutral atoms and molecules of cometary origin that are released by the sublimation of dusty/icy mix of the upper several centimeter layers at the surface of the cometary nucleus move along ballistic trajectories and become ionized with a characteristic ionization lifetime of 10^6–10^7 s. Freshly born ions are accelerated by the motional electric field of the high-speed solar wind flow. The ion trajectory is cycloidal, resulting from the superposition of gyration and $\mathbf{E} \times \mathbf{B}$ drift. The corresponding velocity-space distribution is a ring-beam distribution, which has large velocity space gradients and is unstable to the generation of low frequency transverse waves. The combination of ambient and self-generated

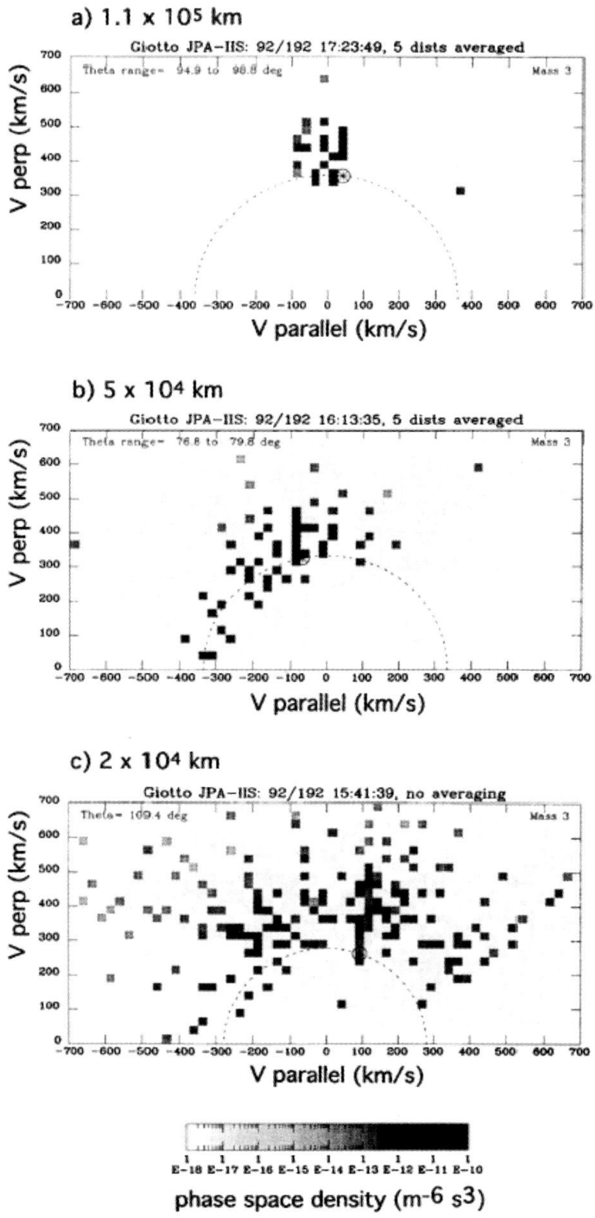

Fig. 14 Examples of the water group ion distributions observed by the Giotto IIS at distances of (**a**) 1.1×10^4 km, (**b**) 5×10^4 km, and (**c**) 2×10^4 km on the G-S outbound pass. Measured zeros are represented by a *light shade of grey*. Areas left white are outside the instrument field of view. Reprinted from Huddleston et al. (1993)

magnetic field turbulence pitch-angle scatters the newborn ions from the pickup ring. As a result of this process the pitch angles of the pickup-ring particles are scattered on the spherical velocity space shell around the local solar wind velocity (e.g. Galeev et al. 1991; Neugebauer et al. 1991; Motschmann and Glassmeier 1993; Coates et al. 1993). Water group ion distribution functions were measured by the Giotto spacecraft during the flyby of comet Grigg-Skjellerup at a range of distances from the nucleus showing the evolution of the pickup ring distribution. See Fig. 14 from Huddleston et al. (1993).

Conservation of momentum and energy causes the solar wind to be decelerated as newly born charged particles are picked up by the magnetized plasma flow. Continuous deceleration of the solar wind flow by mass loading occurs until a weak shock forms and the flow becomes subsonic. The cometosheath is located between the cometary shock and the magnetic field free region in the innermost coma. The plasma population in the cometosheath is a changing mixture of ambient solar wind and particles picked up upstream and downstream of the shock. Near the nucleus ion–neutral chemistry and recombination starts to become more and more important. In the inner coma, an "ionopause" separates the solar wind controlled magnetized plasma from the magnetic field-free cometary ionosphere, the diamagnetic cavity. Deep inside the diamagnetic cavity the cometary plasma and the neutral gas are very strongly coupled by ion–neutral collisions (Cravens et al. 1991a, 1991b), and they move radially outward with the same expansion velocity.

Weaker Jupiter family comets, like the Rosetta mission target comet 67P/Churyumov-Gerasimenko, will only have a small diamagnetic cavity and only for heliocentric distances less than ~2 AU (Hansen et al. 2007). For Halley-class comets and brighter at moderate to small heliocentric distances an extensive set of ion-molecule reactions are responsible for determining the ion composition in comets beginning with an initial set of primary photoionization products (Häberli et al. 1997; Rodgers et al. 2004). The main photoionization product in the cometary coma is H_2O^+. These water ions transfer a proton to other neutral molecules, which have a higher proton affinity than the OH radical. The prime example is $H_2O^+ + H_2O \rightarrow H_3O^+ + OH$, but there are other similar reactions that result in the creation of species like HCO^+ that was detected in comet 1995 O1 Hale-Bopp (Lovell et al. 1999) owing to the large diamagnetic cavity of 30,000 km (Gombosi et al. 1997) that results from its huge gas production rate. The proton affinity of the parent molecules determines the reaction rate; therefore protonated ions of molecules with high proton affinity are more abundant (relative to other ions) than ions of their parent molecules (relative to other neutrals) (Häberli et al. 1997). Therefore, the most important reaction in the innermost coma is proton transfer, e.g.

$$H_3O^+ + M \rightarrow H_2O + MH^+ \tag{6}$$

which leads to typical ion mass spectra where the uneven mass numbers contain the highest peaks, due to the fact that C and O containing molecules can be found at even mass numbers, e.g. mass 31 amu/e (CH_2OH^+, protonated formaldehyde) or mass 33 amu/e ($CH_3OH_2^+$, protonated methanol) (Fig. 15).

Further out or in weaker comets, ion–molecule reactions become less important relative to photo-reactions and the most important ion is then the water ion H_2O^+. Even further away from the nucleus molecular ions are slowly replaced by atomic ions. Inside the so-called magnetic cavity ions and neutrals are in a local chemical equilibrium. The ion population therefore strongly follows the neutral composition. Outside of the cavity this local chemical equilibrium is no longer maintained for all species. The neutral and the ion composition decouple. A good example for this behavior is the sulfur ion. Whereas the densities of most of the other ion species decrease in a monotonous fashion over the magnetic cavity boundary following the decrease in the neutral density, sulfur ions are piled up right at the boundary (see Fig. 16). This is due to the fact that the sulfur ion has a very long lifetime. Sulfur ions produced far away upstream from the magnetic cavity are flushed back by the solar wind towards the comet nucleus and pile up at the magnetic cavity boundary, the contact surface. They are no longer in local chemical equilibrium with the neutral parent species.

In situ measurements of the immediate cometary plasma environs are limited to the ICE flyby of comet 21P/Giacobini-Zinner, the Giotto flybys of comet 1P/Halley and 26P/Grigg-

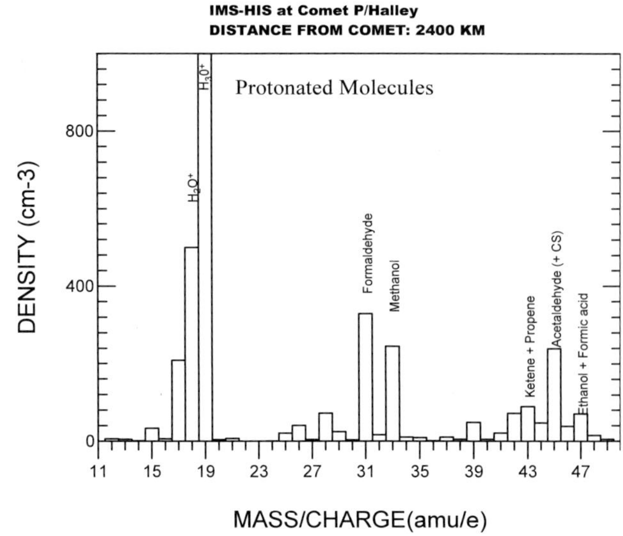

Fig. 15 Ion mass spectrum from the Giotto IMS-instrument

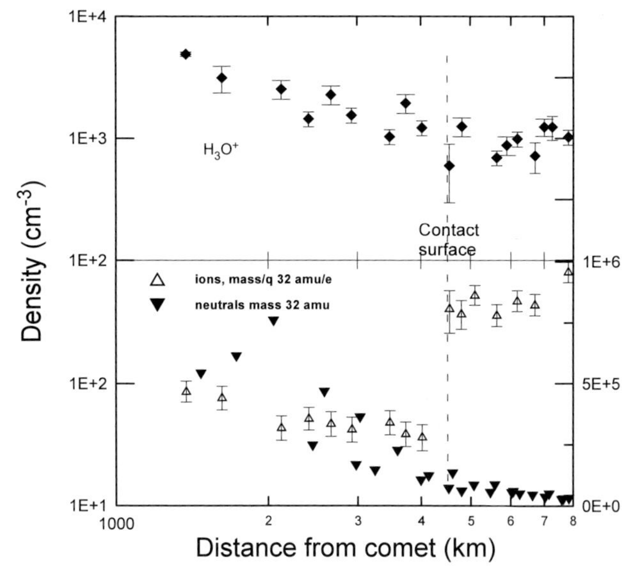

Fig. 16 Data from Giotto during the 1P/Halley flyby: Ion data from the IMS-instrument and neutral data from the NMS-instrument. *Upper panel*: mass/charge 19 amu/ which is due to the hydronium ion; lower panel: mass/charge 32 amu/charge (*left axis*) and neutral mass 32 amu (*right axis*) respectively. The discontinuity of the sulfur ions at the contact surface is due to the long lifetime of these ions

Skjellerup and the flyby of comet 19P/Borrelly by the Deep Space 1 spacecraft. The general features seen in these measurements have been shown to be reasonably reproduced by magnetohydrodynamic (MHD) models of the interaction of the solar wind with the cometary atmosphere (Wegmann 2000; Gombosi et al. 1997; Benna and Mahaffy 2007; Jia et al. 2007). Figure 17 shows steady-state model results for a moderately bright comet at 1 AU from the Sun and showing the classical field-line draping as well as all the normal predicted features, including the upstream bow shock and the inner contact surface and diamagnetic cavity. This result by Jia et al. (2007) is from the latest generation of MHD models implemented on massively parallel computers with a highly refinable computational mesh that can simultaneously resolve small scale phenomena (e.g. shocks and small distances within

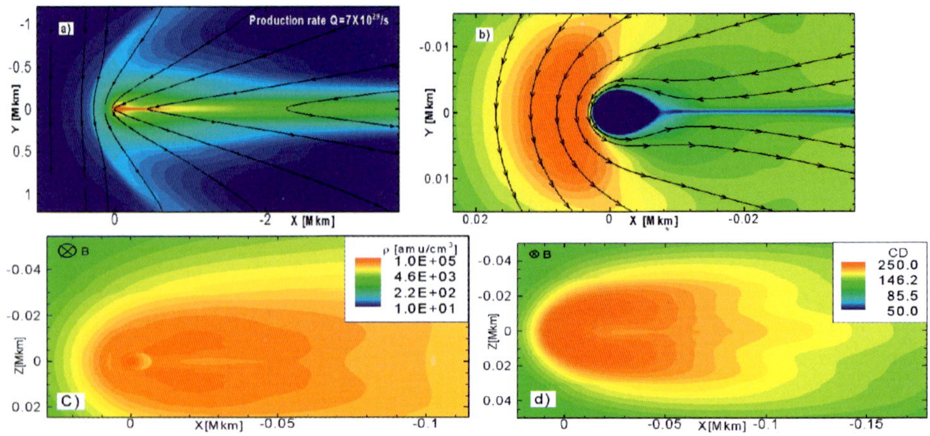

Fig. 17 A comet interacting with a steady-state solar wind. The results of Jia et al. (2007) for the various plasma structures for a moderately bright comet near 1 AU are shown. Magnetic field lines are the *black lines* projected onto the plane containing the interplanetary magnetic field (IMF). **a**) Density contour showing the position of a mass-loaded bow shock. **b**) Magnetic field strength contours in the ecliptic plane showing the structure of the contact surface. **c**) Total ion density contours in the plane perpendicular to the IMF showing the structure of the contact surface and thick tail. **d**) Total ion column density contours in the plane perpendicular to the IMF showing the broad tail. (From Jia et al. 2007)

the contact surface) as well as the global interaction. The more distant plasma tails of comets 1996 B2 Hyakutake (Gloeckler et al. 2000; Jones et al. 2000) and 2006 P McNaught (Neugebauer et al. 2007) have been serendipitously intercepted by the Ulysses spacecraft in 1996 and 2007, respectively. These *in situ* measurements show singly-ionized cometary pickup ions as well as multiply charged C^{++}, O^{++} and O^{3+} which could be cascaded intermediate charge-state heavy solar wind ions, which produce cometary X-rays (Lisse et al. 1996; Cravens 1997b), as well as the surprisingly continuing disturbed solar wind flow even more than 2 AU down stream of the comets.

When comet Hyakutake (C/1996 B2) was observed with the Röntgen Satellite (ROSAT) surprisingly strong X-ray emission was observed (Lisse et al. 1996). The emission morphology was symmetric with respect to a vector from the comet's nucleus toward the sun, but not symmetric around the direction of motion of the comet with respect to interplanetary dust. Observations of other comets in this wavelength range indicate that X-rays are produced by almost all comets. Theoretical and observational work has demonstrated that charge-exchange collisions of highly charged solar wind ions with cometary neutral species can explain this emission. The solar wind contains a large number of minor/heavy ion species with a range of charge states, such as O^{6+}, C^{5+}, N^{5+}, and Si^{10+}. These ions will readily charge transfer with cometary neutrals, producing ions that can be highly excited and consequently emit photons in the extreme ultraviolet and X-ray part of the spectrum. X-ray observations of comets and other solar system objects should be able to provide information on the structure and dynamics of the solar wind flow around these objects (Cravens 1997b).

The ion gyroradius for low activity comet is large compared with the length scale of the comet as an obstacle in the solar wind. In this situation the MHD approximation breaks down and the comet solar wind interaction needs to be considered as a kinetic problem rather than a fluid problem. In such cases hybrid kinetic simulations of comets have been made (Lipatov et al. 1997, 2002; Bagdonat and Motschmann 2002a, 2002b). A careful study of MHD and kinetic simulations for the Rosetta mission target comet 67P/Churyumov-Gerasimenko over the wide range of production rates during its orbit has been done by Hansen et al. (2007). When the comet is nearer perihelion, the hybrid kinetic and MHD simulations give very similar results and a more or less conventional comet/solar wind interaction is present. When the comet is at larger heliocentric distances and has a correspondingly lower production rate, the interaction totally changes character. By 3 AU from the Sun the antisunward directed ion tail totally disappears, and cometary ions simply travel on individual ion-cycloid pickup rings about the direction of the interplanetary magnetic field.

The general aspects of the cometary phenomenon, the production of its neutral dusty-gas atmosphere and its importance to broader planetary studies, specifically the chemistry and physics of the origin and early evolution of the solar system, are discussed in an accompanying papers/chapters by Müller-Wordarg et al. (2008) and Johnson et al. (2008). Ip (2004) presents the most recent up-to-date comprehensive review of the cometary plasma environment. Recent work in modeling the cometary plasma environment has been undertaken to address several important new avenues. Numerical and computational advances have enabled modeling the time dependent phenomena in cometary plasmas such as testing for the causes of the long observed cometary tail disconnection events. Voelzke (2005) presents a review of the past work in this area (Schmidt-Voigt 1989; Wegmann 2000; Yi et al. 1996). Konz et al. (2004) have also explored reconnection events in the context of a 2D resistive MHD calculation. Jia et al. (2007) have tested various types of the tangential solar wind discontinuities that would be expected of a heliospheric current sheet crossing were the trigger for disconnection events as suggested by Brandt and coworkers (Niedner and Brandt 1978; Brandt et al. 1999). They find that the passage of various tangential discontinuities produce cometary tail rays, but only field-flips of greater than 90° when viewed from the appropriate direction with respect to the ambient interplanetary magnetic field direction produce both detectable tail disconnections as well as cometary tail rays. Furthermore they are able to reproduce the recession speeds of the disconnected tail. Figure 18 shows results of Jia et al. (2007) for a time-dependent calculation of a 180° tangential discontinuity passing by a moderately bright comet. In this case the observational geometry with respect to the upstream IMF is not favorable for actually seeing the disconnection, however the appearance of tail rays is quite similar to observations of comet Austin by Bonev and Jockers (1994). The figure shows the event in various stages of the interaction using the disconnection phase nomenclature of Brandt (1982) and Brandt and Chapman (2004).

Benna and Mahaffy (2007) have performed 3D multi-fluid MHD model calculations for the interaction of the solar wind with comet 1P/Halley corresponding to Giotto spacecraft flyby conditions. They include ions, neutrals and electrons as separate interacting fluids. Their comparison with both various Giotto spacecraft data as well as derived quantities compares quite favorably, as had the earlier single-fluid MHD model of Gombosi et al. (1996). In the earlier work the electron temperature was determined separately and imposed on the MHD simulation, whereas in the newer work similar physical assumptions are used to calculate the electron temperature more self-consistently within the MHD calculation itself.

4.2 Tangential Discontinuity and Diamagnetic Cavity; Ionospheric Acceleration and Deceleration

The plasma in the inner coma of an active comet like Halley becomes rather slow (speeds of a few km per second or less) and the cometary plasma density builds up to higher values ($n_e \approx 10^3$–10^4 cm^{-3}). Collisional processes, including chemistry become important in this "cometary ionosphere" (see accompanying paper by Witasse et al. 2008; Cravens 1991a, 1991b). For example, the main ion species for distance less than 2×10^4 km is mass 19

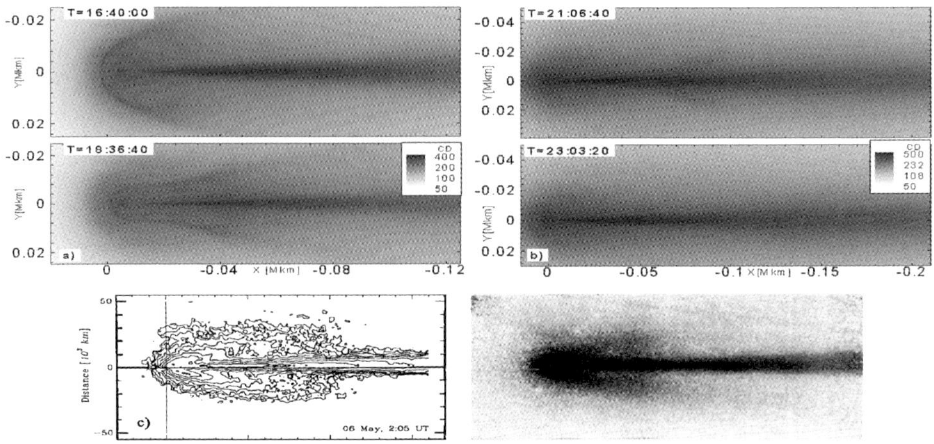

Fig. 18 Comparison of time-dependent MHD model results with observations. Images looking along the direction of the upwind interplanetary magnetic field: Column densities showing tail rays with no observable disconnection in units of 6×10^{10} amu/cm^2. **a**) Phase I. **b**) Phase II–IV, tail rays folded into the tail, but no disconnections can be observed. **c**) Column density of water ions in comet Austin, May 6, 1990, the distances are 10^4 km. Figures **a** and **b** from (Jia et al. 2007), figure **c** from (Bonev and Jockers 1994)

(H_3O^+) rather than H_2O^+ (the main ion produced by photoionization) due to a reaction of H_2O^+ with neutral water (Altwegg et al. 1999). Dissociative recombination is the main loss of plasma in the inner coma, and not transport, with a photochemical density variation that varies inversely with radial distance.

The magnetometer onboard the Giotto spacecraft observed a diamagnetic cavity surrounding the nucleus of Halley at a distance of about 5000 km (Neubauer 1987). The existence of this cavity was successfully explained in papers by Ip and Axford (1987), Cravens (1986, 1989), Puhl-Quinn and Cravens (1995), and Gombosi et al. (1997). Inside the field-free cavity the plasma is coupled with the neutral gas by ion-neutral collisions. Both the ions and the neutrals flow radially away from the nucleus at speeds of about $u_n \approx 1$ km/s. However, in the magnetic barrier outside the diamagnetic cavity boundary the ions are "tied" to the field lines and the neutrals continue to flow outward. In this region an outward ion-neutral friction force of neutrals flowing past stagnated ions is exerted on the plasma. This outward force is balanced by an inward $\mathbf{J} \times \mathbf{B}$ force associated with the magnetic barrier as represented by the following approximate force balance relation:

$$\mathbf{J} \times \mathbf{B} \cong -\nabla(B^2/2\mu_0) = -n_e m_i k_{in} n_n \mathbf{u}_n \tag{7}$$

where n_e is the electron density, m_i is the average ion mass ($\approx$19 amu), n_n is the neutral density, k_{in} is an ion neutral collision coefficient, and u_n is the neutral flow velocity. The $\mathbf{J} \times \mathbf{B}$ force is approximately given by the magnetic pressure gradient force right near the cavity boundary and if the pressure gradient is approximated by the derivative with respect to the radial distance, r, and for a simple $1/r^2$ and $1/r$ (photochemical equilibrium) variations of the neutral and electron densities, respectively, then a simple solution results from integrating this equation is (Cravens 1986):

$$\begin{aligned} B &= B_0\sqrt{[1-(r_{cs}^2/r^2)]} \quad \text{for } r > r_{cs}, \text{ and} \\ B &= 0 \quad \text{for } r < r_{cs} \end{aligned} \tag{8}$$

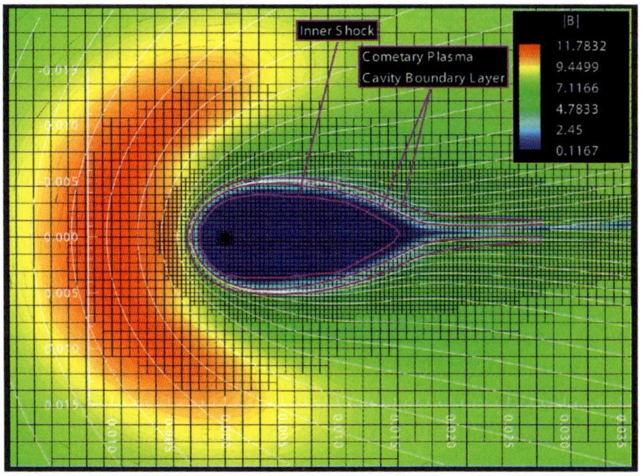

Fig. 19 Magnetic field strength (*color scale* shown in figure) and field lines in the inner coma of comet Halley from a numerical global 3-dimensional MHD model. (From Gombosi et al. 1996)

where B_0 is the field in the magnetic barrier and the radial distance to the diamagnetic cavity boundary for an active comet was found to be given by $r_{cs} = 1.07 \times 10^{-13} Q^{3/4} / B_0$ where Q is the total gas production rate of the comet in units of s^{-1} and B_0 is in units of tesla.

Global MHD models of the cometary plasma environment have reproduced the diamagnetic cavity feature of active comets (e.g., Huebner et al. 1991; Gombosi et al. 1996; Lindgren et al. 1997). Figure 19 shows some magnetic field strengths from a MHD model indicating that the cavity is stretched in the tailward direction, merging with the cometary plasma tail.

5 Ionosphere–Magnetosphere (I-M) coupling

5.1 Terrestrial I-M Interactions

The near-planetary environment, within magnetospheric systems with an ionosphere, can become quite complex. This is because the exchange of mass, momentum and energy between the two is highly nonlinear. It is often extremely difficult to study the magnetosphere-ionosphere coupling because of its complex nature. At many planetary systems, the ionospheric conductivity is uncertain to factors of as much as ten (Eviatar and Richardson 1986; Cheng and Waite 1988; Bunce et al. 2003). Even at Earth, there are many studies that are just attempting to understand the magnitude of the ionospheric conductivity, especially in the dark ionosphere, where the electron density can be difficult to measure. Typically, the conductivity is calculated through the measurement of the electron density and an assumption about the neutral density, or measurement of both, if possible (e.g., Brekke et al. 1974).

One of the best-known processes that give rise to atmospheric plasma flows on Earth is the Dungey (1961) cycle. This process gives rise to a two-cell structure across the auroral/polar regions, in which ionospheric plasma flows in the anti-sunward direction from noon to midnight, and then returns sunward along the dawn and dusk flanks. The process is best understood by considering the reconnection of the Interplanetary Magnetic Field (IMF) lines with the terrestrial field lines at the front of the magnetosphere. The result is that terrestrial field lines that were closed at the front of the magnetosphere are now open and connected to the solar wind. These reconnected field lines are then dragged along by the on-flowing solar wind, and the ionospheric plasma at their footprints in the atmosphere is dragged along with them (cf. Kelley 1989; Schunk and Nagy 2009). Were this process to continue indefinitely, the entire terrestrial field would be connected to the IMF. But as the now-open field lines are swept back into the magnetotail, they move inwards towards the mid-plane where they can reconnect once more, and drift from the midnight–dawn or midnight-dusk sector back towards noon. For Earth, this whole cycle lasts about 3 hours, or 0.125 of a planetary rotation.

One of the most important considerations, from the magnetospheric stand point, is the closure of the currents generated by this reconnection cycle through the ionosphere. The dense ionospheric plasma, mixed with a neutral gas that is typically many orders of magnitude more dense, allows large conductivities to form, serving the purpose of allowing magnetospheric currents to close. This, in fact, serves an additional role—namely the horizontal current allows a diffusion region to form in the ionosphere, such that the field-lines below the ionosphere (which are frozen to the planet) slip away from field-lines above the ionosphere, which are driven strongly by electric fields from the external medium (i.e., the solar wind in most cases). The ionospheric conductance, therefore, allows there to be convection in the magnetosphere. Simplistically, it regulates the speed of the convection—a large conductance slows the fields down, while a smaller conductance allows very large electric fields and therefore very fast flows (Fedder and Lyon 1987; Ridley et al. 2004).

The coupling between a planetary magnetosphere and its ionosphere is not simply just field-aligned currents. Figure 20 illustrates the complex interaction that occurs between the two regions. This figure illustrates what would pass back and forth between different models of the regions. There are typically three feedbacks between the ionosphere and magnetosphere that are studied: (1) the effect of the ionospheric conductivity on the generation of electric fields within the ionosphere and magnetosphere; (2) the neutral wind dynamo effect on magnetospheric convection; and (3) the effect of ionospheric outflow on the magnetosphere.

As stated above, the ionospheric conductivity has a very large uncertainty in is value. Further, the conductivity actually is a tensor that has three-dimensional structure. Lastly, the conductivity cannot be directly measured, and so it must be inferred by different techniques. This makes setting the ionospheric conductivity in models (and validating them) quite difficult. The structure of the ionospheric conductivity strongly affects where currents will flow and how the ionosphere will interact with the thermosphere. The neutral winds are strongly influenced by ion drag, which is directly related to the conductivity, and the frictional heating in the thermosphere is also controlled by the conductivity. Various studies have examined how the global conductivity can affect the flow patterns in the magnetosphere.

One of the most difficult couplings to understand between the ionosphere and magnetosphere is in the particle precipitation, which strongly controls the conductance in the ionosphere. If one simply considers the electrons, there are many different sources for the precipitation: (1) the central plasmasheet that causes the diffuse aurora; (2) the reconnection site that causes the discrete aurora; (3) the ring current interaction with the plasmasphere which can drive pitch angle scattering into the loss cone and a diffuse aurora to form equatorward of the nominal oval; (4) the radiation belts can cause high energy electrons to precipitate at low latitudes; (5) the cusp allows direct entry of magnetosheath plasma into the ionosphere; and (6) polar rain on open field-lines can be caused by streaming electrons on the interplanetary field-lines. Each of these different source populations has different dynamics that are controlled by different processes, although many of the processes may be interlinked. For example, different models of substorm physics may be dependent upon the underlying conductivity in the ionosphere, caused by the diffuse aurora. If this conductivity is too low, then reconnection in the tail is inhibited and,

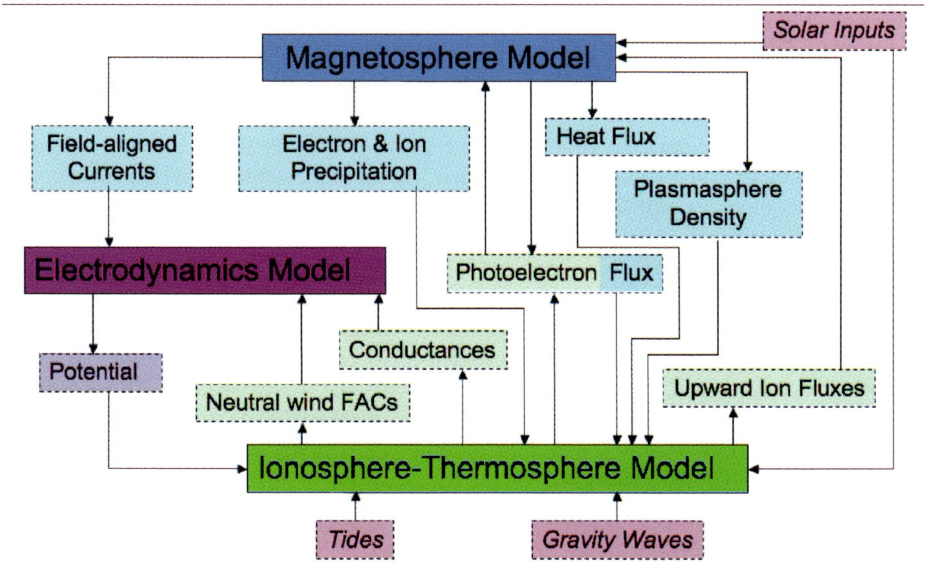

Fig. 20 The flow of information from different models describing the magnetosphere–ionosphere system. The *dark blue*, *purple* and *green boxes* represent models, while the *lighter colors* represent what the model would provide (for example, the *light blue boxes* are provided by the magnetospheric model, which is *shaded dark blue*). The models are driven by external forcing by the Sun (i.e. solar inputs) and the lower atmosphere (i.e. tides and gravity waves)

therefore, the discrete aurora may be extremely weak. If this continues for a long time, the tail may stretch; change the characteristics of the diffuse aurora, and the ionospheric conductivity. Once a certain threshold is reached, reconnection in the tail may be allowed, and a discrete aurora may grow. This concept has been shown in simple models, but the system is even more complex than this, so it is difficult to know whether this type of feedback is of primary importance within the magnetosphere–ionosphere system. Many studies have recently been conducted in which the feedback between the high-latitude ionospheric conductivity and the magnetospheric processes have been investigated (e.g., Fedder and Lyon 1987; Ridley et al. 2004), and in the mid-latitude region, where the conductivity can strongly influence the inner magnetospheric dynamics (e.g., Ebihara et al. 2004; Liemohn et al. 2005)

The neutral wind dynamo has been known to have a strong influence on the ionospheric density structure in the equatorial and low latitude regions of the ionosphere. The neutral winds drag ions (but not electrons) across the magnetic fields, which force a closure current to form (Richmond 1995). This closure current drives electric fields (Fejer and Scherliess 1997) that can lift or push down the ionosphere, dramatically changing the vertical structure (e.g., Kelley 1989; Millward 2001; Immel et al. 2006). In the high latitudes, the same physics occurs, but the electric fields can be communicated to the entire magnetosphere. Therefore, when the high-latitude forcing is constant for many hours (ramping the neutral winds up to significant speeds), and then suddenly changes, the neutral winds may influence the magnetospheric topology through the dynamo electric fields (e.g., Deng et al. 1993; Ridley et al. 2003; Peymirat et al. 2002). This is most important for slower rotating planets, since the neutral winds are tied to a geographic coordinate system and not a sun-fixed coordinate system. On fast rotating planets, the neutrals may not have enough time to accelerate up to appreciable speeds. On slower rotating planets (such as Earth), the neutral wind ramp-up time is approximately 3 hours (e.g., Deng and Ridley 2006), which is small enough that it allows the neutrals to stay in a region in which the acceleration is roughly constant for the full ramp-up time period. On faster rotating planets, the neutral winds may not have a constant acceleration for long enough to reach speeds in which they would strongly influence the magnetospheric convection. On the other hand, because the neutrals and ions would have very different velocities in the high latitudes on fast rotating planets, the Joule heating may be significantly larger than on Earth (e.g., Thayer and Vickrey 1992). This may energize the ions more, which may cause more ion outflow and would create more high-mass particles in the magnetospheric system.

Ion outflow at the Earth is an important source of particles in different regions of the magnetosphere. For example, the plasmaspheric density is controlled by ion outflow at Earth. Further, it has been shown that during strong activity, the plasmasheet can have a larger oxygen density, which is due to ion outflow (Young et al. 1982). At the outer planets, it is unknown what the influence of ion outflow is compared with mass loading from the moons. The direct influence of the ion outflow on the magnetospheric processes is a current subject of many studies (e.g., Winglee et al. 2005).

5.2 I-M Interaction at the Giant Planets

Generally speaking, the giant planets can interact with plasma flows from a very large region of interplanetary space, and these flows map magnetically into the ionosphere. To a first approximation, the solar wind influences plasma on magnetic field lines at the very front of the magnetosphere and on the open fieldlines in the tail and flanks. Other plasma sources and flows occur in the closed field line regions of the magnetosphere. If it is possible to study the flow of plasma in the ionosphere it is possible to derive much information about processes and mechanisms occurring in the magnetosphere itself. For Earth, such measurements can be made on a regular basis essentially *in situ*, making use of a combination of spacecraft and ground-based measurements. Spacecraft measurements of magnetospheric plasma flows are "routine" for the giant planets, but no spacecraft has so far been equipped with instruments to measure plasma flows in the ionosphere. For Jupiter, Saturn and Uranus, however, remote sensing has been developed over the past two decades making use of infrared emission from the H_3^+ molecular ion, a major constituent of giant planet ionospheres (Drossart et al. 1989; Geballe et al. 1993; Trafton et al. 1989, 1993; see reviews by Bhardwaj and Gladstone 2000; Miller et al. 2000, 2006). (So far no H_3^+ emission has not been observed from Neptune.)

Two processes modify the Dungey cycle for Jupiter and Saturn, compared with Earth. The first is the simple observation that these two planets not only have much larger magnetospheres than Earth, but that they are also spinning more than twice as fast—Jupiter rotates on its axis once every 0.41 days and Saturn every 0.44. By the time the solar wind has reached Saturn, it is traveling at $\sim$450 km/s. At this speed, Cowley et al. (2004) have calculated that it takes $\sim$70 hrs for the solar wind to traverse the Saturnian magnetosphere and return reconnected field lines from the front of the magnetosphere—where they open and connect to the IMF—back to a closed condition and returning from midnight to noon. This is approximately 6 Saturnian days as opposed the Earth's Dungey cycle which is approximately $1/8$ of an Earth day.

The second effect is one that is important for both Jupiter and Saturn, known as the Vasyliunas Cycle (see Vasyliunas 1983, for published version of this process). For Jupiter and Saturn, much of the plasma is derived from sources internal to the magnetosphere—from Io's volcanos, in the case of Jupiter, and from the rings and icy moons, in the case of Saturn.

Closed field lines, rotating rapidly with the planet, are heavily mass loaded. In the dawn–noon–dusk sectors, the loaded field lines are generally sufficiently compressed to remain intact. But as they rotate into the dusk-midnight sector, they can stretch out downtail. The mass-loaded field lines then—periodically—"pinch off" in the night-time sector, releasing "plasmoids" downtail. This is an important mechanism for ensuring that the Jovian and Saturnian magnetospheres do not become overloaded with plasma. Like the Dungey cycle, the Vasyliunas cycle is also expected to create a plasma flow signature in the ionosphere.

There is a further process that creates differences between Jupiter and Earth, and between Jupiter and (probably) Saturn, too. This is the breakdown of corotation in the equatorial plasmasheet, which produces very profound effects on Jupiter, creating its main auroral oval. Hill (1979) first described corotation breakdown at the time of Voyager. He firstly explained that Jupiter's magnetic field would drag ions produced by Io's volcanos into corotation with the planet accelerating them from 4.1×10^{-5} radians/s (the angular velocity of Io) to 1.75×10^{-4} rad/s. The momentum required to do this is supplied by the rotation of Jupiter itself: field lines passing through the plasmasheet were kept in corotation by collisions between the neutral atmosphere and ionospheric ions at the feet of the field lines. As centrifugal forces extended the plasmasheet outwards, a point is gradually reached where this mechanism is no longer able to supply sufficient momentum to maintain the magnetospheric plasma in corotation.

Two decades later came the realization that the electric fields and currents that are being generated in the middle magnetosphere, as a result of the lag to corotation in the plasmasheet, are supplying the energetic precipitating particles (mainly keV to MeV electrons, cf. Gerard et al. 1994) responsible for producing aurorae at high latitudes (Cowley and Bunce 2001; Hill 2001). The lag to corotation in the middle magnetosphere not only produces auroral emission from the upper atmosphere, but also has its own signature plasma flow. Essentially, field lines that are lagging to corotation in the middle magnetosphere drag ionospheric ions westward in the planetary frame of reference around the auroral oval. The auroral emissions are due to Lyman and Werner band transitions from molecular hydrogen with total luminosity of about 10^{13} to 10^{14} W (cf., Clarke et al. 2004).

Based on their understanding of the prevailing flows in the magnetosphere, Cowley and co-workers have produced projected ionospheric flows for both Jupiter (Cowley et al. 2003) and Saturn (Cowley et al. 2004). These show ionospheric ions corotating with the planet at lower latitudes where neutral–ion collisions are sufficient to keep closed field lines passing through the equatorial plasmasheet—and the plasmasheet itself—corotating. At larger distances out in the equatorial plane of the magnetosphere, mapping to higher latitudes, however, corotation begins to break down. Although corotation breakdown may take place over a distance of several planetary radii in the plasmasheet, the (approximately dipolar) field lines that cross this region map to high latitudes, where their footprints span a relatively narrow range. Thus, significant planetary westwards ion winds are generated in a relatively narrow (in latitude) region of the atmosphere. Note that for Saturn, Cowley et al. (2004) consider that this breakdown is not sufficient to power the main auroral oval observed on Saturn, although this is disputed (Hill 2005).

The signatures of the Dungey and Vasyliunas cycles are poleward of the region of westward winds generated by corotation breakdown. The addition of the Vasyliunas cycle in this region confines ionospheric flows due to the Dungey cycle to the dawn half of the polar cap. For Saturn, in a sector of the polar region that ranges from ~09:00 local time, through dusk, round to 03:00 local times, the Dungey and Vasyliunas flows combine to produce an anti-sunward ion drift across the whole region. As a result, the return flows are confined to a rather narrow region on the dawn flank. Ishbell et al. (1984) proposed that the polar cap should also be subject to a lag from corotation due to the restraining influence of the solar wind on the open fieldlines that connected into it. Following Ishbell et al. (1984) and Cowley et al. (2004), in the Sun-planet reference frame the angular velocity of the polar cap region may be given by:

$$\Omega_{\text{ion}} = \Omega_{\text{planet}} \mu_0 \Sigma_P{}^* V_{SW} [1 + \mu_0 \Sigma_P{}^* V_{SW}]^{-1} \tag{9}$$

where Ω_{planet} is the angular velocity of the planet, μ_0 is the permeability of free space, Σ_P^* is the *effective*, height-integrated Pedersen conductivity of the ionosphere, and V_{SW} is the solar wind velocity. The *effective* Pedersen conductivity is given by $(1 - K)\Sigma_P$, where Σ_P is the height-integrated Pedersen conductivity of the ionosphere. K is defined (see Cowley and Bunce 2001) as $(\Omega_{\text{planet}} - \Omega_{\text{neut}})/(\Omega_{\text{planet}} - \Omega_{\text{ion}})$, with Ω_{neut} is the angular velocity of the thermospheric neutrals that coexist with the ionospheric ions; K is therefore a height-integrated parameter representing the fraction of ion angular velocity, in the frame of reference that rotates with planet, that is attained by the neutrals as a result of ion-neutral collisions. The consequences of this process for planetary energetics are discussed below.

The M-I currents associated with the Dungey cycle at Jupiter might be manifested in the auroral emissions poleward of the main oval. Sporadic, flare-like UV emissions have been observed in the polar cap (Waite et al. 2001). Auroral X-ray emission has been observed from the polar regions of Jupiter with total X-ray luminosities of $\approx 10^9$ W (cf., Gladstone et al. 2002; Elsner et al. 2005) (also see chapter 7 in this book). The most likely explanation for this polar X-ray aurora is magnetospheric ions accelerated to MeV energies (Cravens et al. 2003) and originating in the magnetopause/cusp region of Jupiter's magnetoshere (Bunce et al. 2004).

Plasma flows have been measured in the ionospheres of Jupiter (Rego et al. 1999) and Saturn (Stallard et al. 2004) from the Doppler shifting of infrared emission from the H_3^+ molecular ion. The regrettable lack of spacecraft data—no mission has or is planned to fly a spectrometer of sufficiently high spectral resolution to measure the ion Doppler shifts—can lead to some ambiguity, as it is only possible to measure velocities in the Earth-planet line of sight, and then make reasonable assumptions about the most likely flow directions. Another issue is which reference frame is the more suitable to represent measured flows: for magnetosphere–ionosphere coupling, and the tracing of magnetospheric plasma flows onto the planet itself, the inertial (Sun-planet) frame of reference is usually used; energy considerations are often better understood by making use of the frame of reference corotating with (the magnetic field of) the planet itself.

The first measurements for Jupiter (Rego et al. 1999) identified an electrojet flowing around the auroral oval, which was associated with the lag to corotation of field lines crossing the equatorial plasmasheet in the middle magnetosphere. Subsequent work by Stallard et al. (2001, 2003) identified auroral oval winds of 0.5 km/s to 1.5 km/s. Poleward of the main Jovian oval, they found near corotation in the polar region, known as the swirl region from UV measurements (Clarke et al. 2004) or Bright Polar Region from IR measurements (Stallard et al. 2001). In the region known as the Dark Polar Region from the IR measurements, a sector that was essential stagnant with respect to the magnetic pole was identified (Cowley et al. 2003; Stallard et al. 2003), which also turned out to be nearly collocated with the polar cap identified from UV measurements (Pallier and Prangé 2004).

The Saturn, measurements carried out in 2003 found a significant lag to corotation right across the auroral/polar region, such that the ion angular velocity, Ω_{ion}, was $\sim 1/3$ that of the planet, Ω_{Sat}, in the inertial reference frame (Stallard et al. 2004). On average, the value of $\Omega_{\text{ion}}/\Omega_{\text{Sat}}$ appears to be ~ 0.43 (Stallard et al. 2007a), compared with the value of 0.25, estimated theoretically by Cowley et al. (2004). From Ishbell et al. (1984), one might deduce

$V_{SW} = 600$ km/s, if the effective Pedersen conductivity is 1 mho. Later work has shown that the velocity profile may be subdivided into regions of greater or less lag to corotation, and that the polar cap itself often has a central region that (nearly) corotates with the planet. This corotating central region has been interpreted as that part of the polar cap that is connected to field lines that stretch so far down tail before reconnecting to the solar wind that they are virtually free to twist as the planet rotates (Stallard et al. 2007b). So far, it has not been possible to observe the flows that would be associated with the Dungey or Vasyliunas cycles either for Jupiter or Saturn.

As described above, current systems in the magnetosphere, generated by plasma drifts, are carried by the magnetic field lines into the upper atmosphere, where they close through the ionosphere. These ionospheric currents can heat the whole upper atmosphere via Joule heating (Waite et al. 1983). In the auroral/polar regions, where the planet's magnetic field is nearly vertical, electric fields may be estimated from the measured ion flows. For Jupiter, the magnetic field is $\sim 10^{-3}$ Tesla. Thus the measured westward (in the planetary reference frame) ion winds of between 0.5 km/s and 1.5 km/s imply an equatorward electric field given by:

$$E_{\mathrm{eq}} = B_{\mathrm{aur}} v_{\mathrm{ion}} \sim 0.5\text{–}1.5 \text{ V/m}. \tag{10}$$

Joule heating is then given simply by:

$$H_{\mathrm{J}} = {E_{\mathrm{eq}}}^2 \Sigma_{\mathrm{P}} \tag{11}$$

where Σ_{P} is the height integrated Pederson conductivity. Millward et al. (2002) have shown that the Jovian Pedersen conductivity, Σ_{P}, is $\sim$1–8 mho for realistic fluxes of precipitating electrons, which implies local Joule heating rates between 0.25 W/m^2 to 18 W/m^2. However, the flow of ions around the auroral oval generates a westwards neutral wind (Miller et al. 2000), which can reach velocities of between 0.5 and 0.7 of the ion velocity (Millward et al. 2005), i.e. the K parameter defined above is between 0.5 and 0.7. In this case, the effective electric field, in the frame of reference fixed to the neutral atmosphere in the auroral regions, is given by $(1 - K)E_{\mathrm{eq}}$, where K is the parameter defined by Cowley and Bunce (2001), equivalent to $v_{\mathrm{neut}}/v_{\mathrm{ion}}$ in the planetary reference frame. Thus the net Joule heating becomes:

$$H_{\mathrm{J}} = [(1 - K)E_{\mathrm{eq}}]^2 \Sigma_{\mathrm{P}}. \tag{12}$$

This reduction in Joule heating, as a result of steady ion-neutral coupling, has also been noted for Earth (Thayer 1998). However, the ion drag generated by the flow of neutrals in the auroral oval against the rest of the neutral atmosphere also contributes to heating, given by (Smith et al. 2005):

$$H_{\mathrm{D}} = K(1 - K){E_{\mathrm{eq}}}^2 \Sigma_{\mathrm{P}}. \tag{13}$$

Thus the total heating generated by the ions drifts and magnetospherically generated electric field is:

$$H_{\mathrm{Elec}} = (1 - K){E_{\mathrm{eq}}}^2 \Sigma_{\mathrm{P}}. \tag{14}$$

Integrated across Jupiter's auroral/polar regions, this gives rise to heating of several times 10^{14} W, at least an order of magnitude greater than the heating due to particle precipitation, which is—in turn—an order of magnitude greater than the total solar EUV radiation absorbed planetwide. For Saturn, the total energy generated is less—by several TW—but this is still greater than heating due to particle precipitation or solar EUV.

More than two decades ago, Waite et al. (1983) questioned whether Joule (and ion drag) heating from the auroral/polar regions of the giant planets could be transferred equatorward, helping to explain the high exospheric temperatures (see Strobel and Smith 1973; Yelle and Miller 2004). Modeling carried out for Jupiter (Bougher et al. 2005) suggests that this may be possible. However, a recent study by Smith et al. (2007) for Saturn indicates that the Coriolis forces generated by the westward flowing neutral atmosphere in the auroral/polar regions turn the heated upper atmosphere poleward, rather than equatorward, and that the heat generated by Joule heating and ion drag is then mainly deposited in the lower atmosphere, rather than further heating the upper atmosphere as a whole.

6 Summary

Understanding the processes involved in the interaction of solar system bodies (planets or satellites) with plasma flows (solar wind/magnetospheric plasma flow) is fundamental to the entire field of space physics. The interaction of solar wind/magnetospheric plasma flow with weakly magnetized or unmagnetized planet/satellite is primarily an ionospheric-atmospheric interaction. Observations as well as various kinds of models have greatly enhanced our knowledge of the properties of such plasma interaction. However, there are still significant uncertainties concerning the nature of the interaction, especially when the obstacle is weakly magnetized or unmagnetized. More detailed analysis of observational data combined with improved model calculations will help to increase our current understanding of the important, controlling interaction processes.

References

M.H. Acuna et al., Science **279**, 1676–1680 (1998). doi:10.1126/science.279.5357.1676 Medline
M.H. Acuna, J.E.P. Connerney, P. Wasilewskii et al., Science **284**, 794–798 (1999). doi:10.1126/science.284.5415.794 Medline
K. Altwegg, H. Balsiger, J. Geiss, Space Sci. Rev. **90**, 3 (1999). doi:10.1023/A:1005256607402
C.S. Arridge et al., J. Geophys. Res. **111**, A11227 (2006). doi:10.1029/2005JA011574
C.S. Arridge et al., Warping of Saturn's magnetospheric and magnetotail current sheets. J. Geophys. Res. (2008, in press)
T. Bagdonat, U. Motschmann, J. Comput. Phys. **183**, 470–485 (2002a). doi:10.1006/jcph.2002.7203
T. Bagdonat, U. Motschmann, Earth Moon Planets **90**, 305–321 (2002b). doi:10.1023/A:1021578232282
H. Backes et al., Science **308**, 992 (2005). doi:10.1126/science.1109763 Medline
S. Barabash et al., Nature **450**, 650–653 (2007a). doi:10.1038/nature06434
S. Barabash et al., Planet. Space Sci. **55**, 1772–1792 (2007b). doi:10.1016/j.pss.2007.01.014
M. Benna, P. Mahaffy, Planet. Space Sci. **55**, 1031–1043 (2007). doi:10.1016/j.pss.2006.11.019
C. Bertucci et al., Geophys. Res. Lett. **30**(2), 1099 (2003). doi:10.1029/2002GL015713
C. Bertucci et al., Geophys. Res. Lett. **34**, L24S02 (2007). doi:10.1029/2007GL030865
A. Bhardwaj, G.R. Gladstone, Rev. Geophys. **38**, 295–353 (2000). doi:10.1029/1998RG000046
T. Bonev, K. Jockers, Icarus **107**(2), 335–357 (1994). doi:10.1006/icar.1994.1028
A. Bösswetter et al., Ann. Geophys. **22**, 4363–4379 (2004)
S.W. Bougher et al., J. Geophys. Res. **110**, E04008 (2005). doi:10.1029/2003JE002230
L.H. Brace et al., J. Geophys. Res. **85**, 7663–7678 (1980). doi:10.1029/JA085iA13p07663
L.H. Brace, A.J. Kliore, Space Sci. Rev. **55**, 81 (1991). doi:10.1007/BF00177136
L.H. Brace, R.E. Hartle, R.F. Theis, Adv. Space Res. **16**, 99 (1995). doi:10.1016/0273-1177(95)00255-D
J.C. Brandt, Observations and dynamics of plasma tails, in *Comets*, ed. by L.L. Wilkening (The University of Arizona Press, Tucson, 1982), pp. 519–537
J.C. Brandt et al., Icarus **137**, 69–83 (1999). doi:10.1006/icar.1998.6030
J.C. Brandt, R.D. Chapman, *Introduction to Comets*, 2nd ed. (Cambridge University Press, Cambridge, UK, 2004), Chapter 4
A. Brekke, J.R. Doupnik, P.M. Banks, J. Geophys. Res. **79**, 3773 (1974). doi:10.1029/JA079i025p03773

E.J. Bunce, S.W.H. Cowley, J.A. Wild, Ann. Geophys. **21**, 1709–1722 (2003)
E.J. Bunce, S.W.H. Cowley, T.K. Yeoman, J. Geophys. Res. **109**, A09S13 (2004). doi:10.1029/2003JA010280
R.W. Carlson, Science **283**, 820–821 (1999). doi:10.1126/science.283.5403.820 Medline
J.Y. Chaufray et al., J. Geophys. Res. **112**, E09009 (2007). doi:10.1029/2007JE002915
A.F. Cheng, J.H. Waite, J. Geophys. Res. **93**, 4107–4109 (1988). doi:10.1029/JA093iA05p04107
J.T. Clarke et al., Jupiter's aurora, in *Jupiter: The Planet, Satellites and Magnetosphere*, ed. by F. Bagenal, T. Dowling, W. McKinnon (Cambridge University Press, Cambridge, 2004), pp. 639–670
P.A. Cloutier et al., in *Venus*, ed. by D.M. Hunten, L. Colin, T.M. Donahue, V.I. Moroz (University of Arizona Press, Tuscon, 1983), p. 941
A.J. Coates et al., J. Geophys. Res. **98**, 20,985–20,994 (1993). doi:10.1029/93JA02535
A.J. Coates et al., Geophys. Res. Lett. **34**, L24S05 (2007). doi:10.1029/2007GL030919
J. Connerney et al., Science **284**, 794–798 (1999). doi:10.1126/science.284.5415.794 Medline
S.W.H. Cowley, E.J. Bunce, Planet. Space Sci. **49**, 1067–1088 (2001). doi:10.1016/S0032-0633(00)00167-7
S.W.H. Cowley et al., Geophys. Res. Lett. **30**, L1220 (2003). doi:10.1029/2002GL016030
S.W.H. Cowley, E.J. Bunce, R. Prangé, Ann. Geophys. **22**, 1379–1394 (2004)
T.E. Cravens, The physics of the cometary contact surface, in Proc. of the 20th ESLAB Symposium on the Exploration of Halley's Comet, eds. B. Battrick, E.J. Rolfe, R. Reinhard. ESA SP-250, **1**, 241 (1986)
T.E. Cravens, A magnetohydrodynamical model of the inner coma of comet Halley. J. Geophys. Res. **94**, 15025 (1989). doi:10.1029/JA094iA11p15025
T.E. Cravens, Plasma processes in the inner coma, in *Comets in the Post-Halley Era*, ed. by R.L. Newburn Jr., M. Neugebauer, J. Rahe (Kluwer, Dordrecht, 1991a), pp. 1211–1258
T.E. Cravens, Collisional processes in cometary plasmas, in *Cometary Plasma Processes*. American Geophysical Union Monograph, vol. 61 (AGU, Washington, 1991b), p. 27
T.E. Cravens, *Physics of Solar System Plasmas* (Cambridge University Press, Cambridge, 1997a)
T.E. Cravens, Geophys. Res. Lett. **24**, 105 (1997b). doi:10.1029/96GL03780
T.E. Cravens et al., J. Geophys. Res. **108**(A12), 1465 (2003). doi:10.1029/2003JA010050
T.E. Cravens et al., Composition of Titan's ionosphere. Geophys. Res. Lett. **33**, L07105 (2006). doi:10.1029/2005GL025575
T.E. Cravens et al., Geophys. Res. Lett. **35**, L03103 (2008). doi:10.1029/2007GL032451
G.K. Crawford et al., VLF imaging of the Venus foreshock. Geophys. Res. Lett. **20**, 2801–2804 (1993). doi:10.1029/93GL01258
D.H. Crider et al., Geophys. Res. Lett. **27**(1), 45 (2000). doi:10.1029/1999GL003625
D.H. Crider et al., J. Geophys. Res. **108**(A12), 1461 (2003). doi:10.1029/2003JA009875
W. Deng et al., J. Geophys. Res. **98**(A5), 7775–7790 (1993). doi:10.1029/92JA02268
Y. Deng, A.J. Ridley, J. Geophys. Res. **111**, A09306 (2006). doi:10.1029/2005JA011368
A.J. Dessler, *Physics of the Jovian Magnetosphere* (Cambridge University Press, Cambridge, 1983)
S.S. Dolginov, V.N. Zhuzgov, Planet. Space Sci. **39**, 1493–1510 (1991). doi:10.1016/0032-0633(91)90077-N
S.S. Dolginov, V.N. Zhuzgov, V.A. Sharova, V.B. Buzin, Kosm. Issled. **16**, 827–863 (1978)
P. Drossart et al., Nature **340**, 539–541 (1989). doi:10.1038/340539a0
J.W. Dungey, Phys. Rev. Lett. **6**, 47 (1961). doi:10.1103/PhysRevLett.6.47
Edberg et al., J. Geophys. Res. (2008, in press). doi:10.1029/2008JA013096
Y. Ebihara et al., J. Geophys. Res. **109**, A08205 (2004). doi:10.1029/2003JA010351
R.C. Elphic et al., J. Geophys. Res. **86**, 11430 (1981). doi:10.1029/JA086iA13p11430
R.F. Elsner et al., J. Geophys. Res. **110**, A01207 (2005). doi:10.1029/2004JA010717
A. Eviatar et al., J. Geophys. Res. **87**, 8091 (1982). doi:10.1029/JA087iA10p08091
A. Eviatar, J.D. Richardson, J. Geophys. Res. **91**, 3299–3303 (1986). doi:10.1029/JA091iA03p03299
A. Eviatar et al., Astrophys. J. **555**, 1013–1019 (2001). doi:10.1086/321510
J.A. Fedder, J.G. Lyon, Geophys. Res. Lett. **14**, 880 (1987). doi:10.1029/GL014i008p00880
B.G. Fejer, L. Scherliess, J. Geophys. Res. **102**(A11), 24,047–24,056 (1997). doi:10.1029/97JA02164
J.L. Fox et al., Space Sci. Rev. (2008, this issue)
A.A. Galeev et al., Quasi linear theory of the ion cyclotron instability and its application to the cometary plasma, in *Cometary Plasma Processes*, 2nd edn., ed. by A.D. Johnstone. Geophys. Monogr. Ser., vol. 61 (AGU, Washington, 1991), pp. 223–240
T.R. Geballe, M.-F. Jagod, T. Oka, Astrophys. J. **408**, L109–L112 (1993). doi:10.1086/186843
J.C. Gerard et al., Science **266**(5191), 1675–1678 (1994). doi:10.1126/science.266.5191.1675 Medline
G.R. Gladstone et al., A pulsating auroral X-ray hot spot on Jupiter. Nature **415**, 1000 (2002). doi:10.1038/4151000a Medline
T.I. Gombosi et al., J. Geophys. Res. **101**, 15233–15252 (1996). doi:10.1029/96JA01075
T.I. Gombosi et al., MHD simulation of comets: The plasma environment of comet Hale-Bopp. Earth Moon Planets **79**, 179 (1997). doi:10.1023/A:1006289418660

G. Gloeckler et al., Nature **404**, 576–578 (2000). doi:10.1038/35007015 Medline
Grebowsky et al., Geophys. Res. Lett. **20**, 2735–2738 (1993). doi:10.1029/93GL02239
R.M. Häberli et al., Icarus **130**(2), 373–386 (1997). doi:10.1006/icar.1997.5835
K.C. Hansen et al., Space Sci. Rev. **128**, 133–166 (2007). doi:10.1007/s11214-006-9142-6
E.M. Harnett, R.M. Winglee, Three-dimensional multifluid simulations of ionospheric loss at Mars from nominal solar wind conditions to magnetic cloud events. J. Geophys. Res. **111**, A09213 (2006). doi:10.1029/2006JA011724
R.E. Hartle et al., J. Geophys. Res. **87**, 1383 (1982). doi:10.1029/JA087iA03p01383
R.E. Hartle et al., Geophys. Res. Lett. **33**, L08201 (2006a). doi:10.1029/2005GL024817
R.E. Hartle et al., Planet. Space Sci. **54**, 1211 (2006b). doi:10.1016/j.pss.2006.05.029
T.W. Hill, J. Geophys. Res. **84**, 6554–6558 (1979). doi:10.1029/JA084iA11p06554
T.W. Hill, J. Geophys. Res. **106**, 8101–8108 (2001). doi:10.1029/2000JA000302
T.W. Hill, Rotationally driven dynamics in the magnetospheres of Jupiter and Saturn, in *Magnetospheres of the Outer Planets*, Leicester, 12 August, 2005
D.P. Hinson et al., J. Geophys. Res. **103**, 29,343–29,357 (1998). doi:10.1029/98JA02659
R.R. Hodges, J. Geophys. Res. **105**, 6971–6981 (2000). doi:10.1029/1999JE001138
R.R. Hodges, Geophys. Res. Lett. **29**(3), 1038 (2002). doi:10.1029/2001GL013853
D.E. Huddleston et al., Mass loading and velocity diffusion models for heavy pickup ions at comet Grigg-Skjellerup. J. Geophys. Res. **98**(A12), 20,995–21,002 (1993). doi:10.1029/93JA02531
W.F. Huebner et al., in *Comets in the Post-Halley Era*, ed. by R.L. Newburn Jr., M. Neugebauer, J. Rahe (Kluwer, Dordrecht, 1991), pp. 907–936
T.J. Immel et al., Geophys. Res. Lett. **33**, L15108 (2006). doi:10.1029/2006GL026161
W.-H. Ip, W.I. Axford, Nature **325**, 418 (1987). doi:10.1038/325418a0
W.-H. Ip, in *Comets II*, ed. by M.C. Festou, H.U. Keller, H.A. Weaver (Univ. of Arizona Press, Tucson, 2004), pp. 605–629
J. Ishbell, A.J. Dessler, J.H. Waite Jr., J. Geophys. Res. **89**, 10716 (1984). doi:10.1029/JA089iA12p10716
Y.-D. Jia et al., J. Geophys. Res. **112**, A05223 (2007). doi:10.1029/2006JA012175
R.E. Johnson, Energetic Charged-Particle Interactions with Atmospheres and Surfaces, 1990
R.E. Johnson et al., Space Sci. Rev. (2008, this issue)
G.H. Jones, A. Balogh, T.S. Horbury, Nature **404**, 574 (2000). doi:10.1038/35007011 Medline
E. Kallio, P. Janhunen, J. Geophys. Res. **107**(A3) (2002). doi:10.1029/2001JA000090
E.I. Kallio et al., Geophys. Res. Lett. **34**, L24S09 (2007). doi:10.1029/2007GL030827
W.T. Kasprzak, H.B. Niemann, Planet. Space Sci. **30**, 1107 (1982). doi:10.1016/0032-0633(82)90121-0
W.T. Kasprzak et al., J. Geophys. Res. **96**, 11,175 (1991). doi:10.1029/91JA00677
M.C. Kelley, *The Earth's Ionosphere: Plasma Physics and Electrodynamic* (Academic Press, San Diego, 1989)
K.K. Khurana, The configuration of Jupiter's magnetosphere, in *Jupiter*, ed. by F. Bagenal, T. Dowling, W. Mckinnon (Cambridge University Press, Cambridge, 2004)
J. Kim et al., J. Geophys. Res. **103**, 29339–29342 (1998). doi:10.1029/98JA02727
A.J. Kliore, Radio occultation observations of the ionospheres of Mars and Venus, in *Venus and Mars: Atmospheres, Ionospheres and Solar Wind Interactions*. Geophys. Monograph, vol. 66 (American Geophysical Union, Washington, 1992), p. 265
A.J. Kliore et al., Science **277**, 355 (1997). doi:10.1126/science.277.5324.355 Medline
A.J. Kliore et al., J. Geophys. Res. **107**, pp. SIA 19–1 (2002)
A.J. Kliore et al., American Geophysical Union, Fall Meeting 2007, abstract #P51D-04 (2007)
W.C. Knudsen et al., J. Geophys. Res. **85**, 7803–7810 (1980). doi:10.1029/JA085iA13p07803
C. Konz, G.T. Birk, H. Lesch, Astron. Astrophys. **415**, 791–802 (2004). doi:10.1051/0004-6361:20031695
S.A. Ledvina et al., Adv. Space Res. **26**, 1691 (2000). doi:10.1016/S0273-1177(00)00075-2
S.A. Ledvina et al., J. Geophys. Res. **110**, A06211 (2005). doi:10.1029/2004JA010771
S.A. Ledvina et al., Space Sci. Rev. (2008, this issue)
M.W. Liemohn et al., J. Geophys. Res. **110**, A12S22 (2005). doi:10.1029/2005JA011109
C.J. Lindgren Jr., T.E. Cravens, S.A. Ledvina, J. Geophys. Res. **102**, 17395 (1997). doi:10.1029/97JA01117
A.S. Lipatov, K. Sauer, K. Baumgärtel, Adv. Space Res. **20**, 279–282 (1997). doi:10.1016/S0273-1177(97)00547-4
A.S. Lipatov, U. Motschmann, T. Bagdonat, Planet. Space Sci. **50**, 403–411 (2002). doi:10.1016/S0032-0633(02)00004-1
C.M. Lisse et al., Science **274**, 205–209 (1996). doi:10.1126/science.274.5285.205
A.J. Lovell et al., Earth Moon Planets **77**, 253–258 (1999). doi:10.1023/A:1006285703521
J.G. Luhmann et al., Geophys. Res. Lett. **10**, 655 (1983). doi:10.1029/GL010i008p00655
J.G. Luhmann, Space Sci. Rev. **44**, 241 (1986). doi:10.1007/BF00200818
J.G. Luhmann, T.E. Cravens, Space Sci. Rev. **55**, 201–274 (1991). doi:10.1007/BF00177138

J.G. Luhmann, J.U. Kozyra, Dayside pickup oxygen ion precipitation at Venus and Mars: Spatial distributions, energy deposition and consequences. J. Geophys. Res. **96**(A4), 5457–5467 (1991). doi:10.1029/90JA01753
J.G. Luhmann, S.A. Ledvina, C.T. Russell, Induced magnetospheres. Adv. Space Res. **33**(11), 1905–1912 (2004). doi:10.1016/j.asr.2003.03.031
J.G. Luhmann, W.T. Kasprzak, C.T. Russell, J. Geophys. Res. **112**(E04S10) (2007). doi:10.1029/2006JE002820
E.T. Lundberg, AGU, Fall Meeting, abstract #P43A-0957 (2005)
R. Lundin et al., Nature **341**, 609–612 (1989). doi:10.1038/341609a0
R. Lundin et al., Geophys. Res. Lett. **17**, 873 (1990). doi:10.1029/GL017i006p00873
Y. Ma et al., J. Geophys. Res. **107**(A10), 1282 (2002). doi:10.1029/2002JA009293
Y. Ma et al., J. Geophys. Res. **109**, A07211 (2004). doi:10.1029/2003JA010367
Y. Ma et al., J. Geophys. Res. **111**, A05207 (2006). doi:10.1029/2005JA011481
Y.J. Ma, A.F. Nagy, Geophys. Res. Lett. **34**, L08201 (2007). doi:10.1029/2006GL029208
Y.J. Ma et al., Geophys. Res. Lett. **34** (2007a). doi:10.1029/2007GL031627
Y. Ma et al., AGU Fall Meeting, abstract #P51D-03 (2007b)
D.J. McComas et al., J. Geophys. Res. **91**, 7939 (1986). doi:10.1029/JA091iA07p07939
M.A. McGrath et al., in *Satellite Atmospheres in Jupiter: The Planet, Satellites, and Magnetosphere*, ed. by F. Bagenal, T. Dowling, W. McKinnon (Cambridge University Press, Cambridge, 2004), pp. 457–483
M. Michael, R.E. Johnson, Planet. Space Sci. **53**, 1510–1514 (2005). doi:10.1016/j.pss.2005.08.001
S. Miller et al., Philos. Trans. R. Soc. **358**, 2485–2502 (2000). doi:10.1098/rsta.2000.0662
S. Miller et al., Philos. Trans. R. Soc. **364**, 3121–3137 (2006). doi:10.1098/rsta.2006.1877
G.H. Millward, J. Geophys. Res. **106**(A11), 24,733–24,744 (2001). doi:10.1029/2000JA000342
G.H. Millward et al., Icarus **160**, 75–107 (2002). doi:10.1006/icar.2002.6951
G.H. Millward et al., Icarus **173**, 200–211 (2005). doi:10.1016/j.icarus.2004.07.027
D.L. Mitchell et al., Geophys. Res. Lett. **27**, 1871 (2000). doi:10.1029/1999GL010754
D.L. Mitchell et al., J. Geophys. Res. **106**, 23419–23427 (2001). doi:10.1029/2000JE001435
R. Modolo, G.M. Chanteur, E. Dubinin, A.P. Matthews, Ann. Geophys. **23**, 433 (2005)
R. Modolo et al., Geophys. Res. Lett. **34**, L24S07 (2007a). doi:10.1029/2007GL030489
R. Modolo et al., Geophys. Res. Lett. **34**, L24S04 (2007b). doi:10.1029/2007GL030482
U. Motschmann, K. Glassmeier, J. Geophys. Res. **98**, 20,977–20,983 (1993). doi:10.1029/93JA02533
I. Müller-Wordarg et al., Space Sci. Rev. (2008, this issue)
A.F. Nagy et al., Space Sci. Rev. **111**(1–2), 33–114 (2004). doi:10.1023/B:SPAC.0000032718.47512.92
A.F. Nagy, T.E. Cravens, Planet. Space Sci. **46**(9–10), 1149 (1998). doi:10.1016/S0032-0633(98)00049-X
N.F. Ness et al., J. Geophys. Res. **87**, 1369 (1982). doi:10.1029/JA087iA03p01369
N.F. Ness et al., Adv. Space Res. **23**(11), 1879–1886 (1999). doi:10.1016/S0273-1177(99)00271-9
F.M. Neubauer, D.A. Gurnett, J.D. Scudder, R.E. Hartle, Titan's magnetospheric interaction, in *Saturn*, ed. by T. Gehrels, M.S. Matthews (Univ. Arizona Press, Tucson, 1984), pp. 760–787
F.M. Neubauer, Astron. Astrophys. **187**, 73 (1987)
F.M. Neubauer et al., J. Geophys. Res. **111**, A10220 (2006). doi:10.1029/2006JA011676
M. Neugebauer et al., Astrophys. J. **372**, 291–300 (1991). doi:10.1086/169975
M. Neugebauer et al., Astrophys. J. **673**, 629–636 (2007)
M.B. Niedner Jr., J.C. Brandt, Astrophys. J. **223**, 655–670 (1978). doi:10.1086/156299
M. Ong et al., Geophys. Res. **96**, 11,133 (1991). doi:10.1029/91JA01100
L. Pallier, R. Prangé, Geophys. Res. Lett. **31**, L06701 (2004). doi:10.1029/2003GL018041
C. Peymirat, A.D. Richmond, R.G. Roble, J. Geophys. Res. **107**(A1), 1006 (2002). doi:10.1029/2001JA900106
J.L. Phillips, J.G. Luhmann, C.T. Rusell, J. Geophys. Res. **89**, 10676 (1984). doi:10.1029/JA089iA12p10676
J.L. Phillips et al., J. Geophys. Res. **93**, 3927–3941 (1988). doi:10.1029/JA093iA05p03927
J.L. Phillips, D.J. McComas, Space Sci. Rev. **55**, 1–80 (1991). doi:10.1007/BF00177135
P. Puhl-Quinn, T.E. Cravens, J. Geophys. Res. **100**, 21631 (1995). doi:10.1029/95JA01820
D. Rego et al., Nature **399**, 121–124 (1999). doi:10.1038/20121
A.D. Richmond, J. Geomag. Geoelectr. **47**, 191 (1995)
W.D. Riedler et al., Nature **341**, 604 (1989). doi:10.1038/341604a0
A.J. Ridley et al., J. Geophys. Res. **108**(A8), 1328 (2003). doi:10.1029/2002JA009464
A.J. Ridley, T.I. Gombosi, D.L. De Zeeuw, Ann. Geophys. **22**(2), 567–584 (2004)
S.D. Rodgers et al., Physical processes and chemical reactions in cometary comae, in *Comets II*, ed. by M.C. Festou, H.U. Keller, H.A. Weaver (University of Arizona Press, Tucson, 2004), pp. 505–522
H. Rosenbauer et al., Nature **341**, 612–614 (1989). doi:10.1038/341612a0
C.T. Russell, Geophys. Res. Lett. **4**, 387–390 (1977). doi:10.1029/GL004i010p00387
C.T. Russell et al., IEEEE Trans. Geophys. Sens. **GE-18**, 32 (1980)

C.T. Russell et al., J. Geophys. Res. **93**, 5461–5469 (1988). doi:10.1029/JA093iA06p05461

C.T. Russell, Planetary magnetospheres. Sci. Prog. **75**, 93 (1991)

C.T. Russell, J.G. Luhmann, R.J. Strangeway, Planet. Space Sci. **54**, 1482–1495 (2006). doi:10.1016/j.pss.2006.04.025

C.T. Russell et al., Venus upper atmosphere and plasma environment: Critical issues for future exploration, in *Exploring Venus as a Terrestrial Planet*. Geophys. Monograph Series, vol. 175 (AGU, Washington, 2007), pp. 139–156. doi:10.1029/176GM09

J. Saur et al., J. Geophys. Res. **104**, 25,105–25,126 (1999). doi:10.1029/1999JA900304

J. Saur et al., Geophys. Res. Lett. **27**, 2893–2896 (2000). doi:10.1029/2000GL003824

M. Schmidt-Voigt, Astron. Astrophys. **210**, 433–454 (1989)

R.W. Schunk, A.F. Nagy, *Ionospheres*, 2nd edn. (Cambridge University Press, Cambridge, 2009)

H. Shinagawa, T.E. Cravens, J. Geophys. Res. **93**, 11263 (1988). doi:10.1029/JA093iA10p11263

I. Sillanpää et al., Adv. Space Res. **38**, 799–805 (2006). doi:10.1016/j.asr.2006.01.005

I. Sillanpää et al., J. Geophys. Res. **112**, A12205 (2007). doi:10.1029/2007JA012348

S. Simon et al., Geophys. Res. Lett. **34**, L24S08 (2007). doi:10.1029/2007GL029967

J.A. Slavin et al., J. Geophys. Res. **85**, 7625–7641 (1980). doi:10.1029/JA085iA13p07625

J.A. Slavin et al., J. Geophys. Res. **96**, 11,235–11,241 (1991). doi:10.1029/91JA00439

C.G.A. Smith, S. Miller, A.D. Aylward, Ann. Geophys. **23**, 1943–1947 (2005)

C.G.A. Smith et al., Nature **445**, 399–401 (2007). doi:10.1038/nature05518 Medline

W.H. Smyth, M.L. Marconi, Icarus **181**, 510–526 (2006). doi:10.1016/j.icarus.2005.10.019

T. Stallard et al., Icarus **154**, 475–491 (2001). doi:10.1006/icar.2001.6681

T.S. Stallard et al., Geophys. Res. Lett. **30**, L1221 (2003). doi:10.1029/2002GL016031

T.S. Stallard et al., Icarus **167**, 204–211 (2004). doi:10.1016/j.icarus.2003.09.006

T. Stallard et al., Icarus **189**, 1–13 (2007a). doi:10.1016/j.icarus.2006.12.027

T. Stallard et al., Icarus **191**, 678–680 (2007b). doi:10.1016/j.icarus.2007.05.016

D.F. Strobel, G.R. Smith, J. Atmos. Sci. **30**, 718–725 (1973). doi:10.1175/1520-0469(1973)030<0718:OTTOTJ>2.0.CO;2

D.F. Strobel, *Comparative Planetary Atmospheres of the Galilean Satellites, Highlights of Astronomy*, vol. 13, ed. by O. Engvold. San Francisco, CA (Astronomical Society of the Pacific, 2005) ISBN 1-58381-189-3. XXIX, p. 894

K. Szego et al., Geophys. Res. Lett. **32**, L20S05 (2005). doi:10.1029/2005GL022646

K. Szego et al., Geophys. Res. Lett. **34**, L24S03 (2007). doi:10.1029/2007GL030677

N. Terada, S. Machida, H. Shinagawa, J. Geophys. Res. **107**(A12), 1471 (2002). doi:10.1029/2001JA009224

J.P. Thayer, J. Geophys. Res. **103**, 471–487 (1998). doi:10.1029/97JA02536

J.P. Thayer, J.F. Vickrey, Geophys. Res. Lett. **19**, 265–268 (1992). doi:10.1029/91GL02868

D.V. Titov et al., Planet. Space Sci. **54**, 1336–1343 (2006). doi:10.1016/j.pss.2006.04.017

L.M. Trafton, D.F. Lester, K.L. Thompson, Astrophys. J. **343**, L73–L76 (1989). doi:10.1086/185514

L.M. Trafton et al., Astrophys. J. **405**, 761–766 (1993). doi:10.1086/172404

J.G. Trotignon et al., J. Geophys. Res. **101**(A11), 24,965–24,977 (1996). doi:10.1029/96JA01898

J.G. Trotignon et al., Planet. Space Sci. **54**, 357 (2006). doi:10.1016/j.pss.2006.01.003

W.-L. Tseng, W.-H. Ip, A. Kopp, Adv. Space Res. **42**(1), 54–60 (2008). doi:10.1016/j.asr.2008.03.009

V.M. Vasyliunas, Plasma distribution and flow, in *Physics of the Jovian magnetosphere*, ed. by A.J. Dessler (Cambridge University Press, Cambridge, 1983), pp. 395–453

D. Vignes et al., Geophys. Res. Lett. **27**, 49–52 (2000). doi:10.1029/1999GL010703

D. Vignes et al., Geophys. Res. Lett. **29**(9), 1328 (2002). doi:10.1029/2001GL014513

M.R. Voelzke et al., Earth Moon Planets **97**(3–4), 399–409 (2005). doi:10.1007/s11038-006-9073-y

J.-E. Wahlund et al., Science **308**, 986–989 (2005). doi:10.1126/science.1109807 Medline

H.Y. Wei et al., Geophys. Res. Lett. **34**, L24S06 (2007). doi:10.1029/2007GL030701

J.H. Waite Jr. et al., J. Geophys. Res. **88**, 6143–6163 (1983). doi:10.1029/JA088iA08p06143

J.H. Waite Jr. et al., Nature **410**, 787–789 (2001). doi:10.1038/35071018 Medline

J.H. Waite et al., Ion neutral mass spectrometer results from the first flyby of Titan. Science **308**, 982–986 (2005). doi:10.1126/science.1110652 Medline

R. Wegmann, Astron. Astrophys. **358**, 759–775 (2000)

R.M. Winglee, W. Lewis, G. Lu, J. Geophys. Res. **110**, A12S24 (2005). doi:10.1029/2004JA010909

O. Witasse et al., Space Sci. Rev. (2008, this issue)

R. Yelle, S. Miller, Jupiter's Ionosphere and Thermosphere, in *Jupiter: The Planet, Satellites, and Magnetosphere*, ed. by F. Bagenal, T. Dowling, W. McKinnon (Cambridge University Press, Cambridge, 2004), pp. 185–218

Y.G. Yeroshenko, Kosmicheskie Issledovania **1**, 92–105 (1979), in Russian

Y.G. Yeroshenko et al., Geophys. Res. Lett. **17**(6), 885–888 (1990). doi:10.1029/GL017i006p00885

Y. Yi et al., J. Geophys. Res. **101**(A12), 27585–27601 (1996). doi:10.1029/96JA02235

D.T. Young, H. Balsiger, J. Geiss, J. Geophys. Res. **87**, 9077–9096 (1982). doi:10.1029/JA087iA11p09077
T.-L. Zhang et al., Geophys. Res. Lett. **18**, 127–129 (1991a). doi:10.1029/90GL02723
T.-L. Zhang, J.G. Luhmann, C.T. Russell, J. Geophys. Res. **96**, 11145–11153 (1991b). doi:10.1029/91JA00088
T.-L. Zhang et al., Planet. Space Sci. **54**, 1279–1297 (2006). doi:10.1016/j.pss.2006.04.018
T.-L. Zhang et al., Nature **450**, 654–656 (2007). doi:10.1038/nature06026 Medline

Exospheres and Atmospheric Escape

R.E. Johnson · M.R. Combi · J.L. Fox · W.-H. Ip ·
F. Leblanc · M.A. McGrath · V.I. Shematovich ·
D.F. Strobel · J.H. Waite Jr.

Originally published in the journal Space Science Reviews, Volume 139, Nos 1–4.
DOI: 10.1007/s11214-008-9415-3

Abstract Aeronomy is a description of the physics and chemistry of the upper atmospheres and ionospheres of planetary bodies. In this chapter we consider those processes occurring

R.E. Johnson (✉)
Engineering Physics, University of Virginia, Charlottesville, VA 22904, USA
e-mail: rej@virginia.edu

R.E. Johnson
Physics Department, New York University, 4 Washington Pl., New York, NY 10003, USA

M.R. Combi
Department of Atmospheric, Oceanic and Space Sciences, University of Michigan, Ann Arbor, MI 48109, USA
e-mail: mcombi@umich.edu

J.L. Fox
Physics Department, Wright State University, Dayton, OH 45435, USA
e-mail: Jane.Fox@wright.edu

W.-H. Ip
Institute of Astronomy, National Central University, Chung Li 32054, Taiwan
e-mail: wingip@astro.ncu.edu.tw

F. Leblanc
Service d'Aeronomie du CNRS, Paris, France
e-mail: francois.leblanc@aerov.jussieu.fr

F. Leblanc
Osservatorio Astronomico di Trieste, 34131 Trieste, France

M.A. McGrath
Marshall Space Flight Center, Huntsville, AL 35899, USA
e-mail: melissa.a.mcgrath@nasa.gov

V.I. Shematovich
Institute of Astronomy RAS, 48 Pyantnitskaya str., Moscow 119017, Russian Federation
e-mail: shematov@inasan.ru

in the upper atmosphere that determine the structure of the corona and lead to molecular escape.

Keywords Atmospheric escape · Planetary corona · Exobase · Exosphere

1 Introduction

Those processes occurring in the upper atmosphere of a planet or planetary satellite that determine the structure of the corona and lead to molecular escape are considered in this paper. A planet's atmosphere decreases in density with increasing altitude, thus an altitude is eventually reached above which molecules can travel planetary scale distances with a very small probability of making a collision. At such altitudes atoms or molecules that have energies greater than their gravitational binding energy can escape to space if their radial velocity is outward. This region of the atmosphere is called the exosphere or the planetary corona. The lower boundary for this region, called the exobase, is defined as that altitude where the atmospheric scale height, H, is about the same size as the mean free path for collisions, l_c. The Knudsen number, Kn, is the ratio between the mean free path and the density scale. It defines the transition region from a gas that is dominated by collisions and behaves like a fluid to a gas that should be modeled stochastically. Therefore, the exobase is also defined to occur where $\mathrm{Kn} \sim 1$. In a gas of randomly moving molecules the mean free path for collisions is $l_c \approx 1/(\sqrt{2}\sigma n)$, where n is the molecular number density and σ is the *hard sphere cross section* which is independent of energy and assumes isotropic scattering. Therefore, setting $[\sigma n_{\mathrm{exo}} H] \sim 1$ is often used to define the exobase. In this estimate the exobase occurs at a column density: $N(r_{\mathrm{exo}}) \sim [n_{\mathrm{exo}} H] \sim 1/\sigma$, where r_{exo} is the exobase radius; $\sigma \sim 1\text{–}3 \times 10^{-15}\ \mathrm{cm}^2$ is often used giving $N(r_{\mathrm{exo}}) \sim 1\text{–}0.3 \times 10^{15}/\mathrm{cm}^2$ (Chamberlain and Hunten 1987). On Mercury, the Moon, and many of the outer solar system icy bodies, the column of bound gas is less than $1/\sigma$, so that escape can occur from the physical surface.

The exobase is often used as the average altitude from which molecules escape to space. This is a crude approximation since hot atoms and molecules escape from depths well below the nominal exobase as seen in Fig. 1 for Titan's atmosphere. In addition, collisions involving atoms and molecules are energy dependent and dominated by forward scattering, so that hard sphere cross sections are a poor approximation and coronas often have molecular constituents. Finally, many of the processes that energize the exobase region reach a maximum well below the exobase, so that the structure of the exosphere is determined by the aeronomy occurring deep in the thermosphere requiring a description of the transport processes.

To more carefully define the nominal exobase, one needs to know the mean free path prior to a significant momentum transfer collision in an atmosphere that is not necessarily exponential. That is, an escaping hot particle can not suffer a momentum transfer collision that is sufficient to reduce its energy below the escape energy or to scatter it so that it is no longer moving upward. Using realistic potentials, a hot particle of energy E moving through

D.F. Strobel
Johns Hopkins University, Baltimore, MD 21218, USA
e-mail: dstrobe1@jhu.edu

J.H. Waite Jr.
Southwest Research Institute, P.O. Drawer 28510, San Antonio, TX 78228, USA
e-mail: hwaite@swri.edu

a background of identical atoms or molecules [e.g., ~eV O in an O thermosphere (Johnson et al. 2000)] is significantly deflected in an average distance $l_c \sim [bn\sigma_d]^{-1}$, where σ_d is the momentum transfer (diffusion) cross section. Here b is a number that depends weakly on the potential; for a steeply varying potential with forward scattering, $b \sim 0.5$ (Johnson 1990, 1994). Assuming hot recoils, or hot particles in the tail of a thermal distribution, moving upward at random angles, the escape probability from a depth r is

$$P_{es} = \exp\left(-\int_r^\infty ds/l_c \cos\theta\right) \approx \exp[-bN(r)\sigma_d/\cos\theta]. \tag{1a}$$

Here the integral is along the particle's trajectory, ds, θ is the direction of motion, assumed upward, and N is the integrated column density. Since the corona is often not a simple exponential, the average column for escape, N_{es}, should be used:

$$N_{es} \approx \left[\int_0^\infty dN \int_0^1 d\cos\theta N P_{es}\right] \Big/ \left[\int_0^\infty dN \int_0^1 d\cos\theta P_{es}\right] \approx c/\sigma_d{}^{-1}. \tag{1b}$$

Here $c \approx (2/3b)$ and the exobase altitude is obtained from $N(r_{exo}) \approx N_{es}$. Simulations and transport equations using realistic potentials give mean escape depths $(N_{es}\sigma_d) \sim 1.3$ (Johnson 1994, Appendix A). Detailed calculations of the O + O collision, averaging over the ground state multiplet (Kharchenko et al. 2000; Tully and Johnson 2001), give $\sigma_d = 1.3 \times 10^{-15}$ cm^2 for a 2 eV O moving in an O thermosphere or $N_{es} \sim 10^{15}$ O/cm^2. Since σ_d decreases slowly with increasing energy, N_{es} should be averaged over the hot particle energy distribution. It is usually evaluated at the escape energy, slightly underestimating N_{es}. Therefore, the exobase is not unique as it depends on the escaping molecule and its energy.

In the Chamberlain approximation for escape described in Sect. 2.1 below, only those atoms that are produced at or above the exobase with energies exceeding the escape energy and oriented upward are considered. However, using a single exobase altitude in a realistic atmosphere is problematic and misleading because it is different for different molecules and is an extended region. Therefore, this model provides only a rough prediction and computer simulations are now common, as described in Sects. 2.3 and 2.4. In the following we use models for the processes considered in the other papers in this issue to describe the physics and chemistry leading to the formation of a planetary corona and molecular escape. Results of simulations are then reviewed for a number of solar system bodies. Bodies with relatively robust atmospheres, for which escape is a relatively small effect at present, are first considered. This is followed by descriptions of icy bodies in the outer solar system for which escape can play a dominant role. The escaping molecules invariably leave a trail of neutral gas as is well known for comets. For a satellite orbiting a planet, this gas can remain gravitationally bound to the planet forming a nearly toroidal atmosphere. Since these atmospheres are extensions of the satellite exospheres and populate the planet's magnetosphere, they are discussed along with their source body.

2 Modeling of the Exosphere and Escape

2.1 Chamberlain Exobase Model

Chamberlain (1963) developed a widely used analytic model of the density of an exosphere. It is based on the assumptions that the speed distribution is Maxwellian at the exobase,

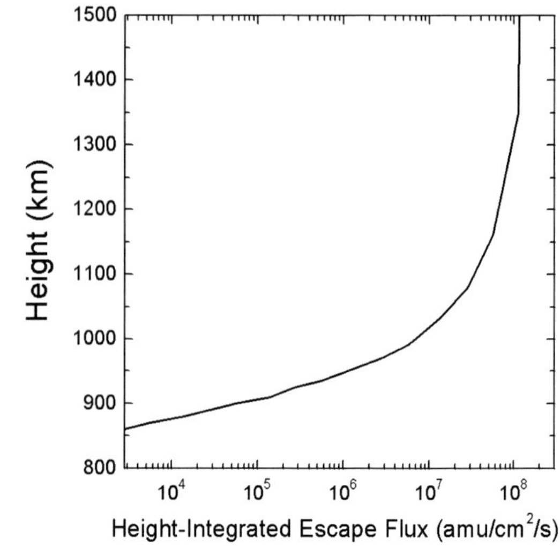

Fig. 1 Height-integrated mass escape flux due to photochemistry and atmospheric sputtering in the transition region of the Titan's upper atmosphere: most escaping molecules originate at altitudes below the nominal exobase ($\sim$ 1450 km) (adapted from: Shematovich et al. 2003). This flux at 1500 km ($\sim 1.2 \times 10^8$ amu/cm^2/s) is about an order of magnitude smaller than recent estimates based on Cassini data (see Sect. 4.4), but the source region of hot particles is the same

that collisions can be ignored above the exobase, that the only force acting is the gravity, and that the speed distribution vs. altitude is obtained from the Liouville equation. Three populations of particles were considered: ballistic, satellite, and escaping. Using the Liouville equation, an analytical formula was derived for the density of each population above the exobase. This approach was generalized for a non-uniform exobase (Vidal-Madjar and Bertaux 1972) and a rotating planet (Hartle 1973; Kim and Son 2000; Kim et al. 2001). Velocity distributions produced by photo-dissociation and atmospheric sputtering have also been used. For instance, in order to fit the recent INMS/CASSINI measurement of Titan's exospheric density, De La Haye et al. (2007a) required an enhanced high energy tail to represent the hot particles produced below the exobase. The full distribution function was described by a kappa-distribution or a Maxwellian combined with a power-law distribution. A best fit was found for the corona densities over a narrow range of altitudes using a kappa distribution to account for both hot and thermal components.

Exospheres like Titan's, having a non-thermal component, require a description of the energizing processes below the exobase where collisions can not be neglected. Deviations from local thermal equilibrium start at Kn ~ 0.1, which occurs at 650 km in Titan's atmosphere. Therefore, production and transport of hot particles into the corona requires solving a Boltzmann equation or a Monte Carlo simulation as discussed below.

2.2 Boltzmann Transport

In the two stream approximation, the flux of hot particles in a background thermal atmosphere is derived from the full Boltzmann transport equation (Schunk and Nagy 2000). Dividing the flux into upward $\Phi^+(E, z)$ and downward $\Phi^-(E, z)$ components, where E is the kinetic energy and z the altitude, results in coupled linear equations that are solved for a range of altitudes up to the exobase. The velocity distribution for the hot particles at the exobase is $\{f(z, v) = [\Phi^+(E, z) + \Phi^-(E, z)]/v(E)\}$, where $v(E)$ is the speed corresponding to the kinetic energy E (Nagy and Banks 1970). The exospheric structure and escape is then determined using the Liouville equation. Nagy et al. (1981) modeled the first detection of hot O in Venus' exosphere produced by dissociative recombination of O_2^+. They used branching ratios from Rohrbaugh and Nisbet (1973) to reproduce the O scale height but over estimated the density. This was subsequently corrected using better input parameters (Nagy and Cravens 1988; Nagy et al. 1990). Measured branching ratios for dissociative recombination (Kella et al. 1997) and the vibrational distribution of the O_2^+ were used by Kim et al. (1998) for solar minimum and maximum conditions at Mars. The role of CO^+ dissociative recombination and photo-dissociation of the CO molecules at Mars' was calculated similarly (Nagy et al. 2001).

Solutions have been developed for a linearized Boltzmann equation in order to describe gas flow in the transition region (Shizgal and Lindenfeld 1982; Shizgal and Blackmore 1986; Shizgal and Arkos 1996; Shizgal 1999). This approximation is used to describe the flow of hot or minor species through the background atmosphere assuming their effect is negligible (Shematovich et al. 1994). Kabin and Shizgal (2002) calculated the Boltzmann collisional term for endogenic or exothermic reactions between species that both have Maxwellian distributions and scatter isotropically. Their results were successfully compared with Monte Carlo test particle simulations described below. They derived the distribution of hot O in the Venusian and Martian atmospheres produced by dissociative recombination of O_2^+ including collisions with thermal O. Pierrard (2003) also solved a linearized Boltzmann equation using a spectral method to determine the flux of H and He through an atmospheric background.

2.3 Monte Carlo Simulations

The Direct Simulation Monte Carlo (DSMC) model is a stochastic method used to describe a rarefied gas and is equivalent to solving the Boltzmann equation. It treats both the dynamic and stochastic nature of the gas (Bird 1994) and is valid if the collisions are statistically independent, multi-particle collisions are negligible, and the collision time is short compared to the time between collisions. All types of collisions can be accounted for including those between atmospheric particles, so that the non-linear processes can be included (Shematovich et al. 1994, 1999, 2005a, 2005b; Krestyanikova and Shematovich 2005, 2006). Each species is described in term of its phase space distribution. The motions of representative particles, each assigned a weight, are followed taking into account collisions and forces. Source distributions and collisions are typically treated using Monte Carlo models. DSMC is time consuming when the domain is highly collisional, but is useful for describing the transition from the collisional to collisionless regime (Marconi et al. 1996). It has been used to simulate heating of the exobase region, the coronal structure, and escape using knowledge of the processes that produce hot atoms and molecules. DSMC simulations have been applied to the exobase region of Mars (Leblanc and Johnson 2001; Krestyanikova and Shematovich 2005, 2006), Titan (Shematovich et al. 2003; Michael and Johnson 2005), and Europa (Shematovich et al. 2005b) as well as to a comet's coma (Combi 1996).

Simplifications are often used, because large numbers of particles are needed to obtain an accurate speed distribution. The velocity space is often divided into a hot particle population (the tail of the distribution) and a thermal background. Test particle simulations track the hot component or trace species allowing collisions only with the background gas and are equivalent to solving a linearized Boltzmann equation. Such simulations have been extensively used at the Earth to model the escape and exospheric formation of minor species (Chamberlain and Campbell 1967; Chamberlain 1969; Barakat and Lemaire 1990; Hodges 1994). These simulations have also been tested against a multi-moment solution (Demars et al. 1993) or a direct solution of the linearized Boltzmann equation (Shizgal and Blackmore 1986). Since the hot particles heat the thermosphere, adding a hot tail to a Maxwellian can be incorrect (Johnson et al. 2000).

1D test particle simulations were also used to describe the O exosphere at Mars produced by dissociative recombination of O_2^+ for different solar conditions (Ip 1988, 1990; Lammer and Bauer 1991). Instead of the standard hard sphere cross sections, Johnson et al. (2000) used the, so-called, universal interaction potential (Ziegler et al. 1985) to describe O + O collisions in an O thermosphere and corona. They showed the escape rates obtained using test particle simulations compared favorably with DSMC results using the same potentials, but differed from hard sphere results. Hodges (2000) developed the first 3D model of the hot O component at Venus and Mars using realistic atmospheric and ionospheric models, but described the thermalization of the hot O by hard sphere collisions. 3D test particle simulations including realistic collision models were used to describe the structure of the exosphere induced by pick-up ion sputtering of Mars (Leblanc and Johnson 2001) and Titan (Michael et al. 2005). In a 1D simulation, (Leblanc and Johnson 2002a) described the effect on atmospheric sputtering of having molecular rather than atomic species at the exobase. They introduced a molecular dynamics collision code which accurately calculated the production of hot neutrals by collisional dissociation (Johnson et al. 2002). Chaufray et al. (2007) used a simpler model for Mars atmosphere, but took into account the dependence of the recombination rate on the solar zenith angle *and* self-consistently coupled the atmosphere simulation to a 3D hybrid simulation of the interaction with the solar wind (Modolo et al. 2005).

Monte Carlo simulations have rapidly increased in complexity and can be more useful than the Boltzmann equation when there are multiple species. The simpler test particle methods are sufficient unless the response of the atmosphere is needed. In addition, test particle methods below the exobase can be coupled to DSMC models in the exobase region and above. However, when the heating rate near the exobase is large accurate representation of the full speed distribution throughout the transition region requires an extensive, often intractable, computational effort in both the Monte Carlo and Boltzmann models. In addition, accurate cross sections are often not available, especially for collisions involving hot particles. Results of simulations for a number of planetary bodies will be described below.

2.4 Exosphere–Plasma Interactions

Since the incident plasma can be an important source of heating and atmospheric loss (Johnson and Luhmann 1998), the principal uncertainty in describing the atmosphere near the exobase is the description of the flow of the ambient and locally produced plasma through the corona.

The effect of a neutral exosphere on the formation of the magnetospheric bow shock and magnetic barrier has been discussed since Pioneer Venus and Mars Global Surveyor observations. The bow shock position at Venus was observed to move away from the planet with increasing solar activity (Alexander and Russell 1985), whereas no significant variation of the bow shock and magnetic pile-up boundary position was observed at Mars (Vignes et al. 2000). Bauske et al. (1998) and Kallio et al. (1998) used 3D single-fluid MHD models, in which a source term was introduced to account for the planetary ions, in order to evaluate the role of the exosphere on the bow shock position, and, hence, on the plasma flow (Ledvina et al. 2008). They concluded that the variations in the bow shock in going from solar minimum to maximum could be reproduced. Ma et al. (2002) and Modolo et al. (2005) used 3D MHD and hybrid models respectively, but did not find any significant dependence of the bow shock position on the exosphere structure at Mars even if an unrealistically high value of the photoionization rate was used. Unlike at Venus, these results emphasize the difficulty at Mars of inferring exospheric structure from bow shock position, contrary to the conclusion of Kotova et al. (1997).

Pérez-de-Tejada (1987, 1998), Lundin et al. (1991), Lundin and Dubinin (1992) calculated the atmospheric escape rate from weakly magnetized planets using the upstream and downstream characteristics of the solar wind. They described the momentum transfer between the incident solar wind and the accelerated plasma by a simple relation using a coefficient for the scavenging efficiency (Lundin and Dubinin 1992). The cross section of the interaction region is defined to be between the mass loading boundary and the magnetopause (Lundin et al. 1991; Lundin and Dubinin 1992). The main difficulty is the lack of constraints on both the efficiency of the momentum transfer and the size of the transfer region. Luhmann et al. (1992) pointed out that this does not allow the use of present-day results to describe the history of the planet-solar wind interaction. Discussion of the interaction for the individual objects is given below.

3 Escape processes

3.1 Thermal Escape

Planetary and satellite atmospheres are confined by gravity. This is often characterized by the so-called Jeans parameter, λ: $\lambda = v^2{}_{\rm esc}/U^2$, where $U = (2kT/m)^{0.5}$ is the most probable speed at temperature T, and $v_{\rm esc} = (2GM/r_{\rm exo})^{0.5}$ is the escape speed at the exobase. Larger values of λ imply a more tightly constrained atmosphere, as is evident by writing the exobase value of λ

$$\lambda \equiv \frac{GMm}{kTr_{\rm exo}} \left[= \frac{\text{gravitational potential energy}}{\text{random kinetic energy}}\right] = \frac{r_{\rm exo}}{H}.$$

As written, m is the mean molecular mass, $g = GM/r^2$ is the gravitational acceleration of an object of mass M, G is the gravitational constant, and the atmospheric scale height is $H = (kT/mg)$ all defined at the exobase. As discussed above, in region where l_c is much less than the local scale height, H ($\mathrm{Kn} \approx l_c/H \ll 1$), the atmosphere can be treated as a fluid and the exobase is at $\mathrm{Kn} \approx 1$. The transition from collision dominated ($\mathrm{Kn} \ll 1$) to the quasi-collisionless exosphere ($\mathrm{Kn} > 1$) is gradual, not abrupt, as discussed. For example, at Titan the 'exobase region' may be ~ 1000 km thick as the atmosphere transitions from N_2 to H_2 domination. While it is often stated that the fluid equations break down at the exobase, they are the first three moments of Boltzmann's equation, so that, with care, they can be applied above the nominal exobase.

3.1.1 Jeans Escape—Thermal Evaporation

An ideal gravitationally bound atmosphere is one in which the escape rate is zero: $\lambda \to \infty$. Although this limit is unattainable, it is instructive. The classic model with no *bulk* outflow is that of Chamberlain, described above, in which escape is due to the fraction of upward moving atoms/molecules with velocities exceeding $v_{\rm esc}$ at the exobase. This gives the Jeans formula for thermal evaporation

$$F_{\rm Jeans}(r_{\rm exo}) = \frac{n(r_{\rm exo})U}{2\sqrt{\pi}}(\lambda + 1)e^{-\lambda} \tag{1}$$

where λ is evaluated at the exobase. Since the tail of the Maxwellian distribution is depleted by escaping particles, it assumed to be rapidly replenished by collisions. In the Jeans limit, $\lambda \to 0$, the atmosphere is no longer bound and blows away with a flux $= n(r_{\rm exo})v_{\rm ther}/4$ where $v_{\rm ther} = (8kT/\pi m)^{0.5}$. This is the upward directed thermally driven flux at the exobase and is used to describe the coma of a comet discussed later in this chapter. In the limit of large λ, the atmosphere is gravitationally retained and Jeans escape is negligible. In the solar system, bodies with values of $\lambda \lesssim 50$ at the exobase have extended atmospheres and large values of $r_{exo}/r_p \geq 1.2$, where r_p is the planet radius (Strobel 2002). Some representative values of λ are: Pluto, $\lambda \sim 20$–25 at $r = 1450$ km and ~ 10 at the exobase ~ 4400 km, Titan, $\lambda \sim 55$–60 at $r = 3300$ km and ~ 45 at the exobase ~ 4300 km, Earth, $\lambda \sim 1000$ at $z = 85$ km and ~ 130 at the exobase ~ 500 km (but highly variable with solar activity), and Jupiter, $\lambda \sim 1860$ at $z = 350$ km and ~ 430 at the exobase $z \sim 2300$ km. The respective values of λ for Venus, Mars, Saturn, Uranus, and Neptune are 1600, 490, 1300, 200, and 450 at the 1 μbar level, whereas at their respective exobases: $\lambda \sim 350$ at $z = 140$ km, 200 at 160 km, 420 at 2500 km, 50 at 4700 km, 120 at 2200 km (cf. Strobel 2002). For comparison, the solar wind with a coronal temperature of 2×10^6 K, has $\lambda \sim 4$.

3.1.2 Hydrodynamic Escape

In contrast to Jeans escape, hydrodynamic escape is an organized outflow with a bulk velocity driven by the heating of the atmosphere. In steady state, with no net production or loss, the atmosphere below the exobase is described by the Euler equations for a pressure driven fluid (e.g., Landau and Lifshitz 1987). For 1D radial flow with velocity, v, the continuity, momentum and energy conservation equations can be written:

$$\frac{1}{r^2}\frac{\partial}{\partial r}(r^2 \rho v) = 0 \tag{2}$$

$$\frac{\partial}{\partial r}\left(\frac{1}{2}v^2\right) + \frac{1}{\rho}\frac{\partial p}{\partial r} + \frac{GM}{r^2} = 0 \tag{3}$$

$$\frac{1}{r^2}\frac{\partial}{\partial r}\left[r^2\left\{\rho v\left(\frac{1}{2}v^2 + \tilde{c}_p T + \Phi_g\right) - \kappa\frac{\partial T}{\partial r}\right\}\right] = Q \tag{4}$$

where the viscous terms, which are important for substantial velocities, are ignored. Here ρ $(= mn)$ is the mass density, v the bulk velocity, $[\tilde{c}_p T]$ the enthalpy with $\tilde{c}_p$ the specific heat at constant p, Q the heating/cooling, κ $[= \kappa_0 (T/T_0)^s]$ the thermal conductivity, and gravity is written as a potential Φ_g: $-\nabla\Phi_g = -GM\vec{r}/r^3$. Equation (2) gives the mass loss rate, written as $4\pi(mF) = [4\pi\rho v r^2]$.

Although one can analyze the importance of the various terms directly, these equations are often written using non-dimensional variables (e.g., Watson et al. 1981):

$$\lambda = \frac{GMm}{rkT_0}, \qquad \tau = \frac{T}{T_0}, \qquad \psi = \frac{mv^2}{kT_0} = \left(\frac{v}{c}\right)^2, \qquad \zeta = \frac{Fk}{\kappa_0 r_0 \lambda_0}, \qquad c_p = \frac{\tilde{c}_p}{k/m} \tag{5}$$

Here $c = (kT_0/m)^{1/2}$ is the speed of sound, $s = 0$ in the expression for $\kappa(T)$, and subscript "0" denotes values at the lower boundary. The steady-state equations are then

$$\frac{d}{d\lambda}\left(\frac{1}{2}\psi\right) + \frac{1}{n}\frac{d(n\tau)}{d\lambda} - 1 = 0 \tag{6}$$

$$\frac{d}{d\lambda}\left(\lambda - c_p\tau - \frac{1}{2}\psi - \frac{\tau^s}{\zeta}\frac{d\tau}{d\lambda}\right) = \frac{(r_0\lambda_0)^3}{\lambda^4}\frac{Q(n(\lambda,\mu_i))}{FkT_0} \tag{7}$$

where Q is a function of n and the mixing ratios of radiatively active constituents, μ_i.

Hydrodynamic escape is formulated in the high density approximation (Parker 1964; McNutt 1989) in which the gravitational potential is deep. Below the exobase the hydrodynamic expansion is slow, $v \ll c$, such that the gravitational energy is greater than the thermal energy and much larger than the flow energy: i.e., $\lambda \geq c_p \tau \gg \psi$. In this case the equations reduce to

$$\frac{d[\ln(n\tau)]}{d\lambda} - \frac{1}{\tau} = 0 \tag{8}$$

$$\frac{d}{d\lambda}\left(\lambda - c_p\tau - \frac{1}{\zeta}\frac{d\tau}{d\lambda}\right) = \frac{(r_0\lambda_0)^3}{\lambda^4}\frac{Q(n(\lambda))}{FkT_0}. \tag{9}$$

Equation (8) is equivalent to (3) neglecting the flow term, resulting in an atmosphere in hydrostatic equilibrium. Equation (9), the scaled energy equation, also neglects the flow term in (4). Therefore, v must be small up to the exobase of the major constituent. The best measurements applicable to planetary escape are the Cassini INMS data for Titan's upper atmosphere and exosphere. Fluid models calibrated to this data assume the flow goes supersonic for the light species, H_2 and CH_4, in an extended region above the exobase of the major constituent N_2 (Cui et al. 2008; Yelle et al. 2008; Strobel 2008a). The relevant asymptotic boundary conditions are n and $\tau \to 0$ as $r \to \infty$ (Parker 1964). Integration of (9) from the lower boundary to $r \to \infty$ ($\lambda \to 0$) yields

$$\lambda_0 = \int_0^{\lambda_0} \frac{(r_0\lambda_0)^3}{\lambda^4}\frac{Q(n(\tau,\lambda))}{FkT_0} d\lambda + \left[c_p\tau + \frac{\tau^s}{\zeta}\frac{d\tau}{d\lambda}\right]_{\lambda_0}. \tag{10}$$

In (10) three processes drive the expansion. If the atmosphere is optically thin to EUV and UV radiation and the interaction with the plasma is negligible, the first term on the right is small. If also the heat conduction is small, then the dominant term on the right is $c_p\tau$ (the internal heat in (4)), plus $\frac{1}{2}\psi$ from (7) which was neglected in (10). Hence, $\lambda_0 \sim [c_p\tau + \frac{1}{2}\psi]$ {i.e., $mgr_0 \sim \tilde{c}_p T + mv^2/2$} with $\tau \approx 1$ {i.e., $T \sim T_0$} due to negligible heating. Ignoring the enthalpy, the atmosphere is traditionally defined to be in a blowoff if $\lambda < \frac{1}{2}\psi$ at the exobase [i.e., $v(r_{\text{exo}}) > v_{\text{esc}}$]. However, $c_p \sim 5/2$ and $7/2$ for atoms and diatomic molecules, respectively, and cannot be ignored. In this limit v is not negligible compared to c, so that ψ must be included in (10). For a vanishingly small total energy flux at infinity, $\lambda_0 \sim c_p + \psi/2$. More likely, the energy flux at infinity is finite and (10) does not apply. This blowoff condition is similar to the jets emitted from the surface of a comet, where $\lambda \ll 1$, and represents chaotic escape.

In contrast, slow hydrodynamic escape refers to an organized, controlled expansion of the atmosphere driven by net heating and upward thermal conduction through some boundary below which heating occurs. Hydrodynamic expansion can be driven by solar heating and escape is limited by the net heating rate (power absorbed times heating efficiency minus radiative cooling) (Hunten and Watson 1982). Therefore, the first term on the right of (10) dominates, so that, in the absence of thermal conduction, $\{[GMm/r_0](4\pi F)\} \approx \int_{r_0}^{\infty} Q 4\pi r^2\, dr$. This gives an upper limit to the mass escape flux by equating the gravitational energy per unit time that would be carried off by the escaping particles to the heating rate. To compare to Jeans escape, assume that the sonic point occurs at the exobase so that the outflow velocity, v, equals c (the isothermal speed of sound). Then $v = v_{\text{esc}}$ if $\lambda = 1/2$ at the exobase. This can be compared with the mean velocity of Jeans escape, $0.36v_{\text{esc}}$ [from (1) with $\lambda = 1/2$], giving the well known discrepancy of about a factor of three between bulk outflow escape and Jeans escape.

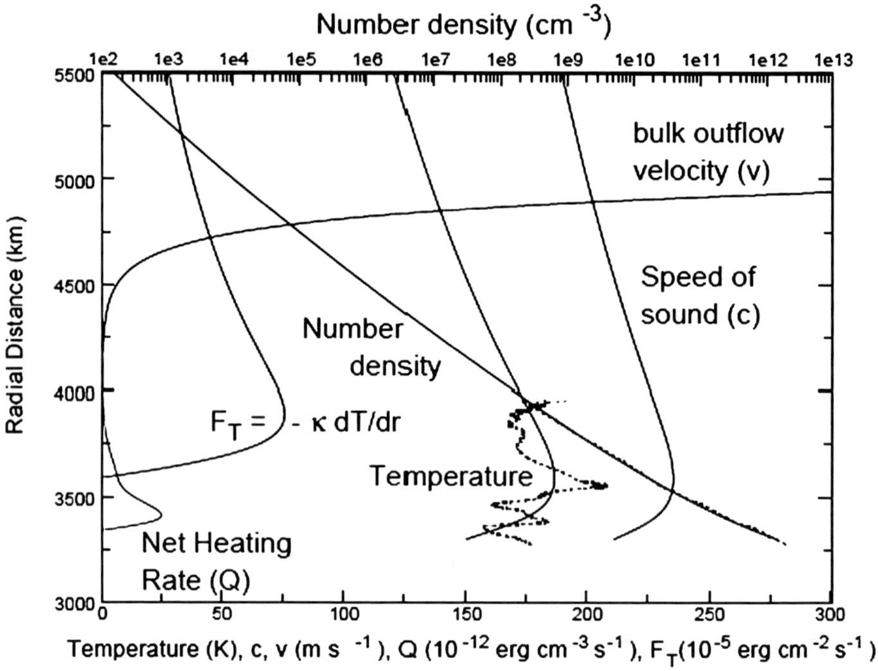

Fig. 2 A model for Titan's atmosphere (Strobel 2008a): n, T, c, v, and upward thermal heat conduction flux (*solid lines*) for an N_2 atmosphere with solar medium conditions: net heating due mostly to CH_4 UV heating above the lower boundary at $r_0 = 3300$ km, $T_0 = 158$ K, and downward thermal heat conduction flux at lower boundary $= 4.0 \times 10^{-3}$ erg/cm^2/s. Comparison with the HASI measurements of total number density and temperature (*dashed lines*) at $\sim 10.3°$S latitude. Assuming the model applies above the nominal exobase (here taken to be 4300 km), the low speed solution (Mach number < 0.3) is valid only to ~ 4750 km and the mass escape rate is 4.5×10^{28} amu/s

Equations (8) and (9) can be solved quasi-analytically if solar heating is represented by a δ function (McNutt 1989; Krasnopolsky 1999). They can also be solved numerically with distributed heating functions (Strobel 2008a, 2008b; for Titan and Pluto discussed below). These solutions were restricted to $\psi \ll 1$ ($v \ll c$) below the exobase. Tian and Toon (2005) solved Euler's equations using distributed heating for Pluto's atmosphere. Although they only included EUV absorption by N_2, omitting the larger UV component absorbed by CH_4, they obtained an order of magnitude larger escape rate than Krasnopolsky (1999) and Strobel (2008b).

For slow hydrodynamic expansion, the physics is as follows. Adiabatic cooling is assumed to occur throughout the atmosphere associated with the bulk outflow velocity in Fig. 2. This outflow can be driven by EUV/UV heating as in the model for Titan. There are also regions of net cooling due to IR emissions that locally exceed solar heating: above 3440 km in Fig. 2. Thermal heat conduction redistributes the heat, producing a temperature maximum at some altitude, below which a downward thermal heat conduction flux adjusts to both adiabatic and radiative cooling. Above the maximum, upward thermal conduction delivers power to sustain an expansion. At levels much above the temperature maximum, heat conduction must be about equal to what is needed to power the upward flow of mass in the model in Fig. 2. Solar heating near the exobase contributes only $\sim 10^{-3}$ of the heat conduction flux and plasma-induced heating is ignored.

Solutions below the exobase must be matched to an escape model, about which there is still controversy. In the model in Fig. 2 it is essential that heat conduction continues to collisionally power the expansion above the exobase until the flow speed exceeds the escape speed. The description of upward heat conduction in the exobase region has been estimated by a 13-moment expansion of the Boltzmann equation, since the fluid is not Maxwellian. In such a model heat conduction is thought to redistribute velocities into an upwardly directed, enhanced tail augmenting escape, as proposed for H_2 escape from a nitrogen atmosphere (Cui et al. 2008). Thus there can be an intimate link between a perturbed velocity distribution and hydrodynamic escape. The expansion below is powered by the delivery of heat from the region of maximum heating to higher altitudes by thermal heat conduction, which requires a decreasing temperature profile. In the above model, hydrodynamic expansion with an upward thermal heat flux and the collisional production of a high speed tail beyond the escape velocity due to heat conduction into the exosphere are two descriptions of the same phenomenon. However, whether or not, for a given λ, the distortion in the tail of the velocity distribution produced by thermal conduction is sufficient to produce the required escape flux has not been demonstrated computationally.

3.2 Photochemical-Induced Escape

Photochemical escape has its origin in the interaction of solar photons with neutrals in the upper thermosphere and exosphere. This includes direct interactions of photons and photoelectrons with thermospheric molecules, as well as chemical reactions of ions with neutrals and electrons. Direct excitation includes photodissociation, e.g.,

$$\mathrm{CO} + h\nu \rightarrow \mathrm{C}^* + \mathrm{O}^* \tag{11}$$

(Brinkmann 1971; Fox and Bakalian 2001); photodissociative ionization, e.g.,

$$\mathrm{N}_2 + \hbar\nu \rightarrow \mathrm{N}^{+*} + \mathrm{N}^* + \mathrm{e} \tag{12}$$

(McElroy et al. 1977; Fox and Dalgarno 1983); photoelectron-impact dissociation, e.g.,

$$\mathrm{N}_2 + \mathrm{e}^* \rightarrow \mathrm{N}^* + \mathrm{N}^* + \mathrm{e}, \tag{13}$$

(McElroy et al. 1977; Fox 1993); and photoelectron-impact dissociative ionization, e.g.,

$$\mathrm{N}_2 + \mathrm{e}^* \rightarrow \mathrm{N}^+ + \mathrm{N}^* + 2\mathrm{e} \tag{14}$$

(Fox and Dalgarno 1983). The asterisks denote electronic excitation or a particle with excess kinetic energy, for neutrals the so-called hot particles. The energy released in (11) is the difference between the photon energy and the molecular dissociation energy, which can be reduced by electronic excitation of the products. Electron impact dissociation of CO primarily produces two ground state atoms (Cosby 1993). Because the interaction of photons and electrons with neutrals is not the same, this assumption should only be adopted when there is no detailed information about the individual process. Thus our ability to determine the energy released depends on knowledge of the branching ratios of the various channels, information that is often not available.

The computation of escape rates due to photodissociation is also affected by the lack of cross sections, which are usually assumed to be equal to the difference between the photoabsorption and photoionization cross sections. Unfortunately, these cross sections are often not measured simultaneously and much of the time they are taken from different sources (Huestis et al. 2008). In addition, photodissociation of many atmospherically important molecules, such as CO, CO_2, H_2 and N_2, proceeds by absorption into discrete states followed by predissociation (Fox and Black 1989; Fox 2007). In such cases, the photoabsorption cross sections must be measured at very high resolution, on the order of 10^{-4} nm, and combined with similarly high resolution solar fluxes in order to compute the dissociation rates and product energies. Fox and Black (1989) constructed such cross sections and solar fluxes for CO and applied the results to Venus. Measurements and calculations of very high resolution cross sections for N_2 (e.g., Stark et al. 2007; Sprengers et al. 2005; Lewis et al. 2005), and CO_2 have been carried out by several groups (Huestis et al. 2008), but are generally not available over the entire energy range of interest.

In photoelectron impact dissociation (13), the excess energy is usually carried away by the electron. The energies of the products are estimated by measurements of appearance potentials or time-of-flight measurements (e.g., Armenante et al. 1985), or by analysis of Doppler-broadened lineshapes. Ajello and Ciocca (1996) analyzed the line shapes of the NI emissions at 1200 Å due to 30 eV electrons on N_2 and found that the excited $N(^4P)$ atoms produced were translationally hot, with average energies ~ 1 eV. Prokop and Zipf (1982) found that the average energy of the N produced in electron impact dissociation of N_2 was ~ 0.45 eV. Fox and Dalgarno (1979) concluded that in electron impact dissociation of CO_2 the mean energy released was ~ 1 eV. Dissociative ionization, processes (12) and (14), are known to produce very energetic products (e.g., Locht and Davister 1995; Tian and Vidal 1998). The energetic ions and neutrals are mostly due to predissociation of electronically excited states of the ions. This is also the case for the energy release when a fast proton produces dissociative charge exchange in N_2 (Luna et al. 2003) and O_2 (Luna et al. 2005). The fraction of atoms with the escape energy must be estimated from such measurements.

Exothermic reactions among the ions, neutrals and electrons near the exobase produce translationally hot species. Dissociative recombination (DR) of molecular ions is one of the most exothermic reactions. DR of O_2^+ is known to be the photochemical source of escaping O from Mars as discussed below. It also is a major source of the hot O in coronas on Earth (Shematovich et al. 1994; Bisikalo et al. 1995), and Venus, also described below. Starting with O_2^+ vibrational ground state, the reaction proceeds via five channels with different exothermicity:

$$O_2^+ + e \rightarrow O(^3P) + O(^3P) + 6.98 \text{ eV} \tag{15a}$$

$$\rightarrow O(^1D) + O(^3P) + 502 \text{ eV} \tag{15b}$$

$$\rightarrow O(^1S) + O(^3P) + 2.79 \text{ eV} \tag{15c}$$

$$\rightarrow O(^1D) + O(^1D) + 3.05 \text{ eV} \tag{15d}$$

$$\rightarrow O(^1D) + O(^1S) + 0.83 \text{ eV}. \tag{15e}$$

The branching ratios have been measured in ion storage rings and are found to be 0.22, 0.42, 0.0, 0.31, and 0.05, respectively (Kella et al. 1997). Peverall et al. (2001) have also measured branching ratios for the channels of DR of O_2^+ as a function of energy and a negligible yield for (15c) is predicted (Guberman and Giusti-Suzor 1991). Since the escape energy for O at the Martian exobase is ~ 1.98 eV, only O produced via (15a) and (15b) can escape. The exothermicities are increased by vibrational excitation of the O_2^+ ($v > 0$). Vibrational distributions of O_2^+ in the atmospheres of Earth and Venus have been calculated (Fox 1985, 1986) as have the O energy spectra at the exobases of the terrestrial planets (Fox and Hać

1997a, 1997b). O_2^+ is predicted to be significantly vibrationally excited at the exobases on all three planets.

Branching ratios for the various energetically allowed channels for dissociative recombination have been measured for N_2^+ (Kella et al. 1997; Peterson et al. 1998) and for CO^+ (Rosen et al. 1998). For DR of CO_2^+

$$CO_2^+ + e \rightarrow CO + O \tag{16a}$$

$$CO_2^+ + e \rightarrow O_2 + C^*. \tag{16b}$$

Siersen et al. (2003) found the branching ratio for (16a) was about 9%. This was questioned by Viggiano et al. (2005) who found that nearly all the reactions proceeded by (16a).

Other exothermic ion–molecule reactions can also produce escape: e.g.

$$N_2^+(v') + O \rightarrow NO^+(v'') + N + 3.05\ \text{eV} \tag{17}$$

(Fox and Dalgarno 1983; Fox 1993). In the center of mass frame, 68% of the exothermicity is carried away by the N atom. The absolute value of the exothermicity is increased by vibrational excitation of N_2^+ and reduced by vibrational excitation of the NO^+. The radiation rate by vibrational-rotational transitions of N_2^+ is small, so that it will be vibrationally excited near and above the Martian exobase (Fox and Hać 1997b). The vibrational distribution was computed for the ionospheres of the Earth (Fox and Dalgarno 1985) and Venus. If we assume that the energy released in (17) is distributed statistically among the vibrational, rotational, and translational modes, with N_2^+ $(v = 0)$ in the vibrational ground state, 31% of the product energy appears as vibrational energy, correspondingly reducing the escape probability.

The non-thermal escape of N_2, CH_4, H, H_2, N, NH, HCN, CN and small hydrocarbons has been studied by Cravens et al. (1997) and by De La Haye et al. (2007b). Cravens et al. computed the escape rate for 19 species and 47 processes, including dissociative recombination of ions, ion-molecule reactions, electron-impact dissociative and dissociative ionization of molecular ions. To obtain the escape fluxes, they integrated the production profiles from the exobase to 2500 km. De La Haye et al. (2007b) included only 12 species, but used the 2-stream approximation to compute the escape rates of these ions through a background mixture of N_2, CH_4, and H_2. They estimated the photochemical escape rates of N and C ions in all forms as $8.3 \times 10^{-24}\ s^{-1}$ and $7.2 \times 10^{24}\ s^{-1}$ respectively.

3.3 Plasma-Induced Escape

For a magnetized planet, the intrinsic field is a natural shield for the atmosphere, and part of the apparent ion loss is recycled by the fields as shown for the Earth (Seki et al. 2001). If the planetary body has a relatively strong magnetic field that can interact with either the solar interplanetary magnetic field or the field of a parent body, such as the magnetospheres of Saturn and Jupiter interacting with the solar wind and the field of Ganymede interacting with the Jovian plasma, then magnetic reconnection processes occur leading to the loss of a body's ionosphere. If the body is weakly or non-magnetized, the morphology of the induced fields and the position exosphere along with the scale lengths for interaction, such as the mean free path and the ion gyroradius, will determine the nature of loss processes.

Ionization in the exosphere and upper atmosphere can lead to *ion loss* as well as the formation of energetic neutral *atoms* (ENAs) by charge exchange. The polar wind (Axford 1968) at the Earth is an example of an ion loss process. The interaction with the solar wind modifies the magnetic field configuration close to the Earth, compressing the field on the sunward side and forming a tail on the anti-sunward side. At high latitudes thermal plasma from the polar ionosphere can flow into the magnetosphere and down the tail resulting in the loss of H^+, He^+ and O^+. Processes similar to this occur on non-magnetized planetary bodies Venus and Mars, as discussed later, resulting in day to night flow down the tail (e.g., Shinagawa and Cravens 1989; Ma and Nagy 2007).

The principal ionization processes are photo-ionization, ionization by electronic impact, and charge exchange. Charge exchange has been studied as a principal source of non-thermal escape of H atoms at the Earth (Shizgal and Arkos 1996), Venus (Hodges 1993) and Mars (Nagy et al. 1990) leading to the high D/H ratio measured in Venus' atmosphere (Donahue et al. 1982). Charge exchange production of ENAs is usually considered as a nearly resonant collision, equivalent to the exchange of an electron without significant momentum transfer (Shizgal and Lindenfeld 1982; Shizgal and Arkos 1996). Kallio et al. (1997) showed that at Mars the region where ENAs are formed by charge exchange between solar wind H^+ and exospheric O occurs on the dayside in a thin layer whose size depends on solar activity. This has been directly observed by ASPERA 3 on Mars Express (Gunell et al. 2006). Zhang et al. (1993) studied the roles of photo-ionization, charge exchange and impact ionization at Mars and Venus, and concluded that the dominant ionization mechanism is electron impact, but was criticized by Krymskii and Breus (1996, see reply by Luhmann 1996). Using their electric and magnetic field model, Kallio and Koskinen (1999) developed a test particle simulation based on an exospheric model (Nagy et al. 1990) in order to describe the ionization of exospheric O and the escape of O^+ ions at Mars. They concluded that ions formed in the exosphere can produce the loss rate measured by Lundin et al. (1989). Ma and Nagy (2007) found that the O^+ escape rate decreases by a factor 3 when photo-ionization and electron impact ionization rates are set to zero, confirming that the escaping O^+ are formed above the Martian exobase. Cravens et al. (2002) used a test particle MHD model to simulate pick-up ion trajectories and reproduced the 55–72 keV ion population reported by Phobos 2 (McKenna-Lawlor et al. 1993).

A fraction of the pick-up ions and ENAs can re-impact the atmosphere with enough energy to induce heating and atmospheric sputtering. This is particularly true when the pick-up ion gyroradius is of the order of the planet radius, as at Mars and Venus for O^+. The sputtering efficiency is given by a yield, Y. It is the ratio between the number of escaping particles and the number of incident particles and varies inversely with the planet's gravitational energy (Johnson 1990). Sputtering of an atmosphere can occur by direct scattering of atmospheric molecules, also called knock-on, which dominates at grazing incidence by incident light ions or ENAs with energies ~keV: $Y \lesssim 1$. For heavy incident ions, a cascade of recoils is set in motion with some having sufficient energies and the appropriate direction of motion to escape: $Y \gtrsim 1$ (Johnson 1994). This occurs when keV to few 100 keV incident O^+ pickup ions and ENAs impact the atmospheres of Mars and Venus (Johnson et al. 2000) or when molecular pick-up ions re-impact Titan's atmosphere (Michael et al. 2005).

Watson et al. (1980) showed that atmospheric sputtering of Mars and Venus by impacting solar wind protons was inefficient. Subsequently Luhmann and Kozyra (1991) calculated the flux of re-impacting pick-up ions and ENAs using a 1D exospheric model of the O density and the solar wind flow derived from a gas-dynamic model. They concluded that a significant number of sputtered O should escape from Mars and Venus. Because the sputtered products add to the exosphere they can in turn be ionized and accelerated back into the exobase resulting in a complicated feedback process (Johnson and Luhmann 1998). Since an expanded corona can push the solar wind interaction region outward, they also pointed out that this could reduce the flux of re-impacting ions, so that there may be negative feedback. Heating of thermospheres and exospheres induced by the deflected ambient plasma, pick-up ions and re-impacting neutrals have been simulated for a number of planetary bodies, as discussed below.

4 Exospheres and Escape from Venus, Mars and Titan

4.1 Overview

Theories of planetary exospheres have been based on ground-based and space observations of emission features such as the 121.6 nm Ly-α and 102.6 nm Ly-β hydrogen lines, the 58.4 nm helium line, and the 130.4 and 135.6 nm atomic oxygen lines. The Mariner observations indicated the presence of hot H (Fig. 3), and the Pioneer Venus UV spectrometer data established the presence of hot O and C at Venus (Nagy et al. 1981; Paxton 1985). Such observations, together with *in situ* mass-spectrometer measurements, as at Titan, allow the density and temperature height profiles of the exospheric components to be constructed. As described above, the exospheres contain hot (suprathermal) neutrals that are a manifestation of the non-thermal processes: dissociative recombination, dissociation by ultraviolet photons and electrons, and exothermic chemical reactions. These are accompanied by the release of energies on the order of several eV, part of which is stored as the internal excitation of the products (Wayne 1991). Charge exchange and atmospheric sputtering induced by energetic plasma ions also produce hot neutrals but with energies up to several hundred eV s (Johnson 1990). These hot neutral sources produce escape, determine the coronal structure, and produce nonthermal emissions. They can also affect the chemistry since non-equilibrium rate coefficients, especially for reactions with high activation energies, can be large.

4.2 Venus

The neutral and plasma environments of Venus are strongly coupled since the atmosphere is not protected by an intrinsic field (Bougher et al. 2008). Pioneer Venus Orbiter (PVO) found a robust magnetic barrier and an almost complete solar wind deflection around the exobase,

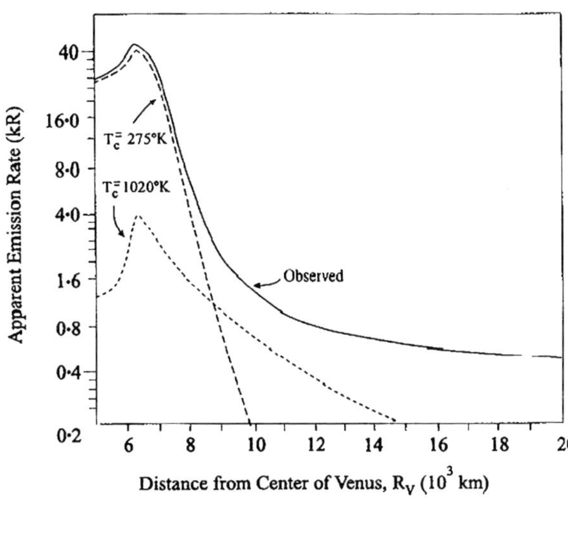

Fig. 3 H corona at Venus discovered by Mariner 5 (Anderson 1976). The emission was fit to a thermal fraction at 275 K and a non-thermal fraction at 1020 K

and the ionopause, where the solar wind and the ionospheric pressures are balanced, was located at a few 100 km altitude. The abrupt drop in ionospheric density at the ionopause is the most striking evidence that the solar wind is scavenging the upper ionosphere. Just above the ionopause is the mantle; between the mantle and the bow shock is the magnetosheath, whose properties are affected by the presence of the hot neutrals.

The exosphere alters the incoming plasma by both mass loading it with ionized gases and charge exchange producing H and O ENAs. Recent numerical estimates (Gunell et al. 2006) indicate that solar wind penetrates fairly deep into the atmosphere at solar maximum, resulting in acceleration and outflow of mainly O^+ with energies up to a several keV. On the other hand, magnetic field measurements from Venus Express (Zhang et al. 2007) show that little solar wind plasma enters the Venus ionosphere at low solar activity.

The creation of hot O is due to exothermic chemistry and to photon and electron impact, as discussed earlier, but loss to space is due mainly to sputtering. Additional loss, due to the ion escape and ionospheric outflow (Terada et al. 2004), occurs when ions produced in the corona are accelerated by the convective electric field and dragged along by solar magnetic field lines wrapping the planet.

The hot O corona at Venus is produced primarily by dissociative recombination of O_2^+ ions [(15a)–(15e)]. Simulations of the hot O have been compared with direct observations (e.g., Fig. 4a). Most simulations use hard sphere models for collisions, and heating of the atmosphere by hot atoms is not considered. Shematovich et al. (2005a), Krestyanikova and Shematovich (2006) and Shematovich and Johnson (2006) recalculated the energy distributions of hot O in thermospheres of terrestrial planets using realistic differential cross sections (e.g., Kharchenko et al. 2000). Results are shown in Fig. 4b, showing that the corona contains a significant hot O component. Using realistic scattering angle distributions resulted in a lower rate of energy loss and, consequently, a higher hot O fraction than when hard sphere models were used (Nagy and Cravens 1988; Hodges 2000).The creation and escape of a hot C corona has also been studied (Fox and Paxton 2005; Liemohn et al. 2004). They concluded that both photodissociation of CO and dissociative recombination of CO^+ and CO_2^+ are the main sources of hot C. Coupled ionosphere and thermosphere models provide important input parameters (Fox and Paxton 2005) allowing one to calculate the distribution of hot H, C, and O in the thermosphere and to estimate escape rates.

Precipitating O^+ pickup ions of exospheric origin follow helical trajectories along interplanetary magnetic field lines draped across Venus and are either be swept away or re-impact the atmosphere with significant energies (up to 1 keV) causing sputtering and population of the corona. Luhmann and Kozyra (1991) suggested that 90% of the O^+ pickup ions ($\sim 1.25 \times 10^6$ cm^{-2} s^{-1}) re-impact producing an escape flux $\sim 2.5 \times 10^6$ O cm^{-2} s^{-1}. Assuming these O^+ ions impact a half sphere, the average sputter loss rate is $\sim 6 \times 10^{24}$ O s^{-1}. This is comparable with the pick-up ion loss and is the only process that produces neutral O escape at Venus.

There are two main sources for pickup: ionization of neutrals inside the corona, producing mainly of O^+, H^+ and C^+, and an ionospheric wind, the outflow of ions produced above the photochemical equilibrium region and below the ionopause (Terada et al. 2004). Relative escape rates depend on the composition of the ionosphere, which is determined by ion-neutral chemistry involving O_2^+, N_2^+, CO^+, and NO^+. About 10^{25} O^+/s are lost through ionization of the hot oxygen exosphere in the magnetosheath and solar wind followed by magnetic drag and subsequent acceleration of the tail plasma (Luhmann et al. 2006b; Russell et al. 2006; Lammer et al. 2006). Measurements by Venus Express show that H^+, He^+, and O^+ are the dominant escaping ions, through the plasma sheet and in boundary

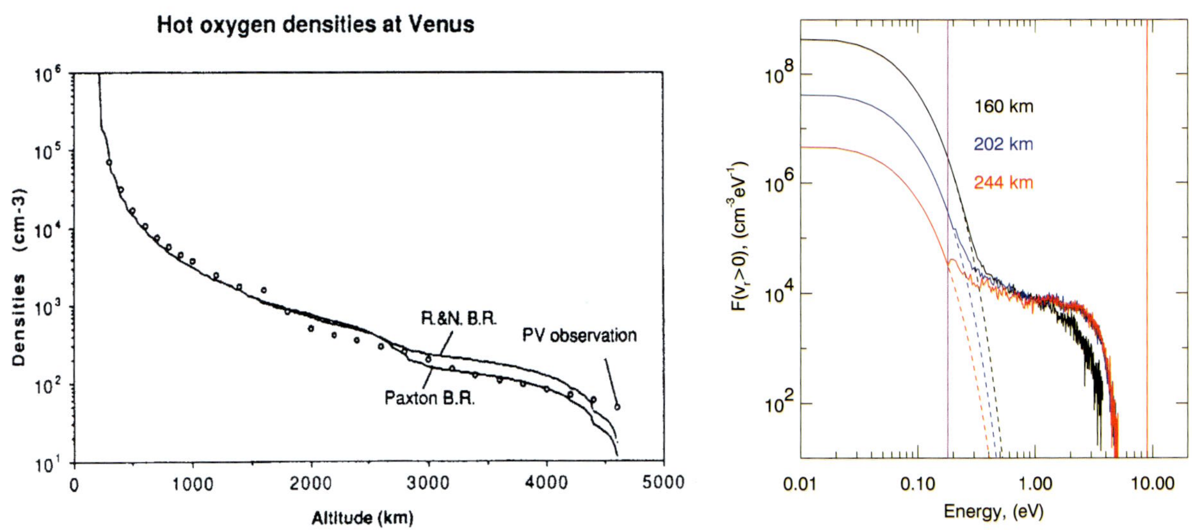

Fig. 4 (*Left*) Hot O at Venus. Calculated and deduced from the PVO UV spectrometer measurements for solar cycle maximum conditions (Nagy et al. 1990). (*Right*) Calculated kinetic energy distributions F ($v_{\mathrm{r}} > 0$) of upward moving O in Venus's upper atmosphere: *solid lines*. Maxwellian: *dashed lines*. *Vertical line* is the escape energy (~ 9 eV) (Shematovich and Johnson 2006)

layer of the induced magnetosphere, with a ratio for H^+ to O^+ of ~ 1.9 (Barabash et al. 2007b). Absolute escape rates are not yet determined.

4.3 Mars

The first measure of escape at Mars was made by Phobos 2 during solar maximum (Lundin et al. 1989). The best estimate was an ion escape rate $\sim$ 1–3 $\times$ 10^{25} s^{-1} of heavy ions tentatively identified as O^+ and O_2^+ (Rosenbauer et al. 1989; Lundin et al. 1989; Lundin and Dubinin 1992). However, this is very uncertain because the position and size of the plasma sheet through which the heavy ions were seen to escape and the solar conditions were not well characterized. Twenty years later, during solar minimum conditions, ASPERA 3 on Mars Express fully covered Mars' magnetotail and obtained a better estimate of ion escape: 1.6×10^{23} O^+ s^{-1}, 1.5×10^{23} O_2^+ s^{-1} and 8×10^{22} CO_2^+ s^{-1} with an uncertainty of less than 50% (Barabash et al. 2007a). This is in agreement with 3D hybrid simulation of Mars' interaction with the solar wind (Modolo et al. 2005) as well as with 3D MHD simulations (Ma et al. 2004; Ma and Nagy 2007). It suggests either the Phobos 2 escape rates (Lundin et al. 1989; Rosenbauer et al. 1989) were significantly overestimated or the rate varies by up to two orders of magnitude from solar minimum to solar maximum.

At Venus, the global day to night motion of the ionospheric plasma is due to the horizontal pressure gradient in the subsolar region which accelerates plasma towards the nightside (Shinagawa 1996). Therefore, a global upward motion of ionospheric ions on the dayside is associated with global downward motion of ions on the nightside, a process that should occur also at Mars. Shinagawa and Cravens (1989) used a 1D multi-species magneto-hydrodynamic simulation between 100 and 480 km to model Mars' ion chemistry and ionosphere. They concluded that large scale horizontal plasma convection needs to occur in the upper atmosphere in order for their model ion profiles to fit the observations of Viking, and the solar wind appears to penetrate the upper atmosphere (Hanson et al. 1977; Johnson 1978). Fox (1997) compared her ionosphere model to Viking measurements and concluded that the main ion escaping from Mars was O_2^+ and the flux might be 4 times larger than that observed by Lundin et al. (1989) at high solar activity and around 3 times lower at low solar activity. The recent measurement by ASPERA 3 on board Mars Express indicate that the solar wind does penetrate deeply into Mars' atmosphere (Lundin and Barabash 2004) but that the tailward flux is significantly different from that in Fox (1997). This suggests the upward dayside ionospheric outflow might not be lost at Mars (Carlsson et al. 2006; Barabash et al. 2007a). In a more recent study using a global 3D multispecies MHD model, Ma and Nagy (2007) concluded that the tailward flow of ionospheric ions is a significant fraction of the ion flux measured by Lundin et al. (1989); they also successfully reproduced the Viking ion densities in a self consistent manner, without the need of ad hoc velocity assumptions.

Most of the models of Mars' interaction with the solar wind indicate that atmospheric escape is dominated by loss of neutrals (Chassefière and Leblanc 2004; Chassefière et al. 2007). To constrain escape the spatial structure, composition and size of the exosphere, and its variability with respect to solar EUV and solar wind activity need to be characterized. In fact, the presence of a hot component in Mars' exosphere is still debated. Lichtenegger et al. (2007) pointed out the discrepancy between estimates of exospheric temperature of 350 ± 100 K based on Lyman α airglow observation (Anderson and Hord 1971) and estimates of 225 K by Viking 1 and 2 probes (Nier and McElroy 1977) at low solar activity, of 220 K for moderate solar conditions by Mars Global Surveyor (Keating et al. 1998), or of 200 ± 10 K for solar minimum conditions from day-glow measurements by SPICAM/Mars

Express (Leblanc et al. 2006, 2007). This discrepancy might be explained by the presence of a significant hot H component in the exosphere like that observed at Venus (Anderson 1976; Bertaux et al. 1978). Chaufray et al. (2008) recently analyzed Mars Express observations of the Lyman α airglow and concluded that, as at Venus, a two component H exosphere fits the observed profiles: a hot component ($T > 500$ K) and a component with $T > 200$ K.

Nagy et al. (1981) and Ip (1988) were the first to describe the production of hot O by dissociative recombination of O_2^+ (15a–15e) releasing excess energies between 0.8 and 6.99 eV depending on the vibrational state of the O_2^+, as discussed earlier. Using a two stream model (Nagy and Cravens 1988) or a test particle simulation (Ip 1988) the partial thermalization of the hot O was described by collisions with atmospheric particles. Both works noted the presence of a substantial hot O component above 500 km at solar minimum (Nagy et al. 1990; Ip 1990; Lammer and Bauer 1991). Kim et al. (1998) developed a model for the hot O exosphere taking into account the role of the vibrational state of the O_2^+ atoms, for both solar minimum and maximum with densities between 2×10^3 O/cm^3 and 6×10^3 O/cm^3 at 1000 km. Nagy et al. (2001) used a two stream model to conclude that photodissociation of CO is the main source of the C exosphere with densities $\sim$ 10–100 C/cm^3 at 1000 km (Fox and Bakalian 2001). Krestyanikova and Shematovich (2005, 2006) used a 1D DSMC simulation with accurate low energy cross-sections and found a significantly hotter O exosphere than in these models.

The sputter contribution to the exosphere (Johnson and Luhmann 1998), although negligible at low solar activity, might be of the same order as the dissociative recombination contribution for higher solar activity and even dominant in early epochs. Recently Cipriani et al. (2007) used a multi-species model of the exosphere of Mars to simulate both dissociative recombination and sputtering. They confirmed the range of densities previously predicted for hot C and O. They also concluded that the sputter contribution to the O exosphere remains significantly smaller than that due to dissociative recombination, but that sputtering populates the exosphere with CO and CO_2 molecules. The Martian hot O corona due to dissociative recombination was a factor of 10 lower at midnight than at noon (Hodges 2000) and the sputter component showed a similar variation (Leblanc and Johnson 2001). Chaufray et al. (2007) estimated the incident flux of pick-up ions incorporating the Modolo et al. (2005) 3D hybrid simulation. Accounting only for the change in the EUV activity, they concluded that at both solar minimum and maximum the contribution to the hot O exosphere by sputtering was one to two orders smaller than that due to dissociative recombination.

Extrapolation to earlier epochs (Lammer et al. 2008) remains uncertain, not only due to uncertainties in the history of the solar activity and solar wind pressure, but also because of the lack of knowledge of the dependency of the escape rate on solar conditions. Modolo et al. (2005) found a variation in the total pick-up ion loss rate by a factor 4 to 5 from solar minimum to solar maximum taking into account *only* the variation of the EUV/UV flux on the ionization rate. Ma and Nagy (2007) found a variation of a factor 2.5 in the ion escape rates due to the EUV/UV flux in going from EUV/UV minimum to maximum, whereas Harnett and Winglee (2006) found the ion loss rates between quiet solar wind conditions and fast solar wind conditions varied by a factor 1.8. Chaufray et al. (2007) examined solar minimum and maximum conditions combining an O exosphere model with the Modolo et al. (2005) hybrid model. They found a global variation of a factor 4 in the neutral escape rates and of a factor 10 in the pick-up ion flux. At Venus, Luhmann et al. (2007) correlated periods of high escape flux with high solar dynamic pressure during PVO observations. Ma and Nagy (2007) modeled this effect at Mars and found an increase in the ion escape rate of an order of magnitude at solar maximum. Therefore, it remains critical at Mars to accurately model and measure escape over a solar cycle.

4.4 Titan

Titan has a thick and extended atmosphere, which consists of over 95% N_2, about 2–3% CH_4, with H_2 and other minor species. Understanding the evolution of Titan's atmosphere provides a critical end point for understanding of the evolution of the atmospheres of the terrestrial planets (Lammer et al. 2008) and the other natural satellites (Johnson 2004). The measured D/H and $^{15}N/^{14}N$ ratios from Cassini-Huygens (Waite et al. 2005a; Niemann et al. 2005) indicate that considerable escape has occurred. The physics of the exosphere is interesting in that thermal escape (Cui et al. 2008), chemical-induced escape (De La Haye et al. 2007b), slow hydrodynamic escape (Strobel 2008b), pick-up and ionospheric ion loss (Ledvina et al. 2005; Ma et al. 2006), and atmospheric sputtering (Michael et al. 2005) have all been proposed as processes that are active at present (Johnson 2008). It is also interesting that atmospheric sputtering varies considerably depending on whether Titan is in Saturn's magnetosphere or exposed to the solar wind flux (Penz et al. 2005).

Heating effects induced by pick-up ions and energetic re-impacting neutrals have been estimated. Initial estimates for N^+ magnetospheric ions penetrating Titan's atmosphere were large (Lammer et al. 2000). However, a DSMC model of the sputtering and heating using a model plasma flux consisting of magnetospheric and pick-up ions led to an increase of the exobase temperature of only a few K and modest loss rates (Michael and Johnson 2005). However, when Titan was within Saturn's magnetosphere, atmospheric sputtering appeared to dominate photon and electron-induced loss processes. Based on such pre-Cassini estimates, it was concluded that the present mass loss rate was small (Shematovich et al. 2003). If that was the case, processes responsible for the isotope ratios must have occurred in an earlier period when the escape processes were more robust. Following the Voyager flybys it was also assumed that ionization of the neutrals escaping from Titan would be the dominant process for supplying Saturn's magnetosphere with heavy ions (Barbosa 1987). However, such ions are rapidly lost down Saturn's magnetotail, so that the dominant source of nitrogen ions to Saturn's magnetosphere is Enceladus (Smith et al. 2007; Johnson et al. 2008). In spite of this, Titan remains an important source of mass loading and plasma in the outer magnetosphere.

With the many transits of Titan's exobase by Cassini, the escape processes can now be characterized. At this writing, data from the Cassini ion neutral mass spectrometer (INMS) have been used to obtain new escape rates. De La Haye et al. (2007a, 2007b) examined the INMS data for a number of early passes and showed that the energy spectra of the molecules in the corona have, not surprisingly, a hot component. In order to simulate the densities of N_2 and CH_4 in a region extending $\sim$ 500 km above the exobase, the molecular energy spectrum was best represented by a kappa distribution. More importantly, for four of the five exobase crossings examined, they could not account for the observed corona structure by assuming that the hot component was only populated by photon and electron induced processes, and concluded that plasma-induced heating associated with the magnetosphere ionosphere interaction must be important. Assuming that is the case, then scaling the required energy deposition rate to a model plasma flux (Michael et al. 2005), a net escape flux is obtained of ~ 0.3–1×10^{10} amu/cm^2/s measured with respect to Titan's surface (De La Haye et al. 2007a). Using the analytic recoil distribution for the hot component gave a rough upper bound of about 5×10^{10} amu/cm^2/s normalized with respect to Titan's surface. More recently Cui et al. (2008) found an H_2 escape rate $\sim 10^{10}$ amu/cm^2/s. In addition a globally average value of $\sim 5 \times 10^{10}$ amu/cm^2/s (4.5×10^{28} amu/s divided by $4\pi R_T^2$) was estimated by assuming slow hydrodynamic escape, as described in Sect. 3.1.2 (Strobel 2008a). Finally, modeling diffusion separation and flow, Yelle et al. (2008) inferred ~ 4–5×10^{10} amu/cm^2/s for CH_4 loss. Remarkably, such rates are equivalent to losing a mass equivalent to the present atmosphere in ~ 4 Gyr. Therefore, loss rates may be much larger than pre-Cassini predictions, but this large loss rate and the mechanism are not agreed upon Johnson (2008). Since plasma-induced heating in the exobase region appears to be a major source of energy, understanding the plasma flow through the transition region is critical for describing the escape rates and structure of the thermosphere and the corona.

5 Exospheres and Escape from Small Bodies

5.1 Io Atmosphere and Torus

The Galilean satellites of Jupiter are instructive for studies of the escape processes in Sect. 3. The most extensively studied is Io, where the first indications of its tenuous atmosphere were the detection of an ionosphere by the Pioneer 10 radio occultation and detection of escaping Na by Brown (1974). SO_2 was subsequently detected by the Voyager IRIS (Pearl et al. 1979) and has been observed since 1990. It was initially unclear whether Io's atmosphere was global or confined to the dayside, where the sublimation of SO_2 frost is orders of magnitude higher than on the night side, and whether the exobase is at or above the surface (McGrath and Johnson 1987). The atmosphere is now understood to be global, although highly non-uniform, and the exobase is variable and above the surface across most of the satellite.

Global characterization of the atmosphere was made by imaging Io using the HST Imaging Spectrograph at the Lyman-α wavelength (1215.67 Å) where SO_2 absorbs strongly, decreasing the surface reflectivity where it is densest. The dayside map in Fig. 5 shows SO_2 is densest at the equator and on the anti-Jovian hemisphere. The sources are active volcanism, sublimation, and, to a lesser degree, sputtering by the Jovian magnetospheric ions that penetrate the atmosphere and reach the surface. Dissociation products, S, O_2, and SO are present at lower abundances, as are minor species (Na, K, Cl, NaCl) which, along with S_2, originate in active volcanoes.

The largest volcanic plumes, such as Pele and Tvashtar, can attain heights of several hundred kilometers. Nonetheless, the best models indicate that plume gas velocities are

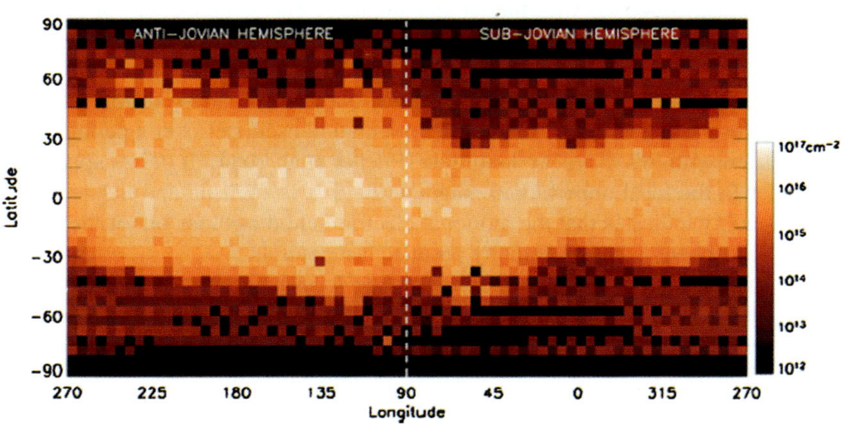

Fig. 5 Map of the SO_2 column density for Io's dayside atmosphere (Feaga et al. 2007)

well below the escape velocity. Therefore, volcanic gas does not escape directly, but contributes to the atmosphere and condenses on the surface. The characteristics of Io's nightside atmosphere are essentially not measured, but active volcanoes will maintain an SO_2 component. The pressure varies significantly from day to night, driving winds that transport the volatile dissociation products, O_2 and, possibly, SO, to the night side (Wong and Johnson 1996; Smyth and Wong 2004). Adding to the complexity is the fact that the magnetospheric plasma continually bombards the atmosphere providing an additional heat source, sputtering. The complicated environment and the spatial and temporal variability of the sources make development of self-consistent models challenging. Nevertheless, it is generally agreed that the exobase altitude varies considerably across the surface and molecular species are present at the exobase.

The interaction between Io's atmosphere and Jupiter's magnetospheric plasma results in mass loss of about one ton per second. As with the other bodies discussed, some of this material is lost from Io in the form of ions, while the principal fraction escapes as neutrals. These neutrals attain enough energy to escape from Io, but, for the most part, not from Jupiter. Therefore, they orbit Jupiter forming large neutral clouds whose morphology reflects the ejection and loss mechanisms. Ejected neutrals are eventually ionized, primarily by electron impact or charge exchange. They are then picked-up and accelerated to the velocity of the rotating magnetic field and swept into a plasma torus surrounding Jupiter at the orbital distance of Io. The plasma torus settles at the centrifugal equator, which is tilted $\sim 7^\circ$ with respect to Io's orbital plane, while Jupiter's magnetic field is tilted by 10° with respect to this plane. The plasma overtakes Io and continually re-impacts the atmosphere ejecting neutrals while the accompanying fields pick-up and remove newly formed ions.

The plasma-induced escape processes have been traced by atomic sodium observations. Although its density is low, Na efficiently fluoresces, with an emission intensity 30 times brighter than any other species at visible and near infrared wavelengths. Therefore, it has been readily observed both near and far from Io since the early 1980s (McGrath et al. 2004). Escape processes for sulfur and oxygen species are less visible, but are modeled by analogy with sodium. Fig. 6 shows images of Na emission at several different scales, illustrating the operative escape mechanisms. The "jet" and "stream" features in the left panel are associated with fast escape, while the "banana" feature is associated with slow escape. The directionality of these features relative to the background magnetic field provides the key to their origins.

Fast Na atoms, associated with the jet and stream features, are a result of Io being immersed in the Jovian magnetic field and plasma. The inclination of the plasma torus relative to Io's orbital plane means that in the course of a ~ 10 hr Jupiter rotation, Io encounters the densest region of the plasma torus twice. The Na jet is a narrow feature that extends away from Io in the anti-Jupiter direction. Its orientation oscillates over a period of several hours correlated with Io's magnetic longitude (Pilcher et al. 1984). It points approximately perpendicular to the local unperturbed magnetic field at Io (Wilson and Schneider 1999). This directionality indicates that the motional electric field drags ions out of the top of Io's anti-Jovian ionosphere, which eventually produce fast neutrals.

The Na stream is a long, narrow feature leading Io in its orbit; it undulates above and below the plasma torus equator with the same period as the jet. The difference between the stream and the jet is the timescale for neutralization of the fresh, approximately corotating, pickup ions. The ions which form the stream recombine in the torus a few hours after leaving Io, whereas those in the jet take only minutes or less to recombine close to Io. In order to produce the observations, the Na in the stream must be produced in ≤ 10 hours. This is incompatible with Na^+ recombination, so the stream must be formed by recombination of

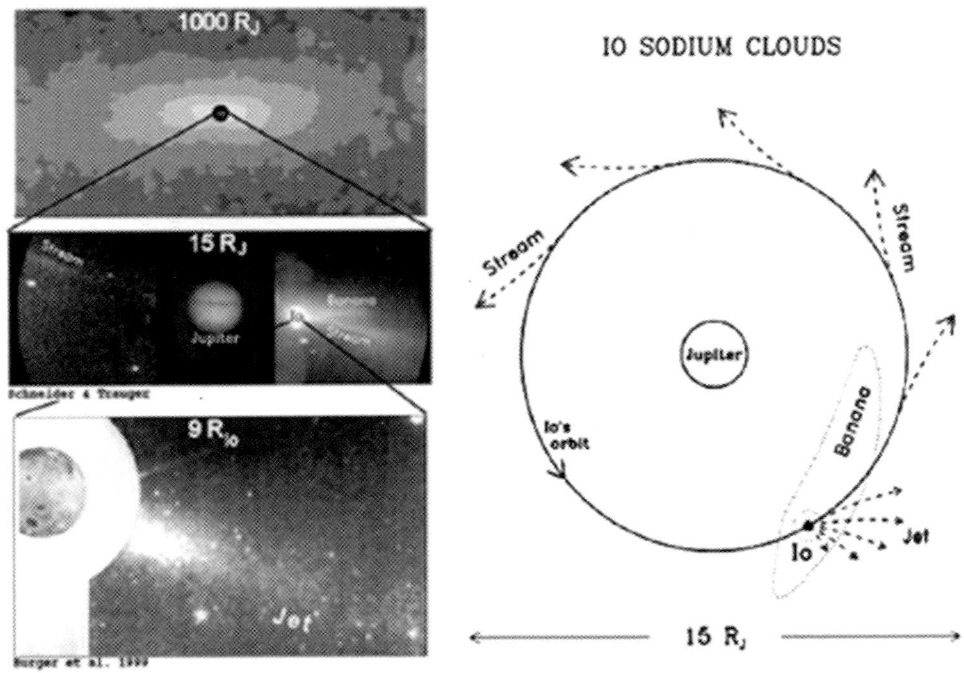

Fig. 6 *Left*: Images at three different scales illustrating the Na escape processes at Io. The width of each image is given by the number at the *top* ($1R_J = 7.14 \times 10^4$ km). *Top*: Na emission from a cloud that is visible to distances of $500R_J$ (Mendillo et al. 1990). *Middle*: Image encompassing Io's orbit showing the Na stream feature (Schneider and Trauger 1995). *Bottom*: The Na jet in the near-Io environment (Burger et al. 1999). The diagram at *right* illustrates the different features schematically (Thomas et al. 2004)

NaX^+ from the ionosphere (Schneider et al. 1991). It is possible that the jet is also produced by escaping NaX^+ ions. The jet may simply represent higher dissociative recombination rates in the part of the stream closest to Io where plasma densities are greatest. The parent, NaX^+, has not been identified, but molecular ions of S and O are obvious candidates (Johnson 1994).

Specific reactions at work in producing the fast features are: charge exchange ($Na^+_{iono} + Na_{atmo} \rightarrow Na_{fast} + Na^+$); dissociative recombination ($NaX^+_{iono} + e^- \rightarrow Na_{fast} + X_{fast}$); impact dissociation ($NaX^+_{iono} + e^- \rightarrow Na_{fast} + X^+ + e^-$). It is uncertain to what extent similar S and O streams and jets exist. Photochemical equilibrium in a collisionally thick and static ionosphere would be dominated by Na^+ and K^+ because of their relatively low ionization potentials (Kumar 1985; Moses et al. 2002), suggesting that if the jet is produced by escaping atomic ions, it may be unique to Na and K. However, the atmosphere is not static and in-situ measurements by the Galileo spacecraft detected signatures near Io at frequencies close to the gyrofrequencies of SO_2^+ and SO^+ (Kivelson et al. 1996; Warnecke et al. 1997), implying that streams or jets of S and O are probably present. The Cassini spacecraft found evidence of extended clouds of S, O, and SO_2 escaping from the Jupiter system via detection of fresh pickup ions (including SO_2^+) well upstream of Jupiter (Krimigis et al. 2002). The parent neutrals most likely result from charge exchange of S^+, O^+ and SO_2^+ with the neutrals in the Io torus, a process which is less important for sodium.

The so-called Na "banana feature" in Fig. 6 is produced by atmospheric sputtering induced by plasma torus or escaping ionospheric ions. Many of the hot neutrals produced do not escape but heat and expand Io's atmosphere (Pospieszalska and Johnson 1996; McGrath et al. 2004). Elastic collisions primarily generate low-energy recoils so that most ejected neutrals have speeds of about a few km/s. These form large clouds extending both ahead of and behind Io, approximately along its orbit. Neutral clouds of sulfur (Durrance et al. 1983, 1995), oxygen (Brown 1981; Durrance et al. 1983; Thomas 1996), and potassium (Trafton 1981) have been detected in addition to Na. These clouds are confined close to the orbital equator, while the plasma torus is confined to the centrifugal equator. The clouds intersect the densest regions of the torus every 6.5 hours, where they are subject to elastic collisions, electron impact dissociation, and ionization, which limit their extent both ahead of and behind the satellite. The complex interplay of ejection speed and direction, orbital motion, and ionization lifetime gives the Na cloud its banana-like shape.

Oxygen and sulfur, due to their higher abundances in Io's atmosphere and proportionally higher sputtering rates, combined with their different ionization lifetimes, will have different morphologies as modeled for O by Smyth and Marconi (2000). The lifetimes against electron impact ionization for sodium, potassium and sulfur are relatively short (2–5 hours) in the densest regions of the torus. The rate coefficient for O, however, is at least an order of magnitude smaller than for S so that charge-exchange loss with torus ions is dominant. The minimum lifetime for O is around 20 hours (Thomas 1992) resulting in significant neutral densities remote from Io and a much more extended, nearly toroidal cloud. Around 180° away from Io itself, densities of neutral O and S are $\sim 29 \pm 16\ cm^{-3}$ and $6 \pm 3\ cm^{-3}$ respectively (Skinner and Durrance 1986). Lagg et al. (1998) derived an average density of 35 oxygen atoms cm^{-3} using measurements from the Galileo energetic particle detector.

5.2 Icy Galilean Satellites

Much less is known about the atmospheres/exospheres and escape for Europa, Ganymede and Callisto (see recent review: Johnson et al. 2008). Molecular oxygen, predicted to be the dominant component and produced by the plasma-induced decomposition of the surface ice

(Johnson et al. 1982; Johnson 1990), was confirmed to be present at Europa and Ganymede from the ratio of atomic oxygen emission lines by Hall et al. (1995, 1998). The derived column abundances (a few $\times 10^{14}$ cm^{-2} disk average value) was close to the predicted value at Europa (Johnson et al. 1982) and indicates that the exobase is at or near the surface for both satellites. Laboratory data showed that O_2 and H_2 are directly produced and ejected from water ice by the incident plasma (Brown et al. 1982; Johnson et al. 2003), a process called radiolysis. Because H_2 escapes more readily, forming a neutral torus (Shematovich et al. 2005b; Smyth and Marconi 2006), and the O_2 does not stick efficiently or escape efficiently, the atmosphere is dominated by O_2, even though the sputtered flux of H_2O dominates. Both Europa and Ganymede also possess ionospheres, detected by Galileo radio occultations (Kliore et al. 1997), which also provided estimates of the neutral column densities.

However, follow up observations of Europa with HST (McGrath et al. 2004, 2008) showed a complex morphology for the O emission, inconsistent with the picture of a uniformly distributed exosphere. Saur et al. (1998) assumed a model O_2 atmosphere which differed significantly from those produced in simulation (e.g., Shematovich et al. 2005b) but described the plasma interaction and production of the O_2. They were able to reproduce the disk averaged HST-GHRS intensities and the nearly uniform O limb UV emission, but not the bright spot emission morphology observed with HST-STIS (cf. McGrath et al. 2004), which may be due to nonuniform sources, nonuniform surface reactivity, or nonuniform plasma excitation (Cassidy et al. 2007). Cassini Observations (Hansen et al. 2005) showed that the OI 1304 emission is more extended than the OI 1356 emission at Europa, which they interpreted as the oxygen exosphere being more extended than the O_2 component consistent with simulations (Shematovich et al. 2005b). Both follow up HST observations and the Cassini observations show emissions from Europa vary significantly in time.

Na and K have both been detected far from Europa (Brown and Hill 1996; Brown 2001; Leblanc et al. 2005). The sodium is also produced by surface sputtering with $\sim 40\%$ of the ejected Na having sufficient energy to escape and with the returning Na redistributed across Europa's surface (Johnson 2000; Leblanc et al. 2002b). The sputtered sodium is ionized primarily by electron impact; photoionization and charge exchange with Io plasma torus ions are negligible (Burger and Johnson 2004; Smyth and Combi 1997). The lifetime is a function of distance from Jupiter, distance from the centrifugal equator, magnetic longitude and local time and varies between 18 and 34 hours at Europa's orbit. Europa's sodium cloud is predominantly a trailing cloud, opposite to the sodium cloud at Io (Burger and Johnson 2004). Mauk et al. (2003) detected energetic neutrals, resulting from charge exchange between protons and the Europa neutral cloud, most likely due to H and H_2 produced by radiolysis and escape (Shematovich et al. 2005b; Smyth and Marconi 2006). Lagg et al. (2003) also reported the depletion of protons with pitch angle of 90°, which are consistent with the presence of an equatorially confined cloud of neutral hydrogen near Europa.

Only a handful of observations exist for the atmospheres of Ganymede and Callisto (Hall et al. 1998; Feldman et al. 2000; Carlson 1999), none of which includes escaping neutrals. Like Europa, Ganymede's atmosphere is dominated by molecular oxygen, but its excitation and morphology are very different because of Ganymede's internal magnetic field. The UV oxygen emissions do not exhibit the features of a globally distributed exosphere. Instead they exhibit a morphology more analogous to auroral emissions on Earth caused by the precipitation of plasma in Ganymede's polar regions. Visible emissions are confined to the equatorial regions, inconsistent with the UV emissions. Neither Na or K have been detected at Ganymede.

The component of Callisto's atmosphere that has been detected is CO_2 (Carlson 1999). A denser molecular oxygen component has been inferred from the ionosphere detections

(Kliore et al. 1997), but attempts to detect O_2, if present, via UV emissions have been unsuccessful (Strobel 2002). However, these observations set an upper limit for CO.

5.3 Saturn Satellite and Ring Atmospheres and Tori

5.3.1 Saturn's Ring Atmosphere

Due to its large area the Saturnian ring system should have an extended gaseous envelope, produced by micrometeoroid impact, photosputtering or energetic ion sputtering. However, because the main rings efficiently absorb energetic particles, the energetic ion flux from the magnetosphere is negligible and is dominated by a low flux of very energetic ions produced by cosmic ray impacts (i.e., the CRAND process; Cooper 1983). Using the interplanetary meteoroid flux of 3×10^{-17} g cm^{-2} s^{-1} (Cook and Franklin 1970), Haff et al. (1983) estimated that the water vapor production rate from micrometeoroid-impacts could be $\sim 5 \times 10^6$ H_2O/(cm^2 s) ($\sim 10^{27}$ H_2O s^{-1} averaged over the ring system; see also Ip 1984a; Pospieszalska and Johnson 1991). The emitted H_2O molecules, however, re-condense on ring particles, which have temperatures $\sim$ 80–100 K, resulting in an average column density that is small ($\sim 10^{11}/\text{cm}^2$). Therefore, Ip (1995) suggested that the O_2, which does not re-condense at these temperatures, is created from the photodissociative products (O and OH) and might have a long lifetime in the ring system leading to the formation of a tenuous, but possibly detectable, atmosphere with average density $\sim 3 \times 10^3$ O_2/cm^3. However, as at Europa, O_2 is also directly produced by charged particle and photo-induced decomposition of ice (Johnson and Quickenden 1997; Johnson et al. 2003). Using laboratory experiments in which ice is exposed to a Lyman-alpha photo-flux (Westley et al. 1995), O_2 was estimated to be produced at a rate $\gtrsim 10^6$ cm^{-2} s^{-1} by the solar EUV/UV flux (Johnson et al. 2006). The interplanetary meteoroid flux at the rings is still very uncertain. If a rate as high as 5×10^{-14} g cm^{-2} s^{-1}, given by Cuzzi et al. (2002) is used, the water vapor production from the ring place would be on the order of 3×10^9 H_2O cm^{-2} s^{-1} ($\sim 5 \times 10^{29}$ H_2O/s averaged over the ring system) thus making meteoroid impact a significant contributor to a ring atmosphere (Ip 2005).

During the Saturn Orbital Insertion (SOI) on July 1, 2004, the Cassini spacecraft flew over the ring plane permitting in-situ observations by the plasma instruments. The INMS (Ion Neutral Mass Spectrometer) and the CAPS (Cassini Plasma Spectrometer) experiments detected the presence of O^+ and O_2^+ ions (Tokar et al. 2005; Waite et al. 2005b). A flux of thermal electrons ($\sim$ 0.6–100 eV) was also detected and found to be in anti-correlation with the optical depth of the rings (Coates et al. 2005), but the neutral atmosphere was below the INMS detection threshold ($\sim 10^5$ cm^{-3}). These results triggered an avalanche of theoretical studies on the formation and structure of the ring atmosphere (Johnson et al. 2006; Bouhram et al. 2006; Luhmann et al. 2006a; Ip 2005). Figure 7 summarizes such simulations showing the spatial distributions of O_2, O_2^+, and O^+ above and below the ring plane.

During SOI the solar zenith angle was about 66° *below* the ring plane. Therefore, only the southern side of the ring system was exposed to the solar flux emitting O_2. The ejected O_2 do not condense out, but thermally equilibrate with the ring particle surfaces. Therefore, equilibrated O_2 exists above and below the ring plane with a scale height $\sim 0.013 R_S$ (780 km). The net column density is determined by the destruction rate due to photo-dissociation and ionization.

The O_2^+ and O^+ ions formed by photoionization will be picked-up and accelerated by the Saturn convective electric field and characterized by their gyromotion giving flat pitch angle distributions. Given the small northward shift ($\sim 0.04 R_S$) of the magnetic dipole center,

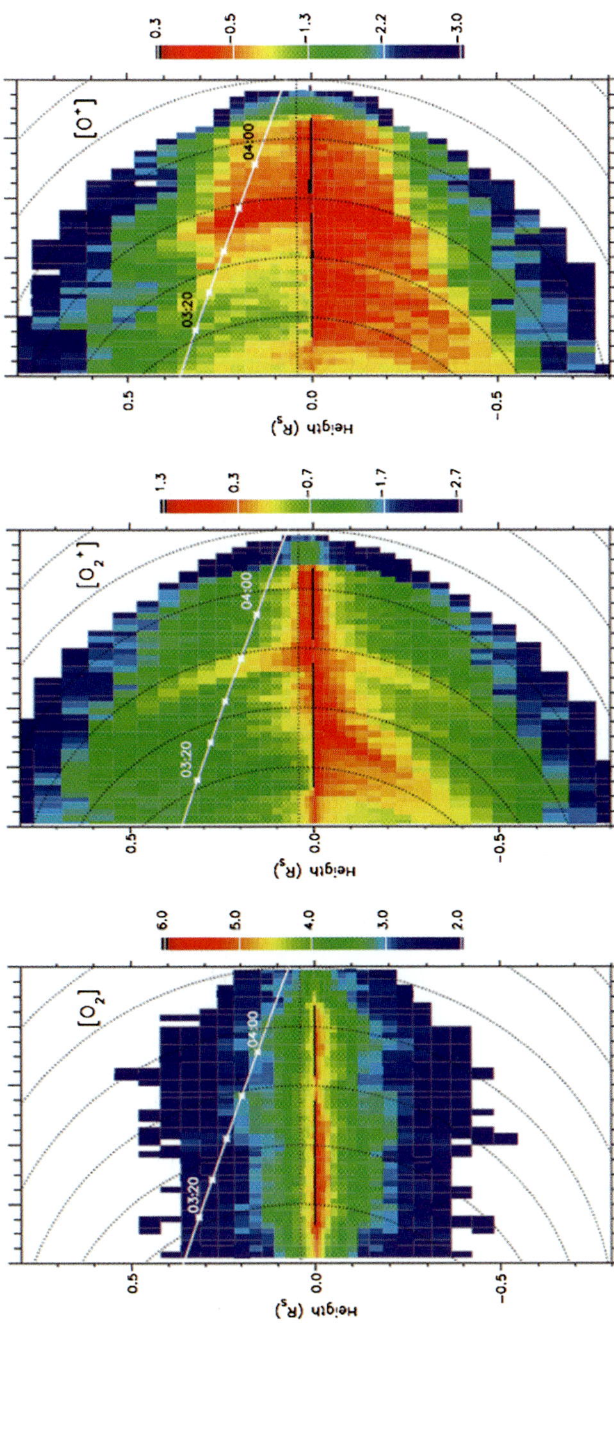

Fig. 7 Density maps from a test-particle simulation of O_2 molecules, O_2^+ and O^+ ions in the vicinity of the Saturning rings: *white line* is the Cassini trajectory (Bouhram et al. 2006)

O_2^+ ions will be formed in a disc-like region below the magnetic equator. The speed of co-rotation becomes smaller than the neutral orbit speed within $\sim 1.86 R_S$, therefore pick-up ions formed at smaller distances from Saturn can be pulled into the southern hemisphere of Saturn (Luhmann et al. 2006a; see also Northrop and Hill 1983; Ip 1983a, 1984b) as suggested by the 2nd and 3rd panels in Fig. 7. Therefore, a steep drop in the ring-ion density was seen by CAPS at $\sim 1.86 R_S$. This pickup ion motion instability limit is almost the same as the sharp boundary between the B ring and the C ring, thus the erosion of the rings might be closely related to the production of the ring atmosphere and the injection of oxygen species ions into the Saturn atmosphere at low latitudes (Connerney and Waite 1984; Moses and Bass 2000; Moses et al. 2000; Moore et al. 2006; Moore and Mendillo 2007).

The O^+ and O_2^+ are subject to ring absorption as they move along the magnetic field lines threading through the ring plane. This loss is indicated by the depletion of the ion densities above the ring plane in Fig. 7. Electrostatic charging of the ring particles can also modify the spatial distribution of these ions (Ip 1984b). The presence of a small electrostatic potential (Φ) can determine whether these low energy ions will be repelled ($\Phi > 0$) or absorbed ($\Phi < 0$) by ring particles. Due to the diurnal variations of the Saturnian ionosphere and the ring shadow, plus the corresponding seasonal changes of the incidence angle of the solar photons on the ring plane, the coupling between the ring atmosphere and the ring plasma system can be highly complex. Farmer and Goldreich (2007) examined the collisional interaction between the neutral ring gas and the ion component to see if there is a strong electrodynamic effect, as suggested by the ring spoke phenomenon (Goertz and Morfill 1983; Farmer and Goldreich 2005; Morfill and Thomas 2005). These authors derived an upper limit of $N < 2 \times 10^{15}$ cm^{-2} for the column density of the oxygen molecules consistent with the models (Johnson et al. 2006).

The Cassini spacecraft flew well above the magnetic equator, therefore the detected ions came from a population that had a significant pitch angle distribution. Whereas the O^+ are formed from O_2 with an additional, randomly oriented, kinetic energy, as discussed earlier, the O_2^+ is formed with no additional kinetic energy. Therefore, the O_2^+ detected at altitudes $\gtrsim 0.1 R_S$ were scattered by charge exchange collisions with O_2 molecules (Johnson et al. 2006). This process also injects O_2 into the Saturnian magnetosphere contributing to O_2^+ detected outside of the ring system (Tokar et al. 2005; Young et al. 2005; Krimigis et al. 2005). Therefore, the aeronomy occurring in Saturn's tenuous ring atmosphere could be the key to the understanding of an array of fundamental issues in the Saturnian system ranging from the large-scale structure of the ring system, the spoke phenomenon, the aeronomy of the Saturnian atmosphere, and the magnetospheric composition and dynamics.

5.3.2 Enceladus and the E-ring

HST observations (Shemansky et al. 1992) showed that the Saturnian system is immersed in a cloud of H_2O and its dissociation products that is much more robust than initially predicted (Johnson et al. 1989). The OH density was estimated to be ~ 160 cm^{-3} at $L \sim 4.5 R_S$, but subsequent HST observations (Hall et al. 1996; Jurac et al. 2002) showed this density could be as much as ~ 1000 cm^{-3} at this radial distance. The detected OH cloud is the product of dissociation of H_2O molecules ejected from the icy bodies that orbit at such distances from Saturn. Therefore, the key question became how to account for the source of the circum-planetary water cloud. After a long series of modeling efforts (Johnson et al. 1989; Pospieszalska and Johnson 1991; Richardson 1998; Ip 1997, 2000), Jurac et al. (2002) showed that the principal source region was near the orbit of Enceladus with a strength of $\sim 0.4 \times 10^{28}$ H_2O/s. Treating the plasma and neutrals self-consistently, Jurac and Richardson

(2005) confirmed the source region, but increased the rate to $\sim 10^{28}$ H_2O/s. Such source rates could not be due to sputtering or meteoroid impact processes, but to some more robust process occurring near Enceladus' orbital radius.

On its way in towards the Saturn system, the Ultraviolet Imaging Spectrometer (UVIS) experiment on the Cassini spacecraft found high temporal variability in O emission over an extended region as seen in Fig. 8a (Esposito et al. 2005). The emission represented a minimum production and loss of $\sim 4 \times 10^{34}$ O over a time interval of three months (an average $\sim 0.5 \times 10^{28}$/s). Although Haff et al. (1983) conjectured that the E-ring grains, which co-orbit with the neutrals seen by HST, might be emitted from geysers on Enceladus, not much attention was paid to this suggestion since no adequate heat source could maintain a subsurface liquid reservoir on this small satellite (Squyres et al. 1983). The situation has now completely changed because of the surprising discovery by the Cassini spacecraft of outgassing at the south pole of Enceladus often referred to as geysers or jets, Fig. 8b.

The magnetometer on Cassini detected perturbations in Saturn's field which suggested that ion formation was occurring close to Enceladus (Dougherty et al. 2006), On July 14, 2005, the spacecraft flew by Enceladus at a closest approach distance of 175 km from its surface. A system of linear fractures (called "Tiger Stripes") was observed by Cassini at the south pole of Enceladus (Porco et al. 2006). In addition, the infrared instrument (CIRS) showed that the temperature was anomalously high indicating the existence of an interior heat source (Spencer et al. 2006). The ISS experiment and the Cassini Dust Analyzer (CDA) also detected the existence of a system of dust jets emanating from the south pole (Fig. 8b), with the "Tiger Stripes" the most likely source. The INMS experiment (Waite et al. 2006) measured an extended distribution of gas cloud composed of H_2O (91%), CO_2 (3.2%), CO or N_2 (3.3%) and CH_4 (1.7%). The radial density distribution of the water molecules suggested the plume could be characterized by two components, one that decayed as $\sim r^{-2}$ from the south polar region plus a minor component, which varied more steeply and appeared to be emitted uniformly over the surface. From the UVIS observation of the occultation of lambda Scorpii by the Enceladus plumes, Hansen et al. (2006) found the production rate that was consistent with that predicted by Jurac and co-workers (Jurac et al. 2002; Jurac and Richardson 2005). Monte Carlo modeling of the ejecta were carried out, assuming a speed distribution and a source rate consistent with rapid sublimation at ~ 180 K (i.e., a thermal surface flux with an energy distribution $[E \exp(-E/kT)/(kT)^2]$ and a cosine distribution corresponding to mean flow speed at the surface of 0.36 km/s on a body with an escape speed 0.24 km/s). The ejecta produced a narrow neutral torus ($\lesssim 0.5 R_S$), and the much larger OH torus, seen earlier by HST, was produced by charge exchange scattering of neutrals from the narrow Enceladus torus (Johnson et al. 2006). In a three-dimensional Monte Carlo model, Burger et al. (2007) showed that the local neutral distribution close to Enceladus could be accounted for by a global source $\sim 8 \times 10^{25}$ H_2O s^{-1} and a localized source at the south pole $\sim 10^{28}$ H_2O s^{-1}. Modeling of the formation and flow of plasma in Enceladus' escaping atmosphere is underway and has much in common with modeling of comet atmospheres discussed below.

5.4 Triton and Pluto

Pluto and Triton, a satellite of Neptune, are twin Kuiper-belt objects. Both have N_2 dominated atmospheres with Pluto having significantly more CH_4. Methane absorbs near-IR solar radiation generating a quasi-isothermal, ~ 100 K, stratosphere that is hotter than Triton's but both have thermospheres with T ~ 100 K. In the absence of data for escape from Pluto, models favor a hydrodynamically escaping atmosphere, as described earlier, with limiting diffusive fluxes for CH_4, H_2, H and total mass loss rates limited by the net heating rate.

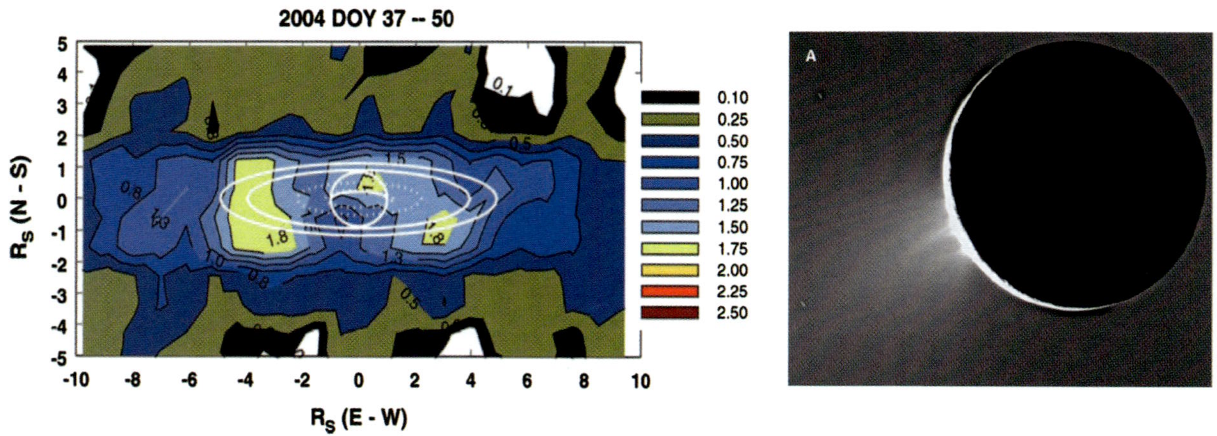

Fig. 8 (**a**) O emission from the Saturn system obtained by Cassini UVIS experiment from $\sim 1000 R_S$ (Esposito et al. 2005); (**b**) ISS NAC clear-filter image of Enceladus' near-surface plumes taken on 27 November 2005 at a phase angle of 161.4° (Porco et al. 2006)

The actual N_2 escape rate is not settled. Tian and Toon (2005) obtained $\sim 1.5 \times 10^{28}$ N_2/s for solar minimum at 30 AU, which is an order of magnitude larger than the estimate by Krasnopolsky (1999), even though they had a factor of 5 lower heating rate due to neglect of CH_4 heating. Strobel (2008b), using a similar model, found a rate in essential agreement with Krasnopolsky (1999), but $\sim 20\%$ larger. The limiting CH_4 escape rate is proportional to its mixing ratio: e.g., for 3%, CH_4, the loss rate would be $\sim 2 \times 10^{27}$ s^{-1}. The calculated N_2 loss rates are power limited and, hence, so are mass escape rates, which include CH_4, H_2, etc. Thus higher CH_4, H_2, etc., mixing ratios and escape rates imply lower N_2 escape rates. The encounter with New Horizon spacecraft will occur in 2015 to study the atmospheric loss rates and the surface-atmosphere interactions.

Voyager 2 solar occultation measurements for Triton's upper atmosphere were used to evaluate the Jeans escape rate. Nitrogen atom densities were extrapolated from 550 km to the 930 km N_2 exobase. Escape rates for H, H_2, C, and O are based on photochemical calculations assuming Jeans escape. The values of λ at the exobase for H, H_2, N, and N_2 are 0.75, 1.5, 10, and 21, respectively. Calculated rates for total H atom ($H + 2H_2$) and N atoms are $\sim$ (7–9) $\times 10^{25}$ and (6–7) $\times 10^{24}$ s^{-1} (Strobel et al. 1995; Kotova et al. 1995). The latter also estimated rates for C and O: $\sim 1.1 \times 10^{24}$ and 4.4×10^{22} s^{-1}. For Pluto, λ values for H, H_2, N, and N_2 are 0.27, 0.54, 3.8, and 7.6 at the N_2 exobase, about a factor of 3 lower than at Triton. However, given the recent evidence for a nearly limiting CH_4 escape flux on Titan, where $\lambda \sim 25$ at the exobase, the possibility of slow hydrodynamic expansion at Triton, which has intermediate values of λ, needs to be examined.

5.5 Comets

The astronomical feature called a coma is the escaping exosphere of a small primitive icy/rocky body (Combi et al. 2004). The exosphere orbits the sun along with the body (the nucleus) and is composed of a head of gas and dust (the coma) and is often accompanied by dust and plasma tails. Beginning with the work of Whipple nearly 60 years ago, we know that the coma and tail of a comet, which can extend from thousands to millions of kilometers in size, are produced by solar heating of a relatively small dark nucleus a few kilometers across. The study of their composition is important because comets are thought to be the least processed remnants from the formation of the solar system. Observations of the coma and tail are important even today in the age of comet missions. While critical information is gained from in situ spacecraft measurements, only a handful of comets have been or will be studied directly even into the next several decades. On the other hand, the compositions of the coma and tail of hundreds of objects are observed remotely and can place the spacecraft measurements in context. Except when comets are observed far from the sun, their nuclei are always shrouded by the coma. Therefore, gaining an understanding of the structure and composition of cometary comae (or exospheres) remains an important task

The nucleus is composed of frozen volatiles that are seen in the coma, as well as the organics and refractory materials that are released in the form of small dust particles during sublimation. Given that gravity is negligible, all but the centimeter sized or larger particles are lifted by the drag force of the expanding gas. As the density of the expanding gas decreases, the dust particles continue to move outward. The ejected molecules and dust follow ballistic orbits under the influence of solar gravity and solar radiation pressure forming the dust tail and the coma. Eventually the parent molecules are dissociated and ionized producing a suite of radicals, atoms and ions. The ions interact with the magnetized solar wind and are dragged into a tail that is often seen via the fluorescence emissions of, mostly, CO^+ and H_2O^+ ions.

Table 1

Comet	S (s^{-1})	R_{coll} (km)
1995 O1 Hale-Bopp (1 AU)	1×10^{31}	240,000
1P/Halley (0.9 AU)	7×10^{29}	16,700
1996 B2 Hyakutake (1 AU)	2×10^{29}	4800
19P/Borrelly-81P/Wild2-9P/Tempel 1 (perihelion)	2×10^{28}	480
67P/ Churyumov-Gerasimenko (perihelion)	1×10^{28}	240
67P/ Churyumov-Gerasimenko (3 AU)	6×10^{25}	surface

The main scientific goal of the Deep Impact mission was to hit the nucleus of comet 9P/Tempel 1 with a high speed projectile and observe the ejecta and potentially the crater both with remote sensing instruments on the main space craft and from ground-based and satellite telescopes in order to get information about the composition, strength and vertical structure below the surface (A'Hearn et al. 2005). In addition, important observations were made of the nucleus prior to and after the impact of a more general nature, independent of the impact from optical and infrared imaging of the nucleus and nearby gas and dust coma. Evidence from the Deep Impact spacecraft (Groussin et al. 2007) indicates that the nucleus has an extremely low thermal conductivity and the nearly black-body surface temperature penetrates only a few centimeters. Therefore, although subsolar temperatures are in excess of 300 K, the exiting water vapor temperature is close to the vacuum sublimation temperature $\sim$ 190–200 K. Measurements and models are coming to a consistent picture. The 200 K sublimation temperature originates a few centimeters below a highly porous and dark refractory layer but does not thermally accommodate to it, explaining the $\sim$ 700 m/s initial gas velocity (Gombosi et al. 1985; Crifo 1987; Combi et al. 2004), consistent with a porous rotating nucleus (Davidsson and Skorov 2002, 2004).

Comet atmospheres are complex since the sublimation is not uniform, gas velocities are not constant, and the production rates can vary by 6–8 orders of magnitude. The simplest model for characterizing the various regimes is a spherical source expanding at a constant velocity for which the gas density is:

$$n = S/(4\pi v r^2),$$

where S is the global production rate, v is the outflow speed and r the distance from the center of the nucleus. Using a collision cross section σ, an expression, analogous to the exobase, is obtained by setting the mean free path equal to the distance to the center of the nucleus (Whipple and Huebner 1976). Therefore, the exobase distance, or collision zone radius, R_{coll}, is: $R_{coll} = S\sigma/4\pi v$. Gas production rates vary widely from comet to comet and with heliocentric distance, and thus the exobase also varies widely, as shown in Table 1.

The structure of the atmosphere depends on the relative size of R_{coll} and the length scales for photodissociation and photoionization. These length scales vary inversely as the square of the comet's heliocentric distance, except near the nucleus where optical depth effects can be important. They also depend on velocities that vary with species, S and r (Bockelée-Morvan and Crovisier 1987; Combi 1987, 1989; Ip 1989). Species produced below R_{coll} can be collisionally quenched while those produced above R_{coll} can escape, similar to the hot coronae around the planets described earlier. The latter are described by Monte Carlo simulations as discussed earlier for the planets and satellites. Even in moderately bright comets the abundance of hot species is large enough to be the principal heat source for the extended atmosphere.

An important complication is suprathermal H. As discussed earlier, heavy products from photodissociation have excess energies $\lesssim$ 2–4 eV, yielding velocities in the range of 1–3 km/s. One to three collisions with the ambient molecules can accommodate a heavy hot species to the local flow. However, H atoms from the photodissociation of H_2O and OH have $\sim$ 1.5 to 2.5 eV, yielding velocities $\sim$ 8 to 20 km/s, requiring $\sim$ 10 or more collisions to accommodate. Slow thermalization of hot H is important for the spatial distribution in the outer exosphere, and is heating the nearly collisionally thick inner coma, increasing its outflow speed. While heavy molecule outflow speeds are $\sim$ 0.7 km/s by 1–2 radii from the nucleus, the outflow speed can vary up to $\sim$ 1 km/s for a Halley-class comet when it is at $\sim$ 1 AU (Lämmerzahl et al. 1987) and up to $\gtrsim$ 2 km/s when the production rate is high, as for 1995 O1 Hale-Bopp, or for 1996 B2 Hyakutake and 2006 P McNaught near perihelion. The criterion is the ratio between the dissociation scale length and R_{coll}.

Outflow speeds have been obtained using radio measurements of OH Doppler-broadened line profiles (Tseng et al. 2007). Figure 9a gives these speeds as a function of distance from the nucleus estimated from the effective aperture size. The larger expansion speeds correspond to higher production rates and small heliocentric distances, where the dissociation scale lengths become small compared with R_{coll}. A comparison of hybrid fluid/kinetic models for the expansion speeds of the productive comet C/1995 OI Hale-Bopp is shown in Fig. 9b for different heliocentric distances (Combi et al. 1999). The variations are reasonably well reproduced, but with the calculated velocities and temperatures systematically larger. The IR cooling, optical trapping and photodissociation branching ratios are the most uncertain aspects (Combi et al. 2004).

The complementary part of this problem is seen in Fig. 10 via the Lyman-alpha line profile in comet C/1996 B2 Hyakutake (Combi et al. 1998). The heavy line shows the model convolved with the GHRS spectral function and the observations (triangles). The model, binned at 1 km/s intervals, is the thin line on which the separate components are seen. The central core comes from H produced in the inner coma merging with the suprathermal H produced from the OH and H_2O outside the collision region.

A hydrodynamic model including corrections for the free expansion of light species (Marconi and Mendis 1983; Ip 1983b; Huebner and Keady 1984) and a hybrid kinetic/hydrodynamic model were used in which the light species were modeled by a test particle method and the heavy-molecule coma is modeled by hydrodynamics (Bockelée-Morvan and Crovisier 1987; Combi 1987; Ip 1989). DSMC models were subsequently carried out for a multispecies coma (Combi 1996) and for transitional and non-spherical flows near weak comets (Crifo et al. 2002, 2005; Skorov et al. 2004, 2006; Tenishev et al. 2008).

The large H coma of comet Hale-Bopp at perihelion (FOV $\sim 40^\circ$), obtained using the SWAN all-sky Ly-α camera on the SOHO (Combi et al. 2000), is due to the large production rate ($> 10^{31}$ s^{-1}). A larger fraction of the H are slowed and thermalized than in bright comets, resulting in slower expansion of the H coma but faster expansion of the heavy atom coma (Tseng et al. 2007). Therefore C and O have also been observed in wide-field images and imaging spectrometers (Feldman et al. 2004). Figure 11 shows the large coma in the forbidden emission from the O(^{1}D) that are produced by photodissociation of H_2O and OH giving a direct measure of the spatial distributions of these parents. Such observations have been useful in determining water production rates (Biermann and Trefftz 1964; Delsemme and Combi 1979; Festou and Feldman 1981; DiSanti and Fink 1991).

In spite of considerable progress, certain branching ratios and energetics remain to be explained (Morgenthaler et al. 2001; Cochran and Cochran 2001; Cochran 2007). In the

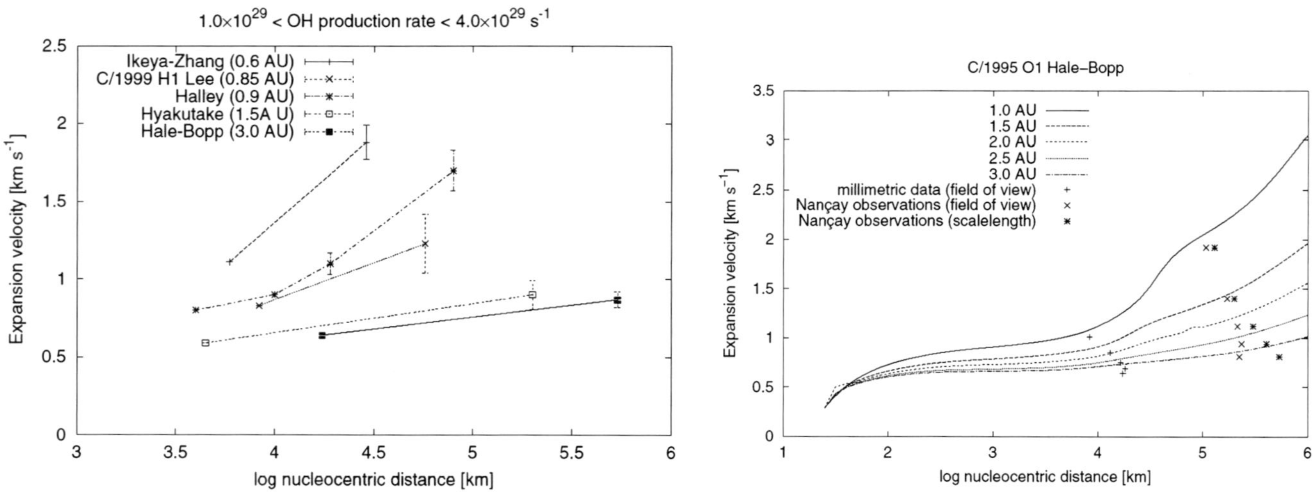

Fig. 9 (**a**) Expansion speeds in comets from radio observations of Doppler broadened OH (Tseng et al. 2007). *Lines* toward the *upper left* tend to larger production rates and smaller heliocentric distances. (**b**) Expansion speeds in C/1995 O1 (Hale-Bopp) for different heliocentric and nucleocentric distances compared with hybrid kinetic/hydrodynamics models (Combi et al. 2000). Points from Tseng et al. (2007): nucleocentric distance in km is estimated from the telescope aperture size projected on the sky plane

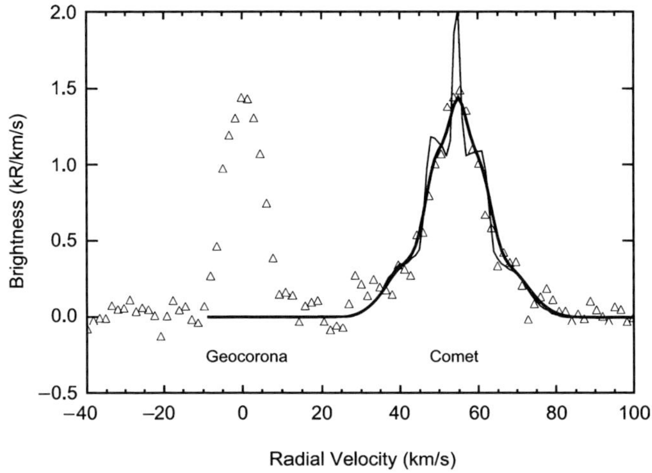

Fig. 10 H Lyman-alpha line profile in C/1996 B2 (Hyakutake). *Triangles*: line profile of the optically thin region of the coma obtained with the GHRS instrument on HST with the small science aperture located 111,000 km sunward of the nucleus. *Thin line*: intrinsic emission at high spectral resolution (1 km/s); *thick line*: model convolved with the instrument spectral function ($\sim$ 4 km/s resolution). H emission in the geocorona: at 0 km/s to the left; comet's emission is Doppler shifted to 55 km/s relative geocentric velocity (Feldman et al. 2004; Combi et al. 1998)

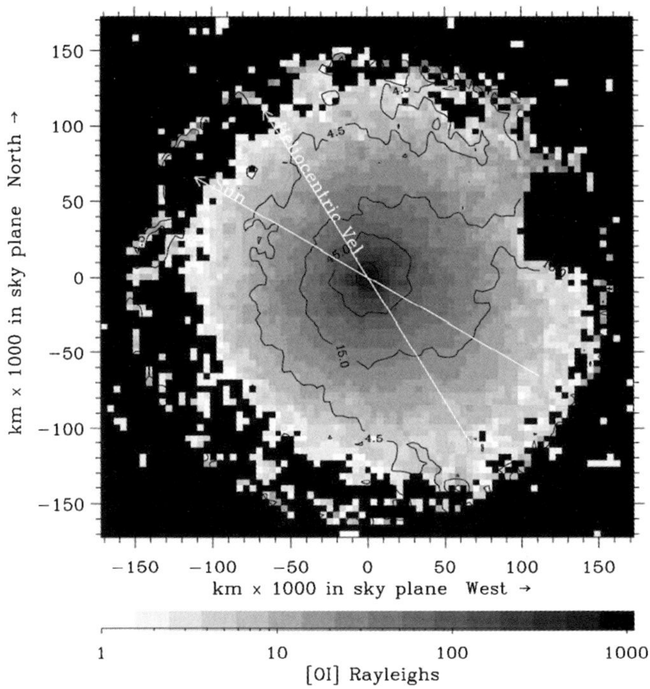

Fig. 11 Wide-field images of the [O I] 6300 Å emission plus continuum from comet C 1996 B2 Hyakutake. *Gray scale*: [O I] emission; *contours*: dust continuum. Emission is from $O(^1D)$ produced by dissociation of H_2O ($\lesssim 10^5$ km) and OH ($\gtrsim 10^5$ km) (Morgenthaler et al. 2007)

absence of better molecular data, it is still problematic to obtain consistent production rates using observations of H_2O (Dello Russo et al. 2006), OH in the near UV from the ground (Schleicher et al. 1998) or from space (Weaver et al. 1999), in the radio from the ground (Gérard et al. 1998) or microwave from space (Bensch et al. 2004), H from the ground (Smyth et al. 1993) or from space (Combi et al. 2000), and [O I] (Fink and Hicks 1996; Fink et al. 1991; Morgenthaler et al. 2001, 2007; Cochran 2007).

6 Summary

The physics and chemistry that describe the formation of a corona and the escape rate from a planetary body are critical aspects of aeronomy. They are of interest in understanding present day observations of upper atmospheres, exospheres, extended neutral clouds and pick-up ions source rates. Accurate descriptions of the corona and the present rate of escape are also absolutely necessary in order to attempt to extrapolate back in time and describe the evolution of a planet's atmosphere.

The fundamental physical and chemical processes that determine planetary escape are reasonably well understood with the possible exception of slow hydrodynamic escape. It is described in detail here, since describing the transition region of an atmosphere is still problematic and there is controversy as to when heating by thermal conduction from below becomes an effective escape process. In addition, the data base for cross sections and branching ratios is inadequate for most of the bodies discussed above (Huestis et al. 2008).

Although enormous progress has been made, escape and corona formation are not yet well described for any of the bodies discussed. This is the case for a number of reasons. A principal reason is that on all non-magnetized bodies, the flow through the transition region of the ambient ions, the pick-up ions and the ionospheric ions is critical, but fully self-consistent models of the plasma flow and the interaction of the ionosphere with the ambient fields are just becoming available. What is especially of interest for extrapolating to earlier epochs at Mars are descriptions of the plasma flow and the ion and neutral escape rates vs. the various solar wind conditions. Some progress has been made recently. For instance, the ambitious model of Chaufray et al. (2007) for Mars does treat the interaction and the atmosphere iteratively, and calculates both neutral and ion loss for solar minimum and maximum, but it uses a simplified model atmosphere and the hybrid interaction model does not have sufficient spatial resolution in the exobase region. More importantly, although the solar EUV was varied, the effect of changes in the solar wind pressure that are applicable to earlier solar conditions was not treated. At Titan, although the interaction of the upper atmosphere with the magnetosphere is not well modeled, a good description of the structure of the thermosphere and corona near the exobase is evolving due to the availability of extensive Cassini data (e.g., De La Haye et al. 2007a, 2007b; Cui et al. 2008). Simulations of this structure are being carried out that will be able to test models for escape.

For all the bodies discussed, the principal focus should be on improving the description of the plasma flow around and through the exobase region of each planetary body. Since iterative models are now being developed, and new in-situ spacecraft data is becoming available, extensive progress is expected in the next few years. However, confidence in the modeling will require a concomitant improvement in the molecular data base.

Acknowledgements Support is acknowledged support from NASA's Planetary Atmospheres Program [REJ grant NNG06GC09G; MRC grant NNG06GF51G; JLF grant NN606GF216], by VIS from RFFI Grant 08-02-00263, W.-H. Ip from NSC 96-2752-M-008-011-PAE and NSC 96-2111-M-008-010.

References

M.F. A'Hearn et al., Science **310**, 258–264 (2005)
J.M. Ajello, M. Ciocca, J. Geophys. Res. **101**, 18,953–18,960 (1996)
C.J. Alexander, C.T. Russell, Geophys. Res. Lett. **12**, 369 (1985)
D.E. Anderson, C.W. Hord, J. Geophys. Res. **76**, 6666–6673 (1971)
D.E. Anderson Jr., J. Geophys. Res. **81**, 1213–1216 (1976)
M. Armenante, V. Santoro, N. Spinelli, F. Vanoli, Int. J. Mass Spectrom. Ion Proc. **64**, 265–273 (1985)
W.I. Axford, J. Geophys. Res. **73**, 6855 (1968)
S. Barabash, A. Fedorov, R. Lundin, J.-A. Sauvaud, Science **315**, 501 (2007a)
S. Barabash et al., Nature **450**, 650–653 (2007b)
D.D. Barbosa, Icarus **72**, 53–56 (1987)
A.R. Barakat, J. Lemaire, Phys. Rev. A **42**(6), 3291–3302 (1990)
R. Bauske, A.F. Nagy, T.I. Gombosi, D.L. De Zeeuw, K.G. Powell, J.G. Luhmann, J. Geophys. Res. **103**(A10), 23625–23638 (1998)
F. Bensch, E.A. Bergin, D. Bockelée-Morvan, G.J. Melnick, N. Biver, Astrophys. J. **609**, 1164–1169 (2004)
J.-L. Bertaux, J. Blamont, M. Marcelin, Planet. Space Sci. **26**, 817–831 (1978)
L. Biermann, D. Trefftz, Z. Astrophys. **59**, 1–28 (1964)
G.A. Bird, *Molecular Gas Dynamics and the Direct Simulation of Gas Flows* (Clarendon, Oxford, 1994)
D.V. Bisikalo, V.I. Shematovich, J.-C. Gerard, J. Geophys. Res. **100**, 3715–3720 (1995)
D. Bockelée-Morvan, J. Crovisier, in *Proceedings, Symposium on The Diversity and Similarity of Comets*. ESA SP-278 (1987)
S. Bougher et al., Space Sci. Rev. (2008, this issue)
M. Bouhram, R.W. Johns, J.-J. Berthelier, J.-M. Illiano, R.L. Tokar, D.T. Young, F.J. Crary, Geophys. Res. Lett. **33**, L05106 (2006). doi:10.1029/2005GL025011
R.T. Brinkmann, Science **174**, 944–945 (1971)
R.A. Brown, in *IAU Symp. 65: Exploration of the Planetary System* (1974), pp. 527–531
R.A. Brown, Astrophys. J. **244**, 1072 (1981)
W.L. Brown et al., Nucl. Instrum. Methods **198**, 1 (1982)
M.E. Brown, R.E. Hill, Nature **380**, 229–231 (1996)
M.E. Brown, Icarus **151**, 190–195 (2001)
M.H. Burger, N.M. Schneider, J.K. Wilson, Geophys. Res. Lett. **26**, 3333–3336 (1999)
M.H. Burger, R.E. Johnson, Icarus **171**, 557–560 (2004)
M.H. Burger, E.C. Sittler Jr., R.E. Johnson, H.T. Smith, O.J. Tucker, V.I. Shematovich, J. Geophys. Res. **112**, A06219 (2007). doi:10.1029/2006JA012086
R.W. Carlson, Science **283**, 821–821 (1999)
E. Carlsson et al., Icarus **182**, 320–328 (2006)
T.A. Cassidy, R.E. Johnson, M.A. McGrath, M.C. Wong, J.F. Cooper, Icarus **191**, 755–764 (2007)
J.W. Chamberlain, Planet. Space Sci. **11**, 901 (1963)
J.W. Chamberlain, Astrophys. J. **155**, 711 (1969)
J.W. Chamberlain, F.J. Campbell, Astrophys. J. **149**, 687–705 (1967)
J.W. Chamberlain, D.M. Hunten, *Theory of Planetary Atmospheres* (Academic Press, New York, 1987)
E. Chassefière, F. Leblanc, Planet. Space Sci. **52**, 1039 (2004)
E. Chassefière, F. Leblanc, B. Langlais, Planet. Space Sci. **55**, 343 (2007)
J.-Y. Chaufray, R. Modolo, F. Leblanc, G.M. Chanteur, R.E. Johnson, J.G. Luhmann, J. Geophys. Res. **112** (2007). doi:10.1029/2007JE002915
J.-Y. Chaufray, J.-L. Bertaux, E. Quémerais, F. Leblanc, Icarus **195**, 598–613 (2008).
F. Cipriani, F. Leblanc, J.J. Berthelier, J. Geophys. Res. E07001 (2007). doi:10.1029/2006JE002818
A.J. Coates et al., Geophys. Res. Lett. **32**, L14S09 (2005). doi:10.1029/2005GL022694
A.L. Cochran, American Astronomical Society, DPS Meeting, #39, #53.09 (2007)
A.L. Cochran, W.D. Cochran, Icarus **154**, 381–390 (2001)
M.R. Combi, Icarus **71**, 178–191 (1987)
M.R. Combi, Icarus **81**, 41–50 (1989)
M.R. Combi, Icarus **123**, 207–226 (1996)
M.R. Combi, M.E. Brown, P.D. Feldman, H.U. Keller, R.R. Meier, W.H. Smyth, Astrophys. J. **494**, 816–821 (1998)
M.R. Combi, A.L. Cochran, W.D. Cochran, D.L. Lambert, C.M. Johns-Krull, Astrophys. J. **512**, 961–968 (1999)
M.R. Combi, A.A. Reinard, J.-L. Bertaux, E. Quemerais, T. Mäkinen, Icarus **144**, 191–202 (2000)
M.R. Combi, W.M. Harris, W.H. Smyth, in *Comets II*, ed. by M.C. Festou et al. (University of Arizona Press, Tucson, 2004), pp. 523–552

J.E.P. Connerney, J.H. Waite, Nature **312**, 136 (1984)
A.F. Cook, A.F. Franklin, Astrophys. J. **75**, 195 (1970)
J.F. Cooper, J. Geophys. Res. **88**, 3945–3954 (1983)
P.C. Cosby, J. Chem. Phys. **98**, 9560–9569 (1993)
T.E. Cravens, C.N. Keller, B. Ray, Planet. Space Sci. **45**, 889 (1997)
T.E. Cravens, A. Hoppe, S.A. Ledvina, S. McKenna-Lawlor, J. Geophys. Res. **107**, 7-1–7-10 (2002). doi:10.1029/2001JA000125
J.F. Crifo, Astron. Astrophys. **187**, 438–450 (1987)
J.F. Crifo, G.A. Loukianov, A.V. Rodionov, G.O. Khanlarov, V.V. Zakharov, Icarus **156**, 249–268 (2002)
J.F. Crifo, G.A. Loukianov, A.V. Rodionov, G.O. Khanlarov, V.V. Zakharov, Icarus **176**, 192–219 (2005)
J. Cui, R.V. Yelle, K. Volk, J. Geophys. Res. (2008, in press)
J.N. Cuzzi et al., Space Sci. Rev. **104**, 209 (2002)
B.J.R. Davidsson, Y.V. Skorov, Icarus **159**, 239–258 (2002)
B.J.R. Davidsson, Y.V. Skorov, Icarus **168**, 163–185 (2004)
V. De La Haye, J.H. Waite Jr., R.E. Johnson, R.V. Yelle, T.E. Cravens, J.G. Luhmann, W.T. Kasprzak, D.A. Gell, B. Magee, F. Leblanc, M. Michael, S. Jurac, I.P. Robertson, J. Geophys. Res. **112**(A7), A07309 (2007a). doi:10.1029/2006JA012222
V. De La Haye, J.H. Waite Jr., T.E. Cravens, A.F. Nagy, R.V. Yelle, R.E. Johnson, S. Lebonnois, I.P. Robertson, Icarus **191**, 236–250 (2007b)
N. Dello Russo, M.J. Mumma, M.A. DiSanti, K. Magee-Sauer, E.L. Gibb, B.P. Bonev, I.S. McLean, L.-H. Xu, Icarus **184**, 255–276 (2006)
A.H. Delsemme, M.R. Combi, Astrophys. J. **228**, 330–337 (1979)
H.G. Demars, A.R. Barakat, R.W. Schunk, J. Atmos. Sol. Terr. Phys. **55**, 1583–1598 (1993)
M.A. DiSanti, U. Fink, Icarus **91**, 105–111 (1991)
T.M. Donahue, J.H. Hoffman, R.R. Hodges, A.J. Watson, Science **216**, 630–633 (1982)
M.K. Dougherty, K.K. Khurana, F.M. Neubauer, C.T. Russell, J. Saur, J.S. Leisner, M.E. Burton, Science **311**, 1406–1409 (2006)
S.T. Durrance, P.D. Feldman, H.A. Weaver, Astrophys. J. **267**, L125 (1983)
S.T. Durrance, P.D. Feldman, W.P. Blair, A.F. Davidsen, G.A. Kriss, J.W. Kruk, K.S. Long, H.W. Moos, Astrophys. J. **447**, 408 (1995)
L.W. Esposito, J.E. Colwell, K. Larsen et al., Science **307**, 1251 (2005)
A.J. Farmer, P. Goldreich, Icarus **179**, 535 (2005)
A.J. Farmer, P. Goldreich, Icarus **188**, 108–119 (2007)
L.M. Feaga, M.F. A'Hearn, J.M. Sunshine, O. Groussin, T.L. Farnham, Icarus **190**, 345 (2007)
P.D. Feldman, M.A. McGrath, D.F. Strobel, H.W. Moos, K.D. Retherford, B.C. Wolven, Astrophys. J. **535**, 1085–1090 (2000)
P.D. Feldman, A.L. Cochran, M.R. Combi, in *Comets II*, ed. by M.C. Festou, H.U. Keller, H.A. Weaver (University of Arizona Press, Tucson, 2004), pp. 425–447
M.C. Festou, P.D. Feldman, Astron. Astrophys. **103**, 154–159 (1981)
U. Fink, M.R. Combi, M.A. DiSanti, Astrophys. J. **383**, 356–371 (1991)
U. Fink, M.D. Hicks, Astrophys. J. **459**, 729–743 (1996)
J.L. Fox, Adv. Space Res. **5**, 165 (1985)
J.L. Fox, Planet. Space Sci. **34**, 1252 (1986)
J.L. Fox, J. Geophys. Res. **98**, 3297–3310 (1993)
J.L. Fox, Geophys. Res. Lett. **24**, 2901 (1997)
J.L. Fox, Icarus **192**, 296–301 (2007)
J.L. Fox, F.M. Bakalian, J. Geophys. Res. **106**, 28785 (2001)
J.L. Fox, J.H. Black, Geophys. Res. Lett. **16**, 291 (1989)
J.L. Fox, A. Dalgarno, J. Geophys. Res. **84**, 7315–7333 (1979)
J.L. Fox, A. Dalgarno, J. Geophys. Res. **88**, 9027–9032 (1983)
J.L. Fox, A. Dalgarno, J. Geophys. Res. **90**, 7557–7567 (1985)
J.L. Fox, A. Hać, J. Geophys. Res. **102**, 24,005–24,011 (1997a)
J.L. Fox, A. Hać, J. Geophys. Res. **102**, 9191–9204 (1997b)
J.L. Fox, L.J. Paxton, J. Geophys. Res. **110**, A01311 (2005). doi:10.1029/2004JA010813
E. Gérard, J.P. Crovisier, N. Colom, D. Biver, D. Bockelée-Morvan, H. Rauer, Planet. Space Sci. **46**, 569–577 (1998)
C.K. Goertz, G. Morfill, Icarus **43**, 219 (1983)
T.I. Gombosi, T.E. Cravens, A.F. Nagy, Astrophys. J. **293**, 328–341 (1985)
O. Groussin, M.F. A-Hearn, J.-Y. Li, P.C. Thomas, J.M. Sunshine, C.M. Lisse, K.J. Meech, T.L. Farnham, L.M. Feaga, W.A. Delamere, Icarus **187**, 16–25 (2007)
S.L. Guberman, A. Giusti-Suzor, J. Chem. Phys. **95**, 2602–2613 (1991)

H. Gunell et al., Icarus **182**(2), 431–438 (2006)
P.K. Haff, A. Eviatar, G.L. Siscoe, Icarus **56**, 426 (1983)
D.T. Hall, D.F. Strobel, P.D. Feldman, M.A. McGrath, H.A. Weaver, Nature **373**, 677–679 (1995)
D.T. Hall, P.D. Feldman, J.B. Holberg, M.A. McGrath, Science **272**, 516 (1996)
D.T. Hall, P.D. Feldman, M.A. McGrath, D.F. Strobel, Astrophys. J. **499**, 475–481 (1998)
C.J. Hansen, D. Shemansky, A. Hendrix, Icarus **176**, 305–315 (2005)
C.J. Hansen, L. Esposito, A.I.F. Stewart, J. Colwell, A. Hendrix, W. Pryor, D.E. Shemansky, R. West, Science **311**, 1422 (2006)
W.B. Hanson, S. Sanatani, D.R. Zuccaro, J. Geophys. Res. **82**, 4351 (1977)
E.M. Harnett, R.M. Winglee, J. Geophys. Res. **111**, A09213 (2006). doi:10.1029/2006JA011724
R.E. Hartle, Planet. Space Sci. **21**, 12123–12137 (1973)
R.R. Hodges, J. Geophys. Res. **98**, 10,833–10,838 (1993)
R.R. Hodges, J. Geophys. Res. **99**, 23,229–23,247 (1994)
R.R. Hodges, J. Geophys. Res. **105**, 6971 (2000)
W.F. Huebner, J.J. Keady, in *Cometary Exploration*, ed. by T. Gombosi (Budapest, 1984), pp. 165–183
D.L. Huestis et al., Space Sci. Rev. (2008, this issue)
D.M. Hunten, A.J. Watson, Icarus **51**, 665–667 (1982)
W.-H. Ip, J. Geophys. Res. **88**, 819 (1983a)
W.-H. Ip, Astrophys. J. **264**, 726–732 (1983b)
W.-H. Ip, J. Geophys. Res. **89**, 395 (1984a)
W.-H. Ip, J. Geophys. Res. **89**, 3829 (1984b)
W.-H. Ip, Icarus **76**, 135–145 (1988)
W.-H. Ip, Astrophys. J. **343**, 946–952 (1989)
W.-H. Ip, J. Geophys. Res. **17**, 2289–2292 (1990)
W.-H. Ip, Icarus **115**, 295–303 (1995)
W.-H. Ip, Icarus **126**, 42 (1997)
W.-H. Ip, Planet. Space Sci. **48**, 775 (2000)
W.-H. Ip, Geophys. Res. Lett. **32**, L13204 (2005). doi:10.1029/2004GL022217
R.E. Johnson, J. Geophys. Res. Lett. **5**, 989 (1978)
R.E. Johnson, *Energetic Charged Particle Interaction with Atmospheres and Surfaces* (Springer, New York, 1990)
R.E. Johnson, Icarus **111**, 65–72 (1994)
R.E. Johnson, Icarus **143**, 429–433 (2000)
R.E. Johnson, Astrophys. J. **609**, L99–L102 (2004)
R.E. Johnson, Proc. R. Soc. (Lond.) (2008, in press)
R.E. Johnson, J.G. Luhmann, J. Geophys. Res. **103**, 3649–3653 (1998)
R.E. Johnson, T.I. Quickenden, J. Geophys. Res. **102**, 10985 (1997)
R.E. Johnson, L.J. Lanzerotti, W.L. Brown, Nuclear Instrum. Methods **198**, 147 (1982)
R.E. Johnson, M.K. Pospieszalska, E.C. Sitler Jr., A.F. Cheng, L.J. Lanzerotti, E.M. Sieveka, Icarus **77**, 311 (1989)
R.E. Johnson, D. Schnellenberger, M.C. Wong, J. Geophys. Res. **105**, 1659 (2000)
R.E. Johnson, M. Liu, C. Tully, Planet. Space Sci. **50**, 123–128 (2002)
R.E. Johnson, T.I. Quickenden, P.D. Cooper, A.J. McKinley, C. Freeman, Astrobiology **3**, 823–850 (2003)
R.E. Johnson et al., Icarus **180**, 393 (2006)
R.E. Johnson et al., in *Europa*, ed. by R. Pappalardo et al. (2008, in press)
S. Jurac, J.D. Richardson, J. Geophys. Res. **110**, A09220 (2005). doi:10.1029/2004JA010635
S. Jurac, M.A. McGrath, R.E. Johnson, J.D. Richardson, V.M. Vasyliunas, A. Eviatar, Geophys. Res. Lett. **29**(24), 2172 (2002). doi:10.1029/2002GL015855
K. Kabin, B.D. Shizgal, J. Geophys. Res. **107**(E7), 5053 (2002). doi:10.1029/2000JE001479
E. Kallio, H. Koskinen, J. Geophys. Res. **104**, 557 (1999)
E. Kallio, J.G. Luhmann, S. Barabash, J. Geophys. Res. **102**, 22,183–22,197 (1997)
E. Kallio, J.G. Luhmann, J.G. Lyon, J. Geophys. Res. **103**(A3), 4753–4754 (1998)
G.M. Keating et al., Science **279**, 1672–1676 (1998)
D. Kella, P.J. Johnson, H.B. Pedersen, L. Vejby-Christensen, L.H. Andersen, Science **276**, 1530–1533 (1997)
V. Kharchenko, A. Dalgarno, B. Zygelman, J.-H. Yee, J. Geophys. Res. **103**, 24899–24906 (2000)
J. Kim, A.F. Nagy, J.L. Fox, T.E. Cravens, J. Geophys. Res. **103**, 29339 (1998)
Y.H. Kim, S. Son, J. Korean Astron. Soc. **33**, 127–135 (2000)
Y.H. Kim, S. Son, J. Kim, J. Korean Astron. Soc. **34**, 25–29 (2001)
M.G. Kivelson, K.K. Khurana, R.J. Walker, J. Warnecke, C.T. Russell, J.A. Linker, D.J. Southwood, C. Polanskey, Science **274**, 396–398 (1996)
A.J. Kliore, D.P. Hinson, F.M. Flasar, A.F. Nagy, Science **277**, 355–358 (1997)

G. Kotova, M. Verigin, A. Remizov, N. Shutte, J. Slavin, K. Szego, M. Tatrallyay, H. Rosenbauer, V.A. Krasnopolsky, D.P.J. Cruikshank, Geophys. Res. **100**, 21,271–21,286 (1995)
G. Kotova, M. Verigin, A. Remizov, N. Shutte, H. Rosenbauer, S. Livi, K. Szego, M. Tatrallyay, J. Slavin, J. Lemaire, K. Schwingenschuh, T.-L. Zhang, Geophys. Res. **102**, 2165–2174 (1997)
V.A. Krasnopolsky, J. Geophys. Res. **104**, 5955–5962 (1999)
M.A. Krestyanikova, V.I. Shematovich, Sol. Syst. Res. **39**, 22 (2005)
M.A. Krestyanikova, V.I. Shematovich, Sol. Syst. Res. **40**, 384–392 (2006)
S.M. Krimigis et al., Nature **415**, 994–996 (2002)
S.M. Krimigis et al., Science **307**, 1270 (2005)
A.M. Krymskii, T.K. Breus, J. Geophys. Res. **101**, 7599 (1996)
S. Kumar, Icarus **61**, 101–123 (1985)
A. Lagg, N. Krupp, J. Woch, S. Livi, B. Wilken, D.J. Williams, Geophys. Res. Lett. **25**, 4039–4042 (1998)
A. Lagg, N. Krupp, J. Woch, D.J. Williams, Geophys. Res. Lett. **30** (2003). doi:10.1029/2003GL017214. 1556
H. Lammer, S.J. Bauer, J. Geophys. Res. **96**, 1819–1825 (1991)
H. Lammer, W. Stumptner, G.J. Molina-Cuberos, S.J. Bauer, T. Owen, Planet. Space Sci. **48**, 529–543 (2000)
H. Lammer, H.I.M. Lichtenegger, H.K. Biernat, N.V. Erkaev, I.L. Arshukova, C. Kolb, H. Gunell, A. Lukyanov, M. Holmstrom, S. Barabash, T.L. Zhang, W. Baumjohann, Planet. Space Sci. **54**, 1445–1456 (2006)
H. Lammer et al., Space Sci. Rev. (2008, this issue)
P. Lämmerzahl, D. Krankowsky, R.R. Hodges, U. Stubbemann, J. Woweries, I. Herrwerth, J.J. Berthelier, J.M. Illiano, P. Eberhardt, U. Dolder, W. Schulte, J.H. Hoffman, Astron. Astrophys. **187**, 169–173 (1987)
L.D. Landau, E.M. Lifshitz, *Fluid Mechanics*, 2nd ed., vol. 6 (Pergamon Press, 1987), p. 539
F. Leblanc, R.E. Johnson, Planet. Space Sci. **49**, 645 (2001)
F. Leblanc, R.E. Johnson, J. Geophys. Res **107** (2002a). doi:10.1029/2000JE001473
F. Leblanc, R.E. Johnson, M.E. Brown, Icarus **159**, 132–144 (2002b)
F. Leblanc, A.E. Potter, R.M. Killen, R.E. Johnson, Icarus **178**, 367–385 (2005)
F. Leblanc, J.Y. Chaufray, O. Witasse, J. Lilensten, J.-L. Bertaux, J. Geophys. Res. **111**, E09S11 (2006). doi:10.1029/2005JE002664
F. Leblanc, J.Y. Chaufray, J.L. Bertaux, Geophys. Res. Lett. **34**, L02206 (2007). doi:10.1029/2006GL028437
S.A. Ledvina, T.E. Cravens, K. Kecskeméty, J. Geophys. Res. **110** (2005). CiteID A06211
S.A. Ledvina et al., Space Sci. Rev. (2008, this issue)
B.R. Lewis, S.T. Gibson, W. Zhang, H. Lefebvre-Brion, J.-M. Robbe, J. Chem. Phys. **122**, 144302 (2005)
H.I.M. Lichtenegger, H. Lammer, Y.N. Kulikov, S. Kazeminejad, G.H. Molina-Cuberos, R. Rodrigo, B. Kazeminejad, G. Kirchengast, Space Sci. Rev. (2007, in press). doi:10.1007/s11214-006-9082-1
M.W. Liemohn, J.L. Fox, A.F. Nagy, X. Fang, J. Geophys. Res. **109**, A10307 (2004). doi:10.1029/2004JA010643
R.W. Locht, M. Davister, Int. J. Mass Spectrom. Ion Proc. **144**, 105–129 (1995)
J.G. Luhmann, J. Geophys. Res. **101**, 7603 (1996)
J.G. Luhmann, J.U. Kozyra, J. Geophys. Res. **96**, 5457–5468 (1991)
J.G. Luhmann, R.E. Johnson, M.H.G. Zhang, Geophys. Res. Lett. **19**, 2151 (1992)
J.G. Luhmann, R.E. Johnson, R.L. Tokar, S.A. Ledvina, T.W. Cravens, Icarus **181**, 465 (2006a)
J.G. Luhmann, S.A. Ledvina, J.G. Lyon, C.T. Russell, Planet. Space Sci. **54**, 1482–1495 (2006b)
J.G. Luhmann, W.T. Kasprzak, C.T. Russell, J. Geophys. Res. **112** (2007). doi:10.1029/2006JE002820
H. Luna, M. Michael, M.B. Shah, R.E. Johnson, C.J. Latimer, J.W. McConkey, J. Geophys. Res. **108**(E4), 5033 (2003). doi:10.1029/2002JE001950
H. Luna, C. McGrath, M.B. Shah, R.E. Johnson, M. Liu, C.J. Latimer, E.C. Montenegro, Astrophys. J. **628**, 1086–1096 (2005)
R. Lundin, S. Barabash, Planet. Space Sci. **52**, 1059–1071 (2004)
R. Lundin, E.M. Dubinin, Adv. Space Res. **12**(9), 255 (1992)
R. Lundin et al., Nature **341**, 609 (1989)
R. Lundin, E.M. Dubinin, S.V. Barabash, H. Koskinen, O. Norberg, N. Pissarenko, A.V. Zakharov, Geophys. Res. Lett. **18**, 1059 (1991)
Y. Ma, A.F. Nagy, Geophys. Res. Lett. **34** (2007). doi:10.1029/2006GL029208
Y. Ma, A.F. Nagy, K.C. Hansen, D.L. DeZeeuw, T.I. Gombosi, K.G. Powell, J. Geophys. Res. **107**, 1282 (2002). doi:10.1029/2002JA009293
Y. Ma, A.F. Nagy, I.V. Sokolov, C.H. Kenneth, J. Geophys. Res. **109**, A07211 (2004). doi:10.1029/2003JA010367
Y. Ma et al., J. Geophys. Res. **111**, A05207 (2006). doi:10.1029/2006JA011481
M.L. Marconi, D.A. Mendis, Astrophys. J. **273**, 381–396 (1983)

M.L. Marconi, L. Dagum, W.H. Smyth, Astrophys. J. **469**, 393–401 (1996)
B.H. Mauk, D.G. Mitchell, C. Paranicas, S.M. Krimigis, Nature **421**, 920–922 (2003)
M.B. McElroy, T.Y. Kong, Y.L. Yung, J. Geophys. Res. **82**, 4379–4388 (1977)
M.A. McGrath, R.E. Johnson, Icarus **69**, 519 (1987)
M.A. McGrath, E. Lellouch, D.F. Strobel, P.D. Feldman, R.E. Johnson, in *Jupiter: Satellites, Atmosphere, Magnetosphere*, ed. by F. Bagenal et al. (Cambridge Univ. Press, Cambridge, 2004), pp. 457–483
M.A. McGrath et al., in *Europa*, ed. by R. Pappalardo et al. (2008, in press)
S.M.P. McKenna-Lawlor, V. Afonin, Y. Yeroshenko, E. Keppler, E. Kirsch, K. Schwingenschuh, Planet. Space Sci. **41**, 373 (1993)
R.L. McNutt, Geophys. Res. Lett. **16**, 1225–1228 (1989)
M. Mendillo, J. Baumgardner, B. Flynn, J.W. Hughes, Nature **348**, 312–314 (1990)
M. Michael, R.E. Johnson, F. Leblanc, M. Liu, J.G. Luhmann, V. Shematovich, Icarus **175**, 263–267 (2005)
M. Michael, R.E. Johnson, Planet. Space Sci. **53**, 1510–1514 (2005)
R. Modolo, G.M. Chanteur, E. Dubinin, A.P. Matthews, Ann. Geophys. **23**, 433 (2005)
L. Moore, M. Mendillo, Geophys. Res. Lett. **34**, L12202 (2007). doi:10.1029/2007GL029381
L. Moore, A.F. Nagy, A.J. Kliore, I. Mueller-Wodarg, J. Richardson, M. Mendillo, Geophys. Res. Lett. **33**, L22202 (2006). doi:10.1029/2006GL027375
G.E. Morfill, H.M. Thomas, Icarus **179**, 539 (2005)
J. Morgenthaler, W.M. Harris, F. Scherb, C.M. Anderson, R.J. Oliversen, N.E. Doane, M.R. Combi, M.L. Marconi, W.H. Smyth, Astrophys. J. **563**, 451–461 (2001)
J. Morgenthaler, W.M. Harris, M.R. Combi, Astrophys. J. **657**, 1162–1171 (2007)
J.I. Moses, S.F. Bass, J. Geophys. Res. **105**, 7013 (2000)
J.I. Moses, B. Bezard, E. Lellouch, G.R. Gladstone, H. Feuchtgruber, M. Allen, J. Geophys. Res. **105**, 7013 (2000)
J.I. Moses, M.Y. Zolotov, B. Fegley, Icarus **156**, 107–135 (2002)
A.F. Nagy, P.M. Banks, J. Geophys. Res. **75**, 6260 (1970)
A.F. Nagy, T.E. Cravens, Geophys. Res. Lett. **15**, 433 (1988)
A.F. Nagy, T.E. Cravens, J.-H. Yee, A.I.F. Stewart, Geophys. Res. Lett. **8**, 629–632 (1981)
A.F. Nagy, J. Kim, T.E. Cravens, Ann. Geophys. **8**, 251 (1990)
A.F. Nagy, M.W. Liemohn, J.L. Fox, J. Kim, J. Geophys. Res. **106**(15), 21,565–21,568 (2001). doi:10.1029/2001JA000007
H.B. Niemann et al., Nature **438**, 779–784 (2005)
A.O. Nier, M.B. McElroy, J. Geophys. Res. **82**, 4341–4349 (1977)
T.G. Northrop, J.R. Hill, J. Geophys. Res. **88**, 6102 (1983)
E.N. Parker, Astrophys. J. **139**, 93–122 (1964)
L.J. Paxton, J. Geophys. Res. **90**, 5089–5096 (1985)
J. Pearl, R. Hanel, V. Kunde, W. Maguire, K. Fox, S. Gupta, C. Ponnamperuma, F. Raulin, Nature **280**, 755–758 (1979)
T. Penz, I.L. Arshukova, N. Terada, H. Shinagawa, N.V. Erkaev, H.K. Biernat, H. Lammer, Adv. Space Res. **36**, 2049–2056 (2005)
H. Pérez-de-Tejada, J. Geophys. Res. **92**, 4713 (1987)
H. Pérez-de-Tejada, J. Geophys. Res. **103**, 31499–31508 (1998)
J.R. Peterson, A. Le Padellec, H. Danared, G.H. Dunn, M. Larsson, A. Larson, R. Peverall, C. Strömholm, S. Rosén, M. Af Ugglas, W.J. van der Zande, J. Chem. Phys. **108**, 1978–1988 (1998)
R. Peverall, S. Rosen, J.R. Peterson, M. Larsson, A. Al-Khalili, L. Vikor, J. Semaniak, R. Bobbenkamp, A. LePadellec, A.N. Maurellis, W.J. van der Zande, J. Chem. Phys. **114**, 6679–6689 (2001)
V. Pierrard, Planet. Space Sci. **51**, 319 (2003)
C.B. Pilcher, J.H. Fertel, W.H. Smyth, M.R. Combi, Astrophys. J. **287**, 427–444 (1984)
C.C. Porco et al., Science **311**, 1393–1401 (2006)
M.K. Pospieszalska, R.E. Johnson, Icarus **93**, 45 (1991)
M.K. Pospieszalska, R.E. Johnson, J. Geophys. Res. **101**, 7565–7573 (1996)
T.M. Prokop, E.C. Zipf, EOS Trans. AGU **63**, 392 (1982)
J.D. Richardson, J. Geophys. Res. **103**, 20,245 (1998)
R.P. Rohrbaugh, J.S. Nisbet, J. Geophys. Res. **78**, 6768–6772 (1973)
S. Rosen et al., Phys. Rev. A **57**, 4462–4471 (1998)
H. Rosenbauer, N. Shutte, I. Apathy, A. Galeev, K. Gringauz, H. Gruenwaldt, P. Hemmerich, K. Jockers, P. Kiraly, G. Kotova, S. Livi, E. Marsh, A. Richter, W. Riedler, A. Remizov, R. Schwenn, K. Schwingenschuh, M. Steller, K. Szego, M. Verigin, M. Witte, Nature **341**, 612 (1989)
C.T. Russell, J.G. Luhmann, R.J. Strangeway, Planet. Space Sci. **54**, 1482–1495 (2006)
J. Saur, D.F. Strobel, F.M. Neubauer, J. Geophys. Res. **103**, 19947–19962 (1998)
D.G. Schleicher, R.L. Millis, P.V. Birch, Icarus **132**, 397–417 (1998)

N.M. Schneider, J.K. Wilson, J.T. Trauger, D.I. Brown, R.W. Evans, D.E. Shemansky, Science **253**, 1394–1397 (1991)
N.M. Schneider, J.T. Trauger, Astrophys. J. **450**, 450 (1995)
R.W. Schunk, A.F. Nagy, *Ionospheres: Physics, Plasma Physics, and Chemistry* (Cambridge Univ. Press, New York, 2000)
K. Seki, R.C. Elphic, M. Hirahara, T. Terasawa, T. Mukai, Science **291**(5510), 1939–1941 (2001)
K. Siersen, A. Al-Khalili, M.J. Jensen, I.B. Nielsen, H.B. Petersen, C.P. Safvan, L.H. Andersen, Phys. Rev. A **68**, 022708 (2003)
D.E. Shemansky, P. Matheson, D.T. Hall, H.-T. Hu, T.M. Tripp, Nature **363**, 329 (1992)
V.I. Shematovich, R.E. Johnson, Hot oxygen coronae at terrestrial planets. Talk at 1st European Congress of Planetary Sciences, Berlin, 2006
V.I. Shematovich, D.V. Bisikalo, J.-C. Gérard, J. Geophys. Res. **99**, 23,217 (1994)
V.I. Shematovich, J.-C. Gerard, D.V. Bisikalo, B. Hubert, J. Geophys. Res. **104**, 4287–4295 (1999)
V.I. Shematovich, R.E. Johnson, M. Michael, J.G. Luhmann, J. Geophys. Res. **108**(E8), 5087 (2003). doi:10.1029/2003JE002094
V.I. Shematovich, D.V. Bisikalo, J.-C. Gerard, Geophys. Res. Lett. **32**, L02105 (2005a)
V.I. Shematovich, R.E. Johnson, J.F. Cooper, M.C. Wong, Icarus **173**, 480–498 (2005b)
H. Shinagawa, J. Geophys. Res. **101**, 26911–26920 (1996)
H. Shinagawa, T.E. Cravens, J. Geophys. Res. **94**, 6506 (1989)
B.D. Shizgal, M.J. Lindenfeld, J. Geophys. Res. **87**, 853 (1982)
B. Shizgal, R. Blackmore, Planet. Space Sci. **34**, 279–291 (1986)
B.D. Shizgal, G.G. Arkos, Rev. Geophys. **34**, 483 (1996)
B.D. Shizgal, J. Geophys. Res. **104**, 14,833 (1999)
T.E. Skinner, S.T. Durrance, Astrophys. J. **310**, 966–971 (1986)
Y.V. Skorov, G.N. Markelov, H.U. Keller, Sol. Syst. Res. **38**, 455–475 (2004)
Y.V. Skorov, G.N. Markelov, H.U. Keller, Sol. Syst. Res. **40**, 219–229 (2006)
H.T. Smith, R.E. Johnson, E.C. Sittler, M. Shappirio, O.J. Tucker, M. Burger, F.J. Crary, D.J. McComas, D.T. Young, Icarus **188**, 356–366 (2007)
W.H. Smyth, M.R. Combi, Icarus **126**, 58–77 (1997)
W.H. Smyth, M.L. Marconi, J. Geophys. Res. **105**, 7783–7792 (2000)
W.H. Smyth, M.L. Marconi, Icarus **181**, 510–526 (2006)
W.H. Smyth, M. Wong, Icarus **171**, 171–182 (2004)
W.H. Smyth, M.L. Marconi, F. Scherb, F.L. Roesler, Astrophys. J. **413**, 756–763 (1993)
J. Spencer et al., Science **311**, 1401–1405 (2006)
J.P. Sprengers, W. Ubachs, K.G.H. Baldwin, J. Chem. Phys. **122**, 144301 (2005)
S.W. Squyres, R.T. Reynolds, P.M. Cassen, Icarus **53**, 319 (1983)
G. Stark, K. Yoshino, P.L. Smith, K. Ito, J. Quant. Spectrosc. Radiat. Transf. **103**, 67–73 (2007)
D.F. Strobel, X. Zhu, M.E. Summers, M.H. Stevens, Icarus **120**, 266–289 (1995)
D.F. Strobel, in *Comparative Aeronomy in the Solar System*, ed. by M. Mendillo, A. Nagy, H. Waite, Geophys. Monogr. Ser. (Am. Geophys. U., 2002), pp. 7–22
D.F. Strobel, Icarus **193**, 588–594 (2008a)
D.F. Strobel, Icarus **193**, 612–619 (2008b)
W.-L. Tseng, D. Bockelee-Morvan, J. Crovisier, P. Colom, W.-H. Ip, Astron. Astrophys. **467**, 729–735 (2007)
V. Tenishev et al., (2008, in press)
N. Terada, H. Shinagawa, S. Machida, Adv. Space Res. **33**, 161–166 (2004)
N. Thomas, Surv. Geophys. **13**, 91–164 (1992)
N. Thomas, Astron. Astrophys. **313**, 306–314 (1996)
N. Thomas, F. Bagenal, T.W. Hill, J.K. Wilson, in *Jupiter. The Planet, Satellites and Magnetosphere*, ed. by F. Bagenal (Cambridge Univ. Press, Cambridge, 2004), pp. 561–591
C. Tian, C.R. Vidal, J. Chem. Phys. **108**, 927–936 (1998)
F. Tian, O.B. Toon, Geophys. Res. Lett. **32**, L18201 (2005). doi:10.1029/2005GL023510
R.L. Tokar et al., Geophys. Res. Lett. **32**, L14S04 (2005). doi:10.1029/2005GL022690
L. Trafton, Astrophys. J. **247**, 1125–1140 (1981)
C. Tully, R.E. Johnson, Planet. Space Sci. **49**, 533–537 (2001)
A. Vidal-Madjar, J.-L. Bertaux, Planet. Space Sci. **20**, 1147 (1972)
A.A. Viggiano, A. Ehlerding, F. Hellberg, R.D. Thomas, V. Zhaunerchyk, W.D. Geppart, H. Montaigne, M. Larsson, M. Kaminska, F. Osterdahl, J. Chem. Phys. **122**(22), 226101 (2005)
D. Vignes et al., Geophys. Res. Lett. **27**, 49 (2000)
J.H. Waite et al., Science **308**, 982–986 (2005a)
J.H. Waite et al., Science **307**, 1260–1262 (2005b)
J.H. Waite et al., Science **311**, 1419–1422 (2006)

J. Warnecke, M.G. Kivelson, K.K. Khurana, D.E. Huddleston, C.T. Russell, Geophys. Res. Lett. **24**, 2139 (1997)
C.C. Watson, P.K. Haff, T.A. Tombrello, in *Proc. Lunar Planet Sci. Conf. 11th* (1980), pp. 2479–2485
A.J. Watson, T.M. Donahue, J.C.G. Walker, Icarus **48**, 150–166 (1981)
R.P. Wayne, *Chemistry of Atmospheres* (Clarendon, Oxford, 1991)
H.A. Weaver, P.D. Feldman, M.F. A'Hearn, C. Arpigny, J.C. Brandt, S.A. Stern, Icarus **141**, 1–12 (1999)
M.S. Westley, R.A. Baragiola, R.E. Johnson, G.A. Barratta, Planet. Space Sci. **43**, 1311–1315 (1995)
F.L. Whipple, W.F. Huebner, Annu. Rev. Astron. Astrophys. **14**, 143–172 (1976)
J.K. Wilson, N.M. Schneider, J. Geophys. Res. **104**, 16567–16584 (1999)
M.C. Wong, R.E. Johnson, J. Geophys. Res. **101**, 23243–23254 (1996)
R.V. Yelle, J. Cui, I.C.F. Muller-Wodarg, J. Geophys. Res. (2008, in press)
D.T. Young et al., Science **307**, 1262–1266 (2005)
M.H.G. Zhang, J. Luhmann, S.W. Bougher, A. Nagy, J. Geophys. Res. **98**, 10,915 (1993)
T.L. Zhang et al., Nature **450**, 654–656 (2007)
J.F. Ziegler, J.P. Biersak, V. Littmark, *The Stopping and Range of Ions in Solids* (Pergamon, New York, 1985)

Atmospheric Escape and Evolution of Terrestrial Planets and Satellites

Helmut Lammer · James F. Kasting · Eric Chassefière · Robert E. Johnson · Yuri N. Kulikov · Feng Tian

Originally published in the journal Space Science Reviews, Volume 139, Nos 1–4.
DOI: 10.1007/s11214-008-9413-5

Abstract The origin and evolution of Venus', Earth's, Mars' and Titan's atmospheres are discussed from the time when the active young Sun arrived at the Zero-Age-Main-Sequence. We show that the high EUV flux of the young Sun, depending on the thermospheric composition, the amount of IR-coolers and the mass and size of the planet, could have been responsible that hydrostatic equilibrium was not always maintained and hydrodynamic flow and expansion of the upper atmosphere resulting in adiabatic cooling of the exobase temperature could develop. Furthermore, thermal and various nonthermal atmospheric escape processes influenced the evolution and isotope fractionation of the atmospheres and water inventories of the terrestrial planets and Saturn's large satellite Titan efficiently.

Keywords Atmosphere evolution · Young Sun/stars · Isotope anomalies · Escape · Magnetic protection · Terrestrial planets

H. Lammer (✉)
Space Research Institute, Austrian Academy of Sciences, Schmiedlstraße 6, 8042 Graz, Austria
e-mail: helmut.lammer@oeaw.ac.at

J.F. Kasting
Department of Geosciences, Penn State University, 443 Deike Building, University Park 16802, USA

E. Chassefière
SA/IPSL Université P. & M. Curie, Boite 102, 4 Place Jussieu, 75252 Paris Cedex 05, France

R.E. Johnson
Engineering Physics, University of Virginia, Charlottesville, 22904, USA

Y.N. Kulikov
Polar Geophysical Institute (PGI), Russian Academy of Sciences, Khalturina Str. 15, 183010 Murmansk, Russia

F. Tian
National Center for Atmospheric Research (NCAR), High Altitude Observatory (HAO), Boulder, CO, USA

1 Introduction

In order to understand the evolution of the planetary atmospheres of Venus, Earth, Mars and Saturn's satellite Titan and the principles that generated Earth's present atmosphere and those of the other terrestrial bodies in the Solar System and possible Earth-type exoplanets, one has to understand the evolutionary influence of the solar/stellar radiation and particle environment on planetary atmospheres. Besides these effects which can modify and fractionate planetary atmospheres over long time spans, surface-atmosphere interaction processes such as the carbon-silicate cycle that controls the CO_2 partial pressure, oxidation processes on the soil, the generation of magnetic dynamos and the influence of life forms and the modification of atmospheres by them, have also taken into account.

The eight major Sections of this Chapter cover a wide range of topics that are connected to the evolution of the atmospheres of terrestrial planetary bodies. In Sect. 2 we discuss the observed isotope anomalies in the atmospheres of Venus, Earth, Mars and Titan and their relevance for atmospheric evolution. In Sect. 3 we focus on the present knowledge of the radiation and particle environment of the young Sun inferred from solar proxies with different ages. After discussing the initial solar and atmospheric conditions we focus in Sect. 4 on questions related to the loss of the initial water inventory from early Venus. In this section we discuss and review in detail the runaway greenhouse effect, and thermal and non-thermal atmospheric escape of Venus' initial H_2O inventory. Section 5 focuses on the evolution of Earth's atmosphere, from its formation, loss processes, magnetospheric protection, to its modification after the origin of primitive life forms. In Sect. 6 we review and discuss the formation, evolution and loss of the initial Martian atmosphere and its water inventory. Finally, in Sect. 7 the evolution of Titan's dense nitrogen atmosphere and its alteration by atmospheric loss processes, the contribution of sputtering, and its relevance to the escape from other satellite atmospheres is reviewed and discussed.

2 Isotope Anomalies in the Atmospheres of Venus, Earth, Mars, and Titan

After the establishment of atmospheric and internal volatile reservoirs during the accretionary and early post-accretionary phases of planet formation, further modifications of isotopic ratios might still occur over long periods of time as a result of thermal and non-thermal escape processes (e.g., Pepin 1991; Becker et al. 2003; Lammer and Bauer 2003). For example, Hutchins and Jakosky (1996) estimated that about $90 \pm 5\%$ of ^{36}Ar and about $80 \pm 10\%$ of ^{40}Ar have been lost by atmospheric sputtering from the martian atmosphere after the intrinsic magnetic field vanished about 4 Gyr ago (Acuña et al. 1998). In this context isotopic fractionation in planetary atmospheres may result from the diffusive separation by mass of isotopic and elemental species and occurs between the homopause, the level above which diffusion rather than turbulent mixing is the controlling process, and the exobase, above which collisions are rare. The lighter particles are more abundant at the exobase and exosphere than the heavier species.

When particles are removed from a planetary exosphere by atmospheric loss processes, the lighter isotopes are preferentially lost and the heavier ones become enriched in the residual gas. The diffusive separation effect leads to enrichment of the lighter isotope in the exosphere, depending on the homopause altitude (Lammer and Bauer 2003). This effect enhances the importance of all atmospheric escape processes that occur at the exobase level. Atmospheric escape mechanisms that can lead to isotope fractionation in a planetary atmosphere are high Jeans escape rates, dissociative recombination, impact

Table 1 Hydrogen, oxygen, carbon and nitrogen Isotope ratios observed in the atmospheres of the three terrestrial planets (Kallenbach et al. 2003; Lammer and Bauer 2003 and references therein) and Titan (Niemann et al. 2005)

Isotope ratios	Venus	Earth	Mars	Titan
D/H	$1.6–2.2 \times 10^{-2}$	1.5×10^{-4}	$8.1 \pm 0.3 \times 10^{-4}$	$2.3 \pm 0.3 \times 10^{-4}$
$^{18}O/^{16}O$	$\sim 2 \times 10^{-3}$	2.04×10^{-3}	$1.89 \pm 0.2 \times 10^{-3}$	
$^{13}C/^{12}C$	$1.14 \pm 0.02 \times 10^{-2}$	1.12×10^{-2}	$1.18 \pm 0.12 \times 10^{-2}$	1.21×10^{-2}
$^{15}N/^{14}N$	$\sim 3.5 \times 10^{-3}$	3.7×10^{-3}	$6.4–5.0 \times 10^{-3}$	$5.46 \pm 0.2 \times 10^{-3}$

dissociation of molecules by energetic electrons, charge exchange, atmospheric sputtering, and ion pick up by the solar wind (Chamberlain and Hunten 1987; Johnson 1990; Lammer and Bauer 1993).

The volatile isotopic compositions in planetary environments were initially established at the time of the formation. For Earth, we have abundant samples of crustal and upper mantle rocks to study, and a well-determined atmosphere. For example from $^{40}Ar/^{36}Ar$ isotope fractionation in the present Earth mantle and the ^{40}Ar degassing rate from the crust it is found that only an early catastrophic degassing model is compatible with the atmospheric $^{40}Ar/^{36}Ar$ ratio (e.g., Hamano and Ozima 1978). The Earth was formed from large planetesimals, therefore, the most likely cause for catastrophic degassing is linked to impacts (e.g., Lange and Ahrens 1982; Matsui and Abe 1986a, 1986b).

Isotopic composition reflects the various reservoirs that went into making up the planets. It is expected that the primary reservoir for oxygen, nitrogen, and probably carbon would have been solid objects, representatives of which may still exist in the various meteorite populations (e.g., Clayton 2003; Grady and Wright 2003). In the case of noble gases and hydrogen the initial reservoirs (Kallenbach et al. 2003; and references therein) were most likely dominated by nebular gases of solar composition, very cold condensates, and solar wind implantation. One does not know how much of any specific reservoir was incorporated in any specific planet and by how much the initial planetary composition was then fractionated by addition of further material or by removal of material from the planet. In addition to infall of micrometeoritic or cometary material, fractionation processes may have occurred during the early stages of the Solar System, caused by high thermal escape rates or rapid non-thermal loss processes from more expanded upper atmospheres which were heated by intense EUV flux from the young Sun.

What we know about the present-day isotopic composition of the planets is limited by observations that have thus far been carried out. Table 1 shows the hydrogen, nitrogen, oxygen and carbon isotope ratios in terrestrial-like planetary atmospheres. For Venus, we have atmospheric data only, with significant uncertainties for many of the isotopic ratios. Interpretations of the mass spectrometry data of Pioneer Venus regarding the D/H ratio suggest that Venus once may have had more water, corresponding to at least 0.3% of an Earth-like ocean. Unfortunately, the D/H ratio on Venus of about $1.6–2.2 \times 10^{-2}$ can be explained two ways: impacts by H_2O-rich planetesimals with similar water abundance as Earth and Mars (Dayhoff et al. 1967; Walker et al. 1970; Donahue and Pollack 1983; Kasting and Pollack 1983; Morbidelli et al. 2000; Raymond et al. 2004), or Venus was formed from condensates in the solar nebula that contained little or no water (Lewis 1970, 1974). The supply of water to the Venus' atmosphere by comets was studied by Lewis (1974), Grinspoon and Lewis (1988) and more recently by Chyba et al. (1990).

However, Grinspoon and Lewis (1988) have also argued that present Venus' water content may be in a steady state where the loss of hydrogen to space is balanced by a continuous input of water from comets or from delayed juvenile outgassing. In case the external water delivery occurs, then no increase of Venus' past water inventory is required to explain the observed D/H ratio. However, recent models of solar system formation (e.g., Morbidelli et al. 2000; Raymond et al. 2004) suggest a wet early Venus (e.g., Dayhoff et al. 1967; Walker et al. 1970; Donahue and Pollack 1983; Kasting and Pollack 1983) because the suggest that most of Earth's water came from the asteroid belt region, not from 1 AU. If so, then Venus must have been hit with H_2O-rich planetesimals as well. The process is stochastic, as it involves large planetesimal impacts, but still it is highly unlikely that Venus ended up with $\leq 10\%$ of Earth's water inventory. This is in agreement with earlier suggestions that the initial water content on early Venus should have been larger (e.g., Shimazu and Urabe 1968; Rasool and DeBergh 1970; Donahue et al. 1982, 1997; Kasting and Pollack 1983; Chassefière 1996a, 1996b).

If Venus was wet, the planet must have lost most of its water during its history. As can be seen in Table 1, besides the enrichment of D in Venus' atmosphere compared to Earth, mass spectrometer measurements of the isotope ratios of $^{15}N/^{14}N$, $^{18}O/^{16}O$ and $^{13}C/^{12}C$ show that these ratios are close to that on Earth (Lammer and Bauer 2003; Kallenbach et al. 2003; and references therein).

Venus' high noble gas abundances and solar-like elemental ratios, except for Ne/Ar, suggest that at least the heavier noble gases in the Venusian atmosphere are not greatly evolved from their primordial states (e.g. Cameron 1983; Pepin 1991, 1997). Neon and Ar isotope ratios also appear to be biased toward solar values compared to their terrestrial counterparts. Venus, therefore, seems to be in a unique position in that its atmosphere may have been altered from its initial composition by a planet specific fractionating loss mechanism to a much smaller extent than the highly processed atmospheres of Earth and Mars. Sekiya et al. (1980, 1981) and Pepin (1997) suggested that hydrodynamic escape from early Venus could have generated Ne and Ar isotope ratios close to the observed values in its present time atmosphere and noble gas ratios similar to those derived for Earth's initial atmosphere.

For Mars, as for Earth, we have data for the atmosphere as well as for some mantle-derived rocks in the form of the martian meteorites. The D/H isotope ratio in the present martian atmospheric H_2O vapor is $8.1 \pm 0.3 \times 10^{-4}$ which is greater than the terrestrial value by a factor of 5.2 ± 0.2 (e.g. Owen et al. 1988; Yung et al. 1988; Krasnopolsky et al. 1997). Modeling the atmospheric D/H ratio by using different methods results in a total H_2O loss of a 3.6–50 m global layer of H_2O from Mars during the past 3.5 Ga (e.g. Yung et al. 1988; Lammer et al. 1996, 2003a; Kass and Yung 1999; Krasnopolsky and Feldman 2001; Bertaux and Montmessin 2001). One should also note that the amplitude and the chronology of water exchange between the atmosphere and the polar caps may also influence the atmospheric D/H ratio. At the present total hydrogen (neutrals and ions) escape rate of about $1.5–2 \times 10^{26}$ s^{-1} (e.g., Anderson and Hord 1971; Krasnopolsky and Feldman 2001; Lammer et al. 2003a), the atmospheric water vapor (10 μm pr.) is completely lost in about 10,000 years. This is a short time; therefore, one cannot exclude that atmosphere-polar caps exchanges, driven by orbital parameter variations and other mechanisms, have an impact on the atmospheric D/H ratio, in addition to escape. Thus, one can see from the wide range of model results and the possible influence of atmosphere-polar cap interactions, that constraining water loss from D/H ratios can result in large uncertainties.

From the mass spectrometer measurements on board of Viking an anomalous $^{15}N/^{14}N$ ratio equal to 1.62 ± 0.16 times the Earth value was observed (Nier 1976; Nier et al. 1976). The $^{15}N/^{14}N$ anomaly on Mars is an important indicator for escape related fractionation processes during the evolution of the Martian atmosphere (Fox and Hać 1997; Manning et al. 2007). In contrast to the nitrogen isotopes, the relative abundances of O and C isotopes on Mars appear to be similar to the observed values on Earth and seem, therefore, to be buffered by surface reservoirs. The atmospheric evolution on Mars can be separated into an early and late period. The early evolutionary epoch can be characterized by a higher CO_2 surface pressure and a possible greenhouse effect, while the second later epoch is related to a low surface pressure and a polar-cap regolith buffered system initiated by polar CO_2 condensation after the late heavy bombardment period about 3.8 Ga ago (Pepin 1994).

During the early evolution period heavy noble gasses were most likely fractionated to their present value by the interplay between solar EUV-driven diffusion-limited hydrogen escape from a steam atmosphere toward the end of accretion (Zahnle et al. 1990) and atmospheric escape and fractionation due to large impacts (Pepin 1997). During this early extreme period in Mars' history, the isotope fractionation the CO_2 surface pressure, and the isotopic history were dictated by an interplay of losses to erosion, sputtering, and carbonate precipitation, additions by outgassing and carbonate recycling, and perhaps also by feedback stabilization under greenhouse conditions.

The atmospheric collapse after the late heavy bombardment period led to an abrupt increase in the mixing ratios of pre-existing Ar, Ne, and N_2 at the exobase and their fast escape by sputtering and pick up loss. Current abundances and isotopic compositions of these species are therefore entirely determined by the action of sputtering and photochemical escape on gases supplied by outgassing during the late evolutionary epoch (Jakosky et al. 1994; Becker et al. 2003; Kallenbach et al. 2003). The present atmospheric Kr inventory on Mars derives almost completely from solar-like Kr degassed during this period (Pepin 1994). Consequently, among current observables, only the Xe and ^{13}C isotopes survive as isotopic tracers of atmospheric history prior to its transition to low surface CO_2 pressure values. The values of the $^{40}Ar/^{36}Ar$ ratio and Ar abundance in the martian atmosphere measured by Viking lead to the conclusion that the martian atmosphere was also generated by secondary degassing from the martian interior (e.g., Hamano and Ozima 1978).

For Titan, recent observations by the Cassini Ion Neutral Mass Spectrometer (INMS) measured in situ at 1250 km altitude an enrichment of ^{15}N of about 1.27 ± 1.58 compared to the terrestrial ratio (Waite et al. 2005). Furthermore, the Huygens probe measured during its decent with the Gas Chromatograph and Mass Spectrometer (GCMS) a similar enrichment of ^{15}N compared to ^{14}N of about 1.47. As on Mars, this enrichment of $^{15}N/^{14}N$ compared to Earth indicates that Titan's atmosphere experienced high escape rates and associated isotope fractionation during its early evolution.

A recent study by Nixon et al. (2008) investigated the $^{12}C/^{13}C$ isotopic ratio in Titan hydrocarbons using Cassini/CIRS infrared spectra. They found that Titan's $^{12}C/^{13}C$ ratio (80.8 ± 2.0) is about 8% lower on Titan than at the Earth and lower than the typical value for outer planets $(88.0 \pm 7.0$; Sada et al. 1996). Because Titan's enrichment in ^{13}C is anomalous in the outer solar system, they suggested that preferential escape of the lighter isotope and isotope dependent chemical reaction rates may have favored the gradual partitioning of ^{12}C into heavier hydrocarbons, so that ^{13}C was left behind in CH_4.

3 Activity of the Young Sun and Stars and Its Relevance to Planetary Atmosphere Evolution

3.1 Evolution of the Solar Radiation Environment

One can only understand the evolution of planetary atmospheres and their water inventories if the evolution of the radiation and particle environment of the Sun is known. Solar luminosity has increased from the time when the young Sun arrived at the Zero-Age-Main-Sequence

Table 2 Wavelength range $\Delta\lambda$ and corresponding flux values in units of erg s^{-1} cm^{-2} normalized to a distance of 1 AU and to the radius of the Sun (Ribas et al. 2005)

$\Delta\lambda$ [nm]	EK Dra	π^1 UMa	κ^1 Cet	β Com	Sun
	[0.1 Gyr]	[0.3 Gyr]	[0.65 Gyr]	[1.6 Gyr]	[4.56 Gyr]
0.1–2	180.2	21.5	7.76	0.97	0.15
2–10	82.4	15.8	10.7	2.8	0.7
10–36	187.2	69.4	22.7	7.7	2.05
36–92	45.6	15.2	7.0	2.85	1.0
92–118	18.1	8.38	2.9	1.7	0.74

(ZAMS) $\sim$ 4.6 Gyr ago up to the present, and its effect on Earth's climate evolution has been studied by various researchers (e.g., Sagan and Mullen 1972; Owen 1979; Guinan and Ribas 2002). The total radiation of the young Sun was about 30% less than today. Solar evolution models show that the luminosity of the Sun will increase in the future and will be 10% about 1 Gyr from now. At that time the Earth's oceans may start to evaporate (e.g., Caldeira and Kasting 1992; Guinan and Ribas 2002), unless negative cloud feedback—not included in the published models—delays the expected surface warming.

Although the total radiation flux of the young Sun was lower than today, observations of young solar-like stars (solar proxies) indicate that the early Sun was a much more active source of energetic particles and electromagnetic radiation in the X-ray and EUV spectral range ($\lambda < 100$ nm) (Newkirk 1980; Skumanich and Eddy 1981; Zahnle and Walker 1982; Ayres et al. 2000; Guinan and Ribas 2002; Ribas et al. 2005). The short wavelength radiation is of particular interest because it can ionize and dissociate atmospheric species, thereby initiating photochemistry that can change atmospheric composition. Additionally, the soft X-rays and EUV radiation is absorbed in a planetary thermosphere, whereby it can heat and expand it significantly (e.g., Lammer et al. 2006a, 2007; Kulikov et al. 2006, 2007; Tian et al. 2008). This results in high predicted atmospheric escape rates from primitive atmospheres (e.g., Sekiya et al. 1980, 1981; Watson et al. 1981; Zahnle et al. 1990; Kulikov et al. 2007; Zahnle et al. 2007).

The active phase of the young Sun lasted about 0.5–1.0 Gyr and included continuous flare events. The period where the particle and radiation environment was up to 100 times, or even more intense than today lasted about 0.15 Gyr after the Sun arrived at the ZAMS (Keppens et al. 1995; Guinan and Ribas 2002; Ribas et al. 2005). This is comparable to, but slightly longer than, the expected time scale for terrestrial planet accretion, 10–100 million years (see, e.g., Morbidelli et al. 2000). The "Sun in Time" observational program was established by Dorren and Guinan (1994) to study the magnetic evolution of the Sun using a homogeneous sample of single nearby G0-V main sequence stars which have known rotation periods and well-determined physical properties, including temperatures, luminosities, metal abundances and ages.

Observations at various wavelength ranges were carried out by the following satellites: ASCA ($\Delta\lambda = 0.1$–2 nm), ROSAT ($\Delta\lambda = 2$–10 nm), EUVE ($\Delta\lambda = 10$–36 nm), FUSE ($\Delta\lambda = 92$–118 nm). The data gap between 36–92 nm is caused by strong interstellar medium absorption. To overcome this problem Ribas et al. (2005) inferred the total integrated flux in that interval by comparison with the flux evolution in the other wavelength ranges. Details of the data sets and the flux calibration procedure employed are provided in Ribas et al. (2005). Table 1 shows a sample of solar proxies that contains stars with ages from 0.1 Gyr up to the age of the Sun. These authors estimated the total irradiance in the wavelength range between 0.1–120 nm and obtained a power law fit for the flux $\Phi(t) = 29.7 \times t^{-1.23}$ in units of [erg s^{-1} cm^{-2}] as a function of stellar age t in units of Gyr (Ribas et al. 2005). From this relation it follows that the fluxes normalized to the present time solar value as a function of time are: ~ 6 times [$t = 3.5$ Gyr ago], ~ 10 times [$t = 3.8$ Gyr ago], ~ 20 times [$t = 4.13$ Gyr ago], ~ 30 times [$t = 4.24$ Gyr ago], ~ 50 times [$t = 4.33$ Gyr ago], ~ 70 times [$t = 4.37$ Gyr ago], and ~ 100 times 4.467 Gyr ago. One should note that during the first 0.1 Gyr the soft X-ray and EUV flux were saturated to these high values and hard X-ray fluxes were even higher (Ribas et al. 2005). It is reasonable to suggest that much stronger high-energy radiation flux of the young Sun should have had a critical impact on ionization, photochemistry, and evolution of the early atmospheres of the terrestrial planets.

3.2 The Solar Wind of the Young Sun

Besides the much higher radiation, which was related to frequent flaring of the young Sun, one should also expect a more powerful stellar wind. HST high-resolution spectroscopic observations of the hydrogen Lyman-α feature of several nearby main-sequence G and K stars by Wood et al. (2002, 2005) have revealed the absorption of neutral hydrogen associated with the interaction between the stars' fully ionized coronal winds with the partially ionized local interstellar medium. These absorption features formed in the astrospheres of the observed stars provided the first empirically-estimated coronal mass loss rates for solar-like G and K main sequence stars.

Wood et al. (2002, 2005) estimated the mass loss rates from the system geometry and hydrodynamics and found from their small sample of stars, where astrospheres can be observed, that mass loss rates increase with stellar activity. The correlation between the mass loss rate and X-ray surface flux follows a power law relationship, which indicates a total plasma density in the early solar wind and Coronal Mass Ejections (CMEs) of about ≥ 100–1000 times higher than today during the first 100 Myr after the Sun reached the ZAMS. The total ejected plasma density decreases as the solar activity subsides and may have been ≥ 30–100 times higher than today at 3.5 Ga ago (e.g., Lundin et al. 2007). However, the present stellar sample analyzed by Wood et al. (2002, 2005), Lundin et al. (2007) is not large enough; therefore, many uncertainties regarding the early solar wind remain, and more observations of young solar-like G and K stars are needed to enhance our knowledge of stellar winds during periods of high coronal activity.

4 Loss of Water from Early Venus

4.1 The Runaway Greenhouse

Venus presents an especially interesting problem for the field of planetary aeronomy. As mentioned in the Introduction, Venus shows clear evidence of having lost substantial amounts of water during its history. The process by which this occurred is typically referred to as a runaway greenhouse, although as we shall see, this term can be defined in different ways that have different physical implications for Venus' history.

The basic concept of the runaway greenhouse has been understood for many years (Ingersoll 1969; Rasool and DeBergh 1970; Walker et al. 1970). Venus' mean orbital distance is 0.72 AU, and so it receives roughly 1.9 times as much sunlight as does Earth. Suppose, following Rasool and DeBergh (1970), that Venus started off with no atmosphere whatsoever, and that it outgassed a mixture of CO_2 and H_2O from volcanoes. If we neglect the

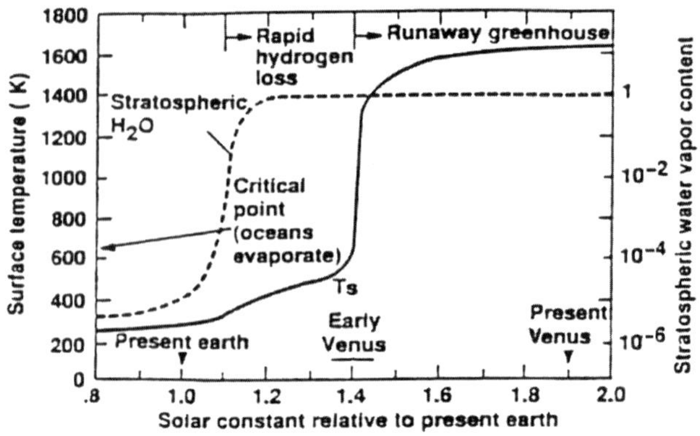

Fig. 1 Diagram illustrating what would happen to Earth if it were slowly pushed inwards towards the Sun. The *horizontal scale* is the solar flux relative to its value at 1 AU. The *solid curve* represents mean global surface temperature (*left-hand scale*). The *dashed curve* represents stratospheric H_2O mixing ratio (*right-hand scale*). The solar flux at Venus today and at 4.5 billion years ago is indicated (adapted from Kasting 1988)

change in solar luminosity over time, as they did, Venus' initial mean surface temperature would have been about 320 K. As its atmosphere grew thicker, however, the surface temperature would have increased as a consequence of the greenhouse effect of CO_2 and H_2O. If one tracks the subsequent evolution, one finds that the surface is always too hot for liquid water to condense, and so all of the outgassed H_2O ends up in the atmosphere as steam. Importantly, even the upper atmosphere would have been H_2O-rich. At these levels, H_2O could have been photodissociated by solar ultraviolet radiation. The hydrogen would have escaped to space by processes described below; the oxygen could either have been dragged along with it, if the escape was fast enough, or it could have reacted with reduced gases (e.g., CO) in the atmosphere and with reduced materials (e.g., ferrous iron) in the planet's crust. Eventually, all of the water would have been lost, and Venus would have been left with the dense CO_2 atmosphere that we observe today.

Although this story sounded satisfactory at the time when it was first proposed, later advances in our understanding of how planets form created problems for this model. The final stages of terrestrial planet accretion are now thought to involve impacts of planetesimals that were Moon-sized or larger. Some of these planetesimals should have originated from the asteroid belt region or beyond (see, e.g., Raymond et al. 2004), and so they would have been rich in H_2O and other volatiles. When they collided with a growing planet, either Venus or Earth, most of these volatiles should have been injected directly into its atmosphere in a process termed impact degassing (Lange and Ahrens 1982). Hence, the atmosphere and ocean, if it was stable, should have formed as the planet itself formed. This process has been simulated using numerical models that include both the atmosphere and the growing solid planet (Matsui and Abe 1986a, 1986b; Zahnle et al. 1988). These calculations indicate that the planet's entire surface should have been molten during the main part of the accretion period, creating a magma ocean, and that it should have been overlain by a dense ($\sim$ 100 bar), steam atmosphere that was in quasi-equilibrium with the magma. For Venus, it is uncertain whether this steam atmosphere would have condensed out when the accretion process ended or whether it would have remained as vapor. In any case, as we will see, its fate should have been similar to that predicted by the earlier models: loss by photodissociation, followed by escape of hydrogen to space.

It is easier to understand how this process works by examining a somewhat simpler calculation described by Kasting (1988), the results of which are summarized in Fig. 1. In this numerical simulation, a planet resembling modern Earth was "pushed" closer to the Sun by gradually increasing the incident solar flux. (The horizontal axis, S_{eff}, represents the solar flux relative to its value at modern Earth, $\sim$ 1365 W/m^2.) The solid curve in the figure shows the evolution of the planet's mean surface temperature, T_s. T_s increases slowly at first, then "runs away" to extremely high values when S_{eff} reaches ~ 1.4. At this point, all remaining water vaporizes, leaving the Earth with a dense, 270 bar steam atmosphere that is in every sense a true runaway greenhouse.

Figure 1 also shows something else, however: the dashed curve, which goes with the scale on the right, represents the mixing ratio of H_2O in the planet's stratosphere, $f(H_2O)$. At low surface temperatures (corresponding to low S_{eff}), $f(H_2O)$ is very low—only a few times 10^{-6}, i.e., a few parts per million by volume (ppmv). This corresponds to the situation on modern Earth, for which $f(H_2O)$ is about 3–5 ppmv. But for surface temperatures exceeding ~340 K, or 70°C, $f(H_2O)$ rises quickly to values near unity, and the stratosphere becomes water-dominated. In this model, this phenomenon occurs at $S_{eff} \geq 1.1$. This should lead to photodissociation of H_2O and escape of H to space, as before, with the difference being that liquid water remains present on the planet's surface until the very last part of the escape process.

The model calculation described here is heuristic and may not apply directly to early Venus because its atmosphere and initial water inventory were almost certainly different from modern Earth. The results of the calculation nevertheless suggest what may have happened on Venus. The Sun was about 30 percent less luminous when it formed (Gough 1981), so the solar flux on early Venus was approximately 1.34 times the value for modern Earth, or ~ 1825 W/m^2. This value is right near the "runaway greenhouse" threshold in this model, when the oceans actually vaporize, and it is well above the critical solar flux for water loss. If clouds—which were not explicitly included in the model shown here—act to cool the surface, and if Venus' initial water endowment was a substantial fraction of Earth's, then early Venus could well have had liquid oceans on its surface. This hypothesis may be testable at some time in the future when we have the technology to sample Venus' surface and subsurface.

4.2 Thermal Escape of Light Species

The theory of thermal escape from an atmosphere was developed in the 1960s (Chamberlain 1961; Öpik and Singer 1963). Because the density of the atmosphere decreases with altitude the atmosphere becomes collisionless above a certain level, called the exobase. The exobase distance, where the atmospheric scale height is equal to the collisional mean free path, is ≈ 200 km on present Venus. Present thermal escape, or "Jeans" escape, consists of the (small) upward flux of atoms whose velocity is larger than the escape velocity (10.4 km s^{-1}) at the exobase. Because of the low exospheric temperature of Venus (≈ 275 K), which is caused by the large abundance of CO_2, a strong infrared emitter, present thermal escape of hydrogen on Venus is almost negligible. But at epochs in the past when the water abundance in Venus' atmosphere was higher and when the Sun was a more powerful EUV emitter, the exospheric temperature was probably much higher and thermal escape could have taken the form of a "hydrodynamic" escape. Hydrodynamic escape is a global, cometary-like, expansion of the atmosphere. It requires the deposition of a large flux of EUV energy into the atmosphere to allow species to overcome gravity. Such conditions may have been reached in H- or He-rich thermospheres heated by the strong EUV flux of the young Sun (Sekiya et al. 1980, 1981; Watson et al. 1981; Zahnle and Walker 1982; Yelle 2004, 2006; Tian et al. 2005; Munoz 2007; Penz et al. 2008), e.g. in the following cases:

(i) primordial H_2/He atmospheres;
(ii) an outgassed H_2O-rich atmosphere during an episode of runaway and/or wet greenhouse.

The theory of hydrodynamic escape was developed by Parker (1963) for the solar wind plasma and was first applied to study hydrodynamic escape of hydrogen-rich early atmospheres of terrestrial planets by Sekiya et al. (1980, 1981), Watson et al. (1981), Kasting and Pollack (1983), and later by many other authors. Öpik and Singer (1963) defined the state of an expanding atmosphere when its outflow velocity, v_{exo}, at the exobase is equal to or exceeds the escape velocity, v_{esc}, from a planet at that altitude ($v_{exo} \geq v_{esc}$) as blow-off. This corresponds to a Jeans escape parameter, $\lambda(= GMm/r_{exo}kT_{exo})$, of < 1.5. This condition may occur if an atmosphere is sufficiently heated and if the flow of the main escaping species is not diffusion limited.

Hydrodynamic models were also applied to mass fractionation of planetary atmospheres (Zahnle and Kasting 1986; Hunten et al. 1987; Chassefière and Leblanc 2004). However, the models of hydrodynamic escape of atomic hydrogen from water-rich early atmospheres of terrestrial type planets were not quite satisfactory (e.g. Chassefière 1996a). The main reason for this is the fact that these models did not take into account the transition from the fluid regime to the collisionless regime in the upper planetary corona. Once collisions become infrequent, solar EUV energy cannot be readily converted into bulk translational kinetic energy (Chassefière 1996a).

Sekiya et al. (1980, 1981) and Watson et al. (1981) in their pioneering work studied hydrodynamic escape of an atomic hydrogen rich atmosphere from a terrestrial planet due to solar EUV heating by applying idealized hydrodynamic equations. From their thermospheric model of the Earth Watson et al. (1981) obtained supersonic flow solutions for which the sonic point was reached at a distance of about 2×10^5 km or some 30 planetary radii, r_0. These authors argued that supersonic hydrodynamic escape of atomic hydrogen was possible from hydrogen dominated Earth's atmosphere even if it were exposed to the present time solar EUV flux. However, as pointed out above, these authors assumed that the fluid equations applied above the exobase, which is not internally self-consistent. So, there is some question as to whether their supersonic solutions could really be achieved. Indeed, the flow at the exobase ($r_{exo} \approx 7.5r_0$) in their model is subsonic, and its velocity of ~ 100 m s^{-1} is an order of magnitude lower than the escape velocity of 1.5 km s^{-1}. As the conversion of internal thermal energy of the neutral gas into kinetic energy of the flow is retarded by the lack of collisions above the exobase, the flow of neutral particles cannot be accelerated anymore and it is not clear that either the sonic or even the escape velocity can be reached.

These negative considerations should be tempered by the realization that H_2- or H-dominated upper atmospheres on rocky planets are not likely to remain hydrostatic if some appreciable stellar EUV heating is present. As pointed out by Kasting and Pollack (1983), their more H_2O-rich early Venus atmospheres would remain collisional out to all distances if the hydrostatic assumption was adopted. Application of the barometric law would then imply that the atmospheric mass was infinite a result that cannot be physically correct (Chamberlain and Hunten 1987; Walker 1977). Hence, such atmospheres must be expanding hydrodynamically into space, albeit perhaps at somewhat less than the escape rate that corresponds to transonic outflow. Accurately calculating the escape rate in such cases could in principle be accomplished by using a hybrid approach similar to that employed by Chassefière (1996a), in which a fluid dynamical solution was joined to a modified Jeans' solution at the exobase (see below for more details). Alternatively, a "moment" type of approach (e.g., Schunk and Watkins 1979), in which the particle velocity distribution is calculated self-consistently, could be applied at all altitudes. In carrying out such modeling efforts it should be borne in mind that the real escape problem is inherently 2- or 3-dimensional as a consequence of interactions of the escaping gas with the impinging stellar wind. Hence, any 1-D approximation, regardless of its level of sophistication, is just that—an approximation.

If we ignore these complications for the moment, and simply acknowledge that the published hydrodynamic solutions represent upper limits on the hydrogen escape rate, we can see that thermal escape rates from hydrogen-rich terrestrial planets could have been large in the past, especially during the earliest epochs when the solar EUV radiation was much more intense than today (Ribas et al. 2005).

In a recent study Tian et al. (2008), like Chassefière (1996a), matched subsonic outflow solutions to Jeans escape boundary conditions at the exobase. They showed that heavier species like C, O or N atoms can be incorporated in the hydrodynamic flow if the heating is strong enough. Adiabatic cooling associated with the hydrodynamic flow results in reduced exobase temperatures and thus controls the escape rates.

It is thought that the thermosphere of Venus was rich in water vapor at the time when a runaway greenhouse occurred (Fig. 1), theoretically allowing hydrodynamic escape to develop, although there is no clear evidence that such an episode of intense hydrodynamic escape ever occurred on terrestrial planets.

Kasting and Pollack (1983), following Watson et al. (1981), developed a coupled photochemical-dynamic model of hydrodynamic escape on Venus. In their model, the vertical thermal structure of the thermosphere up to the exosphere and its chemical composition were calculated self-consistently. The temperature at the cold trap, that is the bottom of the thermosphere, was assumed to be 170 K. The altitude of the cold trap, which controls the mixing ratio of water vapor in the thermosphere, is presently 90 km but was probably larger at primitive epochs, when the atmosphere was hotter. Accordingly, several cases with H_2O mass mixing ratios at the cold trap in the range from $\approx 10^{-3}$ up to ≈ 0.5 were studied.

Hydrodynamic expansion, starting at a level of about 200 km altitude, results in a flow where the bulk velocity increases with altitude (up to ≈ 1 km s^{-1} at ≈ 10 planetary radii), and the temperature moderately increases up to the distance ≈ 1 planetary radius (≈ 500 K), and decreases above this height due to adiabatic cooling. A hydrogen escape flux up to $\approx 3 \times 10^{11}$ cm^{-2} s^{-1} was found for a large H_2O mixing ratio and present solar EUV conditions. This value has to be multiplied by 10 or even higher values for relevant primitive solar EUV conditions. At this rate, the hydrogen of an Earth-type ocean could be removed in a few hundred million years.

As mentioned above, the Kasting and Pollack calculation assumed collisional flow up to infinity, although the exobase level was reached at an altitude of ≈ 1 planetary radius. Because the temperature of the flow at the exobase level is only a few hundred Kelvins, the transition from the collisional to the non-collisional regime is expected to inhibit expansion. But the expansion cannot be stopped entirely; otherwise, the atmosphere would again be collisional at all altitudes.

In order to study the possible effect of this transition, Chassefière (1996a) proposed a hybrid formulation, using both a dynamic model for the inner fluid region and a Jeans approach for the upper, collisionless region. The conservation equations were solved from the base of the expanding flow up to the exobase using a complete scheme of solar EUV energy deposition. An additional source of energy was introduced at the top of the dynamic model (exobase level), representing the collisional deposition of the kinetic energy of energetic neutral atoms (ENAs) created by charge exchange between escaping H atoms and solar wind protons. This energy diffuses inward, throughout the (subsonic) expanding flow, and heats the expanding medium in addition to solar EUV. The solar wind energy deposition controls the temperature gradient below the exobase, which is taken as a boundary condition of the model. The upward flux at the exobase was calculated using the classical Jeans theory and compared to the flux below the exobase, as provided by the dynamic model. Self-consistent solutions, for which the upward flux was continuous across the exobase,

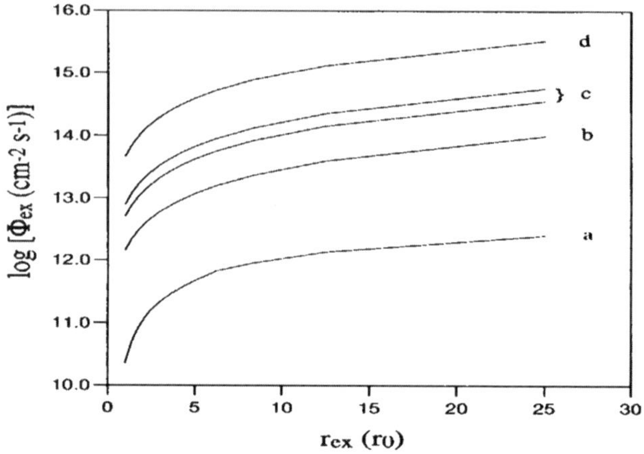

Fig. 2 Escape flux as a function of the planetocentric altitude of the exobase (in units of planetary radius r_0) for a solar wind enhancement factor with respect to present of: (**a**) 1, (**b**) 30, (**c**) 500–2000 and (**d**) 100,000 (from Chassefière (1997)

were exhibited. Calculations done for present solar EUV conditions are in agreement with the values found by Kasting and Pollack and showed that the additional contribution of energy from particle heating by solar wind-produced ENAs may be quite substantial. It was noted that for an exobase altitude of one planetary radius, any planetary magnetic field pushing away the obstacle up to an altitude larger than ≈ 3 planetary radii inhibits the solar wind energy source.

In a follow-up paper, Chassefière (1997) used a simplified approach to quantify the effect of an enhanced solar wind on the hydrodynamic escape flux from a hydrogen-rich upper Venus' atmosphere. Numerical simulations using the hybrid model showed that at high solar EUV flux the altitude of the exobase might reach ≈ 10 planetary radii, although numerical instabilities did not allow him to obtain firm, self-consistent solutions. The goal of the simplified approach presented in the 1997 paper was to calculate the Jeans escape flux as a function of the exobase altitude, assuming energy balance between incoming energetic neutrals and outgoing escaping atoms. The results are displayed in Fig. 2.

Assuming that the exobase was at 10 planetary radii altitude and that the solar wind density was larger by one order of magnitude at primitive epochs, an escape flux of 10^{13} cm^{-2} s^{-1} or more was derived, sufficient to remove all the hydrogen contained in an Earth-type ocean in less than ten million years. It was emphasized that the escape rate in this case might be limited by diffusion at the cold trap and be possibly below the energetically possible value.

This would necessarily have become true once the bulk of Venus' water had been lost and water vapor became a minor constituent of the lower atmosphere. Interestingly, energetic neutrals are formed at ≈ 20 planetary radii from the planet (assuming the exobase is at ≈ 10 radii altitude), and this mechanism would work even in the presence of a magnetosphere of the size of the terrestrial magnetosphere.

Although the EUV-powered hydrodynamic escape is of thermal nature, the interaction with the solar wind may result in an additional source of energy. The process described above is only one possible mechanism, although energetically representative of the maximum possible contribution of the solar wind, as all the kinetic energy carried by the solar wind beam intercepted by the exobase is assumed to be deposited. However, one should note that recent studies and observations of present Venus and Mars indicate that the main production region of these ENAs occurs at solar zenith angles > 30 degrees and, because the ENAs carry the energy and momentum of the solar wind protons, they essentially follow the streamlines of the flow past the planet (e.g., Kallio et al. 1997; Holmström et al. 2002;

Lichtenegger et al. 2002; Futaana et al. 2006; Galli et al. 2007). Therefore, only a smaller fraction of these ENAs may contribute to the heating of the upper thermosphere. Other interactions, like sputtering, where ions originate in the atmosphere itself (Luhmann and Kozyra 1991), are examined in a following section.

Because of the lack of observational constraints, it is difficult to assess the reliability of the existing approaches, but depending on the young Sun radiation and particle conditions it appears plausible that hydrodynamic escape was able to remove all the hydrogen contained in an Earth-sized ocean from the primitive Venus' atmosphere within a few tens to a hundred million years. A (still missing) precise measurement of the noble gas isotopic ratios in the Venus' atmosphere and a detailed comparative study in reference to the Earth case are necessary to better understand the evolution of the primitive atmospheres of the two planets and would provide a diagnostic tool for estimating the role of hydrodynamic escape.

4.3 Thermal Loss of Oxygen from an H_2O-Rich Early Venus

The absence of molecular oxygen at a substantial level in the atmosphere of Venus is still poorly understood. If all the hydrogen contained in the initial water of Venus has been removed by hydrodynamic escape, as previously described, what was the fate of the oxygen atoms contained in water molecules and released by photodissociation in the high atmosphere? If oxygen has remained in the atmosphere, this process would provide a way for a planet to form a massive abiotic oxygen atmosphere (Zahnle and Kasting 1986). This possibility, as pointed out by Kasting (1997), deserves to be seriously studied in order to interpret future observations of the chemical composition of extrasolar planets from space (DARWIN, TPF). Studying Venus offers an opportunity to understand what is the fate of oxygen on a planet that loses its water by early massive hydrogen escape.

A first possibility is oxidation of the crust. Assuming FeO represents 5% in mass of the mantle, it may be calculated that an extrusion rate of $\approx 20\ \text{km}^3\ \text{yr}^{-1}$, similar to the present terrestrial rate, averaged over 4.5 Ga is required to provide the chemical reservoir able to absorb the amount of oxygen contained in an Earth-type ocean (Lewis and Prinn 1984). Independent estimates of the present volcanic activity on Venus, based on geophysical, geological, and geochemical data, generally suggest maximum extrusion rates of approximately $0.4\ \text{km}^3\ \text{yr}^{-1}$ (Bullock and Grinspoon 1993).

Considering that extrusions are assumed to account for only 5–10% of the total crust production, the upper limit of the crustal growth rate including extrusions may be about 4 $\text{km}^3\ \text{yr}^{-1}$ (D. Breuer, personal communication, 2007), too small to account for the removal of the oxygen content of a full Earth-type ocean. Similar conclusions were reached by Lewis and Prinn (1984, p. 190). However, crustal overturn on Venus may be highly episodic (Turcotte 1993), and so the oxygen consumption rate averaged over time could be larger than estimated here.

Escape to space provides an alternative, and/or complementary, potential sink for oxygen (Zahnle and Kasting 1986; Chassefière 1996a, 1996b). We will examine in this section the hypothesis of thermal (hydrodynamic) escape, whereas possible non-thermal mechanisms are described later. Indeed, in the case of an intense hydrodynamic escape of atomic hydrogen, the theory predicts that heavy atoms can be dragged off along with escaping H atoms (Hunten et al. 1987). A heavy constituent "2", of mass m_2 and mixing ratio X_2, is dragged off along with a light escaping constituent "1" (H or H_2), of mass m_1 and mixing ratio X_1, according to the following law:

$$F_2 = \frac{X_2}{X_1} F_1 \left[\frac{(m_c - m_2)}{(m_c - m_1)} \right], \tag{1}$$

where F_i are the fluxes and

$$m_c = m_1 + \left(\frac{kTF_1}{bgX_1} \right), \tag{2}$$

is the "crossover mass" (b is the product of the density by the diffusion coefficient of "2" in "1"). If $m_2 < m_c$, "2" can escape with "1" (the flux F_2 is proportional to the difference $m_c - m_2$).

The possibility that oxygen atoms produced by H_2O photodissociation could be dragged off along with hydrogen atoms may be assessed by using Hunten's theory, with $m_1 = 1$ uma (H) and $m_2 = 16$ uma (O). The crossover mass m_c may be estimated for Venus, assuming present solar EUV conditions. Assuming escape is limited by energy (EUV only), with a typical efficiency factor of 0.25 (the fraction of incident EUV energy converted into escape energy), and taking into account the geometrical amplification of the intercepted EUV flux due to the enhanced altitude of the exobase, m_c is in the range from 1.4 uma to 7.2 uma for present EUV conditions (Chassefière 1996b), with a most likely value of 2.8 uma. Since m_c is (nearly) proportional to the amplitude of the EUV flux, and assuming that this flux varies with time t as $(t_0/t)^{5/6}$, where t_0 is the present time (4.6 Gyr), m_c falls below 16 at ≈ 600 Myr, with a large uncertainty (between 200 Myr and 1.8 Gyr). This means that, theoretically, oxygen could escape together with hydrogen during the first hundreds million years. But, if oxygen was massively dragged off with hydrogen (and therefore is not a minor species like in the theory of Hunten), the EUV energy required for removing a 2:1 stoichiometric mixture of H and O (2 H atoms for 1 O atom) is 9 times larger than for hydrogen alone (ratio of 18 for H_2O to 2 for H_2). Thus, if Venus' atmosphere lost most of its oxygen with the hydrogen, the "effective" crossover mass would have been $2.8 \times 9 = 25$ uma, pushing the end of the hydrodynamic escape phase of oxygen back to 40 Myr (between 30 Myr and 130 Myr). Through an analytical rigorous theory derived from Hunten's theory, Chassefière (1996b) has shown that no more than 30% of the oxygen content of a Venusian Earth-sized ocean might have been lost by EUV-driven hydrodynamic escape over the period from 100 Myr to 1 Gyr.

Finally, assuming that the solar wind was more intense at primitive epochs, and applying the simplified treatment previously described to estimate the energy deposited at the exobase by energetic neutrals formed through charge exchange between escaping atoms and solar wind protons (Chassefière 1997), it has been found that, if the solar wind was enhanced by three orders of magnitude at primitive stages, it is theoretically possible to remove most of the oxygen of an Earth-sized ocean in ten million years by hydrodynamic escape. However, early planetary intrinsic or induced magnetic fields could have reduced this heating process and the resulting loss rates. The fate of oxygen originating from water released by impacting bodies at a later stage could be high thermal and non-thermal loss rates and/or oxidation of the crust.

As a conclusion, in the case of a purely EUV-driven hydrodynamic escape, the removal of all (or most of) the oxygen contained in an Earth-sized ocean was possible only at very early times ($t < 30$–40 Myr). Such a removal could have occurred later ($t > 100$ Myr) only if there was a substantial additional source of energy such as the (possibly) enhanced primitive solar wind. An enhancement factor of $\approx 10^3$ with respect to the present value is theoretically able to remove the oxygen in ≈ 10 Myr. Another possible loss mechanism caused by solar wind interaction with an upper atmosphere is non-thermal escape, which is described in the following section. It may be concluded that an extended period of water delivery by impacting bodies, until ≈ 300 Myr (Weissman 1989) or even later, resulting in the progressive building of an ocean, would be difficult to reconcile with the hypothesis of massive hydrodynamic oxygen escape, except if a very strong solar wind (three orders of magnitude above the present value) survived for a few hundred million years after the formation of the Sun. On the other hand, if most of the water was delivered to Venus at the very beginning, during accretion, EUV-SW-powered hydrodynamic escape was potentially able to remove large amounts of water from a primitive atmosphere.

4.4 Non-Thermal Oxygen Loss During Venus History

The flow of the solar wind around non-magnetized planets like Venus and Mars has been studied extensively by using gas dynamic and convection magnetic field models (e.g., Spreiter et al. 1966; Spreiter and Stahara 1980), semi-analytical magnetohydrodynamic (MHD) flow models (e.g. Shinagawa et al. 1991; Biernat et al. 2001), and by hybrid models (e.g., Terada et al. 2002; Kallio et al. 2006). The solar X-ray and EUV radiation produces an ionized region in the upper atmosphere where large concentrations of ions and free electrons can exist. This region, where the solar wind generates a magnetic field and interacts with the ionospheric plasma of a non-magnetized planet, builds up an atmospheric obstacle, over which the stellar wind plasma is deflected. For the non-thermal loss processes, like ion pick-up from un-magnetized or weakly magnetized planets, the solar activity dependence of the ionopause altitude becomes a controlling factor. The atmosphere below the ionopause is protected against erosion by the solar wind, while neutral gas above can be ionized and picked up by it. As a result, the ion escape rate during a planet's history would have depended on the early solar X-ray, EUV, and solar particle flux conditions.

If early Venus had no intrinsic planetary magnetic field that was strong enough to shield the solar wind of the young Sun, the solar plasma flow should have been blocked like today by the ionospheric plasma pressure. This pressure balance occurs in the collision-free regime above the exobase level because the Interplanetary Magnetic Field (IMF) is enhanced above the ionosphere by the ionospheric induction current (e.g., Alfvén and Fälthammar 1963), by which the shocked solar wind is deflected.

Neutral atoms and molecules above the ionopause can be transformed to ions by charge exchange with solar or stellar-wind particles, EUV radiation or electron impact. These newly generated planetary ions are accelerated to higher altitudes and energies by the interplanetary electric field and are guided by the solar- or stellar wind plasma flow around the planetary obstacle to space, where they are lost from the planet (e.g., Spreiter and Stahara 1980; Lundin et al. 1989, 1990; 2007; Lichtenegger and Dubinin 1998; Biernat et al. 2001; Lammer et al. 2006b; Terada et al. 2002).

Another important effect of the ions pick up process is that a part of neutral atoms above the ionopause can be directed back to the upper atmosphere of the planet where they collide with the background gas so that the collision partners can be accelerated by sputtering to energies above the escape energy. As can be seen in Fig. 3, atmospheric sputtering refers to a mechanism by which incident energetic particles (mostly charged particles) interact with a planetary atmosphere or surface and produce the ejection of planetary material.

Sputtering has been recognized as an important source of atmospheric non-thermal loss in the case of Mars, but of less importance for larger planets like Venus (Luhmann and Kozyra 1991). For planets with the mass of Venus or Earth, sputtering accelerates atmospheric particles to high altitudes from where they can also be lost by ionization and stellar wind via the pick up process. On present Venus, sputtering yields O loss rates of the order of 5×10^{24} s^{-1} which is about 2 times lower than the ion pick up rate. However, it is difficult to say how efficient sputtering by an enhanced solar wind from an extended upper atmosphere compared with ion pick up is. As mentioned before, the extreme plasma interaction with early Venus might have induced a strong magnetic field which could have a reverse

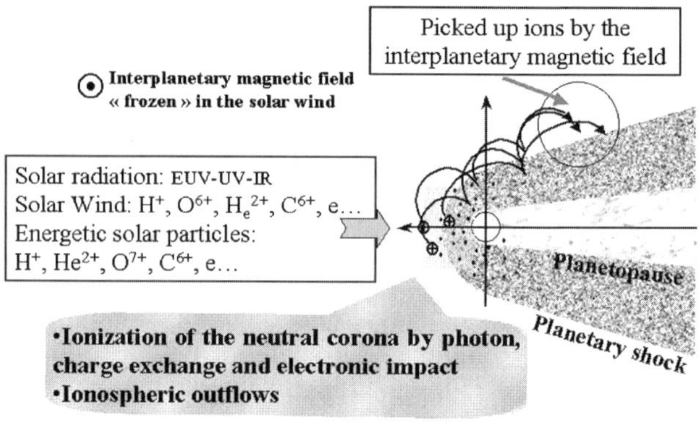

Fig. 3 Illustration of picked up planetary ions, directed backwards to a planetary atmosphere, which is not protected by a strong magnetic field. These ions can act together with solar wind particles as sputter agents (courtesy of F. Leblanc)

effect on the sputter loss between 4–4.5 Gyr ago. But to be sure how efficient sputtering is compared with other non-thermal loss processes, model calculations under extreme early Venus conditions have to be carried out in the future.

Barabash et al. (2007) find from the analysis of direct measurements by the Venus Express plasma instrument package that the dominant escaping ions from Venus are O^+, He^+, and H^+, which leave Venus through the plasma sheet, a central portion of the wake, and a boundary layer of the induced magnetosphere. They reported that the cool O^+ ion outflow triggered by the solar wind interaction through the plasma tail is of the order of $\leq 10^{26}$ s^{-1}.

In addition to ion pick up and cool ion escape, plasma clouds are observed above the ionopause, primarily near the terminator and further downstream. The detailed analysis of several detached plasma clouds has shown that the ions within the clouds themselves are ionosphere-like in electron temperature and density (Brace et al. 1982; Russell et al. 1982). In the magnetic barrier, plasma is accelerated by a strong magnetic tension directed perpendicular to the magnetic field lines. This magnetic tension forms specific types of plasma flow stream lines near the ionopause, which are orthogonal to the magnetic field lines. This process favors the appearance of Kelvin-Helmholtz and interchange instabilities that can detach ionospheric plasma in the form of detached ion clouds from a planet. One can model the Kelvin-Helmholtz instability at a planetary obstacle by applying the one-fluid, incompressible magnetohydrodynamic (MHD) equations.

For studying the ion loss due to the Kelvin Helmholtz instability, Terada et al. (2002) applied a global hybrid model to present Venus. They found that the dynamic ion removal process associated with this plasma instability plays a significant role additionally to other ion loss processes. Terada et al. (2002) obtained a loss rate for O^+ ions of the order of $\sim 5 \times 10^{25}$ s^{-1}. Table 3 summarize the present time escape rates from Venus. One can see that thermal escape of hydrogen is negligible at present Venus.

Kulikov et al. (2006) studied the expected O^+ ion pick up loss rates over Venus' history by using the X-ray and EUV satellite data discussed in Sect. 3.1, as well as a range of solar wind plasma densities and velocities expected for the young active Sun and discussed in Sect. 3. For modeling the Venusian thermosphere over the planetary history, Kulikov et al. (2006) used a diffusive-gravitational equilibrium and thermal balance model which was applied for a study of the heating of the early thermosphere by photodissociation and ionization processes, exothermic chemical reactions, and cooling by CO_2 IR emission in the 15 μm band. As can be seen in Fig. 4, their model simulations resulted in expanded thermospheres with exobase altitudes between about 200 km for present EUV flux values and about 1700 km for 100 times higher EUV fluxes after the Sun arrived at the Zero-Age-Main Sequence.

Table 3 Thermal and non-thermal loss rates of oxygen and hydrogen from present Venus

Escape process	Loss [s^{-1}] 1 EUV
Jeans: H	2.5×10^{16} (1)
Photochemical reactions: H	3.8×10^{25} (1)
Electric field force: H^+	$\leq 7 \times 10^{25}$ (2)
Solar wind ion pick up: H^+	1×10^{25} (1)
Solar wind ion pick up: H_2^+	$< 10^{25}$ (1)
Solar wind ion pick up: O^+	1.5×10^{25} (1)
Detached plasma clouds: O^+	5×10^{24}–10^{25} (1, 3)
Sputtering: O	6×10^{24} (4)
Cool plasma outflow: O^+	$\leq 10^{26}$ (5)

(1) Lammer et al. (2006a); (2) Hartle and Grebowsky (1993); (3) Terada et al. (2002); (4) Luhmann and Kozyra (1991); (5) Barabash et al. (2007)

Kulikov et al. found that exospheric temperatures during the active phase of the young Sun could have reached about 8000 K if the atmosphere had a similar composition as that observed on present Venus after the Sun arrived at the ZAMS (see Fig. 3). Kulikov et al. (2006) applied a numerical test particle model for the simulation of the O^+ pick up ion loss from non-magnetized Venus over its history and found a total loss of about 180–280 bar ($\sim$ 70–110% TO: Terrestrial Ocean) for the maximum solar wind estimated by Wood et al. (2002), about 40–60 bar ($\sim$15–25% TO) for the average solar wind, and about 10–15 bar (4–6% TO) for the minimum solar wind.

From our knowledge of Earth, Venus, Mars and Titan, Yamauchi and Wahlund (2007) point out that the ionopause builds up above the exobase no matter what the solar wind conditions are. In that case the lower range of ion pick up loss rates modeled by Kulikov et al. (2006), corresponding to the planetary obstacle boundaries located near the exobase, may be more realistic. They obtain O^+ pick up loss rates at 4 Gyr ago (15 EUV) of about 1.5–5 $\times$ 10^{27} s^{-1} for minimum and average early solar wind flux conditions as estimated by Wood et al. (2002). These O^+ pick up loss rates for a 100 EUV CO_2 atmosphere (4.5–4.6 Gyr ago) correspond to loss rates of about 0.35-1.5 $\times$ 10^{30} s^{-1} for minimum and average solar wind conditions expected for the young Sun.

Thus, if one considers uncertainties in observations of stellar mass loss from young active solar-like stars (Wood et al. 2005), early Venus may have lost during its history an amount of oxygen, via the ion pick up process, equivalent to an atmosphere loss of about 5–50 bar. One should also note that the ion pick up loss rates would be different if Venus' early atmosphere had a different composition than today. This was most likely the case during the evaporation of the Venusian water ocean, as discussed in Sect. 4.3. Furthermore, the expected shift in exobase altitude shown in Fig. 4 will affect the D/H fractionation estimates of Donahue et al. (1997) and, the homopause-exobase distance will increase enhancing isotope fractionation.

In a hydrogen-rich thermosphere the exobase moves too a much larger distance compared with that calculated for the CO_2-rich thermosphere by Kulikov et al. (2006). In such a case it may be possible that oxygen and heavier species may be protected by the dense hydrogen corona until the hydrogen inventory is lost by thermal and non-thermal escape processes.

Even though the cool ion outflow and Kelvin-Helmholtz instability induced plasma clouds are more efficient ion escape processes from present Venus compared with ion pick up, it is difficult to estimate their contribution to atmospheric loss over Venus' past. While conservative $\mathbf{O^+}$ pick up estimates indicate that the planet could have lost the oxygen from

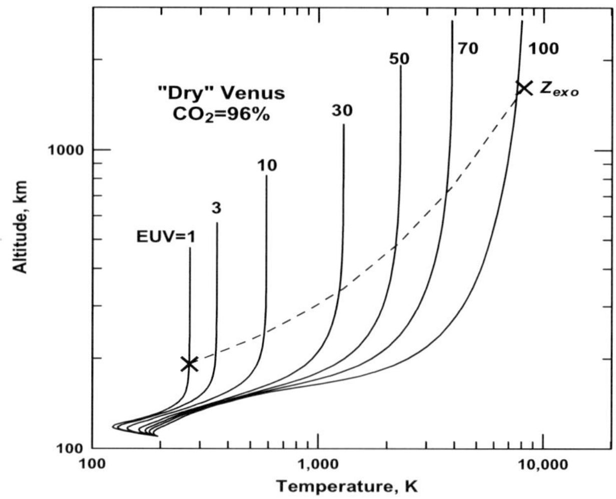

Fig. 4 Temperature profile of a "dry" CO_2-rich Venus atmosphere as a function of solar EUV flux. The *dashed line* corresponds to the exobase distance where the mean free path equals the scale height

an evaporated ocean equivalent to about 5–50 bar over Venus' history, it is possible that cool ion outflow and plasma clouds may enhance this loss up to a factor of 2–5. Hence, it is important to estimate contribution of these ion loss processes to the total loss over the solar cycle by analyzing spacecraft data (PVO, VEX, etc.), so that MHD and hybrid models could be adjusted for higher solar activity and atmospheric conditions expected during Venus' early history.

5 Early Evolution of Earth's Atmosphere

5.1 Formation of the Atmosphere

Earth's atmosphere is thought to have formed in much the same way as did Venus' atmosphere, by impact degassing of large, volatile-rich planetesimals. So, the first part of the discussion in the previous section applies here as well. The big difference, of course, is that Earth is farther from the Sun than is Venus; hence, once the main phase of accretion had stopped and the molten surface had solidified ($\sim$ 100 million years), liquid oceans should have definitely formed. This prediction has now been spectacularly confirmed by studies of oxygen isotopes in zirconium silicate minerals, or zircons, with ages as old as 4.4 Gyr (Valley et al. 2002). The $^{18}O/^{16}O$ ratio in these zircons, which is different from that in Earth's mantle, can only be explained if these minerals crystallized from magmas formed from high-^{18}O rocks that had interacted with liquid water at or near Earth's surface. The actual upper limit on surface temperature from these measurements is 200°C, which is still quite warm, but is well below the expected 1500°C temperature of a steam atmosphere (Zahnle et al. 1988).

What happened next is highly uncertain. It depends, in part, on how rapidly Earth formed relative to the lifetime of the solar nebula. If the nebula was entirely gone by the time Earth's formation was complete, then the early atmosphere may have been a weakly reduced mixture of CO_2 and N_2 (Rubey 1951; Walker 1977). If, however, the nebula was still present during the latter stages of accretion, as planetary scientists from the Japanese school have long argued (Hayashi et al. 1985), then Earth's earliest atmosphere may have been rich in

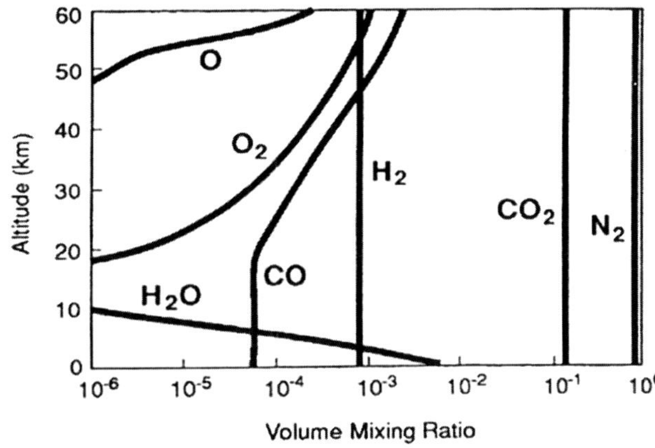

Fig. 5 An example of a typical, weakly reduced atmosphere, as simulated using a 1-D photochemical model. A surface pressure of 1 bar has been assumed. The CO_2 partial pressure, 0.2 bars, is approximately the amount needed to offset 30 percent reduced solar luminosity. The O_2 in the middle atmosphere is produced from CO_2 photolysis (from Kasting 1993)

H_2 and/or CH_4. Alternatively, an atmosphere rich in these highly reduced gases could have been produced by impacts, especially those that occurred during the earlier stages of accretion when elemental iron-rich impactors were still abundant (Schaefer and Fegley 2007; Hashimoto et al. 2007). Hence, the nature of Earth's earliest atmosphere should be viewed as an unresolved question.

Regardless of which planetary formation model is correct, the early atmosphere should have contained a substantial amount of H_2—enough to make the upper atmosphere hydrogen-rich. As can be seen in Fig. 5, even a weakly reduced lower atmosphere should have had an H_2 mixing ratio of the order of 10^{-3} (1000 ppmv) or greater (Kasting 1993; Holland 2002). This estimate is obtained by balancing the outgassing of reduced species from volcanoes with escape of hydrogen to space, assuming that the escape takes place at the diffusion-limited rate. If the escape rate was slower, as some researchers have suggested (Tian et al. 2005), then the atmospheric H_2 mixing ratio should have been even higher.

The concerns about the rapidity of hydrodynamic escape, expressed in earlier sections, could conceivably raise estimated H_2 concentrations still more. Much of the interest in this question results from its relevance to the origin of life (Chyba 2005). If the atmosphere was more reduced, then Miller-Urey type synthesis (from lightning) of prebiotic organic compounds is much more efficient (Miller and Schlesinger 1984). This is one motivation for the discussion of hydrogen escape that follows.

Once life had evolved, the composition of Earth's atmosphere would almost certainly have changed. One of the first things to happen may have been the conversion of much of the existing H_2 into CH_4 (Walker 1977; Kharecha et al. 2005). This reaction is carried out by methanogenic bacteria, or methanogens, which are thought to be amongst the earliest organism to have evolved (Woese and Fox 1977). Methanogens are anaerobic bacteria that are poisoned by free O_2 and that therefore live today in restricted habitats such as the intestines of cows and other ruminants and in the mud beneath rice paddies. On the early Earth, with its lack of atmospheric O_2, methanogens should have been ubiquitous.

Methanogens can produce methane by a number of different pathways, the most direct being the reaction

$$CO_2 + 4H_2 \rightarrow CH_4 + 2H_2O. \tag{3}$$

But they can also start from organic compounds, e.g. acetate (CH_3COOH), produced by the fermentation of more complex forms of organic matter. This process would have continued within the oceans and in sediments even after the origin of oxygenic photosynthesis sometime before 2.7 Gyr (Brocks et al. 1999). Indeed, methane is generated at depth within marine sediments today; however, nearly all of it is consumed by other, methanotrophic bacteria before it can make its way into the atmosphere. CO should also have been consumed by such an ecosystem, either by direct uptake by acetogens (Kharecha et al. 2005) or by the photochemically catalyzed water-gas reaction: $CO + H_2O \rightarrow CO_2 + H_2$.

A weakly reduced atmosphere is believed to have persisted until about 2.4 Ga, at which time it was replaced by one rich in O_2, like today's atmosphere (Holland 1994; Farquhar et al. 2000). So, hydrogen escape to space was probably extremely important for at least the first half of Earth's history. Indeed, the escape of hydrogen to space may have played a critical role in causing the rise of O_2 (Kasting et al. 1993; Catling et al. 2001; Claire et al. 2006). Because most of the hydrogen arrived initially in the form of H_2O, its escape left large amounts of oxygen behind. In the Kasting et al. (1993) model, this O_2 was mostly taken up by Earth's mantle, where it could conceivably have caused a change in mantle redox state. Although mantle redox change now appears unlikely, based on various petrologic indicators (Li and Lee 2004), the mantle may indeed have absorbed much of this O_2. Some of it, though, appears to have been taken up by oxidation of rocks on the continents, and this may have helped set up the O_2 rise at 2.4 Ga (Catling et al. 2001; Clair et al. Claire et al.).

Surprisingly, hydrogen may have continued to escape rapidly even following the rise of atmospheric O_2. Pavlov et al. (2003) have suggested that CH_4 concentrations may have remained relatively high, 50–100 ppmv, during the early- to mid-Proterozoic Eon, 2.5–0.8 Ga. Their argument assumes that atmospheric O_2 concentrations remained somewhat lower than today and that the deep oceans remained largely anoxic, as others have suggested previously (Canfield 1998). The recent modeling study by Goldblatt et al. (2006) supports this hypothesis. In their model, CH_4 decreased dramatically just prior to the rise of O_2, but then it increased again soon afterwards.

Indeed, high Proterozoic CH_4 levels and rapid hydrogen escape may have been required in order to balance Earth's redox budget at that time. According to this argument, hydrogen was escaping rapidly prior to the rise of O_2; hence, it must have continued to escape rapidly following the rise of O_2; otherwise, an equivalent amount of reducing power would have had to be lost as organic matter in sediments. But the relative constancy of the carbon isotope record, averaged over long time periods, indicates that no such change in organic carbon burial took place (Goldblatt et al. 2006). This last argument is speculative, but it suggests that hydrogen escape could have played a fundamental role in Earth's atmospheric evolution throughout a large fraction of the planet's history.

5.2 Thermal and Non-thermal Escape from Present and Early Earth's Atmosphere

The main problem for modeling atmospheric escape from early Earth is that there are many unknown parameters on which it depends. Besides the uncertainties in the solar wind conditions, atmospheric composition, internal, and surface heating and outgassing sources, such as volcanic activity, we do not know if early Earth was magnetized or non-magnetized at the time when life emerged. There is no magnetic record in the Earth's crust before 3.5 Gyr ago (e.g., Hale and Dunlop 1984; Sumita et al. 2001; Yoshihara and Hamano 2004; Ozima et al. 2005). A palaeointensity measurement on the Komati formation which has an age of about 3.5 Gyr may imply that the Earth's dynamo might not be very strong before the solid-state inner core was formed (Hale and Dunlop 1984). On the other hand, new paleomagnetic data (Tarduno et al. 2007) suggest that the Earth's magnetic field at about 3.2 Ga could be as strong as that of today, implying that the differentiation of the Earth's inner core began no later than 3.2 Gyr.

Depending on atmospheric composition and the exobase temperature, the observed non-thermal loss rate from the Earth-mass and size planets is much faster than Jeans escape, except for light species like H, H_2 and He (e.g., Lundin and Dubinin 1992; Cully et al. 2003; Wahlund et al. 2005 and references therein; Yamauchi and Wahlund 2007). The observed non-thermal loss rate of hydrogen from the Earth's upper atmosphere/ionosphere from non-thermal ion heating processes is of the order of about 1–10 kg s^{-1} (6×10^{26}–$\sim 10^{27}$ s^{-1}) (e.g., Moore et al. 1999; Cully et al. 2003; Yamauchi and Wahlund 2007). One should note that these ion loss rates can even be higher than the diffusion limited escape rate of neutral hydrogen. The amount of up-welling ions is connected to the solar wind pressure and activity. When for instance a magnetic cloud or a CME collide, it squeezes Earth's magnetic field, squirting particles stored in the magnetotail up the field lines towards the poles. Jeans escape of neutral H atoms is estimated to be larger at solar maximum but smaller than the non-thermal escape rate of protons during solar minimum. The upper limit of the loss rate of H atoms, which is diffusion limited, is about 10^{27} s^{-1} (Vidal-Madjar 1978; Kasting and Catling 2003, and references therein).

For present Earth the main escaping ion is O^+ which originates in the ionosphere, and the O^+ loss rate is larger than the H^+ loss rate, even during the solar maximum. The escape rate related to non-thermal ion heating strongly depends on the magnetospheric activity, with the largest source located in the dayside polar region (e.g., Kondo et al. 1990; Norqvist et al. 1998; Yamauchi and Wahlund 2007), where the solar wind can directly penetrate to the ionosphere through the magnetosphere. What is important for early Earth is that the escape rate of heavy ions like O^+ and N^+ increases to higher values compared with that for H^+ during high solar activity periods and major magnetic storms (Chappell et al. 1982; Cully et al. 2003). For instance, the non-thermal O^+ loss rate from the ionosphere increases by a factor of 100, while the non-thermal H^+ loss rate increases only by a factor of 2–3 when the solar $F_{10.7}$ flux increases by a factor of about 3 (Cully et al. 2003; Yamauchi and Wahlund 2007).

In a recent study Tian et al. (2008) investigated the response of the Earth's atmosphere to extreme solar EUV conditions and found that the upper atmosphere of an Earth-mass planet with the present Earth's atmospheric composition would start to rapidly expand if thermospheric temperatures exceeded 7000–8000 K.

In such a case the thermosphere is cooled adiabatically due to the outflow of the dominant species (O, N, etc.). From Fig. 6 it is seen that exobase moves upward as a consequence of the outflow. It can in fact exceed the present subsolar average magnetopause stand-off distance of about 10 Earth-radii. Kulikov et al. (2006) showed that even a "dry Venus" with the present 96% CO_2 could have reached temperature values around 8000 K during the first 100 Myr after the Sun arrived at the ZAMS. Of course, the very early Venus' atmosphere had a very different composition which would result in a different thermal structure than that modeled by Kulikov et al. (2006).

Depending on the solar EUV flux and planetary and atmospheric parameters, one can see from Fig. 6 that the exosphere could expand beyond the magnetopause. Therefore, the constituents beyond the magnetopause could be ionized and picked up by the solar wind plasma. Furthermore, other ion loss processes similar than at Venus and discussed in Sect. 4.4 would have contributed to the loss of the early water inventory. The expanded thermosphere-exosphere region, therefore, will result in high non-thermal atmospheric loss rates (Lundin et al. 2007).

It is also seen in Fig. 6 is that high amounts of CO_2, like on present Venus and Mars, can cool the thermosphere much better than Earth-like nitrogen/oxygen atmospheres, so that the exobase level remains much closer to the planetary surface. In such a case the atmosphere would be protected against erosion by the solar wind. Therefore, one can expect,

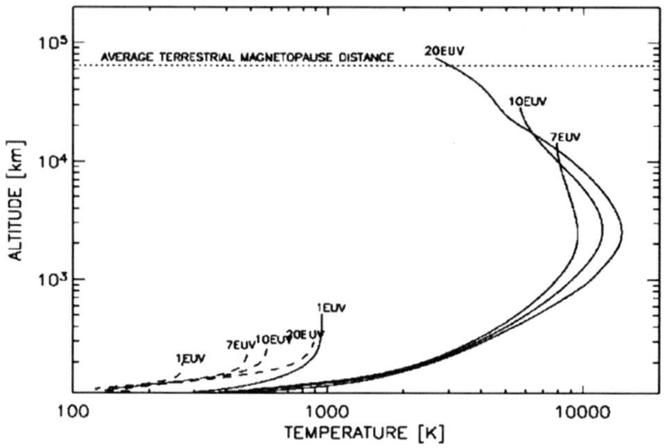

Fig. 6 Thermospheric temperature profiles between 100 km and the corresponding exobase levels for present (=1), 7, 10 and 20 times higher EUV solar fluxes than today, applied to Venus (Kulikov et al. 2006) and Earth (Tian et al. 2008) with the present time atmospheric composition. The efficient IR-cooling due to large amount (96%) of CO_2 in the hydrostatic thermosphere of Venus yields much lower exobase temperatures and atmospheric expansion compared with an Earth-like atmospheric composition

in agreement with Kulikov et al. (2007) that the atmosphere of the early Earth may have had during its first 500 Myr a higher amount of CO_2 in its thermosphere, which resulted in a less expanded upper atmosphere and exobase levels below the magnetopause. Otherwise early Earth's atmosphere would have been hot and unstable. By contrast an early CO_2-poor Earth's atmosphere may have experienced high nonthermal loss rates. In case that the early Earth's upper atmosphere was hydrogen-rich, as suggested by Tian et al. (2005), most of the expanded hydrogen exosphere would be ionized and lost from the planet by nonthermal loss processes like ion pick up, even if the thermal loss rate was lower due to a cooler exosphere as suggested by these authors. To investigate if early Earth could have kept its atmosphere, ion-loss test particle and MHD models have to be applied to extended atmospheres.

6 Evolution of Mars' Atmosphere

6.1 Early Mars' Climate: Was There a Dense CO_2 Atmosphere?

Mars, as one of the terrestrial planets, probably formed in much the same way as did Venus and Earth. So, volatiles should have been delivered to its surface by impact degassing of planetesimals originating from the asteroid belt or beyond. Mars, however, is different from Earth and Venus in one important respect: its mass is just slightly over $1/10^{\text{th}}$ of Earth's mass. Mars' small mass has likely had a huge impact on its initial retention of volatiles and on its subsequent evolution.

Consider the retention issue first. As discussed earlier, impact degassing of incoming planetesimals is widely accepted as a source of planetary volatiles. However, impact erosion has also been widely discussed as a loss mechanism for volatiles (see, e.g., Walker 1986; Melosh and Vickery 1989). It should be noted that there is no generally accepted theory that describes how this process works, and so the two references given differ widely in their predictions. The efficiency of impact erosion is, not surprisingly, highly dependent on the mass of the growing planet. Large planets are better able to hold onto their atmospheres because their escape velocities are higher relative to the expected impact velocities of incoming planetesimals. In a pioneering study, Melosh and Vickery (1989) concluded that if Mars had simply been given a 1-bar CO_2 atmosphere initially at 4.5 Ga, it could have lost nearly all of it by 3.8 Gyr as a consequence of impacts that occurred during the heavy bombardment period of Solar System history. This process could conceivably explain why Mars has such a thin atmosphere ($\sim$ 6 mbar surface pressure) today.

This hypothesis raises several issues that require further discussion. First, how could Mars first accumulate an atmosphere and then lose it by essentially the same process, i.e., impacts? A possible answer is that the presumed impact velocities of the incident planetesimals were different at different times in Mars' history. During the early phases of accretion, planetesimals were small, and they should also have been on nearly circular orbits because collisions with other small bodies were relatively frequent. Hence, the relative velocity between the planetesimals and the growing protoplanet should have been smaller. By contrast, the bodies that arrived several hundred million years later are assumed to have been perturbed (by Jupiter) from initial orbits in the asteroid belt. They would have had higher eccentricities and would thus have hit Mars at higher relative velocities. Hence, the planetesimals that arrived early added to Mars' atmosphere, while those that arrived later may have removed it. That said, it seems unlikely that Mars could have lost its entire initial atmosphere in this way, as the impact of even one large, slow-moving body during the latter stages of accretion would have left an appreciable amount of volatiles behind. Such an explanation has been offered to account for the thick atmosphere on Saturn's moon, Titan (Griffith and Zahnle 1995).

The heavy bombardment period is itself a matter of contention. The idea that the inner solar system was subjected to an intense bombardment by late-arriving planetesimals grew out of the analysis of Moon rocks brought back by the Apollo missions between 1969 and 1973 (see, e.g., Hartmann 1973; Neukum and Wise 1976). These rocks had radiometric age dates that clustered near 3.8–3.9 Gyr. Although some researchers interpreted this as a "pulse" of impacts at about this time (Ryder 2003, and references therein), others suggested that the impacts that formed these rocks represented the tail end of an extended period of heavy bombardment. The latter view has prevailed until just recently. However, a new dynamical model for Solar System formation (Tsiganis et al. 2005; Gomes et al. 2005) suggests that the "pulse" hypothesis may indeed have been correct. In this model—which has been termed the "Nice model" because several of its authors are from the vicinity of the city of Nice in southern France—Jupiter and Saturn began their lives closer to each other than they are now. Jupiter migrated inward and Saturn migrated outward as a result of interactions with planetesimals in the disk. After some elapsed time ($\sim$ 700 million years if one chooses parameters properly), they crossed the 2:1 mean motion resonance, where Saturn's orbital period was exactly twice that of Jupiter.

At this point, all hell broke loose from a dynamical standpoint. Uranus and Neptune, which were formed close to Saturn in this model, were thrown into the outer Solar System where they perturbed the remaining population of planetesimals. These icy planetesimals from the outer Solar System were then responsible for causing a great pulse of bombardment on both the Moon and the terrestrial planets. Because they would have arrived with high relative velocities, these impacts would almost certainly have caused extensive atmospheric erosion.

Returning now to the question of Mars' early atmospheric evolution, we can see that from a theoretical standpoint it is highly uncertain. The Nice model is just that—a model—and it may or may not be correct. Hence, we cannot be sure at this time whether Mars (or Earth) was subjected to an extended heavy bombardment, and we should therefore have little confidence in our ability to predict how its atmosphere should have formed and evolved.

What we do have for Mars is lots of observations of its surface, both from spacecraft that have orbited the planet and from landers and rovers that have sampled the surface directly. The heavily cratered southern highlands of Mars are covered by fluvial features, such as the ones seen in Fig. 7. So, a flowing liquid—almost certainly water—was present on Mars' surface at some time prior to 3.8 Gyr ago. By contrast, the less heavily cratered northern plains are essentially devoid of such features, suggesting that the planet dried up and became much colder soon after this time. This last conclusion is reinforced by geochemical data from instruments such as TES (the Thermal Emission Spectrometer) that flew aboard the Mars Global Surveyor spacecraft. Such studies have revealed the widespread presence of minerals such as olivine that react readily with liquid water (Hoefen et al. 2003). So, Mars' surface has evidently been dry throughout most of its history.

Adding further to our confusion about Mars' early history is the fact that we do not understand how the fluvial features were formed. Some researchers (e.g., Segura et al. 2002) have suggested that they could have been created in the aftermath of large impacts, even if the early Martian climate was quite cold. Others (Pollack et al. 1987) have argued for a warm, almost Earth-like, early Mars. But the warm early Mars theory has problems because climate models (Kasting 1991) suggest that it is difficult to bring Mars' average global surface temperature above freezing using the greenhouse effect of a dense CO_2 atmosphere. At high CO_2 partial pressures, the increase in albedo caused by Rayleigh scattering outweighs the increased greenhouse effect from infrared absorption. CO_2 ice clouds may have helped to warm the surface (Forget and Pierrehumbert 1997), but this mechanism only works well for nearly 100 percent cloud cover. Furthermore, despite intensive spectroscopic searches from a series of orbiting spacecraft, no outcrops of carbonate rocks have ever been found [although carbonate minerals have been identified in Martian dust (Bandfield et al. 2003)].

If CO_2 was abundant, and if liquid water was present, why didn't they form? One suggestion is that the surface was too acidic, and that the CO_2 was lost from the upper atmosphere (Fairen et al. 2004). If so, it is obviously important to understand how this processes work. So, our theories about how Mars' atmosphere has evolved are strongly shaped by our knowledge of atmospheric escape processes.

6.2 Loss of Water and Other Volatiles from Early Mars

The evolution of the martian atmosphere and the evidence of the existence of an early hydrosphere are of great interest for studies regarding the evolution of the planet's water inventory and the search for life by current and future Mars missions. As shown in Fig. 7 the history of the martian atmosphere can be divided into early and late evolutionary periods (e.g., Carr 1987; Zahnle et al. 1990; Carr 1996; Pepin 1994; Hutchins and Jakosky 1996; Chassefière and Leblanc 2004; Donahue 2004; Chassefière and Leblanc 2004). Although the martian climate is at present too cold and the atmosphere too thin to allow liquid water to be stable on the surface, there are many indications that the situation was different during the Noachian epoch.

Besides geological evidence of outflow channels, river beds, possible shorelines (e.g., Head III et al. 1999; Clifford and Parker 2001) and evidence of standing bodies of water, an observed large deuterium (D) enrichment in the atmospheric water vapor (e.g., Zahnle et al. 1990; Owen et al. 1988) indicates that significant amount of water has been lost from the surface by atmospheric escape processes over the planet's history.

After the young Sun arrived at the ZAMS, heavy noble gases, including nonradiogenic Xe isotopes, may have been hydrodynamically fractionated during the accretion phase of the planet, with corresponding depletions and fractionations of lighter primordial atmospheric

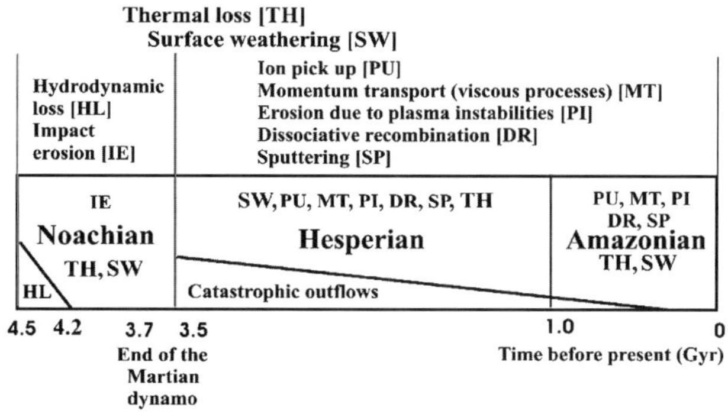

Fig. 7 Schematic illustration of various atmospheric escape processes and their expected relevance during the martian history from the early Noachian to the Hesperian and Amazonian epochs

species like deuterium (D) or H atoms (Hunten et al. 1987; Zahnle et al. 1990; Pepin 1994; Donahue 2004). Subsequently the CO_2 pressure history and the isotopic evolution of atmospheric species during this early period were determined by the interplay between impact erosion (Melosh and Vickery 1989; Chyba et al. 1990; Brain and Jakosky 1998) and impact delivery, carbonate precipitation and oxidation, by outgassing and carbonate recycling, and perhaps also by feedback stabilization under greenhouse conditions (Carr 1987, 1996; Pepin 1994). This period was also influenced by thermal and non-thermal atmospheric loss processes (e.g. Zahnle et al. 1990; Donahue 2004; Kulikov et al. 2007, and references therein). This in turn depended partly on the time of the onset of the martian magnetic dynamo, the field strength and the decrease-time of the magnetic moment, and the radiation and particle environment of the young Sun.

Carr and Head (2003) estimated the potential early martian water reservoirs from geomorphological analysis of possible shorelines of the post-Noachian epoch with the help of Mars Global Surveyor (MGS) images and altimeter data. They suggested that an amount of water equivalent to a global martian ocean with the depth of about 150–200 m could explain the observed geological surface features. However, early Mars could have had more water than this because erosional processes may have obscured and erased the geological signatures of hydrological activity during the Noachian epoch.

The second period of martian atmospheric evolution, from the Hesperian to the present Amazonian epoch, is characterized by uniform atmospheric loss enhanced by the vanished intrinsic magnetic field and various non-thermal atmospheric escape processes that have resulted in the present surface pressure of about 7–10 mbar (e.g., Jakosky et al. 1994; Lammer et al. 2003a, 2003b, and references therein).

Table 4 summarizes the most reasonable results of atmospheric escape rate models for three level of solar EUV flux: 1 EUV (present moderate martian solar activity), 2 EUV and 6 EUV (roughly corresponding to the flux about 3.5 Gyr ago (Zahnle and Walker 1982; Ribas et al. 2005) at the beginning of the Hesperian epoch). More results can be found in the literature, but many escape rates were revised after more accurate atmospheric data and plasma data of the martian environment became available. The question marks in Table 4 correspond to species and escape processes for which no escape rates have been modeled.

Carlsson et al. (2006) and Barabash et al. (2007) estimated the present loss rates for molecular O_2^+ and CO_2^+ ions from the analysis of the Mars Express (MEX) Ion Mass Analyzer (IMA) sensor of the ASPERA-3 instrument. Loss rates for moderate solar activity for O_2^+ and CO_2^+ and O^+ related ion loss rates are about 1.8×10^{24}–3.6×10^{24} (O_2^+) and 8.0×10^{23}–2.0×10^{24} (CO_2^+), respectively. Recently Ma and Nagy (2007) reproduced the observed O^+, O_2^+ and CO_2^+ ion escape rates for low solar activity Mars Express mission conditions with a 3D multi-species non-ideal magnetohydrodynamic model. Recent hybrid model results by Chaufray et al. (2007) yield similar ion loss rates.

The ASPERA instrument on board the Phobos 2 spacecraft observed strong interaction between the solar wind plasma and the cold ionospheric plasma in the Martian topside ionosphere. The solar plasma appears to transfer momentum directly to the Martian ionosphere from the dayside transition region to the deep plasma tail (Lundin et al. 1989, 1990). This is in agreement with reported the detection of cold electrons above the Martian ionopause, indicating the presence of detached plasma clouds (Acuña et al. 1998; Cloutier et al. 1999).

Pérez-de-Tejada (1992), Lundin and Dubinin (1992), Pérez-de-Tejada (1998), and Lammer et al. (2003b) found that this momentum transport process is capable of accelerating ionospheric O^+ to velocities > 5 km s^{-1} resulting in energies larger than the martian escape energy. Analytic models (Pérez-de-Tejada 1992; Lammer et al. 2003b) give estimates which are in rough agreement with the observations. As shown in Table 4, cool ion escape from the martian plasma tail can yield O^+ loss rates for moderate solar activity of about 10^{25} s^{-1}.

Assuming the oxygen which was lost from Mars during the Amazonian and Hesperian period originated from H_2O these authors estimated that Mars may have lost the equivalent of a global ocean with a depth of ≤ 15 m over 3.5 Gyr. This is smaller than the $\sim$30–80 m reported in earlier studies (Luhmann et al. 1992; Jakosky et al. 1994; Kass and Yung 1995, 1996, 1999; Krasnopolsky and Feldman 2001), but larger than the estimates of 3 to 5 m obtained by Yung et al. (1988) and Lammer et al. (1996). The models of Leblanc and Johnson (2002), Lammer et al. (2003a, 2003b) and Penz et al. (2004) used atmospheric input parameters for higher the EUV flux obtained form Zhang et al. (1993).

Finally, the results in Table 4 should only be considered rough estimates until accurate thermosphere-ionosphere-hot particle-exosphere models related to the evolution of the solar EUV flux are obtained based on MHD and hybrid simulations.

While there are agreements between different model results and ion escape observations, the dissociative recombination O atoms loss rates for 1 EUV (Luhmann 1997) shown in Table 4 may be larger. A recent study of the martian coronae and related escape by a complex 3 D Monte Carlo model give escape rates of $\sim 10^{25}$ s^{-1} and 4×10^{25} s^{-1} for low and high solar activity conditions respectively (Chaufray et al. 2007). However, we show in Table 4 the values of the Luhmann (1997) model because this author applied the model also to higher EUV values. We note that dissociative recombination related escape of atomic O is important for present Mars, but it is suggested to be less important during earlier periods (Johnson and Luhmann 1998; Lillis et al. 2006).

Lammer et al. (2006a) and Kulikov et al. (2007) applied a thermospheric model to the CO_2 atmosphere of Mars for high EUV radiation levels (10, 50, and 100 times the average present solar value). They found that the average dayside exobase temperature grows on Mars in a 95% CO_2 atmosphere by approximately a factor of 3 from about 355 K to about 1230 K for the EUV flux increasing from 10 to 100 times that of the present Sun. As shown by Zahnle et al. (1990) a H_2-rich early martian atmosphere may have developed hydrodynamic conditions.

It appears that the early evaporation of the martian CO_2 atmosphere by thermal loss processes was very unlikely, and if early Mars had a strong magnetic dynamo, it is unlikely that the planet lost several bars of CO_2, C, nitrogen and oxygen due to non-thermal loss processes (Kulikov et al. 2007). If early Mars lost its main atmosphere and water inventory during the first hundred Myr after the planet's origin, the model results would be in agreement with the observations by the OMEGA instrument on board of Mars Express which found no definite evidence that CO_2 sustained a long-term greenhouse effect enabling liquid water to remain stable for geological time periods on the surface of Mars in

Table 4 Modeled thermal and non-thermal loss rates of atomic and molecular hydrogen, oxygen, nitrogen and carbon species (neutrals and ions) from Mars at present time moderate solar activity conditions (1 EUV), at 2 EUV periods and for 6 EUV (∼ 3.5 Gyr ago)

Species	EUV	Jeans	Photochem.	Sputtering	Pick up	Plasma clouds	Cool ion outflow
H	1	1.5×10^{26} [1]	?	?			
H_2	1	3.3×10^{24} [2]	?	?			
H^+	1				1.2×10^{25} [3]	?	?
H_2^+	1				$\sim 10^{25}$ [3]	?	?
	1		2.8×10^{24} [4]	3.5×10^{23} [5(3)]			
O	2		$3.0{\times}^{25}$ [4]	1.3×10^{23} [5(3)]			
	6		8.0×10^{25} [4]	1.5×10^{27} [5(3)]			
	1				$3.0{\times}^{24}$ [3]	$1.0{\times}^{24}$ [6]	$\sim 10^{25}$ [7]
O^+	2				4.0×10^{25} [3]	$8.0{\times}^{24}$ [7]	$5.0{\times}^{2}$ [7]
	6				8.3×10^{25} [3]	2.0×10^{26} [6]	3.0×10^{27} [7]
N	1		4.5×10^{23} [8]	?	?	?	?
O_2^+	1				1.8×10^{24}– 3.6×10^{24} [9]		
C	1		3.0×10^{24} [10]	?	?	?	?
	1		8.0×10^{23} [11]	3.7×10^{22} [5(3)]	?	?	?
CO	2			$2.0{\times}^{24}$ [5(3)]	?	?	?
	6			2.5×10^{23} [5(3)]	?	?	?
	1			$5.0{\times}^{22}$ [5(3)]			
CO_2	2			2.3×10^{24} [5(3)]			
	6			4.0×10^{25} [5(3)]			
CO_2^+	1				8.0×10^{23}– 2.0×10^{24} [9]		

[1] Anderson and Hord (1971), [2] Krasnopolsky and Feldman (2001), [3] Lammer et al. (2003a), [4] Luhmann (1997) for 1 EUV, 2 EUV, 6 EUV also in agreement with Kim et al. (1998) for 1 EUV, [5] Leblanc and Johnson (2002), [6] Penz et al. (2004), [7] Lammer et al. (2003b), [8] Fox and Dalgarno (1983), [9] molecular ion outflow is estimated (Carlsson et al. 2006), [10] Nagy et al. (2001), [11] Fox and Bakalian (2001)

the post-Noachian terrains (Bibring et al. 2005). Bibring et al. (2005) concluded that the OMEGA observations are consistent with early strong escape of the most of the martian CO_2 atmosphere.

The simulated loss rates discussed in this section are highly model dependent and have to be compared with future observational data and measurements by some martian aeronomy and environmental orbiter. However, the missing data which may help us to understand the evolution of the early martian magnetic dynamo, the atmospheric surface pressure, atmospheric sputtering and photochemical loss processes, etc. over the planet's history can only be procured by using a comprehensive package of instruments during a high solar activity period, such as proposed for the low altitude Mars Magnetic and Environmental Orbiter (MEMO) (Leblanc et al. 2007).

7 Evolution of Titan's Atmosphere

7.1 Origin of Titan's Atmosphere and the Relevance of the $^{15}N/^{14}N$ Isotope Fractionation to Its Evolution

The origin of Titan's atmosphere which contains mainly N_2 and CH_4 was not well understood before the arrival and observations of Cassini/Huygens although thermodynamic models of the solar nebula predicted that C and N_2 were mainly available in the form of CO and N_2. Two possible sources of volatiles have been suggested: comets that condensed outside the Saturnian nebula (e.g. Prinn and Fegley 1989), and b) planetesimals that condensed within a Saturnian subnebula (Griffith and Zahnle 1995). Carbon within cometary matter is mainly concentrated in the form of heavy organics like CO and CO_2, with a small fraction of CH_4. But CO is much less abundant than Titan's CH_4 (e.g., Gautier and Raulin 1997).

One can overcome this problem if Titan was generated in Saturn's subnebula which was warmer than the surrounding solar nebula so that the temperature-pressure conditions favored the conversion of CO to CH_4 as well as the conversion of N_2 into NH_3, respectively. Based on this scenario Lunine and Stevenson (1987) suggested that CH_4 and NH_3 were trapped in the planetesimals which formed Titan as hydrate and clathrate hydrates from where they were outgassed as NH_3 and CH_4 (Atreya et al. 1978; McKay et al. 1988).

Mousis et al. (2002) investigated this hypothesis in more depth and modeled for the first time the formation of clathrate hydrates of CH_4 and of hydrates of NH_3 in an evolutionary solar nebula and found that Titan formed from planetesimals that were relics of those embedded in the feeding zone of Saturn and contained NH_3 hydrate and CH_4 clathrate hydrates. They also found that for plausible abundances of CH_4 and NH_3 in the solar nebula at 10 AU the masses of CH_4 and NH_3 trapped in Titan could even be higher than the estimate of these compounds in Titan's primitive atmosphere.

Data obtained by the Cassini/Huygens spacecraft contributed to the understanding of Titan's atmosphere evolution. Measurements with the Gas Chromatograph Mass Spectrometer (GCMS) aboard the Huygens probe confirmed the low abundance of CO. The abundance of noble gasses like Ar was also found to be very low and Kr and Xe were even below the detection threshold (Niemann et al. 2005). The detected low noble gas abundances are not in agreement with the thermo dynamical calculations which predict solar abundances or even over-solar in Titan (Prinn and Fegley 1989; Mousis et al. 2002).

In a more recent study Alibert and Mousis (2007) calculated Saturn's subnebula consistent with the end phase of Saturn's formation by avoiding the limitations in Mousis et al. (2002) such as "equilibrium of Saturn's subnebula during its cooling phase" and neglecting the fact that Saturn accreted gas and gas coupled material during a substantial fraction of the subnebula lifetime (Lubow et al. 1999; Magni and Coradini 2004). Alibert and Mousis (2007)

Two scenarios were studied, one where Titan is formed in the late cold subnebula from preserved planetesimals produced in Saturn's feeding zone and Titan is formed in an early subnebula. They found that in the first scenario the CO/CH_4 molar mixing ratio would be orders of magnitude larger than that observed in Titan's atmosphere, but the second scenario predicted abundances similar to the observed ones. However, in addition to these scenarios, volatiles delivered by comets could have, modified the initial atmospheric inventory (Griffith and Zahnle 1995).

Recent in situ measurements by the Cassini Ion Neutral Mass Spectrometer (INMS) at 1250 km altitude found an enrichment of ^{15}N that is only about 1.27–1.58 the terrestrial value (Waite et al. 2005). Furthermore, the Huygens probe measured during its decent with the Gas Chromatograph and Mass Spectrometer (GCMS) a similar enrichment of ^{15}N compared to ^{14}N of about 1.47 (Niemann et al. 2005). These $^{15}N/^{14}N$ isotopic ratio observations are an indication that Titan experienced considerable nitrogen escape. Waite et al. (2005) compared the INMS measurements with the model results of Lunine et al. (1999), by assuming that the initial nitrogen ratio was similar to the present terrestrial value and that the temperature between the exosphere and the homopause remained unchanged over the course of atmospheric evolution. By considering these assumptions they found that Titan may have lost 1.7 ± 0.05 to 10 ± 5 times its present atmosphere. The large uncertainty in their estimate is due to the unknown efficiency for dissociative fractionation of the isotopes. Further, Waite et al. (2005) mention that these values correspond to the upper-end of the INMS-measured range. If they use the lower end of the INMS-measured range, the range of atmospheric loss over Titan's history becomes 2.8 ± 0.2 to 100 ± 75.

If one considers the present solar activity and nitrogen loss rates caused by sputtering in the order of about 10^{25}–10^{26} s^{-1} (e.g. Shematovich et al. 2003; Michael et al. 2005) or loss of CH_5^+, $C_2H_5^+$, H_2CN^+, $C_xH_y{}^+$ ions due to ionospheric outflow of about 5×10^{24}–10^{25} s^{-1} (Hartle et al. 1982; Lammer and Bauer 1991; Keller et al. 1994; Keller and Cravens 1994; Keller et al. 1998; Nagy et al. 2001; Sillanpää et al. 2006; Ma et al. 2007) its difficult to understand how Titan could have lost several times the present atmosphere mass (see also Johnson et al. 2008). Even if CH_4 escapes from present Titan in the order of about 4–5×10^{10} amu cm^{-2} s^{-1} (Yelle et al. 2008; Johnson et al. 2008) one can not explain the ^{15}N enrichment.

In a recent study Penz et al. (2005) used astrophysical observations on radiative fluxes and stellar winds of solar-like stars with different ages and lunar and meteorite fossil records (Newkirk 1980). These data indicate that the early Sun underwent indeed a highly active phase resulting in up to about 100 times higher X-ray and EUV radiation fluxes (Zahnle and Walker 1982; Ribas et al. 2005) and much higher solar wind mass fluxes (Wood et al. 2002) 100–500 Myr after it arrived to the Zero-Age-Main-Sequence. The results of Penz et al. (2005) indicate, in agreement with Johnson (2004), that atmospheric sputtering even with a strong early solar wind cannot be responsible for the observed enrichment in ^{15}N isotopes in Titan's atmosphere. The estimated non-thermal nitrogen loss rates during the young Sun epoch after Titan's origin are 100–1000 times higher ($\leq 10^{28}$ s^{-1}) than that of today but the time period was too short to have lost several bar of atmosphere (Penz et al. 2005).

But they suggest that Titan's early atmosphere may have been in a state of nitrogen blow-off due to EUV enhanced heating and exobase expansion of the upper atmosphere. These authors suggested that, because of Titan's low gravity and an expanded exobase level the dynamically driven nitrogen flow could overcome the escape velocity at the exobase level, so that more than 30 times of its present atmospheric mass may have escaped (Penz et al. 2005). Such an expected rise in exobase altitude would result in a larger homopause-exobase distance Δz and, hence, in a strong effect of mass-driven diffusive separation (Lunine et al. 1999; Lammer et al. 2000), where the diffusive separation factor

$$f = \mathrm{EXP}\left(\frac{\Delta z}{H_d}\right) - 1, \tag{4}$$

with

$$H_d = \frac{kT(r)}{(m_2 - m_1)\, g(r)}, \tag{5}$$

where H_d is the diffusive scale height, k the Boltzmann's constant, m_2 and m_1 is the mass of the heavier ^{15}N and lighter ^{14}N isotope, respectively. T and g are the temperature and gravitational acceleration halfway between the homopause and the exobase levels.

By assuming that nitrogen was the main species, as it is today, and the mass fractionation during escape is the Rayleigh process, the original atmospheric mass relative to the present one can be written as (Lunine et al. 1999)

$$\frac{n_1^0}{n_1} = \left(\frac{n_2}{n_1} \Big/ \frac{n_2^0}{n_1^0}\right)^{\frac{(1+f)}{f}}. \tag{6}$$

The ratio n_2/n_1 is the measured isotope fractionation and n_2^0/n_1^0 is the initial value prior to atmospheric enrichment and can be assumed to be the terrestrial value.

Figure 8 shows the initial nitrogen reservoir of Titan needed to reproduce the measured average ^{15}N isotope enrichment of about 1.47 (Waite et al. 2005; Niemann et al. 2005) as a function of exobase levels above the surface and different temperatures in (5) and resulting different diffusive scale heights. The homopause position in Fig. 8 corresponds to the observed altitude of 1195 km (Waite et al. 2005). Because, of enhanced thermosphere heating by the young Sun, and concomitant exobase expansion the temperature between the homopause and exobase might rise rather than remain close to 150 K as assumed by Lunine et al. (1999) and Waite et al. (2005). As a result, the diffusive scale height in (5) would be larger, resulting in a decrease of the diffusive separation factor f in (4).

As one can see from Fig. 8, it is hard to constrain the amount of atmospheric loss over Titan's history. The uncertainties are largely due to our imprecise knowledge of the position of the homopause and exobase levels as well as due to the unknown temperature value between the homopause and exobase levels. Correspondingly the measured nitrogen isotope anomaly is an indication that Titan's atmosphere was at least several times denser than today.

If one considers reasonable temperatures of ~150–500 K between the homopause and exobase one can see from Fig. 8 that for exobase levels at altitudes $\geq$ 3000 km above Titan's surface the satellite may have lost 2–10 times of its present atmospheric mass. Whereas the nitrogen isotope measurements suggest considerable atmospheric loss, the carbon isotope ratios, remarkably, do not. Prior to the Cassini observations it had been suggested that photo-absorption by methane and its photoproducts played an important role in heating the atmosphere. However, if the supply of methane to the atmosphere is episodic, then, the due to the depleted hydrocarbons, the nitrogen atmosphere might cool and could become thin or collapse prior to the next outgassing event (Lorenz et al. 1997; Lunine et al. 1998).

This would clearly affect the estimates of nitrogen loss over time. The carbon isotope ratios from the Cassini measurement confirm that there must be a subsurface source

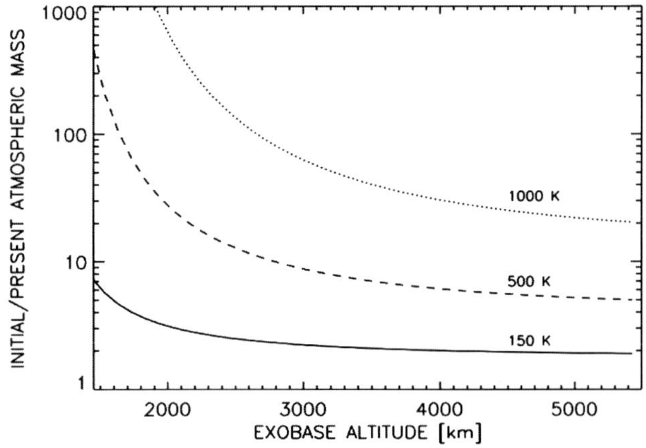

Fig. 8 Titan's initial nitrogen reservoir, normalized to the present nitrogen atmospheric mass as a function of exobase altitude from Titan's surface and average temperature between the homopause and exobase levels

of methane. Cryovolcanic outgassing of methane stored as clathrate hydrates within an icy shell above an ammonia-enriched water ocean has been proposed (Tobie et al. 2006; Atreya et al. 2006). Whether such a source is steady or episodic is not clear. Therefore, in future atmospheric evolution studies, the effect of cryovolcanism on the atmosphere structure needs to be considered.

In addition self consistent hydrodynamic models of the thermosphere are needed which examine adiabatic cooling due to dynamic expansion caused by a rise in thermospheric temperature as well as cooling as a function of the change in mixing ratios of minor atmospheric species like HCN. Such studies are important for finding out, to which altitude the exobase level could expand due to EUV heating by the young Sun and if Titan's exosphere could reach hydrodynamic blow off conditions, and, if so, over which time periods such conditions may have been active. An explanation of the nitrogen isotope anomaly is important for enabling us to estimate the nitrogen reservoir required to produce the present Titan atmosphere. It is also of importance for understanding the formation, evolution, and escape of atmospheres around other satellites like Callisto, Ganymede, Europa, Triton and small planetary bodies like Pluto because their early atmosphere environments should have also experienced an enhanced EUV flux. Below we consider one aspect of this, the role of the incident plasma in driving escape.

7.2 Contribution of Atmospheric Sputtering to Titan's Isotope Fractionation

Estimates of the magnetospheric ion and the pick-up ion flux onto Titan's exobase were made using a hybrid calculation based on the ambient ion fluxes from Voyager (see Bretch et al.; Ledvina Chapter). These fluxes were used in a number of Monte Carlo simulations of Titan's exobase region in order to describe the plasma heating (Michael and Johnson 2005) and sputtering of Titan's atmosphere (Shematovich et al. 2003; Michael et al. 2005). Such simulations showed that, using present atmospheric sputtering rates, the fraction of Titan's atmosphere that would be lost over its lifetime is only about 0.5% of the present atmospheric molecular nitrogen inventory. If the exobase region was populated by NH_3 instead of N_2 over a significant fraction of its history, then the net loss would be, very roughly, about twice that, which is still too small to affect the isotope ratios.

Lammer, Bauer, and co-workers (Lammer et al. 2000; Lammer and Bauer 2003) obtained similar results, but also considered the fact that an early more robust solar wind would have compressed Saturn's magnetosphere and, possibly, sputtered the atmosphere more efficiently. Early estimates of the net loss assuming a T-tauri phase suggested that such a process might explain the isotope ratios. That was subsequently re-examined (Penz et al. 2005), as described above.

With the large number of passes of Cassini through Titan's exobase region, we are now in a position to re-examine this process in more detail. That is, rather than use model fluxes, the corona structure and escape rates can be linked to actual plasma fluxes. For instance, Cassini INMS measurements show that the structure of Titan's corona above the nominal exobase differs from that produced thermally (De La Haye et al. 2007a) and this structure and exobase temperature appear to vary spatially and/or with local time. The non-thermal component, however, cannot be re-produced by detailed models of the photon and electron induced chemistry in Titan's exobase region (De La Haye et al. 2007b). Therefore, it is suggested that the observation might be explained by atmospheric sputtering. Since the energetic particle flux onto Titan's exobase is not much different from that assumed in earlier simulations (Ledvina et al. 2004), it is suggested to be due to an enhanced flux of low-energy pick-up ions or "hot" out-flowing ionospheric particles associated with fields which penetrate below the exobase (De La Haye et al. 2007a). In addition, estimates made using INMS data suggest that the loss rates for hydrogen and methane may be larger than earlier estimates (Yelle et al. 2008; Strobel 2007). Therefore, present Titan's loss rates are not easy to explain, although they are not likely to be large enough to account for the observed isotope ratios.

7.3 Relevance of Sputter-Loss from Titan to Loss from Other Satellite Atmospheres

Although it has a very thick atmosphere, Titan is similar in size to the other large moon's of the giant planets that do not have thick atmosphere's. For example, Triton is sufficiently far from the Sun, so that much of its atmosphere could be frozen out on the surface. This is not the case for the large Jovian moons, suggesting that they possibly lost their dense gravitationally bound atmospheres by some atmospheric erosion process. Whereas Io's relatively thin atmosphere is produced by present volcanism, there is no evidence for volatiles associated with nitrogen or carbon. In addition, Europa, Ganymede, and Callisto have thin atmospheres which appear to be formed by sublimation and radiation-induced decomposition of water ice containing some trapped volatiles and, possibly, trace minerals (Johnson et al. 2004; McGrath et al. 2004).

Scaled by the parent planet radius, Callisto is farther from Jupiter, in Jupiter radii, than Titan is from Saturn, in Saturn radii, but Titan has retained a large atmosphere and Callisto has not. This has been attributed to differences in solar driven escape rates and impact erosion rates (Griffith and Zahnle 1995). However, we also note that all three icy Galilean satellites orbit much deeper in Jupiter's magnetosphere than Titan does in Saturn's magnetosphere. That is, they reside a considerable distance from the magnetopause, in a region of much higher field strength. At present, they also experience plasma pressures that are, going from Callisto to Io, 10 to 10^4 times that experienced by Titan when it is in Saturn's magnetosphere. Although the calculation of accurate atmospheric loss rates requires detailed consideration of the molecular physics, this pressure is a measure of the ability to remove an atmosphere and to retain the ions formed, allowing plasma to build up. Therefore, estimates of present atmospheric sputtering rates were used to show that Io and Europa would have rapidly lost a Titan-like atmosphere, whereas Ganymede and Callisto would have lost $\sim$30% and 3% respectively of a Titan-like atmosphere at present plasma bombardment rates. Assuming a more dense plasma torus when Io and Europa were being stripped, atmospheric sputtering alone might be able to account for the lack of a primordial atmosphere on the Jovian satellites, although Callisto with its more copious CO_2 inventory may be an interesting intermediate case.

7.4 Relevance of Loss from Titan to Loss from Other Satellite Atmospheres

Although it has a very thick atmosphere, Titan is similar in size to the other large moon's of the giant planets that do not have thick atmosphere's. For example, Triton is sufficiently far from the Sun, so that much of its atmosphere could be frozen out on the surface. This is not the case for the large Jovian moons, suggesting that they possibly lost their dense gravitationally bound atmospheres by some atmospheric erosion process. Whereas Io's relatively thin atmosphere is produced by present volcanism, there is no evidence for volatiles associated with nitrogen or carbon. In addition, Europa, Ganymede, and Callisto have thin atmospheres which appear to be formed by sublimation and radiation-induced decomposition of water ice containing some trapped volatiles and, possibly, trace minerals (Johnson et al. 2004; McGrath et al. 2004).

Scaled by the parent planet radius, Callisto is farther from Jupiter, in Jupiter radii, than Titan is from Saturn, in Saturn radii, but Titan has retained a large atmosphere and Callisto has not. However, all three icy Galilean satellites orbit much deeper in Jupiter's magnetosphere than Titan does in Saturn's magnetosphere. That is, they reside a considerable distance from the magnetopause, in a region of much higher field strength. At present, they also experience plasma pressures that are, going from Callisto to Io, 10 to 10^4 times that experienced by Titan when it is in Saturn's magnetosphere. Although the calculation of accurate atmospheric loss rates requires detailed consideration of the molecular physics, this pressure is a measure of the ability to remove an atmosphere and to retain the ions formed, allowing plasma to build up. Therefore, estimates of present atmospheric sputtering rates were used to show that Io and Europa would have rapidly lost a Titan-like atmosphere, even at present atmospheric rates, whereas Ganymede and Callisto would have lost $\sim 30\%$ and 3% respectively of a Titan-like atmosphere. Assuming a more dense plasma torus when Io and Europa were being stripped, atmospheric sputtering alone might be able to account for the lack of a primordial atmosphere on the Jovian satellites, although Callisto with its more copious CO_2 inventory may be an interesting intermediate case.

8 Conclusion

The origin and evolution of the atmospheres of the terrestrial planets in the solar system and Saturn's large satellite Titan were discussed. Due to the extreme radiation (X-ray, soft X-ray and EUV) and plasma (solar wind mass flux) environment of the young Sun we expect that the atmospheres and planetary water inventories were strongly affected by thermal and various nonthermal escape processes mainly during the first Gyr after the Sun arrived at the Zero-Age-Main-Sequence. Due to the heating of the much higher solar EUV flux the thermosphere and exobase levels extended to higher altitudes than at present time, which resulted in larger solar wind—atmosphere interaction areas and higher nonthermal loss rates. The extended exobase levels and resulting larger homopause-exobase distances were also responsible for the enrichment of heavy isotopes in the present atmospheres. Under certain activity conditions of the young Sun hydrostatic equilibrium could not kept resulting in large thermal escape rates.

Acknowledgements Helmut Lammer, James F. Kasting, Eric Chassefière and Yuri N. Kulikov thank the Helmholtz-Gemeinschaft as this research has been supported by the Helmholtz Association through the research alliance "Planetary Evolution and Life". Yu. Kulikov and H. Lammer acknowledge also support by the Austrian Academy of Sciences, "Verwaltungsstelle für Auslandsbeziehungen", by the Russian Academy of Sciences (RAS), for supporting working visits to the PGI/RAS in Murmansk, Russian Federation. H. Lammer and Yu. N. Kulikov also acknowledge the International Space Science Institute (ISSI; Bern, Switzerland) and the ISSI team "Evolution of Exoplanet Atmospheres and their Characterization". R.E Johnson acknowledges the support of NASA's Planetary Atmospheres Program.

References

M.H. Acuña, J.E.P. Connerney, P. Wasilewski et al., Science **279**, 1676–1680 (1998)
H. Alfvén, C.G. Fälthammar, *Cosmical Electrodynamics. Fundamental Principles* (Clarendon, Oxford, 1963)
Y. Alibert, O. Mousis, Astron. Astrophys. **465**, 1051–1060 (2007)
D.E. Anderson Jr., C.W. Hord, J. Geophys. Res. **76**, 6666–6673 (1971)
S.K. Atreya, T.M. Donahue, W.R. Kuhn, Science **201**, 611–613 (1978)
S.K. Atreya, E. Sushil, Y. Adams et al., Planet. Space Sci. **54**, 1177–1187 (2006)
T.R. Ayres, B. Alexander, R.A. Osten et al., Astrophys. J. **549**, 554–577 (2000)
J.L. Bandfield, T.D. Christensen, R. Philip, Science **301**, 1084–1087 (2003)
S. Barabash, A. Fedorov, J.A. Sauvaud et al., Nature (2007). doi:10.1038/nature06434
R.H. Becker, R.N. Clayton, E.M. Galimov, H. Lammer, B. Marty, R.O. Pepin, R. Wieler, Space Sci. Rev. **106**, 377–410 (2003)
J.-P. Bertaux, F. Montmessin, J. Geophys. Res. **106**, 32,879–32,884 (2001)
J.P. Bibring, Y. Langevin, A. Gendrin, the OMEGA team, Science **307**, 1576–1581 (2005)
H.K. Biernat, N.V. Erkaev, C.J. Farrugia, Adv. Space Res. **28**, 833–839 (2001)
L.H. Brace, R.F. Theis, W.R. Hoegy, Planet. Space Sci. **30**, 29–37 (1982)
D.A. Brain, B.M. Jakosky, J. Geophys. Res. **103**, 22,689–22,694 (1998)
J.J. Brocks, G.A. Logan, R. Buick, E.R. Summons, Science **285**, 1033–1036 (1999)
M.A. Bullock, D.H. Grinspoon, Geophys. Res. Lett. **20**, 2147–2150 (1993)
K. Caldeira, J.F. Kasting, Nature **359**, 226–228 (1992)
A.G.W. Cameron, Icarus **56**, 195–201 (1983)
D.E. Canfield, Nature **396**, 450–453 (1998)
E. Carlsson, A. Fedorov, S. Barabash et al., Icarus **182**, 320–328 (2006)
M.H. Carr, Nature **326**, 30–34 (1987)
M.H. Carr, *Water on Mars* (Oxford Univ. Press, New York, 1996)
M.H. Carr, J.W. Head III, J. Geophys. Res. **108**, 5042 (2003). doi:10.1029/2002JE001963
D.C. Catling, K.J. Zahnle, C.P. McKay, Science **293**, 839–843 (2001)
J.W. Chamberlain, Astophys. J. **133**, 675–687 (1961)
J.W. Chamberlain, D.M. Hunten, *Theory of Planetary Atmospheres* (Academc Press, Arizona, 1987)
C.R. Chappell, R.C. Olsen, J.L. Green, J.F.E. Johnson, J.H. Waite Jr., Geophys. Res. Lett. **9**, 937–940 (1982)
E. Chassefière, J. Geophys. Res. **101**, 26039–26056 (1996a)
E. Chassefière, Icarus **124**, 537–552 (1996b)
E. Chassefière, Icarus **126**, 229–232 (1997)
E. Chassefière, F. Leblanc, Planet. Space Sci. **52**, 1039–1058 (2004)
J.Y. Chaufray, R. Modolo, F. Leblanc, G. Chanteur, R.E. Johnson, J. Geophys. Res. **112** (2007). doi:10.1029/2007JE002915
C.F. Chyba, Science **308**, 962–963 (2005)
C.F. Chyba, P.J. Thomas, L. Brookshaw, C. Sagan, Science **249**, 366–373 (1990)
M.W. Claire, D.C. Catling, K.J. Zahnle, Geobiology **4**, 239–269 (2006)
R.N. Clayton, Space Sci. Rev. **106**, 19–33 (2003)
S.M. Clifford, T.J. Parker, Icarus **154**, 40–79 (2001)
P. Cloutier, C.C. Law, D.H. Crider et al., Geophys. Res. Lett. **26**, 2685–2688 (1999)
C.M. Cully, E.F. Donovan, A.W. Yau, G.G. Arkos, J. Geophys. Res. **108**, 1093 (2003). doi:10.1029/2001JA009200
M.O. Dayhoff, R.V. Eck, F.R. Lippincott, C. Sagan, Science **155**, 556–558 (1967)
V. De La Haye, J.H. Waite Jr., R.E. Johnson et al., J. Geophys. Res. **112**, A07309 (2007a). doi:10.1029/2006JA012222
V. De La Haye, J.W. Waite Jr., T.E. Cravens et al., Icarus **191**, 236–250 (2007b)

T.M. Donahue, Icarus **167**, 225–227 (2004)
T.M. Donahue, J.B. Pollack, in *Venus*, ed. by D.M. Hunten, L. Colin, T.M. Donahue, V.I. Moroz (University of Arizona Press, Tucson, 1983), p. 1003
T. Donahue, J.H. Hoffman, A.J. Watson, Science **216**, 630–633 (1982)
T.M. Donahue, D.H. Grinspoon, R.E. Hartle et al., in *Venus II*, ed. by S.W. Bougher, D.M. Hunten, R.J. Phillips (The University of Arizona Press, Tucson, 1997), pp. 385–414
J.D. Dorren, E.F. Guinan, in *The Sun as a Variable Star*, ed. by J.M. Pap, C. Frölich, H.S. Hudson, S.K. Solanki (Cambridge University Press, Cambridge, 1994), p. 206
A.G. Fairen, D. Fernandez-Remolar, J.M. Dohm, V.R. Baker, R. Amils, Nature **431**, 423–426 (2004)
J. Farquhar, J. Savarino, T.L. Jackson, M.H. Thiemens, Nature **404**, 50–52 (2000)
F. Forget, R.T. Pierrehumbert, Science **278**, 1273–1276 (1997)
J.L. Fox, F.M. Bakalian, J. Geophys. Res. **106**, 28,785–28,795, 2001)
J.L. Fox, A. Dalgarno, J. Geophys. Res. **88**, 9027–9032 (1983)
J.L. Fox, A. Hać, J. Geophys. Res. **102**, 24,005–24,011 (1997)
Y. Futaana, S. Barabash, A. Grigoriev et al., Icarus **182**, 424–430 (2006)
A. Galli, P. Wurz, H. Lammer et al., Space Sci. Rev. **126**, 447–467 (2007)
D. Gautier, F. Raulin, in *Hygens: Science, Payload and Mission*, vol. SP-1177 (ESA, Nordwijk, 1997), pp. 359–364
C. Goldblatt, T.M. Lenton, A.J. Watson, Nature **443**, 683–686 (2006)
R. Gomes, H.F. Levison, K. Tsiganis, A. Morbidelli, Nature **435**, 466–469 (2005)
D.O. Gough, Sol. Phys. **74**, 21–34 (1981)
M.M. Grady, I.P. Wright, Space Sci. Rev. **106**, 211–131 (2003)
C.A. Griffith, K. Zahnle, J. Geophys. Rev. **100**, 16,907–16,922 (1995)
D.H. Grinspoon, J.S. Lewis, Icarus **74**, 21–35 (1988)
E.F. Guinan, I. Ribas, in *The Evolving Sun and its Influence on Planetary Environments*, vol. 269, ed. by B. Montesinos, A. Giménez, E.F. Guinan (ASP, San Francisco, 2002), p. 85
C.J. Hale, D. Dunlop, Geophys. Res. Lett. **11**, 97–100 (1984)
Y. Hamano, M. Ozima, in *Terrestrial Rare Gases*, ed. by E.C. Alexander Jr., L. Ozima (Japan Scientific Societies Press, Tokyo, 1978), p. 155–177
G.L. Hashimoto, Y. Abe, S. Sugita, J. Geophys. Res. **112**, E05010 (2007). doi:10.1029/2006JE002844
R.E. Hartle, J.M. Grebowsky, J. Geophys. Res. **98**, 7437–7445 (1993)
R.E. Hartle, E.C. Sittler, K.W. Oglivie et al., J. Geophys. Res. **87**, 1383–1394 (1982)
W.K. Hartmann, J. Geophys. Res. **78**, 4096–4116 (1973)
C. Hayashi, K. Nakazawa, Y. Nakagawa, in *Protostars and Planets II*, ed. by D.C. Black, M.S. Mathews (University of Arizona Press, Tucson, 1985), p. 1100
J.W. Head III, H. Hiesinger, M.A. Ivanov et al., Science **286**, 2134–2137 (1999)
T.M. Hoefen, R.N. Clark, J.L. Bandfield et al., Science **302**, 627–630 (2003)
H.D. Holland, in *Early Life on Earth*, ed. by S. Bengtsson (Columbia Univ. Press, New York, 1994), p. 237
H.D. Holland, Geochim. Cosmochim. Acta **66**, 3811–3826 (2002)
M. Holmström, S. Barabash, E. Kallio, J. Geophys. Res. **107**, SSH 4-1 (2002). CiteID 1277, doi:10.1029/2001JA000325
D.M. Hunten, R.O. Pepin, J.C.G. Walker, Icarus **69**, 532–549 (1987)
K.S. Hutchins, B.M. Jakosky, J. Geophys. Res. **101**, 14,933–14,950, (1996)
A.P. Ingersoll, J. Atmos. Sci. **26**, 1191–1198 (1969)
B.M. Jakosky, R.O. Pepin, R.E. Johnson, J.L. Fox, Icarus **111**, 271–288 (1994)
R.E. Johnson, *Energetic Charged Particle Interactions with Atmospheres and Surfaces* (Springer, Heidelberg, 1990)
R.E. Johnson, Astrophys. J. **609**, L99–L102 (2004)
R.E. Johnson, J.G. Luhmann, J. Geophys. Res. **103**, 3649–3653 (1998)
R.E. Johnson, R.W. Carlson, J.F. Cooper et al., in *Jupiter-The Planet, Satellites and Magnetosphere*, ed. by F. Bagenal, T. Dowling, W.B. McKinnon (Cambridge University, Cambridge, 2004), p. 485
R.E. Johnson, M.R. Combi, J.L. Fox et al., Space Sci. Rev. (2008, this issue)
R. Kallenbach, T. Encrenaz, J. Geiss, K. Mauersberger, T. Owen, F. Roberts, *Solar System History from Isotope Signatures of Volatile Elements* (Kluwer, Dordrecht, 2003)
E. Kallio, J.G. Luhmann, S. Barabash, J. Geophys. Res. **102**, 22,183–22,198 (1997)
E. Kallio, R. Jarvinen, P. Janhunen, Planet. Space Sci. **54**, 1472–1481 (2006)
D.M. Kass, Y.L. Yung, Science **268**, 697–699 (1995)
D.M. Kass, Y.L. Yung, Science **274**, 1932–1933 (1996)
D.M. Kass, Y.L. Yung, Geophys. Res. Lett. **26**, 3653–3656 (1999)
J.F. Kasting, Icarus **74**, 472–494 (1988)
J.F. Kasting, Icarus **94**, 1–13 (1991)

J.F. Kasting, Science **259**, 920–926 (1993)
J.F. Kasting, Orig. Life **27**, 291–307 (1997)
J.F. Kasting, D. Catling, Ann. Rev. Astron. Astrophys. **41**, 429–463 (2003)
J.F. Kasting, J.B. Pollack, Icarus **53**, 479–508 (1983)
J.F. Kasting, D.H. Eggler, S.R. Raeburn, J. Geol. **101**, 245–257 (1993)
C.N. Keller, T.E. Cravens, J. Geophys. Res. **99**, 6527–6536 (1994)
C.N. Keller, T.E. Cravens, L. Gan, J. Geophys. Res. **99**, 6511–6525 (1994)
C.N. Keller, V.G. Anicich, T.E. Cravens, Planet. Space Sci. **46**, 1157–1174 (1998)
R. Keppens, K.B. MacGregor, P. Charbonneau, Astron. Astrophys. **294**, 469–487 (1995)
P. Kharecha, J.F. Kasting, J.L. Siefertet, Geobiology **3**, 53–76 (2005)
J. Kim, A.F. Nagy, J.L. Fox, T.J. Cravens, Geophys. Res. **103**, 29,339–29,342 (1998)
T. Kondo, B.A. Whalen, A.W. Yau, W.K. Peterson, J. Geophys. Res. **95**, 12,091–12,102 (1990)
V.A. Krasnopolsky, P.D. Feldman, Science **294**, 1914–1917 (2001)
V.A. Krasnopolsky, G.L. Bjoraker, M.J. Mumma, D.E. Jennings, J. Geophys. Res. **102**, 6524–6534 (1997)
Y.N. Kulikov, H. Lammer, H.I.M. Lichtenegger et al., Planet. Space Sci. **54**, 1425–1444 (2006)
Y.N. Kulikov, H. Lammer, H.I.M. Lichtenegger et al., Space Sci. Rev. **129**, 207–244 (2007)
H. Lammer, S.J. Bauer, J. Geophys. Res. **96**, 1819–1825 (1991)
H. Lammer, S.J. Bauer, Planet. Space Sci. **41**, 657–663 (1993)
H. Lammer, S.J. Bauer, Space Sci. Rev. **106**, 281–292 (2003)
H. Lammer, W. Stumptner, S.J. Bauer, Geophys. Res. Lett. **23**, 3353–3356 (1996)
H. Lammer, W. Stumptner, G.J. Molina-Cuberos, S.J. Bauer, T. Owen, Planet. Space Sci. **48**, 529–543 (2000)
H. Lammer, C. Kolb, T. Penz et al., Int. J. Astrobiol. **2**, 1–8 (2003a)
H. Lammer, H.I.M. Lichtenegger, C. Kolb et al., Icarus **106**, 9–25 (2003b)
H. Lammer, Y.N. Kulikov, H.I.M. Lichtenegger, Space Sci. Rev. **122**, 189–196 (2006a)
H. Lammer, H.I.M. Lichtenegger, H.K. Biernat et al., Planet. Space Sci. **54**, 1445–1456 (2006b)
H. Lammer, H.I.M. Lichtenegger, Yu.N. Kulikov et al., Astrobiology **7**, 185–207 (2007)
M.A. Lange, T.J. Ahrens, Icarus **51**, 96–120 (1982)
F. Leblanc, R.E. Johnson, J. Geophys. Res. **107** (2002). doi:10.1029/2000JE001473
F. Leblanc, B. Langlais, T. Fouchet, Astrobiology (2007, submitted)
S. Ledvina, J.G. Luhmann, S.H. Brecht, T.E. Cravens, Adv. Space Res. **33**, 2092–2102 (2004)
J.S. Lewis, Earth Planet. Sci. Lett. **10**, 73–80 (1970)
J.S. Lewis, Science **186**, 440–443 (1974)
J.S. Lewis, R.G. Prinn, *Planets and Their Atmospheres: Origin and Evolution* (Academic Press, Orlando, 1984)
Z.X.A. Li, C.T.A. Lee, Earth Planet. Sci. Lett. **228**, 483–493 (2004)
H.I.M. Lichtenegger, E.M. Dubinin, Earth Planets Space **50**, 445–452 (1998)
H.I.M. Lichtenegger, H. Lammer, W. Stumptner, J. Geophys. Res. **107**, SSH 6-1 (2002). CiteID 1279. doi:10.1029/2001JA000322
R.J. Lillis, M. Manga, D.L. Mitchell, R.P. Lin, M.H. Acuña, Geophys. Res. Lett. **33** (2006). doi:10.1029/2005GL024905
R.D. Lorenz, C.P. McKay, J.I. Lunine, Science **275**, 642–644 (1997)
S.H. Lubow, M. Seibert, P. Artymowicz, Astrophys. J. **526**, 1001–1012 (1999)
J.G. Luhmann, J. Geophys. Res. **102**, 1637 (1997)
J.G. Luhmann, J.U. Kozyra, J. Geophys. Res. **96**, 5457–5468 (1991)
J.G. Luhmann, R.E. Johnson, M.G.H. Zhang, Geophys. Res. Lett. **19**, 2151–2154 (1992)
R. Lundin, E.M. Dubinin, Adv. Space Res. **12**, 255–263 (1992)
R. Lundin, A. Zakharov, R. Pellinen et al., Nature **341**, 609–612 (1989)
R. Lundin, A. Zakharov, R. Pellinen et al., Geophys. Res. Lett. **17**, 873–876 (1990)
R. Lundin, H. Lammer, I. Ribas, Space Sci. Rev. **129**, 245–278 (2007)
J.I. Lunine, D.J. Stevenson, Astrophys. J. Suppl. **58**, 493–531 (1987)
J.I. Lunine, R.D. Lorenz, W.K. Hartmann, Planet. Space Sci. **46**, 1099–1107 (1998)
J.I. Lunine, Y.L. Yung, R.D. Lorenz, Planet. Space Sci. **47**, 1291–1303 (1999)
Y. Ma, A.F. Nagy, Geophys. Res. Lett. **34**, 8 (2007). CiteID L08201
Y. Ma, A.F. Nagy, G. Toth et al., Geophys. Res. Lett. **34**, L24S10 (2007). doi:10.1029/2007GL031627
G. Magni, A. Coradini, Planet. Space Sci. **52**, 343–360 (2004)
C.V. Manning, C.P. McKay, K.J. Zahnle, *American Geophys. Fall Meeting*, Abstract #P13D-1556 (2007)
T. Matsui, Y. Abe, Nature **319**, 303–305 (1986a)
T. Matsui, Y. Abe, Nature **322**, 526–528 (1986b)
M.A. McGrath, E. Lellouch, D.F. Strobel, P.D. Feldman, R.E. Johnson, in *Jupiter—The Planet, Satellites and Magnetosphere*, ed. by F. Bagenal, T. Dowling, W.B. McKinnon (Cambridge University Press, Cambridge, 2004), p. 457

C.P. McKay, T.W. Scattergood, J.B. Pollack, W.J. Borucki, H.T. van Ghyseghem, Nature **332**, 520–522 (1988)
H.J. Melosh, A.M. Vickery, Nature **338**, 487–489 (1989)
M. Michael, R.E. Johnson, Planet. Space Sci. **53**, 1510–1514 (2005)
M. Michael, R.E. Johnson, F. Leblanc et al., Icarus **175**, 263–267 (2005)
S.L. Miller, G. Schlesinger, Orig. Life **14**, 83–90 (1984)
A. Morbidelli, J. Chambers, J.I. Lunine et al., Meteorit. Planet. Sci. **35**, 1309–1320 (2000)
T.E. Moore, R. Lundin, D. Alcayde et al., Space Sci. Rev. **88**, 7–84 (1999)
O. Mousis, D. Gautier, D. Bockelée-Morvan, Icarus **156**, 162–175 (2002)
A.G. Munoz, Planet. Space Sci. **55**, 1426–1455 (2007)
A.F. Nagy, M.W. Liemohn, J.L. Fox et al., J. Geophys. Res. **106**, 21,565–21568 (2001)
G. Neukum, D.U. Wise, Science **194**, 1381–1387 (1976)
G. Newkirk Jr., Geochim. Cosmochim. Acta Suppl. **13**, 293–301 (1980)
H.B. Niemann, S.K. Atreya, S.J. Bauer et al., Nature **438**, 779–784 (2005)
A.O. Nier, Science **194**, 70–72 (1976)
A.O. Nier, M.B. McElroy, Y.L. Yung, Science **194**, 68–70 (1976)
C.A. Nixon, R.K. Achterberg, S. Vinatier et al., Icarus **195**, 778–791 (2008)
P. Norqvist, M. Andre, M. Tyrland, J. Geophys. Res. **103**, 23,459–23,474 (1998)
T. Owen, in *Evolution of Planetary Atmospheres and Climatology of the Earth* (CNRS, Toulouse, 1979), p. 1
T. Owen, J.P. Maillard, C. DeBergh, B.L. Lutz, Science **240**, 1767–1770 (1988)
E.J. Öpik, S.F. Singer, Phys. Fluids **4**, 221–233 (1963)
M. Ozima, K. Seki, N. Terada, Y.N. Miura, F.A. Podosek, H. Shinagawa, Nature **436**, 655–659 (2005)
E.N. Parker, *Interplanetary Dynamical Processes* (Interscience, New York, 1963)
A.A. Pavlov, M.T. Hurtgen, J.F. Kasting, M.A. Arthur, Geology **31**, 87–90 (2003)
T. Penz, N.V. Erkaev, H.K. Biernat et al., Planet. Space Sci. **52**, 1157–1167 (2004)
T. Penz, H. Lammer, Y.N. Kulikov, H.K. Biernat, Adv. Space Res. **36**, 241–250 (2005)
T. Penz, N.V. Erkaev, Y.N. Kulikov et al., Planet. Space Sci. (2008). doi:10.1016/j.pss.2008.04.005
R.O. Pepin, Icarus **92**, 2–79 (1991)
R.O. Pepin, Icarus **111**, 289–304 (1994)
R.O. Pepin, Icarus **126**, 148–156 (1997)
H. Pérez-de-Tejada, J. Geophys. Res. **97**, 3159–3167 (1992)
H. Pérez-de-Tejada, J. Geophys. Res. **103**, 31499–31508 (1998)
J.B. Pollack, J.F. Kasting, S.M. Richardson et al., Icarus **71**, 203–224 (1987)
R.G. Prinn, B. Fegley Jr., in *Origin and Evolution of Planetary and Satellite Atmospheres* (University of Arizona Press, Tucson, 1989), pp. 78–136
S.I. Rasool, C. DeBergh, Nature **226**, 1037–1039 (1970)
S.N. Raymond, T. Quinn, J.I. Lunine, Icarus **168**, 1–17 (2004)
I. Ribas, E.F. Guinan, M. Güdel, M. Audard, Astophys. J. **622**, 680–694 (2005)
W.W. Rubey, Geol. Soc. Am. Bull. **62**, 1111–1148 (1951)
C.T. Russell, J.G. Luhmann, R.C. Elphic et al., Geophys. Res. Lett. **9**, 45–48 (1982)
G. Ryder, Astrobiology **3**, 3–6 (2003)
P.V. Sada, G.H. McCabe, G.L. Bjoraker et al., Astrophys. J. **472**, 903–907 (1996)
C. Sagan, G. Mullen, Science **177**, 52–56 (1972)
L. Schaefer, J.B. Fegley, Icarus **186**, 462–483 (2007)
R.W. Schunk, D.S. Watkins, Planet. Space Sci. **27**, 433–444 (1979)
T.L. Segura, O.B. Toon, A. Colaprete, K. Zahnle, Science **298**, 1977–1980 (2002)
M. Sekiya, K. Nakazawa, C. Hayashi, Earth Planet. Sci. Lett. **50**, 197–201 (1980)
M. Sekiya, C. Hayashi, K. Nakazawa, Prog. Theor. Phys. **66**, 1301–1316 (1981)
V.I. Shematovich, R.E. Johnson, M. Michael, J.G. Luhmann, J. Geophys. Res. **108**, 5087 (2003). doi:10.1029/2003JE002094
Y. Shimazu, T. Urabe, Icarus **9**, 498–506 (1968)
H. Shinagawa, J. Kim, A.F. Nagy, T.E. Cravens, J. Geophys. Res. **96**, 11,083–11,095 (1991)
I. Sillanpää, E. Kallio, R. Jarvinen et al., Adv. Space Res. **38**, 799–805 (2006). doi:10.1016/j.asr.2006.01.005
J.A. Skumanich, Eddy, in *Solar phenomena in stars and stellar systems, Proceedings of the Advanced Study Institute* (Reidel, Dordrecht, 1981), p. 349
J.R. Spreiter, S.S. Stahara, J. Geophys. Res. **98**, 17,251–17,262 (1980)
J.R. Spreiter, A.L. Summers, A.Y. Alksne, Planet. Space Sci. **14**, 223–253 (1966)
D.F. Strobel, Icarus (2007, in press)
I. Sumita, T. Hatakeyama, A. Yoshihara, Y. Hamano, Phys. Earth Planet. Int. **128**, 223–241 (2001)
J.A. Tarduno, R.D. Cottrell, M.K. Watkeys, D. Bauch, Nature **446**, 657–660 (2007)
N. Terada, S. Machida, H. Shinagawa, J. Geophys. Res. **107**, 1471–1490 (2002)

F. Tian, O.B. Toon, A.A. Pavlov, H. DeSterck, Science **308**, 1014–1017 (2005)
F. Tian, J. Kasting, H. Liu, R.G. Roble, J. Geophys. Res. (2008, accepted)
K. Tsiganis, R. Gomes, A. Morbidelli, H.F. Levison, Nature **435**, 459–461 (2005)
G. Tobie, J.I. Lunine, C. Sotin, Nature **440**, 61–64 (2006)
D.L. Turcotte, An episodic hypothesis for Venusian tectonics. J. Geophys. Res. **98**, 17,061–17,068 (1993)
J.W. Valley, W.H. Peck, E.M. King, S.A. Wilde, Geology **30**, 351–354 (2002)
A. Vidal-Madjar, Geophys. Res. Lett. **5**, 29–32 (1978)
J.H. Waite Jr., H. Nieman, R.V. Yelle et al., Science **308**, 982–985 (2005)
J.-E. Wahlund, R. Bostrom, G. Gustafsson et al., Science **308**, 986–989 (2005)
J.C.G. Walker, *Evolution of the Atmosphere* (Macmillan, New York, 1977)
J.C.G. Walker, Icarus **68**, 87–98 (1986)
J.C.G. Walker, K.K. Turekian, D.M. Hunten, J. Geophys. Res. **75**, 3558–3561 (1970)
A.J. Watson, T.M. Donahue, J.C.G. Walker, Icarus **48**, 150–166 (1981)
P.R. Weissman, in *Origin and Evolution of Planetary and Satellite Atmospheres*, ed. by S.K. Atreya, J.B. Pollack, M.S. Matthews (University of Arizona Press, Tucson, 1989), p. 230
C.R. Woese, G.E. Fox, Proc. Natl. Acad. Sci. USA **74**, 5088–5090 (1977)
B.E. Wood, H.-R. Müller, G. Zank, J.L. Linsky, Astrophys. J. **574**, 412–425 (2002)
B.E. Wood, H.-R. Müller, G.P. Zank, J.L. Linsky, S. Redfield, Astrophys. J. **628**, L143–L146 (2005)
M. Yamauchi, J.-E. Wahlund, Astrobiology **7**, 783–800 (2007)
R.V. Yelle, Icarus **170**, 167—179 (2004)
R.V. Yelle, Icarus **183**, 508 (2006)
R.V. Yelle, J. Cui, I.C.F. Muller-Wodarg, J. Geophys. Res. (2008, in press)
A. Yoshihara, Y. Hamano, Precambr. Res. **131**, 111–142 (2004)
Y.L. Yung, J.-S. Chen, J.P. Pinto, M. Allen, S. Paulsen, Icarus **76**, 146–159 (1988)
K.J. Zahnle, J.F. Kasting, Icarus **68**, 462–480 (1986)
K.J. Zahnle, J.C.G. Walker, Rev. Geophys. Space Phys. **20**, 280–292 (1982)
K.J. Zahnle, J.F. Kasting, J.B. Pollack, Icarus **74**, 62–97 (1988)
K.J. Zahnle, J.B. Pollack, J.F. Kasting, Icarus **84**, 503–527 (1990)
K.J. Zahnle, N. Arndt, C. Cockell et al., Space Sci. Rev. **129**, 35–78 (2007)
M.G.H. Zhang, J.G. Luhmann, A.F. Nagy et al., J. Geophys. Res. **98**, 10915–10923 (1993)

Aeronomy of Extra-Solar Giant Planets

Roger Yelle · Helmut Lammer · Wing-Huen Ip

Originally published in the journal Space Science Reviews, Volume 139, Nos 1–4.
DOI: 10.1007/s11214-008-9420-6 © Springer Science+Business Media B.V. 2008

Abstract The intense stellar UV radiation field incident upon extra-solar giant planets causes profound changes to their upper atmospheres. Upper atmospheric temperatures can be tens of thousands of kelvins, causing thermal dissociation of H_2 to H. The stellar ionizing flux converts H to H^+. The high temperatures also drive large escape rates of H, but for all but the planets with the smallest orbits, this flux is not large enough to affect planet evolution. The escape rate is large enough to drag off heavier atoms such as C and O. For very small orbits, when the hill sphere is inside the atmosphere, escape is unfettered and can affect planet evolution.

Keywords Extra-solar planets · Aeronomy · Atmospheric escape

1 Introduction

Approximately 250 extrasolar planets have been detected at the time of writing, about 40% of which have orbits with semi-major axes less than ≤ 0.1 AU. Detections have been made largely by the identification of periodic variations in radial velocities of the star. Most of these newly discovered planets have relatively large masses and therefore are similar to Jupiter or Saturn in our own solar system; consequently, we referred to these objects as Extra-solar Giant Planets (EGPs).

Their proximity to the central star implies that EGPs are subjected to intense radiation and plasma fluxes, and that strong tidal forces could distort their atmospheres. Short wavelength stellar radiation and charged particle fluxes deposited in the upper atmospheres of the

R. Yelle (✉)
Department of Planetary Sciences, University of Arizona, Tucson, AZ 85721, USA
e-mail: yelle@lpl.arizona.edu

H. Lammer
Space Research Institute, Austrian Academy of Sciences, Schmiedlstraße 6, 8042 Graz, Austria

W.H. Ip
Institute of Astronomy, National Central University, Jhongli City, Taiwan

EGPs can heat the atmospheres to high levels and cause drastic chemical changes. Particularly important is the possibility that high temperatures in the upper atmospheres of EGPs could lead to escape fluxes that may even be large enough to alter the evolution of the planet. Observations of EGP atmospheres are needed to understand the aeronomical effects of this hostile environment.

Direct observations of the atmospheres of EGPs are rare because of the difficulty of separating the planetary signal from the nearby, much brighter, stellar signal. Fortunately, for some EGPs, the orbital plane lies along the line-of-sight to Earth, enabling studies of the atmosphere through measurements of the absorption of starlight as it passes in front of its star. Charbonneau et al. (2002) reported the first results: detection of the absorption signature of Na resonance lines in the atmosphere of HD209458b. Subsequently, Tinetti et al. (2007) observed a transit of HD189733b and found that absorption by H_2O vapour is the most likely cause of the wavelength-dependent variations in the effective radius of the planet at the infrared wavelengths 3.6 μm, 5.8 μm and 8 μm. Nevertheless, the most interesting results have been in the UV (Vidal-Madjar et al. 2003).

In this report we give an overview and update on the aeronomy of extrasolar Jupiter-type planets. We focus mainly on the interaction between the stellar radiation and particle environment with these planets and discuss the chemistry of their thermospheres, the thermospheric heating and cooling mechanisms, the related atmospheric expansion and thermal and non-thermal atmospheric loss rates, the role of the Roche lobe and effects related to the plasma flow of the host star and the magnetospheres of extrasolar planets.

2 Observations

Vidal-Madjar et al. (2003) made the first observations of an EGP upper atmosphere through measurement of the absorption of stellar Lyman α as HD209458b passed in front of HD209458a (Fig. 1). The observations were made with the Space Telescope Imaging Spectrograph (STIS) using the G140M grating and were characterized by a spectral resolution of 0.08 Å. From analysis of these data Vidal-Madjar et al. (2003) determined that the planet absorbed $15 \pm 4\%$ of the stellar Lyman α between wavelengths of 1215.15 and 1216.1 Å. HD209458b obscures only 1.5% of the stellar flux in the visible (Ballester et al. 2007), thus the Lyman α signature implies an atomic hydrogen cloud around the planet of roughly 3.3 planetary radii (R_P). The Hill radius (also called the Roche sphere), which defines the region of space dominated by the planet's gravity, is, for HD209458b, at $4.08 R_P$, thus the extended upper atmosphere of HD209458b is comparable to the size of the Hill sphere. Atmospheric atoms or molecules that reach the Hill sphere have escaped the planet; thus the large extent of the atmosphere implies large escape rates. We discuss this in detail below. Moreover, the width of the absorption signature measured by Vidal-Madjar et al. (2003) is quite broad, corresponding to velocities of ~ 100 km/s both towards and away from the star. The velocities are far above the escape velocity from the planet and the blue-shifted observations can be interpreted as a direct measurement of escape molecules.

Subsequently Vidal-Madjar et al. (2004) observed the same system with the STIS G140L grating that supplies lower spectral resolution but greater spectral coverage. The goal was to search for other absorption features and Vidal-Madjar et al. (2004) detected absorption in HI, OI, and CII lines of $5 \pm 2\%$, $13 \pm 4.5\%$ and $7.5 \pm 3.5\%$. The smaller HI absorption depth is a consistent with the earlier results because the lower spectral resolution implies a smaller absorption signature. The large OI and CII absorptions indicate that these species, like H, must be escaping the atmosphere at a rapid rate.

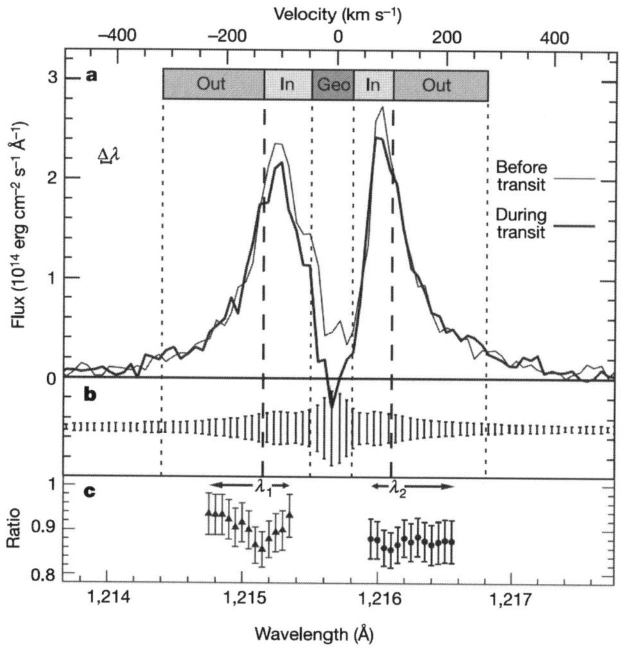

Fig. 1 From Vidal-Madjar et al. (2003). The absorption signature of HD209458b on the Lyman α line profile

Ben-Jaffel (2007) has reanalyzed the Vidal-Madjar et al. (2003) observations. Their data reduction approach differs in a number of ways from Vidal-Madjar et al. (2003) as they made extensive use of the time-tagged nature of the HST measurements. Ben-Jaffel (2007) derived a absorption depth of $8.9 \pm 2.1\%$, corresponding to an obscuring cloud with a radius of $2.47 \pm 0.30 R_P$. This absorption is an average over the 1214.83–1215.36 Å and 1215.89–1216.43 Å wavelength regions. Ben-Jaffel (2007) do not find any evidence for absorption in the line wings, corresponding to velocities of $\sim$ 100 km/s, observed by Vidal-Madjar et al. (2003). Instead, Ben-Jaffel (2007) suggests that the apparent absorption in the Vidal-Madjar et al. (2003) analysis was due to stellar variability. At present, the issue is unresolved.

Vidal-Madjar et al. (2003) present a forceful response to Ben-Jaffel's criticisms. Vidal-Madjar et al. (2003) point out that Ben-Jaffel (2007) calculates the absorption depth over a broader wavelength region than Vidal-Madjar et al. (2003). The larger wavelength range dilutes the absorption signature, Ben-Jaffel's $8.9 \pm 2.1\%$ is consistent with the original value of $15 \pm 4\%$ (Vidal-Madjar et al. 2003). Vidal-Madjar et al. (2003) also argue that the difference in opinion about absorption in the far wings of the line are related to different normalization procedures adopted in the two studies and argues in favor of the Vidal-Madjar et al. (2003) approach, partly because it accounts for possibly variations in the stellar Lyman alpha signal.

For our purposes, the most important difference between the analyzes of Vidal-Madjar et al. (2003) and Ben-Jaffel (2007) concerns the interpretation of the data. Ben-Jaffel (2007) claims that his analysis implies that the atmosphere is HD209458B is not escaping rapidly. This is based on the absence of absorption in the line wings due to atoms with velocities above the escape velocity and the fact that the inferred extent of the optically thick upper atmosphere is smaller than the Hill sphere. These arguments, however, do not survive careful examination. The existence of high velocity atoms is interesting from a physical point of view, and may provide insight into the interaction between the star and planetary atmospheres, but the these atoms do not represent the bulk of atmospheric escape. Most escaping atom have much lower velocities and cause absorption in the core of the Lyman alpha line. The second problem with Vidal-Madjar et al. (2003)'s argument is a confusion over the optical obstacle presented by the H cloud and the Hill sphere. These are two different concepts and cannot be compared directly. It is not necessary for the optically thick atmosphere to extend to the Hill sphere for substantial escape; it is only necessary for there to be some molecules that make it to the Hill sphere. The important quantity is the flux of H atoms incident upon the Hill sphere, not the H density that makes the atmosphere optically thick. All models for the upper atmospheric escape (discussed below) have implicitly recognized this. In fact, as discussed below, there is little doubt that an atmosphere as extended as that of HD209458b will be escaping rapidly. Finally, Vidal-Madjar et al. (2003) point out the Hill sphere is not spherical but elongated in the direction of the star and a more careful analysis shows that the H cloud should extend beyond the Hill sphere.

Ballester et al. (2007) conducted broadband UV and visible observations of HD209458b with the STIS 430L grating and discovered a new absorption feature starting at 3650 Å and extending to 3900 Å. They interpret this feature as absorption in the Balmer continuum of HI. The measured absorption is only 0.03% but, because of the high signal levels in this spectral region, this is larger the uncertainty in the observations. Absorption in the Balmer continuum indicates the presence of hot H in the upper atmosphere of HD209458b. As discussed below, this is also expected based on aeronomical considerations.

3 Composition and Thermal Structure of an EGP

The thermal structure in the upper atmosphere, and the thermal escape rate of an EGP depends on the composition of the atmosphere. Unlike the atmospheres of planets in our own solar system whose main constituents are usually stable over geological time scales, the atmospheric composition of extra-solar giant planets is altered by the intense stellar UV radiation field from the central star. Moreover, the thermal structure and composition are closely coupled through the dependence of heating efficiency and cooling rates on composition (Yelle 2004, 2006; Muñoz 2007).

When the temperature reaches a few thousand kelvins and the density $\sim 10^{10}$ cm^{-3}, thermal decomposition

$$\mathrm{H_2} + M \rightarrow \mathrm{H} + \mathrm{H} + M$$

causes the atmosphere to change from predominantly H_2 to predominantly H. Both the temperature and density of H are consistent with the observations of Ballester et al. (2007). The balance between photo-ionization and recombination determines the charge state of the atmosphere. In the lower and middle thermosphere the balance favors neutral H but at high altitudes where the stellar radiation field is most intense and the densities are low, H^+ should be the dominant constituent. Thus, as altitude increases in the thermosphere, the composition changes from H_2 to H to H^+. This change in composition causes the atmosphere to become more extended. The transition from H_2 to H causes the scale height of the atmosphere to double and the transition from H to H^+ causes the scale height to increase again by a factor of two or more, depending on the electron and ion temperature.

Photo-ionization also leads to production of some minor species, the most important of which is H_3^+, which is created primarily by

$$\mathrm{H_2^+} + \mathrm{H_2} \rightarrow \mathrm{H_3^+} + \mathrm{H}$$

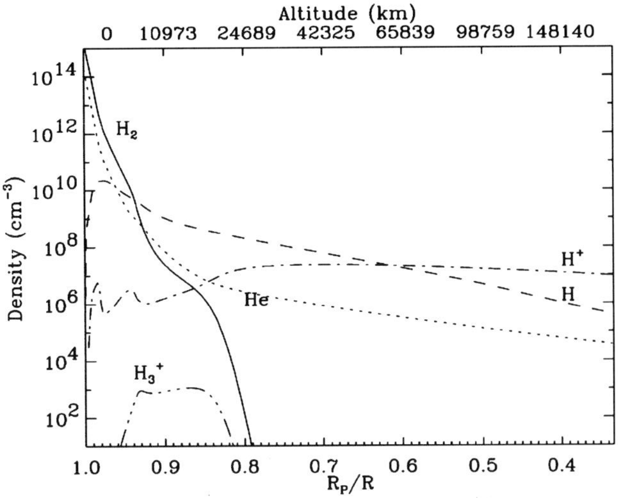

Fig. 2 The calculated composition of HD209458B (from Yelle 2004)

The predicted density of H_3^+ is fairly low, compared with Jupiter for example, because the high electron density on HD209458b favors H^+ as the primary ion. Nevertheless, H_3^+ plays an important role in the thermal structure, as discussed below. An example of the calculated composition is shown in Fig. 2.

The models presented in Yelle (2004) assumed only H and He in the upper atmosphere based on analogy with Jupiter where diffusive separation sequesters heavier species in the lower atmosphere. The detection of CII and OI in the vicinity of HD209458b suggests that this assumption should be re-examined. O is present in the deeper atmosphere of HD209458b in the form of H_2O and C as either CH_4 or CO. If transported to the upper atmosphere these species could be broken down into atomic and ionic form through photochemistry. Escape might then lead to the presence of O and C^+ in the vicinity of HD209458b as seen by Vidal-Madjar et al. (2008).

There are two physical processes that could bring heavy species into the thermosphere of an extra-solar giant planet. A very large vertical mixing rate could offset diffusive separation and cause the relative density of C and O species to remain constant. If the vertical mixing rate is parameterized in terms of an eddy diffusion coefficient, then the existence of C and O in the upper atmosphere requires that the eddy coefficient be equal to or larger than the molecular diffusion coefficient. The eddy coefficient can be expressed as $K \sim vH$ where H is the atmospheric scale height and v the characteristic velocity of the turbulence in the atmosphere, whereas the molecular diffusion coefficient is approximately $D \sim cL$, where c is the sound speed and L the mean free path. An eddy coefficient comparable to the molecular diffusion coefficient then implies that the turbulent velocity is comparable to the sound speed, because the atmospheric scale height becomes comparable to the mean free path near the top of the atmosphere. This is not impossible, but requires extreme forcing. Unfortunately, we have too little information of the dynamics of EGP atmospheres to permit a more quantitative estimate.

Diffusive separation can also be overcome by the vigorous atmospheric escape. If the escape flux of the light constituent is large enough, collisions between a light primary constituent and a heavy minor constituent can increase the scale height of the minor constituent, counteracting the effects of diffusive separation (Hunten et al. 1987). The necessary flux depends on the mass ratio of the light to heavy species. Mathematically, we have

$$m_c = m\,(1 + HF/b) \tag{1}$$

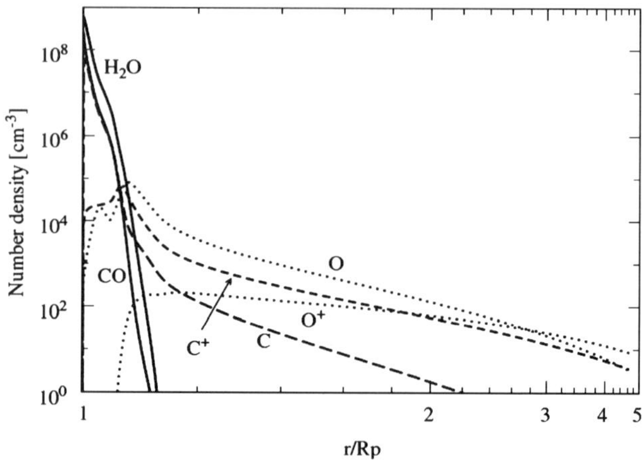

Fig. 3 Model for the density of O and C species (from Muñoz 2007)

where m is the mass of the background atmosphere, H is the scale height, F is the escape flux of the background atmosphere and b is the binary collision parameter. The quantity, m_c is known as the cross-over mass. For a given escape flux F, species lighter than m_c will be carried along with the ambient flow and escape the atmosphere; species heavier than m_c will still experience diffusive separation, will be kept at lower altitudes, and will not escape the atmosphere (Hunten et al. 1987). Adopting an H escape flux for HD209458b of $F = 3 \times 10^{13}$ cm^{-2} s^{-1}, a scale height of $H = 920$ km, a binary diffusion parameter $b = 2 \times 10^{20}$ cm^{-1} s^{-1} and assuming mostly H escape, implies a cross-over mass of 152 amu, which is larger than the mass of C or O. Thus, it is likely that drag forces could cause C and O species to escape the atmosphere along with H.

Muñoz (2007) presents sophisticated calculations of the distribution of heavy species in the atmosphere of HD209458b. The results from one model are shown in Fig. 3. The eddy diffusion coefficient for this model is 10^8 cm^2 s^{-1}, implying that diffusive separation should occur. The fact that the O and C^+ densities are roughly parallel to the total atmospheric density, suggest that drag forces overwhelm diffusive separation, but the issue is not discussed in this context in Muñoz (2007). The predicted densities of O and C^+ are of the right order to explain the absorption measured by Vidal-Madjar et al. (2004) observations, but the line observed absorption line widths of 0.065 Å and 0.11 Å for the OI and CII lines is significantly larger than predicted for even for the elevated temperatures in the upper atmosphere of HD209458b. Muñoz (2007) speculates that the large absorption line widths might be explained velocity fluctuations associated with turbulence in the escaping flow, but the issue is unresolved at present.

The thermal balance, like the chemical balance, varies strongly with altitude through an EGP thermosphere in response partly to the changing composition. Typical results are shown in Fig. 4. At the lowest altitudes heating by stellar radiation is balance primarily by radiative cooling by H_3^+. At higher levels the balance is primarily between heating by stellar radiation and adiabatic cooling. The same is true for more distant EGPs, but for closer-in EGPs radiative cooling due to $e + H^+$ recombination becomes more important than adiabatic cooling (Yelle 2004). Unlike our Jovian planets, thermal conduction never plays a dominant role, despite the very large temperature increase in the thermosphere. This is a consequence of the extended nature of the atmosphere and the large scale heights that minimize the consequences of diffusive effects.

The balance between stellar heating and adiabatic cooling indicates that the escape rate from and EGP is limited by the rate of absorption of stellar energy. This can be seen clearly

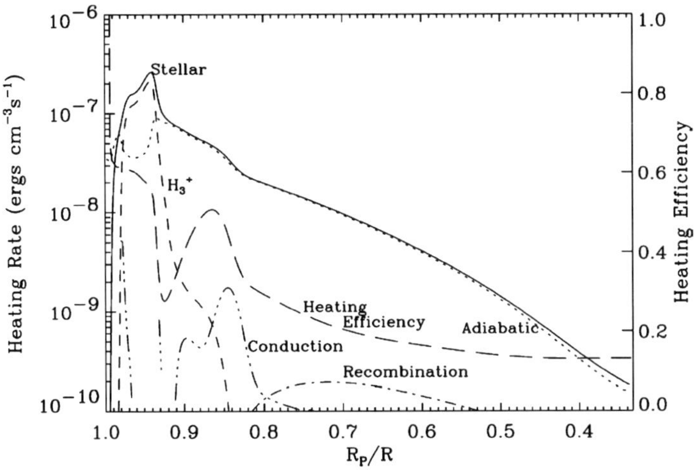

Fig. 4 Calculated heating rates for HD209458b (from Yelle 2004)

Fig. 5 Escape rates versus semi-major axis (from Muñoz 2007)

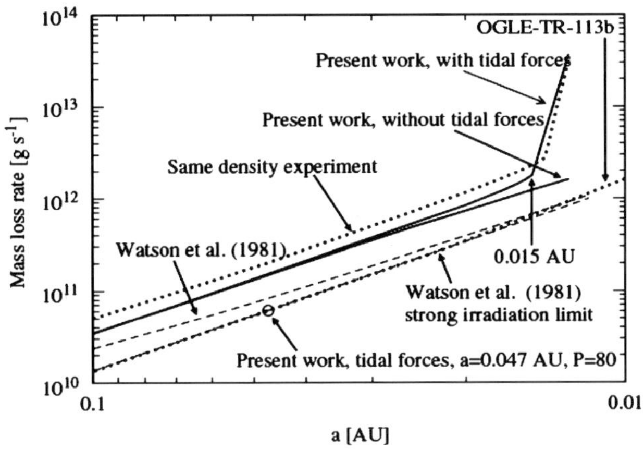

in Fig. 5 (from Muñoz 2007) that shows that calculated escape rates vary as the inverse square of the semi-major axis except for very small values of where tidal forces become important. Yelle (2004) found a similar result. This may explain the good agreement between various estimates of the escape rate. Table 1 presents the published calculations of the escape rate from HD209458b. Except for the earliest rough estimates, the calculated values for HD209458b are all in the range $3–7 \times 10^{10}$ g s^{-1}. Despite this agreement, it remains important to verify the models through continued and improved observations.

4 Tidal Effects at Close-in Extrasolar Planets

Lecavelier des Etangs et al. (2004), Jaritz et al. (2005) and Lecavelier des Etangs (2007) argued that for some close-in gas giants, due to expected high exospheric temperatures, the exobase level R_c can reach the Hill sphere R_{HS} before classical hydrodynamic blow-off conditions may develop and, thus, affect their atmospheric loss rates by one or two orders of magnitude.

Table 1 Calculated escape rates

Escape rate ($g\,s^{-1}$)	Author	Comments
5×10^{11}	Lecavelier des Etangs et al. (2004)	Assumed T, used Jeans formula
$\sim 10^{12}$	Lammer et al. (2003)	Watson et al. (1981) approximation
4.5×10^{10}	Yelle (2004)	Navier-Stokes equations with variable composition
3.5×10^{10}	Tian et al. (2005)	Navier-Stokes equations with constant composition
4.8×10^{10}	Muñoz (2007)	Navier-Stokes equations with variable composition
3.5×10^{10}	Penz et al. (2008)	Navier-Stokes equations with constant composition

Muñoz (2007) took an interesting approach to assess the importance of tides on the escape rate. The problem is that to be treated rigorously tides require a 3 dimensional calculation of atmospheric structure but all aeronomical models to date are one-dimensional. Muñoz (2007) avoided the problem by calculating the effect of tides only along the line between the planet and star. This approach probably overestimates the effect of tides but the calculations are likely to be of the correct order of magnitude. Figure 5 presents results for OGLE-TR-113b. According to these calculations the influence of tides is drastic once the Hill sphere is inside the exobase, but of minor importance otherwise. The transition between these two regimes is quite sharp.

Erkaev et al. (2007) derived the gravitational potential difference between the base of the thermosphere, and the Roche lobe boundary, R_{Rl}

$$\Delta\Phi = \Phi_0 \left(1 - \frac{3}{2\xi} + \frac{1}{2\xi^3}\right) = \Phi_0 K(\xi), \tag{2}$$

where

$$\xi = \frac{R_{\mathrm{HS}}}{R_P} \approx \left(\frac{M_P}{3M_*}\right)^{\frac{1}{3}} \lambda. \tag{3}$$

$K(\xi)$ or its inverse $1/K(\xi)$ can be considered as the potential energy reduction or atmospheric mass loss enhancement factor due to the stellar tidal forces, respectively. By using appropriate stellar and planetary parameters the mass loss enhancement factor $1/K(\xi)$ is for example about 1.97 for the extrasolar planet TresS-1 at 0.023 AU and about 1.53 for HD209458b at 0.045 AU (Erkaev et al. 2007). This is significantly larger than calculated by Muñoz (2007); thus, it appears that the influence of tides requires further investigation.

The sharp increase in escape rate at 0.015 AU shown in Fig. 5 may indicate the onset of true blow-off where the mean kinetic energy of atmospheric molecules in comparable to the gravitational binding energy, more similar to a cometary atmosphere than a bound planetary atmosphere. Erkaev et al. (2007) reformulated the blow-off condition to include tidal effects from the central star. Ff $R_c \leq R_{\mathrm{HS}}$. We have,

$$\frac{3}{2} k T_c = m \Delta\Phi, \tag{4}$$

where m the molecular mass, and $\Delta\Phi$ the perturbed gravitational potential difference between the exobase level and the Hill sphere. Solving for T_c gives

$$T_c = \frac{2 m M_{pl} G K(\xi)}{3k}. \tag{5}$$

If the blow-off temperature for an extrasolar planet without the effect of the tides is, for example, 10,000 K, a similar extrasolar planet, which is closer to its host star, may start to evaporate hydrodynamically due to the tidal effect at about 5000 K if $K(\xi) = 0.5$ like it is for TresS-1. This indicates that the host stars tidal forces enhance the possibility that close-in extrasolar gas giants may reach hydrodynamic blow-off conditions more easily and stay much longer in this regime compared to similar planets at locations where the stellar tides can be neglected (Erkaev et al. 2007).

5 Stellar Plasma Interaction With Close-in Extrasolar Planets

Because EGPs orbit very close to their host stars their magnetospheres and atmospheres are exposed to much stronger stellar wind conditions and stellar plasma conditions than the planets in our solar system. The characteristics of stellar winds and CMEs from other stars are generally poorly known but they are likely to play an important role in various star-planet interaction processes. Preusse et al. (2005) studied the stellar wind regimes of close-in extrasolar planets on the basis of a solar wind model by Weber and Davis (1967) and found in agreement with Erkaev et al. (2005) that in contrast to Solar System planets most extrasolar planets which orbit their host stars less than about 0.1 AU may represent obstacles in a sub-Alfénic stellar wind plasma flow. This condition occurs if the magnetosonic Mach number of the stellar wind $M_s < 1$

$$M_s = M_A \left(1 + \frac{\gamma\beta}{2}\right)^{-\frac{1}{2}}, \tag{6}$$

where γ is the adiabatic coefficient, β the plasma parameter and M_A the Alfén Mach number

$$M_A = \frac{v_{sw}}{v_A} = \frac{v_{sw}\sqrt{\mu_0 \rho}}{B}, \tag{7}$$

with the magnetic permeability μ_0, the stellar wind particle density ρ, the interplanetary magnetic field strength B, the stellar wind velocity v_{sw} and v_A the Alfén velocity. Under this condition the stellar plasma interaction with extrasolar planets does not lead to the formation of a bow shock (Erkaev et al. 2005; Preusse et al. 2005) and only Alfén waves and slow shocks will be produced by this plasma flow regime. In such a case, the stellar wind magnetic pressure is comparable or even larger than the dynamic flow pressure. We note that the plasma flow regime of low magnetosonic Mach numbers and its plasma interaction with planetary obstacles may have a similarity with the interaction of the Saturnian or Jovian subsonic corotating magnetospheric plasma flow with Titan and Io.

If a close-in extrasolar planet collides with a stellar CME the interaction regime is different (Erkaev et al. 1995). The average CME velocity at orbital distances of close-in exoplanets is about 500 km s^{-1}, which is comparable to the present solar wind velocity at 1 AU, a bow shock should form like that observed on Solar System planets, and after the shock the CME plasma would be deflected around the planetary obstacle (Erkaev et al. 1995; Lammer et al. 2007).

Grießmeier et al. (2004) studied the effect of tidal-locking to close-in extrasolar planets and found that due to the slow rotation one may expect weaker magnetic moments compared to fast rotating Solar System gas giant like Jupiter. This effect ma lead to weaker magnetospheric protection of the EUV heated and expanded upper atmospheres of close-in extrasolar planets against the dense stellar plasma flow (Erkaev et al. 2005;

Khodachenko et al. 2007). The various atmospheric effects expected due to direct stellar wind or CME plasma interaction with the ionized part of the upper atmosphere are discussed in more detail in the following subsections.

5.1 Magnetospheres and Non-Thermal Atmospheric Loss of Close-in Extrasolar Planets

Additionally to the high thermal atmospheric escape rates in the orders of about $\geq 10^{10}$–10^{11} g s^{-1} of hydrogen-rich gas giants in orbit locations between 0.02–0.05 AU non-thermal escape processes should contribute to the total loss rates. In these loss processes mainly ionized neutrals are incorporated within the stellar plasma flow around a non- (Venus-like) or magnetic (Earth-like) planetary obstacle. The efficiency of this additional atmospheric loss rates depend on the strength of the magnetic dynamo and hence the atmosphere protecting magnetosphere of the extrasolar planet. The magnetosphere of the planet is generated due to the interaction of the stellar plasma flow with the internal magnetic field.

Due to the close orbit location of many extra-solar planets they should be tidal locked due to gravitational dissipation (e.g. Goldreich and Soter 1966). Grießmeier et al. (2004) found that the expected reduced rotation rate should have implications for the planetary magnetic moment M which is proportional to the velocity ϖ of rotation of the planet around its axis, the core radius r_c, the mass density ρ_c and conductivity σ_c of the dynamo region (Busse 1976; Stevenson 1983; Mizutani et al. 1992; Sano 1983). Because the internal core properties like r_c, ρ_c and σ_c are unknown on extrasolar planets one can apply empirical scaling laws developed for Solar System planets where $r_c \propto M_{pl}^{\alpha} R_{pl}^{\beta}$. If one normalizes r_c, M_{pl} and R_{pl} to the Jovian values and by using the best fits for Saturn, Uranus and Neptune one obtains for factor $\alpha = 0.75$ and for $\beta = -0.96$ (Grießmeier et al. 2004). If the radius and the mass and ϖ of an extrasolar planet is known one can estimate the magnetic moment which was estimated for the tidal locked Jovian-type extrasolar planet HD209458b at 0.045 AU with $\varpi = 0.12\ \varpi_{\mathrm{Jup}}$ of about $\leq 0.1\ M_{\mathrm{Jup}}$ (where $M_{\mathrm{Jup}} = 1.5 \times 10^{27}$ Am^2).

Grießmeier et al. (2004) studied the expected magnetic topology of extrasolar planets based on magnetospheric models which were developed for Solar System planets (e.g. Voigt 1981; Jordan 1994; Song et al. 2001; Stadelmann 2004). A magnetic obstacle (magnetopause) R_{MP} forms when the dynamic pressure of the stellar plasma flow is in equilibrium with the magnetic field and atmospheric pressure

$$R_{\mathrm{MP}} = \left(\frac{\mu_0 f_0^2 M^2}{8\pi^2 n m v^2} \right)^{\frac{1}{6}}, \tag{8}$$

where n, m and v correspond to the stellar plasma density, proton mass and velocity, M is the magnetic moment, μ_0 is the magnetic permeability, and f_0 is a magnetopause shape function of about 1.5 for spherical magnetospheres or 1.16 for more realistic non-spherical shapes (Voigt 1995). Neutrals which reach planetary distances beyond R_{MP} can be ionized due to charge exchange with stellar plasma particles, electron impact or soft X-rays and EUV radiation. After ionization the particles will be incorporated in the stellar plasma flow so that most of them are lost from the planetary atmosphere.

One can see from (8) that due to much higher stellar plasma densities at close orbital distances nmv^2 will be much larger and due to tidal locking induced lower magnetic moments M, one can expect much weaker and compressed magnetopause boundaries $R_{\mathrm{MP}} < 3R_{\mathrm{PL}}$ compared to that of Jupiter in 5 AU (Grießmeier et al. 2004). Erkaev et al. (2005) calculated the stellar wind ion pick up loss rates of HD209458bs extended neutral upper atmosphere based on the maximum magnetic moment estimations of Grießmeier et al. (2004) for present time stellar activity with a numerical test particle model and obtained H^+ ion pick up loss rates of about 1.5 $\times 10^9$ g s^{-1} for a neutral density of about 10^5 cm^{-3} at an assumed magnetopause location at about $3.5R_{\rm Jup}$. Their loss rates are more or less in agreement with the simple ion loss estimations by Guillot et al. (1996) of about 10^{10} g s^{-1}. Depending on the atmospheric density and the strength of the magnetic moment the non-thermal loss rates can be lower or even higher than the thermal loss rates.

Khodachenko et al. (2007) studied the H^+ pick up loss rates of HD209458b by taking into account charge exchange, electron impact and EUV ionization processes during the time period when the planet collides with dense stellar CME plasma. Their results obtained with a numerical test particle model indicate that short-periodic slow rotating hydrogen-rich gas giants at orbital distances ≤ 0.05 AU around solar-like stars should experience strong atmospheric loss during CME collision. They found that if HD209458b is protected by a magnetic moment $M = 0.1M_{\rm Jup}$ the dense CME plasma flow would be deflected around the planet at planetocentric distances of about 1.54–$2.33R_{\rm PL}$ which is deeper in the thermosphere compared to the ordinary stellar wind interaction. In such a case the EUV heated and dynamically expanded upper atmosphere may interact with the CME plasma flow in a Venus-type plasma interaction regime resulting in H^+ loss rates in the order of about 1.5×10^{11}–2.0×10^{13} g s^{-1}.

Calculations of the CME-caused atmospheric H^+ ion pick-up loss rate of HD209458b over its lifetime, by assuming the observed solar CME occurrence rate indicate that this extrasolar planet could have lost during its life time up to 20% of its present mass. By assuming a weaker magnetic moment where the CME plasma flux is deflected at distances $\sim 1.3R_{\rm PL}$, then the planet may be eroded during a 5 Gyr period down to its core-size. Because HD209458b at 0.045 AU or OGLE-TR-56b at about 0.0225 AU have masses of 0.69 and 1.45 $M_{\rm Jup}$, Khodachenko et al. (2007) concluded that both extrasolar planets and also all the others observed at such close orbital distances, might have strong enough magnetic moments which are able to keep the magnetopause at high enough altitudes where the interaction of the dense CME plasma with the upper atmosphere is not so efficient.

One can see from these studies that non-thermal ion loss processes from weakly extrasolar planets within very close orbital distances to their host stars may have important implications for the evolution of these planets. One should also note that the numerical test particle models applied by Erkaev et al. (2005) and Khodachenko et al. (2007) are not able to calculate the atmospheric altitude where dense plasma fluxes are deflected around the planet in a self consistent way. Therefore, it is important that future studies have to apply magnetohydrodynamic models, which can self consistently calculate the formation of the planetary obstacles in the upper atmospheres of close-in extrasolar planets and their related ion erosion rates.

5.2 Star Planetary Magnetospheric Interaction

In analogy with the solar wind interaction of planetary magnetospheres, it was first proposed by Rubenstein and Schaefer (2000) that strong interaction between the magnetic fields of a central star and its near-by extrasolar planet could lead to energy dissipation in the stellar atmosphere via magnetic field reconnection. In turn, the heating of the stellar chromosphere and corona could produce visible features such as enhancement in the Ca II H and K line emission (Cuntz et al. 2000) and radio emission (Zarka et al. 2001). The diagnostic signature is that the induced emission should be correlated to the orbital phase of the exoplanets. Along this line of thinking, Ip et al. (2004) produced a simple model depicting the interaction of the dipole field of the central star with a hot Jupiter with a surface field comparable

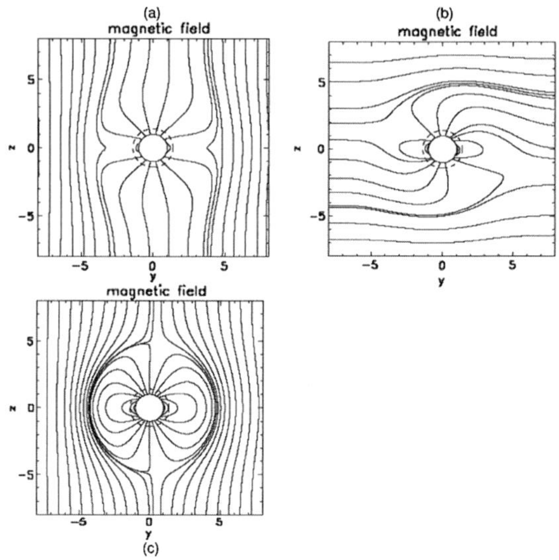

Fig. 6 Schematic views of magnetospheric configurations of close-in hot Jupiter's with different orientations of the stellar coronal magnetic field under sub-Alfvenic condition, from Ip et al. (2004)

to that of Jupiter (Ip et al. 2004). It was pointed out in this study that the close orbital distances of the hot Jupiter's tend to make the magnetospheric interaction highly sub-Alfvenic. No bow shock will form upstream. It was estimated that the magnetic energy input could be comparable to a large solar flare. In particular, it was pointed out by these authors that the upper atmosphere of a hot Jupiter could be subject to additional heating because of such star-magnetosphere interaction. The expectation is therefore when a hot Jupiter is located between its host star and the observer, the chromospheric hot spot generated by star-magnetosphere interaction should be most visible.

However, there are several lines of arguments which indicate that the situation might be more complicated. First, the continuous Ca II H and K spectroscopy of several close-in giant exoplanets by Shkolnik et al. (2005) showed that some of them (e.g., Tau Boo) has no periodic appearance of chromospheric Ca II, H and K emission.

And for those which periodic (or quasi-periodic) emission features have been detected, the correlations exhibit significant phase lags. For example, the repeatable enhanced emission of HD 179949 was found to peak at phase $\varphi = 0.83$ when the planet is approaching inferior conjunction when the extrasolar planet is between the host star and observer. In the case of υ Andromeda, the emission peaks occurred at $\varphi \sim 0.2$ when the planet was nearly behind the host star. Recently, non-dipolar magnetic field and multipolar magnetic field configurations have been invoked to explain these behaviors (McIvor et al. 2006; Cranmer and Saar 2007). Because of this puzzling phenomena, new studies have been initiated to study the stellar wind outflows of low-mass main-sequence stars (Cranmer 2007) and to measure the surface magnetic field structures of some of the host stars of close-in extrasolar planets (Catala et al. 2007; Moutou et al. 2007). Obviously, not for too long we might be able to understand better the physical processes involved in star-magnetosphere interaction.

5.3 The Role of Stellar Plasma Conditions to Expected Radio Emission from Extrasolar Planets

All strongly magnetized planets in the Solar System generate non-thermal radio emission caused by the cyclotron maser instability (e.g. Zarka 1998).The power emitted in Jupiter's hectometric emission and Saturn's kilometric emission are strongly correlated with solar wind plasma parameters (e.g. Gallagher and Dangelo 1981; Desch and Rucker 1983). Theoretical studies indicate that for Jovian-like extrasolar planets, which are exposed to denser stellar plasma flows at small orbital distances, the radio emission should be much stronger compared to Solar System planets (Zarka 2007; Zarka et al. 2001; Farrell et al. 1999; 2004; Lazio et al. 2004; Stevens 2005; Grießmeier et al. 2005, 2004). If the decametric radio emission from an extrasolar planet exceeds the emission of its host star it may be detected with ground-based radio telescopes in the near future (Grießmeier et al. 2005; Grießmeier et al. 2007b). Such a discovery can be used to extract information about the magnetic fields of extrasolar planets. The power $P_{\rm rad}$ which is emitted by radio waves can be written as (Grießmeier et al. 2007a)

$$P_{\rm rad} \propto P_{\rm in} \propto R_{\rm MP} n v_{\rm eff}^3, \tag{9}$$

where $P_{\rm in}$ is the power input, $R_{\rm MP}$ the magnetopause distance (see (8)), n the stellar proton density and

$$v_{\rm eff} = \left(v_{\rm Kepler}^2 + v^2\right)^{\frac{1}{2}}, \tag{10}$$

is the effective velocity resulting from the orbital velocity $v_{\rm Kepler}$ and the stellar plasma velocity v. The radio flux at the location of the observer can be written as (Grießmeier et al. 2007a)

$$\Phi_s = \frac{4\pi^2 m_e R_{\rm MP}^3 P_{\rm rad}}{e\mu_0 \Omega s^2 M}, \tag{x11}$$

where Ω is the solid angle of the beam, m_e and e are the electron mass and charge, s is the distance of the observer, μ_0 the magnetic permeability and M the magnetic moment. By using appropriate stellar and planetary parameters Grießmeier et al. (2007a) estimated the radio flux from the 4.4 Jupiter-mass extrasolar planet τ Bootes b which orbits in 15.6 pc a 2.4 Gyr old, 1.42 solar mass star (Fuhrmann et al. 1998) at 0.0489 AU (Leigh et al. 2003) depended on stellar plasma parameters. By assuming that τ Bootes b is exposed very often by CMEs they obtained a radio flux Φ_s at the Earth of about 2×10^{-2} Jy compared to about 10^5–10^6 Jy observed from Jupiter (Grießmeier et al. 2007a). The flux from the extrasolar planet would even be larger if one uses the ordinary solar wind flux as an input. However, the expected radio flux from extrasolar planets in the order of about 10^{-2}–5×10^{-2} Jy may be reached within the "French-Austrian-Ukrainian" decametre radio astronomy cooperation where a radio receiver (Robin 2) operates at the UTR-2 antenna array in Charkov (Rucker 2002). Furthermore, in the near future radio emission and a better knowledge of magnetospheres of extrasolar planets may be obtained after the Low Frequency Array (LOFAR) with its sensitivity of about 10^{-3} Jy in the decametre range will be go in operation.

Acknowledgements R.V. Yelle acknowledges support from NASA Planetary Atmospheres Program through grant NNG05GF47G. H. Lammer thanks the Austrian Ministry for Science, Education and Culture (bm:bwk) and ASA for funding the CoRoT project.

References

G.E. Ballester, D.K. Sing, F. Herbert, Nature **445**, 511–514 (2007)
L. Ben-Jaffel, Astrophys. J. Lett. **671**, L61–L64 (2007)
F.H. Busse, Phys. Earth Planet. Int. **12**, 350–358 (1976)
C. Catala, J.-F. Donati, E. Shkolnik, D. Bohkebderm, E. Alecian, Mon. Not. R. Astron. Soc. **374**, L42 (2007)

D. Charbonneau, T.M. Brown, R.W. Noyes, R.L. Gilland, Astrophys. J. **568**, 377–384 (2002)
S.R. Cranmer, arXiv:astro-ph/0701561v1 (2007)
S.R. Cranmer, S.H. Saar, astro-ph/0702530v1 (2007)
M. Cuntz, S.H. Saar, Z.E. Musielak, Astrophys. J. **533**, L151–L154 (2000)
M.D. Desch, H.O. Rucker, J. Geophys. Res. **88**, 8999 (1983)
N.V. Erkaev, C.J. Farrugia, H.K. Biernat, L.F. Burlaga, G.A. Bachmaier, J. Geophys. Res. **100**(A10), 19919–19931 (1995)
N.V. Erkaev, T. Penz, H. Lammer, H.I.M. Lichtenegger, P. Wurz, H.K. Biernat, J.-M. Griessmeier, W.W. Weiss, Astrophys. J. Suppl. Ser. **157**, 396–401 (2005)
N.V. Erkaev, Yu.N. Kulikov, H. Lammer, F. Selsis, D. Langmayr, G.F. Jaritz, H.K. Biernat, Astron. Astrophys. **472**, 329–334 (2007)
W.M. Farrell, M.D. Desch, P. Zarka, J. Geophys, Res. **104**(E6), 14025–14032 (1999)
W.M. Farrell, T.J.W. Lazio, P. Zarka, T.J. Bastian, M.D. Desch, B.P. Ryabov, Planet. Space Sci. **52**(15), 1469–1478 (2004)
K. Fuhrmann, M.J. Pfeiffer, J. Bernkopf, Astron. Astrophys. **336**, 942–952 (1998)
D.L. Gallagher, N. Dangelo, Geophys. Res. Lett. **8**, 1087 (1981)
J.M. Grießmeier, A. Stadelmann, T. Penz, H. Lammer, F. Selsis, I. Ribas, E.F. Guinan, U. Motschmann, H.K. Biernat, W.W. Weiss, Astron. Astrophys. **425**, 753–762 (2004)
J.-M. Grießmeier, U. Motschmann, G. Mann, H.O. Rucker, Astron. Astrophys. **439**, 771–775 (2005)
J.-M. Grießmeier, S. Preusse, M.L. Khodachenko, U. Motschmann, G. Mann, H.O. Rucker, Planet. Space Sci. **55**, 618–630 (2007a)
J.-M. Grießmeier, P. Zarka, H. Spreeuw, Astron. Astrophys. **475**, 359–368 (2007b)
P. Goldreich, S. Soter, Icarus **5**, 375–389 (1966)
T. Guillot, A. Burrows, W.B. Hubbard, J.I. Lunine, D. Saumon, Astrophys. J. **459**, L35 (1996)
D.M. Hunten, R.O. Pepin, J.C.G. Walker, Icarus **69**, 532–549 (1987)
W.-H. Ip, A. Kopp, J.-H. Hu, Astrophys. J. **602**, L53–L56 (2004)
G. Jaritz, S. Endler, D. Langmayr, H. Lammer, J.-M. Grießmeier, N.V. Erkaev, H.K. Biernat, Astron. Astrophys. **439**, 771–775 (2005)
C.E. Jordan, Rev. Geophys. **32**, 139 (1994)
M.L. Khodachenko, H. Lammer, H.I.M. Lichtenegger, D. Langmayr, N.V. Erkaev, J.-M. Grießmeier, M. Leitner, T. Penz, H.K. Biernat, U. Motschmann, H.O. Rucker, Planet. Space Sci. **55**, 631–642 (2007)
H. Lammer, F. Selsis, I. Ribas, E.F. Guinan, S.J. Bauer, Astrophys. J. **598**, L121–L124 (2003)
H. Lammer, H.I.M. Lichtenegger, Y.N. Kulikov, J.-M. Grießmeier, N. Terada, N.V. Erkaev, H.K. Biernat, M.L. Khodachenko, I. Ribas, T. Penz, F. Selsis, Astrobiology **7**, 185–207 (2007)
T.J.W. Lazio, W.M. Farrell, J. Dietrick, E. Greenless, E. Hogan, C. Jones, L.A. Hennig, Astrophys. J. **612**, 511–518 (2004)
A. Lecavelier des Etangs, Astron. Astrophys. **461**, 1185–1193 (2007)
A. Lecavelier des Etangs, A. Vidal-Madjar, J.C. McConnell, G. Hébrard, Astron. Astrophys. **418**, L1–L4 (2004)
C. Leigh, A. Collier Cameron, K. Horne, A. Penny, D. James, Mon. Not. R. Astron. Soc. **344**, 1271–1282 (2003)
T. McIvor, M. Jardine, V. Holzwarth, Mon. Not. R. Astron. Soc. **367**, L1 (2006)
H. Mizutani, T. Yamamoto, A. Fujimura, Adv. Space Res. **12**, 265–279 (1992)
C. Moutou, J.-F. Donati, R. Savalle, G. Hussain, E. Alecian, F. Bouchy, C. Catala, A. Collier Cameron, S. Udry, A. Vidal-Madjar, Astron. Astrophys. **473**, 651 (2007)
A.G. Muñoz, Planet. Space Sci. 2007. doi:10.1016/j.pss.2007.03.007
T. Penz, N.V. Erkaev, Y.N. Kulikov, D. Langmayr, H. Lammer, G. Micela, C. Cecchi-Pestellini, H.K. Biernat, F. Selsis, P. Barge, M. Deleuil, A. Léger, Planet. Space Sci. **56**, 1260 (2008)
S. Preusse, A. Kopp, J. Büchner, U. Motschmann, Astron. Astrophys. **434**, 1191–1200 (2005)
E.P. Rubenstein, R.E. Schaefer, Astrophys. J. **529**, 1031 (2000)
H.O. Rucker, ESA **SP-518**, 421–424 (2002)
Y. Sano, J. Geomagn. Geoelectr. **45**, 65–77 (1983)
E. Shkolnik, A.H. Walker, D.A. Bohlender, P.-G. Gu, M. Kuerster, Astrophys. J. **622**, 1075 (2005)
P. Song, H.J. Singer, G.L. Siscoe, in *Space Weather (Geophysical Monograph)*. Geophys. Monograph Series, vol. 125 (AGU, Washington, 2001), p. 440
A. Stadelmann, Ph.D. Thesis, Technische Universität Braunschweig, Braunschweig, Germany, 2004
I.R. Stevens, Mon. Not. R. Astron. Soc. **356**, 1053–1063 (2005)
D.J. Stevenson, Rep. Prog. Phys. **46**, 555–620 (1983)
F. Tian, O.B. Toon, A.A. Pavlov, H. De Sterck, Astrophys. J. **621**, 1049–1060 (2005)
G. Tinetti, A. Vidal-Madjar, M.-C. Liang, J.-P. Beaulieu, Y. Yung, S. Carey, R.J. Barber, J. Tennyson, I. Ribas, N. Allard, G.E. Ballester, D.K. Sing, F. Selsis, Nature **448**, 169–171 (2007)

A. Vidal-Madjar, A. Lecavalier des Etangs, J.-M. Désert, G.E. Ballester, R. Ferlet, G. Hébrand, M. Mayor, Nature **422**, 143–146 (2003)
A. Vidal-Madjar, J.-M. Désert, A. Lecavelier des Etangs, G. Hébrard, G.E. Ballester, D. Ehrenreich, R. Ferlet, J.C. McConnell, M. Mayor, C.D. Parkinson, Astrophys. J. **604**, L69–L72 (2004)
A. Vidal-Madjar, A. Lecavelier des Etangs, J.-M. Désert, G.E. Ballester, R. Ferlet, G. Hébrard, M. Mayor, Astrophys. J. **676**, L57–L60 (2008)
G.-H. Voigt, Planet. Space Sci. **29**, 1–20 (1981)
G.-H. Voigt, in *Handbook of Atmospheric Electrodynamics*, ed. by H. Volland, vol. II (CRC Press, Boca Raton, 1995), p. 333
A.J. Watson, T.M. Donahue, J.C.G. Walker, Icarus **48**, 150 (1981)
E.J. Weber, L.J. Davis, Astrophys. J. **148**, 217–227 (1967)
R. Yelle, Icarus **170**, 167–179 (2004)
R. Yelle, Icarus **183**, 508 (2006)
P. Zarka, J. Geophys. Res. **103**, 20159 (1998)
P. Zarka, Planet. Space Sci. **55**, 598 (2007)
P. Zarka, R.A. Treumann, B.P. Ryabov, V.B. Ryabov, Astrophys. Space Sci. **277**, 293–300 (2001)

Space Science Series of ISSI

1. R. von Steiger, R. Lallement and M.A. Lee (eds.): *The Heliosphere in the Local Interstellar Medium*. 1996 ISBN 0-7923-4320-4
2. B. Hultqvist and M. Øieroset (eds.): *Transport Across the Boundaries of the Magnetosphere*. 1997 ISBN 0-7923-4788-9
3. L.A. Fisk, J.R. Jokipii, G.M. Simnett, R. von Steiger and K.-P. Wenzel (eds.): *Cosmic Rays in the Heliosphere*. 1998 ISBN 0-7923-5069-3
4. N. Prantzos, M. Tosi and R. von Steiger (eds.): *Primordial Nuclei and Their Galactic Evolution*. 1998 ISBN 0-7923-5114-2
5. C. Fröhlich, M.C.E. Huber, S.K. Solanki and R. von Steiger (eds.): *Solar Composition and its Evolution – From Core to Corona*. 1998 ISBN 0-7923-5496-6
6. B. Hultqvist, M. Øieroset, Goetz Paschmann and R. Treumann (eds.): *Magnetospheric Plasma Sources and Losses*. 1999 ISBN 0-7923-5846-5
7. A. Balogh, J.T. Gosling, J.R. Jokipii, R. Kallenbach and H. Kunow (eds.): *Co-rotating Interaction Regions*. 1999 ISBN 0-7923-6080-X
8. K. Altwegg, P. Ehrenfreund, J. Geiss and W. Huebner (eds.): *Composition and Origin of Cometary Materials*. 1999 ISBN 0-7923-6154-7
9. W. Benz, R. Kallenbach and G.W. Lugmair (eds.): *From Dust to Terrestrial Planets*. 2000 ISBN 0-7923-6467-8
10. J.W. Bieber, E. Eroshenko, P. Evenson, E.O. Flückiger and R. Kallenbach (eds.): *Cosmic Rays and Earth*. 2000 ISBN 0-7923-6712-X
11. E. Friis-Christensen, C. Fröhlich, J.D. Haigh, M. Schüssler and R. von Steiger (eds.): *Solar Variability and Climate*. 2000 ISBN 0-7923-6741-3
12. R. Kallenbach, J. Geiss and W.K. Hartmann (eds.): *Chronology and Evolution of Mars*. 2001 ISBN 0-7923-7051-1
13. R. Diehl, E. Parizot, R. Kallenbach and R. von Steiger (eds.): *The Astrophysics of Galactic Cosmic Rays*. 2001 ISBN 0-7923-7051-1
14. Ph. Jetzer, K. Pretzl and R. von Steiger (eds.): *Matter in the Universe*. 2001 ISBN 1-4020-0666-7
15. G. Paschmann, S. Haaland and R. Treumann (eds.): *Auroral Plasma Physics*. 2002 ISBN 1-4020-0963-1
16. R. Kallenbach, T. Encrenaz, J. Geiss, K. Mauersberger, T.C. Owen and F. Robert (eds.): *Solar System History from Isotopic Signatures of Volatile Elements*. 2003 ISBN 1-4020-1177-6
17. G. Beutler, M.R. Drinkwater, R. Rummel and R. von Steiger (eds.): *Earth Gravity Field from Space – from Sensors to Earth Sciences*. 2003 ISBN 1-4020-1408-2
18. D. Winterhalter, M. Acuña and A. Zakharov (eds.): *"Mars" Magnetism and its Interaction with the Solar Wind*. 2004 ISBN 1-4020-2048-1
19. T. Encrenaz, R. Kallenbach, T.C. Owen and C. Sotin: *The Outer Planets and their Moons* ISBN 1-4020-3362-1
20. G. Paschmann, S.J. Schwartz, C.P. Escoubet and S. Haaland (eds.): *Outer Magnetospheric Boundaries: Cluster Results* ISBN 1-4020-3488-1
21. H. Kunow, N.U. Crooker, J.A. Linker, R. Schwenn and R. von Steiger (eds.): *Coronal Mass Ejections* ISBN 978-0-387-45086-5

22. D.N. Baker, B. Klecker, S.J. Schwartz, R. Schwenn and R. von Steiger (eds.): *Solar Dynamics and its Effects on the Heliosphere and Earth* ISBN 978-0-387-69531-0
23. Y. Calisesi, R.-M. Bonnet, L. Gray, J. Langen and M. Lockwood (eds.): *Solar Variability and Planetary Climates* ISBN 978-0-387-48339-9
24. K.E. Fishbaugh, P. Lognonné, F. Raulin, D.J. Des Marais, O. Korablev (eds.): *Geology and Habitability of Terrestrial Planets* ISBN 978-0-387-74287-8
25. O. Botta, J.L. Bada, J. Gomez-Elvira, E. Javaux, F. Selsis, R. Summons (eds.): *Strategies of Life Detection* ISBN 978-0-387-77515-9
26. A. Balogh, L. Ksanfomality, R. von Steiger (eds.): *Mercury* ISBN 978-0-387-77538-8
27. R. von Steiger, G. Gloeckler, G.M. Mason (eds.): *The Composition of Matter* ISBN 978-0-387-74183-3
28. H. Balsiger, K. Altwegg, W. Huebner, T.C. Owen, R. Schulz (eds.): *Origin and Early Evolution of Comet Nuclei, Workshop honouring Johannes Geiss on the occasion of his 80th birthday* ISBN 978-0-387-85454-0
29. A.F. Nagy, A. Balogh, T.E. Cravens, M. Mendillo, I. Mueller-Wodarg (eds.): *Comparative Aeronomy* ISBN 978-0-387-87824-9
30. F. Leblanc, K.L. Aplin, Y. Yair, R.G. Harrison, J.P. Lebreton and M. Blanc (eds.): *Planetary Atmospheric Electricity* ISBN 987-0-387-87663-4

Springer – Dordrecht / Boston / London

9